Geometric Analysis
and Nonlinear Partial Differential Equations

Springer-Verlag Berlin Heidelberg GmbH

Stefan Hildebrandt
Hermann Karcher
Editors

Geometric Analysis and Nonlinear Partial Differential Equations

With 100 Figures, 19 in Color

Springer

Editors

Stefan Hildebrandt
Hermann Karcher

University of Bonn
Department of Mathematics
Beringstraße 1
53115 Bonn, Germany

anke@iam.uni-bonn.de

Cataloging-in-Publication Data applied for

Bibliographic information published by Die Deutsche Bibliothek

Die Deutsche Bibliothek lists this publication in the Deutsche Nationalbibliografie; detailed bibliographic data is available in the Internet at <http://dnb.ddb.de>.

Mathematics Subject Classification (2000): 35-xx, 49-xx

ISBN 978-3-642-62887-0 ISBN 978-3-642-55627-2 (eBook)
DOI 10.1007/978-3-642-55627-2

http://www.springer.de

Originally published by Springer-Verlag Berlin Heidelberg New York in 2003

Typeset by the authors using a Springer TEX macro package. Edited by PTP-Berlin.
Cover design: *design & production* GmbH, Heidelberg

SPIN: 10993071 41/3111LK - 5 4 3 2 1 - Printed on acid-free paper

Preface

This book is not a textbook, but rather a coherent collection of papers from the field of partial differential equations. Nevertheless we believe that it may very well serve as a good introduction into some topics of this classical field of analysis which, despite of its long history, is highly modern and well prospering. Richard Courant wrote in 1950: *"It has always been a temptation for mathematicians to present the crystallized product of their thought as a deductive general theory and to relegate the individual mathematical phenomenon into the role of an example. The reader who submits to the dogmatic form will be easily indoctrinated. Enlightenment, however, must come from an understanding of motives; live mathematical development springs from specific natural problems which can be easily understood, but whose solutions are difficult and demand new methods or more general significance."*

We think that many, if not all, papers of this book are written in this spirit and will give the reader access to an important branch of analysis by exhibiting interesting problems worth to be studied. Most of the collected articles have an extensive introductory part describing the history of the presented problems as well as the state of the art and offer a well chosen guide to the literature. This way the papers became lengthier than customary these days, but the level of presentation is such that an advanced graduate student should find the various articles both readable and stimulating.

All contributions to this book grew out of work carried out at the Sonderforschungsbereich 256 *Nichtlineare partielle Differentialgleichungen* at Bonn University which was supported by the *Deutsche Forschungsgemeinschaft* (DFG) during the years 1986–2001. Most of the articles presented here reflect the very recent research at the SFB 256; yet the connection to the preceding development is well documented. Clearly these papers can represent only a small part of the work which profited from the DFG support at Bonn, and they certainly cannot do justice to the wealth of discoveries by analysts all over the world. We hope, however, that this book offers significant samples of the mathematical research that presently is going on.

To structure the material of this book we have arranged the topics that are dealt with in two sections. This division is in some cases not compelling, which is no surprise as it was one aim of the SFB to encourage collaboration of the participating mathematicians. In any case it is a natural tendency in mathematics that, sooner or later, separated currents will unite, thereby multiplying their strength. The ordering of the papers is not arbitrary or artificial, it corresponds to an inherent connection.

Section I contains the contributions of geometric nature: Spectral theory on regular and singular spaces, parametrization of Teichmüller space and its Thurston boundary, CR-manifolds, global estimates of conformal mappings, complete minimal surfaces, numerical methods for constructing minimal surfaces with prescribed boundaries, singular minimal surfaces solving an isoperimetric problem generated by partitioning of bodies, mean curvature flows. The final three papers treat the regularity of solutions of variational problems.

Section II consists of articles on partial differential equations, mostly of nonlinear type, which originate from problems in physics, particularly fluid mechanics and elasticity theory, biology and stochastics. A study of nonlinear differential equations through the geometry of matrix space is followed by a survey on the Korteweg–de Vries equation and by KAM theory for partial differential equations. Approximations of weak solutions of conservation laws are discussed. Also of hyperbolic nature is a nonlinear system describing motion in biological systems. Another paper from biology deals with orientational patterns in the arrangement of lengthy proteins. Two further contributions treat stochastic differential equations for processes in infinite dimensional spaces and a Markov semigroup approach to harmonic maps. Two papers on elliptic problems follow, dealing with a transmission problem for nonlinear elliptic systems and the numerical treatment of elliptic boundary value problems, respectively. The final seven articles are concerned with equations which are essentially parabolic in nature: Cahn-Hilliard equation and the influence of elastic effects, uniqueness and regularity for solutions to the Navier–Stokes equations, the no–slip boundary condition in fluid dynamics, finite volume schemes and Galerkin methods, electrorheological fluids, waiting time phenomena and "Ostwald ripening".

We thank all colleagues for their contributions to this book and for the smooth cooperation in making it more coherent and attractive. We are very much indebted to Martin Rumpf for the design of the graphic material presented in this book, and to Anke Thiedemann for her editorial work.

On behalf of everyone involved in the making of this book we are expressing our gratitude to the *Deutsche Forschungsgemeinschaft* which, for fifteen years, supported the work of the SFB 256 so substantially. We are grateful to the experts from other universities who accompanied our work, and to Bonn University for lasting support and encouragement.

We dedicate this book to Olga A. Ladyzhenskaya, whose work has influenced so many of us, as a token of our admiration.

On May 13, 2002, Professor Ladyzhenskaya has received the degree of a doctor honoris causa of the Rheinische Friedrich-Wilhelms-Universität Bonn. It is a great pleasure to include the celebration address given by Michael Struwe at this occasion.

Bonn, May 2002

Stefan Hildebrandt
Hermann Karcher

Table of Contents

Olga Ladyzhenskaya
A Life-Long Devotion to Mathematics

Michael Struwe

ETH Zürich, Zürich, Schweiz
struwe@math.ethz.ch

Summary. On May 13, 2002, Professor Ladyzhenskaya was awarded the degree of "Doctor honoris causa" by the University of Bonn. The following is the text of the Laudatio in honor of Professor Ladyzhenskaya read at this occasion.

1 The Beginning of a Mathematical Career

Olga Alexandrovna Ladyzhenskaya was born on March 7, 1922, in the town of Kologriv in a family of old Russian nobility. Her father Alexander Ivanovich Ladyzhenski was teaching mathematics at the local school. He transmitted his passion for mathematics not only to his students but also to his daughter Olga who from early childhood showed a strong talent for logical thinking. In 1939 she was admitted to the Leningrad Teachers' Training College and from 1941 to 1943 she taught mathematics to the senior classes at Kologriv Secondary School, the same school where her father had worked. From 1943 to 1947 she then studied mathematics at the University of Moscow. Among her teachers in Moscow were Kourosh, Stepanov, Petrovsky, and Gelfand. In 1947 she graduated and was recommended for a postgraduate fellowship. In the same year she married Andrei Alexevich Kiselev, specialist in the Theory of Numbers in the city of Leningrad/St. Petersburg, and moved to St. Petersburg where she became a postgraduate student of St. Petersburg University. Formally her scientific supervisor was Sobolev, but in reality her advisor was Smirnov. She defended her Ph.D. in the spring of 1949. In the same year already, she became assistant professor of St. Petersburg University. In the spring of 1953, at Moscow University, she handed in her thesis for the D. Sc. degree, comparable to the German "Habilitation". Not much later, in 1954, she was elected associate professor and in 1956 she became full professor of St. Petersburg University. At the same time, from 1954 to 1961 she also held a position as a Leading Scientific Researcher at Steklov Mathematical Institute (Leningrad branch). Finally, in 1961, she was appointed head of the newly created Laboratory of Mathematical Physics. Her mathematical successes soon brought her wide recognition, both in the Soviet Union and abroad.

This is what we can read about Professor Ladyzhenskaya's youth and the beginning of her mathematical career in the account of A. D. Alexandrov, A. P. Oskolkov, N. N. Ural'tseva, and L. D. Faddeev [1] on the occasion of her 60th birthday. We get the impression of an uneventful youth passed in rural tranquility and economic security in the family of a state official and a mathematical gift that matured in a creative free atmosphere.

However, the truth is very far from this, and it could only be told after communist rule of Russia had ended. Those were difficult days for an intellectual, in particular, for a descendent of the Russian noble class. In 1937 Professor Ladyzhenskaya's father was arrested by Stalin's men. In fact, as Alexander Solschenizyn recalls in his epic account of the "Gulag", Alexander I. Ladyzhenski had been warned by a peasant that his name was in "their" lists; but he stayed, he would not leave the students who depended on him; see [31], p. 23-24. In a show trial he was convicted as an "enemy of the Russian people" and sentenced to death. Olga Alexandrovna Ladyzhenskaya was lucky enough to be allowed to finish high school - unlike her elder sister who had to leave the university. In 1939 Olga Alexandrovna applied to enter prestigious Leningrad University, but as the daughter of a "class enemy" she was not admitted. When she finally was allowed to enter Moscow University in 1943, it was only because state policy had changed in the difficult period of wartime. Although she had completed her second thesis as early as 1951 she was not allowed to defend it before 1953, after Stalin's death.

There is only one explanation why in spite of such adversity Olga Alexandrovna Ladyzhenskaya was able to rise to the top of renowned Steklov Institute and to become the uncontested head of the Leningrad school of Partial Differential Equations, and this is her work.

2 Work

Professor Ladyzhenskaya has written more than 250 mathematical papers; her work covers the whole spectrum of partial differential equations, ranging from hyperbolic equations to differential equations generated by symmetric functions of the eigenvalues of the Hessian, and discussing topics ranging from uniqueness to convergence of Fourier series or finite-difference approximation of solutions. She developed the functional analytic treatment of nonlinear stationary problems by Leray-Schauder degree theory and pioneered the theory of attractors for dissipative equations.

She is author of six monographs, three of which have greatly influenced the development of the field of partial differential equations throughout the second half of the last century; in fact, her book on "Linear and quasilinear elliptic equations" [21], that she wrote jointly with her former student Nina Ural'tseva, was one of the first mathematical monographs I ever bought. I only regret that I did not buy her book with Ural'tseva and Solonnikov [20] on parabolic equations at the same time. It has been reprinted now, but only with a soft cover which may not be entirely appropriate for a book on "hard" analysis.

In these books, starting from the work of DeGiorgi [6] and Nash [26], she and her co-authors provide a regularity theory for elliptic and parabolic equations in divergence form under the most general assumptions on the coefficients of the equations and prove the exact dependence of the regularity of the solution on the regularity of the data, thereby giving a complete answer to Hilbert's 19th problem for a large class of equations.

With Ural'tseva [22], [23], using the work of Krylov and Safonov [28], she later extended these results to equations in non-divergence form with bounded measurable coefficients. Previously, as an important intermediate step towards a full existence and regularity theory for such problems, in her book [21] she already derived closedness in L^2 for the corresponding elliptic operators in the case of continuous coefficients. Remarkably, in two dimensions she only needed to assume uniform ellipticity and boundedness of the coefficients, thus recovering a result of Bernstein [2].

But the problems closest to her heart, I believe, always have been the equations of hydrodynamics, in particular, the Navier-Stokes equations, to which she has made deep and lasting contributions, some of which are summarized in her third milestone monograph on "The mathematical theory of viscous incompressible flow" [13].

I would like to recall some of these achievements here.

3 An Interpolation Inequality and Applications

Right at the beginning, on the second page of Chapter 1 of her book [13] we find the inequality

$$\int_{\mathbb{R}^2} |u|^4 dx \leq 4 \left(\int_{\mathbb{R}^2} |u|^2 dx \right) \left(\int_{\mathbb{R}^2} |\nabla u|^2 dx \right) \tag{1}$$

for any smooth function $u : \mathbb{R}^2 \to \mathbb{R}$ of compact support. The beauty of this inequality lies in its simplicity, and also the proof, taken from Ladyzhenskaya's paper [15] of 1959, is particularly instructive and elegant. We reproduce it here with minor modifications.

Labelling space coordinates as $x = (x_1, x_2)$, for any function u of compact support for $i = 1, 2$ we obtain

$$u^2(x_1, x_2) = 2 \int_{-\infty}^{x_i} u \frac{\partial u}{\partial x_i} dx_i \leq 2 \int_{-\infty}^{\infty} |u||\nabla u| dx_i .$$

Hence by Fubini's theorem and Hölder's inequality we find

$$\begin{aligned}
\int_{\mathbb{R}^2} u^4(x)\, dx &= \int_{-\infty}^{\infty} \int_{-\infty}^{\infty} u^4(x_1, x_2)\, dx_1\, dx_2 \\
&\leq 4 \int_{-\infty}^{\infty} \int_{-\infty}^{\infty} \Big(\int_{-\infty}^{\infty} |u||\nabla u|\, dx_1 \Big) \Big(\int_{-\infty}^{\infty} |u||\nabla u|\, dx_2 \Big)\, dx_1\, dx_2 \\
&= 4 \Big(\int_{-\infty}^{\infty} \int_{-\infty}^{\infty} |u||\nabla u|)\, dx_1\, dx_2 \Big)^2 \\
&= 4 \Big(\int_{\mathbb{R}^2} |u||\nabla u|\, dx \Big)^2 \leq 4 \Big(\int_{\mathbb{R}^2} |u|^2\, dx \Big) \Big(\int_{\mathbb{R}^2} |\nabla u|^2\, dx \Big) ,
\end{aligned}$$

as claimed. The constant may be improved from 4 to 2; what is important is that the constant is independent of the size of the support of u.

Inequality (1) also appears as a special case of the interpolation inequalities that were derived by Gagliardo [9], [10] and Nirenberg [27] in the late 50's, that is, around the same time as Ladyzhenskaya. Their motivation was functional analytic and inspired by the study of linear elliptic equations, while Professor Ladyzhenskaya arrived at inequality (1) by studying the (nonlinear) Navier Stokes system; see [13], p. 203.

In fact, the analytic treatment of nonlinear equations often relies on inequalities like (1). The particular estimate (1) also played a key role in an early paper of mine, dealing with the heat flow of harmonic maps of surfaces [32]. It turned out that it was precisely the inequality (1) which was needed to conclude the argument; but I was not sure if such a "borderline" estimate could be valid. In search of help, in [20] I first came across a very general statement of interpolation inequalities in Sobolev spaces. After checking again and again all the values of n, p, q, etc., I could hardly believe my good fortune that the desired estimate, indeed, should be true. Only when I later saw the above argument in [13] was I really convinced that I had not made a mistake.

To see how the estimate (1) may be applied in hydrodynamics we need to take a look at the Navier-Stokes equations.

4 Navier-Stokes Equations

The mathematical problems in hydrodynamics received special attention recently, when, in May 2000, the regularity problem for the Navier-Stokes equations was named as one of the seven millenium prize problems of the Clay Institute. An award of 1 million US Dollars is promised to whoever is able to show - in one version of this problem - that for any given smooth initial velocity $u_0 = (u_0^1, u_0^2, u_0^3) : \mathbb{R}^3 \to \mathbb{R}^3$, periodic in each of the spatial variables and of vanishing divergence

$$\operatorname{div} u_0 = \sum_{i=1}^{3} \frac{\partial u_0^i}{\partial x_i} = 0,$$

there exists a global smooth velocity field $u = (u^1, u^2, u^3) : \mathbb{R}^3 \times [0, \infty[\to \mathbb{R}^3$ and an associated smooth pressure distribution $p : \mathbb{R}^3 \times [0, \infty[\to \mathbb{R}$, both also periodic in the space variables, satisfying the Navier-Stokes equations

$$\frac{\partial u^i}{\partial t} + u \cdot \nabla u^i = \Delta u^i - \frac{\partial p}{\partial x^i},\ i = 1, \ldots, 3, \tag{2}$$

on $\mathbb{R}^3 \times [0, \infty[$ together with the divergence condition

$$\operatorname{div} u = 0, \tag{3}$$

and with initial data

$$u = u_0 \text{ at } t = 0. \tag{4}$$

Note that we have scaled the viscosity parameter ν usually present on the right of (2) to unity.

Alternatively, the prize will be awarded to whoever presents an example of smooth, divergence-free initial data u_0 such that the problem (2),(3), and (4) does *not* admit a smooth solution (u, p) on $\mathbb{R}^3 \times [0, \infty[$, where one is allowed to add a smooth forcing term on the right of (2), or for the answer - affirmative or negative - to the corresponding regularity question in the non-periodic case but with suitable decay conditions near spatial infinity; see the problem description by Fefferman [7] for further details.

Professor Ladyzhenskaya probably would not have agreed with this formulation of the problem. To her, I believe, as to many other people working in the field, rather than existence and *regularity* the main problem is existence and *uniqueness*.

Uniqueness as a central issue already had played a key role in her diploma thesis and again in her "habilitation" thesis in 1953. As she observed, for each of the classical initial value problems one can investigate the question of well-posedness on a whole scale of solution spaces. For the problem of determining the "correct" solution space the issues of existence and uniqueness are in conflict with one another: If one broadens the concept of solution by allowing also distribution solutions with very "bad" regularity properties it may be easy to show existence while it may be hard - if not impossible - to show uniqueness. Conversely, if one narrows the concept of solution by allowing only solutions with "good" regularity properties, it may be easy to prove uniqueness but it may be difficult to obtain the existence of a solution in the restricted class. The question then becomes whether for some choice of solution space both requirements can simultaneously be fulfilled.

For the Navier-Stokes system (2) - (4) the corresponding question was posed and answered in the affirmative - at least in two space dimensions - in the celebrated work of Kiselev-Ladyzhenskaya [12] from 1957. Estimates for the L^4-norm of a solution play a key role; in fact, uniqueness may be most easily obtained by using Ladyzhenskaya's inequality (1), as was later done by Lions-Prodi [25].

For convenience, let us look at the problem on the whole space. Our point of departure is the classical energy identity for (sufficiently smooth) solutions of (2) - (4), valid in any space dimension n.

As observed by Leray [24] and Hopf [11], when we multiply equation (2) with the test function u and take account of condition (3), the nonlinear term and the pressure term yield a term of divergence type

$$\sum_{i=1}^{n} \left(u \cdot \nabla u^i + \frac{\partial p}{\partial x^i} \right) u^i = u \cdot \nabla(\frac{1}{2}|u|^2 + p) = \operatorname{div} \left(u(p + \frac{1}{2}|u|^2) \right)$$

that vanishes after integration. From the remaining terms, upon integrating by parts, we obtain the identity

$$\frac{1}{2} \frac{d}{dt} \int_{\mathbb{R}^n} |u|^2 \, dx + \int_{\mathbb{R}^n} |\nabla u|^2 \, dx = 0 \, .$$

After another integration in time there results the energy identity

$$\int_{\mathbb{R}^n} |u(t)|^2\, dx + 2\int_0^t \int_{\mathbb{R}^n} |\nabla u|^2\, dx\, dt = \int_{\mathbb{R}^n} |u_0|^2\, dx =: E(u_0) \qquad (5)$$

valid for (sufficiently smooth) solutions of (2). By a Galerkin approximation, and using similar arguments as leading to (5) above, we obtain a distribution solution (u,p) to (2)-(4) satisfying (5) in a weak sense, that is, with left hand side not exceeding the quantity $E(u_0)$ for any t. This is the class of Hopf solutions to (2)-(4).

Now suppose that $n = 2$. Let u and v be Hopf solutions on some time interval $[0,T]$ for the same data $u = v = u_0$ at $t = 0$, and set $w = v - u$. We intend to show that $w = 0$, thus establishing existence and uniqueness within the Hopf class of solutions. (For the following argument we need not worry about the pressure p.) Taking the difference of equations (2) for v and u and multiplying by w, then similar to our derivation of (5) above, upon integrating by parts we obtain the identity (in the distribution sense on $[0,T]$ and with $(a,b) = \sum_i a^i b^i$ for $a = (a^1, a^2)$, $b = (b^1, b^2) \in \mathbb{R}^2$)

$$\frac{1}{2}\frac{d}{dt}\int_{\mathbb{R}^2} |w|^2\, dx + \int_{\mathbb{R}^2} |\nabla w|^2\, dx = -\int_{\mathbb{R}^2} (w, w\cdot\nabla u)\, dx.$$

The remaining terms vanish because u and v, and therefore also w, are divergence-free. By Hölder's inequality we can estimate the term on the right

$$\left|\int_{\mathbb{R}^2} (w, w\cdot\nabla u)\, dx\right| \le \Big(\int_{\mathbb{R}^2} |\nabla u|^2\, dx\Big)^{1/2}\Big(\int_{\mathbb{R}^2} |w|^4\, dx\Big)^{1/2},$$

which yields the estimate

$$\frac{1}{2}\frac{d}{dt}||w||^2_{L^2} + ||\nabla w||^2_{L^2} \le \left|\int_{\mathbb{R}^2} (w, w\cdot\nabla u)\, dx\right| \le ||w||^2_{L^4}||\nabla u||_{L^2}. \qquad (6)$$

Observing that we are in $n = 2$ space dimensions, we may now use (1) to bound

$$||w||^2_{L^4} \le 2||w||_{L^2}||\nabla w||_{L^2}.$$

With the familiar estimate $2|ab| \le a^2 + b^2$ we proceed to estimate the right hand side in (6) above

$$||w||^2_{L^4}||\nabla u||_{L^2} \le 2||w||_{L^2}||\nabla w||_{L^2}||\nabla u||_{L^2} \le ||w||^2_{L^2}||\nabla u||^2_{L^2} + ||\nabla w||^2_{L^2}.$$

From (6) then we conclude the differential inequality

$$\frac{d}{dt}||w||^2_{L^2} \le 2||w||^2_{L^2}||\nabla u||^2_{L^2}.$$

Recalling (5), we therefore obtain the uniform bound

$$||w(t)||^2_{L^2} \le ||w(0)||^2_{L^2}\exp\Big(2\int_0^t ||\nabla u(s)||^2_{L^2}\, ds\Big) \le ||w(0)||^2_{L^2}\exp(||u_0||^2_{L^2})$$

for any $t \leq T$ and therefore $w = 0$, as desired. The above formal argument may be justified by mollifying the function w in time. Inequality (1) ensures that all integrals are well-defined.

The solution class originally considered by Kiselev-Ladyzhenskaya was slightly more narrow, involving also square integrable time and mixed second space-time derivatives. This illustrates the rich variety of possible choices between the class of Leray-Hopf solutions on the one extreme and smooth, classical solutions on the other.

In fact, in two space dimensions, as was shown by Serrin [29], [30], for smooth initial data u_0 all these choices are equivalent as any Hopf solution is smooth and smooth solutions are easily seen to be unique. The previously made distinction between the question of regularity and the question of uniqueness thus may seem artificial.

However, there are parabolic equations with a structure not too much different from that of the Navier-Stokes equations where this distinction *is* meaningful. For example, in my result [32] that I already briefly mentioned in relation to inequality (1) above, in 1985 I showed the existence of a unique solution to the heat flow of harmonic maps of surfaces whose energy is monotonically decreasing and which is regular away from finitely many points in space-time. Many people believed that the solution would actually be smooth until Chang-Ding-Ye [5] in 1992 gave an example where the solution indeed blows up from smooth data after a finite time.

It may be instructive to develop this example still a bit further. Freire [8] in 1995 showed that my solution is unique even in the class of distribution solutions with monotonically decreasing energy. On the other hand, very recently Topping [33] and Bertsch et al. [3] independently constructed examples where the heat flow admits infinitely many distinct weak solutions that all satisfy a weak version of the energy inequality similar to the Hopf solutions that we considered above.

Could something similar be the case for the Navier-Stokes system in three or more space dimensions? We do not know the answer; the right notion of "solution" remains a mystery. The class of smooth solutions may be too small. However, since it seems to be physically impossible to observe, for instance, isolated ("point") singularities of a flow, there seems to be no reason to insist that solutions be smooth but in an average sense.

On the other hand, the (weak) energy inequality alone may not be sufficient to guarantee uniqueness of weak solutions. In fact, in 1969 Professor Ladyzhenskaya gave an example showing nonuniqueness within the Hopf class of weak solutions in 3 space dimensions, however, on a space-time domain $Q_T \subset [0, T] \times \mathbb{R}^3$ which is singular at the origin [16]. Further energy or entropy constraints, that derive, for instance, from the second principle of thermodynamics, seem to be needed in order to distinguish physically realistic weak solutions from non-physical, purely mathematical ones.

A natural choice for a solution space thus could be the class of "suitable weak solutions" constructed by Caffarelli-Kohn-Nirenberg [4]. These are Hopf solutions satisfying, in addition, a local form of the energy inequality which implies a constraint similar to the monotonicity of the energy required in Freire's uniqueness result for the harmonic map heat flow. Very recently, in joint work with Seregin [19], Ladyzhenskaya has also contributed to the analysis of this class of solutions.

Professor Ladyzhenskaya considered hydrodynamics not only within the frame defined by the Navier-Stokes equations, but she has also considered alternative models to explain the multiple phenomena that we observe in fluid mechanics and she explored their mathematical properties. Likewise she developed new and mathematically extremely fruitful ideas about the nature of turbulence, that, in particular, led her to consider the notion of attractor for infinite dimensional dynamical systems [17]; see also her monograph [14].

Thus, we see that since the early 1950's Professor Ladyzhenskaya continues to work on the forefront of research on hydrodynamics, often providing the initial stimulus also for the fruitful investigations of others. Her ideas and the questions she raised remain relevant today, in particular, for the Navier-Stokes system.

5 Ladyzhenskaya and the Steklov Institute

With her impressive mathematical achievements, helped by her cultured and charming personality, Professor Ladyzhenskaya has attracted a large number of excellent students to work with her at the Laboratory of Mathematical Physics of Steklov Institute and at Leningrad University, among them L. Faddeev, K. Golovkin, A. Ivanov, V. Rivkind, V. Solonnikov, and N. Ural'tseva. Much of this activity is documented in Ladyzhenskaya's account [18].

Professor Ladyzhenskaya already was famous for her work worldwide when, in 1981, she was elected corresponding member of the Russian Academy of Science, and, in 1990, full member. She also is foreign member of numerous academies abroad, among them the Leopoldina, the oldest German academy. Until 1998 she was President of the Mathematical Society of St. Petersburg, thus a successor of Leonhard Euler in this office.

The year 1989 brought about the end of communist rule and a turn towards democracy and market economy in Russia. Russian mathematicians were allowed to travel more freely; some of them were able to visit Western countries for the first time. At the same time their economic situation deteriorated at the rate with which their salaries decreased in value in comparison with market prizes for standard goods.

Thus we can easily sympathize with those scientists, among them many leading Russian mathematicians, who accepted offers from abroad and left their country to find more favorable working conditions and a secure future for their families elsewhere. Professor Ladyzhenskaya, however, stayed and helped steer the Steklov Institute through these years of economic change and foster the careers of her coworkers there, thus remaining faithful to the legacy of her father.

She only retired officially from her position at Steklov Institute in 2000. To a large extent she uses her new freedom to travel to make contacts for the benefit of her students and for the better of the institution which has been her scientific home for over 50 years. The University of Bonn stands out prominently among the institutions with whom she has built lasting relations which have proved highly valuable for both sides.

I am grateful to be able to share this moment when these efforts and the mathematical work on which this cooperation is founded will be recognized by awarding the degree of a "Doctor honoris causa" to Professor Ladyzhenskaya.

Acknowledgement. I am indebted to Denis Labutin, Gregory Seregin, and the Mathematics Department of the University of Bonn for their help in providing background information and gathering references for this survey.

References

1. A. D. Aleksandrov, A. P. Oskolkov, N. N. Ural'tseva, L. D. Faddeev: Ol'ga Aleksandrovna Ladyzhenskaya (on her sixtieth birthday). Russian Math. Surv. 38.5 (1983), 171-181
2. S. N. Bernstein: Sur la généralization du problème de Dirichlet. Math. Ann. 69 (1910), 82-136
3. M. Bertsch, R. dal Passo, R. van der Hout: Nonuniqueness for the heat flow of harmonic maps on the disk. To appear in Arch. Rat. Mech. Analysis
4. L. Caffarelli, R. Kohn, L. Nirenberg: Partial regularity of suitable weak solutions of the Navier-Stokes equations. Comm. Pure Appl. Math. 35 (1982), 771-831
5. K.-C. Chang, W.-Y. Ding, R. Ye: Finite-time blow-up of the heat flow of harmonic maps from surfaces. J. Diff. Geom. 36, No.2 (1992), 507-515
6. E. DeGiorgi: Sulla differenziabilità e l'analiticità delle estremali degli integrali multipli regolari. Mem. Accad. Sci. Torino, P. I., III. Ser. 3 (1957), 25-43
7. C. L. Fefferman: Existence and smoothness of the Navier-Stokes equations. Clay Prize Description, May 1, 2000
8. A. Freire: Uniqueness for the harmonic map flow from surfaces to general targets. Comment. Math. Helv. 70, No.2, 310-338 (1995). Correction: Comment. Math. Helv. 71, No.2, 330-337 (1996)
9. E. Gagliardo: Proprietà di alcune classi di funzioni in più variabili. Ricerche Mat. 7 (1958), 102-137
10. E. Gagliardo: Ulteriori proprietà di alcune classi di funzioni in più variabili. Ricerche Mat. 8 (1959), 24-51
11. E. Hopf: Über die Anfangswertaufabe für die hydrodynamischen Grundgleichungen. Math. Nachr. 4 (1951), 213-231
12. A. A. Kiselev, O. A. Ladyzhenskaya: On the existence and uniqueness of solutions of the non-stationary problems for flows of non-compressible fluids. Am. Math. Soc., Transl., II. Ser. 24, 79-106 (1963); translation from Izv. Akad. Nauk SSSR. Ser. Mat. 21, 655-680 (1957)
13. O. A. Ladyzhenskaya: The mathematical theory of viscous incompressible flow. Gordon and Breach, rev. second edition, 1969
14. O. A. Ladyzhenskaya: Attractors for semigroups and evolution equations. Lezioni Lincee, Cambridge University Press, 1991
15. O. A. Ladyzhenskaya: Solution "in the large" of the nonstationary boundary value problem for the Navier-Stokes system in two space variables. Comm. Pure Appl. Math. 12 (1959), 427-433

16. O. A. Ladyzhenskaya: Example of nonuniqueness in the Hopf class of weak solutions for the Navier-Stokes equations. Math. USSR - Izvestija 3 (1969), No. 1, 229-236; translated from Izv. Akad. Nauk. SSSR, Ser. Mat. Tom 33 (1969)
17. O. A. Ladyzhenskaya: A dynamical system generated by the Navier-Stokes equations. J. Soviet Math. 3, No. 4 (1975), 458-479
18. O. A. Ladyzhenskaya: On some trends of investigations carried out in the laboratory of mathematical physics at the Leningrad branch of the mathematical institute. Proc. Steklov Inst. Math. 2 (1988), 229-258
19. O. A. Ladyzhenskaya, G. A. Seregin: On partial regularity of suitable weak solutions to the three-dimensional Navier-Stokes equations. J. Math. Fluid Mech. 1 (1999), 356-387
20. O. A. Ladyzhenskaya, V. A. Solonnikov, N. N. Ural'tseva: Linear and quasilinear equations of parabolic type. Amer. Math. Soc., Transl. Math. Monographs 23 (1968, reprinted 1988), Providence, Rhode Island; translated from the Russian edition of 1967
21. O. A. Ladyzhenskaya, N. N. Ural'tseva: Linear and quasilinear elliptic equations. Academic Press, New York and London, 1968; translated from the Russian edition of 1964
22. O. A. Ladyzhenskaya, N. N. Ural'tseva: Estimate of the Hölder norm of the solutions of second-order quasilinear elliptic equations of the general form. J. Sov. Math. 21 (1983), 762-768
23. O. A. Ladyzhenskaya, N. N. Ural'tseva: Estimates of Hölder constants for bounded solutions of second-order quasi-linear parabolic equations of nondivergent form. Adv. Math., Suppl. Stud. 10 (1986), 1-22
24. J. Leray: Sur le mouvement d'un liquide visqueux emplissant l'espace. Acta Math. 63 (1934), 193-248
25. J.-L. Lions, G. Prodi: Un théorème d'existence et unicité dans les équations de Navier-Stokes en dimension 2. C. R. Acad. Sci., Paris 248 (1959), 3519-3521
26. J. Nash: Continuity of solutions of parabolic and elliptic equations. Am. J. Math. 80 (1958), 931-954
27. L. Nirenberg: On elliptic partial differential equations. Ann. Scuola Norm. Sup. Pisa, Sci. fis. Mat., III. Ser. 123 (1959), 115-162
28. M. V. Safonov: Harnack's inequality for elliptic equations and the Hölder property of their solutions. J. Sov. Math. 21 (1983), 851-863; translated from Zap. Nauchn. Semin. Leningr. Otd. Mat. Inst. Steklova 96 (1980), 272-287
29. J. Serrin: On the interior regularity of weak solutions of the Navier-Stokes equations. Arch. Rat. Mech. Anal. 9 (1962), 187-195
30. J. Serrin: The initial value problem for the Navier-Stokes equations. In: R. E. Langer (ed) Nonlinear problems. Univ. Wisc. Press, Madison (1963), 69-98
31. A. Solschenizyn: Der Archipel Gulag. Scherz-Verlag, Bern, 1974 (translated from the Russian)
32. M. Struwe: On the evolution of harmonic mappings of Riemannian surfaces. Comment. Math. Helv. 60 (1985), 558-581
33. P. Topping: Reverse bubbling and nonuniqueness in the harmonic map flow. IMRN 2002 No. 10, 505-520

I. Geometric Analysis and Calculus of Variations

On the Spectral Theory of Surfaces with Cusps

Werner Ballmann[1] and Jochen Brüning[2*]

[1] Mathematisches Institut, Universität Bonn,
Beringstrasse 1, 53115 Bonn, Germany ballmann@math.uni-bonn.de

[2] Institut für Mathematik, Humboldt–Universität,
Rudower Chaussee 5, 12489 Berlin, Germany bruening@mathematik.hu-berlin.de

Summary. We are interested in the spectral properties of Dirac operators on noncompact surfaces. Under the assumption that 1) the ends of the given surface M are cusps as in the case of finite area surfaces of negative curvature and 2) the geometry of the Dirac bundle in question is closely related to the geometry of M we investigate the essential spectrum of the corresponding Dirac operator D and discuss its Fredholm index.

1 Introduction

For a compact Riemannian manifold, M, one has a fairly good understanding of the spectral theory of self-adjoint elliptic operators. To illustrate the main results in a specific and very well known case we consider the operator

$$D := d + d^*,$$

acting on the space $\Omega(M)$ of smooth differential forms on M, i.e., on the space of smooth sections of the bundle ΛT^*M. Then D is a first order elliptic differential operator, and is symmetric in the Hilbert space $L^2(\Lambda T^*M)$ with domain $\Omega(M)$. Its square is the Laplacian on forms which restricts to forms of a fixed degree and hence decomposes in a direct sum, contrary to D.

The above operator D is not closed, but it is closable, and its closure is a self-adjoint operator in $L^2(\Lambda T^*M)$; such operators are called essentially self-adjoint. One may think of this as expressing the fact that M has no boundary and hence presents no boundary conditions for D. Thus we may consider the spectrum $\operatorname{spec} D$ of D, and the main theorem here tells us that D is discrete, i.e., $\operatorname{spec} D$ is a discrete subset of $\mathbb{R}$ such that each spectral value is an eigenvalue of finite multiplicity.

In particular, D is a Fredholm operator, but necessarily with index 0. If we break the symmetry by a self-adjoint involution, α, which anticommutes with D, then we may split D in a possibly nonsymmetric way and obtain nonzero indices. This procedure produces the Gauss-Bonnet Operator, with index the Euler characteristic $\chi(M)$, and the Hirzebruch Signature Operator, with index the signature of M.

These facts indicate already that $\operatorname{spec} D$ contains important information about the structure of M. An extensive literature is devoted to investigations of how much

* Both authors were partially supported by SFB 256 (U Bonn) and SFB 288 (HU Berlin).

of the structure and, in particular, of the geometry of M is actually determined by the spectrum. In doing so it is natural to admit spectral data from a larger class of geometric operators, the so called generalized Dirac operators, which we introduce below. The main new object is the Dirac operator associated with a spinor bundle on M, introduced by Atiyah and Singer in their work on the Index Theorem.

The whole situation becomes quite different if we drop the assumption of compactness. Then D may not be essentially self-adjoint any more, but this difficulty does not arise if the manifold is assumed to be complete. The new phenomenon we have to expect, whether the underlying manifold is complete or not, consists in the possible presence of spectral points which are not isolated eigenvalues of finite multiplicity; the union of all such points is called the essential spectrum, $\mathrm{spec}_e\, D$. In particular, D may not be Fredholm any more; indeed, this is easily seen to be equivalent to the fact that 0 is not a point in the essential spectrum.

If one wants to extend the theory mentioned above to certain classes of complete noncompact manifolds, then, consequently, one has to focus on the determination of the essential spectrum. If one restricts attention to the Fredholm case, then Fredholm criteria are needed; moreover, the index formulas have to be derived.

In the cases we consider, the underlying manifold M has finitely many ends, and each end U of M is diffeomorphic to a product of the form $S \times (0, \infty)$, where S is compact and connected [1], and the metric on U is of the form

$$f^2(s,t)ds^2 + dt^2.$$

If $f = \text{const.} > 0$, then such an end is called *cylindrical*. This case was investigated by Atiyah, Patodi, and Singer in their discussion of the index formula associated with boundary value problems for generalized Dirac operators, see [1]. Another case which has been considered is the case of locally symmetric spaces of finite volume; in this case, a much more detailed description of the essential spectrum is possible, see for example [21, 12, 18, 19]. Index formulas have also been obtained for manifolds with ends which are *warped products* with suitable warping functions ([22, 5]) and for certain more general spaces ([6, 8, 9]); the essential spectrum of the Laplacian has been described in great generality by Lott ([15]). One of the points of the present investigation is to keep the assumptions on the function f general enough to include surfaces of pinched negative curvature and finite area. However, we do not want to assume a priori bounds on the derivative of the curvature of M; in fact, under our assumptions, the derivative of the curvature may explode. This restricts the regularity of f severely and is one of the main technical problems we have to overcome. In this article, we actually examine the most simple class of such manifolds, namely complete noncompact surfaces of finite area with pinched negative curvature near infinity. It is somewhat surprising that already this rather narrow family displays a great variety of phenomena. In particular, even on the same surface (but for different operators, of course), $\mathrm{spec}_e\, D$ may be empty or equal to all of $\mathbb{R}$ (in this context, see also [2]). On the other hand, the classical index formulas extend to the Fredholm case, but contain additional terms coming from the singularity at infinity.

---

[1] Actually, since we consider the case of surfaces, S will be the unit circle.

It does not seem possible to determine $\operatorname{spec}_e D$ in general but we can derive some criteria which are useful and essentially reduce the problem to an ordinary differential operator.

We now proceed to a more detailed presentation of our results which makes it necessary to introduce some terminology.

Let M be a noncompact surface with a complete Riemannian metric. Assume that M has finitely many ends and that each end U of M is a cusp. By this we mean that U is diffeomorphic to $S \times (0, \infty)$, where $S = \mathbb{R}/\mathbb{Z}$ is the circle, the metric on U is of the form $f^2 ds^2 + dt^2$ as above and has finite area, and the Gauss curvature K satisfies

$$-b^2 \leq K = -\partial_t^2 f / f \leq -a^2$$

on U, where a, b are some appropriate positive constants. These assumptions imply that M is conformally equivalent to a closed surface $\bar{M}$ with finitely many points deleted.

The most important examples are complete surfaces with finite area and globally pinched negative Gauss curvature. For that class to be included in our discussion, we have to be somewhat modest in the regularity assumption since then the above diffeomorphism need only be C^2, cf. [11], [13]. More precisely, we assume that f is C^1 and that the second partial derivative of f in the t–direction exists and is continuous. When necessary we label the ends by an index k.

Let E be a graded Dirac bundle over M in the sense of Gromov and Lawson, see [16]. That is, E is a bundle of left modules over the Clifford bundle $\mathrm{Cl}\, M$ of M with compatible Hermitian metric and connection together with a parallel field α of unitary involutions which anticommutes with Clifford multiplication by vector fields. These data determine a Dirac operator D and a decomposition $E = E^+ \oplus E^-$ into the eigenspaces of α with eigenvalue ± 1. We are interested in the spectral properties of D as an unbounded and essentially self–adjoint operator in the space $L^2(M, E)$ of square integrable sections of E.

There is no reason to expect close relations between the geometry of M and the spectral properties of D if the underlying Dirac bundle is not closely related to the geometry of M. To establish such a relation we propose the following notion. We say that a vector bundle E over M with Hermitian metric and connection is *geometric* if there is a parallel twoform S with values in the bundle of skew Hermitian endomorphisms of E such that the curvature R of E satisfies

$$R(X, Y)u = KS(X, Y)u$$

for all vector fields X, Y of M and sections u of E. The class of geometric bundles over M admits all the standard operations on vector bundles and contains all flat bundles and all bundles which are associated to the Riemannian structure or a spin structure of M.

Now let E be a fixed graded Dirac bundle over M and assume that E is geometric. Then for any local orthonormal frame X, Y of TM, $C := XYS(X, Y)$ is a parallel field of Hermitian endomorphisms of E independent of the choice of X and Y and hence is globally defined. We have

$$XY \cdot R(X,Y)u = KCu$$

for any section u of E. Now the field α of involutions of E is parallel and anti-commutes with Clifford multiplication by vector fields, hence $C(E^{\pm}) \subset E^{\pm}$. We set

$$C^{\pm} = C|E^{\pm} : E^{\pm} \to E^{\pm} .$$

Let $U = S \times (0, \infty)$ be an end of M. Let $T = \partial_t$ and $X = f^{-1}\partial_s$ and orient U by this frame; if M is oriented, then we assume that the orientations of U and M coincide. The oriented orthonormal frame T, X is a section of the $SO(2)$–principal bundle $SO(U)$ of oriented orthonormal frames on U and determines a trivialization $SO(U) = U \times SO(2)$. Let $\widetilde{SO}(U) = U \times \mathbb{R} \to U \times SO(2)$ be the corresponding lift, where we map $r \in \mathbb{R}$ to the counterclockwise rotation by the angle $2\pi r$. Then $\widetilde{SO}(U)$ is an $\mathbb{R}$–principal bundle over U and any bundle associated to $\widetilde{SO}(U)$ via a unitary representation of $\mathbb{R}$ comes with a Hermitian metric and connection. Now for $c \in \mathbb{R}$ and $w \in \mathbb{C}$ with $|w| = 1$ we let $L(c, w)$ be the complex line bundle associated to $\widetilde{SO}(U)$ via the unitary representation $\exp(2\pi icr)$ of $\widetilde{SO}(2)$, and with atlas given by the holonomy requirement $[s + 1, t, 0, u] = [s, t, 0, wu]$ along the circles $S \times \{t\}$. Note that $L(1,1)$ is the tangent bundle of U with its natural complex structure determined by the chosen orientation.

The complex area element $\omega_{\mathbb{C}} = iTX$ defines a parallel field of unitary involutions of E over U which commutes with α and C. We set

$$E^{\pm}(c, \varepsilon) = \{u \in E^{\pm}|U \mid Cu = cu \, , \omega_{\mathbb{C}}u = \varepsilon u\} \, ,$$

where $c \in \mathbb{R}$ and $\varepsilon \in \{+1, -1\}$. We will show that $E^+(c, \varepsilon)$ is isomorphic to an orthogonal sum of line bundles $L(-\varepsilon c, w)$, and we denote by $m(c, \varepsilon, w)$ the number of times $L(-\varepsilon c, w)$ occurs in this sum. The spectral characteristics which we investigate in this paper are given in terms of C^+ and the integers $m(c, \varepsilon, w)$. All data associated to an end $U = U_k$ will be indexed by k if necessary.

Suppose $E = \mathrm{Cl}\, M$ is the Clifford bundle over M with the even–odd decomposition. Then $E^+|U = E^+(0,1) \oplus E^+(0,-1)$, $m(0, \varepsilon, 1) = 1$ and $m(0, \varepsilon, w) = 0$ for $w \neq 1$. If M is oriented and E is the spinor bundle associated to the corresponding spin^c structure, then $E^+|U = E^+(0,1)$, $m(0,1,1) = 1$ and $m(0,1,w) = 0$ for $w \neq 1$. If E is a spinor bundle associated to a spin structure of M, then $E^+|U = E^+(1/2, 1)$. If the spin structure is the restriction of a spin structure of the closed surface $\bar{M}$, then $m(1/2, 1, -1) = 1$ and $m(1/2, 1, w) = 0$ for $w \neq -1$. If M has more than one cusp, then there exist spinor bundles on M with $m_k(1/2, 1, 1) = 1$ for some k, see [2].

We first analyze the essential spectrum $\mathrm{spec}_e\, D$ of D which is localized on the ends of M by the decomposition principle. Recall that D is Fredholm if and only if $0 \notin \mathrm{spec}_e\, D$.

Theorem A *1) If* $m_k(c,\varepsilon,1) = 0$ *for all* k, c, ε, *then* $\operatorname{spec}_e D = \emptyset$.
2) If $m_k(1/2,\varepsilon,1) = 0$ *for all* k *and* ε, *then* D *is Fredholm.*
3) If $m_k(1/2,\varepsilon,1) \neq 0$ *for some* k *and* ε, *then* $\operatorname{spec}_e D = \mathbb{R}$.

Note that when M has more than one cusp, then depending on the spin structure, the essential spectrum of the spin Dirac operator on M is empty or equal to $\mathbb{R}$, and both cases occur.

Set $D^+ = D|L^2(M,E^+) \cap C^1(M,E^+)$. In the case when D is Fredholm, we also determine the index of D^+ explicitly.

Theorem B *Let* $m = \operatorname{rank} E^+$ *and suppose that* $m_k(1/2,\varepsilon,1) = 0$ *for all* k *and* ε. *Then*

$$\begin{aligned}\operatorname{ind} D^+ = (m/2 - \operatorname{tr} C^+)\,\chi(M) &+ \frac{1}{2}\sum_{k,c,\varepsilon} \operatorname{sign}(1/2 - c)\, m_k(c,\varepsilon,1) \\ &- \frac{1}{2}\sum_{k,c,\varepsilon,w} \varepsilon\,\eta(w)\, m_k(c,\varepsilon,w)\,,\end{aligned}$$

where $\eta(1) = 0$ *and* $\eta(\exp(2\pi i \rho)) = 1 - 2\rho$ *for* $0 < \rho < 1$.

We emphasize that in our theorems, we do not need assumptions on derivatives of the curvature or, respectively, assumptions on the asymptotic behaviour of third or higher derivatives of the metric.

Special cases of the above theorems were obtained in [10], [22], [5], and [2]. In recent work, John Lott deals with the Laplacian, D^2, on forms, for manifolds of pinched negative curvature and finite volume in arbitrary dimensions [15]. He proves the finiteness of the space of L^2–harmonic forms and shows that the essential spectrum of D^2 coincides with the essential spectrum of an ordinary differential operator which is our $D^2_{\infty 0}$ introduced in (22) below. In the preprint version [3] of this paper we showed that $\operatorname{spec}_e D \subset \operatorname{spec}_e D_{\infty 0}$ for all Dirac operators, which is the main point of the Fredholm analysis. Motivated by [15], we now also prove equality by an elementary argument using Weyl sequences for Dirac operators.

In the case when E is the Clifford bundle with the even–odd decomposition, then D^+ is called the *Gauss–Bonnet operator*. We get

$$\operatorname{ind} D^+ = \chi(M) + \#\{\text{ends}\} = \chi(\bar{M})\,.$$

When M is oriented and E is the spinor bundles associated to the spin^c structure of M, then D^+ is called the *Riemann–Roch operator*. In this case

$$\operatorname{ind} D^+ = \big(\chi(M) + \#\{\text{ends}\}\big)/2 = \chi(\bar{M})/2\,.$$

When M is oriented and E is the spinor bundle associated to a a spin structure which extends to a spin structure on $\bar{M}$, then $\operatorname{ind} D^+ = 0$.

Recall that M is conformally equivalent to a closed surface $\bar{M}$ with finitely many points deleted. Hence the Riemannian metric on M is conformally equivalent to a

Riemannian metric with cylindrical ends. The latter were discussed in the work of Atiyah, Patodi and Singer [1], and therefore it is interesting to know whether the dimension of the space $\mathcal{H}^2(M,E)$ of square integrable harmonic sections of E is a conformal invariant. Now it is a rather straightforward consequence of our description of geometric bundles in Section 2 that $\dim \mathcal{H}^2(M, E^+(1))$ and $\dim \mathcal{H}^2(M, E^-(0))$ are conformally invariant, where

$$E^{\pm}(c) = \{u \in E^{\pm} \mid Cu = cu\},$$

but in other cases conformal invariance may fail. Using the conformal invariance of $\dim \mathcal{H}^2(M, E^-(0))$ the Gauss–Bonnet and Riemann–Roch formulas above are easy consequences of the results in [1], but also follow easily by a direct argument for surfaces with cylindrical ends as in Section 4 of [1].

The plan of the paper is as follows. In Section 1 we discuss the geometry of the ends in more detail. In Section 2 we characterize geometric bundles and determine their structure over the ends of M. In Section 3 we introduce operators which model the Dirac operator along the ends. In Section 4 we prove our main analytical lemmas. In Section 5 we investigate the essential spectrum and prove Theorem A. In Section 6 we compute the index formula from Theorem B.

The first author would like to thank the Department of Mathematics at the University of Pennsylvania for its hospitality during the spring of 1999, when a large part of the present work was completed. The first author presented the state of our work at the geometry festival, which took place in Philadelphia that year. At the time, we did have our main results on the essential spectrum. However, we were still missing some arguments in the discussion of the index and could only establish the index formulas under the additional assumption that the derivative of the Gauss curvature of M is uniformly bounded.

After we finished the first version of this article, which appeared as a preprint of the SFB 256 in Bonn in 1999 [3], we received a preprint of the article [15] by John Lott where related problems are discussed. One of his arguments lead to an improvement of the second assertion of Theorem 6.5, where now an equality of different spectra is claimed, replacing the inclusion relation of [3]. Except for this change and some additional comments in the text, the present version of the article is identical with [3]. We would like to thank Hermann Karcher for pointing out the occasionally hermetic presentation and for identifying some places, where additional remarks seemed helpful to clarify the arguments as well as the results. We hope that these additions will make the article more accessible.

We would also like to thank Charles Epstein for helpful discussions.

2 The Ends

By assumption each end U is diffeomorphic to $S \times (0,\infty)$, where $S = \mathbb{R}/\mathbb{Z}$ is the circle, and the metric is of the form $f^2ds^2 + dt^2$, where $f = f(s,t)$ is continuously differentiable with respect to s and twice continuously differentiable with respect to t. We recall that we assume that the area of U is finite. We have $f'' + Kf = 0$, where the *Gauss curvature* $K = K(s,t)$ is continuous and bounded by $-b^2 \leq K \leq -a^2 < 0$ for appropriate positive constants a, b. Here and below the prime $'$ denotes differentiation with respect to the variable t.

The curves $c_s(.) = (s, .)$ are unit speed geodesics. The circles $S_t = S \times \{t\}$ are perpendicular to the geodesics c_s. The length and geodesic curvature of S_t are given by

$$\bar{f}(t) = \int_0^1 f(s,t)ds \tag{1}$$

and

$$\kappa(s,t) = -f'(s,t)/f(s,t)\,, \tag{2}$$

respectively. Recall that κ satisfies the *Riccati equation*

$$\kappa' = \kappa^2 + K\,. \tag{3}$$

By the assumption on finite area f is the *stable solution* of the Jacobi equation. Hence

$$\begin{gathered} f(s,0)\exp(-bt) \leq f(s,t) \leq f(s,0)\exp(-at)\,, \\ a \leq \kappa(s,t) \leq b\,. \end{gathered} \tag{4}$$

We also introduce the *averaged geodesic curvature*

$$\bar{\kappa}(t) = \frac{1}{\bar{f}(t)} \int_0^1 \kappa(s,t) f(s,t)ds\,. \tag{5}$$

By (2),

$$\bar{\kappa}(t) = -\bar{f}'(t)/\bar{f}(t)\,. \tag{6}$$

Furthermore,

$$\begin{gathered} \bar{f}(0)\exp(-bt) \leq \bar{f}(t) \leq \bar{f}(0)\exp(-at)\,, \\ a \leq \bar{\kappa}(t) \leq b\,. \end{gathered} \tag{7}$$

Later it will be important to consider the *arc length* along S_t,

$$r(s,t) = \frac{1}{\bar{f}(t)} \int_0^s f(s,t)ds\,. \tag{8}$$

Then

$$r(0,t) = 0 \quad \text{and} \quad r(1,t) = 1\,. \tag{9}$$

Let ∇ be the Levi–Civita connection and define the orthonormal frame T, X as in the introduction. This frame is parallel in the T–direction and we have

$$\nabla_X T = -\kappa X \quad \text{and} \quad \nabla_X X = \kappa T\,. \tag{10}$$

3 The Bundles

In the surface case, the structure of the geometric operators under study is to a large extent determined by the structure of the bundles they are connected with; therefore, we examine the geometric bundles first and show that they can be written as direct sums of certain line bundles, depending only on two parameters. This is possible in view of the simple structure of the ends and the endomorphism $C = TXS(T, X)$ as in the introduction.

We fix an end $U = S \times (0, \infty)$of M and let $SO(U)$ be the $SO(2)$–principal bundle of oriented orthonormal frames over U. Our preferred frame $F = (T, X)$ is a global section of $SO(U)$ and determines a trivialization

$$U \times SO(2) \to SO(U)\,, (s,t,A) \mapsto F(s,t)A\,,$$

and a lift

$$\widetilde{SO}(U) := U \times \mathbb{R} \xrightarrow[\mathrm{id}\times\pi]{} U \times SO(2) = SO(U)$$

of $SO(U)$ to an $\mathbb{R}$–principal bundle over U, where $\pi(r)$ is counterclockwise rotation by the angle $2\pi r$.

The tangent bundle with its natural parallel complex structure — counterclockwise rotation by a right angle — is the complex line bundle associated to the representation $\exp(2\pi i r)$ of $\mathbb{R}$ on $\mathbb{C}$.

The section $\tilde{F}(s,t) = (s,t,0)$ is a lift of the frame $F = F(s,t)$ and determines lifts $\tilde{T} = \tilde{F}_* T$ of T and $\tilde{X} = \tilde{F}_* X$ of X along the image of $\tilde{F}$. In terms of $\tilde{T}$ and $\tilde{X}$, the connection form ω and curvature form Ω of the Levi-Civita connection are given by

$$\omega(\tilde{T}) = 0\,, \quad \omega(\tilde{X}) = -\kappa i \quad \text{and} \quad \Omega(\tilde{T},\tilde{X}) = -Ki\,.$$

Fix $c \in \mathbb{R}$ and $w \in \mathbb{C}$ with $|w| = 1$. Let $L(c, w)$ be the complex line bundle associated to $\widetilde{SO}(U)$ via the representation $\exp(2\pi i c r)$ of $\mathbb{R}$ on $\mathbb{C}$ with an atlas given by the holonomy relation $[s+1, t, 0, u] = [s, t, 0, wu]$. Since the representation of $\mathbb{R}$ is unitary, $L(c, w)$ has a natural Hermitian metric. Furthermore, ω determines a unitary connection on $L(c, w)$. It is easy to see that $L(c, w)$ is a geometric bundle; we prove a more precise statement in the lemma below. We also have

$$L(c,w) \otimes L(c',w') = L(c+c', ww')\,. \tag{11}$$

Lemma 3.1 *The line bundle $L(c, w)$ has a section Φ of norm 1 such that*

$$\Phi(s+1,t) = w\Phi(s,t)\,, \quad \nabla_T\Phi = 0\,, \quad \nabla_X\Phi = -\kappa i c\Phi\,, \quad R(T,X)\Phi = -Kic\Phi\,.$$

Proof. In the standard notation for bundles associated to principal bundles let $\Phi = \Phi(s,t) = [s,t,0,1]$. Then Φ has norm 1 and the asserted holonomy. Now under the representation $\exp(2\pi i c r)$, the connection form ω and curvature form Ω are multiplied by c. Hence Φ also satisfies the remaining equations.

We now describe the structure of geometric bundles over U. A similar description is also valid over all of M. However, for ease of presentation we stick to geometric bundles over U.

Lemma 3.2 *Let E be a geometric bundle over U. Then E is isomorphic to an orthogonal sum of line bundles $L(c, w)$.*

Proof. Let S be the parallel tensor with $R(T, X) = KS(T, X)$ as in the definition of geometric bundles. Since S is parallel and skew symmetric in T and X, $S(T, X)$ is a parallel field of endomorphisms of E over U. Hence E is the direct sum of the parallel and pairwise perpendicular subbundles

$$E(c) = \{u \in E \mid S(T, X)u = -icu\}, \quad c \in \mathbb{R}.$$

Hence we may assume that $E = E(c)$ for some fixed c.

Let $t > 0$ and Φ, Ψ be sections of $E|S_t$ solving

$$\nabla_X \Phi = -\kappa i c \Phi, \quad \nabla_X \Psi = -\kappa i c \Psi.$$

Then

$$\begin{aligned} X\langle \Phi, \Psi \rangle &= \langle \nabla_X \Phi, \Psi \rangle + \langle \Phi, \nabla_X \Psi \rangle \\ &= \langle -\kappa i c \Phi, \Psi \rangle + \langle \Phi, -\kappa i c \Psi \rangle = 0. \end{aligned}$$

Now extend Φ by parallel translation in the T-direction and recall that $[T, X] = \kappa X$, see (10). We get

$$\nabla_T(\nabla_X \Phi) = \nabla_{[T,X]}\Phi + R(T, X)\Phi = -\kappa^2 i c \Phi - K i c \Phi.$$

Using (3) we also have

$$\nabla_T(-\kappa i c \Phi) = -\kappa' i c \Phi = -(\kappa^2 + K) i c \Phi.$$

Hence $\nabla_X \Phi$ and $-\kappa i c \Phi$ solve the same differential equation. Hence $\nabla_X \Phi = -\kappa i c \Phi$ on all of U.

We conclude that there is an orthonormal frame $\Phi_1, \ldots, \Phi_m$ of E, where $m = \operatorname{rank} E$, such that $\nabla_T \Phi_j = 0$ and $\nabla_X \Phi_j = -\kappa i c \Phi_j$ for all j. It follows that the holonomy $H = (h_{ij})$, defined by

$$\Phi_j(s+1, t) = \sum h_{ij} \Phi_i(s, t),$$

is independent of (s, t) and unitary. Now an appropriate unitary change of frame with constant coefficients diagonalizes H.

We now turn to geometric Dirac bundles. Fix $c \in \mathbb{R}$, $\varepsilon \in \{-1, +1\}$ and $w \in \mathbb{C}$ with $|w| = 1$. Let

$$L(c, \varepsilon, w) = L(-\varepsilon c, w) \oplus L(\varepsilon(1 - c), w).$$

Viewing the Clifford bundle as the bundle associated to $\widetilde{SO}(U)$ via the representation $\exp(2\pi i r)$ on $\mathrm{Cl}\,\mathbb{R}^2$, we describe Clifford multiplication on $L(c, \varepsilon, w)$ by the following representation of $\mathrm{Cl}\,\mathbb{R}^2$ on $\mathbb{C}^2 = \mathbb{C} \oplus \mathbb{C}$,

$$\begin{pmatrix} 0 & -1 \\ 1 & 0 \end{pmatrix}, \quad \begin{pmatrix} 0 & \varepsilon i \\ \varepsilon i & 0 \end{pmatrix}, \quad \begin{pmatrix} -\varepsilon i & 0 \\ 0 & \varepsilon i \end{pmatrix}.$$

Here the matrices describe Clifford multiplication by $e_1, e_2, e_1 e_2 \in \mathrm{Cl}\,\mathbb{R}^2$ respectively. Furthermore, let $\alpha = \varepsilon\omega_{\mathbb{C}}$, where $\omega_{\mathbb{C}} = iTX$ is the complex area element. It is easy to see that $L(c, \varepsilon, w)$ is a graded Dirac bundle with $L^+(c, \varepsilon, w) = L(-\varepsilon c, w)$ and $L^-(c, \varepsilon, w) = L(\varepsilon(1-c), w)$. Moreover, it is immediate that

$$L(c, \varepsilon, w) \otimes L(c', w') = L(c - \varepsilon c', \varepsilon, ww'). \tag{12}$$

Lemma 3.3 *There are sections Φ^+ of $L^+(c, \varepsilon, w)$ and Φ^- of $L^-(c, \varepsilon, w)$ of norm 1 such that*

$$\begin{gathered} \Phi^\pm(s+1, t) = w\Phi^\pm(s, t), \quad T\Phi^\pm = \pm\Phi^\mp, \quad X\Phi^\pm = \varepsilon i \Phi^\mp, \\ \nabla_T \Phi^\pm = 0, \quad X\nabla_X \Phi^+ = -\kappa c \Phi^-, \quad X\nabla_X \Phi^- = \kappa(1-c)\Phi^+, \\ TXR(T, X)\Phi^+ = Kc\Phi^+, \quad TXR(T, X)\Phi^- = K(1-c)\Phi^-. \end{gathered}$$

Proof. Let Φ^+ be a section of $L(-\varepsilon c, w)$ as in Lemma 3.1 and set $\Phi^- = T\Phi^+$. It is immediate from the definition of Clifford multiplication that these sections have the asserted properties.

We now describe the structure of graded geometric Dirac bundles over U. As in the case of geometric bundles, there is a similar description for graded geometric Dirac bundles over all of M.

Lemma 3.4 *Let E be a graded geometric Dirac bundle over U with involution α. Then E is isomorphic to an orthogonal sum of bundles $L(c, \varepsilon, w)$.*

It is clear that the numbers $m(c, \varepsilon, w)$ from the introduction are equal to the number of times that the bundles $L(c, \varepsilon, w)$ occur in the decomposition of E over U in Lemma 3.4.

Proof (of Lemma 3.4). As in the introduction, let $TXR(T, X) = KC$. We have

$$C(v \cdot u) = v \cdot (1 - C)u,$$

for any $v \in TM$ and $u \in E$ with the same foot point [2]. Note that E^+ and E^- are invariant under R and Clifford multiplication by TX, hence $C(E^\pm) \subset E^\pm$.

The complex area element $\omega_{\mathbb{C}} = iTX$ commutes with the involution α of E and C and hence $E^\pm$ is the direct sum of the pairwise perpendicular and parallel subbundles

[2] For this conclusion we need that $K(s, t) \neq 0$ for some $(s, t) \in U$.

$$E^{\pm}(c,\varepsilon) = \{u \in E^{\pm} \mid C(u) = cu,\ \omega_{\mathbb{C}}u = \varepsilon u\}\,,$$

where $c \in \mathbb{R}$ and $\varepsilon \in \{+1,-1\}$. Hence E is the direct sum of the pairwise perpendicular and parallel graded Dirac subbundles

$$E(c,\varepsilon) = E^{+}(c,\varepsilon) \oplus E^{-}(1-c,-\varepsilon)\,.$$

Hence we may assume $E = E(c,\varepsilon)$. But then by Lemma 3.2, E^+ is isomorphic to the orthogonal sum of line bundles $L(-\varepsilon c, w)$. It follows easily that E is isomorphic to the orthogonal sum of the corresponding graded Dirac bundles $L(c,\varepsilon,w)$.

Examples 3.5 *1) The Clifford bundle* $\mathrm{Cl}\,U$ *together with the even–odd decomposition is equal to* $L(0,1,1) \oplus L(0,-1,1)$.

2) The chosen orientation determines a spin^c *structure on* U*. The corresponding spinor bundle is equal to* $L(0,1,1)$.

3) There are two spin structures on U*. The* trivial spin structure *is* $\widetilde{SO}(U)/2\mathbb{Z}$*. The spinor bundle associated to the trivial spin structure is* $L(1/2,1,1)$.

The correspondence $(s,t) \leftrightarrow \exp(-(t+is))$ *identifies* U *with the unit ball* B *with midpoint removed. The restriction of the unique spin structure of* B *to* U *is called the* nontrivial spin structure. *The spinor bundle associated to the nontrivial spin structure of* U *is equal to* $L(1/2,1,-1)$.

It follows from the above description of spinor bundles and (12) that we can factor graded geometric Dirac bundles with involution $\alpha = \omega_{\mathbb{C}}$ by spinor bundles,

$$L(c,1,w) = L(1/2,1,\pm 1) \otimes L(1/2-c,\pm w)\,. \tag{13}$$

4 The Model Operators

The simple structure of the geometric bundles described in the previous section implies a correspondingly simple structure of the geometric operators. But we emphasize that the model operators do not have constant coefficients; the whole difficulty of the analysis lies with this variability. A particular phenomenon due to the finiteness of the volume is the "explosion" of the spectrum of a certain coefficient operator, as explained in Proposition 4.1 below; this fact seems to be quite important.

Lemma 3.4 implies that the properties of the Dirac operator on a graded Dirac bundle E over an end $U = S \times (0,\infty)$ are determined by the properties of the Dirac operators on the bundles $L(c,\varepsilon,w)$.

For $t > 0$ write

$$C^1(S_t)_w = \{u \in C^1([0,1] \times \{t\}) \mid u(1,t) = u(0,t)/w\}\,.$$

For an interval $I \subset (0,\infty)$ write correspondingly

$$C^1(S \times I)_w = \{u \in C^1([0,1] \times I) \mid u(1,t) = u(0,t)/w\}\,.$$

We let $C_0^1(S \times I)_w$ be the subspace of $u \in C^1(S \times I)_w$ which have compact support in $[0,1] \times I$.

We now transform the Dirac operators on the bundles $L(c, \varepsilon, w)$ to operators which are convenient for our purposes. Define matrices

$$\gamma = \begin{pmatrix} 0 & -1 \\ 1 & 0 \end{pmatrix} \quad \text{and} \quad \tau = \begin{pmatrix} 1 & 0 \\ 0 & -1 \end{pmatrix} . \tag{14}$$

Then the map

$$\begin{aligned} \Psi : C_0^1\big(S \times (0,\infty)\big)_w \otimes \mathbb{C}^2 &\to C_0^1(U, L(c,\varepsilon,w)) , \\ u = (u^+, u^-) &\mapsto u^+ \Phi^+ + u^- \Phi^- , \end{aligned}$$

transforms the Dirac operator to the *model operator*

$$\begin{aligned} D_\infty := \Psi^{-1} D \Psi &= \gamma\big((\partial_t - \kappa/2) + (\varepsilon i X + \kappa(1/2 - c))\tau\big) \\ &= \gamma\big((\partial_t - \kappa/2) + A + \kappa B\big) , \end{aligned} \tag{15}$$

where

$$A = A(t) = \begin{pmatrix} A^+(t) & 0 \\ 0 & -A^+(t) \end{pmatrix} = \begin{pmatrix} \varepsilon i X & 0 \\ 0 & -\varepsilon i X \end{pmatrix} \tag{16}$$

and

$$B = (1/2 - c) \begin{pmatrix} 1 & 0 \\ 0 & -1 \end{pmatrix} = \begin{pmatrix} 1/2 - c & 0 \\ 0 & c - 1/2 \end{pmatrix} . \tag{17}$$

The operator D_∞ depends on the Riemannian metric on $U = U_k$ and on the parameters c, ε and w.

We recall that the area element on U is $f ds dt$. For $t > 0$ and functions $u, v \in C^1(S_t)_w \otimes \mathbb{C}^2$ we let

$$(u,v)_t = \int_S \langle u(s,t), v(s,t) \rangle f(s,t) ds , \tag{18}$$

where $\langle .\,,.\rangle$ denotes the inner product in $\mathbb{C}^2$. For an interval $I \subset (0,\infty)$ and functions $u, v \in C^1(S \times I)_w \otimes \mathbb{C}^2$ we let

$$(u,v)_I = \int_I (u,v)_t dt = \int_I \int_S \langle u(s,t), v(s,t) \rangle f(s,t) ds dt , \tag{19}$$

denote the L^2–inner product of u and v. We use analogous notation for the corresponding norms. Note that Ψ extends to a unitary map

$$\Psi : L^2\big((S \times (0,\infty); f ds dt\big) \otimes \mathbb{C}^2 \to L^2(U, L(c,\varepsilon,w)) .$$

We now turn to the family of operators $A = A(t)$ from (15) which splits into operators $A^+(t) = \varepsilon i X$ and $A^-(t) = -\varepsilon i X$ according to (16). We recall that the length of S_t is $\bar{f}(t)$ and that X has norm one. Hence we have the following lemma.

Proposition 4.1 *The operator* iX *is essentially self–adjoint as an operator in* $L^2(S_t)$ *with domain* $C^1(S_t)_w$. *Its spectrum is discrete and consists of the eigenvalues*

$$-2\pi(\rho + k)/\bar{f}(t)\,, \quad k \in \mathbb{Z}\,,$$

where we write $w = \exp(2\pi i \rho)$. *The corresponding eigenspaces are spanned by the functions* $\exp(2\pi i(\rho + k)r/\bar{f}(t))$, *where* $r = r(s,t)$ *denotes arc length along* S_t *as in (8).*

We note that 0 is in the spectrum of $A(t)$ if and only if $w = 1$ and that the nonzero eigenvalues grow like $1/\bar{f}(t) \to \infty$. Thus except for the possible eigenvalue 0, there is a larger and larger spectral gap for $A(t)$ about 0. This observation will be of central importance below.

We let $P_0 = P_0(t)$ be the orthogonal projection onto the kernel of $A(t)$. If $w \neq 1$, then $P_0 = 0$ and if $w = 1$, then

$$P_0 u(s,t) = \bar{u}(t) = \frac{1}{\bar{f}(t)} \int_S u(s,t) f(s,t) ds\,. \tag{20}$$

Note that $\bar{u}$ is continuously differentiable if u is and that

$$\operatorname{supp} \bar{u} \subset \{t \mid \operatorname{supp} u \cap S_t \neq \emptyset\}\,. \tag{21}$$

In the case $w = 1$ we will also consider the following symmetric *model operator*

$$D_{\infty 0} := P_0 D_\infty P_0 = \gamma\big((\partial_t - \bar{\kappa}/2) + \bar{\kappa}(1/2 - c)\tau\big)\,, \tag{22}$$

with domain $C_0^1(0,\infty) \otimes \mathbb{C}^2$ in $L^2\big((0,\infty), \bar{f}dt\big) \otimes \mathbb{C}^2$. Here we identify functions in $C_0^1(0,\infty) \otimes \mathbb{C}^2$ with functions on $S \times (0,\infty)$ which do not depend on s, compare (20). Note that the substitution $\tilde{u} = \sqrt{\bar{f}}u$ defines a unitary equivalence of $L^2\big((0,\infty), \bar{f}dt\big) \otimes \mathbb{C}^2$ and $L^2\big((0,\infty), dt\big) \otimes \mathbb{C}^2$ and transforms $D_{\infty 0}$ into the Dirac system

$$\gamma\big(\partial_t + \bar{\kappa}(1/2 - c)\tau\big) \tag{23}$$

with domain $C_0^1(0,\infty) \otimes \mathbb{C}^2$.

5 The Weapons

Now we are ready to derive the estimates which are needed for our main results. Besides an argument based on integration by parts (in Lemma 5.1) they take the form of a priori estimates for the model operator, as one might expect. Basically, they tell us that the model operator behaves well under reduction by P_0 (Lemma 5.3), that the "high energy part" (in the language of [15]) $(I - P_0)u$ of a solution u is well controlled in norm by the graph norm (Lemma 5.4), and that the "low energy part" of the solution is also controlled by the graph norm if $c \neq \frac{1}{2}$ (Lemma 5.5).

Of these facts, we will make a twofold use: they will be used in the next section to derive information on the essential spectrum from the study of $D_{\infty 0}$, and to derive an index formula in the last section, using an appropriate vanishing theorem.

We fix an end $U = S \times (0, \infty)$ and consider a model operator D_∞ as in (15). We recall that the area element on U is $f ds dt$. The following lemma is now immediate from the definition of the L^2–inner product (19) and partial integration.

Lemma 5.1 *Let $I = [t_0, t_1] \subset (0, \infty)$ and assume that $u, v \in C^1(S \times I)_w \otimes \mathbb{C}^2$. Then*

$$(\partial_t u, v)_I = (u, -\partial_t v + \kappa v)_I + \{(u, v)_{t_1} - (u, v)_{t_0}\}.$$

In particular, $\gamma(\partial_t - \kappa/2)$ is symmetric on $C^1_0(S \times (0, \infty))_w \otimes \mathbb{C}^2$.

Now P_0 is self–adjoint, see (20). Hence the following formulas are immediate from Lemma 5.1.

Lemma 5.2 *On $C^1(S \times (0, \infty))_w \otimes \mathbb{C}^2$ we have*

$$P_0 \partial_t P_0 = \partial_t P_0 \quad \text{and} \quad P_0 \kappa P_0 = \bar{\kappa} P_0 .$$

Therefore we also get

$$(1 - P_0)\partial_t P_0 = 0, \qquad (1 - P_0)\kappa P_0 = (\kappa - \bar{\kappa}) P_0,$$
$$P_0 \partial_t (1 - P_0) = P_0(\kappa - \bar{\kappa}), \qquad P_0 \kappa (1 - P_0) = P_0 (\kappa - \bar{\kappa}).$$

Our first weapon is the following lemma.

Lemma 5.3 *On $C^1(S \times (0, \infty))_w \otimes \mathbb{C}^2$ we have*

$$P_0 D_\infty P_0 = \gamma\big((\partial_t - \bar{\kappa}/2) + \bar{\kappa} B\big) P_0,$$
$$(1 - P_0) D_\infty P_0 = \gamma(\kappa - \bar{\kappa})(B - 1/2) P_0,$$
$$P_0 D_\infty (1 - P_0) = \gamma P_0 (\kappa - \bar{\kappa})(B + 1/2)(1 - P_0).$$

Proof. By definition, $A P_0 = P_0 A = 0$. Furthermore, $[P_0, \gamma] = [P_0, \tau] = 0$ and $B\gamma + \gamma B = 0$. The formulas are now immediate from Lemma 5.2.

By $P_>(t)$ and $P_<(t)$ we denote the spectral projection in $L^2(S_t)_w \otimes \mathbb{C}^2$ corresponding to the eigenvalues of $A(t)$ which are > 0 and < 0, respectively. The following lemma is our second weapon.

Lemma 5.4 *Let $I = [t_0, t_1] \subset (0, \infty)$ be a compact interval and suppose $u \in C^1(S \times I)_w \otimes \mathbb{C}^2$ satisfies $P_>(t_0) u_{t_0} = 0$ and $P_<(t_1) u_{t_1} = 0$. Then*

$$\|Au\|_I \le \|D_\infty u\|_I + C_1 \|u\|_I,$$

where $C_1 = b \max(|c|, |1 - c|)$.

Since the nonzero eigenvalues of $A(t)$ get larger and larger as t increases, Lemma 5.4 tells us that the "high energy part" of u is very small if the right hand side is a priori bounded and t_0 is very large. The first instance, where we use this phenomenon, is the proof of Theorem 6.5, where we discuss bounds on the essential spectrum.

Proof (of Lemma 5.4). By approximation we may assume that u is C^2. Now by (10), $[\partial_t, X] = \kappa X$. Therefore $\partial_t Au = A\partial_t u + \kappa Au$ and hence, by Lemma 5.1,

$$(\partial_t u, Au)_I = (u, -A\partial_t u)_I + \{(u, Au)_{t_1} - (u, Au)_{t_0}\}.$$

Now A is symmetric. Therefore our assumption on u implies

$$2\,\mathrm{Re}(\partial_t u, Au)_I = \{(u, Au)_{t_1} - (u, Au)_{t_0}\} \geq 0.$$

Hence

$$\begin{aligned}\|D_\infty u\|_I \|Au\|_I &\geq \mathrm{Re}(D_\infty u, \gamma Au)_I \\ &= \mathrm{Re}(\partial_t u, Au)_I + \|Au\|_I^2 + \mathrm{Re}(\kappa(B - 1/2)u, Au)_I \\ &\geq \|Au\|_I^2 - b\max(|c|, |1-c|)\|u\|_I \|Au\|_I .\end{aligned}$$

Recall that $P_0 = 0$ if $w \neq 1$. We assume now that $w = 1$. Then for $u = (u^+, u^-) \in C_0^1(S \times (0,\infty)) \otimes \mathbb{C}^2$ we have $P_0 u = (\bar{u}^+, \bar{u}^-)$, where $\bar{u}^\pm$ is defined as in (20). We set $P_0^+ u = (\bar{u}^+, 0)$ and $P_0^- u = (0, \bar{u}^-)$ so that $P_0 = P_0^+ + P_0^-$. Now our third and last weapon reads as follows.

Lemma 5.5 *Assume $w = 1$ and $c \neq 1/2$. Let $I = [t_0, t_1] \subset (0,\infty)$ be a compact interval and suppose $u \in C^1(S \times I) \otimes \mathbb{C}^2$ satisfies*

$$\begin{aligned} P_0^-(t_0)u_{t_0} &= 0\,, P_0^+(t_1)u_{t_1} = 0 \quad \text{if } 1/2 < c\,, \\ P_0^+(t_0)u_{t_0} &= 0\,, P_0^-(t_1)u_{t_1} = 0 \quad \text{if } 1/2 > c\,. \end{aligned}$$

Then

$$a|1/2 - c|\,\|P_0 u\|_I \leq \|P_0 D_\infty u\|_I + C_2\,\|(1-P_0)u\|_I\,,$$

where $C_2 = b|1/2 - c| + (b-a)/2$.

We recall that the boundary conditions of Atiyah, Patodi and Singer are nonlocal and actually concern spectral projections of the operators $A(0)$. In our discussion of the index, we will need a variant of their boundary condition. Lemma 5.5 gives precisely the estimates we need, and for the correct boundary conditions.

Proof (of Lemma 5.5). We set $\tilde{\tau} = \mathrm{sign}(1/2 - c)\,\tau$, with τ from (14). Now we have $P_0 A = AP_0 = 0$ by the definition of P_0. Hence

$$\begin{aligned}(P_0 D_\infty u, \gamma\tilde{\tau}P_0 u)_I &= ((\partial_t - \kappa/2)P_0 u, \tilde{\tau}P_0 u)_I + ((\partial_t - \kappa/2)(1-P_0)u, \tilde{\tau}P_0 u)_I \\ &\quad + (\kappa BP_0 u, \tilde{\tau}P_0 u)_I + (\kappa B(1-P_0)u, \tilde{\tau}P_0 u)_I\,.\end{aligned}$$

Next we estimate the four terms on the right hand side. By Lemma 5.1, the first term is

$$((\partial_t - \kappa/2)P_0 u, \tilde{\tau} P_0 u)_I = -(P_0 u, (\partial_t - \kappa/2)\tilde{\tau} P_0 u)_I + (P_0 u, \tilde{\tau} P_0 u)_{t_1} - (P_0 u, \tilde{\tau} P_0 u)_{t_0} .$$

By our assumption

$$(P_0 u, \tilde{\tau} P_0 u)_{t_1} - (P_0 u, \tilde{\tau} P_0 u)_{t_0} = \begin{cases} \|\bar{u}^-\|_{t_1} + \|\bar{u}^+\|_{t_0} & \text{if } 1/2 < c, \\ \|\bar{u}^+\|_{t_1} + \|\bar{u}^-\|_{t_0} & \text{if } 1/2 > c, \end{cases}$$

so the above computation gives

$$\mathrm{Re}((\partial_t - \kappa/2)P_0 u, \tilde{\tau} P_0 u)_I \geq 0 .$$

By Lemma 5.2, the second term can be estimated as follows,

$$\begin{aligned} |((\partial_t - \kappa/2)(1 - P_0)u, \tilde{\tau} P_0 u)_I| &= |(P_0 \frac{\kappa - \bar{\kappa}}{2}(1 - P_0)u, \tilde{\tau} P_0 u)_I| \\ &\leq \frac{b - a}{2} \|(1 - P_0)u\|_I \|P_0 u\|_I . \end{aligned}$$

Recall that $B = (1/2 - c)\tau$. Hence the absolute value of the fourth term can be estimated by

$$|(\kappa B(1 - P_0)u, \tilde{\tau} P_0 u)_I| \leq b|1/2 - c| \|(1 - P_0)u\|_I \|P_0 u\|_I .$$

The crucial estimate involves the third term, but this is immediate from the definition of B and $\tilde{\tau}$:

$$(\kappa B P_0 u, \tilde{\tau} P_0 u)_I \geq a|1/2 - c| \|P_0 u\|_I^2 .$$

We conclude that

$$\begin{aligned} &\|P_0 D_\infty u\|_I \|P_0 u\|_I \geq \\ &\qquad a|1/2 - c| \|P_0 u\|_I^2 - (b|1/2 - c| + (b - a)/2) \|(1 - P_0)u\|_I \|P_0 u\|_I . \end{aligned}$$

This finishes the proof of the lemma.

6 The Essential Spectrum

Now we are going to use our estimates to characterize the essential spectrum of the Dirac operator under study. We first remark that the essential spectrum "lives at infinity" - as one should expect - hence is determined by the end structure. Next we exploit the a priori estimates to show that the exploding spectral subspaces on the cross section (the circle) do not contribute to the essential spectrum; hence we are left with an ordinary differential operator.

Since M is complete, the Dirac operator D with domain $C_0^1(M, E)$ is essentially self–adjoint in $L^2(M, E)$, see [16, p.117]. We denote by $\bar{D}$ the closure of D and recall that the *spectrum* spec $\bar{D}$ of $\bar{D}$ is contained in $\mathbb{R}$.

Definition 6.1 *A real number* λ *belongs to the* discrete spectrum $\operatorname{spec}_d \bar{D}$ *of* $\bar{D}$ *if* λ *is an eigenvalue of finite multiplicity of* $\bar{D}$ *and an isolated point of* $\operatorname{spec} \bar{D}$*. The complement* $\operatorname{spec}_e \bar{D} = \operatorname{spec} \bar{D} \setminus \operatorname{spec}_d \bar{D}$ *is called the* essential spectrum *of* $\bar{D}$.

In particular, $\bar{D}$ is Fredholm if and only if $0 \notin \operatorname{spec}_e \bar{D}$. An easy consequence of Weyl's Criterion and the Rellich Lemma, the so–called Decomposition Principle, says that $\operatorname{spec}_e \bar{D}$ does not depend on compact parts of M, see below.

We say that a sequence (u_n) in $C_0^1(M, E)$ is a *special Weyl sequence (for E)* if

$$\lim_{n \to \infty} \|u_n\| = 1 \quad \text{and} \quad \operatorname{supp} u_n \to \infty \quad \text{as } n \to \infty . \tag{24}$$

Now the following well known characterization of $\operatorname{spec}_e \bar{D}$ by special Weyl sequences is a version of the Decomposition Principle.

Lemma 6.2 $\lambda \in \operatorname{spec}_e \bar{D}$ *if and only if there is a special Weyl sequence* (u_n) *with*

$$\lim_{n \to \infty} \|(D - \lambda)u_n\| = 0 .$$

Corollary 6.3 *Assume that there is a compact subset K of M and a constant* $c_K > 0$ *such that* $\|Du\| \geq c_K \|u\|$ *for all* $u \in C_0^1(M \setminus K, E)$*. Then*

$$\operatorname{spec}_e \bar{D} \cap (-c_K, c_K) = \emptyset .$$

It is convenient to introduce the following notation. Let $\tilde{E}$ be a Hermitian vector bundle over a complete Riemannian manifold $\tilde{M}$ with compact boundary. Let $\tilde{D}$ be a symmetric first order differential operator on $C_0^1(\tilde{M}, \tilde{E})$. Set

$$\begin{aligned} \operatorname{spec}_\infty \tilde{D} = \{\lambda \in \mathbb{R} \mid & \text{ there is a special Weyl sequence } (u_n) \\ & \text{for } \tilde{E} \text{ with } \lim_{n \to \infty} \|(\tilde{D} - \lambda)u_n\| = 0\} . \end{aligned} \tag{25}$$

Then $\operatorname{spec}_e \bar{D} = \operatorname{spec}_\infty D$ by Lemma 6.2, but, in general, (25) ignores possible contributions from the boundary. The following result is now obvious from Lemma 3.4.

Lemma 6.4 *We have* $\operatorname{spec}_e \bar{D} = \cup \operatorname{spec}_\infty D_\infty$*, where the union is over all* k, c, ε*, and* w *with* $m_k(c, \varepsilon, w) > 0$.

It remains to discuss $\operatorname{spec}_\infty D_\infty$. In the case $w = 1$, the model operator $D_{\infty 0}$ from (22) comes into play.

Theorem 6.5 *Fix an end* U *and constants* c, ε, *and* w*. Then we have:*
1) If $w \neq 1$, *then* $\operatorname{spec}_\infty D_\infty = \emptyset$.
2) If $w = 1$, *then* $\operatorname{spec}_\infty D_\infty = \operatorname{spec}_\infty D_{\infty 0}$*. Furthermore,*

$$\operatorname{spec}_\infty D_\infty \cap (-a|1/2 - c|, a|1/2 - c|) = \emptyset .$$

Proof. Let $\lambda \in \operatorname{spec}_\infty D_\infty$ and (u_n) be a special Weyl sequence for D_∞ with $(D_\infty - \lambda)u_n \to 0$. Now for any $N > 0$ there is a constant $t_0 > 0$ such that the nonzero eigenvalues of $A(t)$ are of absolute value $\geq N$ for all $t \geq t_0$. Hence $(1 - P_0)u_n \to 0$, by Lemma 5.4. This completes the proof of the first assertion.

Assume now that $w = 1$. By what we just said we have $(1 - P_0)u_n \to 0$, hence $\|P_0 u_n\| \to 1$ and therefore, by Lemma 5.5,

$$\liminf_{t\to\infty} \|D_\infty u_n\| \geq \liminf_{t\to\infty} \|P_0 D_\infty u_n\| \geq a|1/2 - c|\,.$$

This establishes the inequality in the second assertion. Furthermore, by Lemma 5.3,

$$P_0(D_\infty - \lambda)(1 - P_0)u_n = \gamma P_0(\kappa - \bar{\kappa})(B + 1/2)(1 - P_0)u_n \to 0\,.$$

Therefore

$$(P_0 D_\infty P_0 - \lambda)P_0 u_n = P_0(D_\infty - \lambda)P_0 u_n \to 0\,.$$

But then $(P_0 u_n)$ is a special Weyl sequence with $(D_{\infty 0} - \lambda)P_0 u_n \to 0$. Hence λ is in $\operatorname{spec}_\infty D_{\infty 0}$, hence $\operatorname{spec}_\infty D_\infty \subset \operatorname{spec}_\infty D_{\infty 0}$.

Conversely, let $\lambda \in \operatorname{spec}_\infty D_{\infty 0}$ and let (u_n) be a special Weyl sequence with $\lim_{n\to\infty}(D_{\infty 0} - \lambda)u_n = 0$. By passing to a subsequence if necessary we may assume that $\operatorname{supp} u_n \subset S \times (n+1, \infty)$. We decompose

$$\begin{aligned} D_\infty &= (1 - P_0)D_\infty(1 - P_0) + D_{\infty 0} + P_0 D_\infty(1 - P_0) + (1 - P_0)D_\infty P_0 \\ &=: D_{\infty 1} + D_{\infty 0} + D_{\infty 01} + D_{\infty 10}\,. \end{aligned}$$

Note that, by Lemma 5.3, $D_{\infty 01}$ and $D_{\infty 10}$ are bounded multiplication operators. Now we want to construct a special Weyl sequence (w_n) with $w_n = u_n + v_n$ and $P_0 v_n = 0$, such that the following properties hold as $n \to \infty$:

$$\operatorname{supp} v_n \to \infty\,, \tag{26}$$

$$\|v_n\| \to 0\,, \tag{27}$$

$$(D_\infty - \lambda)(w_n) \to 0\,. \tag{28}$$

These properties clearly imply that $\lambda \in \operatorname{spec}_\infty D_\infty$, hence we obtain $\operatorname{spec}_\infty D_{\infty 0} \subset \operatorname{spec}_\infty D_\infty$, as desired. Next we compute

$$(D_\infty - \lambda)(u_n + v_n) = (D_{\infty 0} - \lambda)u_n + (D_{\infty 1} - \lambda)v_n + D_{\infty 01}v_n + D_{\infty 10}u_n\,. \tag{29}$$

To verify (5.10), we select a sequence of smooth functions ϕ_n with $0 \leq \phi_n \leq 1$, $\phi_n(x) = 1$ if $0 \leq x \leq n$, and $\phi_n(x) = 0$ if $x \geq n+1$, and satisfying $|d\phi_n(x)| \leq 2$. Next we denote by $D^n_{\infty 1}$ the self-adjoint extension of $D_{\infty 1}$ in $L^2(S \times (n, \infty))_w \otimes \mathbb{C}^2$ defined by Atiyah-Patodi-Singer boundary conditions at $S \times \{n\}$. Then it follows from Proposition 4.1 and Lemma 5.4 that

$$\lim_{n\to\infty} \|(D^n_{\infty 1} - \lambda)^{-1}\| = 0\,. \tag{30}$$

Now we let

$$v_n := (\phi_n - 1)(D_{\infty 1}^n - \lambda)^{-1}(D_{\infty 10} u_n) \,. \tag{31}$$

Then the condition (5.8) is clearly satisfied, and it follows from (5.12) that (5.9) holds, too. Finally, (5.11) implies (5.10) since ϕ_n has uniformly bounded gradient and $\operatorname{supp} u_n$ and $\operatorname{supp} \phi_n$ are disjoint.

The lower bound on the essential spectrum in part two of the theorem follows in certain cases from the work of McKean [17].

7 The Index

To prove the desired index formulas in the Fredholm case, we now proceed as follows. We separate the ends from the rest of the surface by introducing boundary conditions at a cross section S, sufficiently close to infinity; these conditions are defined by spectral projections of the coefficient operator and determine two elliptic problems, one in the interior and one in the exterior domain of M defined by S. The interior and exterior boundary conditions are related in the natural way, and one will expect additivity in the sense that the indices of the two boundary value problems add up to the index of the operator on M. This fact is certainly known to the experts, for a proof we refer to [7]. The index of the interior problem can then be computed using the already mentioned fundamental result of [1] plus the addendum on the variation of boundary conditions due to Agranovic and Dynin, whereas the exterior problem is more difficult in general. But our estimates will imply a vanishing theorem for the right boundary conditions; it then only remains to compute the limit of all terms in the index formula, as S tends to infinity.

We are ready to discuss the calculation of the index of D^+ in detail. As usual, the ends of M come with an index k. For each end $U = U_k$, we also enumerate the line bundles $L(c, \varepsilon, w)$ which occur in the decomposition of $E|U$ by an additional index j, $L_{kj} = L(c_{kj}, \varepsilon_{kj}, w_{kj})$, where $1 \le j \le m = \dim E^+$. These indices will also be used for other objects attached to L_{kj} if necessary.

Throughout this section we assume that $m_k(1/2, \varepsilon, 1) = 0$ for all k. In other words, we assume that $w_{kj} \neq 1$ whenever $c_{kj} = 1/2$. Then the Dirac operator $\bar{D}$ is Fredholm, by Theorem A. Consider an end $U = U_k = S \times (0, \infty)$ and set

$$U_t = U_{k,t} = S \times (t, \infty) \,, \quad S_t = S_{k,t} = S \times \{t\} \,. \tag{32}$$

Furthermore set

$$V_t = \cup_k U_{k,t} \,, \quad M_t = M \setminus V_t \,. \tag{33}$$

Then M_t is a compact surface with smooth boundary ∂M_t, consisting of the pairwise disjoint circles $S_{k,t}$. We let $\bar{U}_{k,t}$ and $\bar{V}_t$ denote the closures of $U_{k,t}$ and V_t, respectively.

Our aim is to reduce the computation of the index of D^+ to a boundary value problem on M_t, where $t > 0$ is sufficiently large.

By an *elliptic boundary value problem* we mean a pseudodifferential boundary condition which is well–posed in the sense of Seeley [20].

Lemma 7.1 *Let $t > 0$. Then any self–adjoint extension of D over M_t or $\bar{V}_t$, defined by an elliptic boundary value problem at ∂M_t is Fredholm.*

Proof. In Theorem 4.1 of [7] Brüning and Lesch characterize elliptic boundary value problems in terms of orthogonal projections and show that they are Fredholm. This implies the result for M_t.

The local analysis used for the result just mentioned implies a 'decomposition principle' for self–adjoint extensions of D over $\bar{V}_t$, hence the second assertion since $\bar{D}$ is Fredholm, by Theorem 6.5.

We choose $t_0 >> 0$ and introduce boundary conditions at t_0. To that end we write a section u over U_k as

$$u = \sum_j \left(u_{kj}^+ \Phi_{kj}^+ + u_{kj}^- \Phi_{kj}^- \right) , \tag{34}$$

where Φ_{kj}^+, Φ_{kj}^- is a frame of L_{kj} as in Lemma 3.3. We recall that $P_{kj,>}(t_0)$ and $P_{kj,<}(t_0)$ denote the spectral projections in $L^2(S_{k,t_0})_{w_{kj}} \otimes \mathbb{C}^2$ corresponding to the eigenvalues of $A_{kj}(t_0)$ which are > 0 and < 0, respectively. Since A_{kj} decomposes into two operators according to the decomposition $u = (u^+, u^-)$, we have a corresponding decomposition $P_{kj,>} = P_{kj,>}^+ + P_{kj,>}^-$ and similarly for $P_{kj,<}$. We also recall the orthogonal projection $P_{kj,0}(t_0)$ onto the kernel of $A_{kj}(t_0)$ in $L^2(S_{k,t_0})_{w_{kj}} \otimes \mathbb{C}^2$ and the decomposition $P_{kj,0} = P_{kj,0}^+ + P_{kj,0}^-$ as in Lemma 5.5.

In their discussion of the index problem for Dirac operators on compact manifolds with boundary, Atiyah, Patodi, and Singer used the projections

$$P_{kj,APS} = P_{kj,<} + P_{kj,0}^+ \tag{35}$$

on the component S_{kt_0} of ∂M_{t_0}, assuming in addition that the metric is a product near the boundary, i.e., $f(s,t) = \bar{f}(t_0)$ for $s \in [0,1]$ and t near t_0, see [1]. In our discussion we need a variation of this. We set

$$P_{kj}(t_0) = \begin{cases} P_{kj,<}(t_0) + P_{kj,0}^+(t_0) \text{ if } 1/2 < c\,, \\ P_{kj,<}(t_0) + P_{kj,0}^-(t_0) \text{ if } 1/2 > c\,. \end{cases} \tag{36}$$

The identity

$$\gamma P_{kj}(t_0) = (1 - P_{kj}(t_0))\gamma \tag{37}$$

where γ is as in (14), implies that D with each of the following two domains

$$\mathcal{D}_{\text{int}} = \{ u \in C^1(M_{t_0}, E) \mid P_{kj}(t_0)(u_{kj}|S_{k,t_0}) = 0 \text{ for all } k, j \}$$
$$\mathcal{D}_{\text{ext}} = \{ u \in C_0^1(\bar{V}_{t_0}, E) \mid (1 - P_{kj}(t_0))(u_{kj}|S_{k,t_0}) = 0 \text{ for all } k, j \}$$

is symmetric. We denote the corresponding operators by D_{int} and D_{ext}, respectively. Then the boundary conditions are elliptic. To see this, we invoke again [7, Theorem 4.1]. This requires a localization in the standard form used in [7, (4.2)]: the map

$$u \mapsto (f_{t_0}/f_t)^{1/2} u$$

transforms D_{kj} into the operator

$$D_{kj} = \gamma(\partial_t + \tilde{A}(t)), \tag{38}$$

on the Hilbert space $L^2((S \times (0,\infty), f_{t_0} ds dt)$. Here

$$\tilde{A}(t_0) = A(t_0) + \kappa(t_0)B =: \tilde{A} = \begin{pmatrix} \tilde{A}_+ & 0 \\ 0 & -\tilde{A}_+ \end{pmatrix}.$$

Next we need to determine a projection $P_+(\tilde{A})$ with the properties listed in (3.10) of [BL], and we have to show that $(P_+(\tilde{A}), P(t_0))$ forms a Fredholm pair. Now it is clear that the projection

$$P_+(\tilde{A}) = \begin{pmatrix} P_{\geq 0}(\tilde{A}_+) & 0 \\ 0 & P_{<0}(\tilde{A}_+) \end{pmatrix}$$

of Atiyah, Singer and Patodi is an admissible choice. Since B is bounded and $P(t_0)$ differs by a finite rank projection from the analogous projection $P_+(A(t_0))$, the Fredholm property follows from standard perturbation theory. It follows also from Theorem 4.1 in [7] (with an obvious modification for D_{ext}) that both D_{int} and D_{ext} are essentially self–adjoint. By Lemma 7.1, they are both Fredholm.

The boundary projections respect the decomposition $E = E^+ \oplus E^-$, hence we have the subdomains $\mathcal{D}^+_{\text{int}}$ and $\mathcal{D}^+_{\text{ext}}$ and the corresponding restrictions

$$D^+_{\text{int}} : \mathcal{D}^+_{\text{int}} \to L^2(M_{t_0}, E^-),$$
$$D^+_{\text{ext}} : \mathcal{D}^+_{\text{ext}} \to L^2(\bar{V}_{t_0}, E^-).$$

The proof of our index formula below rests on the following result.

Theorem 7.2 (see [7]) $\operatorname{ind} D^+ = \operatorname{ind} D^+_{\text{int}} + \operatorname{ind} D^+_{\text{ext}}$.

The proof of this formula consists in showing that the transmission condition at ∂M_{t_0} for sections in $H^1(M, E^+)$ can be deformed into the above boundary conditions in the definition of D_{int} and D_{ext} without affecting the index. The important point is that the interior boundary condition above is independent of the exterior one, so that they can be handled separately. Another important feature of our choice of boundary condition is that the exterior contribution to the index can be computed explicitly: it is zero if t_0 is sufficiently large.

Lemma 7.3 *For t_0 sufficiently large,* $\ker D_{\text{ext}} = 0$. *In particular,* $\operatorname{ind} D^+_{\text{ext}} = 0$.

Proof. It suffices to consider the (self–adjoint) closure of the operator $D_{kj,\infty} =: D_{kj}$ on $\mathcal{D}_{kj,\text{ext}}$, that is, on $C_0^1(S \times [t_0, \infty))_{w_{kj}} \otimes \mathbb{C}^2$ with the boundary condition $(1 - P_{kj}(t_0))$. Let $C_3 > 0$ be a lower bound for the absolute value of the nonzero eigenvalues of the operator $A_{kj}(1)$. Then $C_3\bar{f}(1)/\bar{f}(t_0)$ is a lower bound for the nonzero eigenvalues of $A_{kj}(t)$ for all $t \geq t_0$, by Proposition 4.1, hence for any $u \in C_0^1([t_0, \infty))_{w_{kj}} \otimes \mathbb{C}^2$ satisfying the boundary condition we have

$$C_3 \frac{\bar{f}(1)}{\bar{f}(t_0)} \|(1 - P_0)u\|_{[t_0,\infty)} \leq \|Au\|_{[t_0,\infty)}$$
$$\leq C_4 \left(\|D_{kj}u\|_{[t_0,\infty)} + \|(1 - P_0)u\|_{[t_0,\infty)} \right),$$

by Lemmas 5.4 and 5.5, where C_4 is a constant independent of k, j, and t_0. Since $\mathcal{D}_{kj,\text{ext}}$ is dense in the domain of the closure of D_{kj}, the above inequality and the inequality in Lemma 5.5 persist to hold in that domain. On the other hand, $\bar{f}(t) \to 0$ as $t \to \infty$ so

$$C_3\bar{f}(1) > C_4\bar{f}(t_0)$$

if t_0 is sufficiently large. Thus if $u \in \ker D_{kj}$, then $(1 - P_0)u = 0$ and so, by Lemma 5.5, also $u = 0$.

We now come to the proof of the asserted index formula. By Theorem 7.2 and Lemma 7.3 we have

$$\operatorname{ind} D^+ = \operatorname{ind} D^+_{\text{int}}, \tag{39}$$

and hence it remains to compute $\operatorname{ind} D^+_{\text{int}}$. Now in the standard versions of the index theorem for manifolds with boundary, it is assumed that near the boundary the manifold is a product. To arrive at such a situation, we deform the given metric in a small neighborhood of t_0 by replacing f with

$$f^\alpha = (1 - \alpha)f + \alpha(\phi\bar{f}(t_0) + (1 - \phi)f), \quad \alpha \in [0, 1], \tag{40}$$

where ϕ is a suitable cut–off function near t_0. Using the representation (38), we see that the family D^α_{int} is graph continuous on $\mathcal{D}_{\text{int}}$, hence has constant index. Hence we can, and will, assume from now on that near the boundary, the metric of M_{t_0} is a product. Thus we are finally in the situation considered in [1], except for a variation in the boundary condition. This is taken care of by a result of Agranovich-Dynin, see Theorem 23.1 in [4] or Theorem 4.2 in [7].

Theorem 7.4 *Denote by ω_{ind} the index form of D. Then*

$$\operatorname{ind} D^+_{\text{int}} = \int_{M_{t_0}} \omega_{\text{ind}} - \frac{1}{2}\sum_{k,j} \dim\ker A^+_{kj}(t_0) + \frac{1}{2}\sum_{k,j} \eta_{kj}(t_0)$$
$$+ \sum_{k,j} \operatorname{ind}(P^+_{kj}(t_0) : \operatorname{im} P^+_{APS}(t_0) \to \operatorname{im} P^+_{kj}(t_0)),$$

where $\eta_{kj}(t_0)$ denotes the η–invariant of $A^+_{kj}(t_0)$ and where P_{APS} denotes the Atiyah–Singer–Patodi spectral projection.

It remains to explain and evaluate the different terms on the right hand side of this formula.

We start with the index form ω_{ind}. We could compute ω_{ind} by using the local structure of graded geometric Dirac bundles as in Section 2. However, there is also the following way: By the Local Index Theorem, we have

$$\omega_{\mathrm{ind}}(p) = \lim_{t\to 0} \mathrm{tr}_E[\omega_{\mathbb{C}} e^{-tD^2}(p,p)]\,,$$

hence the computation of ω_{ind} is a local problem. Therefore we may consider an open contractible subset $W \subset M$, for which we choose an orientation. If M is oriented, we assume that the orientations of W and M coincide. The field C is parallel, hence the subbundles $E^{\pm}$ split over W as a direct sum of the pairwise orthogonal and parallel subbundles

$$E^{\pm}(c,\varepsilon) = \{u \in E^{\pm} \mid Cu = cu\,,\ \omega_{\mathbb{C}} u = \varepsilon u\}\,,$$

where $c \in \mathbb{R}$ and $\varepsilon \in \{-1,+1\}$. Now the subbundles

$$E(c,\varepsilon) = E^+(c,\varepsilon) \oplus E^-(1-c,-\varepsilon)\,.$$

are graded Dirac subbundles of E, hence we may assume that $E = E(c,\varepsilon)$. We discuss the case $\varepsilon = +1$ first. Then

$$E^+ = \{u \in E \mid \omega_{\mathbb{C}} u = u\}\,.$$

Let Σ be the spinor bundle associated to the spin structure of W. Then we have

$$E = \Sigma \otimes F \quad \text{with} \quad F = \mathrm{Hom}_{\mathrm{Cliff}}(\Sigma, E)\,,$$

where $\mathrm{Hom}_{\mathrm{Cliff}}(\Sigma, E)$ is the bundle of homomorphisms from Σ to E which are linear over Clifford multiplication, a parallel subbundle of $\mathrm{Hom}(\Sigma, E)$ with the canonical connection. Now the curvature endomorphism C_Σ on $\Sigma^{\pm}$ is multiplication by $K/2$. By our assumption on E, the curvature endomorphism C_E on $E^{\pm}$ is multiplication by Kc and $K(1-c)$, respectively. It follows easily that for any oriented orthonormal frame X, Y of W, $R_F(X,Y)$ is multiplication by $Ki(c-1/2)$. Hence

$$\omega_{\mathrm{ind}} = \frac{1}{2\pi}(1/2 - c)\dim F\, KdA = \frac{1}{2\pi}(1/2-c)\dim E^+ KdA\,,$$

by the formula for the index form of twisted Dirac operators as explained in [1].

In the case where $\varepsilon = -1$ we reverse the roles of E^+ and E^-, then we are back in the previous case but ω_{ind} changes sign. We get

$$-\omega_{\mathrm{ind}} = \frac{1}{2\pi}(1/2-(1-c))\dim E^- KdA = -\frac{1}{2\pi}(1/2-c)\dim E^+ KdA\,.$$

In conclusion,

$$\omega_{\text{ind}} = \frac{1}{2\pi}\{m/2 - \text{tr}(C^+)\}KdA\,,$$

and this gives the first term of the claimed index formula.

Now the second term in the index formula 7.4 is obviously equal to $-1/2$ times $\#\{(k,j) \mid w_{kj} = 1\}$. Furthermore, note that for each pair (k,j) the projection $P^+_{kj} : \text{im}\, P^+_{APS} \to \text{im}\, P^+_{kj}$ is surjective, by (35) and (36). It has trivial kernel if $w_{kj} \neq 1$ or if $w_{kj} = 1$ and $c_{kj} > 1/2$. The kernel is of dimension 1 if $w_{kj} = 1$ and $c_{kj} < 1/2$. This gives the second term of the claimed index formula.

If $w_{kj} = 1$, then the spectrum of $A^+_{kj}(t_0)$ is symmetric about 0 and hence $\eta_{kj}(t_0) = 0$ in this case. If $w_{kj} \neq 1$, we write $w_{kj} = \exp(2\pi i\rho)$ with $0 < \rho = \rho_{kj} < 1$. Then if $\varepsilon_{kj} = 1$, we get from Proposition 4.1

$$\eta_{kj}(s) = -\rho^{-s} - \sum_{k\geq 1}\{(k+\rho)^{-s} - (k-\rho)^{-s}\}\,.$$

If $\varepsilon_{kj} = -1$, we get the corresponding negative of the right hand side. Now

$$\lim_{s\to 0}\sum_{k\geq 1}\{(k+\rho)^{-s} - (k-\rho)^{-s}\} = -2\rho\,.$$

This concludes the proof of Theorem B.

Remark 7.5 *The proof of Theorem B gives a somewhat more general result. We only need to assume that the given graded Dirac bundle E is geometric along the ends of M. Then the same index formula holds, except that the first term on the right hand side has to be replaced by the integral over the corresponding index form.*

References

1. Atiyah, M., Patodi, V., Singer, I. (1975): Spectral asymmetry and Riemannian geometry. I. Math. Proc. Cambridge Philos. Soc., **77**, 43–69
2. Bär, C. (2000): The Dirac operator on hyperbolic manifolds of finite volume. J. Differential Geometry, **54**, 439–488
3. Ballmann, W., Brüning, J. (1999): On the spectral theory of surfaces with cusps. Preprint 616, SFB 256, Bonn
4. Booß–Bavnbek, B., Wojciechowski, K. (1993): Elliptic boundary problems for Dirac operators. Birkhäuser, Basel
5. Brüning, J. (1990): L^2-index theorems on certain complete manifolds. J. Differential Geometry, **32**, 491–532
6. Brüning, J. (1992): On L^2-index theorems on complete manifolds of rank-one-type. Duke Math. J., **66**, 257–309
7. Brüning, J., Lesch, M. (1999): Spectral theory of boundary value problems for Dirac type operators. Geometric aspects of partial differential equations (Roskilde, 1998). Contemporary Math., **242**, 203–215

8. Carron, G. (2001): Théorèmes de l'indice sur les variétés non-compactes. J. Reine Angew. Math., **541**, 81–115
9. Carron, G. (2001): Un théorème de l'indice relatif. Pacific J. Math., **198**, 81–107
10. Donnelly, H., Xavier, F. (1984): On the differential form spectrum of negatively curved Riemannian manifolds. Amer. J. Math., **106**, 169–185
11. Eberlein, P. (1980): Lattices in manifolds of nonpositive curvature. Annals of Math., **111**, 435–476
12. Faddeev, L. (1967): The eigenfunction expansion of Laplace's operator on the fundamental domain of a discrete group on the Lobačevskiĭ plane. (Russian) Trudy Moskov. Mat. Obšč., **17**, 323–350
13. Heintze, E., Im Hof, H.C. (1977): Geometry of horospheres. J. Differential Geometry, **12**, 481–491
14. Hopf, E. (1940): Statistik der Lösungen geodätischer Probleme vom unstabilen Typus. II. Math. Annalen, **117**, 590–608
15. Lott, J. (2001): On the spectrum of a finite-volume negatively-curved manifold. Amer. J. of Math., **123**, 185–205
16. Lawson, H.B., Michelsohn, M.L. (1989): Spin geometry. Princeton University Press, Princeton
17. McKean, H.P. (1970): An upper bound to the spectrum of Δ on a manifold of negative curvature. J. Differential Geometry, **4**, 359–360.
18. Müller, W. (1987): Manifolds with cusps of rank one, spectral theory and L^2 index theorem. Lecture Notes in Math., **1244**, Springer-Verlag, Berlin
19. Müller, W (1992): Spectral geometry and scattering theory for certain complete surfaces of finite volume. Inventiones math., **109**, 265–305
20. Seeley, R.T. (1969): Topics in pseudo–differential operators. In: Conference on pseudo–differential operators 1968, C.I.M.E., Edizione Cremonese, Roma
21. Selberg, A. (1962): Discontinuous groups and harmonic analysis. Proceedings ICM, Stockholm, 177–189
22. Stern, M. (1984): L^2–index theorems on warped products. PhD Thesis, Princeton University, Princeton

The Dirac Determinant of Spherical Space Forms

Christian Bär and Sven Schopka

Universität Hamburg, FB Mathematik, Bundesstr. 55, 20146 Hamburg
{baer, schopka}@math.uni-hamburg.de

Summary. The ζ-regularized determinants of the Dirac operator and of its square are computed on spherical space forms. On S^2 the determinant of Dirac operators twisted by a complex line bundle is also calculated.

1 Introduction

In classical field theory the physical fields φ have to satisfy certain field equations or, equivalently, have to be critical points of some action functional $\mathcal{S}[\varphi]$. When passing to quantum field theory this requirement is discarded and one instead looks at partition functions defined as a functional integral

$$Z = \int \exp(-\mathcal{S}[\varphi])\mathcal{D}\varphi$$

over the space of all fields. It is now a serious problem that in most cases the space of fields is infinite dimensional and the "measure" $\mathcal{D}\varphi$ does not exist. If the space of fields is a Hilbert space and the action is of the form $\mathcal{S}[\varphi] = \frac{1}{2}(L\,\varphi, \varphi)$ where L is a positive self-adjoint operator, then one can *define* the functional integral by

$$Z = (\det L)^{-1/2}.$$

This is motivated by the fact that if the Hilbert space is of finite dimension N, then

$$\int_{\mathbb{R}^N} \exp(-\frac{1}{2}(L\,x, x))dx^1 \cdots dx^N = (2\pi)^{N/2} \det(L)^{-1/2}.$$

One can then regard $\mathcal{D}x = \frac{dx^1}{\sqrt{2\pi}} \cdots \frac{dx^N}{\sqrt{2\pi}}$ as a renormalized Lebesgue measure. At first it may seem that one has simply shifted the problem since in the physical case the operator L will typically have unbounded spectrum and the product of its eigenvalues will diverge. So one has to find a reasonable definition for the determinant of L. We will look at two closely related definitions for the determinant, the one most commonly used is coming from the ζ-function of L while the proper time regularized determinant makes essential use of the asymptotic heat kernel expansion. This will be explained in detail in the next section. In order to distinguish these regularized determinants from the usual ones in finite dimensions we will denote them by DET and by $\mathrm{DET}_{\mathrm{p.t.}}$.

Even though these concepts are standard in quantum field theory not many of these determinants have been computed explicitly. Due to the somewhat involved nature of their definition an explicit computation can be achieved only in cases of high symmetry. In this paper we provide such explicit computation for the Dirac operator and its square on spherical space forms. The square of the Dirac operator D^2 is a non-negative self-adjoint elliptic differential operator acting on spinor fields. In Theorem 4.7 we present a formula for the determinant of D^2 on the n-dimensional sphere, $n \geq 2$, with its standard Riemannian metric of constant curvature 1. The determinant is given by a linear combination of the Riemann ζ-function and its first derivative evaluated at certain non-positive integers. The proper time regularized determinant is given in Corollary 4.10.

Since the Dirac operator D itself has a spectrum unbounded from above and from below one has to find a reasonable definition for its determinant. Since regularized determinants are in general not multiplicative,

$$\mathrm{DET}(A\,B) \neq \mathrm{DET}(A)\,\mathrm{DET}(B),$$

it is not sufficient to simply set $\mathrm{DET}(D) := \sqrt{\mathrm{DET}(D^2)}$. In (3) we define $\mathrm{DET}(D)$ to be of the form $\exp(i\varphi)\sqrt{\mathrm{DET}(D^2)}$ where the ζ-invariant of D^2 and the η-invariant of D enter into the phase φ. This definition is motivated in Section 2. We give a simple expression for the *multiplicative anomaly* $\exp(i\varphi)$ in Theorem 4.9. It turns out to be trivial for odd-dimensional spheres. The case of the 1-dimensional sphere needs special treatment since S^1 has two different spin structures. The results for S^1 are contained in Theorem 4.11 and turn out to depend on the choice of spin structure. The numerical values strongly suggest that the Dirac determinant on the n-dimensional sphere tends to 1 as the dimension n goes to ∞,

$$\lim_{n\to\infty} \mathrm{DET}(D; S^n) = 1.$$

We have no rigorous proof for this conjecture.

We then pass to spherical space forms, i. e. to quotients of the sphere. We first look at the determinant of D^2 and compute the *covering anomaly*

$$\frac{\mathrm{DET}(D^2; G\backslash S^n)}{\mathrm{DET}(D^2; S^n)^{\frac{1}{|G|}}}$$

in Theorem 5.12. Here only odd dimensions n have to be considered. For the example of real-projective space it turns out that the covering anomaly is trivial,

$$\mathrm{DET}(D^2; \mathbb{RP}^n) = \sqrt{\mathrm{DET}(D^2; S^n)}.$$

In Theorem 5.15 we give the formula for the determinant of D itself on spherical space forms. For the computation of the multiplicative anomaly we need to recall a formula for the η-invariant of spherical space forms which we do in Theorem 3.4.

In the final section we restrict our attention to S^2 but we twist the Dirac operator with a complex line bundle. The Chern number of such a bundle can be interpreted as a topological charge. In Theorem 6.16 we compute the eigenvalues of these twisted Dirac operators and in Theorem 6.17 we compute their determinants. As a special case this includes the determinant of the Laplace-Beltrami operator acting on functions.

It is always understood that the spherical space forms are equipped with their standard metrics of constant curvature 1. On S^2, S^4, and S^6 the standard metric is known to have an interesting criticality property for the determinant of D^2 [4, 5, 13, 15].

Acknowledgement. The first author has been partially supported by the Research and Training Networks "EDGE" and "Geometric Analysis" funded by the European Commission.

2 Determinant of the Dirac Operator

Throughout this paper let L be a non-negative self-adjoint elliptic differential operator of second order acting on sections of a Riemannian or Hermitian vector bundle over a compact n-dimensional Riemannian manifold M. See [10] and [14] for the basics on spin and spectral geometry. The most common regularization scheme for making sense of the determinant of L is the definition via its *ζ-function* which reads

$$\mathrm{DET}(L; M) := \exp\left(-\zeta_L'(0)\right) \tag{1}$$

with

$$\zeta_L(s) := \sum_{\lambda \in \mathrm{Spec}(L)-\{0\}} \frac{1}{\lambda^s}.$$

This way of defining the determinant is motivated by the observation that for positive numbers λ_k we have $\ln \lambda_k = -\frac{d}{ds}\lambda_k^{-s}|_{s=0}$ and $\exp\left(\sum_k \ln \lambda_k\right) = \prod_k \lambda_k$. The series defining $\zeta_L(s)$ converges for $\Re(s) > n/2$. It can be shown that ζ_L extends meromorphically to $\mathbb{C}$ and that it has no pole at $s = 0$.

Note that this definition of the determinant excludes all information on the possible eigenvalue zero. It has to be taken into account separately.

There is the following expression of $\zeta_L(s)$ in terms of its Mellin transform:

$$\zeta_L(s) = \frac{1}{\Gamma(s)} \int_0^\infty t^{s-1}\, \mathrm{tr}(e^{-tL} - P_0) dt \tag{2}$$

where P_0 denotes the projector onto the kernel of L.

So far we have a definition of the determinant only in case the operator L is non-negative. This can be applied to the square of the Dirac operator, $L = D^2$, but not to the Dirac operator itself. We adopt the following convention for the determinant of D [20]:

$$\text{DET}(D;M) := \exp\left(i\frac{\pi}{2}(\zeta_{D^2}(0) - \eta_D(0))\right) \cdot \exp\left(-\frac{\zeta'_{D^2}(0)}{2}\right) \tag{3}$$

where the *η-function* is given by

$$\eta_D(s) := \sum_{\lambda \in \text{Spec}(D)-\{0\}} \frac{\text{sgn}\lambda}{|\lambda|^s}$$

for $\Re(s) >> 0$. What enters into the phase of the determinant is the *η-invariant* $\eta(M) := \eta_D(0)$. It measures the spectral asymmetry and is zero in case of a symmetric spectrum.

Why is the definition in (3) reasonable? Denote the positive eigenvalues of D by λ_k and the negative ones by $-\nu_k$. We can formally write down a ζ-function for D itself as follows:

$$\begin{aligned}\zeta_D(s) &= \sum_k \lambda_k^{-s} + \sum_k (-1)^{-s}\nu_k^{-s} \\ &= \sum_k \left(\frac{\lambda_k^{-s}+\nu_k^{-s}}{2} + \frac{\lambda_k^{-s}-\nu_k^{-s}}{2}\right) + (-1)^{-s}\sum_k \left(\frac{\lambda_k^{-s}+\nu_k^{-s}}{2} - \frac{\lambda_k^{-s}-\nu_k^{-s}}{2}\right) \\ &= \frac{\zeta_{D^2}(\frac{s}{2}) + \eta_D(s)}{2} + (-1)^{-s}\frac{\zeta_{D^2}(\frac{s}{2}) - \eta_D(s)}{2}.\end{aligned}$$

This is a meromorphic function well-defined up to the sign ambiguity in $(-1)^{-s} = e^{\mp i\pi s}$. Choosing $(-1)^{-s} = e^{-i\pi s}$ we get

$$\exp\left(-\zeta'_D(0)\right) = \exp\left(i\frac{\pi}{2}(\zeta_{D^2}(0) - \eta_D(0))\right) \cdot \exp\left(-\frac{\zeta'_{D^2}(0)}{2}\right)$$

thus yielding (3).

We want to compare the above considerations with another way of defining a regularized determinant. For $\varepsilon > 0$ define

$$\ln \text{DET}_\varepsilon(L;M) := -\int_\varepsilon^\infty t^{-1}\,\text{tr}(e^{-tL} - P_0)dt. \tag{4}$$

Since the integrand decays exponentially fast the integral is finite for every positive ε. One says that $\text{DET}_\varepsilon(L;M)$ is obtained from the (infinite) determinant of L by cutoff in proper time [17, p. 170]. To motivate this definition note that (2) gives for $\Re(s) >> 0$

$$\begin{aligned}\zeta'_L(s) = &-\frac{\Gamma'(s)}{\Gamma(s)^2}\int_0^\infty t^{s-1}\text{tr}(e^{-tL} - P_0)dt \\ &+\frac{1}{\Gamma(s)}\int_0^\infty \ln(t)t^{s-1}\text{tr}(e^{-tL} - P_0)dt.\end{aligned}$$

After replacing the lower integral boundary 0 by $\varepsilon > 0$ we can take the limit $s \to 0$. The Laurent expansion

$$\Gamma(s) = \frac{1}{s} + \mathrm{O}(1) \tag{5}$$

gives $\lim_{s\to 0} \frac{1}{\Gamma(s)} = 0$ and $\lim_{s\to 0}\left(-\frac{\Gamma'(s)}{\Gamma(s)^2}\right) = \frac{d}{ds}\Gamma(s)^{-1}|_{s=0} = 1$. This way we obtain the right hand side of (4).

The proper time regularized determinant is now defined as the "finite part" of $\mathrm{DET}_\varepsilon(L;M)$ for $\varepsilon \searrow 0$. To make this more precise we look at the asymptotic expansion of the heat kernel of $L = D^2$ for $t \searrow 0$

$$\operatorname{tr} e^{-tL} \sim \sum_{k=0}^{\infty} \Phi_{k-\frac{n}{2}}(L)\, t^{k-\frac{n}{2}}.$$

In particular, we can plug

$$\operatorname{tr} e^{-tL} = \sum_{k=0}^{[n/2]} \Phi_{k-\frac{n}{2}}(L) t^{k-\frac{n}{2}} + R(t) \tag{6}$$

into (4) where the remainder term $R(t)$ is of order $\mathrm{O}(t)$ if n is even and of order $\mathrm{O}(\sqrt{t})$ if n is odd. After splitting the integral in (4) into one over $[\varepsilon,1]$ and one over $[1,\infty]$ we get

$$\begin{aligned}
\ln \mathrm{DET}_\varepsilon(L;M) &= -\int_\varepsilon^1 \left(\sum_{k=0}^{[n/2]} \Phi_{k-\frac{n}{2}}(L) t^{k-\frac{n}{2}-1} + R(t)t^{-1} - \dim\ker(L)t^{-1}\right) dt \\
&\quad - \int_1^\infty t^{-1}\operatorname{tr}(e^{-tL} - P_0)dt \\
&= -\sum_{k=0}^{[(n-1)/2]} \frac{\Phi_{k-\frac{n}{2}}(L)}{k-\frac{n}{2}} + \sum_{k=0}^{[(n-1)/2]} \frac{\Phi_{k-\frac{n}{2}}(L)\varepsilon^{k-\frac{n}{2}}}{k-\frac{n}{2}} + (\Phi_0(L) \\
&\quad - \dim\ker(L))\ln\varepsilon - \int_\varepsilon^1 R(t)t^{-1}dt - \int_1^\infty t^{-1}\operatorname{tr}(e^{-tL} - P_0)dt.
\end{aligned}$$

Here we use the convention $\Phi_0(L) = 0$ if n is odd. The terms $\frac{\Phi_{k-\frac{n}{2}}(L)}{k-\frac{n}{2}}\varepsilon^{k-\frac{n}{2}}$, $k \le [(n-1)/2]$, and $(\Phi_0(L) - \dim\ker(L))\ln\varepsilon$ explode for $\varepsilon \searrow 0$ unless they vanish. We now abandon the terms divergent in the limit $\varepsilon \searrow 0$ and we are led to

Definition 2.1 *The* proper time regularized determinant $\mathrm{DET}_{\mathrm{p.t.}}(L;M)$ *is defined by*

$$\begin{aligned}
\ln \mathrm{DET}_{\mathrm{p.t.}}(L;M) := \sum_{k=0}^{[(n-1)/2]} \frac{\Phi_{k-\frac{n}{2}}(L)}{\frac{n}{2}-k} - \int_0^1 R(t)t^{-1}dt \\
- \int_1^\infty t^{-1} tr(e^{-tL} - P_0)dt
\end{aligned} \tag{7}$$

where $\Phi_{k-\frac{n}{2}}(L)$ and $R(t)$ are as in (6).

The following relation (cf. [17, (28.11), p. 171]) shows that the two regularizations do not differ significantly.

Proposition 2.2 *The ζ-regularized determinant* $\mathrm{DET}(L; M)$ *and the proper time regularized determinant* $\mathrm{DET}_{\mathrm{p.t.}}(L; M)$ *are related by*

$$\ln \mathrm{DET}(L; M) - \ln \mathrm{DET}_{\mathrm{p.t.}}(L; M) = \Gamma'(1)(\Phi_0(L) - \dim \ker L) = \Gamma'(1)\zeta_L(0).$$

Proof. In order to compare (7) with the ζ-determinant we insert the heat kernel expansion (6) into (2) and do the corresponding integral splitting into one over $[0, 1]$ and one over $[1, \infty]$.

$$\zeta_L(s) = \frac{1}{\Gamma(s)}\left(\sum_{k=0}^{[(n-1)/2]} \frac{\Phi_{k-\frac{n}{2}}(L)}{s - k + \frac{n}{2}} + \frac{\Phi_0(L) - \dim \ker L}{s} + \int_0^1 R(t)t^{s-1}dt \right.$$
$$\left. + \int_1^\infty t^{s-1}\mathrm{tr}(e^{-tL} - P_0)dt\right). \tag{8}$$

Using (5) we find

$$\zeta_L(0) = \Phi_0(L) - \dim \ker L.$$

Equation (8) gives us for the determinant

$$\ln \mathrm{DET}(L; M) = -\zeta_L'(0) = \sum_{k=0}^{[(n-1)/2]} \frac{\Phi_{k-\frac{n}{2}}(L)}{\frac{n}{2} - k} + \Gamma'(1)\zeta_L(0) - \int_0^1 R(t)t^{-1}dt$$
$$- \int_1^\infty t^{-1}\mathrm{tr}(e^{-tL} - P_0)dt \tag{9}$$

Comparing (7) and (9) we get Proposition 2.2.

3 Dirac Spectrum of Spherical Space Forms

The Dirac spectrum of the sphere S^n, $n \geq 2$, with constant curvature 1 has been computed by different methods in [2, 7, 18, 19]. The eigenvalues are

$$\pm\left(\frac{n}{2} + k\right), \tag{10}$$

$k \in \mathbb{N}_0$, with multiplicity $2^{[n/2]}\begin{pmatrix} k+n-1 \\ k \end{pmatrix}$. For $n \geq 2$ the sphere is simply connected, hence has only one spin structure. It is given by the standard projection $\mathrm{Spin}(n+1) \to \mathrm{Spin}(n+1)/\mathrm{Spin}(n) = S^n$.

We now look at spherical space forms $M = G\backslash S^n$ where G is a finite fixed point free subgroup of $\mathrm{SO}(n+1)$. Spin structures correspond to homomorphisms $\varepsilon : G \to \mathrm{Spin}(n+1)$ such that the diagram

$$\begin{array}{ccc} & & \mathrm{Spin}(n+1) \\ & \overset{\varepsilon}{\nearrow} & \downarrow \\ G & \longrightarrow & \mathrm{SO}(n+1) \end{array}$$

commutes. The corresponding spin structure is given by $\varepsilon(G)\backslash\mathrm{Spin}(n+1) \to G\backslash(\mathrm{Spin}(n+1)/\mathrm{Spin}(n)) = M$. Thus spinors on M correspond to $\varepsilon(G)$-invariant spinors on S^n.

Since any eigenspinor on M can be lifted to S^n all eigenvalues of M are also eigenvalues of S^n, hence of the form (10). To know the spectrum of M one must compute the multiplicities μ_k of $\frac{n}{2}+k$ and μ_{-k} of $-(\frac{n}{2}+k)$. It is convenient to encode them into two power series, so-called *Poincaré series*

$$F_+(z) := \sum_{k=0}^{\infty} \mu_k z^k,$$
$$F_-(z) := \sum_{k=0}^{\infty} \mu_{-k} z^k.$$

To formulate the result recall that in even dimension $2m$ the spinor representation is reducible and can be decomposed into two half spinor representations

$$\mathrm{Spin}(2m) \to \mathrm{Aut}(\Sigma_{2m}^{\pm}),$$

$\Sigma_{2m} = \Sigma_{2m}^+ \oplus \Sigma_{2m}^-$. Denote their characters by $\chi^{\pm} : \mathrm{Spin}(2m) \to \mathbb{C}$.

Theorem 3.3 ([2]). *Let $M = G\backslash S^n$, $n = 2m-1$, be a spherical space form with spin structure given by $\varepsilon : G \to \mathrm{Spin}(2m)$. Then the eigenvalues of the Dirac operator are $\pm(\frac{n}{2}+k), k \geq 0$, with multiplicities determined by*

$$F_+(z) = \frac{1}{|G|}\sum_{g\in G} \frac{\chi^-(\varepsilon(g)) - z\cdot\chi^+(\varepsilon(g))}{\det(1_{2m} - z\cdot g)},$$
$$F_-(z) = \frac{1}{|G|}\sum_{g\in G} \frac{\chi^+(\varepsilon(g)) - z\cdot\chi^-(\varepsilon(g))}{\det(1_{2m} - z\cdot g)}.$$

Note that only odd-dimensional spherical space forms are of interest because in even dimensions real projective space is the only quotient and in this case it is not even orientable.

This kind of encoding the spectrum of M is in fact well-suited for computation of the η-invariant.

Theorem 3.4 ([3]). *Let $M = G\backslash S^{2m-1}$ be a spherical space form with spin structure given by $\varepsilon : G \to \mathrm{Spin}(2m)$. Then the η-invariant of M is given by*

$$\eta(G\backslash S^{2m-1}) = \frac{2}{|G|} \sum_{g\in G-\{1_{2m}\}} \frac{(\chi^- - \chi^+)(\varepsilon(g))}{\det(1_{2m} - g)}.$$

For the convenience of the reader and since we will need the arguments again we briefly sketch the proof. We define the *θ-functions*

$$\begin{aligned}\theta_\pm(t) &:= e^{-\frac{n}{2}t} \cdot F_\pm(e^{-t}) \\ &= \frac{e^{-(m+\frac{1}{2})t}}{|G|} \sum_{g\in G} \frac{\chi^\mp(\varepsilon(g)) - e^{-t}\cdot\chi^\pm(\varepsilon(g))}{\det(1_{2m} - e^{-t}\cdot g)}.\end{aligned} \tag{11}$$

Since G acts freely the non-trivial elements $g \in G$ do not have 1 as an eigenvalue and thus $\det(1_{2m} - g) \neq 0$. Therefore only the summand for $g = 1_{2m}$ contributes to the pole of $\theta_\pm$ at $t = 0$. Hence for $\theta := \theta_+ - \theta_-$ the poles at $t = 0$ cancel, θ is holomorphic at $t = 0$ with

$$\theta(0) = \frac{2}{|G|} \sum_{g\in G-\{1_{2m}\}} \frac{(\chi^- - \chi^+)(\varepsilon(g))}{\det(1_{2m} - g)}. \tag{12}$$

Next we observe that

$$\theta_+(t) = \sum_{k=0}^{\infty} \mu_k e^{-(n/2+k)t} = \sum_{\lambda>0} e^{-\lambda t},$$

where the last sum is taken over all positive eigenvalues. Similarly for θ_-,

$$\theta_-(t) = \sum_{k=0}^{\infty} \mu_{-k} e^{-(n/2+k)t} = \sum_{\lambda<0} e^{\lambda t}.$$

Application of the Mellin transformation yields

$$\eta(s) = \frac{1}{\Gamma(s)} \int_0^\infty \theta(t) t^{s-1} dt.$$

Therefore

$$\eta(G\backslash S^{2m-1}) = \lim_{s\to 0} \frac{1}{\Gamma(s)} \int_0^\infty \theta(t)t^{s-1}dt = \mathrm{Res}_{s=0}\left(\int_0^\infty \theta(t)t^{s-1}dt\right).$$

Since θ decays exponentially fast for $t \to \infty$ the function $s \mapsto \int_1^\infty \theta(t)t^{s-1}dt$ is holomorphic at $s = 0$. Thus

$$\eta(G\backslash S^{2m-1}) = \mathrm{Res}_{s=0}\left(\int_0^1 \theta(t)t^{s-1}dt\right) = \theta(0)$$

which together with (12) proves Theorem 3.4.

Compare this to Goette's computation of the *equivariant* η-invariant of spheres in [11, Satz 6.10]. See also [9, 10] where the η-invariant of all twisted signature operators on spherical space forms is determined and used to compute their K-theory. In [8] the η-invariant of all twisted Dirac operators on 3-dimensional spherical space forms is computed.

Using Theorem 3.4 it is easy to discuss real projective space.

Corollary 3.5 ([3]) *For $n \geq 2$ real projective space $\mathbb{RP}^n$ is spin if and only if $n \equiv 3$ mod 4, in which case it has exactly two spin structures. The η-invariant for the Dirac operator is given by*

$$\eta(\mathbb{RP}^n) = \pm 2^{-m}, \quad n = 2m - 1,$$

where the sign depends on the spin structure chosen.

Let us now look at lens spaces. For $\alpha \in \mathbb{R}$ define the rotation matrix

$$R(\alpha) := \begin{pmatrix} \cos(2\pi\alpha) & -\sin(2\pi\alpha) \\ \sin(2\pi\alpha) & \cos(2\pi\alpha) \end{pmatrix}.$$

For natural numbers $q, p_1, \ldots, p_m$ where p_j and q are coprime for all j, let G be the cyclic subgroup of $\mathrm{SO}(2m)$ generated by

$$\begin{pmatrix} R(p_1/q) & & 0 \\ & \ddots & \\ 0 & & R(p_m/q) \end{pmatrix}.$$

We denote the resulting lens space by $L(q, p_1, \ldots, p_m) := G\backslash S^{2m-1}$. Now Theorem 3.3 takes the form (compare [1, Satz 5.3]):

a) If q is odd, then $L(q, p_1, \ldots, p_m)$ has exactly one spin structure and the Poincaré series are given by

$$F_+(z) = \frac{1}{q}\sum_{k=0}^{q-1} \frac{\sum\limits_{\substack{\varepsilon_1\cdots\varepsilon_m\\=(-1)^{m+1}}} \exp(\pi i k(q+1)\sum_j \varepsilon_j p_j/q) - z \cdot \sum\limits_{\substack{\varepsilon_1\cdots\varepsilon_m\\=(-1)^{m}}} \exp(\pi i k(q+1)\sum_j \varepsilon_j p_j/q)}{\prod\limits_{j=1}^{m} (\exp(2\pi i k p_j/q) - z)(\exp(-2\pi i k p_j/q) - z)},$$

$$F_-(z) = \frac{1}{q}\sum_{k=0}^{q-1} \frac{\sum\limits_{\substack{\varepsilon_1\cdots\varepsilon_m\\=(-1)^{m}}} \exp(\pi i k(q+1)\sum_j \varepsilon_j p_j/q) - z \cdot \sum\limits_{\substack{\varepsilon_1\cdots\varepsilon_m\\=(-1)^{m+1}}} \exp(\pi i k(q+1)\sum_j \varepsilon_j p_j/q)}{\prod\limits_{j=1}^{m} (\exp(2\pi i k p_j/q) - z)(\exp(-2\pi i k p_j/q) - z)}.$$

b) If q is even and $p_1 + \ldots + p_m$ is odd (i. e. if m is odd), then $L(q, p_1, \ldots, p_m)$ has no spin structures.

c) If q is even and $p_1 + \ldots + p_m$ is even (i. e. if m is even), then $L(q, p_1, \ldots, p_m)$ has two different spin structures. The Poincaré series for the first spin structure are given by

$$F_+(z) = \frac{1}{q}\sum_{k=0}^{q-1} \frac{\sum\limits_{\substack{\varepsilon_1\cdots\varepsilon_m\\=(-1)^{m+1}}} \exp(\pi i k \sum\limits_j \varepsilon_j p_j/q) - z \cdot \sum\limits_{\substack{\varepsilon_1\cdots\varepsilon_m\\=(-1)^{m}}} \exp(\pi i k \sum\limits_j \varepsilon_j p_j/q)}{\prod\limits_{j=1}^{m} (\exp(2\pi i k p_j/q) - z)(\exp(-2\pi i k p_j/q) - z)},$$

$$F_-(z) = \frac{1}{q}\sum_{k=0}^{q-1} \frac{\sum\limits_{\substack{\varepsilon_1\cdots\varepsilon_m\\=(-1)^{m}}} \exp(\pi i k \sum\limits_j \varepsilon_j p_j/q) - z \cdot \sum\limits_{\substack{\varepsilon_1\cdots\varepsilon_m\\=(-1)^{m+1}}} \exp(\pi i k \sum\limits_j \varepsilon_j p_j/q)}{\prod\limits_{j=1}^{m} (\exp(2\pi i k p_j/q) - z)(\exp(-2\pi i k p_j/q) - z)},$$

while for the second spin structure they are given by

$$F_+(z) = \frac{1}{q}\sum_{k=0}^{q-1} (-1)^k \frac{\sum\limits_{\substack{\varepsilon_1\cdots\varepsilon_m\\=(-1)^{m+1}}} \exp(\pi i k \sum\limits_j \varepsilon_j p_j/q) - z \cdot \sum\limits_{\substack{\varepsilon_1\cdots\varepsilon_m\\=(-1)^{m}}} \exp(\pi i k \sum\limits_j \varepsilon_j p_j/q)}{\prod\limits_{j=1}^{m} (\exp(2\pi i k p_j/q) - z)(\exp(-2\pi i k p_j/q) - z)},$$

$$F_-(z) = \frac{1}{q}\sum_{k=0}^{q-1} (-1)^k \frac{\sum\limits_{\substack{\varepsilon_1\cdots\varepsilon_m\\=(-1)^{m}}} \exp(\pi i k \sum\limits_j \varepsilon_j p_j/q) - z \cdot \sum\limits_{\substack{\varepsilon_1\cdots\varepsilon_m\\=(-1)^{m+1}}} \exp(\pi i k \sum\limits_j \varepsilon_j p_j/q)}{\prod\limits_{j=1}^{m} (\exp(2\pi i k p_j/q) - z)(\exp(-2\pi i k p_j/q) - z)}.$$

Here the notation $\sum\limits_{\substack{\varepsilon_1\cdots\varepsilon_m\\=(-1)^{m+1}}}$ indicates that the sum is taken over all choices of $\varepsilon_j = \pm 1$ for which $\varepsilon_1 \cdots \varepsilon_m = (-1)^{m+1}$ and similarly for $(-1)^m$. For the η-invariant this yields in case a)

$$\eta = \theta(0)$$

$$= \frac{2}{q}\sum_{k=1}^{q-1} \frac{\sum\limits_{\substack{\varepsilon_1\cdots\varepsilon_m\\=(-1)^{m+1}}} \exp(\pi i k(q+1)\sum_j \varepsilon_j p_j/q) - \sum\limits_{\substack{\varepsilon_1\cdots\varepsilon_m\\=(-1)^{m}}} \exp(\pi i k(q+1)\sum_j \varepsilon_j p_j/q)}{\prod\limits_{j=1}^{m}(\exp(2\pi i k p_j/q)-1)(\exp(-2\pi i k p_j/q)-1)}$$

$$= (-1)^{m+1}\frac{2}{q}\sum_{k=1}^{q-1}\frac{\sum\limits_{\varepsilon_1,\ldots,\varepsilon_m}\prod\limits_{j=1}^{m}\varepsilon_j\exp(\pi i k(q+1)\varepsilon_j p_j/q)}{\prod\limits_{j=1}^{m}(2-2\cos(2\pi k p_j/q))}$$

$$= (-1)^{m+1}\frac{2}{q}\sum_{k=1}^{q-1}\frac{\prod\limits_{j=1}^{m}(\exp(\pi i k(q+1)p_j/q)-\exp(-\pi i k(q+1)p_j/q))}{2^m\prod\limits_{j=1}^{m}(1-\cos(2\pi k p_j/q))}$$

$$= (-1)^{m+1} i^m \frac{2}{q}\sum_{k=1}^{q-1}\prod_{j=1}^{m}\frac{\sin(\pi k(q+1)p_j/q)}{1-\cos(2\pi k p_j/q)}.$$

If m is odd we see that η is imaginary, but on the other hand the η-invariant is always real. Hence it must vanish in this case. In fact, the η-invariant of the Dirac operator always vanishes unless the dimension of the manifold is $n \equiv 3 \bmod 4$. Case c) is treated similarly. Case b) cannot occur for even m. We summarize:

Corollary 3.6 *Let $m \in \mathbb{N}$ be even and let $q, p_1, \ldots, p_m \in \mathbb{N}$ be such that q and p_j are coprime for all j.*

Then if q is odd $L(q, p_1, \ldots, p_m)$ has exactly one spin structure and the η-invariant is given by

$$\eta(L(q,p_1,\ldots,p_m)) = (-1)^{3m/2+1}\frac{2}{q}\sum_{k=1}^{q-1}\prod_{j=1}^{m}\frac{\sin(\pi k(q+1)p_j/q)}{1-\cos(2\pi k p_j/q)}.$$

If q is even, then $L(q, p_1, \ldots, p_m)$ has two different spin structures. For the first spin structure the η-invariant is given by

$$\eta(L(q,p_1,\ldots,p_m)) = (-1)^{3m/2+1}\frac{2}{q}\sum_{k=1}^{q-1}\prod_{j=1}^{m}\frac{\sin(\pi k p_j/q)}{1-\cos(2\pi k p_j/q)}$$

while for the second it is

$$\eta(L(q,p_1,\ldots,p_m)) = (-1)^{3m/2+1}\frac{2}{q}\sum_{k=1}^{q-1}(-1)^k\prod_{j=1}^{m}\frac{\sin(\pi k p_j/q)}{1-\cos(2\pi k p_j/q)}.$$

The case $q = 2$ and $p_1 = \ldots = p_m = 1$ recovers Corollary 3.5. We collect a few examples in Tables 1 and 2. For even q we denote the η-invariants for the two spin structures by η_1 and η_2.

Table 1.

M	$\dim(M)$	η
$L(3,1,1)$		$\frac{4}{9}$
$L(3,1,2)$		$-\frac{4}{9}$
$L(5,1,1)$	3	$\frac{4}{5}$
$L(5,1,2)$		0
$L(5,2,3)$		$-\frac{4}{5}$
$L(3,1,1,1,1)$		$-\frac{4}{27}$
$L(3,1,1,1,2)$	7	$\frac{4}{27}$
$L(5,1,1,1,1)$		$-\frac{12}{25}$
$L(3,1,1,1,1,1,1)$		$\frac{4}{81}$
$L(3,1,1,1,1,1,2)$	11	$-\frac{4}{81}$
$L(5,1,1,1,1,1,1)$		$\frac{8}{25}$
$L(3,1,1,1,1,1,1,1,1)$		$-\frac{4}{243}$
$L(3,1,1,1,1,1,1,1,2)$	15	$\frac{4}{243}$
$L(5,1,1,1,1,1,1,1,1)$		$-\frac{28}{125}$
$L(3,1,1,1,1,1,1,1,1,1,1)$		$\frac{4}{729}$
$L(3,1,1,1,1,1,1,1,1,1,2)$	19	$-\frac{4}{729}$
$L(5,1,1,1,1,1,1,1,1,1,1)$		$\frac{4}{25}$

4 Dirac Determinant of the Sphere

In this section we compute the determinant of the square of the Dirac operator and of the Dirac operator itself over the sphere of constant sectional curvature 1. For this task one could use the general machinery developed in [6] for homogeneous operators on compact locally symmetric spaces. We found a direct approach more convenient. Recall the *Riemann ζ-function* ζ_R and the *Hurwitz ζ-function* ζ_H defined by meromorphic extension of the series

$$\zeta_R(s) = \sum_{j=1}^{\infty} \frac{1}{j^s} \qquad \text{and} \qquad \zeta_H(s,b) = \sum_{j=0}^{\infty} \frac{1}{(j+b)^s}$$

with $b > 0$. Observe that the binomial coefficient $M_n(x) := \binom{x+n-1}{x} = \frac{(x+n-1)\cdots(x+1)}{(n-1)!}$ is a polynomial in x of degree $n-1$. Hence we can define

$$A(k,n) := \frac{1}{k!} \frac{d^k}{dx^k} M_n(x)|_{x=-\frac{n}{2}} \tag{13}$$

Table 2.

M	$\dim(M)$	η_1	η_2
$L(4,1,1)$		$\frac{5}{8}$	$-\frac{3}{8}$
$L(4,1,3)$		$\frac{3}{8}$	$-\frac{5}{8}$
$L(6,1,1)$		$\frac{35}{36}$	$-\frac{19}{36}$
$L(6,1,5)$		$\frac{19}{36}$	$-\frac{35}{36}$
$L(8,1,1)$	3	$\frac{21}{16}$	$-\frac{11}{16}$
$L(8,1,3)$		$\frac{3}{16}$	$\frac{3}{16}$
$L(8,1,5)$		$-\frac{3}{16}$	$-\frac{3}{16}$
$L(8,3,5)$		$\frac{11}{16}$	$-\frac{21}{16}$
$L(10,1,1)$		$\frac{33}{20}$	$-\frac{17}{20}$
$L(4,1,1,1,1)$		$-\frac{9}{32}$	$\frac{7}{32}$
$L(4,1,1,1,3)$		$-\frac{7}{32}$	$\frac{9}{32}$
$L(6,1,1,1,1)$	7	$-\frac{329}{432}$	$\frac{265}{432}$
$L(8,1,1,1,1)$		$-\frac{105}{64}$	$\frac{87}{64}$
$L(10,1,1,1,1)$		$-\frac{1221}{400}$	$\frac{1029}{400}$
$L(4,1,1,1,1,1,1)$		$\frac{17}{128}$	$-\frac{15}{128}$
$L(6,1,1,1,1,1,1)$	11	$\frac{3611}{5184}$	$-\frac{3355}{5184}$
$L(8,1,1,1,1,1,1)$		$\frac{657}{256}$	$-\frac{623}{256}$

or, equivalently,

$$M_n(x) = \sum_{k=0}^{n-1} A(k,n)\left(x+\frac{n}{2}\right)^k. \tag{14}$$

Moreover, we set

$$C(k,n) := \begin{cases} \sum_{j=1}^{m-1} j^k \ln(j), & \text{if } n = 2m \text{ is even,} \\ \sum_{j=0}^{m-1} (j+\frac{1}{2})^k \ln(j+\frac{1}{2}), & \text{if } n = 2m+1 \text{ is odd.} \end{cases} \tag{15}$$

Now we can formulate the result for D^2 which in different notation is also contained in [4, Section 8].

Theorem 4.7. *The determinant of the square of the Dirac operator on the n-dimensional sphere S^n, $n \geq 2$, with constant sectional curvature 1 is given by the following formulas:*

If $n = 2m$ is even, then

$$\ln \mathrm{DET}(D^2; S^{2m}) = -2^{m+2} \sum_{k=0}^{n-1} A(k,n)\,(\zeta_R'(-k) + C(k,n)), \tag{16}$$

and if $n = 2m+1$ is odd, then

$$\ln \mathrm{DET}\,(D^2; S^{2m+1}) =$$
$$-2^{m+2} \sum_{k=0}^{n-1} A(k,n) \Big(\zeta_R'(-k)(\frac{1}{2^k} - 1) + \frac{\ln 2}{2^k}\,\zeta_R(-k) + C(k,n) \Big) \tag{17}$$

where ζ_R is the Riemann ζ-function, $A(k,n)$ is as in (13) and $C(k,n)$ is as in (15).

Proof. From (10) we see that on S^n

$$\zeta_{D^2}(s) = 2^{[n/2]+1} \sum_{j=0}^{\infty} \frac{\binom{j+n-1}{j}}{(j+\frac{n}{2})^{2s}}.$$

By (14) we obtain

$$\begin{aligned} \zeta_{D^2}(s) &= 2^{[n/2]+1} \sum_{k=0}^{n-1} A(k,n) \sum_{j=0}^{\infty} (j+\frac{n}{2})^{k-2s} \\ &= 2^{[n/2]+1} \sum_{k=0}^{n-1} A(k,n)\,\zeta_H(2s-k, \frac{n}{2}). \end{aligned} \tag{18}$$

In case $n = 2m$ is even we use

$$\zeta_H(s,m) = \zeta_R(s) - \sum_{j=1}^{m-1} j^{-s}$$

valid for $m \in \mathbb{N}$ to get

$$\zeta_{D^2}'(0) = 2^{m+2} \sum_{k=0}^{n-1} A(k,n) \left(\zeta_R'(-k) + \sum_{j=1}^{m-1} j^k \ln(j) \right).$$

This proves (16). For $n = 2m+1$ odd we obtain

$$\zeta_{D^2}'(0) = 2^{m+2} \sum_{k=0}^{n-1} A(k,n) \zeta_H'(-k, m+\frac{1}{2}).$$

From

$$\zeta_H(s, m+\frac{1}{2}) = \zeta_H(s, \frac{1}{2}) - \sum_{j=0}^{m-1} (j+\frac{1}{2})^{-s},$$

valid for $m \in \mathbb{N}$ and

$$\zeta_H(s, \frac{1}{2}) = (2^s - 1)\zeta_R(s)$$

(see e.g. [12, 9.535]) we get (17).

Example 4.8 *From this theorem we get*

$$\begin{aligned}
\mathrm{DET}(D^2; S^2) &= \exp(-8\,\zeta_R'(-1)),\\
\mathrm{DET}(D^2; S^3) &= 2^{-\frac{1}{2}} \exp\left(3\,\zeta_R'(-2)\right)),\\
\mathrm{DET}(D^2; S^4) &= \exp\left(\frac{8}{3}\zeta_R'(-1) - \frac{8}{3}\zeta_R'(-3)\right),\\
\mathrm{DET}(D^2; S^5) &= 2^{\frac{3}{16}} \exp\left(-\frac{5}{4}\,\zeta_R'(-2) + \frac{5}{8}\,\zeta_R'(-4)\right),\\
\mathrm{DET}(D^2; S^6) &= \exp\left(-\frac{16}{15}\,\zeta_R'(-1) + \frac{4}{3}\,\zeta_R'(-3) - \frac{4}{15}\,\zeta_R'(-5)\right),\\
\mathrm{DET}(D^2; S^7) &= 2^{-\frac{5}{64}} \exp\left(\frac{259}{480}\,\zeta_R'(-2) - \frac{35}{96}\,\zeta_R'(-4) + \frac{7}{160}\,\zeta_R'(-6)\right),\\
\mathrm{DET}(D^2; S^8) &= \exp\left(\frac{16}{35}\,\zeta_R'(-1) - \frac{28}{45}\,\zeta_R'(-3) + \frac{8}{45}\,\zeta_R'(-5) - \frac{4}{315}\,\zeta_R'(-7)\right),\\
\mathrm{DET}(D^2; S^9) &= 2^{\frac{35}{1024}} \exp\left(-\frac{3229}{13440}\,\zeta_R'(-2) + \frac{47}{256}\,\zeta_R'(-4) - \frac{21}{640}\,\zeta_R'(-6)\right.\\
&\qquad \left. + \frac{17}{10752}\,\zeta_R'(-8)\right),\\
\mathrm{DET}(D^2; S^{10}) &= \exp\left(-\frac{64}{315}\,\zeta_R'(-1) + \frac{164}{567}\,\zeta_R'(-3) - \frac{13}{135}\,\zeta_R'(-5)\right.\\
&\qquad \left. + \frac{2}{189}\,\zeta_R'(-7) - \frac{1}{2835}\,\zeta_R'(-9)\right).
\end{aligned}$$

For $n = 2, 4, 6$ this reproduces the values computed by Branson in [4, Thm. 8.1]. For the Dirac operator itself we obtain

Theorem 4.9. *The determinant of the Dirac operator on the n-dimensional sphere S^n, $n \geq 2$, with constant sectional curvature 1 is given by the following:*

If $n = 2m$ is even, then

$$\mathrm{DET}(D; S^{2m}) = \exp\left(-i\pi 2^m \sum_{k=0}^{2m-1} A(k, 2m)\frac{B_{k+1}(m)}{k+1}\right) \sqrt{\mathrm{DET}(D^2; S^{2m})},$$

and if n is odd, then

$$\mathrm{DET}(D; S^n) = \sqrt{\mathrm{DET}(D^2; S^n)}$$

where $\mathrm{DET}(D^2; S^n)$ is as in Theorem 4.7, $A(k, n)$ as in (13) and $B_k(x)$ are the Bernoulli polynomials.

Proof. By (3) we have

$$\mathrm{DET}(D;S^n) = \exp\left(i\frac{\pi}{2}(\zeta_{D^2}(0) - \eta_D(0))\right)\sqrt{\mathrm{DET}(D^2;S^n)}.$$

The η-invariant vanishes because the Dirac spectrum of S^n is symmetric. It remains to compute $\zeta_{D^2}(0)$. Plugging

$$\zeta_H(-k,b) = -\frac{B_{k+1}(b)}{k+1}$$

(see [12, 9.531 and 9.623(3)]) into (18) we get

$$\zeta_{D^2}(0) = -2^{[n/2]+1}\sum_{k=0}^{n-1} A(k,n)\frac{B_{k+1}(\frac{n}{2})}{k+1}. \tag{19}$$

This proves the even-dimensional case. If n is odd we observe that $\Phi_0(D^2) = 0$ and $\ker D^2 = \{0\}$ and hence

$$\zeta_{D^2}(0) = 0. \tag{20}$$

This concludes the proof.

Using Proposition 2.2 we can combine Theorem 4.7, (19) and (20) to obtain the proper time regularized determinant of D^2 on S^n.

Corollary 4.10 *The proper time regularized determinant of the square of the Dirac operator on the n-dimensional sphere S^n, $n \geq 2$, with constant sectional curvature 1 is given by the following formulas:*
If $n = 2m$ is even, then

$$\ln \mathrm{DET}_{\mathrm{p.t.}}(D^2;S^{2m}) = -2^{m+1}\sum_{k=0}^{2m-1} A(k,2m)\left(2\,\zeta_R'(-k) + 2\,C(k,2m) - \Gamma'(1)\frac{B_{k+1}(m)}{k+1}\right)$$

and if n is odd, then

$$\mathrm{DET}_{\mathrm{p.t.}}(D^2;S^n) = \mathrm{DET}(D^2;S^n)$$

where ζ_R is the Riemann ζ-function, $A(k,n)$ is as in (13), $C(k,n)$ is as in (15) and $B_k(x)$ are the Bernoulli polynomials.

The 1-dimensional case needs to be treated separately because S^1 has two spin structures. Let us call the spin structure which extends to the unique spin structure of the 2-dimensional disk the *bounding* spin structure, the other one the *non-bounding* spin structure. For the first spin structure formula (10) still holds while the non-bounding spin structure has $\mathbb{Z}$ as its Dirac spectrum. We get

Theorem 4.11. *The determinants of the Dirac operator and its square on the* 1*-dimensional sphere* S^1 *with length* 2π *take the values*

$$\mathrm{DET}(D; S^1) = 2 \quad \textit{and} \quad \mathrm{DET}(D^2; S^1) = 4$$

for the bounding spin structure and

$$\mathrm{DET}(D; S^1) = -2\pi i \quad \textit{and} \quad \mathrm{DET}(D^2; S^1) = 4\pi^2$$

for the non-bounding spin structure.

Recall that by our convention for the determinant the eigenvalue 0 is ignored. Otherwise the determinant on S^1 would vanish for the non-bounding spin structure.

Proof. From (10) we get for the bounding spin structure

$$\zeta_{D^2}(s) = 2\sum_{j=0}^{\infty} \frac{1}{(j+\frac{1}{2})^{2s}} = 2\,\zeta_H(2s, \frac{1}{2}) = 2(2^{2s}-1)\zeta_R(2s).$$

Using $\zeta_R(0) = -\frac{1}{2}$ we obtain

$$\zeta'_{D^2}(0) = 4\ln 2 \cdot \zeta_R(0) = -2\ln 2$$

and hence

$$\mathrm{DET}(D^2; S^1) = \exp\left(-\zeta'_{D^2}(0)\right) = 4.$$

The computation for the non-bounding spin structure is even easier. For ζ_{D^2} we get

$$\zeta_{D^2}(s) = 2\sum_{j=1}^{\infty} \frac{1}{j^{2s}} = 2\,\zeta_R(2s).$$

Now $\zeta'_R(0) = -\frac{1}{2}\ln(2\pi)$ yields

$$\zeta'_{D^2}(0) = 4\zeta'_R(0) = -2\ln(2\pi)$$

and therefore

$$\mathrm{DET}(D^2; S^1) = 4\pi^2.$$

This proves the formulas for $\mathrm{DET}(D^2; S^1)$. The η-invariant vanishes in both cases due to spectral symmetry. The result for $\mathrm{DET}(D; S^1)$ follows from (3) and

$$\zeta_{D^2}(0) = \begin{cases} 0 & \text{for the bounding spin structure} \\ -1 & \text{for the non-bounding spin structure.} \end{cases}$$

In the following tables we give some results obtained by numerical approximation. The phase φ is defined by

$$\mathrm{DET}(D; S^n) = |\mathrm{DET}(D; S^n)|e^{i\varphi}.$$

Here are the values of the determinant on the first few even dimensional spheres.

We do not need the $\mathrm{DET}_{\mathrm{p.t.}}(D^2; S^n)$–column in the odd dimensional case because of Corollary 4.10. The phase vanishes according to Theorem 4.9.

Table 3.

n	$\lvert\textrm{DET}(D;S^n)\rvert$	φ	$\textrm{DET}_{\textrm{p.t.}}(D^2;S^n)$
2	1.938054383626833	-0.523598775598299	3.098641928647827
4	0.796336885184459	0.191986217719376	0.680506885334576
6	1.096240639130369	-0.0793709255073612	1.167199468652179
8	0.961240935820022	0.0345879931923003	0.935802870851431
10	1.017761616865532	-0.0155276797389130	1.029945138033507
12	0.992018956602283	0.00710558492613086	0.986674523357944
14	1.003708484448039	-0.00329528874347978	1.006211552565484
16	0.998272959376817	0.00154335558060539	0.997114236587699
18	1.000814152574965	-0.000728325704396029	1.001360932301507
20	0.999614434504206	0.000345772030034635	0.999355987552311

Table 4.

n	$\textrm{DET}(D)$
3	0.803354268824629
5	1.090359845142337
7	0.963796369884191
9	1.016473922384390
11	0.992614518464762
13	1.003422630166412
15	0.998408322304586
17	1.000749343263366
19	0.999645452552308
21	1.000168795852563

A view at the numerical values leads us to

Conjecture 1. The determinant of the Dirac operator on the n-dimensional sphere S^n tends to 1 for $n \to \infty$,

$$\lim_{n\to\infty} \textrm{DET}(D;S^n) = 1.$$

For example, we have

$$\lvert\textrm{DET}(D;S^{200})\rvert \approx 0.99999999999999999999999999999999023251476,$$

and for the phase of $\mathrm{DET}(D; S^{200})$ we have

$$\varphi \approx 0.88944347908781047952551018696458465891 76 \cdot 10^{-31}.$$

We have no explanation for this phenomenon.

5 Dirac Determinant of Spherical Space Forms

Next we determine the Dirac determinant of spherical space forms $M = G\backslash S^{2m-1}$ where $G \subset \mathrm{SO}\,(2m)$ is a fixed point free subgroup. The spin structure is given by a homomorphism $\varepsilon : G \to \mathrm{Spin}\,(2m)$ lifting the inclusion of G into $\mathrm{SO}\,(2m)$ as explained in Section 3. In order to distinguish geometric data for different manifolds we write $\zeta_{D^2}\,(s; M)$ for the ζ-function of the square of the Dirac operator on M.

Theorem 5.12. *Let $M = G\backslash S^{2m-1}$ be a spherical space form with spin structure given by $\varepsilon : G \to \mathrm{Spin}\,(2m)$. Then the square of the Dirac operator on M has the determinant*

$$\begin{aligned}\mathrm{DET}\,(D^2; G\backslash S^{2m-1}) \quad &= \quad \mathrm{DET}\,\left(D^2; S^{2m-1}\right)^{\frac{1}{|G|}} \cdot \\ \prod_{g \in G - \{1_{2m}\}} \exp &\left(-\frac{2\chi(\varepsilon(g))}{|G|} \cdot \int_0^\infty \frac{e^{-(m-\frac{1}{2})t}}{\det(1_{2m} - e^{-t}g)} \frac{1-e^{-t}}{t}\,dt \right)\end{aligned}$$

where $\mathrm{DET}\,(D^2; S^{2m-1})$ is given by Theorem 4.7 and χ is the character of the spinor module for $\mathrm{Spin}\,(2m)$.

Proof. To compute the ζ-function $\zeta_{D^2}\,(s\,; G\backslash S^{2m-1})$ we recall the θ-functions

$$\theta_\pm\,(t) \;=\; e^{-(m-\frac{1}{2})t} \cdot F_\pm\,(e^{-t}) \;=\; \frac{e^{-(m-\frac{1}{2})t}}{|G|} \sum_{g \in G} \frac{\chi^{\mp}(\varepsilon(g)) - e^{-t} \cdot \chi^{\pm}(\varepsilon(g))}{\det(1_{2m} - e^{-t} \cdot g)},$$

see (11). Hence by the following Mellin transformation

$$\begin{aligned}\Gamma\,(2s)\,\zeta_{D^2}\,(s; G\backslash S^{2m-1}) &= \sum_{\lambda \in \mathrm{Spec}(D)} |\lambda|^{-2s} \int_0^\infty u^{2s-1}\,e^{-u}du \\ &= \sum_{\lambda \in \mathrm{Spec}(D)} \int_0^\infty t^{2s-1}\,e^{-|\lambda|t}dt \\ &= \sum_{k=0}^\infty (\mu_k + \mu_{-k}) \int_0^\infty t^{2s-1}\,e^{-(m-\frac{1}{2}+k)t}dt \\ &= \int_0^\infty t^{2s-1}\,(\theta_+\,(t) + \theta_-\,(t))\,dt \\ &= \frac{1}{|G|} \sum_{g \in G} \chi(\varepsilon(g)) \int_0^\infty \frac{t^{2s-1} \cdot e^{-(m-\frac{1}{2})t} \cdot (1-e^{-t})}{\det(1_{2m} - e^{-t} \cdot g)}\,dt\end{aligned}$$

we obtain

$$\zeta_{D^2}\left(s; G\backslash S^{2m-1}\right) = \frac{1}{|G|\cdot\Gamma(2s)}\sum_{g\in G}\chi(\varepsilon(g))\int_0^\infty t^{2s}\,\frac{e^{-(m-\frac{1}{2})t}}{\det(1_{2m}-e^{-t}g)}\,\frac{1-e^{-t}}{t}dt.$$

The special case $G=\{1_{2m}\}$ yields

$$\zeta_{D^2}\left(s; S^{2m-1}\right) = \frac{1}{\Gamma(2s)}\cdot\chi(1)\cdot\int_0^\infty t^{2s}\,\frac{e^{-(m-\frac{1}{2})t}}{\det(1_{2m}-e^{-t}\cdot 1_{2m})}\,\frac{1-e^{-t}}{t}dt$$

and therefore

$$\begin{aligned}\zeta_{D^2}\left(s; G\backslash S^{2m-1}\right) &= \frac{\zeta_{D^2}(s; S^{2m-1})}{|G|}\\ &+ \frac{1}{|G|\cdot\Gamma(2s)}\sum_{g\in G-\{1_{2m}\}}\chi(\varepsilon(g))\int_0^\infty t^{2s}\,\frac{e^{-(m-\frac{1}{2})t}}{\det(1_{2m}-e^{-t}g)}\,\frac{1-e^{-t}}{t}dt\,.\end{aligned}$$

Since $g\in G-\{1_{2m}\}$ does not have any positive real eigenvalues the integral defines a function in s holomorphic at 0. Thus

$$\zeta_{D^2}\left(0; G\backslash S^{2m-1}\right) = \frac{\zeta_{D^2}(0; S^{2m-1})}{|G|} = 0 \tag{21}$$

by (20) and

$$\begin{aligned}\zeta'_{D^2}\left(0; G\backslash S^{2m-1}\right) &= \frac{\zeta'_{D^2}(0; S^{2m-1})}{|G|}\\ &+\frac{2}{|G|}\sum_{g\in G-\{1_{2m}\}}\chi\left(\varepsilon(g)\right)\int_0^\infty \frac{e^{-(m-\frac{1}{2})t}}{\det(1_{2m}-e^{-t}g)}\,\frac{1-e^{-t}}{t}dt\end{aligned}$$

by (5). This implies the formula for $\mathrm{DET}(D^2; G\backslash S^{2m-1})$.

Equality of $\mathrm{DET}\,(D^2; G\backslash S^{2m-1})$ and $\mathrm{DET}_{\mathrm{p.t.}}(D^2; G\backslash S^{2m-1})$ follows from Proposition 2.2 and (21).

Example 5.13 *Let us look at real projective space $M=\mathbb{RP}^{2m-1}$. In this case $G=\{1_{2m}, -1_{2m}\}$. We assume m to be odd so that M is spin and has two spin structures. They are given by $\varepsilon(-1_{2m})=\pm\omega$ where $\omega=e_1\cdots e_{2m}$ is the volume element considered as an element in $\mathrm{Spin}(2m)\subset \mathrm{Cl}(\mathbb{R}^{2m})$. The volume element acts by multiplication by ±1 on the two half spinor spaces so that $\chi^+(\omega)=-\chi^-(\omega)$. Hence for both spin structures we have $\chi(\varepsilon(-1_{2m}))=0$. It follows that*

$$\mathrm{DET}(D^2; \mathbb{RP}^{2m-1}) = \sqrt{\mathrm{DET}(D^2; S^{2m-1})} = \mathrm{DET}(D; S^{2m-1}).$$

Example 5.14 *Let us now look at the lens space $M = L(q, p_1, \dots, p_m)$. We assume that q is odd so that M has exactly one spin structure. For*

$$g = \begin{pmatrix} R(k\, p_1/q) & & 0 \\ & \ddots & \\ 0 & & R(k\, p_m/q) \end{pmatrix}$$

we compute

$$\begin{aligned} \chi(\varepsilon(g)) &= \sum_{\varepsilon_1, \cdots, \varepsilon_m} \exp\left(\pi i k (q+1) \sum_{j=1}^{m} \varepsilon_j p_j / q \right) \\ &= \sum_{\varepsilon_1, \cdots, \varepsilon_m} \prod_{j=1}^{m} \exp(\pi i k (q+1) \varepsilon_j p_j / q) \\ &= 2^m \prod_{j=1}^{m} \cos\left(\frac{\pi\, k\, (q+1)\, p_j}{q} \right) \end{aligned}$$

and

$$\begin{aligned} \det\left(1_{2m} - e^{-t}\, g\right) &= \prod_{j=1}^{m} \det\left(1_{2m} - e^{-t}\, R(k\, p_j/q)\right) \\ &= \prod_{j=1}^{m} \left(1 - 2\, e^{-t}\, \cos(2\,\pi\, k\, p_j/q) + e^{-2t}\right). \end{aligned}$$

Now Theorem 5.12 yields

$$\mathrm{DET}(D^2; L(q, p_1, \dots, p_m)) = \mathrm{DET}(D^2; S^{2m-1})^{1/q} \cdot \exp(A)$$

where the covering anomaly $\exp(A)$ *is given by*

$$A = -\frac{2^{m+1}}{q} \sum_{k=1}^{q-1} \int_0^\infty \frac{(1 - e^{-t})\, e^{-(m-1/2)t}\, dt}{t \prod_j (1 - 2\, e^{-t}\, \cos(2\,\pi\, k\, p_j/q) + e^{-2t})} \prod_{j=1}^{m} \cos\left(\frac{\pi\, k\, (q+1)\, p_j}{q} \right).$$

Theorem 5.15. *Let $M = G \backslash S^{2m-1}$ be a spherical space form with spin structure given by $\varepsilon : G \to$* Spin$(2m)$*. Then the Dirac operator on M has the determinant*

$$\mathrm{DET}\,(D; G\backslash S^{2m-1}) = \exp\left(\frac{i\pi}{|G|} \sum_{g \in G - \{1_{2m}\}} \frac{(\chi^+ - \chi^-)(\varepsilon(g))}{\det(1_{2m} - g)} \right) \cdot \sqrt{\mathrm{DET}(D^2; G\backslash S^{2m-1})}\,,$$

where $\mathrm{DET}\,(D^2; G\backslash S^{2m-1})$ *is given by Theorem 5.12 and $\chi^\pm$ are the characters of the half spinor modules of* Spin$\,(2m)$.

Proof. Follows directly from (3), Theorem 3.4, and (21).

6 Spectrum and Determinant of Twisted Dirac Operators on S^2

In this section we compute the spectrum and determinant of certain twisted Dirac operators on the two-sphere S^2. Recall that complex line bundles over S^2 are in 1-1 correspondence with $\mathbb{Z}$ via their Chern numbers. Hence each complex line bundle is of the form $\mathcal{L}^k$ for some $k \in \mathbb{Z}$ where $c_1(\mathcal{L}) = 1$. The spinor bundle ΣS^2 decomposes into a sum of two complex line bundles, $\Sigma S^2 = \Sigma^+ S^2 \oplus \Sigma^- S^2$, the bundles of spinors of positive and of negative chirality. Since $c_1(\Sigma^\pm S^2) = \pm 1$ we have $\mathcal{L} = \Sigma^+ S^2$. The metric and the Levi-Civita connection on $\Sigma^+ S^2$ induce Hermitian metrics and metric connections on all the line bundles $\mathcal{L}^k$. Using these connections we can form the twisted Dirac operator D_k acting on sections of the bundle $\Sigma S^2 \otimes \mathcal{L}^k$.

Theorem 6.16. *The eigenvalues of the twisted Dirac operator D_k acting on sections of the spinor bundle tensored with the line bundle of Chern number k over S^2 are given by*

$$\pm\sqrt{j(|k|+j)}, \qquad j = 0, 1, 2, \dots$$

with multiplicity

$$2j + |k|.$$

Proof. We start by computing the relevant curvatures. The curvature tensor $R^{\Sigma S^2}$ of the spinor bundle can be expressed in terms of the curvature tensor R^{TS^2} of the tangent bundle by

$$R^{\Sigma S^2}(X,Y) = \frac{1}{4}\sum_{i,j=1}^{2} \left\langle R^{TS^2}(X,Y)e_i, e_j \right\rangle e_i e_j.$$

where e_1, e_2 is an orthonormal tangent frame acting on spinors via Clifford multiplication. Since in our case S^2 is of constant sectional curvature 1 we have

$$R^{TS^2}(X,Y)Z = \langle Y, Z\rangle X - \langle X, Z\rangle Y$$

and therefore

$$R^{\Sigma S^2}(X,Y) = \frac{1}{4}(YX - XY).$$

Hence

$$R^{\Sigma S^2}(e_1, e_2) = -\frac{1}{2} e_1 \cdot e_2.$$

Since the area element $e_1 \cdot e_2$ acts on $\Sigma^\pm S^2$ by multiplication by $\pm i$ we get

$$R^{\mathcal{L}}(e_1, e_2) = R^{\Sigma^+ S^2}(e_1, e_2) = -\frac{i}{2}$$

and thus

$$R^{\mathcal{L}^m}(e_1, e_2) = -\frac{im}{2}.$$

The decomposition $\Sigma S^2 = \mathcal{L} \oplus \mathcal{L}^{-1}$ yields the parallel decomposition $\Sigma S^2 \otimes \mathcal{L}^k = \mathcal{L}^{k+1} \oplus \mathcal{L}^{k-1}$ with respect to which the curvature tensor has the form

$$R^{\Sigma S^2 \otimes \mathcal{L}^k}(e_1, e_2) = -\frac{i}{2} \begin{pmatrix} k+1 & 0 \\ 0 & k-1 \end{pmatrix}.$$

The Dirac operator interchanges the chirality of spinors and hence has the form

$$D_k = \begin{pmatrix} 0 & D_k^+ \\ D_k^- & 0 \end{pmatrix}$$

where $D_k^- = (D_k^+)^*$. Denote the connection Laplace operator on $\mathcal{L}^m$ by $\nabla^*\nabla =: \Delta_m$. The curvature endomorphism in the Weitzenböck formula for D_k^2 is given by

$$\begin{aligned} \mathcal{K}_k &= e_1 \cdot e_2 \cdot R^{\Sigma S^2 \otimes \mathcal{L}^k}(e_1, e_2) \\ &= -\frac{i}{2} \begin{pmatrix} i & 0 \\ 0 & -i \end{pmatrix} \begin{pmatrix} k+1 & 0 \\ 0 & k-1 \end{pmatrix} \\ &= \frac{1}{2} \begin{pmatrix} k+1 & 0 \\ 0 & -k+1 \end{pmatrix}. \end{aligned}$$

The Weitzenböck formula $D_k^2 = \nabla^*\nabla + \mathcal{K}_k$ now says

$$D_k^+ D_k^- = \Delta_{k+1} + \frac{k+1}{2}, \tag{22}$$

$$D_k^- D_k^+ = \Delta_{k-1} + \frac{-k+1}{2}. \tag{23}$$

Taking the difference of (22) and (23) with $k+2$ instead of k we obtain

$$D_k^+ D_k^- = D_{k+2}^- D_{k+2}^+ + k + 1. \tag{24}$$

Since $D_k^+ D_k^- = (D_k^-)^* D_k^-$ and $D_k^- D_k^+ = (D_k^+)^* D_k^+$ are non-negative operators they do not have negative eigenvalues. Denote the positive eigenvalues by $\lambda_j(D_k^+ D_k^-)$ and $\lambda_j(D_k^- D_k^+)$, $j = 1, 2, 3, \ldots$, and their multiplicities by $\mu_j(D_k^+ D_k^-)$ and $\mu_j(D_k^- D_k^+)$ respectively. Write $\lambda_0(D_k^+ D_k^-) = \lambda_0(D_k^- D_k^+) = 0$ and let $\mu_0(D_k^+ D_k^-)$ and $\mu_0(D_k^- D_k^+)$ be the multiplicities of the eigenvalue 0. Here $\mu_0(D_k^+ D_k^-) = 0$ or $\mu_0(D_k^- D_k^+) = 0$ is not excluded.

From now on assume $k \geq 0$. The case of negative k can be treated similarly by interchanging the roles of D_k^+ and D_k^-. For non-negative k equation (22) shows

$$\mu_0(D_k^+ D_k^-) = 0. \tag{25}$$

The Atiyah-Singer index formula yields

$$\mu_0(D_k^- D_k^+) - \mu_0(D_k^+ D_k^-) = \mathrm{ind}(D_k^+) = c_1(\mathcal{L}^k) = k$$

and thus

$$\mu_0(D_k^- D_k^+) = k. \tag{26}$$

Since D_k^+ intertwines the operators $D_k^- D_k^+$ and $D_k^+ D_k^-$ it induces isomorphisms on the eigenspaces for non-zero eigenvalues. Hence for $j \geq 1$

$$\lambda_j(D_k^+ D_k^-) = \lambda_j(D_k^- D_k^+) \quad \text{and} \quad \mu_j(D_k^+ D_k^-) = \mu_j(D_k^- D_k^+). \tag{27}$$

Combining equations (24) and (27) yields for $j \geq 1$

$$\lambda_j(D_k^- D_k^+) = \lambda_{j-1}(D_{k+2}^- D_{k+2}^+) + k + 1 \text{ and } \mu_j(D_k^- D_k^+) = \mu_{j-1}(D_{k+2}^- D_{k+2}^+). \tag{28}$$

The parameter shift from j to $j-1$ is due to the fact that 0 is counted as an eigenvalue (of multiplicity 0) for $D_k^+ D_k^-$. Using (28) inductively j times we get

$$\lambda_j(D_k^- D_k^+) = \lambda_0(D_{k+2j}^- D_{k+2j}^+) + \sum_{m=1}^{j}(k + 2m - 1) = j(k + j)$$

and by (26)

$$\mu_j(D_k^- D_k^+) = \mu_0(D_{k+2j}^- D_{k+2j}^+) = k + 2j.$$

We have computed the spectrum of $D_k^2 = \begin{pmatrix} D_k^+ D_k^- & 0 \\ 0 & D_k^- D_k^+ \end{pmatrix}$, namely 0 is an eigenvalue of multiplicity k and $j(k+j)$ has multiplicity $2(k+2j)$, $j = 1, 2, 3, \ldots$.

It remains to observe that the spectrum of D_k is symmetric about 0 because for each eigenspinor of D_k of the form $\psi = \psi_+ + \psi_-$ with respect to the splitting $\Sigma S^2 \otimes \mathcal{L}^k = \Sigma^+ S^2 \otimes \mathcal{L}^k \oplus \Sigma^- S^2 \otimes \mathcal{L}^k$ the spinor $\hat{\psi} = \psi_+ - \psi_-$ is an eigenspinor for the opposite eigenvalue. Taking square roots proves the theorem.

For $k = 0$ Theorem 6.16 gives the spectrum of the classical Dirac operator which we used in the previous sections. Similarly, for $k = 1$ we recover the spectrum of the Laplace operator acting on functions $\Delta = D_1^- D_1^+$.

Theorem 6.16 can also be proved by trivialising the twisted spinor bundle over S^2 minus a point and then solving the eigenvalue equation explicitly using spin-weighted spherical harmonics. See [16, Sec. 3.1] for this approach.

Knowing the eigenvalues of D_k explicitly we can compute its determinant.

Theorem 6.17. *The determinant of the twisted Dirac operator D_k acting on sections of the spinor bundle tensored with the line bundle of Chern number k over S^2 is given by*

$$\mathrm{DET}(D_k; S^2) = \exp\left(i\frac{\pi}{2}(-|k| - \frac{1}{3})\right) e^{-4\,\zeta_R'(-1)+k^2/2} \cdot \prod_{m=1}^{|k|} m^{|k|-2m},$$

the determinant of its square is

$$\mathrm{DET}(D_k^2; S^2) = e^{-8\,\zeta_R'(-1)+k^2} \cdot \prod_{m=1}^{|k|} m^{2|k|-4m}$$

and the proper time regularized determinant

$$\mathrm{DET}_{\mathrm{p.t.}}(D_k^2; S^2) = e^{-8\,\zeta_R'(-1)+k^2+\Gamma'(1)(|k|+\frac{1}{3})} \cdot \prod_{m=1}^{|k|} m^{2|k|-4m}.$$

Proof. By Theorem 6.16 we obtain for the ζ-function

$$\begin{aligned}\zeta_{D_k^2}(s) &= 2\sum_{j=1}^{\infty} \frac{2j+|k|}{j^s(j+|k|)^s} \\ &= 2\sum_{j=1}^{\infty}(j^{-s}(j+|k|)^{1-s} + j^{1-s}(j+|k|)^{-s}).\end{aligned}$$

For $j > |k|$ we can expand $(1+\frac{|k|}{j})^\alpha$ into a binomial series and we get

$$\begin{aligned}\zeta_{D_k^2}(s) &= \\ &= 2\sum_{j=1}^{|k|} \frac{2j+|k|}{j^s(j+|k|)^s} + 2\sum_{j=|k|+1}^{\infty} j^{1-2s} \sum_{i=0}^{\infty}\left[\binom{1-s}{i} + \binom{-s}{i}\right]\left(\frac{|k|}{j}\right)^i \\ &= 2\sum_{j=1}^{|k|} \frac{2j+|k|}{j^s(j+|k|)^s} + 2\sum_{i=0}^{\infty}\left[\binom{1-s}{i} + \binom{-s}{i}\right] |k|^i \zeta_H(2s+i-1, |k|+1)\end{aligned}$$

Now we partly follow a similar calculation of Weisberger [21, Appendix C]. For the coefficients

$$d_i(s) := \binom{1-s}{i} + \binom{-s}{i}.$$

we have

$$\begin{aligned}d_0(s) &= 2, \\ d_1(s) &= -2s+1, \\ d_2(s) &= s^2, \\ d_i(s) &= (-1)^i \frac{i-2}{i(i-1)} s + \mathrm{O}(s^2), \qquad i \geq 3.\end{aligned}$$

Hence

$$\begin{aligned}\zeta_{D_k^2}(0) &= 2\sum_{j=1}^{|k|}(2j+|k|) + 2\left(2\zeta_H(-1,|k|+1) + |k|\zeta_H(0,|k|+1)\right)\\ &= 4\zeta_R(-1) + 2|k|\zeta_R(0)\\ &= -\frac{1}{3} - |k| \end{aligned} \tag{29}$$

and since $\zeta_H(s,|k|+1)$ has only one pole of first order at $s=1$ with residue 1 we have

$$\begin{aligned}\zeta'_{D_k^2}(0) \quad &= \\ = -2\sum_{j=1}^{|k|}&(2j+|k|)(\ln j + \ln(j+|k|)) + 8\zeta'_H(-1,|k|+1) + 4|k|\zeta'_H(0,|k|+1)\\ &-4|k|\zeta_H(0,|k|+1) + |k|^2 + 2\sum_{i=3}^{\infty}(-1)^i\frac{i-2}{i(i-1)}|k|^i\zeta_H(i-1,|k|+1)\\ = -2\sum_{j=1}^{|k|}&(2j+|k|)(\ln j + \ln(j+|k|)) + 8\zeta'_H(-1,|k|+1) + 4|k|\zeta'_H(0,|k|+1)\\ &+2|k| + 5|k|^2 + 2\sum_{i=3}^{\infty}(-1)^i\frac{i-2}{i(i-1)}|k|^i\zeta_H(i-1,|k|+1). \end{aligned} \tag{30}$$

For the evaluation of the last sum we use the Mellin transform of the Hurwitz ζ-function

$$\zeta_H(s,a) = \frac{1}{\Gamma(s)}\int_0^\infty t^{s-1}e^{-at}(1-e^{-t})^{-1}dt. \tag{31}$$

This yields

$$\begin{aligned} I &:= \sum_{i=3}^{\infty}(-1)^i\frac{i-2}{i(i-1)}|k|^i\zeta_H(i-1,|k|+1)\\ &= \sum_{i=3}^{\infty}(-1)^i\frac{i-2}{i(i-1)}|k|^i\frac{1}{(i-2)!}\int_0^\infty t^{i-2}e^{-(|k|+1)t}(1-e^{-t})^{-1}dt\\ &= |k|\int_0^\infty t^{-1}(-e^{-(2|k|+1)t} + e^{-(|k|+1)t}(1-|k|t))(1-e^{-t})^{-1}dt\\ &\quad -2\int_0^\infty t^{-2}(e^{-(2|k|+1)t} - e^{-(|k|+1)t}(1-|k|t+\frac{1}{2}(|k|t)^2))(1-e^{-t})^{-1}dt. \end{aligned}$$

To compute these integrals we multiply the integrand by t^s, evaluate the integral using (31) and then let $s \to 0$.

$$
\begin{aligned}
I = \lim_{s\to 0}\{&|k|\Gamma(s)(-\zeta_H(s,2|k|+1)+\zeta_H(s,|k|+1))\\
&-|k|^2\Gamma(s+1)\zeta_H(s+1,|k|+1)-2[\Gamma(s-1)(\zeta_H(s-1,2|k|+1)\\
&-\zeta_H(s-1,|k|+1))+|k|\Gamma(s)\zeta_H(s,|k|+1)\\
&-\frac{1}{2}|k|^2\Gamma(s+1)\zeta_H(s+1,|k|+1)]\}\\
= \lim_{s\to 0}&\{|k|\Gamma(s)(-\zeta_H(s,2|k|+1)-\zeta_H(s,|k|+1))+2\Gamma(s-1)h(s-1)\}
\end{aligned}
$$

with the definition

$$
h(s) \quad := \quad \sum_{j=1}^{|k|}(j+|k|)^{-s} \quad = \quad \zeta_H(s,|k|+1)-\zeta_H(s,2|k|+1).
$$

Using $-2\zeta_H(0,|k|+1) = 1+2|k|$, $h(0) = |k|$, and $\mathrm{Res}_{s=-l}\Gamma(s) = \frac{(-1)^l}{l!}$ we conclude

$$
\begin{aligned}
I \quad &= \\
&= \lim_{s\to 0}\left\{|k|s\Gamma(s)\left(-2\frac{\zeta_H(s,|k|+1)-\zeta_H(0,|k|+1)}{s}+\frac{h(s)-h(0)}{s}+\frac{1+3|k|}{s}\right)\right.\\
&\qquad \left. +2\Gamma(s-1)h(s-1)\right\}\\
&= -2|k|\zeta_H'(0,|k|+1)+|k|h'(0)\\
&\qquad +\lim_{s\to 0}\{2\Gamma(s-1)[h(s-1)+(s-1)\frac{|k|}{2}(1+3|k|)]\}\\
&= -2|k|\zeta_H'(0,|k|+1)+|k|h'(0)-2h'(-1)-|k|(1+3|k|)\\
&= -2|k|\zeta_H'(0,|k|+1)-|k|\sum_{j=1}^{|k|}\ln(j+|k|)+2\sum_{j=1}^{|k|}\ln(j+|k|)(j+|k|)\\
&\qquad -|k|(1+3|k|)\\
&= -2|k|\zeta_H'(0,|k|+1)+\sum_{j=1}^{|k|}\ln(j+|k|)(2j+|k|)-|k|(1+3|k|).
\end{aligned}
$$

Inserting this into (30) we get

$$
\begin{aligned}
\zeta_{D_k^2}'(0) &= -2\sum_{j=1}^{|k|}(2j+|k|)\ln j+8\zeta_H'(-1,|k|+1)+2|k|+5|k|^2\\
&\quad -2|k|(1+3|k|)\\
&= 8\zeta_R'(-1)-|k|^2+2\sum_{j=1}^{|k|}(2j-|k|)\ln j,
\end{aligned}
$$

therefore

$$\text{DET}(D_k^2; S^2) = e^{-8\zeta_R'(-1)+k^2} \cdot \prod_{m=1}^{|k|} m^{2|k|-4m}.$$

The formula for $\text{DET}(D_k; S^2)$ now follows from (29) and from the vanishing of the η-invariant. Proposition 2.2 yields the formula for $\text{DET}_{\text{p.t.}}(D_k^2; S^2)$.

Remark 6.18 *The case $k = 0$ yields Theorem 4.7, 4.9 and 4.10 for $n = 2$. Since the Laplace-Beltrami operator acting on functions is given by $\Delta = D_1^- D_1^+$ it has the same spectrum as D_1^2 except that all non-zero eigenvalues have only half the multiplicity. Hence $\text{DET}(\Delta; S^2) = \sqrt{\text{DET}(D_1^2; S^2)}$ and we get (compare [4, Thm. 8.1])*

Corollary 6.19 *The determinant of the Laplace-Beltrami operator Δ acting on functions on S^2 is given by*

$$\text{DET}(\Delta; S^2) = e^{-4\zeta_R'(-1)+\frac{1}{2}}.$$

References

1. Bär, C. (1991): Das Spektrum von Dirac-Operatoren. Bonner Math. Schr., **217**
2. Bär, C. (1996): The Dirac operator on space forms of positive curvature. J. Math. Soc. Japan, **48**, 69–83
3. Bär, C. (2000): Dependence of the Dirac Spectrum on the Spin Structure. Séminaires et Congrès, **4**, 17–33
4. Branson, T. (1993): The Functional Determinant. Global Analysis Research Center Lecture Note Series, Number 4, Seoul National University
5. Branson, T., Ørsted, B. (1991): Explicit functional determinants in four dimensions. Proc. Am. Math. Soc., **113**, 669–682
6. Bunke, U., Olbrich, M. (1995): Selberg Zeta and Theta Functions. Akademie Verlag, Berlin
7. Camporesi, R., Higuchi, A. (1996): On the eigenfunctions of the Dirac operator on spheres and real hyperbolic spaces. J. Geom. Phys., **20**, 1–18
8. Cisneros-Molina, J.L. (2001): The η-invariant of twisted Dirac operators of S^3/Γ. Geom. Dedicata, **84**, 207–228
9. Gilkey, P. (1984): Invariance theory, the heat equation and the Atiyah-Singer index theorem. Publish or Perish, Wilmington, Delaware
10. Gilkey, P. (1989): The geometry of sperical space form groups. World Scientific, Singapore
11. Goette, S. (1997): Äquivariante η-Invarianten homogener Räume. Shaker Verlag, Aachen
12. Gradstein, I.S., Ryshik, I.M. (1981): Tables of series, products and integrals. Harri Deutsch, Frankfurt
13. Gursky, M. (1997): Uniqueness of the Functional Determinant. Commun. Math. Phys., **189**, 655–665
14. Lawson, H.B., Michelsohn, M. -L. (1989): Spin geometry. Princeton University Press, Princeton
15. Onofri, E. (1982): On the positivity of the effective action in a theory of random surfaces. Commun. Math. Phys., **86**, 321–326

16. Schopka, S. (2001): Anomalies in quantum field theory and determinants of elliptic operators. Diplomarbeit, Universität Hamburg, Hamburg
17. Schwarz, A.S. (1993): Quantum field theory and topology. Grundlehren 307, Springer-Verlag, Berlin
18. Sulanke, S. (1979): Berechnung des Spektrums des Quadrates des Dirac-Operators auf der Sphäre. Dissertation, HU Berlin, Berlin
19. Trautman, A. (1995): The Dirac operator on hypersurfaces. Acta Phys. Polon. B, **26**, 1283–1310
20. Wojciechowski, K.P. (1999): Heat equation and spectral geometry. Lecture given at the Summer School on Geometric Methods in Quantum Field Theory, `http://wwwlma.univ-bpclermont.fr/manifestation/vdl/vdl99.html` , Villa de Leyva
21. Weisberger, W.I. (1987): Normalization of the path integral measure and the coupling constant for bosonic strings. Nucl. Phys. **B, 284**, 171–200

Constructing Isospectral Metrics via Principal Connections

Dorothee Schueth

Mathematisches Institut, Universität Bonn, Beringstr. 6, D-53115 Bonn, Germany
schueth@math.uni-bonn.de

1 Recent Developments Concerning Isospectral Manifolds

The spectrum of a closed Riemannian manifold is the eigenvalue spectrum of the associated Laplace operator acting on functions, counted with multiplicities; two manifolds are said to be isospectral if their spectra coincide. Spectral geometry deals with the mutual influences between the spectrum of a Riemannian manifold and its geometry. To which extent does the spectrum determine the geometry?

For example, the spectrum determines a sequence of so-called heat invariants, the first few of which are the dimension, the volume, and the total scalar curvature (see e.g. [4]). Some Riemannian manifolds are known to be completely characterized by their spectra; e.g., flat tori in dimensions two and three [13], or round spheres in dimensions up to six [20].

In order to find geometric properties which are not determined by the spectrum, one needs examples of isospectral manifolds. One method of constructing such manifolds which was very productive in the 1980's is the so-called Sunada method and its generalizations (see [17], [5], [2], [3], [12]). This construction yielded pairs (or families) of isospectral manifolds arising as quotients of a common Riemannian covering manifold by different discrete subgroups of isometries. In particular, these pairs were always locally isometric, and their geometries could be distinguished only by global properties.

The first examples of locally nonisometric manifolds were found in the first half of the 1990's by Zoltan I. Szabó [18] (in the case with boundary) and Carolyn Gordon [6]. In both cases, the manifolds arose from very special constructions; isospectrality was proven, at first, more or less by explicit computation of the spectra, which happened to be possible here. However, in a second paper on her examples [7], Gordon interpreted them as instances of the following general principle:

Theorem 1.1. **[7]** *If a torus acts on two Riemannian manifolds freely and isometrically with totally geodesic orbits, and if the quotients of the manifolds by any subtorus of codimension at most one are isospectral when endowed with the submersion metric, then the original two manifolds are isospectral.*

Several other examples of locally nonisometric isospectral manifolds, all in some sense related to Szabó's and Gordon's original examples, were subsequently found, and proven to be isospectral by the above theorem ([11], [9], [14]). Some of these examples show that neither the range of the scalar curvature function [9] nor certain curvature integrals like $\int_M \mathrm{scal}^2 dvol_g$, $\int_M \|\mathrm{Ric}\|^2 dvol_g$, $\int_M \|R\|^2 dvol_g$ are determined by the spectrum [14]. The latter is particularly remarkable because a certain linear combination of these three terms is a heat invariant, thus spectrally determined.

It is not very obvious from the theorem how to find new applications. A main motive of the author's habilitation thesis [15] which was (just as [9], [14], and [16]) written under partial support of the SFB 256 at Bonn, was to give a specialized version which would be useful for finding new situations in which Gordon's theorem applied. Such a specialized version is the following:

Let T be a torus, endowed with a fixed left invariant metric, and let $P \to B$ be a principal T-bundle, where B is a closed and connected Riemannian manifold. For any principal connection ω on P there is a unique Riemannian metric g_ω on P such that the projection $P \to B$ is a Riemannian submersion, the horizontal distribution is equal to $\ker \omega$, and the induced metric on each fiber is the given metric on T. With respect to this metric g_ω, the fibers are totally geodesic.

Theorem 1.2. [15] ("Connection Technique") *Let ω, ω' be principal connections on P. Suppose that for each μ in the dual of the Lie algebra $\mathfrak{z}$ of T there is a bundle automorphism $F_\mu : P \to P$ which factors over an isometry of B and satisfies $\mu \circ \omega = F_\mu^*(\mu \circ \omega')$. Then (P, g_ω) and $(P, g_{\omega'})$ are isospectral.*

The idea of the proof is to show that if $W < T$ is a subtorus of codimension one and $\mu \in \mathfrak{z}^*$ is such that $\ker \mu = T_e W$, then F_μ induces an isometry from $(P/W, g_\omega^W)$ to $(P/W, g_{\omega'}^W)$, where g_ω^W and $g_{\omega'}^W$ denote the submersion metrics.

Ballmann [1] gave an interesting extension of Theorem 1.2 to other fiber bundles associated with P.

One application of Theorem 1.2 was the construction of isospectral, locally non-isometric metrics on $S^2 \times T^2$ [15]; thereby, dimension four became the lowest dimension of manifolds on which such metrics are so far known to exist. Geometrically, these isospectral metrics on $S^2 \times T^2$ can be distinguished by the dimension of the locus of the maximal scalar curvature. Another application of Theorem 1.2 was the construction of left invariant isospectral metrics on compact Lie groups [15]; see more about this in Section 2.

Soon afterwards, Carolyn Gordon [8] discovered a new version of her original theorem in which the torus actions were no longer required to be free. (Earlier there had already been versions, first by Gordon and Szabó [10], then by the author [15] in which the T-orbits were not anymore required to be totally geodesic.) As an application, Gordon obtained isospectral metrics on spheres, in dimension at least eight; together with certain examples of very different type found independently by Z.I. Szabó [19], these were the first examples of isospectral spheres. The author established a specialized formulation of Gordon's new theorem, in the spirit of the above connection technique:

Let (M, g_0) be a closed and connected Riemannian manifold on which a torus T with Lie algebra $\mathfrak{z}$ acts effectively by isometries. For $Z \in \mathfrak{z}$ we denote by Z^* the induced vectorfield on M. We call a 1-form λ on M *horizontal* if it vanishes on all Z^*. With every $\mathfrak{z}$-valued, T-invariant, horizontal 1-form on M we associate a Riemannian metric $g_\lambda : (X, Y) \mapsto g_0(X + \lambda(X)^*, Y + \lambda(Y)^*)$ on M.

Theorem 1.3. [16] *Let λ, λ' be $\mathfrak{z}$-valued, T-invariant, horizontal 1-forms on M. If for every $\mu \in \mathfrak{z}^*$ there is a T-equivariant isometry F_μ of (M, g_0) such that $\mu \circ \lambda = F_\mu^*(\mu \circ \lambda')$, then (M, g_λ) and $(M, g_{\lambda'})$ are isospectral.*

Remark 1. Note that the union $\hat{M}$ of principal orbits in M is a principal T-bundle again. If we let ω_λ denote the principal connection on $\hat{M}$ whose kernel is the horizontal distribution with respect to g_λ, then $\omega_\lambda = \omega_0 + \lambda$. In particular, the condition $\mu \circ \lambda = F_\mu^*(\mu \circ \lambda')$ from Theorem 1.3 is equivalent to $\mu \circ \omega_\lambda = F_\mu^*(\mu \circ \omega_{\lambda'})$, which is a condition on connection forms just as in Theorem 1.2.

Theorem 1.3 was used in [16] to construct examples of isospectral metrics on spheres in dimension five. A sketch of proof for Theorem 1.3 is as follows: Consider the Sobolev spaces $\mathcal{H} := H^{1,2}(M, g_\lambda)$ and $\mathcal{H}' := H^{1,2}(M, g_{\lambda'})$, and decompose each as an orthogonal sum of subspaces $\mathcal{H}_\mu$, $\mathcal{H}'_\mu$ using Fourier decomposition with respect to the unitary T-action on $\mathcal{H}$ and $\mathcal{H}'$. Here, μ runs through the dual lattice $\mathcal{L}^* \subset \mathfrak{z}^*$ of $T = \mathfrak{z}/\mathcal{L}$. We claim that $F_\mu^* : \mathcal{H}'_\mu \to \mathcal{H}_\mu$ preserves both the L^2- and the $H^{1,2}$-norms, whence the theorem will follow by the variational characterization of eigenvalues via the Rayleigh quotient. Preservation of L^2-norms is trivial since F_μ is a g_0-isometry and g_0, g_λ, $g_{\lambda'}$ all have the same volume element. That $F_\mu^* : \mathcal{H}'_\mu \to \mathcal{H}_\mu$ also preserves the norm of the gradient (hence the $H^{1,2}$-norm) can now be derived using the condition $\mu \circ \lambda = F_\mu^*(\mu \circ \lambda')$.

Actually Theorem 1.3 applies to the case of manifolds with boundary as well. In that case, both Dirichlet and Neumann isospectrality is implied; see [16], [8].

Note that it is crucial that F_μ may depend on μ. If all F_μ are equal, then $\lambda = F_\mu^* \lambda'$ (for any μ), which is easily seen to imply that g_λ and $g_{\lambda'}$ are isometric. In the opposite case we say that λ, λ' satisfy the conditions of Theorem 1.3 *nontrivially*.

At this point, the reader might be wondering how one is able to produce examples of $\mathfrak{z}$-valued 1-forms λ, λ' which do satisfy the conditions of Theorem 1.3 nontrivially. The next section describes one fruitful way of doing this.

2 Almost Conjugate Subspaces of Compact Lie Algebras and Associated Principal Connections

We start this section by describing families of almost conjugate subspaces in certain classical compact Lie algebras. Such families have repeatedly been used (at first in [7] and [11]) as ingredients in earlier constructions of isospectral manifolds which either arose from, or can be viewed as arising from, Theorem 1.3. We will try to explain as generally as possible how these almost conjugate subspaces can be used to produce nontrivial pairs of 1-forms satisfying the conditions of Theorem 1.3. Several known examples will serve as illustration. In Section 3, we will then give some new applications.

Note that the isospectral, locally nonisometric metrics on the low-dimensional manifolds $S^2 \times T^2$ [15] and S^5 or $S^3 \times S^1$ [16] do, however, *not* arise from the approach discussed here.

In the following, let H be a compact connected semisimple Lie group with Lie algebra $\mathfrak{h}$, and let $\mathfrak{z}$ be a euclidean vector space.

Definition 1. (i) *Two linear maps $j, j' : \mathfrak{z} \to \mathfrak{h}$ are called* isospectral *if for each $Z \in \mathfrak{z}$ there is $a_Z \in H$ such that $j'_Z = \mathrm{Ad}_{a_Z}(j_Z)$.*
(ii) *j and j' are called* equivalent *if there is $\Phi \in \mathrm{Aut}(\mathfrak{h})$ and $C \in \mathrm{O}(\mathfrak{z})$ such that $j'_Z = \Phi(j_{C(Z)})$ for all $Z \in \mathfrak{z}$.*

Note that if j, j' are isospectral, then the linear subspaces $j(\mathfrak{z})$ and $j'(\mathfrak{z})$ of $\mathfrak{h}$ are "almost conjugate" in the sense that there is a linear bijection (given by $j_Z \leftrightarrow j'_Z$) between them which preserves adjoint orbits. The following facts turn out to be very pleasant with respect to isospectral purposes:

Lemma 1. *Let $\mathfrak{z} = \mathbb{R}^2$, equipped with the standard metric, and denote by $\mathcal{J}$ the vector space of all linear maps from $\mathfrak{z}$ to $\mathfrak{h}$.*

(i) [11] *If $\mathfrak{h} = \mathfrak{so}(m)$, where m is any positive integer other than $1, 2, 3, 4$, or 6, then there is a Zariski open subset $\mathcal{O}$ of $\mathcal{J}$ such that each $j \in \mathcal{O}$ belongs to a d-parameter family of isospectral, inequivalent elements of $\mathcal{J}$. Here $d \geq m(m-1)/2 - [m/2]([m/2]+2) > 1$. For $m = 6$, there exist at least 1-parameter families in $\mathcal{J}$ with these properties.*
(ii) [15] *If $\mathfrak{h} = \mathfrak{su}(m)$, where $m \geq 3$, then there is a Zariski open subset $\mathcal{O}$ of $\mathcal{J}$ such that each $j \in \mathcal{O}$ belongs to a continuous family of isospectral, inequivalent elements of $\mathcal{J}$.*

The definitions of isospectrality and equivalence which were used in [11] and [15] were different in minor ways from our above definition; however, it is not hard to see that this does not affect the statements of Lemma 1.

Now suppose that our compact Lie group H acts on a compact connected Riemannian manifolds N by isometries. Then each $j_Z \in \mathfrak{h}$ induces a vectorfield j_Z^* on N; by taking the dual with respect to the Riemannian metric on N, we obtain a 1-form λ_Z on N.

Key observation: *If j_Z, j'_Z belong to the same adjoint orbit then the associated 1-forms λ_Z, λ'_Z on N belong to the same $H \subset \mathrm{Isom}(N)$-orbit.*

This observation, and how one may use it systematically, was already mentioned, but not further elaborated on, in the author's Remark 3.4 in [15].

Our first application is the construction of isospectral metrics on $M := N \times T$, where T is a torus with Lie algebra $\mathfrak{z}$. Let T be equipped with a fixed left invariant metric, and denote by g_0 the Riemannian product metric on M. Given any linear map $j : \mathfrak{z} \to \mathfrak{h}$, we consider the associated 1-forms λ_Z ($Z \in \mathfrak{z}$) on N as above. We define a $\mathfrak{z}$-valued 1-form λ on N by $\langle \lambda(X), Z \rangle = \lambda_Z(X)$ for all $Z \in \mathfrak{z}$ and $X \in TN$, where $\langle\ ,\ \rangle$ is the given euclidean inner product on $\mathfrak{z}$. Let the pullback of λ to $M = N \times T$ be denoted λ again. Note that λ is T-invariant and horizontal by construction.

Proposition 1. *Let H act on N by isometries. If $j, j' : \mathfrak{z} \to \mathfrak{h}$ are isospectral, then the associated $\mathfrak{z}$-valued 1-forms λ, λ' on $M = N \times T$ satisfy the conditions of Theorem 1.3 on (M, g_0); in particular, (M, g_λ) and $(M, g_{\lambda'})$ are isospectral.*

Proof. Let $\mu \in \mathfrak{z}^*$, and let $Z \in \mathfrak{z}$ be the dual vector with respect to the inner product on $\mathfrak{z}$. Choose $a_Z \in H$ such that $j'_Z = \mathrm{Ad}_{a_Z}(j_Z)$. Then straightforward calculation shows that the T-equivariant isometry $F_\mu := (a_Z\,, \mathrm{Id})$ of $M = N \times T$ satisfies $\mu \circ \lambda = F_\mu^*(\mu \circ \lambda')$. □

Example 1. The following known examples of isospectral manifolds can be viewed as applications of Proposition 1:

(i) The isospectral families of left invariant metrics on $H \times T^2$ from [15], where H is one of $\mathrm{SO}(m \geq 5)$, $\mathrm{Spin}(m \geq 5)$, or $\mathrm{SU}(m \geq 3)$. Here we have $\mathfrak{z} = \mathbb{R}^2$, $\mathfrak{h} = \mathfrak{so}(m)$ (resp. $\mathfrak{su}(m)$), and the Riemannian manifold N is H, endowed with a biinvariant metric. We consider the action of H on itself from the left and use the isospectral families of maps from Lemma 1. Actually, our present construction leads to *right* invariant vectorfields j_Z and 1-forms λ_Z, hence to right invariant isospectral metrics on $H \times T^2$. However, under the canonical identification of right invariant metrics with left invariant ones via the map $a \mapsto a^{-1}$, these classes of isospectral families of metrics on $H \times T^2$ are in fact the same. The author showed in [15] that, generically, these isospectral homogeneous metrics differ by the norm of the associated Ricci tensors.

(ii) The isospectral families of metrics on $S^{m-1 \geq 4} \times T^2$ from [9]. Here $\mathfrak{z} = \mathbb{R}^2$, $\mathfrak{h} = \mathfrak{so}(m)$, $H = \mathrm{SO}(m)$, and $N = S^{m-1}$ endowed with a round metric and with the canonical action of H from the left. Again one uses the isospectral maps from Lemma 1(i). The original construction of these isospectral families of metrics on $S^{2m-1} \times T^2$ was done in a very different context; the manifolds were viewed as submanifolds of certain Riemannian nilmanifolds. Moreover, it was shown in [9] that the maximum of the scalar curvature is in general nonconstant during these isospectral deformations.

We are now going to give an extension of Proposition 1 which produces isospectral metrics on irreducible manifolds as well:

Let T again be a torus with Lie algebra $\mathfrak{z}$, and suppose that $H \times T$ acts on a compact connected Riemannian manifold (M, g_0) by isometries. With any linear map $j : \mathfrak{z} \to \mathfrak{h}$ we associate vectorfields j_Z^*, 1-forms λ_Z and a $\mathfrak{z}$-valued 1-form λ on M exactly as we did above on N. In short,

$$\langle \lambda(X), Z \rangle = g_0(j_Z^*(p), X)$$

for all $Z \in \mathfrak{z}$ and $X \in T_pM$ ($p \in M$), where $\langle\ ,\ \rangle$ is the given inner product on $\mathfrak{z}$ (not to be confused with the metrics induced by g_0 on T-orbits in M).

Note that λ is T-invariant because the actions of H and T on M commute. However, λ will in general not be horizontal; this is the reason for the additional orthogonality assumption in the following result.

Proposition 2. *Let $H \times T$ act on (M, g_0) by isometries. If $j, j' : \mathfrak{z} \to \mathfrak{h}$ are isospectral, and if $j_Z^*(p), j_Z'^*(p) \perp W^*(p)$ for all $Z, W \in \mathfrak{z}$ and $p \in M$, then the associated $\mathfrak{z}$-valued 1-forms λ, λ' on M satisfy the conditions of Theorem 1.3; in particular, (M, g_λ) and $(M, g_{\lambda'})$ are isospectral.*

Proof. By the orthogonality assumption, λ and λ' are now indeed horizontal. We proceed as in the proof of Proposition 1, this time letting $F_\mu := a_Z \in H$; these maps are T-equivariant because H commutes with T. □

Example 2. The following known examples of isospectral manifolds can be viewed as applications of Proposition 2 (each time, the reservoir of isospectral j-maps from Lemma 1 is used):

(i) Isospectral families of right invariant isospectral metrics on G, corresponding to the left invariant ones from [15], where G is one of SO$(n \geq 9)$, Spin$(n \geq 9)$, or SU$(n \geq 6)$. Here $\mathfrak{z} = \mathbb{R}^2$, $\mathfrak{h}$ is $\mathfrak{so}(m \geq 5)$ or $\mathfrak{su}(m \geq 3)$, H is SO(m) or Spin(m) or SU(m), respectively, and T is a maximal torus in SO(4) or Spin(4) or SU(3), respectively. We consider the left action of $H \times T$ on $M := G$ given by the inclusion $H \times T \subset \mathrm{SO}(m) \times \mathrm{SO}(4) \subset \mathrm{SO}(m+4) = G$ (and similarly for Spin and SU). Let g_0 be a biinvariant metric on G. Then λ, and hence g_λ, is right invariant on G (and left invariant under T). The orthogonality assumption of Proposition 2 is satisfied because H- and T-orbits meet perpendicularly with respect to g_0.

(ii) Gordon's families of isospectral metrics on $S^{n-1 \geq 8}$ from [8]. Here $\mathfrak{z} = \mathbb{R}^2$, $\mathfrak{h} = \mathfrak{so}(m \geq 5)$, $H = \mathrm{SO}(m)$; T is again a maximal torus in SO(4), and $M = S^{m+3} \subset \mathbb{R}^m \oplus \mathbb{R}^4$, endowed with the standard metric g_0. The action of $H \times T \subset \mathrm{SO}(m) \times \mathrm{SO}(4) \subset \mathrm{SO}(m+4) =: G$ on M is the restriction of the canonical action of G on the ambient space.

Remark 2. Further instances of Proposition 2 are the isospectral manifolds constructed in [14] and [10], and the conformally equivalent isospectral metrics on certain products of Lie groups from the last chapter of [15].

How can we get around the orthogonality assumption in Proposition 2? Our last observation in this section is that we can make λ horizontal by "brute force" if necessary: Suppose $H \times T$ acts isometrically on (M, g_0), and let λ be the associated $\mathfrak{z}$-valued, T-invariant 1-form on M as above. Then we define

$$\lambda_h(X) := \|Z_1^* \wedge \ldots \wedge Z_r^*\|^2 \lambda(X) - \sum_{k=1}^{r} \langle Z_1^* \wedge \ldots \wedge Z_{k-1}^* \wedge X \wedge Z_{k+1}^* \wedge \ldots \wedge Z_r^*,\ Z_1^* \wedge \ldots \wedge Z_r^* \rangle \lambda(Z_k^*)$$

for all $X \in TM$, where $\{Z_1, \ldots, Z_r\}$ is a basis of $\mathfrak{z}$, the Z_k^* are the induced vectorfields on M, and on each $\bigwedge^r T_pM$ we use the inner product induced by g_0. Obviously, λ_h is indeed horizontal now, and is again T-invariant.

Proposition 3. *Let $H \times T$ act on (M, g_0) by isometries. If $j, j' : \mathfrak{z} \to \mathfrak{h}$ are isospectral, then the associated 1-forms λ_h, λ_h' on M satisfy the conditions of Theorem 1.3; in particular, (M, g_{λ_h}) and $(M, g_{\lambda_h'})$ are isospectral.*

Proof. The proof is the same as for Proposition 2; in order to show that $F_\mu := a_Z \in H$ satisfies $\mu \circ \lambda_h = F_\mu^*(\mu \circ \lambda'_h)$, one must now also use the invariance of the vecorfields Z_k^* under the g_0-isometry $a_Z \in H$, which is another consequence of the fact that H and T commute. □

Example 3. There is just one instance of previously known examples which can in hindsight be viewed as an application of Proposition 3 (but not of Proposition 2): Certain continuous isospectral families of metrics on S^7 constructed by the author in [16]. In our present notation, we there had $\mathfrak{z} = \mathbb{R}^2$, $\mathfrak{h} = \mathfrak{su}(3)$, $H = \mathrm{SU}(3)$, $T = \{e^{it}\mathrm{Id} \mid t \in \mathbb{R}\} \times \mathrm{U}(1) \subset \mathrm{U}(3) \times \mathrm{U}(1)$, and $M = S^7 \subset \mathbb{C}^3 \oplus \mathbb{C}$, endowed with the standard metric g_0 and the action of $H \times T \subset U(3) \times U(1) \subset \mathrm{U}(4) =: G$ which is the restriction of the canonical action of G on S^7. (To be precise, our present λ_h differ from those used in [16] by multiplication with the $H \times T$-invariant function $\|Z_2^*\|^2$.)

3 Some New Applications

In the earlier examples of isospectral manifolds which served as illustrations for Propositions 1–3 in the previous section, the only occurring actions of the compact simple groups H, or the compact groups $G \supset H \times T$, were the actions of these groups either on themselves, or on some round sphere. Between these two "extremes", one may as well consider their actions on any associated homogeneous space. Continuing to use the notation from Section 2, we obtain:

Corollary 1. *Let $j, j' : \mathfrak{z} \to \mathfrak{h}$ be isospectral.*

(1) *Let $K \subset H$ be a Lie subgroup. For any left invariant metric on H which is right invariant under K, consider the associated H-invariant homogeneous metric on H/K. Applying Proposition 1 to the Riemannian manifold $N := H/K$, we obtain isospectral metrics g_λ, $g_{\lambda'}$ on $(H/K) \times T$.*

(2) *Let G be a compact Lie group containing $H \times T$ as a Lie subgroup, and let K be any Lie subgroup of G. Given a left invariant metric on G which is right invariant under K, let g_0 denote the associated G-invariant homogeneous metric on $M := G/K$. If the orthogonality condition $j_Z^*(p), j_Z'^*(p) \perp W^*(p)$ holds with respect to g_0 for all $Z, W \in \mathfrak{z}$ and $p \in M$, then Proposition 2 yields isospectral metrics g_λ, $g_{\lambda'}$ on $M = G/K$.*

(3) *In the context of* (2), *applying Proposition 3 to $M = G/K$ yields isospectral metrics g_{λ_h}, $g_{\lambda'_h}$ on G/K even if the orthogonality condition is not satisfied.*

Remark 3. (i) In Examples 1–3 of Section 2, we have already seen examples of the above corollary. We had $K = \{e\}$ in Examples 1(i) and 2(i), $K = \mathrm{SO}(m-1) \subset \mathrm{SO}(m) = H$ in Example 1(ii), $K = \mathrm{SO}(m+3) \subset \mathrm{SO}(m+4) = G$ in Example 2(ii), and $K = \mathrm{U}(3) \subset \mathrm{U}(4) = G$ in Example 3. In each case, the metrics g_0 were associated to biinvariant metrics on the groups.

(ii) In case the reader has become worried by our silence on general nontriviality statements for the above constructions, he should at least find it reassuring that in Examples 1–3 above, the isospectral manifolds have indeed been proven to be nonisometric under generic conditions, provided that the isospectral maps $j, j' : \mathfrak{z} \to \mathfrak{h}$ are not equivalent ([9], [15], [8], [16]). So the natural expectation that the isospectral manifolds will generically be nonisometric if there is no apparent reason for them to be otherwise was indeed justified in all earlier examples. In order not to make the present contribution too technical, we restrict ourselves to giving a nonisometry proof only for the first (which is the least complicated one) of the three new families of examples below.

Example 4. Isospectral metrics on $N \times T^2$, where N is one of the real Grassmann manifolds $\mathrm{Gr}_{k,m}$ *of k-planes in $\mathbb{R}^{m\geq 5}$, $1 \leq k \leq m-1$.*

This is a new application of Proposition 1 as specialized in part (1) of Corollary 1. In case $k = 1$ (or $k = m-1$) we get isospectral products of a projective space and a torus; these are Riemannian subcoverings of the manifolds from Example 1(ii), obtained by dividing by the canonical action of $\mathbb{Z}_2$ on the sphere factor.

In the context of part (1) of Corollay 1, we let $\mathfrak{z} = \mathbb{R}^2$, equipped with the standard inner product, $\mathfrak{h} = \mathfrak{so}(m \geq 5)$, $H = \mathrm{SO}(m)$, $K = \mathrm{S}(\mathrm{O}(k) \times \mathrm{O}(m-k)) \subset H$. Consider the biinvariant metric on H given by $\langle X, Y \rangle = \frac{1}{2}\mathrm{tr}({}^t XY)$ and the induced normal homogeneous metric on $N := H/K = \mathrm{Gr}_{k,m}$. Let $T = \mathbb{R}^2/\mathbb{Z}^2$ be the standard twodimensional torus with the canonical metric, and let g_0 be the Riemannian product metric on $(H/K) \times T$. Then for each pair of isospectral linear maps $j, j' : \mathfrak{z} \to \mathfrak{h}$ the associated metrics $g_\lambda, g_{\lambda'}$ on $(H/K) \times T$ are isospectral. Below we will show that if j and j' are not equivalent (in the sense of Definition 1), and if j' is *generic* in the sense that $j'(\mathfrak{z})$ has trivial centralizer in $\mathfrak{h}$, then these two metrics are not isometric. In particular, there exist d-parameter families of pairwise nonisometric isospectral metrics on $(H/K) \times T$, where d is as in Lemma 1(i).

Example 5. Isospectral metrics on the real Stiefel manifolds $W_{k,n}$ of orthonormal k-frames in $\mathbb{R}^{n\geq 9}$, $1 \leq k \leq n-1$.

This is a new application of Proposition 2 as specialized in part (2) of Corollary 1; in case $k = 1$ we obtain Gordon's isospectral spheres from Example 2(ii) again.

Let $\mathfrak{z} = \mathbb{R}^2$, $\mathfrak{h} = \mathfrak{so}(m \geq 5)$, $H = \mathrm{SO}(m)$, and T be a maximal torus in $\mathrm{SO}(4)$; thus $H \times T \subset \mathrm{SO}(m) \times \mathrm{SO}(4) \subset \mathrm{SO}(m+4) =: G$. Let M be the set of orthonormal k-tuples $(p_1, \ldots, p_k)$ with $p_\ell \in \mathbb{R}^{m+4}$ $(1 \leq \ell \leq k)$, viewed as a submanifold of $(\mathbb{R}^{m+4})^k = \mathbb{R}^{(m+4)k}$ and endowed with the metric g_0 which is induced on M by the standard metric of $\mathbb{R}^{(m+4)k}$. Then G, hence $H \times T$, acts by isometries on M. The orthogonality condition from Proposition 2 is satisfied. In fact, for any $j_Z \in j(\mathfrak{z}) \subset \mathfrak{h} = \mathfrak{so}(m) \subset \mathfrak{so}(m) \oplus \mathfrak{so}(4) \subset \mathfrak{g}$ and $W \in \mathfrak{z} = \mathfrak{so}(4) \subset \mathfrak{so}(m) \oplus \mathfrak{so}(4) \subset \mathfrak{g}$ we have

$$\begin{aligned}\langle j_Z^*(p), W^*(p) \rangle &= \langle (j_Z p_1, \ldots, j_Z p_k), (W p_1, \ldots, W p_k) \rangle \\ &= \sum_{\ell=1}^{k} \langle j_Z p_\ell, W p_\ell \rangle = -\sum_{\ell=1}^{k} \langle W j_Z p_\ell, p_\ell \rangle = 0\end{aligned}$$

because $W j_Z = 0$ in $\mathrm{End}(\mathbb{R}^{m+4})$. Thus for each pair of isospectral maps $j, j' : \mathfrak{z} \to \mathfrak{h}$ we obtain a pair of isospectral metrics $g_\lambda, g_{\lambda'}$ on M.

The manifold M can be identified with $G/K = W_{k,n}$, where $n = m + 4$ and $K = \{\mathrm{Id}\} \times \mathrm{SO}(m+4-k) \subset \mathrm{SO}(k) \times \mathrm{SO}(m+4-k) \subset \mathrm{SO}(m+4) = G$. Note, however, that the homogeneous metric on M which we used above is *not* induced by a biinvariant metric on G, except if $k = 1$. Instead, as one can check, it is induced by the left invariant and K-right invariant metric g on G given by

$$g(X,Y) = \langle X,Y\rangle + \langle \mathrm{pr}_k \circ X \circ \mathrm{pr}_k\,,\, \mathrm{pr}_k \circ Y \circ \mathrm{pr}_k\rangle$$

for all $X,Y \in \mathfrak{g} = \mathfrak{so}(m+4) \subset \mathrm{End}(\mathbb{R}^{m+4})$, where $\langle X,Y\rangle = \frac{1}{2}\mathrm{tr}({}^tXY)$, and pr_k denotes the projection of $\mathbb{R}^{m+4}$ to the first k coordinates. The normal homogeneous metric would *not* have satisfied the orthogonality condition of Proposition 2 except if $k = 1$.

Example 6. Isospectral metrics on the real Grassmann manifolds $\mathrm{Gr}_{k,n}$ *of k-planes in* $\mathbb{R}^{n\geq 9}$, $1 \leq k \leq n-1$.

This is a new application of Proposition 3 as specialized in part (3) of Corollary 1. In case $k = 1$ (or $k = n-1$) we obtain isospectral projective spaces which are just $\mathbb{Z}_2$-subcoverings of Gordon's isospectral spheres from Example 2(ii).

We let $\mathfrak{z}$, $\mathfrak{h}$, H, T, G, and n be as in the previous example, but now choose $K := \mathrm{S}(\mathrm{O}(k) \times \mathrm{O}(m+4-k))$. Let $j, j' : \mathfrak{z} \to \mathfrak{h}$ be isospectral linear maps. The normal homogeneous metric g_0 on $M := G/K = \mathrm{Gr}_{k,n}$, induced from the biinvariant metric on G given by $\langle X,Y\rangle = \frac{1}{2}\mathrm{tr}({}^tXY)$, will in general not satisfy the orthogonality condition from Proposition 2; in contrast to the case of Stiefel manifolds in the previous example, there is no remedy for this in choosing a different G-invariant metric on M. Thus the construction from Proposition 3 is our last recourse here. We use the normal homogeneous metric g_0 on M (or any other G-invariant metric) and obtain associated isospectral metrics $g_{\lambda_h}, g_{\lambda'_h}$ on M, where λ_h, λ'_h are constructed as in Proposition 3 from the $\mathfrak{z}$-valued 1-forms λ, λ' on M associated with j and j'.

We conclude this contribution by giving a nonisometry proof for the isospectral manifolds of Example 4. We use the same notation as there.

In order to simplify our subsequent calculations, it is convenient to view the Grassmann manifold N as a submanifold of the space of symmetric matrices $\mathrm{Sym}(\mathbb{R}^m) \subset \mathrm{End}(\mathbb{R}^m)$ via the identification of each $p = aK \in H/K = N$ with the projection ap_0a^{-1}, where p_0 denotes the projection of $\mathbb{R}^m$ to the first k coordinates. This is just the canonical embedding of $\mathrm{Gr}_{k,m}$ given by identifying a k-plane in $\mathbb{R}^m$ with the orthogonal projection onto this k-plane. Using the inner product $\langle X,Y\rangle = \frac{1}{2}\mathrm{tr}({}^tXY)$ on $\mathrm{End}(\mathbb{R}^m) \supset \mathrm{Sym}(\mathbb{R}^m)$, this identification is actually an isometry with respect to our normal homogeneous metric on N and the one induced from the ambient space $\mathrm{Sym}(\mathbb{R}^m)$. On this new copy of N, we have

$$\lambda_Z(X) = \langle [j_Z\,, p], X\rangle$$

for all $Z \in \mathfrak{z}$, $X \in T_pN \subset \mathrm{Sym}(\mathbb{R}^m)$, and hence

$$d\lambda_Z(X,Y) = 2\langle [j_Z\,, X], Y\rangle = -2\langle j_Z\,, [X,Y]\rangle.$$

Our proof follows the strategy used in [16].

We first show that if j is generic (i.e., if $j(\mathfrak{z})$ has trivial centralizer in $\mathfrak{h}$), then T is a maximal torus in $\mathrm{Isom}(N \times T, g_\lambda)$. Let F_t be a 1-parameter family of isometries of $(N \times T, g_\lambda)$ commuting with T. Then the F_t are T-equivariant and preserve the g_λ-horizontal distribution, hence the associated principal connection $\omega_\lambda = \omega_0 + \lambda$, hence they preserve also $d\omega_\lambda = d\omega_0 + d\lambda = d\lambda$; note that $d\omega_0 = 0$ here. In particular, the F_t induce a 1-parameter family of isometries $\bar{F}_t$ of the normal homogeneous metric on N which preserve $d\lambda$, hence $d\lambda_Z$ for all $Z \in \mathfrak{z}$. But each 1-parameter family of isometries of the Grassmann manifold $N = H/K$ belongs to $H = \mathrm{SO}(m)$. Hence we have $\bar{F}_t : p \mapsto a_t p a_t^{-1}$, where a_t is a 1-parameter family in H. Preservation of $d\lambda$ implies, by the above formula, that

$$\langle j_Z , [X,Y] \rangle \equiv \langle j_Z , a_t[X,Y]a_t^{-1} \rangle = \langle a_t^{-1} j_Z a_t , [X,Y] \rangle$$

for all $Z \in \mathfrak{z}$, $t \in \mathbb{R}$, and $X, Y \in T_pN \subset \mathrm{Sym}(\mathbb{R}^m)$. One easily checks that these $[X, Y]$, which are skew-symmetric, span the whole space $\mathfrak{h}$. Thus $j_Z \equiv a_t^{-1} j_Z a_t$ for all $Z \in \mathfrak{z}$, contradicting the genericity of j; this shows that T is indeed a maximal torus in $\mathrm{Isom}(N \times T, g_\lambda)$ if j is generic.

Now suppose that there were an isometry $F : (N \times T, g_\lambda) \to (N \times T, g_{\lambda'})$. By assumption, j' is generic; hence T is a maximal torus in $\mathrm{Isom}(N \times T, g_{\lambda'})$. Since all maximal tori are conjugate, we can assume (after possibly composing F with an isometry of $g_{\lambda'}$) that conjugation by F maps $T \subset \mathrm{Isom}(N \times T, g_\lambda)$ to $T \subset \mathrm{Isom}(N \times T, g_{\lambda'})$. Let Ψ denote the automorphism of $\mathfrak{z} = T_eT$ induced by conjugation by F. Then $F_*(Z^*) = \Psi(Z)^*$ for all $Z \in \mathfrak{z}$. In particular, $\Psi \in \mathrm{O}(\mathfrak{z})$, and F maps T-orbits to T-orbits. By similar arguments as above, F induces an isometry $\bar{F}$ of the normal homogeneous metric on N satisfying $d\lambda = \Psi \circ (\bar{F}^* d\lambda')$. But every isometry $\bar{F}$ of N is induced by conjugation by some $a \in \mathrm{O}(m)$, possibly combined with the orthocomplementation map $\beta : p \mapsto \mathrm{Id} - p$ in case $m = 2k$. From our above formula for $d\lambda$, $d\lambda'$ we see that $d\lambda'$ is invariant under β^* and therefore, again by similar arguments as before, that

$$j_Z = a^{-1} j'_{\Psi^{-1}(Z)} a$$

for all $Z \in \mathfrak{z}$ and some $a \in \mathrm{O}(m)$, contradicting the assumed inequivalence of j and j'. □

References

1. Ballmann, W. (2001): On the construction of isospectral manifolds. (Preprint)
2. Bérard, P. (1992): Transplantation et isospectralité I. Math. Ann., **292**, 547–559
3. Bérard, P. (1993): Transplantation et isospectralité II. J. London Math. Soc., **48**, 565–576
4. Berger, M., Gauduchon, P., Mazet, E. (1971): Le Spectre d'une Variété Riemannienne. Springer Lect. Notes Math., **194**
5. DeTurck, D., Gordon, C.S. (1987): Isospectral Deformations II: Trace formulas, metrics, and potentials. Comm. Pure Appl. Math., **XL**, 1067–1095
6. Gordon, C.S. (1993): Isospectral closed Riemannian manifolds which are not locally isometric. J. Diff. Geom., **37**, 639–649

7. Gordon, C.S. (1994): Isospectral closed Riemannian manifolds which are not locally isometric: II. In: Brooks, R., Gordon, C., Perry, P. (eds) Geometry of the Spectrum. Contemp. Math., **173**, 121–131
8. Gordon, C.S. (2001): Isospectral deformations of metrics on spheres. Invent. Math., **145**, 317–331
9. Gordon, C.S., Gornet, R., Schueth, D., Webb, D., Wilson, E. (1998): Isospectral deformations of closed Riemannian manifolds with different scalar curvature. Ann. Inst. Fourier, **48**, 593–607
10. Gordon, C.S., Szabó, Z.I. (2001): Isospectral deformations of negatively curved Riemannian manifolds with boundary which are not locally isometric. Duke Math. J. (to appear)
11. Gordon, C.S., Wilson, E. (1997): Continuous families of isospectral Riemannian metrics which are not locally isometric. J. Diff. Geom., **47**, 504–529
12. Pesce, H. (1996): Représentations relativement équivalentes et variétés riemanniennes isospectrales. Comment. Math. Helvetici, **71**, 243–268
13. Schiemann, A. (1997): Ternary positive definite quadratic forms are determined by their theta series. Math. Ann., **308**, 507–517
14. Schueth, D. (1999): Continuous families of isospectral metrics on simply connected manifolds. Ann. of Math., **149**, 287–308
15. Schueth, D. (2001): Isospectral manifolds with different local geometries. J. reine angew. Math., **534**, 41–94
16. Schueth, D. (2001): Isospectral metrics on five-dimensional spheres. J. Diff. Geom. (to appear)
17. Sunada, T. (1985): Riemannian coverings and isospectral manifolds. Ann. of Math., **121**, 169–186
18. Szabó, Z.I. (1997): Locally non-isometric yet super isospectral spaces. Geom. Funct. Anal., **9**, 185–214
19. Szabó, Z.I. (2001): Isospectral pairs of metrics on balls, spheres, and other manifolds with different local geometries. Ann. of Math., **154**, 437–475
20. Tanno, S. (1973): Eigenvalues of the Laplacian of Riemannian manifolds. Tôhoku Math. J., **25**, 391–403

Parametrizations of Teichmüller Space and Its Thurston Boundary

Ursula Hamenstädt*

Mathematisches Institut, Universität Bonn, Beringstr. 6, D-53115 Bonn
ursula@math.uni-bonn.de

Summary. For $g \geq 2$ let $\mathcal{T}_g$ be the Teichmüller space of hyperbolic metrics on a closed surface of genus g, and let $\partial\mathcal{T}_g$ be its Thurston boundary. Using intersection with $6g-5$ simple closed geodesics, we construct an embedding of $\partial\mathcal{T}_g$ into the real projective space $\mathbb{R}P^{6g-6}$.

1 Introduction

The easiest way to describe all hyperbolic metrics on a closed surface S_0 of genus $g \geq 2$ is to look at the *Teichmüller space* $\mathcal{T}_g$ of *marked* hyperbolic metrics on S_0. This Teichmüller space is the set of all pairs (f, h) where h is a hyperbolic metric on a surface S and f is the homotopy class of a homeomorphism $F : S_0 \to S$ of S_0 onto S. With respect to a natural topology, the space $\mathcal{T}_g$ is homeomorphic to an open cell of dimension $6g-6$.

Let γ be a closed curve on the base surface S_0 which is not null-homotopic. For every $(f, h) \in \mathcal{T}_g$, the curve $f(\gamma)$ is freely homotopic to a unique closed geodesic with respect to the hyperbolic metric h. The length $\ell_\gamma(f, h)$ of this geodesic defines a smooth (in fact real analytic) function on $\mathcal{T}_g$ which we call the *length function* of γ. Thus for a number $k > 0$ and an ordered k-tuple of fixed nontrivial free homotopy classes $[\gamma_1], \ldots, [\gamma_k]$ on S_0 we obtain a real analytic map of $\mathcal{T}_g$ into the k-dimensional euclidean space $\mathbb{R}^k$ by assigning to (f, h) the nonzero vector $(\ell_{\gamma_1}(f, h), \ldots, \ell_{\gamma_k}(f, h)) \in \mathbb{R}^k$. We call such a map *geometric*, and *special geometric* if it is defined by pairwise distinct free homotopy classes of *simple* closed curves, i.e. curves without self-intersections.

There are finite collections of simple closed curves on S_0 for which the corresponding geometric map is an embedding of Teichmüller space (see [2]). One can obtain rather easily such an embedding with a collection of $9g-9$ simple closed curves. On the other hand it has been known for a long time that there are no such embeddings by $6g-6$ length functions (see the discussion in [6] and the references there). The question on the minimal number of curves needed to obtain an embedding was answered by Schmutz. He showed the following theorem [6].

* Partially supported by the Sonderforschungsbereich 256.

Theorem A *For every $g \geq 2$ the Teichmüller space $\mathcal{T}_g$ can be parametrized by $6g-5$ length functions of simple closed geodesics.*

Below we give a proof of Theorem A which is a modification and simplification of the construction of Schmutz. Another very simple geometric embedding of $\mathcal{T}_g$ into $\mathbb{R}^{6g-5}$ can be found in [4]; however this embedding is not special.

A *geodesic current* for a closed hyperbolic surface S is a locally finite Borel-measure on the space of unoriented geodesics in the hyperbolic plane $\mathbf{H}^2$ which is invariant under the action of the fundamental group $\pi_1(S)$ of S. The space $\mathcal{C}$ of geodesic currents for S only depends on the topological type of S and will be equipped with the weak*-topology. There is a continuous bilinear form i on $\mathcal{C}$, the so called *intersection form*, which extends the geometric intersection number between two closed geodesics on S viewed as geodesic currents [1]. The subset $\mathcal{L}$ of $\mathcal{C}$ of all geodesic currents μ with vanishing self-intersection $i(\mu,\mu) = 0$ is the space of *measured geodesic laminations* and is homeomorphic to $\mathbb{R}^{6g-6}$ [1]. Every weighted simple closed geodesic can naturally be viewed as a geodesic current, and this set of currents is dense in $\mathcal{L}$.

Denote by $\partial\mathcal{T}_g$ the projectivization of $\mathcal{L} - \{0\}$. With respect to the natural quotient topology, $\partial\mathcal{T}_g$ is a compact space. Following Thurston (see [3]), we say that a sequence of points $\{(f_i, h_i)\}_i \subset \mathcal{T}_g$ *converges* in $\mathcal{T}_g \cup \partial\mathcal{T}_g$ to the projective class $[\mu]$ of a nontrivial measured lamination $\mu \in \mathcal{L}$ if and only if there are numbers $a_i > 0$ $(i > 0)$ such that for every k-tuple of free homotopy classes of curves $\gamma_1, \dots, \gamma_k$ on S the sequence $\{a_i(\ell_{\gamma_1}(f_i, h_i), \dots, \ell_{\gamma_k}(f_i, h_i))\}_i$ converges in $\mathbb{R}^k$ to $(i(\gamma_1, \mu), \dots, i(\gamma_k, \mu))$.

Thurston showed that this defines a topology on $\mathcal{T}_g \cup \partial\mathcal{T}_g$ which restricts to the usual topologies on $\mathcal{T}_g$ and $\partial\mathcal{T}_g$. With respect to this topology the space $\mathcal{T}_g \cup \partial\mathcal{T}_g$ is homeomorphic to a closed ball of dimension $6g-6$ (see [3]). In particular, the space of projective measured laminations is homeomorphic to a sphere of dimension $6g-7$. It is called the *Thurston boundary* of $\mathcal{T}_g$.

Recall that a collection $\psi_1, \dots, \psi_k$ of simple closed geodesics on S *fills up* S if every geodesic on S intersects one of the curves ψ_i transversely. This is equivalent to saying that the complement of $\{\psi_1, \dots, \psi_k\}$ in S consists of a finite collection of connected simply connected regions. If the curves $\psi_1, \dots, \psi_k$ fill up S then for every geodesic lamination μ the vector $(i(\psi_1, \mu), \dots, i(\psi_k, \mu))$ is non-zero and therefore it defines a point in the real projective space $\mathbb{R}P^{k-1}$. In particular, the projectivization of the geometric map defined by $\psi_1, \dots, \psi_k$ is a map of $\mathcal{T}_g$ into $\mathbb{R}P^{k-1}$ which extends continuously to the Thurston boundary. We call such a map *projective geometric*, and *special projective geometric* if it is defined by simple closed curves.

As before, it is well known (see [3]) that there are special projective geometric maps which embed $\mathcal{T}_g \cup \partial\mathcal{T}_g$ into a real projective space of dimension not bigger than $9g-9$. More is known for surfaces with cusps where our above discussion is also valid. Namely, the Teichmüller space of all marked complete finite volume hyperbolic structures on a noncompact punctured surface and its Thurston boundary admit projective geometric embeddings onto a compact finite sided polyhedron in a real projective space of minimal dimension [4]. However these embeddings are not special.

For closed surfaces of genus g the existence of projective geometric embeddings of $\mathcal{T}_g \cup \partial\mathcal{T}_g$ into $\mathbb{R}P^{6g-6}$ is an open problem. The only partial result is due to Schmutz [7] who constructed a special geometric embedding of $\mathcal{T}_2$ into $\mathbb{R}P^6$. However this embedding does not extend to an embedding of the Thurston boundary. We show below that the projectivization of the geometric embedding of $\mathcal{T}_g$ into $\mathbb{R}P^{6g-5}$ from Theorem A also does not extend to an injective map on the Thurston boundary.

This leads us to the simpler question whether the Thurston boundary $\partial\mathcal{T}_g$ of $\mathcal{T}_g$ can be parametrized by intersections with $6g-5$ simple closed curves. This question can be answered affirmatively. We show.

Theorem B *For every $g \geq 2$ there is a special projective geometric embedding of $\partial\mathcal{T}_g$ onto the boundary of a finite sided polyhedron in $\mathbb{R}P^{6g-6}$.*

We do not know whether our curve system which we construct in Theorem B defines an embedding of $\mathcal{T}_g$ into $\mathbb{R}P^{6g-6}$. We conjecture however that this is true and that the image of this embedding equals the interior of our polyhedron.

2 A Parametrization of Teichmüller Space

In this section we describe a parametrization of Teichmüller space by $6g-5$ length functions. Our construction is a modification of the construction of Schmutz [6].

First, recall that a *pair of pants* is a surface with boundary which is homeomorphic to a sphere with 3 holes. Let $P_1, \dots, P_{2g-2}$ be a collection of $2g-2$ pairs of pants. For $i \leq g-1$ label the boundary circles of P_{2i-1} by b_{2i-1}, b_{2i}, d_i. For $i \leq g-2$ label the boundary circles of P_{2i} by b_{2i}, b_{2i+1}, d_i. The boundary circles of P_{2g-2} are labelled by b_{2g-2}, b_1, d_{g-1}.

Glue the pairs of pants along the boundary circles with the same label. We obtain a closed surface S of genus g with a decomposition into $2g-2$ pairs of pants which is given by the circles $b_1, \dots, b_{2g-2}$ and the circles $d_1, \dots, d_{g-1}$.

For every hyperbolic structure on S the distinguished circles are freely homotopic to simple closed geodesics which decompose S into $2g-2$ pairs of pants with hyperbolic structures and geodesic boundaries. In the sequel we denote these geodesics by the same letters $b_1, \dots, b_{2g-2}, d_1, \dots, d_{g-1}$.

Recall that a *Dehn twist* along the simple closed curve b_i is a diffeomorphism of the surface S which leaves the complement of an embedded cylinder C in S with core-curve b_i pointwise fixed. It maps a simple arc c in C with endpoints on the different boundary components of C to an arc with the same endpoints which goes around the cylinder one more time.

Let a_0 be a simple closed geodesic which intersects each of the curves b_i transversely in a single point and does not intersect any of the curves d_j. Denote by a_i the geodesic obtained from a_0 by a Dehn twist along b_i. The curves a_0, a_i, b_i then mutually intersect transversely in a unique point.

Similarly, define e_0 to be a simple closed geodesic which intersects each of the curves b_{2j-1}, d_j in a single point and does not intersect any of the curves b_{2j}. Let e_j be the image of e_0 under a Dehn twist along d_j.

Notice that all our above closed geodesics are simple and non-separating and mutually intersect in at most one point. We have (compare [6]).

Proposition 2.1 *The special geometric map defined by the curves*

$$b_1, \ldots, b_{2g-2}, a_0, \ldots, a_{2g-2}, d_1, \ldots, d_{g-2}, e_0, \ldots, e_{g-1}$$

is an embedding of $\mathcal{T}_g$ into $\mathbb{R}P^{6g-5}$.

Proof. Denote by $\mathcal{F}$ our above family of geodesics. We have to show that the hyperbolic structure on our surface S is determined by the lengths of the curves from the collection $\mathcal{F}$. For this it is sufficient to show that the lengths of the boundary circles $b_1, \ldots, b_{2g-2}, d_1, \ldots, d_{g-1}$ of our pants decomposition of S and the *twist parameters* which determine how these pants are glued along their boundaries are determined by the lengths of the curves from $\mathcal{F}$.

Cut S open along the curves $d_1, \ldots, d_{g-1}$. We obtain a bordered torus with $2g-2$ boundary circles. Each of these circles is a copy of one of the curves $d_1, \ldots, d_{g-1}$. Following Schmutz (Proposition 1 and the discussion on p.283 of [S1]), the hyperbolic structure of this torus and the common length of the two boundary circles which correspond to the curve d_{g-1} are determined by the lengths of the geodesics $d_1, \ldots, d_{g-2}$ and the lengths of the inner geodesics $b_1, \ldots, b_{2g-2}, a_0, \ldots, a_{2g-2}$ (see also the discussion below). In particular, for every $i \in \{1, \ldots, 2g-2\}$ the twist parameter for the glueing of P_i, P_{i+1} along b_i is determined.

Consider the bordered torus T with $2g+2$ boundary circles which we obtain by cutting S open along the curves b_{2i} $(i = 1, \ldots, g-1)$. We have to show that the hyperbolic structure on T is determined by the lengths of the inner geodesics $e_0, \ldots, e_{g-1}, d_1, \ldots, d_{g-1}$ and the hyperbolic structures of the *X-pieces* X_i which we obtain by glueing P_{2i} and P_{2i+1} along the geodesic b_{2i+1}.

For this recall that e_i is the image of e_0 under a full Dehn twist along d_i. In particular, the geodesics e_0, e_i, d_i intersect mutually in a single point. The angles between them at these intersection points are determined by their lengths on S [6].

We argue now as in [6] and show that the twist parameters for glueing along d_i are determined. Namely, choose a lift of e_0 to the universal cover $\mathbf{H}^2$. We obtain a geodesic segment α of a given length $\ell = \ell_{e_0}$.

Since the hyperbolic structure on the subsurface X_i of T is determined, we know the length of the shortest arc c_i in X_i connecting the geodesics d_i and d_{i+1} and which is freely homotopic to the arc $e_0 \cap X_i$ relative to the boundary curves d_i, d_{i+1}. The subsegment α_i of α which is the lift of $e_0 \cap X_i$ forms together with segments from lifts of d_i, d_{i+1} and a lift $\tilde{c}_i$ of c_i a geodesic quadrangle Q_i with two right angles at the endpoints of $\tilde{c}_i$. The length of the side $\tilde{c}_i$ of Q_i and all the angles of Q_i are known. Hyperbolic trigonometry shows that Q_i is completely determined by these data. Thus the length of the arc α_i is also known.

Now the twist parameter for the glueing of X_i, X_{i+1} along d_{i+1} can be computed from the signed difference between the lengths of the sides of Q_i and Q_{i+1} which are contained in our lift of d_{i+1}. This then shows that the twist parameters for glueing along the geodesics $d_1, \dots, d_{g-1}$ are determined by the lengths of the curves a_i, b_j, d_k, e_ℓ. **q.e.d.**

We do not know whether the projective geometric map of $\mathcal{T}_g$ into $\mathbb{R}P^{6g-6}$ defined by our above curves is an embedding. However it is easy to see that its extension to the Thurston boundary is not injective.

Lemma 2.2 *The map which assigns to a projective measured lamination* $[\mu]$ *the projective class of the* $6g-5$*-tuple of intersection numbers with our curves*

$$b_1, \dots, b_{2g-2}, a_0, \dots, a_{2g-2}, d_1, \dots, d_{g-1}, e_0, \dots, e_{g-1}$$

is not injective.

Proof. It is sufficient to find two distinct simple closed geodesics γ_1, γ_2 on S which have the same number of intersections with each of our above curves b_i, a_j, d_k, e_ℓ.

For this let c_0 be the image of a_0 under a full Dehn twist along b_2 and a full Dehn twist along b_4. Let c_1 be the image of c_0 under a full Dehn twist along b_2, and let c_2 be the image of e_0 under a full Dehn twist along b_4. By the definition of our curve system we have $i(a_i, c_j) = i(a_i, c_0) + 1$, $i(b_i, c_j) = i(b_i, c_0)$, $i(d_i, c_j) = i(d_i, c_0)$, $i(e_i, c_j) = i(e_i, c_0)$ for all i and $j = 1, 2$ from which our lemma follows. **q.e.d.**

3 A Parametrization of the Thurston Boundary

Let $b_1, \dots, b_{3g-g}$ be any family of pairwise disjoint simple closed geodesics on a closed hyperbolic surface S of genus $g \geq 2$. These curves decompose our surface S into $2g-2$ pairs of pants. By the Dehn-Thurston theorem (see [5]), a simple closed geodesic γ is uniquely determined by its intersection numbers with the curves b_j and its *twist parameters* at b_j. The twist parameter $\tau_j(\gamma)$ of γ at b_j is an integer which measures the signed number of times our curve winds around b_j with respect to some reference arc. Thus if γ is any simple closed geodesic then the image γ' of γ under a positive Dehn twist along b_j satisfies $i(\gamma', b_i) = i(\gamma, b_i)$ for $i = 1, \dots, 3g-3$, $\tau_k(\gamma') = \tau_k(\gamma)$ for $k \neq j$ and $\tau_j(\gamma') = \tau_j(\gamma) + 1$.

Choose a simple closed curve a_0 on S which intersects each of the curves b_i in at least one point and has vanishing twist parameters. For $i = 1, \dots, 3g-3$ let a_i be the image of a_0 under a full Dehn twist along b_i. Moreover, for every $j \in \{1, \dots, 3g-3\}$ choose a simple closed curve c_j with vanishing twist parameters and such that the $(3g-3, 3g-3)$-matrix A with entry $i(b_i, c_j)$ in the i-th line and j-th row has maximal rank.

We can obtain such a pants decomposition and corresponding family of curves c_j as follows. Choose $g-1$ separating simple closed geodesics $b_2, \dots, b_g$ which cut S into $g-2$ tori $T_2, \dots, T_{g-1}$ with two boundary circles and 2 tori T_1, T_g with one boundary circle. Assume that the geodesic b_i separates T_{i-1} from T_i. Extend this system of curves to a pants decomposition of S by simple closed geodesics $b_1, \ldots, b_{3g-3}$.

For each $i \in \{2, \ldots, g\}$ choose a simple closed nonseparating geodesic c_i which intersects b_i in exactly two points and does not intersect b_j for $j \neq i$. Such a curve is contained in the subsurface $T_{i-1} \cup T_i$ of S. Choose moreover a simple closed geodesic c_1, c_{g+1} in T_1, T_g which intersects the curve from our pants decomposition contained in T_1, T_g in a single point.

Let $i \in \{2, \ldots, g-1\}$ and denote by b', b'' the geodesics from our pants decomposition which are contained in T_i. If we cut T_i open along b'' then we obtain a sphere with 4 holes. The simple closed geodesic b' separates two of these holes from the other two. Choose a simple closed geodesic c' which is contained in this sphere and intersects b' in exactly two points. Similarly we can find a simple closed geodesic c'' which intersects b'' in exactly two points and does not intersect b'. In this way we construct successively a family of curves $c_1, \ldots, c_{3g-3}$ with the property that with respect to a suitable numbering we have $i(c_i, b_i) = 2$ for $i = 2, \ldots, g, g+2, \ldots, 3g-3$, $i(c_1, b_1) = i(c_{g+1}, b_g) = 1$ and $i(b_i, c_j) = 0$ otherwise.

We show

Proposition 3.3 *The special projective geometric map defined by the curves $a_0, \ldots, a_{3g-3}, c_1, \ldots, c_{3g-3}$ is an embedding of $\partial\mathcal{T}_g$ into $\mathbb{R}P^{6g-6}$.*

Proof. Recall that a *simple geodesic multicurve* is a collection of mutually disjoint simple closed geodesics with nonnegative multiplicities on our surface S. Simple geodesic multicurves are dense in the space of measured geodesic lamination and therefore it is enough to show that a simple geodesic multicurve is determined by its intersections with our $6g-5$ geodesics a_i, c_j.

For this recall that the curves $b_1, \ldots, b_{3g-3}$ define a pants decomposition of S and therefore a simple geodesic multicurve is uniquely determined by its intersections with the curves b_j and its twist parameters at b_j.

We show first that the intersection numbers $i(\gamma, a_i)$ determine the intersection numbers $i(\gamma, b_j)$ ($j = 1, \ldots, 3g-3$) and the signs of all the twist parameters.

Namely, denote by D_j the Dehn twist along b_j such that $a_j = D_j a_0$. Recall that D_j leaves the complement of a cylinder C_j about b_j pointwise fixed. Assume that $i(\gamma, b_j) \neq 0$ and let m_j be the twist parameter of γ at b_j. If $i(\gamma, b_j) = k$ then the multicurve γ intersects the cylinder C_j in k arcs. Up to homotopy, each of these arcs meets b_j in a unique point, and it has $|m_j| i(b_j, a_0)$ intersection points with the multi-arc $a_0 \cap C_j$ and $|m_j - 1| i(b_j, a_0)$ intersection points with the multi-arc $a_j \cap C_j$. Outside the cylinder C_j the number of intersection points between γ and a_0, a_j coincide. But this just means that $i(\gamma, b_j) = |i(\gamma, a_0) - i(\gamma, a_j)| / i(a_0, b_j)$. Moreover our twist parameter m_j is positive if and only if $i(\gamma, a_j) < i(\gamma, a_0)$, and it is nonpositive otherwise.

We are left with showing that the absolute values of the twist parameters are also determined. For this it is enough to show that for any simple geodesic multicurve and any $3g-3$-tuple $k_1, \ldots, k_{3g-3}$ of integers the intersection numbers with the geodesics a_i, c_j of the multicurves γ and $D_1^{k_1} \circ \cdots \circ D_{3g-3}^{k_{3g-3}}\gamma = \gamma'$ do not all coincide.

Recall that the signs of the twist parameters are determined by the intersections with the curves a_j, so we may assume that the twist parameters of γ and γ' have the same sign. Write $q(i) = i(\gamma, b_i)$ if the twist parameter of γ at b_i is positive, and write $q(i) = -i(\gamma, b_i)$ otherwise. Since the twist parameters of the curve c_j vanish, for each i the Dehn twist $D_i^{k_i}$ applied to γ changes the intersection of γ with the multi-arc $c_j \cap C_i$ in the cylinder C_i with core curve b_i by $q(i)k_i i(b_i, c_j)$.

Thus we have $i(D_i^{k_i}\gamma, c_j) = i(\gamma, c_j) + k_i i(b_i, c_j)q(i)$ and consequently our above statement follows if we can show that the system of $3g-3$ equations

$$\sum_{i=1}^{3g-3} k_i i(b_i, c_j)q(i) = 0 \quad (j = 1, \ldots, 3g-3)$$

only has solutions which satisfy $k_i = 0$ for all i such that $q(i) \neq 0$.

Let E be the $(3g-3, 3g-3)$-matrix whose j-th line is the vector

$$e_j = (i(b_1, c_j)q(1), \ldots, i(b_{3g-3}, c_j)q(3g-3)).$$

After reordering of our indices we may assume that $q(i) \neq 0$ for $i \leq k$ and $q(i) = 0$ otherwise. Then the last $3g-3-k$ rows of the matrix E vanish identically. Since we may multiply a row of E with a nonzero number without changing the rank of E, the rank of C coincides with the rank of a $(3g-3, k)$-submatrix of our matrix A. By our assumption on A, the rank of this submatrix equals k and therefore the proposition follows. **q.e.d.**

The next lemma concludes the proof of our Theorem B.

Lemma 3.4 *The image of $\partial\mathcal{T}_g$ under our projective geometric embedding equals the boundary of a finite sided polyhedron in $\mathbb{R}P^{6g-6}$.*

Proof. We show that the image of the space $\mathcal{L}$ of measured laminations under the map Φ which assigns to a lamination μ its $6g-5$-tuple of intersection numbers with the geodesics a_i, c_j equals a finite union of $6g-6$-dimensional convex polyhedra in $\mathbb{R}^{6g-5}$, each contained in a linear hyperplane, and which intersect along their boundaries.

Consider again the pants decomposition of S by the curves $b_1, \ldots, b_{3g-3}$. There is a finite number of *train tracks* in standard position with respect to this pants decomposition. Each of these train tracks T consists of a finite number of *branches* and 3-*valent switches*. Every measured lamination is *carried* by one of these tracks, i.e. there is such a track T such that the lamination is given by a collection of nonnegative weights on the branches of T which satisfy the algebraic *switch conditions*.

For every given such train track T the set U of laminations which are carried by T and which are defined by positive weights on all branches of T is an open subset of $\mathcal{L}$. Its boundary consists of laminations which are carried by more than one train track in standard position (for all this see [5]).

For every $\mu \in U$ the intersection $i(\mu, b_i)$ is represented by a weight on one of the branches of T and the same is true for the twist parameter $\tau_i(\mu)$ about b_i, where the sign of this twist parameter is determined by the train track T. Since the switch conditions are linear equations in the weights, the set U admits a natural convex linear structure in such a way that the intersection functionals $i(\cdot, b_j)$ and the twist parameters τ_i are linear.

By the proof of Proposition 3.3, for every i there is an integer $p_i \geq 1$ such that the intersection number of any measured geodesic lamination μ with the curve b_i equals $|i(a_0, \mu) - i(a_i, \mu)|/p_i$. Similarly, again by the proof of Proposition 3.3, for any fixed choice of signs for the twist parameters τ_j at the pants curves b_j, the twist parameters $\tau_j(\mu)$ of a lamination μ with twist parameters of these signs are solutions of $3g - 3$ independent linear equations in the intersection numbers of μ with the curves c_j. Therefore there are $6g - 5$ linear equations in the intersection numbers $i(a_i, \mu), i(c_j, \mu)$ which depend on T and determine the intersection numbers $i(\mu, b_j)$ and the twist parameters $\tau_j(\mu)$.

Thus if $(u_1, \ldots, u_{6g-5}) \in \mathbb{R}^{6g-5}$ spans the kernel of the corresponding linear map, then the restriction of our map Φ to U is a linear map of a convex cone in an affine space of dimension $6g - 6$ onto a convex set in the hyperplane

$$\{(x_1, \ldots, x_{6g-5}) \in \mathbb{R}^{6g-5} \mid \sum_{i=1}^{6g-5} u_i x_i = 0\}.$$

But this just means that the image of Φ equals the union of finitely many convex cones in hyperplanes which intersect at their boundaries. Since $\mathcal{L}$ is homeomorphic to $\mathbb{R}^{6g-6}$ this implies that the image of $\partial \mathcal{T}_g$ under the special geometric map which is the projectivization of Φ equals the boundary of a finite sided polyhedron in $\mathbb{R}P^{6g-6}$. This finishes the proof of our lemma. **q.e.d.**

References

1. F. Bonahon: The geometry of Teichmüller space via geodesic currents. Invent. Math. 92 (1988), 139–162
2. P. Buser: Geometry and spectra of compact Riemann surfaces. Birkhäuser, Boston 1992
3. A. Fathi, F. Laudenbach, V. Poénaru: Travaux de Thurston sur les surfaces. Astérisque 66–67 (1979)
4. U. Hamenstädt: Length function and parameterizations of Teichmüller space for surfaces with cusps. preprint 2001
5. R. Harer, J. Penner: Combinatorics of train tracks. Princeton University Press, Princeton 1992
6. P. Schmutz: Die Parametrisierung des Teichmüllerraumes durch geodätische Längenfunktionen. Comment. Math. Helvetici 68 (1993), 278–288
7. P. Schmutz Schaller: Teichmüller space and fundamental domains of Fuchsian groups. L'Enseignement Mathématique 45 (1999), 169–187

Linearization of Isotropic Automorphisms of Non-quadratic Elliptic CR-Manifolds in $\mathbb{C}^4$

Vladimir V. Ežov[1] and Gerd Schmalz[2]

[1] University of South Australia, School of Mathematics, Mawson Lakes, SA 5095, South Australia
vladimir.ejov@unisa.edu.au

[2] Mathematisches Institut der Universität Bonn, Beringstraße 1, D-53115 Bonn
schmalz@math.uni-bonn.de

1 Introduction

In 1974 Chern and Moser [4] constructed normal forms for real-analytic hypersurfaces with non-degenerate Levi-form in $\mathbb{C}^{n+1}$. For a real-analytic hypersurface M in $\mathbb{C}^2$ this means that there are local coordinates z, w centered in $0 \in M$ such that the equation of M takes the form

$$M : \operatorname{Im} w = |z|^2 + \sum_{\substack{k,\ell \geq 2 \\ \max(k,\ell) \geq 4}} n_{k\ell}(\operatorname{Re} w) z^k \bar{z}^\ell. \tag{1}$$

These coordinates are determined up to a rather complicate action of a 5-dimensional Lie group. This Lie group is nothing but the isotropy group of the hyperquadric

$$Q : \operatorname{Im} w = |z|^2,$$

i.e. the group of germs of biholomorphic mappings that preserve 0 and Q. Normal forms proved to be very useful in the study of automorphisms of hypersurfaces since an isotropic automorphism can be viewed as a transformation from normal form to normal form if the hypersurface was given in normal form. This way Kružilin and Loboda [13] established for real-analytic strictly pseudoconvex hypersurfaces that are not locally equivalent to the sphere the existence of normal coordinates such that all isotropic automorphisms are linear mappings simultaneously, i.e. they linearized the isotropy group of M in normal coordinates. The quadric (sphere) has to be excluded because of their automorphisms

$$z^* = \frac{z + aw}{1 - 2\mathrm{i}\,\bar{a}z - (r + \mathrm{i}\,|a|^2)w}$$
$$w^* = \frac{w}{1 - 2\mathrm{i}\,\bar{a}z - (r + \mathrm{i}\,|a|^2)w}$$

with $a \in \mathbb{C}$, $r \in \mathbb{R}$.

For general M we consider the analogous subgroup of "nil-isotropic" automorphisms

$$\mathrm{Aut}_{0,\mathrm{id}} = \left\{ \phi \in \mathrm{Aut}(M) : \phi(0) = 0, \ \frac{\partial z^*}{\partial z} = 1 \right\}.$$

Since isotropic automorphisms of M coincide with quadric automorphisms up to 3rd order, non-triviality of $\mathrm{Aut}_{0,\mathrm{id}}$ is an obstruction against linearizability of the isotropy group. Belošapka [1] and Loboda [14] proved the triviality of $\mathrm{Aut}_{0,\mathrm{id}}$ also for hypersurfaces with non-degenerate indefinite Levi form. However, the isotropy groups of such hypersurfaces cannot be simultaneously linearized in general (see [7]). Though, for each individual isotropic automorphism ϕ there are normal coordinates where ϕ is linear (see [5]).

Besides real hypersurfaces that appear naturally as boundaries of complex domains also real submanifolds of higher codimension are important in the study of properties of complex domains. Their geometry turned out to be much more difficult than the geometry of hypersurfaces. It is impossible to construct a universal normal form even with fixed dimension and codimension. There are only a few cases where the non-degenerate vector-valued Levi forms admit a discrete classification similar to the classification by signature of a non-degenerate scalar hermitian form. The simplest case of 2-codimensional submanifolds in $\mathbb{C}^4$ led to an interesting and rich geometry.

A submanifold M in $\mathbb{C}^N$ is called a CR-manifold if the complex dimension of the spaces $H_x = \{\xi \in T_x M : \mathrm{i}\xi \in T_x M\}$ is constant. This is automatically satisfied for hypersurfaces. For 2-codimensional submanifolds in $\mathbb{C}^4$ we consider the case when this constant (called the CR-dimension) equals 2.

The Levi form is then the algebraic bracket

$$H_x \otimes \bar{H}_x \to T_x M / H_x M$$

defined as the Lie bracket followed by the projection and divided by i. This Levi form can be viewed as a pair of two hermitian forms on $H_x \cong \mathbb{C}^2$ defined up to an action of $\mathrm{GL}(2,\mathbb{C}) \times \mathrm{GL}(2,\mathbb{R})$ related to the choice of coordinates in H_x and $T_x M / H_x M$. We impose the following non-degeneracy condition: the two scalar hermitian forms are linearly independent and they have no common annihilator. The non-degenerate $\mathbb{R}^2$-valued hermitian forms on $\mathbb{C}^2$ comprise 3 classes represented by: $(|z_1|^2, |z_2|^2)$ (called hyperbolic), $(\mathrm{Re}\, z_1 \bar{z}_2, \mathrm{Im}\, z_1 \bar{z}_2)$ (called elliptic) and $(|z_1|^2, \mathrm{Re}\, z_1 \bar{z}_2)$ (parabolic).

The hyperbolic case appears whenever the scalar hermitian forms possess a positive definite linear combination. Then they can be simultaneously diagonalized and the above form can be obtained by a choice of suitable linear combinations. It is clear that hyperbolicity of the Levi form at one point implies hyperbolicity in some neighbourhood.

Similarly, the elliptic Levi forms appear if the two scalar hermitian forms have no degenerate (real) linear combinations. This is also an open condition.

In difference to this parabolicity is not an open condition. Therefore, the study of CR-manifolds in a neighbourhood of a parabolic point is much more difficult.

The linearizability of the isotropy group of hyperbolic CR-manifolds was proved by Schmalz in [17]. Here, in addition to the quadric $\operatorname{Im} w_1 = |z_1|^2$, $\operatorname{Im} w_2 = |z_2|^2$ one has to exclude direct products of two hypersurfaces in $\mathbb{C}^2$ one of them being equivalent to a sphere.

From now on let M be an elliptic CR-manifold.

According to [9] (and further, [8]), M can be written locally in normal coordinates $(z_1, z_2, \omega_1, \omega_2)$ by one complex equation

$$V - z_1 \bar{z}_2 = N(z_1, \bar{z}_1, z_2, \bar{z}_2, \eta, \bar{\eta}),$$

where $V = \frac{\omega_1 - \bar{\omega}_2}{2\mathrm{i}}$, $\eta = \frac{\omega_1 + \bar{\omega}_2}{2}$ and N uniquely splits into the sum

$$N = N^{CM}(z_1, \bar{z}_2, \eta) + N^{tor}(z_1, \bar{z}_1, z_2, \bar{z}_2, \eta, \bar{\eta}).$$

The "Chern-Moser component" N^{CM} is a "polarization" of the right hand side in (1), i.e. its restriction to $z_1 = z_2$ and $\eta \in \mathbb{R}$ is a power series in $(z_1, \bar{z}_1, \eta)$ whose real and imaginary part are in classical Chern-Moser normal form (1).

The "torsion component" is defined by $N^{tor} = \sum_{\substack{k,\ell \geq 1 \\ k+\ell \geq 3}} N^{tor}_{k\ell}$ where the homogeneous polynomials $N^{tor}_{k\ell}$ of degree k in z and ℓ in $\bar{z}$ meet the conditions

$$\begin{aligned} N^{tor}_{k,0} &= 0, \quad k \geq 0, & N^{tor}_{1,1} &= 0 \\ \frac{\partial N^{tor}_{k1}}{\partial \bar{z}_2} &= 0, \quad k \geq 2, & \frac{\partial^2 N^{tor}_{21}}{\partial z_1 \partial z_2} &= 0 \\ \frac{\partial N^{tor}_{1k}}{\partial z_1} &= 0, \quad k \geq 2, & \frac{\partial^2 N^{tor}_{12}}{\partial \bar{z}_1 \partial \bar{z}_2} &= 0 \\ \frac{\partial^4 N^{tor}_{22}}{\partial z_1 \partial z_2 \partial \bar{z}_1 \partial \bar{z}_2} &= 0. \end{aligned}$$

It was proved in [3] that partially integrable abstract elliptic CR-manifolds of CR dimension and codimension 2 without torsion are automatically real-analytic, embeddable and N^{tor} vanishes identically.

In [12] the authors showed that the subgroup $\operatorname{Aut}_{0,\mathrm{id}} M$ of nil-isotropic automorphisms $\psi = (f, g): \quad z^* = f(z, \omega), \quad \omega^* = g(z, \omega)$ with

$$\left. \frac{\partial f}{\partial z} \right|_{\substack{z=0 \\ \omega=0}} = \mathrm{id}$$

is isomorphic to either $\{\mathrm{id}\}$ or $\mathbb{C}$. The case $\operatorname{Aut}_{0,\mathrm{id}} M = \mathbb{C}$ is extremely restrictive and occurs only for a family of exceptional manifolds parameterized by at most one complex and two integer parameters. The existence and explicit description of such exceptional manifolds are yet to be established.

In this paper we prove linearizability of Aut_0 for non-exceptional elliptic manifolds. This provides us with a simple description of non-exceptional manifolds with non-trivial isotropy group in terms of the normal form.

We will use the following notations: Recall that for any $\psi \in \operatorname{Aut}_0 M$ the linear map

$$L_\psi : \quad (z,\omega) \mapsto (Cz, \rho\omega)$$

with

$$C = \left.\frac{\partial f}{\partial z}\right|_0, \qquad \rho = \left.\frac{\partial g}{\partial \omega}\right|_0$$

is an automorphism of the quadric $Q: \quad V - z_1\bar{z}_2 = 0$. By $\operatorname{Aut}_0^0 M$ we denote the subgroup of $\operatorname{Aut}_0 M$ that consists of automorphisms ψ such that L_ψ belongs to the connected component of the identity in $\operatorname{Aut}_0 Q$. We note that for $\psi \in \operatorname{Aut}_0^0 M$ the matrices C and ρ are diagonal.

Let $\psi = (f,g) \in \operatorname{Aut}_0^0 M$. Define $a = (a_1, a_2) \in \mathbb{C}^2$ and $r \in \mathbb{C}$ via

$$\left.\frac{\partial f}{\partial \omega}\right|_0 = C \begin{pmatrix} a_1 & 0 \\ 0 & a_2 \end{pmatrix} \tag{2}$$

$$\frac{1}{2}\left(\left.\frac{\partial^2 g_1}{(\partial \omega_1)^2}\right|_0 + \overline{\left.\frac{\partial^2 g_2}{(\partial \omega_2)^2}\right|_0} \right) = \rho_{11} r.$$

Since M is given in normal form we can consider ψ as a normalization with parameters C, ρ, a, r (see [9]). Therefore, ψ equals up to terms of order 3 in z, ω to a fractional linear automorphism of the elliptic CR-quadric $V - z_1\bar{z}_2 = 0$:

$$\begin{aligned} \tilde{z}_1 &= \frac{z_1 + a_1\omega_1}{1 - 2\,\mathrm{i}\,\bar{a}_2 z_1 - (r + \mathrm{i}\, a_1\bar{a}_2)\omega_1} \\ \tilde{z}_2 &= \frac{z_2 + a_2\omega_2}{1 - 2\,\mathrm{i}\,\bar{a}_1 z_2 - (\bar{r} + \mathrm{i}\, a_2\bar{a}_1)\omega_2} \\ \tilde{\omega}_1 &= \frac{\omega_1}{1 - 2\,\mathrm{i}\,\bar{a}_2 z_1 - (r + \mathrm{i}\, a_1\bar{a}_2)\omega_1} \\ \tilde{z}_2 &= \frac{\omega_2}{1 - 2\,\mathrm{i}\,\bar{a}_1 z_2 - (\bar{r} + \mathrm{i}\, a_2\bar{a}_1)\omega_2}. \end{aligned} \tag{3}$$

If we plug ψ into the equation of M we obtain an identity which is satisfied for all points of M:

$$\frac{g_1 - \bar{g}_2}{2\,\mathrm{i}} - f_1\bar{f}_2 - N\left(f_1, \bar{f}_1, f_2, \bar{f}_2, \frac{g_1 + \bar{g}_2}{2}, \frac{\bar{g}_1 + g_2}{2}\right)\Bigg|_{V = z_1\bar{z}_2 - N} \equiv 0 \tag{4}$$

To analyze this identity we will use the following notion of weight: The weight γ is defined by

$$\gamma(z_j) = 1, \qquad \gamma(\omega_j) = \gamma(\eta) = \gamma(\bar{\eta}) = 2$$

$$\gamma\left(\frac{\partial}{\partial z_j}\right) = -1, \qquad \gamma\left(\frac{\partial}{\partial \omega_j}\right) = \gamma\left(\frac{\partial}{\partial \eta}\right) = \gamma\left(\frac{\partial}{\partial \bar{\eta}}\right) = -2,$$

and the property that the weight of a product is the sum of the weights of its factors.

If $\psi \in \mathrm{Aut}_0 M$ is linear, i.e. $\psi = L_\psi$, identity (4) takes the simple form

$$(z, \bar{z}, \eta, \bar{\eta}) = T_\psi N(z, \bar{z}, \eta, \bar{\eta}).$$

where T_ψ denotes the action of L_ψ on N (or weighted homogeneous polynomials from the expansion of N) induced by the linear mapping L_ψ.

Now we can describe the manifolds with linearizable isotropy group $\mathrm{Aut}_0 M$ *as those manifolds whose equation in suitable normal coordinates consists of polynomials which are invariant with respect to* T_ψ *for* $\psi \in \mathrm{Aut}_0 M$.

For any monomial Π in $z_1, z_2, \bar{z}_1, \bar{z}_2, \eta, \bar{\eta}$ we define the 1-defect $d_1(\Pi)$ as the difference of the degrees with respect to z_1 and with respect to $\bar{z}_2$.

By $\bar{1}$-defect $d_{\bar{1}}(\Pi)$ we will denote the difference of the degrees with respect to $\bar{z}_1$ and with respect to z_2.

The total defect $\delta(\Pi)$ is the sum of d_1 and $d_{\bar{1}}$.

According to these definitions we expand

$$N(z, \bar{z}, \eta, \bar{\eta}) = \sum_{\gamma=\gamma_0}^{\infty} N_\gamma(z, \bar{z}, \eta, \bar{\eta}),$$

where N_γ is the component of weight γ in N and $N_{\gamma_0} \neq 0$. We can further split N_γ into components $N_{\gamma, d_1, d_{\bar{1}}}$ of 1-defect d_1 and $\bar{1}$-defect $d_{\bar{1}}$. We are interested in certain properties of 3 consecutive components $N_{\gamma_0}, N_{\gamma_0+1}, N_{\gamma_0+2}$ that arise from the components of weights $\gamma_0, \gamma_0+1, \gamma_0+2$ in (4).

For N_{γ_0} identity (4) implies

$$\rho_{11} N_{\gamma_0}(z, \bar{z}, \eta, \bar{\eta}) = N_{\gamma_0}(Cz, \bar{C}\bar{z}, \rho_{11}\eta, \rho_{22}\bar{\eta}). \tag{5}$$

Suppose that $a_1 = 0$ and let $N_{\gamma_0,\delta}$ be the component of maximal total defect in N_{γ_0}. Making use of the notation of the vector field

$$\chi_1^1 = \mathrm{i}\,\bar{a}_2 \left(2z_1^2 \frac{\partial}{\partial z_1} + 2(\eta + \mathrm{i}\,V) z_1 \frac{\partial}{\partial \omega_1} - \mathrm{i}(\eta - \mathrm{i}\,V) \frac{\partial}{\partial \bar{z}_2} \right)$$

we derive from the component of weight-defect $(\gamma_0+1, \delta+1)$ in identity (4)

$$\begin{aligned} &\rho_{11} N_{\gamma_0+1,\delta+1}(z, \bar{z}, \eta, \bar{\eta}) = \\ &\qquad N_{\gamma_0+1,\delta+1}(Cz, \bar{C}\bar{z}, \rho_{11}\eta, \rho_{22}\bar{\eta}) + \rho_{11} C_{11} \chi_1^1 N_{\gamma_0,\delta}(z, \bar{z}, \eta, \bar{\eta}). \end{aligned} \tag{6}$$

And, finally, if $a = 0$ then the (γ_0+2, δ)-component of identity (4) provides

$$\rho_{11} N_{\gamma_0+2,\delta}(z, \bar{z}, \eta, \bar{\eta}) = N_{\gamma_0+2,\delta}(Cz, \bar{C}\bar{z}, \rho_{11}\eta, \rho_{22}\bar{\eta}) + 2\rho_{11}^2 (\mathrm{Re}\,\chi_2) N_{\gamma_0,\delta} \tag{7}$$

where χ_2 is the vector field

$$\chi_2 = \sum_{j=1,2} r_j z_j \omega_j \frac{\partial}{\partial z_j} + r_j \omega_j^2 \frac{\partial}{\partial \omega_j},$$

with $r_1 = r = \bar{r}_2$.

Now we are ready to prove the individual linearization of isotropic automorphisms.

2 Linearization of a Single Automorphism

Let M be a non-exceptional manifold. We may assume that M is given in normal form. Then any automorphism

$$\psi : z^* = f(z,\omega) \quad \omega^* = g(z,\omega)$$

from $\mathrm{Aut}_0^0 M$ can be considered as a normalization, and, therefore, it is uniquely determined by the parameters $\lambda \in \mathbb{R}_{>0}, \mu \in \mathbb{R}_{>0}, -\pi < \theta \leq \pi, -\frac{\pi}{2} < \phi \leq \frac{\pi}{2}$ defined by

$$C = \left.\frac{\partial f}{\partial z}\right|_0 = \lambda \,\mathrm{e}^{\mathrm{i}\,\phi} \begin{pmatrix} \mu\,\mathrm{e}^{\mathrm{i}\,\theta} & 0 \\ 0 & \frac{1}{\mu}\,\mathrm{e}^{-\mathrm{i}\,\theta} \end{pmatrix}$$

and by the parameters a, r defined in (2). Though, all these parameters are free for the quadric, this is no longer the case for automorphisms of non-quadratic manifolds.

Lemma 2.1 *Parameters* (a, r, λ) *are uniquely determined by* (μ, ϕ, θ).

Proof. Suppose that two automorphisms in $\mathrm{Aut}_0^0 M$ have the same parameters $\lambda, \mu, \phi, \theta$. Then according to the Main Theorem of [12] automorphism $\psi_0 = \psi_1 \circ \psi_2^{-1} \in \mathrm{Aut}_{0,\mathrm{id}}$ is the identical map. Thus, $\psi_1 = \psi_2$ and the corresponding parameters a and r coincide. To show that λ is also determined by (μ, ϕ, θ) we consider the equation of M in normal form

$$V = z_1 \bar{z}_2 + N(z, \bar{z}, \eta, \bar{\eta}) \tag{8}$$

By T_C we denote the action of C on N (or homogeneous polynomials from the expansion of N) induced by the linear mapping

$$z^* = Cz, \quad \omega^* = \rho\omega,$$

where $\rho = \lambda^2\,\mathrm{e}^{2\,\mathrm{i}\,\theta}\,\mathrm{id}$, i.e. the operator

$$T_C N(z, \bar{z}, \eta, \bar{\eta}) = \frac{\mathrm{e}^{-2\,\mathrm{i}\,\theta}}{\lambda^2} N(Cz, \overline{Cz}, \lambda^2\,\mathrm{e}^{2\,\mathrm{i}\,\theta}\,\eta, \lambda^2\,\mathrm{e}^{-2\,\mathrm{i}\,\theta}\,\bar{\eta}).$$

Let N_{γ_0} be the first non-vanishing weighted component of N. Since ψ preserves the equation (8) then

$$T_C N_{\gamma_0} = N_{\gamma_0}.$$

Any monomial

$$\Pi = c z_1^{\alpha} \bar{z}_1^{\bar{\alpha}} z_2^{\beta} \bar{z}_2^{\bar{\beta}} \eta^{\delta} \bar{\eta}^{\bar{\delta}}$$

in N_{γ_0} is an eigenvector of T_C with the eigenvalue

$$\mathrm{e}^{\mathrm{i}(\alpha+\beta-\bar{\alpha}-\bar{\beta})\phi}\,\mathrm{e}^{\mathrm{i}(\alpha+\bar{\beta}-\bar{\alpha}-\beta+2\delta-2\bar{\delta}-2)\theta}\,\lambda^{\gamma_0-2}\mu^{\alpha+\bar{\alpha}-\beta-\bar{\beta}}.$$

Hence, λ is determined by $\lambda = \mu^{\frac{\alpha+\bar{\alpha}-\beta-\bar{\beta}}{2-\gamma_0}}$. □

Remark 2.2 *If $\alpha + \bar{\beta} + 2\delta = \bar{\alpha} + \beta + 2\bar{\delta} + 2$ then, in general, it is impossible to determine θ via ϕ. If $\alpha + \beta = \bar{\alpha} + \bar{\beta}$ then ϕ cannot be determined by θ.*

Corollary 2.3 *The dimension of* $\mathrm{Aut}_0 M$ *does not exceed 3.*

Since parameters a, r depend on the other parameters in C it will be our first aim to choose normal coordinates so that in these coordinates $a = 0$ and $r = 0$ (this is the simplest possible dependence).

Lemma 2.4 *For each $\psi \in \mathrm{Aut}_0^0 M$ there exist normal coordinates such that $a = 0$.*

Proof. According to Theorem 3 in [9] it suffices to find coordinates such that the chain $\Gamma_0 = \{z = 0, v = 0\}$ is invariant with respect to ψ. Therefore, we will look for an invariant chain Γ and pass to such normal coordinates where Γ will be mapped to Γ_0. Since Γ is uniquely defined by $T_0\Gamma$, ψ-invariance of Γ means that

$$d\psi|_0 T_0\Gamma = T_0\Gamma. \tag{9}$$

Applying Jacobian

$$d\psi|_0 = \left(\begin{array}{cc|cc} \lambda\mu\,\mathrm{e}^{\mathrm{i}(\phi+\theta)} & 0 & \lambda\mu\,\mathrm{e}^{\mathrm{i}(\phi+\theta)}\,a_1 & 0 \\ 0 & \frac{\lambda}{\mu}\,\mathrm{e}^{\mathrm{i}(\phi-\theta)} & 0 & \frac{\lambda}{\mu}\,\mathrm{e}^{\mathrm{i}(\phi-\theta)}\,a_2 \\ \hline 0 & 0 & \lambda^2\,\mathrm{e}^{2\,\mathrm{i}\,\theta} & 0 \\ 0 & 0 & 0 & \lambda^2\,\mathrm{e}^{-2\,\mathrm{i}\,\theta} \end{array}\right)$$

to the basis $\tau_1 = (b_1, 0, 1, 0)^t$, $\tau_2 = (0, b_2, 0, 1)^t$ of $T_0\Gamma$ we obtain

$$d\psi|_0 \begin{pmatrix} b_1 & 0 \\ 0 & b_2 \\ 1 & 0 \\ 0 & 1 \end{pmatrix} = \begin{pmatrix} \lambda\mu\,\mathrm{e}^{\mathrm{i}(\phi+\theta)}(b_1 + a_1) & 0 \\ 0 & \frac{\lambda}{\mu}\,\mathrm{e}^{\mathrm{i}(\phi-\theta)}(b_2 + a_2) \\ \lambda^2\,\mathrm{e}^{2\,\mathrm{i}\,\theta} & 0 \\ 0 & \lambda^2\,\mathrm{e}^{-2\,\mathrm{i}\,\theta} \end{pmatrix}$$

Now, equation (9) implies that the columns in the matrix at the right hand side are linear combinations of the two basis vectors. This can only happen if the first one is τ_1 scaled by $\lambda^2\,\mathrm{e}^{2\,\mathrm{i}\,\theta}$ and the second one is τ_2 scaled by $\lambda^2\,\mathrm{e}^{-2\,\mathrm{i}\,\theta}$ and therefore

$$\begin{aligned} a_1 &= \left(\frac{\lambda}{\mu}\,\mathrm{e}^{\mathrm{i}(\theta-\phi)} - 1\right) b_1 \\ a_2 &= \left(\lambda\mu\,\mathrm{e}^{-\,\mathrm{i}(\theta+\phi)} - 1\right) b_2. \end{aligned}$$

Using the splitting $C = \Lambda U$ with $\Lambda = \lambda \begin{pmatrix} \mathrm{e}^{\mathrm{i}\,\theta} & 0 \\ 0 & \mathrm{e}^{-\,\mathrm{i}\,\theta} \end{pmatrix}$, $U = \mathrm{e}^{\mathrm{i}\,\phi} \begin{pmatrix} \mu & 0 \\ 0 & \frac{1}{\mu} \end{pmatrix}$ this can be written as

$$(\Lambda U^{-1} - \mathrm{id})b = a. \tag{10}$$

This system is always soluble for b if ΛU^{-1} does not have eigenvalue 1. Then, solving (10) for b we find the unique ψ-invariant chain tangent to the span of τ_1, τ_2.

Now, suppose that 1 is an eigenvalue of ΛU^{-1}. Then either $\lambda = \frac{1}{\mu}$ and $\theta = -\phi$ (i.e., $C_{11} = 1$), or $\lambda = \mu$ and $\theta = \phi$ (i.e., $C_{22} = 1$). Without loss of generality we may restrict ourselves to the first case. System (10) will have a (non-unique) solution if and only if $a_2 = 0$. In this situation we find a complex 1-parameter family of ψ-invariant chains.

We will show by contradiction that, in fact, $a_2 = 0$ if ψ is an automorphism of a non-exceptional manifold. Then (5) and (6) take the form

$$T_C N_{\gamma_0} = N_{\gamma_0} \tag{11}$$

$$(T_C - \mathrm{id}) N_{\gamma_0+1,\delta+1} = \# \chi_1^1 N_{\gamma_0,\delta}. \tag{12}$$

Here and everywhere below # means a non-vanishing constant.

By (11) N_{γ_0} belongs to the kernel of $T_C - \mathrm{id}$ which indicates a certain balance in the distribution of the variables in the monomials of N_{γ_0}. The vector field χ_1^1 increases the 1-defect by 1, i.e. it changes the balance in such a way that that T_C would act with an additional factor C_{11}. But, since $C_{11} = 1$, it follows that $\chi_1^1 N_{\gamma_0}$ belongs to the kernel of $T_C - \mathrm{id}$ as well. Therefore, the right hand side of (12) equals $\#\chi_1^1 N_{\gamma_0,\delta}$ and lies in the kernel of $T_C - \mathrm{id}$ whereas the left hand side of (12) lies in the range of $T_C - \mathrm{id}$. As $T_C - \mathrm{id}$ is diagonalizable (the monomials form an eigenbasis) its range is spanned by the eigenvectors with eigenvalues different from 0 and therefore intersects with the kernel by $\{0\}$. Thus, $\chi_1^1 N_{\gamma_0,\delta} = 0$.

On the other hand, in [12] it was proved that for a non-exceptional manifold $\chi_1^1 N_{\gamma_0} \neq 0$ if $a_2 \neq 0$. Contradiction. □

Having chosen coordinates where Γ_0 is an ψ-invariant chain, we can now change the coordinates corresponding to a reparameterization of this chain in order to make $r = 0$:

Lemma 2.5 *Let* $\psi \in \mathrm{Aut}_0^0 M$ *and* $a = 0$ *then in some normal coordinates* $r = 0$.

Proof. Suppose that $\lambda^2 e^{2i\theta} \neq 1$. Consider the transformation

$$\begin{aligned} z^* &\doteq (\mathrm{id} - R)^{-1} z \\ \omega^* &= (\mathrm{id} - R)^{-1} \omega \end{aligned} \tag{13}$$

where

$$R = \begin{pmatrix} r'\omega_1 & 0 \\ 0 & \bar{r}'\omega_2 \end{pmatrix}, \quad z = \begin{pmatrix} z_1 \\ z_2 \end{pmatrix}, \quad \omega = \begin{pmatrix} \omega_1 \\ \omega_2 \end{pmatrix}.$$

One computes that in $*$-coordinates the new parameter

$$r^* = r - (\lambda^2 e^{2i\theta} - 1) r'.$$

Taking

$$r' = \frac{r}{1 - \lambda^2 e^{2i\theta}}$$

we obtain $r^* = 0$.

Suppose now that

$$\lambda^2 e^{2i\theta} = 1, \tag{14}$$

i.e., $\lambda = 1$, and $e^{2i\theta} = 1$. We show that then $r = 0$ automatically. Identity (7) can be written as

$$T_C N_{\gamma_0} = N_{\gamma_0}$$
$$(T_C - \mathrm{id})N_{\gamma_0+2} = \#\,\mathrm{Re}\,\chi_2 N_{\gamma_0}.$$

It is a consequence of (14) that the right hand side of the last equation belongs to the kernel of $T_C - \mathrm{id}$ though the left hand side belongs to the range of $T_C - \mathrm{id}$. By the same argument as in the proof of the previous lemma this implies $\mathrm{Re}\,\chi_2 N_{\gamma_0} = 0$ which is impossible if $r \neq 0$. □

Combining Lemmas 2.4 and 2.5 we easily prove that any $\psi \in \mathrm{Aut}_0^0 M$ can be linearized.

Theorem 3 *For each individual automorphism $\psi \in \mathrm{Aut}_0^0 M$ there exist normal coordinates such that ψ takes the form of a linear transformation*

$$z^* = \Lambda U z$$
$$\omega^* = \Lambda^2 \omega$$

Proof. According to Lemmas 2.4 and 2.5 we can choose normal coordinates such that the parameters a and r of ψ considered as a normalization of M vanish. By Theorem 3 in [9] the vanishing of a means that ψ is a mapping of the form

$$z^* = \Lambda U(\mathrm{id} - R)^{-1} z$$
$$\omega^* = \Lambda^2 (\mathrm{id} - R)^{-1} \omega.$$

Condition $r = 0$ implies that $R = 0$ and therefore ψ has the required form. □

3 Simultaneous Linearization

The goal of this section is to prove

Theorem 4 *For every non-exceptional manifold M there exist normal coordinates (z, ω) such that all the elements of $\mathrm{Aut}_0 M$ take the form*

$$z^* = \Lambda U z \tag{15}$$
$$\omega^* = \Lambda^2 \omega$$

The proof of this Theorem requires several preliminary results.

Lemma 3.6 *The group* $\operatorname{Aut}_0^0 M$ *is Abelian.*

Proof. This is an immediate consequence of the fact that an automorphism from $\operatorname{Aut}_0^0 M$ is uniquely determined by the parameter C which is the restriction of the differential in 0 to the complex tangent space and that the matrices C are diagonal. □

Lemma 3.7 *Every* $\psi \in \operatorname{Aut}_0^0 M$ *splits into the composition of two automorphisms* ψ_1 *with parameter*

$$C_1 = \begin{pmatrix} e^{i(\phi+\theta)} & 0 \\ 0 & e^{i(\phi-\theta)} \end{pmatrix}, \tag{16}$$

referred to as a rotation, and ψ_2 *with parameter*

$$C_2 = \begin{pmatrix} \lambda\mu & 0 \\ 0 & \frac{\lambda}{\mu} \end{pmatrix}$$

referred to as a scaling.

Proof. Let $\psi \in \operatorname{Aut}_0^0 M$ and $(\lambda, \mu, \phi, \theta, a, r)$ be the corresponding set of parameters. According to the Theorem on individual linearization from the previous section we can start with a normal coordinate system such that $a = 0$, $r = 0$ and ψ is a linear map (15).

By analyzing the action of the linear map ψ on the equation of M we observe that ψ splits into the required composition of two automorphisms

$$\psi_1 : \quad z^* = \begin{pmatrix} e^{i(\phi+\theta)} & 0 \\ 0 & e^{i(\phi-\theta)} \end{pmatrix} z, \quad w^* = w \tag{17}$$

and

$$\psi_2 : \quad z^* = \begin{pmatrix} \lambda\mu & 0 \\ 0 & \frac{\lambda}{\mu} \end{pmatrix} z, \quad w^* = \lambda^2 w. \tag{18}$$

□

It follows now that the group $\operatorname{Aut}_0^0 M$ decomposes into the direct product of the subgroup S of all scalings and the subgroup K of all rotations.

Lemma 3.8 *The subgroup* S *is either trivial or one-parametric.*

Proof. Let $\mathrm{id} \neq \psi \in S$. We choose normal coordinates such that ψ is linear of the form (18). Set $\ell := \ln \lambda$ and $m := \ln \mu$. Consider the one-parametric group

$$\psi^t : \quad z^* = \begin{pmatrix} e^{(\ell+m)t} & 0 \\ 0 & e^{(\ell-m)t} \, 0 \end{pmatrix} z, \quad w^* = e^{2\ell t} w.$$

ψ^t acts on a monomial Π of the equation of M by multiplication with e^{ht} where h is an integer that depends only on Π. Since $\psi^1 = \psi$ is an automorphism this integer must vanish. Thus, all ψ^t are automorphisms. Since the dimension of S does not exceed 1 it follows that S is either trivial or one-parametric. □

We prove that all automorphisms from K can be simultaneously linearized.

Lemma 3.9 *There exist normal coordinates such that all automorphisms $\psi \in K$ take the form (17).*

Proof. First we prove that all automorphisms in K have a common invariant chain. If K contains ψ with parameter C such that neither $e^{i(\phi+\theta)} = 1$ nor $e^{i(\phi-\theta)} = 1$ then there is a unique ψ-invariant chain Γ. Since K is Abelian Γ is then also invariant with respect to the other automorphisms from K. If such ψ does not exist then either for all elements in K the parameter ϕ equals θ or for all elements ϕ equals $-\theta$, i.e. the group K is either one-parametric or finite cyclic. In both cases it suffices to find an invariant chain for a generator.

Without loss of generality we may now assume that the standard chain Γ_0 is invariant for all $\psi \in K$. For all automorphisms with $\theta = 0$ the parameter r vanishes automatically, according to Lemma 2.5, regardless to parameterization of Γ_0. These automorphisms form, obviously, a subgroup of K. The factor group of K by this subgroup is either one-parametric or finite cyclic. In both cases we can find coordinates such that the parameter r vanishes for an automorphism from a generating coset. Then $r = 0$ for all elements of K. □

Corollary 3.10 *K is a compact Lie group.*

We consider now the situation when $\mathrm{Aut}_0^0 M$ contains a scaling σ since, otherwise, by the lemma above we know that $\mathrm{Aut}_0^0 M = K$ can be linearized.

Let Γ_K be the family of all K-invariant chains.

Lemma 3.11 *The family Γ_K contains a σ-invariant chain Γ_σ.*

Proof. Assume K is not trivial (otherwise there is nothing to prove). We choose normal coordinates such that K is linearized. Family Γ_K might consist of a single chain Γ_0 (this is the case if for some $\psi \in K$ $\phi \neq \theta$ and $\phi \neq -\theta$) or else

$$T_0\Gamma = \mathrm{span}\left\{\begin{pmatrix}0\\0\\1\\0\end{pmatrix}, \begin{pmatrix}0\\b\\0\\1\end{pmatrix}\right\}, \quad b \in \mathbb{C}, \tag{19}$$

(this is the case if $\phi = \theta$ for every $\psi \in K$), or

$$T_0\Gamma = \mathrm{span}\left\{\begin{pmatrix}b\\0\\1\\0\end{pmatrix}, \begin{pmatrix}0\\0\\0\\1\end{pmatrix}\right\}, \quad b \in \mathbb{C},$$

(this is the case if $\phi = -\theta$ for every $\psi \in K$).

If Γ_0 is the only K-invariant chain we are done because σ commutes with K and, therefore, preserves Γ_0.

Since the two remaining cases are similar to each other we consider only the case $\phi = \theta$ for all $\psi \in K$. The Jacobian $d\sigma|_0$ has the form

$$d\psi|_0 = \begin{pmatrix} \lambda\mu & 0 & \lambda\mu a_1 & 0 \\ 0 & \frac{\lambda}{\mu} & 0 & \frac{\lambda}{\mu} a_2 \\ \hline 0 & 0 & \lambda^2 & 0 \\ 0 & 0 & 0 & \lambda^2 \end{pmatrix}$$

Since σ commutes with K it maps any chain from Γ_K into a chain from Γ_K. From the first component of

$$d\psi|_0 \begin{pmatrix} 0 \\ 0 \\ 1 \\ 0 \end{pmatrix} = \begin{pmatrix} 0 \\ 0 \\ \lambda^2 \\ 0 \end{pmatrix},$$

one reads that $a_1 = 0$. If $a_2 = 0$ then $\Gamma_\sigma = \Gamma_0$ and we are done. If $a_2 \neq 0$ then, as it had been shown in the proof of Lemma 2.4, $\lambda \neq \mu^{-1}$, and Γ_σ is the chain tangent to (19) for $b = \frac{a_2}{\lambda\mu - 1}$. □

Let now Γ_0 be a common invariant chain for $\mathrm{Aut}_0^0 M$.

Lemma 3.12 *There exist normal coordinates such that* $\mathrm{Aut}_0^0 M$ *is linearized.*

Proof. If the parameter λ of the scaling σ is equal to 1 then $r = 0$ automatically. So, it remains then to choose coordinates such that Γ_0 is a common invariant chain and K is linearized.

Let $\lambda \neq 1$. Choose normal coordinates such that Γ_0 remains a common invariant chain and σ is linearized. It turns out that K is then also linearized automatically. Indeed, if for $\psi \in K$ the parameter $r \neq 0$ then N_{γ_0} and N_{γ_0+2} satisfy the conditions:

$$\begin{aligned} T_{C_\sigma} N_{\gamma_0} &= N_{\gamma_0} \\ T_{C_\psi} N_{\gamma_0} &= N_{\gamma_0} \\ T_{C_\sigma} N_{\gamma_0+2} &= N_{\gamma_0+2} \\ T_{C_\psi} N_{\gamma_0+2} - N_{\gamma_0+2} &= \# T_{C_\psi} \operatorname{Re} \chi_2 N_{\gamma_0}. \end{aligned}$$

Since $2 \operatorname{Re} \chi_2 N_{\gamma_0} \neq 0$ the right hand side of the last equation contains monomials Π such that $T_{C_\sigma} \Pi = \lambda^2 \Pi$, though, the left hand side contains only monomials form the kernel of $T_{C_\sigma} - \mathrm{id}$, by the third equation. Contradiction. □

Lemma 3.13 $\mathrm{Aut}_0 M$ *has a common invariant chain.*

Proof. Aut_0 consists at most of two cosets with respect to $\mathrm{Aut}_0^0 M$. Therefore, a chain that is invariant for $\mathrm{Aut}_0^0 M$ and one element of $\psi \in \mathrm{Aut}_0 M \setminus \mathrm{Aut}_0^0 M$ will be invariant for the entire group.

We start with normal coordinates where $\mathrm{Aut}_0^0 M$ is linearized. The Jacobian of ψ has the form

$$d\psi|_0 = \left(\begin{array}{cc|cc} 0 & \lambda\mu\,\mathrm{e}^{\mathrm{i}(\phi+\theta)} & 0 & \lambda\mu\,\mathrm{e}^{\mathrm{i}(\phi+\theta)}\,a_1 \\ \frac{\lambda}{\mu}\,\mathrm{e}^{\mathrm{i}(\phi-\theta)} & 0 & \frac{\lambda}{\mu}\,\mathrm{e}^{\mathrm{i}(\phi-\theta)}\,a_2 & 0 \\ \hline 0 & 0 & 0 & \lambda^2\,\mathrm{e}^{2\,\mathrm{i}\,\theta} \\ 0 & 0 & \lambda^2\,\mathrm{e}^{-2\,\mathrm{i}\,\theta} & 0 \end{array}\right) \tag{20}$$

Notice that ψ^2 already belongs to Aut_0^0 and therefore it is linear. So, the right upper block of the square of (20) must vanish. This gives the equations

$$\begin{aligned} a_1 + \frac{\lambda}{\mu}\,\mathrm{e}^{\mathrm{i}(\theta-\phi)} a_2 &= 0 \\ \lambda\mu\,\mathrm{e}^{-\,\mathrm{i}(\theta+F)} a_1 + \qquad\qquad a_2 &= 0. \end{aligned} \tag{21}$$

If $\lambda \neq 1$ or $\phi \neq 0$ this immediately implies that $a_1 = a_2 = 0$, i.e., the standard chain is invariant for ψ. Assume now $\lambda = 1$ and $\phi = 0$. Then

$$d\psi|_0 = \left(\begin{array}{cc|cc} 0 & \mu\,\mathrm{e}^{\mathrm{i}\,\theta} & 0 & \mu\,\mathrm{e}^{\mathrm{i}\,\theta}\,a_1 \\ \frac{1}{\mu}\,\mathrm{e}^{-\,\mathrm{i}\,\theta} & 0 & e^{-2\,\mathrm{i}\,\theta} a_1 & 0 \\ \hline 0 & 0 & 0 & \mathrm{e}^{2\,\mathrm{i}\,\theta} \\ 0 & 0 & \mathrm{e}^{-2\,\mathrm{i}\,\theta} & 0 \end{array}\right)$$

Since $\mathrm{Aut}_0^0 M$ is a normal subgroup of $\mathrm{Aut}_0 M$ then $d\psi|_0 T_0 \Gamma_0$ is invariant for $\mathrm{Aut}_0 M$. It follows that either $a_1 = 0$ (and therefore Γ_0 is invariant for ψ, as well), or $\mathrm{Aut}_0^0 M = \{\mathrm{id}\}$ (because we would obtain more than a one-parametric family of invariant chains). In this case we find an invariant chain for ψ by solving the equation (10). □

It follows that there exist normal coordinates such that $\mathrm{Aut}_0^0 M$ is linearized and any $\psi \in \mathrm{Aut}_0 M \setminus \mathrm{Aut}_0^0 M$ has the form

$$\begin{aligned} z^* &= J\Lambda U(\mathrm{id} - R)^{-1} z \\ \omega^* &= J\Lambda^2(\mathrm{id} - R)^{-1}\omega \end{aligned} \tag{22}$$

where

$$J = \begin{pmatrix} 0 & 1 \\ 1 & 0 \end{pmatrix}, \quad R = \begin{pmatrix} r\omega_1 & 0 \\ 0 & \bar{r}\omega_2 \end{pmatrix}.$$

It suffices to find a transformation of the form (13) such that $\mathrm{Aut}_0^0 M$ remains linearized and one (and therefore every) element from $\mathrm{Aut}_0 M \setminus \mathrm{Aut}_0^0 M$ becomes linear.

Lemma 3.14 *If for some* $\psi \in \mathrm{Aut}_0 M$ *the parameter* λ *is different from* 1 *then* $\mathrm{Aut}_0 M$ *is automatically linearized if* $\mathrm{Aut}_0^0 M$ *is linearized.*

Proof. Without loss of generality we assume that $\lambda \neq 1$ for some $\psi \in \mathrm{Aut}_0 M \setminus \mathrm{Aut}_0^0 M$. Taking the square of ψ in the form (22) we obtain a fractional linear mapping with parameter

$$r' = r + \lambda^2 \mathrm{e}^{-2\mathrm{i}\theta} \bar{r} \tag{23}$$

On the other hand we know that ψ^2 is already linear, thus $r' = 0$. But this is only possible if r was already 0. □

Lemma 3.15 *If* $\mathrm{e}^{2\mathrm{i}\theta}$ *is different from* 1 *for some* $\psi \in \mathrm{Aut}_0^0 M$ *then* $\mathrm{Aut}_0 M$ *is automatically linearized if* $\mathrm{Aut}_0^0 M$ *is linearized.*

Proof. Let $\psi_1 \in \mathrm{Aut}_0 M \setminus \mathrm{Aut}_0^0 M$ have parameters θ_1 and r. Take $\psi_2 \in \mathrm{Aut}_0^0 M$ with $\mathrm{e}^{2\mathrm{i}\theta_2} \neq 1$. The condition $r' = 0$ in (23) implies that $2(\theta_1 - \arg r) = \pi$, or $r = 0$. Considering (23) for the composition $\psi_1 \circ \psi_2$ leads to $2(\theta_1 - \theta_2 - \arg r) = \pi$, or $r = 0$. By the assumption of the lemma it follows that $r = 0$. □

Theorem 4 is now a consequence of the lemma below.

Lemma 3.16 *If* $\lambda = 1$ *and* $\mathrm{e}^{2\mathrm{i}\theta} = 1$ *for all* $\psi \in \mathrm{Aut}_0^0 M$ *then* $\mathrm{Aut}_0 M$ *can be linearized.*

Proof. According to the proof of Lemma 2.5, any coordinate change of the form (13) preserves the linear form of Aut_0^0. Therefore it suffices to find a transformation (13) that line-arises one $\psi \in \mathrm{Aut}_0 M \setminus \mathrm{Aut}_0^0 M$. This changes the parameter r of ψ into $r + r' + \mathrm{e}^{-2\mathrm{i}\theta} \bar{r}'$. If $r \neq 0$ then (23) ensures that there exists r' such that $r + r' + \mathrm{e}^{-2\mathrm{i}\theta} \bar{r}' = 0$ □

References

1. Belošapka, V.K. (1979): On the dimension of the group of automorphisms of an analytic hypersurface (Russian). Izv. Akad. Nauk SSSR Ser. Mat. **43** no. 2, 243–266, English transl. in Math. USSR Izvestija 14 (1980)
2. Belošapka, V.K. (1990): A uniqueness theorem for automorphisms of a nondegenerate surface in the complex space (Russian). Mat. Zametki **47** no. 3, 17–22, English transl. in Math. Notes 47 (1990)
3. Čap, A., Schmalz, G. (2000): Partially integrable almost CR manifolds of CR dimension and codimension two. Preprint, Ernst-Schrödinger-Institut Wien, ESI 937, electronically available at www.esi.at
4. Chern, S.S., Moser, J.M. (1974): Real hypersurfaces in complex manifolds. Acta Math. **133** no. 3-4, 219–271
5. Ežov, V.V. (1985): On the linearization of automorphisms of a real analytic hypersurface (Russian). Izv. Akad. Nauk SSSR Ser. Mat. **49** no. 4, 731–765, English translation: Math. USSR Izvestia, 27:1 (1985), 53–84.
6. Ežov, V.V. (1998): Triviality of Scalar Linear Type Isotropy Subgroup by Passing to an Alternative Canonical Form of a Hypersurface. Ann. Pol. Math. **70**, 85–97
7. Ežov, V.V. (1992): Cauchy-Riemann automorphism groups that cannot be projectively realized. J. Geom. Anal. **2.5**, 417–427

8. Ežov, V.V., Isaev, A.V., Schmalz, G. (1999): Invariants of elliptic and hyperbolic CR-Structures of codimension 2. Int. J. Math. **10:1** 1–52
9. Ežov, V.V., Schmalz G. (1996): Normal form and 2-dimensional chains of an elliptic CR surface in $\mathbb{C}^4$. J. Geom. Anal. **6.4** 495–529
10. Ežov, V.V., Schmalz, G. (1999): Infinitesimale Starrheit hermitescher Quadriken in allgemeiner Lage. Math. Nachr. **204** 41–60.
11. Ežov, V.V., Schmalz, G. (2001): Holomorphic classification of CR-manifolds with rigid Levi form (Russian). In: Čirka, E.M. (ed.) Complex analysis in modern mathematics. Dedicated to the memory of B.V.Shabat, Fazis, Moscow
12. Ežov V.V., Schmalz, G. (2001): Triviality problem for nil-isotropic automorphisms on CR manifolds. Bonner Math. Schriften **338**. Bonn
13. Kružilin, N.G., Loboda, A.V. (1983): Linearization of local automorphisms of pseudoconvex surfaces (Russian). Dokl. Akad. Nauk SSSR (Ser. Mat.) **271** no. 2, 280–282
14. Loboda, A.V. (1981): On local automorphisms of real analytic hypersurfaces (Russian). Izv. Akad. Nauk. SSSR (Ser. Math) **45** no. 3, 620–645
15. Loboda, A.V. (1988): Generic real analytic manifolds of codimension 2 in $\mathbb{C}^4$ and their biholomorphic mappings (Russian). Izv. Akad. Nauk. SSSR (Ser. Math) **52(5)**, 970–990. Engl. transl. in Math. USSR Izvestiya vol. 33(2), 295–315, (1989)
16. Loboda, A.V. (1990): Linearizability of holomorphic mappings of generating manifolds of codimension 2 in $\mathbb{C}^4$ (Russian). Izv.AN.SSSR **54(3)**, 632–644
17. Schmalz, G. (1998): Über die Automorphismen einer streng pseudokonvexen CR-Mannigfaltigkeit der Kodimension 2 im $\mathbb{C}^4$. Math. Nachr. **196**, 189–229
18. Schmalz, G., Slovák, J. (2000): The geometry of hyperbolic and elliptic CR-manifolds of codimension two. Asian J. Math. **4** no. 3, 565–597
19. Webster, S.M. (1978): On the Moser normal form at a non-umbilic point. Math. Ann. **233(2)**, 97–102

Global $C^{2+\alpha}$-Estimates for Conformal Maps

Friedrich Sauvigny

Institut für Mathematik der Brandenburgischen Technischen Universität Cottbus, Universitätsplatz 3-4, D-03044 Cottbus, Germany

Summary. For conformal maps g on the closed unit disc $\overline{B}$ we shall estimate $\|g\|_{C^{2+\alpha}(\overline{B})}$ by the relevant geometric data from above. Here we bound the modulus of their derivatives $|g'(w)| > 0$, $w \in \overline{B}$, quantitatively from below. Only the maximum principle and Minding's formula for the geodesic curvature are necessary in the proof. For instance, our estimates suffice to construct conformal maps of the class $C^{2+\alpha}(\overline{B})$ approximating $C^{2+\alpha}$-domains by real-analytic Jordan-domains.

1 Introduction

In the theory of partial differential equations one usually constructs the solutions of boundary value problems in sufficiently large classes of functions. The questions how well the solution behaves especially at the boundary and whether singularities occur is normally postponed to the subsequent regularity investigations. The desire to keep the solutions in those small classes adequate to the prescribed data requires strict a priori estimates. One usually begins with interior estimates and then considers the boundary.

In this paper we derive $C^{2+\alpha}$-estimates for conformal maps and establish a positive bound for the modulus of their derivatives from below even up to the boundary. Applying the maximum principle to certain auxiliary functions from differential geometry we proceed from the boundary into the interior of the domain. We hope that similar estimates can be developed for solutions of Plateau's problem.

Let $B := \{w = u + iv \in \mathbb{C} : |w| < 1\}$ be the unit disc and $\Omega \subset \mathbb{C}$ a bounded domain such that $\Gamma = \partial\Omega$ is a Jordan curve. With the aid of Dirichlet's principle (compare [2]) one can construct a conformal map $g : B \to \Omega$ which is topological from $\overline{B}$ onto $\overline{\Omega}$. If Γ is a real-analytic curve, we can extend g across its boundary ∂B as a conformal map applying a generalized version of the Schwarz' reflection principle. Let us now approximate a $C^{2+\alpha}$-domain Ω by real-analytic domains Ω_n $(n = 1, 2, 3, \dots)$ and observe the convergence of their conformal maps $g_n : B \to \Omega_n$ towards the limit map $g : B \to \Omega$. This usually happens in the class $C^2_{loc}(B) \cap C^\circ(\overline{B})$, and the regularity $g \in C^{2+\alpha}(\overline{B})$ as well as the condition $g'(w) \neq 0$ for all $w \in \overline{B}$ are investigated a posteriori. Even in the more general framework of minimal surfaces this concept has been elaborated (see [5] and [3], ch.7).

With the aid of Minding's formula and the maximum principle we shall derive a priori estimates for $\|g\|_{C^{2+\alpha}(\overline{B})}$ and $\|\frac{1}{E}\|_{C^0(\overline{B})}$ from above for conformal maps $g : \overline{B} \to \overline{\Omega} \in C^{2+\alpha}(\overline{B})$ abbreviating $E(w) := |g'(w)|^2$, $w \in \overline{B}$. These estimates decisively depend on the curvature for the boundary of the domain Ω. In this way, we can keep our conformal maps in the class $C^{2+\alpha}(\overline{B})$ differential-geometrically regular on $\overline{B}$ during the approximation. For an excellent lecture on complex analysis we refer the reader to [4].

2 A Global Bound for the First Derivatives

The basis for our estimate of the first derivatives yields the following

Proposition 2.1 *On the disc $B_\varrho(z_1) := \{z \in \mathbb{C} : |z - z_1| < \varrho\}$ with the center $z_1 \in \mathbb{C}$ and the radius $\varrho > 0$ let the holomorphic function*

$$w = f(z) : B_\varrho(z_1) \longrightarrow B \in C^1(\overline{B_\varrho(z_1)}) \tag{1}$$

be given. We require

$$|f(z_1)| \leq 1 - \varepsilon \tag{2}$$

with a positive $\varepsilon > 0$ and $z_2 \in \partial B_\varrho(z_1)$ being a boundary point with

$$|f(z_2)| = 1. \tag{3}$$

Then we have the estimate

$$|f'(z_2)| \geq \frac{\varepsilon^2}{\varrho}. \tag{4}$$

Proof. We consider the linear transformation

$$\ell(w) := z_1 + (z_2 - z_1)w, \quad w \in \overline{B},$$

on the unit disc B satisfying

$$\ell(0) = z_1, \quad \ell(1) = z_2, \quad |\ell'(w)| = |z_2 - z_1| = \varrho \quad \text{for all } w \in \overline{B}. \tag{5}$$

Setting $w_1 = f(z_1) \in B$ and $w_2 = f(z_2) \in \partial B$ we use the Möbius-transformation

$$h(w) := e^{i\vartheta} \frac{w - w_1}{\overline{w_1}w - 1}, \quad w \in \overline{B}, \tag{6}$$

with a suitable $\vartheta \in [0, 2\pi)$.

We obtain

$$h(w_1) = 0, \quad h(w_2) = 1, \tag{7}$$

and we calculate

$$\left.\begin{aligned} |h'(w_2)| &= \frac{|(\overline{w}_1 w_2 - 1) - \overline{w}_1(w_2 - w_1)|}{|\overline{w}_1 w_2 - 1|^2} = \frac{|1 - |w_1|^2|}{|1 - \overline{w}_1 w_2|^2} \\ &\leq \frac{1}{(1 - |\overline{w}_1 w_2|)^2} = \frac{1}{(1 - |w_1|)^2} \leq \frac{1}{(1 - (1 - \varepsilon))^2} = \frac{1}{\varepsilon^2}\,. \end{aligned}\right\} \tag{8}$$

We consider the origin-preserving holomorphic map

$$\Phi(w) := h \circ f \circ \ell(w), \quad w \in \overline{B}\,,$$

of the class $C^1(\overline{B}, \overline{B})$, and the lemma of H.A. Schwarz yields

$$|\Phi(w)| \leq |w|, \quad w \in \overline{B}\,. \tag{9}$$

For all $r \in (0, 1)$ this implies the inequality

$$\left| \frac{\Phi(r) - \Phi(1)}{r - 1} \right| \geq \frac{|\Phi(1)| - |\Phi(r)|}{1 - r} \geq \frac{1 - r}{1 - r} = 1$$

and consequently

$$|\Phi'(1)| \geq 1. \tag{10}$$

Combining (10), (8) and (5) we obtain

$$1 \leq |\Phi'(1)| = |h'(w_2) \cdot f'(z_2) \cdot \ell'(1)| \leq \frac{1}{\varepsilon^2}\, |f'(z_2)| \varrho$$

and finally

$$|f'(z_2)| \geq \frac{\varepsilon^2}{\varrho}\,. \qquad \text{Q.e.d.}$$

Now let $\Omega \subset \mathbb{C}$ be a bounded C^2-domain, whose boundary $\Gamma = \partial\Omega$ is a Jordan-curve of the class C^2, parametrized by

$$\zeta = \zeta(t) \,:\, \mathbb{R} \longrightarrow \Gamma \in C^2 \quad \text{with} \quad |\zeta'(t)| > 0 \quad \text{in } \mathbb{R}$$

in positive orientation. We fix a point $z_0 \in \Omega$. By $|\Omega|$ we denote the area of Ω and $\kappa(z) \in \mathbb{R}$ means the oriented curvature of the boundary point $z \in \Gamma$.

Proposition 2.2 *The domain Ω as above satisfies the restrictions*

$$|\Omega| \leq \omega_0 \quad \textit{and} \quad |\kappa(z)| \leq \kappa_0 \quad \textit{for all } z \in \partial\Omega \tag{11}$$

with the constants $\omega_0, \kappa_0 \in (0, +\infty)$.

Then there exists a Lipschitz constant $L = L(\omega_0, \kappa_0) \in (0, +\infty)$ *with the following property:*
For all conformal maps

$$f : \Omega \longrightarrow B \in C^1(\overline{\Omega}, \overline{B}), \quad f(z_0) = 0, \;\; f'(z_0) > 0 \tag{12}$$

with their inverse maps

$$g : B \longrightarrow \Omega \in C^1(\overline{B}, \overline{\Omega}), \quad g(0) = z_0, \;\; g'(0) > 0 \tag{13}$$

we have the a priori estimate

$$\sup_{w \in B} |g'(w)| \leq L(\omega_0, \kappa_0). \tag{14}$$

Proof. On account of the curvature condition (11) we can find a radius $\varrho = \varrho(\kappa_0) > 0$ such that each boundary point $z_2 \in \Gamma = \partial\Omega$ admits a support circle $B_\varrho(z_1) \subset \Omega$ satisfying $z_2 \in \partial B_\varrho(z_1) \cap \partial\Omega$. Obviously

$$\text{dist}\,(z_1, \Gamma) = \varrho \tag{15}$$

holds true. We easily verify a chord-arc-condition for our C^2-Jordan-curve Γ: For two successive points $\zeta_j = \zeta(t_j) \in \Gamma$ $(j = 1, 2)$ with $t_1 < t_2$ whose arc

$$\overset{\frown}{\zeta_1\zeta_2} := \{\zeta(t) \in \Gamma \, : \, t_1 \leq t \leq t_2\}$$

has the length $|\, \overset{\frown}{\zeta_1\zeta_2} \,|$, we find a constant $\lambda = \lambda(\kappa_0) \in (1, +\infty)$ such that

$$|\, \overset{\frown}{\zeta_1\zeta_2} \,| \leq \lambda |\zeta_1 - \zeta_2|. \tag{16}$$

For an arbitrary point $w_0 \in \partial B$ and $\varepsilon > 0$ we consider the domain

$$\Theta_\varepsilon = \Theta_\varepsilon(w_0) := \{w \in B \, : \, |w - w_0| < \varepsilon\}.$$

Since Dirichlet's integral of the map g satisfies

$$\left.\begin{aligned} D(g) &:= \iint_B \left(|g_u|^2 + |g_v|^2\right) dudv \;=\; 2 \iint_B |g'(w)|^2 \, dudv \\ &= 2|\Omega| \;\leq\; 2\omega_0 \end{aligned}\right\} \tag{17}$$

we infer the following inequality from the Courant-Lebesgue-lemma

$$\int\limits_{\substack{w \in B \\ |w - w_0| = \varepsilon*}} |dg| \leq 2\sqrt{\frac{2\pi\omega_0}{\log\frac{1}{\varepsilon}}} \,. \tag{18}$$

Here $\varepsilon^* \in [\varepsilon, \sqrt{\varepsilon}]$ is an intermediate value.

Together with (16) we can estimate the length of the Jordan-curve $g(\partial\Theta_{\varepsilon^*})$ as follows

$$|g(\partial\Theta_{\varepsilon^*})| \leq 2(1+\lambda)\sqrt{\frac{2\pi\omega_0}{\log\frac{1}{\varepsilon}}} \leq 2\varrho \tag{19}$$

choosing $\varepsilon = \varepsilon(\omega_0, \kappa_0) > 0$ sufficiently small. Each point $z \in g(\Theta_{\varepsilon^*}) \supset g(\Theta_\varepsilon)$ satisfies

$$\text{dist}\,(z, \Gamma) < \varrho \tag{20}$$

on account of (19). Since $w_0 \in \partial B$ has been chosen arbitrarily, each point z in the strip

$$\Sigma := \{z = g(w) \,:\, w \in B,\ 1-\varepsilon < |w| < 1\}$$

satisfies (20). Consequently, the center $z_1 \in \Omega$ of the support circle above is not contained in Σ due to (15) and its preimage $f(z_1) \in B$ fulfills

$$|f(z_1)| \leq 1 - \varepsilon. \tag{21}$$

We now apply Proposition 2.1 and obtain

$$|f'(z_2)| \geq \frac{\varepsilon^2}{\varrho} \quad \text{for all } z_2 \in \partial\Omega. \tag{22}$$

For the inverse map $g : B \to \Omega \in C^1(\overline{B}, \overline{\Omega})$ this implies

$$|g'(w)| \leq \frac{\varrho}{\varepsilon^2} \quad \text{for all } w \in \partial B. \tag{23}$$

Setting $L = L(\omega_0, \kappa_0) := \frac{\varrho}{\varepsilon^2}$ we conclude

$$\sup_{w\in B} |g'(w)| \leq L(\omega_0, \kappa_0) \tag{24}$$

from the maximum principle for holomorphic maps. Q.e.d.

3 A Global Bound for the Second Derivatives

We need the following auxiliary lemma proved by the maximum principle for harmonic functions.

Proposition 3.3 *The holomorphic function*

$$G(w) = x(w) + iy(w) = x(u, v) + iy(u, v) \in C^1(\overline{B}, \mathbb{C})$$

satisfies

$$\left|\frac{d}{dt}\,y(e^{it})\right| \leq \ell \quad \textit{for all } t \in \mathbb{R} \tag{25}$$

with $\ell \in (0, +\infty)$.

Then we have a constant $m = m(\ell) \in (0, +\infty)$ *such that*

$$|G'(w)| \le m \quad \textit{for all } w \in \overline{B} \tag{26}$$

holds true.

Proof. We consider the Jordan-curve

$$\widehat{\Gamma} := \{(\cos t, \sin t, y(e^{it})) \in \mathbb{R}^3 \, : \, 0 \le t \le 2\pi\}.$$

At each point $(u_0, v_0, y_0) = (u_0, v_0, y(u_0, v_0)) \in \widehat{\Gamma}$ we have a superior and inferior support plane

$$y^{\pm}(u, v) := y_0 + \alpha^{\pm}(u - u_0) + \beta^{\pm}(v - v_0), \quad (u, v) \in \mathbb{R}^2, \tag{27}$$

situated totally on one side of the curve $\widehat{\Gamma}$. Due to (25) we have a constant $m = m(\ell) \in (0, +\infty)$ such that the real coefficients $\alpha^{\pm}, \beta^{\pm}$ fulfill

$$\sqrt{(\alpha^{\pm})^2 + (\beta^{\pm})^2} \le m. \tag{28}$$

The maximum principle for harmonic functions yields

$$\left.\begin{array}{l} y^-(u, v) \le y(u, v) \le y^+(u, v) \quad \text{for all } (u, v) \in \overline{B} \\ \text{and} \quad y^-(u_0, v_0) = y_0 = y^+(u_0, v_0). \end{array}\right\} \tag{29}$$

This implies

$$\left| \frac{\partial}{\partial r} y(re^{it}) \right|_{r=1} \le m \quad \text{for all } t \in \mathbb{R} \tag{30}$$

and consequently

$$|y_w(w)| \le m \quad \text{for all } w \in \partial B \tag{31}$$

using (25). The maximum principle for the holomorphic function $y_w = \frac{1}{2}(y_u - i y_v)$ yields

$$|y_w(w)| \le m \quad \text{for all } w \in \overline{B}. \tag{32}$$

Utilizing the Cauchy-Riemann equations for the function

$$G(w) = x(w) + iy(w), \quad w \in \overline{B},$$

we arrive at the estimate (26). Q.e.d.

Proposition 3.4 *For all conformal maps*

$$g : B \longrightarrow \Omega \in C^2(\overline{B}, \overline{\Omega}), \quad g(0) = z_0, \; g'(0) \ge \gamma_0 > 0 \tag{33}$$

onto the domains Ω as in Proposition 2 we have the a priori estimates

$$\inf_{w\in B} |g'(w)| \geq M(\omega_0, \kappa_0, \gamma_0) > 0 \tag{34}$$

and

$$\sup_{w\in B} |g''(w)| \leq N(\omega_0, \kappa_0) < +\infty. \tag{35}$$

Proof. With the holomorphic function

$$g : \overline{B} \longrightarrow \mathbb{C} \in C^2(\overline{B}, \mathbb{C})$$

we associate the Gaussian metric

$$ds^2 = E(w)(du^2 + dv^2) = |g'(w)|^2(du^2 + dv^2).$$

For the geodesic curvature κ of the boundary curve $\Gamma = \partial\Omega$ we have Minding's formula (compare [1] §83)

$$\left.\begin{aligned} &\frac{\partial}{\partial r} \log \sqrt{E(r\cos t, r\sin t)}\Big|_{r=1} = \kappa(g(e^{it}))\sqrt{E(\cos t, \sin t)} - 1 \\ &\text{for all} \quad t \in \mathbb{R}. \end{aligned}\right\} \tag{36}$$

Now we consider the holomorphic function

$$\left.\begin{aligned} G(w) := \log g'(w) &= \log |g'(w)| + i \arg g'(w) \\ &= \log \sqrt{E(w)} + i \arg g'(w) =: x(w) + iy(w), \quad w \in \overline{B}. \end{aligned}\right\} \tag{37}$$

From (36) and Proposition 2.2 we find a constant $\ell = \ell(\omega_0, \kappa_0) \in (0, +\infty)$ such that

$$\left|\frac{d}{dt} y(e^{it})\right| = \left|\frac{d}{dr} x(re^{it})\right|_{r=1} \leq \ell. \tag{38}$$

With the aid of Proposition 3.3 we find a constant $m = m(\ell) = m(\omega_0, \kappa_0) \in (0, +\infty)$ such that

$$|G'(w)| \leq m \quad \text{for all } w \in \overline{B} \tag{39}$$

is valid. This implies

$$\left.\begin{aligned} &|g''(w)| \leq m|g'(w)| \leq mL =: N(\omega_0, \kappa_0) < +\infty \\ &\text{for all } w \in \overline{B}. \end{aligned}\right\} \tag{40}$$

Finally, we observe

$$G(0) = \log |g'(0)| \in [\log \gamma_0, \log L(\omega_0, \kappa_0)]. \tag{41}$$

Together with (39) we find a constant $C(\omega_0, \kappa_0, \gamma_0)$ such that

$$\left.\begin{array}{l} |\log g'(w)| = |G(w)| \leq C(\omega_0, \kappa_0, \gamma_0) < +\infty \\ \text{for all } w \in \overline{B} \end{array}\right\} \tag{42}$$

is valid. This gives us the a priori constant $M(\omega_0, \kappa_0, \gamma_0) > 0$ in (34).

Q.e.d.

4 The Global $C^{2+\alpha}$-Estimate

By potential-theoretic methods, one can prove an analogue of Proposition 3.3 in the Hölder-class.

Proposition 4.5 *The holomorphic function*

$$G(w) = x(u,v) + iy(u,v) \in C^{\alpha}(\overline{B}, \mathbb{C})$$

with $\alpha \in (0,1)$ *satisfies the Hölder-condition*

$$|y(e^{it_1}) - y(e^{it_2})| \leq \beta |t_1 - t_2|^{\alpha} \quad \textit{for all } t_1, t_2 \in \mathbb{R} \tag{43}$$

with $\beta \in (0, +\infty)$.
Then we have a Hölder-constant $h = h(\alpha, \beta) \in (0, +\infty)$ *such that*

$$|G(w_1) - G(w_2)| \leq h(\alpha, \beta)|w_1 - w_2|^{\alpha} \quad \textit{for all } w_1, w_2 \in \overline{B} \tag{44}$$

holds true.

Proof. We refer the reader to [3], 7.1, Lemma 6 or to [5], 10.1, Lemma 10.2. Q.e.d.

We are now able to establish the principal result of this paper, namely the following

Theorem 4.6. *Let* $\Omega \subset \mathbb{C}$ *be a* C^2*-domain bounded by the Jordan-curve* $\Gamma = \partial\Omega$, *whose curvature* κ *satisfies the conditions*

$$\left.\begin{array}{l} |\kappa(z)| \leq \kappa_0 \quad \textit{for all } z \in \Gamma \quad \textit{and} \\ |\kappa(z_1) - \kappa(z_2)| \leq \beta_0 |z_1 - z_2|^{\alpha_0} \quad \textit{for all } z_1, z_2 \in \Gamma\,; \end{array}\right\} \tag{45}$$

with the constants $\kappa_0, \beta_0 \in (0, +\infty)$ *and* $\alpha_0 \in (0,1)$. *Furthermore, we fix* $z_0 \in \Omega$ *and assume* $|\Omega| \leq \omega_0$ *with a constant* $\omega_0 \in (0, +\infty)$.
Then we have a priori constants $M(\omega_0, \kappa_0, \gamma_0) > 0$ *and* $P(\omega_0, \kappa_0, \gamma_0, \alpha_0, \beta_0) \in (0, +\infty)$ *such that the estimates*

$$\|g\|_{C^{2+\alpha}(\overline{B})} \leq P(\omega_0, \kappa_0, \gamma_0, \alpha_0, \beta_0) < +\infty \tag{46}$$

and

$$\inf_{w\in B}|g'(w)| \geq M(\omega_0,\kappa_0,\gamma_0) > 0 \tag{47}$$

hold true for all conformal maps

$$g\,:\,B \longrightarrow \Omega \in C^{2+\alpha_0}(\overline{B},\overline{\Omega}),\quad g(0)=z_0\,,\; g'(0)\geq\gamma_0>0. \tag{48}$$

Proof. Taking Proposition 2.2 and 3.4 into account, we have only to establish a Hölder-estimate for the function $g''(w) : B \to \mathbb{C} \in C^{\alpha_0}(\overline{B})$. As in the proof of Proposition 3.4, we consider the auxiliary function

$$G(w) := \log g'(w) =: x(w)+iy(w),\quad w\in\overline{B}.$$

Its imaginary part satisfies the boundary condition

$$\frac{d}{dt}\,y(e^{it}) = \kappa(g(e^{it}))|g'(e^{it})|-1,\quad t\in\mathbb{R}. \tag{49}$$

Due to (45) and the Propositions 2.2 and 3.4 we can find a Hölder-constant $\beta = \beta(\beta_0,\omega_0,\gamma_0,\kappa_0)\in(0,+\infty)$ such that

$$\left|\frac{d}{dt}\,y(e^{it})\Big|_{t_1} - \frac{d}{dt}\,y(e^{it})\Big|_{t_2}\right| \leq \beta|t_1-t_2|^{\alpha_0}\quad\text{for all } t_1,t_2\in\mathbb{R} \tag{50}$$

holds true. We now apply Proposition 4.5 to the derivative

$$G'(w) = \frac{g''(w)}{g'(w)},\quad w\in\overline{B}, \tag{51}$$

and obtain a Hölder-constant $h = h(\alpha_0,\beta_0,\omega_0,\gamma_0,\kappa_0)\in(0,+\infty)$ such that

$$|G'(w_1)-G'(w_2)| \leq h|w_1-w_2|^{\alpha_0}\quad\text{for all } w_1,w_2\in\overline{B} \tag{52}$$

holds true. Together with (34) and (35) we establish the Hölder-condition for the function $g''(w)$. Q.e.d.

5 Boundary Behaviour of Conformal Maps

We shall utilize Theorem 4.6 to study the boundary behaviour of conformal maps for domains Ω, which are bounded by a regular $C^{2+\alpha}$-Jordan-curve $\Gamma = \partial\Omega$ with $\alpha\in(0,1)$. For the convenience of the reader we supply the fundamental

Proposition 5.7 *Let $\Omega\subset\mathbb{C}$ be a domain bounded by a regular, real-analytic Jordan-curve $\Gamma=\partial\Omega$.*

Then there exists a conformal map $g : B\to\Omega$ which can be extended across ∂B in the following sense: For each $w_1\in\partial B$ we have a convergent power-series

$$\mathfrak{P}(w) := \sum_{k=0}^{\infty} a_k (w - w_1)^k \,, \quad w \in \mathbb{C} \text{ with } |w - w_1| < \varepsilon$$

such that

$$g(w) = \mathfrak{P}(w) \quad \text{for all } w \in B \text{ with } |w - w_1| < \varepsilon \tag{53}$$

holds true. Here $\varepsilon > 0$ *is chosen sufficiently small, and the coefficients satisfy* $a_k \in \mathbb{C}$ $(k \in \mathbb{N} \cup \{0\})$ *and* $a_1 \neq 0$.

Proof. With the aid of Dirichlet's principle, one can construct a conformal map $g : B \to \Omega$ which is a homeomorphism $g : \overline{B} \to \overline{\Omega}$ on the closure of the domains (see [2]). The following arguments can be taken from the lecture [4]. Around the point $z_1 = g(w_1)$, the regular real-analytic boundary Γ is given by

$$\zeta(t) = \sum_{k=0}^{\infty} \frac{1}{k!} \zeta^{(k)}(t_1)(t - t_1)^k \,, \quad t_1 - \varepsilon < t < t_1 + \varepsilon, \tag{54}$$

with $t_1 \in \mathbb{R}$, $\varepsilon > 0$ and $\zeta'(t_1) \neq 0$. Setting $r = t + is \in \mathbb{C}$ we can extend the convergent power-series into the complex field and obtain

$$\left.\begin{aligned} &h(r) := \sum_{k=0}^{\infty} \frac{1}{k!} \zeta^{(k)}(t_1)(r - t_1)^k = z_1 + \zeta'(t_1)(r - t_1) + \cdots \\ &\text{for all} \quad r \in \mathbb{C} \text{ with } |r - t_1| < \varepsilon. \end{aligned}\right\} \tag{55}$$

On account of $\zeta'(t_1) \neq 0$ we have the holomorphic inverse map h^{-1} in a neighbourhood of $z_1 = h(t_1)$. We map the upper half-plane

$$\mathbb{H}^+ := \{z = x + iy \in \mathbb{C} \,:\, y > 0\}$$

conformally onto the unit disc B by the Moebius-transformation

$$\ell \,:\, \mathbb{H}^+ \longrightarrow B, \quad \ell(0) = w_1 \,. \tag{56}$$

We apply the Schwarz' reflection principle on the following holomorphic mapping

$$\psi(z) := h^{-1} \circ g \circ \ell(z), \quad z \in \mathbb{H}^+ \,, \ |z| < \varepsilon \,. \tag{57}$$

We obtain the holomorphic function

$$\psi(z), \quad |z| < \varepsilon, \tag{58}$$

defined in the full disc around the origin. Now, the function

$$h \circ \psi \circ \ell^{-1}(w) = \sum_{k=0}^{\infty} a_k (w - w_1)^k =: \mathfrak{P}(w), \quad |w - w_1| < \varepsilon, \tag{59}$$

is holomorphic by expansion into a convergent power-series around the point $w_1 \in \partial B$. From (57) and (59) we infer

$$\left.\begin{aligned} g(w) &= \sum_{k=0}^{\infty} a_k (w - w_1)^k = \mathfrak{P}(w) \\ \text{for all} \quad & w \in B \text{ with } |w - w_1| < \varepsilon. \end{aligned}\right\} \tag{60}$$

Since $g : \overline{B} \to \overline{\Omega}$ is a homeomorphism, the coefficient a_1 does not vanish.

Q.e.d.

Theorem 5.8. *Let $\Omega \subset \mathbb{C}$ be a domain bounded by a regular Jordan-curve $\Gamma = \partial\Omega$ of the class $C^{2+\alpha}$.*
Then there exists a conformal map $g : \overline{B} \to \overline{\Omega} \in C^{2+\alpha}(\overline{B})$ such that $g'(w) \neq 0$ for all $w \in \overline{B}$ holds true.

Proof: We approximate the domain Ω by real-analytic domains Ω_n $(n = 1, 2, \dots)$ such that their boundary-curvatures κ_n are uniformly bounded and satisfy a Hölder-condition with the exponent α. For the conformal maps

$$g_n : B \longrightarrow \Omega_n \in C^{2+\alpha}(\overline{B}, \overline{\Omega}_n), \quad g_n(0) = z_0\,, \;\; g_n'(0) \geq \gamma_0 > 0, \tag{61}$$

we have the a priori estimate

$$\|g_n\|_{C^{2+\alpha}(\overline{B})} \leq c_1 \;\; (n \in \mathbb{N}); \quad |g_n'(w)| \geq c_2 > 0, \;\; w \in \overline{B}, \;\; n \in \mathbb{N}, \tag{62}$$

with the constants $c_1, c_2 \in (0, +\infty)$. Therefore, we can extract a subsequence converging in the space $C^{2+\alpha}(\overline{B})$ to a conformal map $g : \overline{B} \to \overline{\Omega} \in C^{2+\alpha}(\overline{B})$ with $|g'(w)| \geq c_2 > 0$ for all $w \in \overline{B}$. Q.e.d.

Acknowledgement. Here I would like to express my profound gratitude to all those who created, supported and organized the Sonderforschungsbereich 256 "Nonlinear Partial Differential Equations" at the Universität Bonn. Especially, I sincerely thank Professor Dr. Dr. h. c. mult. Stefan Hildebrandt for his numerous kind invitations to the SFB 256 and for our intensive scientific collaboration in friendship.

References

1. W. Blaschke: Vorlesungen über Differentialgeometrie I. Springer, Berlin, 1945
2. R. Courant: Dirichlet's principle, conformal mapping, and minimal surfaces. Springer, Berlin, Reprint 1977
3. U. Dierkes, S. Hildebrandt, et al.: Minimal surfaces II. Grundlehren der math. Wiss. **296**, Springer, Berlin, 1992
4. H. Grauert: Funktionentheorie I. Vorlesungsausarbeitung am Mathematischen Institut der Universität Göttingen, Wintersemester 1964/65
5. C. Pommerenke: Univalent functions. Vandenhoek und Ruprecht, Göttingen, 1975

On Karcher's Twisted Saddle Towers

Matthias Weber

Indiana University, Math Department, Rawles Hall, Bloomington, IN 47405, USA
weber@indiana.edu

1 Introduction

Progress is usually judged by its achievements, and this also applies to mathematics. Therefore, the simplicity of a new idea is frequently well hidden under technical ballast which is necessary to press new results out of the idea.

I feel this is true in particular for the flat structures on surfaces which Mike Wolf and myself have been using to solve high-dimensional period problems for minimal surfaces ([13, 12, 14, 8, 11]).

I would like to take this opportunity to apply our method to a simple class of examples which deserves more attention: Karcher's twisted saddle towers. We will see that the relatively harmless 1-dimensional period problem which one has to solve here *disappears completely* if one uses the right flat coordinate system.

Moreover, for these examples, we don't need to talk about flat structures at all, euclidean polygons will do.

The only prerequisites for this paper are familiarity with the Weierstraß representation and knowledge of the classical Schwarz-Christoffel formula.

We begin now with an informal description of the original *untwisted surfaces*. The classical singly periodic Scherk surface is invariant under a vertical translation, so it is natural to do all computations and arguments on the quotient surface. This is a sphere with four punctures which represent the four orthogonal cylindrical ends.

There is a simple variation of this: For every integer $k \geq 2$, there is a singly periodic minimal surface S_k, invariant under a vertical translation, such that the quotient surface is a sphere with $2k$ cylindrical ends. Here, consecutive ends make an angle π/k (see figure 1). S_2 is then the original Scherk surface.

One can in fact deform these surfaces by changing the angles between the ends, but this doesn't concern us here, we are interested in another deformation of this *most symmetric* case.

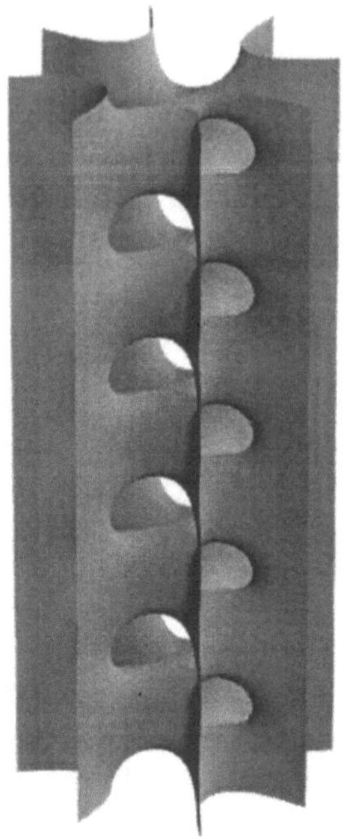

Fig. 1. Singly periodic Scherk surface S_4 with eight ends

To describe this twist deformation, fix some integer $k \geq 2$. For any *twist parameter* $0 < \phi < \pi/k$, there is an embedded singly periodic minimal surface $KS_{k,\phi}$, invariant under a vertical screw motion, whose rotational part is by angle $2 \cdot \phi$, such that the quotient surface is a sphere with $2k$ *helicoidal* ends. If we normalize the vertical translation to be of amount 1, the surfaces $KS_{k,\phi}$ converge to S_k for $\phi \to 0$.

In [4], the existence of these surfaces is shown for any $k \geq 2$ and $0 < \phi < \pi/k$ in two different ways: First, using a Plateau construction with barriers, and secondly using the Weierstraß representation. While the Plateau construction is elegant and non-computational, it uses non-elementary machinery. On the other hand, the Weierstraß representation approach requires the solution of a period problem. Neither method shows that the constructed surfaces come in a smooth family.

We will give a simple proof of this existence result, using the Weierstrass representation enriched by the usage of euclidean geometry. As a byproduct, we will obtain the smoothness of the family $KS_{k,\phi}$ in ϕ.

These surfaces are significant for at least two reasons:

For one, they were the first and still are the simplest examples of screw-motion invariant minimal surfaces. Since then, many other such surfaces have been found ([10, 6, 1, 2, 3, 9, 11, 7]).

In general, such screw-motion invariant surfaces are hard to describe by the usual Weierstraß representation approach, because the Gauß map is not single valued on

Fig. 2. A twisted Karcher-Scherk surface with eight helicoidal ends (cf. color plate 6, Page 669)

the quotient surface. Karcher's way to overcome this was to use the logarithmic derivative of the Gauß map, which is well-defined on the quotient, but this made the period problem a little bit more difficult as it requires an additional integration.

For two, the $KS_{k,\phi}$ surfaces show a surprising phenomenon which has changed from its status as a curiosity into a focus of current research: When the twist parameter ϕ approaches its limit π/k, the minimal surface will degenerate. This degeneration can be made visible in two interesting ways by rescaling: First, one can obtain a single helicoid in k different ways, one for each pair of opposite helicoidal ends of $KS_{k,\phi}$. Secondly, one can obtain a minimal limit foliation by horizontal planes in the sense of Colding and Minicozzi.

This foliation will have k singular vertical lines through the kth roots of unity. These correspond to the k collapsed helicoids from the first case of limits.

This means that close to the limit $\phi = \pi/k$, the surface will look like k copies of a helicoid glued together. Once this is observed, one can obtain many new examples by 'glueing together helicoids', appropriately positioned and in a suitable way, see [7].

2 The Weierstrass Representation a la Euclid

In this section, we review the Weierstraß representation for all relevant surfaces and relate it to the Schwarz-Christoffel formula.

2.1 Singly Periodic Scherk Surfaces with k-Fold Dihedral Symmetry

We will discuss briefly the Weierstraß representation of the singly periodic Scherk surfaces S_k with k-fold dihedral symmetry.

We do this on the quotient surface Σ_k obtained by dividing S_k by a minimal vertical translation. This is a conformal sphere, punctured at $2k$ point at the $(2k)^{\text{th}}$ roots of unity. The points 0 and ∞ are umbilic points with vertical normal.

Furthermore, the $2k$ lines $t \mapsto te^{j\pi i/k}, \quad j = 0, \dots, 2k-1$ are mapped to straight lines on Σ_k.

Here is the Weierstraß representation for S_k, see [5, 4] for a detailed discussion. The choices of the signs makes the formula consistent with later formulas.

$$G(w) = -iw^{k-1} \tag{1}$$

$$dh = i\frac{w^{k-1}dw}{1-w^{2k}} \tag{2}$$

All claims made about the surfaces so far may be checked directly using these formulas. For our purposes it is important to note that the vertical translation of the surfaces is represented by a cycle around one puncture.

More concretely, a curve starting at 0, going once around one puncture and then back to 0, will be mapped to a curve connecting two umbilic points which are above each other. To check this, one has to compute period integrals involving the Weierstraß 1-forms Gdh, $\frac{1}{G}dh$ and dh. For the explicit computation, we again refer to [4], but it should be clear that such a computation is simple, as periods of meromorphic 1-forms on a sphere are just residues.

2.2 The Twisted Karcher-Scherk Surfaces

The twisted Karcher-Scherk surfaces $KS_{k,\phi}$ are invariant under a vertical screw motion with twist angle 2ϕ. We write $\phi = a\pi$.

These surfaces exist for $0 < a < \frac{1}{k}$.

The quotient surface $\Sigma_{k,\phi}$ is again a $2k$-punctured sphere. A slight complication arises because it is not possible to have the punctures at the roots of unity. Using an additional parameter r, one places the punctures at the roots of $w^k = r^k$ and $w^k = -1/r^k$.

The straight lines of S_k survive the deformation as straight lines, this is what made the Plateau construction possible.

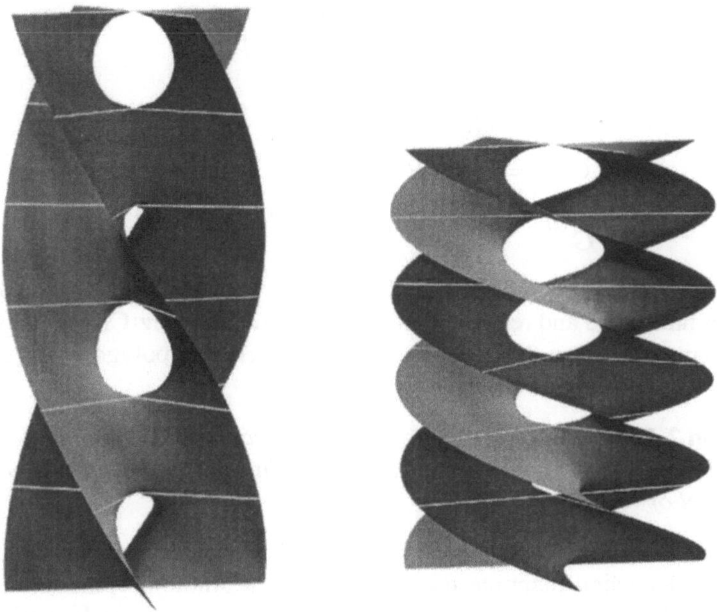

Fig. 3. Twisted Karcher-Scherk surface with four helicoidal ends (cf. color plate 7, Page 669)

The real parameter r is to be determined so that a horizontal period condition is satisfied. This again requires the evaluation of period integrals. While the height differential is still quite simple, the Gauss map is much more complicated then for S_k. Here is the formula from [4]:

$$G(w) = -iw^{k-1} \cdot \left(\frac{r^k - w^k}{r^k w^k + 1} \right)^a \tag{3}$$

$$dh = \frac{iw^{k-1}}{(r^k - w^k)(r^k w^k + 1)} dw \tag{4}$$

For $a = 0$ and $r = 1$, we obtain the classical Scherk surfaces S_k, see (1).

The Gauß map has the unpleasant feature that it is not anymore single valued on the quotient surface.

In fact, the multi-valuedness measures precisely the twist angle: If we analytically continue $G(z)$ around a puncture, the value $G(z)$ changes to $e^{a \cdot 2\pi i}$, so that $2\phi = 2\pi a$ is the twist angle.

2.3 Towards the Schwarz-Christoffel Formula

The first step in simplifying the Weierstraß representation is to exploit the k-fold symmetries. This is done by a simple coordinate change:

Write $z = w^k$ and $R = r^k$. Then, up to constant real factors,

$$\begin{aligned}
G(z) &= -iz^{1-1/k} \cdot \left(\frac{R-z}{Rz+1}\right)^a \\
dh &= \frac{idz}{(R-z)(Rz+1)} \\
Gdh &= z^{1-\frac{1}{k}}(R-z)^{a-1}(Rz+1)^{-a-1}\,dz \\
\frac{1}{G}dh &= -z^{\frac{1}{k}-1}(R-z)^{-a-1}(Rz+1)^{a-1}\,dz
\end{aligned}$$

The key observation now is that the 1-forms Gdh and $\frac{1}{G}dh$ are single-valued in the upper half plane and represent integrands in the Schwarz-Christoffel formula for the Riemann map from the upper half plane to euclidean polygons. Let us pause for a minute to review this formula.

Definition 2.1 *A* euclidean polygon *is a simply connected region in the plane such that all boundary components consist of straight lines and segments, meeting in a discrete set of vertices.*

That is, we allow for unbounded polygons but exclude domains with holes.

By the Riemann mapping theorem, there is a biholomorphic map f from the upper half plane to any such euclidean polygon. This map extends continuously and one-to-one to the (extended) real axes. Let us denote the preimages of the vertices by t_i, this will be a finite subset of the (extended) real line.

Denote by $\alpha_j = (a_j - 1)\pi$ the interior angle of the polygon at $f(t_j)$. Then we have

Theorem 2.2 (Schwarz-Christoffel formula) *Given a euclidean polygon as above, there are constants C_1, C_2 and w_0 such that*

$$f(z) = C_1 + C_2 \int_{w_0}^{w} \prod_j (z - t_j)^{a_j}\,dz$$

Now back to Gdh and $\frac{1}{G}dh$.

We conclude from theorem 2.2 that the map $z \mapsto \int^z Gdh$ maps the upper half plane to an open euclidean quadrilateral such that the points $0, R, -\frac{1}{R}, \infty$ are mapped to vertices with angles $(2-\frac{1}{k})\pi, a\pi, -a\pi, \frac{1}{k}\pi$. Similarly, the map $z \mapsto \int^z Gdh$ maps the upper half plane to an open euclidean quadrilateral such that the points $0, R, -\frac{1}{R}, \infty$ are mapped to vertices with angles $\frac{1}{k}\pi, -a\pi, a\pi, (2-\frac{1}{k})\pi$.

2.4 The Period Condition

This is a quite remarkably simple interpretation of Gdh and $\frac{1}{G}dh$, but what does it help? We still have to determine the parameter $R = r^k$ in the Weierstraß representation such that a certain period condition is satisfied. We will now make this condition explicit:

Consider two closed cycles: γ_1 commencing at 0, and looping around R back to 0, and γ_2 commencing at 0 and around $-1/R$ in the z-plane. These will be mapped to curves on the surface which represent a purely vertical translation. The verticality is taken care of by the definition of dh, but we haven't argued why the two endpoints of image curves lie strictly above each other. This is the content the *horizontal period condition* which can be stated as

$$\int_{\gamma_i} G dh = \overline{\int_{\gamma_i} \frac{1}{G} dh} \tag{5}$$

But, in terms of our understanding of Gdh and $\frac{1}{G}dh$ as integrands of Schwarz-Christoffel integrals, this boils down to a pair of geometric conditions on the euclidean image quadrilaterals. We will exhibit this geometric condition in the next section.

Moreover, if we can find a pair of image quadrilaterals which are conformally equivalent (as they must both be conformal to the upper half plane quadrilateral with vertices at $0, R, \infty, -\frac{1}{R}$), have the correct angles, and satisfy the geometric conditions, we can *define* Gdh and $\frac{1}{G}dh$ as Schwarz-Christoffel integrands for these quadrilaterals, without using any formulas. This is our next goal, so we will forget about minimal surfaces and the Weierstraß representation for a little while.

3 A Geometric Construction of the Weierstraß Data

We begin with an elementary construction:

For $0 < a < \frac{1}{k}$, construct an acute triangle with angles $\frac{1}{k}\pi$, $b\pi = (\frac{1}{2} - \frac{1}{2k} - \frac{a}{2})\pi$ and $c\pi = (\frac{1}{2} - \frac{1}{2k} + \frac{a}{2})\pi$. In honor of the case $k = 2$, we call the edge opposite to the vertex with angle $\frac{1}{k}\pi$ the *hypothenuse* of the triangle.

Observe that this construction is possible if and only if $0 < a < \frac{1}{k}$, because we will need the initial triangle to be acute.

Remark 3.3 *To be precise, the construction is also possible for all a with $0 < a < \frac{1}{k}$ modulo 2π. This leads to* branched *polygons and* overtwisted *Karcher-Scherk surfaces which are not embedded any longer.*

Reflect the triangle at its hypothenuse and extend all edges to infinity. Mark the edges as indicated as *dotted* and *dashed*. Observe that the angle $a\pi$ in the figure is implied by the construction.

Separate the dashed and dotted polygons, and reflect the dashed polygon at the x-axes. These polygons form now two unbounded quadrilaterals Q_1 and Q_2 in two different euclidean planes. Label the vertices (including the vertex at infinity) with the symbols $0, R, -1/R, \infty$ as indicated.

Define maps f_1 and f_2 from the upper half plane to the dashed and dotted domains Q_1 and Q_2 so that labels are mapped to labels. This *defines* uniquely the real value of R as the location of the symbol on the real axes.

That the same R works for both domains follows from the fact that both domains are congruent and have the same modulus. We will see this shortly also by a simple computation.

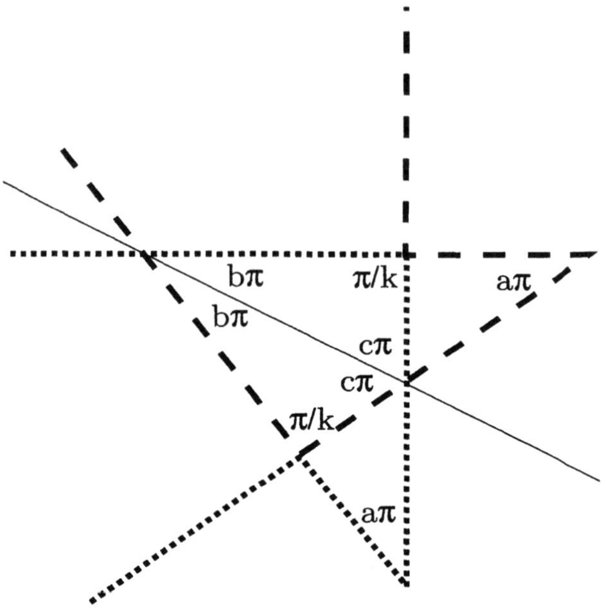

Fig. 4. Construction of the quadrilaterals

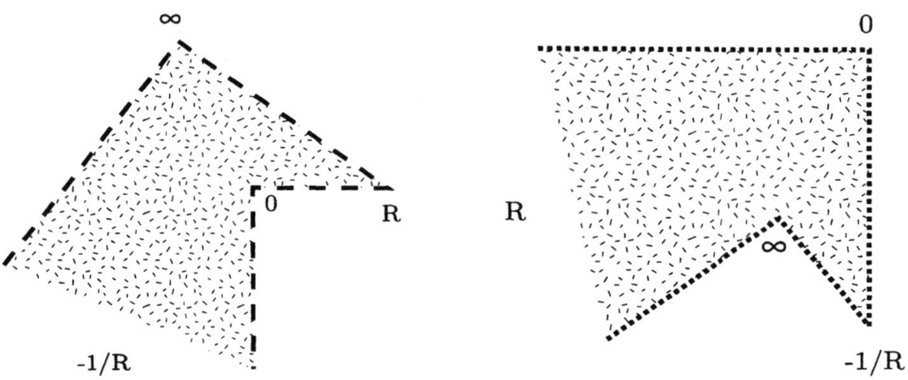

Fig. 5. Separation in dashed and dotted domains Q_1 and Q_2

If we now *define*

$$Gdh = df_1(z) = f_1'(z)dz$$
$$\frac{1}{G}dh = df_2(z) = f_2'(z)dz$$

then the Schwarz-Christoffel formula implies that, up to rotation and scaling,

$$Gdh = z^{1-\frac{1}{k}}(R-z)^{a-1}(Rz+1)^{-a-1}$$
$$\frac{1}{G}dh = -z^{\frac{1}{k}-1}(R-z)^{-a-1}(Rz+1)^{a-1}$$

This is exactly the kind of formula we were aiming for.

To verify that these formulas are correct without rotating or scaling, we need to check that the image domains Q_1 and Q_2 under the Schwarz-Christoffel maps $\int G dh$ and $\int 1/G dh$ are *congruent* and *correctly aligned* as in the figure.

For the correct alignment, observe that the integrand $G dh$ (resp. $\frac{1}{G} dh$) is positive (resp. negative) in the interval $(0, R)$, so that this interval is mapped to horizontal lines with the correct orientation as in the figure.

For the congruence, we use the automorphism $\iota(z) = -\frac{1}{z}$ of the upper half plane (which permutes the vertices $0, R, \infty$ and $-1/R$) and compute that

$$\begin{aligned}\iota^* G dh &= \left(\frac{1}{z}\right)^{1-\frac{1}{k}} \left(R + \frac{1}{z}\right)^{a-1} \cdot \left(1 - \frac{R}{z}\right)^{-1-a} \frac{dz}{z^2} \\ &= e^{i\theta} \frac{1}{G} dh\end{aligned}$$

for some real number θ. This means that Q_1 and Q_2 differ by a rotation (and not by a scale factor).

We now verify that the horizontal period conditions are satisfied. Let's begin with the condition on the cycle γ_1 which encircles R. The condition requires to compute the periods $\int_{\gamma_1} G dh$ and $\int_{\gamma_1} \frac{1}{G} dh$. The cycle γ_1 commences at 0, runs in the upper half plane to some point on the segment (R, ∞), and then continues back (say by reflection at the real axes) in the lower half plane to 0. Thus, the vector $u_1 = \int_{\gamma_1} G dh$ can be made visible in the image quadrilateral Q_1 as the vector from 0 to the mirror image of 0 at the straight segment $[R, \infty]$. The same statement holds for $v_1 = \int_{\gamma_1} \frac{1}{G} dh$ in Q_2.

Similarly, recall that γ_2 is a cycle commencing at 0, crossing the segment $(-\infty, -\frac{1}{R})$ and going back to 0. Thus the period vector $u_2 = \int_{\gamma_2} G dh$ (resp. $v_2 = \int_{\gamma_2} \frac{1}{G} dh$) is the vector from 0 to the mirror image of 0 at the edge $[-\infty, -\frac{1}{R}]$.

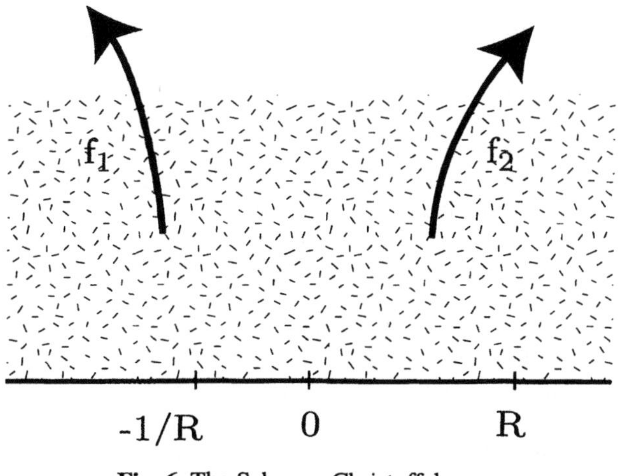

Fig. 6. The Schwarz-Christoffel maps

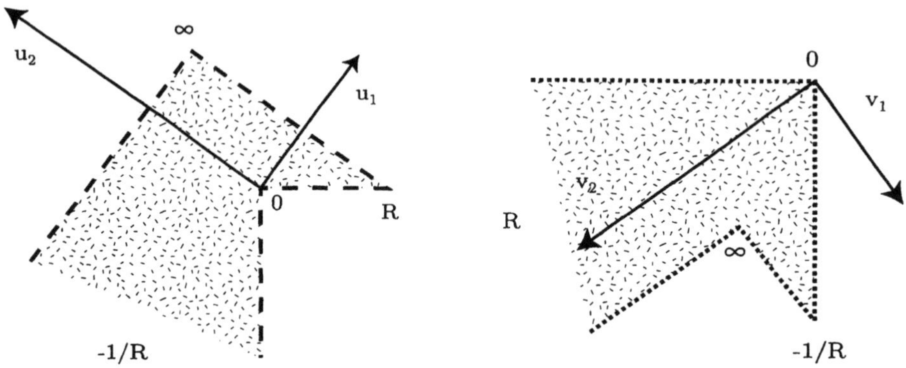

Fig. 7. Conjugacy of the period vectors

By the geometric construction of the quadrilaterals, the period vectors U_1, v_1 and u_2, v_2 are complex conjugate. This shows that that the period condition (5) is satisfied.

This finishes the construction of the twisted Karcher-Scherk surface for any $k \geq 2$ and $0 < \phi < \pi/k$. As the domains Q_1 and Q_2 vary smoothly with ϕ, so do the 1-forms Gdh and $\frac{1}{G}dh$, and we obtain that the family of surfaces is smooth in ϕ as a byproduct.

References

1. M. Callahan, D. Hoffman, H. Karcher: A family of singly-periodic minimal surfaces invariant under a screw motion. Experimental Mathematics, pages 157–182, 1993
2. D. Hoffman, H. Karcher, F. Wei: The genus one helicoid and the minimal surfaces that led to its discovery. In: K. Uhlenbeck (ed) Global Analysis and Modern Mathematics. Publish or Perish Press, 119–170, 1993
3. D. Hoffman, H. Karcher, F. Wei: The singly periodic genus-one helicoid. Commentarii Math. Helv., 248–279, 1999
4. H. Karcher: Embedded minimal surfaces derived from Scherk's examples. Manuscripta Math., 62, 83–114, 1988
5. H. Karcher: Construction of minimal surfaces. Surveys in Geometry, 1–96, 1989. University of Tokyo, 1989, and Lecture Notes No. 12, SFB 256, Bonn, 1989
6. A. Lynker: Einfach periodische elliptische Minimalflächen. Diplomarbeit Bonn, 1993
7. M. Traizet, M. Weber: Embedded minimal surfaces with helicoidal ends. preprint, 2001
8. M. Weber: Period quotient maps of meromorphic 1-forms and minimal surfaces on tori. preprint Bonn, 1999
9. M. Weber: The genus one helicoid is embedded. Habilitationsschrift Bonn, 2000
10. E. Koch, W. Fischer: On 3-periodic minimal surfaces with non-cubic symmetry. Zeitschrift für Kristallographie, 183, 129–152, 1988
11. M. Weber, D. Hoffman, M. Wolf: An embedded helicoid with a handle. Preprint, 2001
12. M. Weber, M. Wolf: Teichmüller theory and handle addition for minimal surfaces. Annals of Math. To appear.

13. M. Weber, M. Wolf: Minimal surfaces of least total curvature and moduli spaces of plane polygonal arcs. Geom. and Funct. Anal., 8, 1129–1170, 1998
14. M. Weber, M. Wolf: Minimal Surfaces and Teichmueller Theory. 2001. In preparation.

Unstable Periodic Discrete Minimal Surfaces

Konrad Polthier

Technische Universität Berlin, Institut für Mathematik
polthier@math.tu-berlin.de

Summary. In this paper we define the new *alignment energy* for non-conforming triangle meshes, and describes its use to compute unstable conforming discrete minimal surfaces. Our algorithm makes use of the duality between conforming and non-conforming discrete minimal surfaces which was observed earlier. In first experiments the new algorithm allows us the computation of unstable periodic discrete minimal surfaces of high numerical precision. The extraordinary precision of the discrete mesh enables us to compute the index of several triply periodic minimal surfaces.

1 Introduction

The numerical computation of unstable discrete minimal surfaces with *excellent numerical qualities* is among the most challenging problems. Good quality means that the numerical solution is sufficiently accurate to allow further investigations of the numerical data set. For example, discrete minimal surfaces are critical points of the discrete area functional and the calculation of the second derivative of the area functional at such a critical point requires that the discrete mesh is of very high numerical quality, i.e. very close to an exact critical point. Away from a critical point the second derivative is of minor value. Currently, good data sets of unstable discrete minimal surfaces are very hard to produce, even worse, they are hardly available to the community. Solutions obtained by integrating the Weierstrass functions are usually interpolating the smooth surface very well at the vertices but have unknown variational properties and are usually not discrete minimal surfaces in our sense.

In this paper we introduce a new algorithm to compute solutions of free-boundary value problems of minimal surfaces where the boundaries are constrained to meet a collection of planes orthogonally, or equivalently, to consist of planar symmetry lines lying on a set of given planes. For example, a patch may be bounded by the faces of a convex polyhedron like the images in Figures 7 and 8. To circumvent the instability of solutions of free boundary value problems one can use an indirect approach when the conjugate minimal surface is stable - this is the so-called conjugate surface construction [3][4]. It is known from our previous work [7][8] that all discrete minimal surfaces come in pairs of a *conforming* and a conjugate *non-conforming* surface. Non-conforming triangle meshes are characterized by the property that adjacent triangles are connected at their edge midpoints rather than having both endpoints of the edge in common.

In order to compute a conforming discrete minimal surface via the conjugate surface construction one computes its non-conforming conjugate minimale surface. But non-conforming meshes are highly unstable with respect to the area functional, such that, previous to this work, non-conforming minimal surfaces could only be obtained by starting from a conforming discrete minimal surface.

The essential ingredient in our algorithm is the introduction of the new *alignment energy* for non-conforming triangle meshes. It turns out that minimizing the alignment energy in the class of *non-conforming* triangle meshes leads to non-conforming discrete minimal surfaces whose discrete conjugate surfaces are then solutions of free-boundary value problems for *conforming* triangle meshes.

In a previous work [9] we computed unstable conforming discrete catenoids and helicoids by deriving explicit formulas for the vertices. Brakke's surface evolver [1] allows usage of the Hessian to reach an unstable critical point if the surface is already near enough the critical point for the Hessian mode to work.

This paper first introduces in Section 2 the different types of discrete minimal surfaces relevant for our discussion, and recalls in Section 3 the duality of conforming and non-conforming discrete minimal surfaces. In Section 4 we introduce the new alignment energy for non-conforming discrete minimal surfaces which is the basis on a new algorithm for unstable conforming minimal surfaces. We conclude with a number of test computations of well-known triply periodic minimal surfaces, and a numerical evaluation of their index.

2 Review of Non-conforming Triangle Meshes

In the theory of discrete minimal surfaces we observed the simultaneous appearance of two different types of triangle meshes, so-called conforming and non-conforming triangle meshes [7][8]. Because of their importance for the present work we recall their definitions.

Definition 2.1 *A* simplicial surface M_h *in* $\mathbb{R}^m$ *is a simplicial complex consisting of a finite set* $\mathfrak{T}$ *of planar triangles such that*

1. *Any vertex* $p \in M_h$ *belongs to at least one triangle* $T \in \mathfrak{T}$.
2. *The star of each vertex* $p \in M_h$ *is a simplicial disk.*

Note, a simplicial surface is continuous across edges and at the vertices of its triangles. We follow the common practice in finite element theory to call a simplicial surface a *conforming surface*.

Definition 2.2 *On a simplicial surface* M_h *we define the function space* S_h *of* conforming functions*:*

$$S_h := \left\{ v : M_h \to \mathbb{R}^d \;\middle|\; v \in C^0(M_h) \text{ and } v \text{ is linear on each triangle} \right\}$$

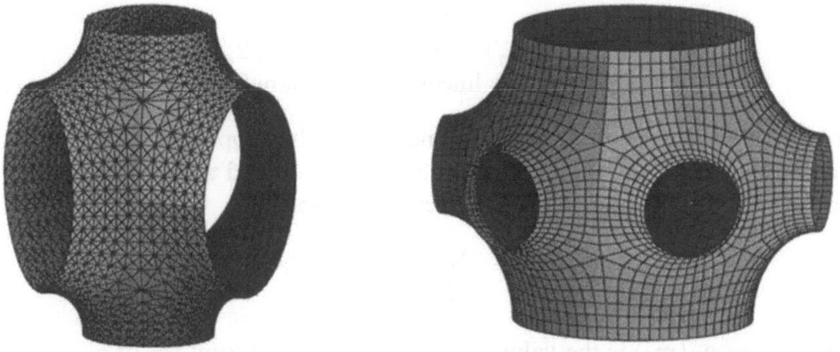

Fig. 1. Discrete H-T surface with fundamental patch in a 3-prism, and differently assembled in a 6-prism. In the right figure sets of four coplanar triangles were joined to emphasize the observed alignment of edges with curvature lines. (cf. color plate 1, Page 667)

S_h is a finite dimensional space spanned by the Lagrange basis functions $\{\varphi_1, .., \varphi_n\}$ corresponding to the set of vertices $\{p_1, ..., p_n\}$ of M_h, that is for each vertex p_i we have a function

$$\begin{gathered} \varphi_i : M_h \to \mathbb{R} \\ \varphi_i(p_j) = \delta_{ij} \quad \forall i, j \in \{1, .., n\} \\ \varphi_i \text{ is linear on each triangle.} \end{gathered} \tag{1}$$

Then each function $u_h \in S_h$ has a unique representation

$$u_h(p) = \sum_{j=1}^{n} u_j \varphi_j(p) \quad \forall\, p \in M_h$$

where $u_j = u_h(p_j) \in \mathbb{R}^d$.

Second, we introduce the following space of non-conforming finite elements. Since this space includes discontinuous functions its use was sometimes called a *variational crime* in the finite element literature [2]. In our setting of discrete minimal surfaces the space of non-conforming functions and surfaces does naturally appear [7][8].

Definition 2.3 *For a simplicial surface M_h, we define the space of* non-conforming functions *by*

$$S_h^* := \left\{ v : M_h \to \mathbb{R}^d \;\middle|\; \begin{array}{l} v_{|T} \textit{ is linear for each } T \in M_h, \textit{ and} \\ v \textit{ is continuous at all edge midpoints} \end{array} \right\}.$$

The space S_h^* is a superset of the space of conforming functions S_h, since S_h^* additionally includes functions which are discontinuous at edges away from the edge midpoint. Let $\{m_i\}$ denote the set of edge midpoints of M_h, then for each midpoint m_i we have the basis function

$$\begin{aligned} \psi_i &: M_h \to \mathbb{R} \\ \psi_i(m_j) &= \delta_{ij} \quad \forall\, i,j \in \{1,2,..\} \\ \psi_i &\text{ is linear on each triangle.} \end{aligned} \tag{2}$$

Note, that ψ_i may be multi-valued along edges. The support of a function ψ_i consists of the (at most two) triangles adjacent to the edge e_i, and ψ_i is usually not continuous on M_h. Each function $v \in S_h^*$ has a representation

$$v_h(p) = \sum_{\text{edges } e_i} v_i \psi_i(p) \quad \forall\, p \in M_h$$

where $v_i = v_h(m_i)$ is the value of v_h at the edge midpoint m_i of e_i.

We will use the term *non-conforming surface* M_h^* to denote the image of a simplicial surface under a non-conforming function.

3 Discrete Minimal Surfaces

The area of a conforming or non-conforming simplicial surface is defined as the sum of the area of all elements. This leads to the following variational characterization of discrete minimal surfaces which applies both to conforming and non-conforming meshes:

Definition 3.4 *A conforming surface M_h respectively non-conforming surface M_h^* is a* discrete minimal surface *iff the discrete area functional of M_h respectively M_h^* is critical with respect to variations of any set of interior vertices respectively interior edge midpoints. To include symmetry properties into this definition we allow a constrained variation of boundary points:*

- *if a boundary segment is a straight line, then its points may vary along the straight line*
- *if a boundary segment is a planar curve, then its points may vary within the plane*
- *in all other cases the boundary points always remain fixed.*

The relaxed boundary constraints allow one to solve free boundary value problems such that a discrete minimal surfaces may be extended by reflection to a discrete minimal surface. We recall the explicit representation of the area gradient given in [6]:

Lemma 3.5 *A simplicial surface M_h is discrete minimal if and only if at each interior vertex p*

$$\frac{d}{dp} \text{area}\, M_h = \frac{1}{2} \sum_j (\cot \alpha_j + \cot \beta_j)(p - q_j) = 0 \tag{3}$$

where $\{q_j\}$ denotes the set of vertices of link p *and α_j,β_j denote the two angles opposite to the edge pq_j. At boundary vertices on symmetry arcs the area gradient is constrained to be tangential to the straight line or to the plane.*

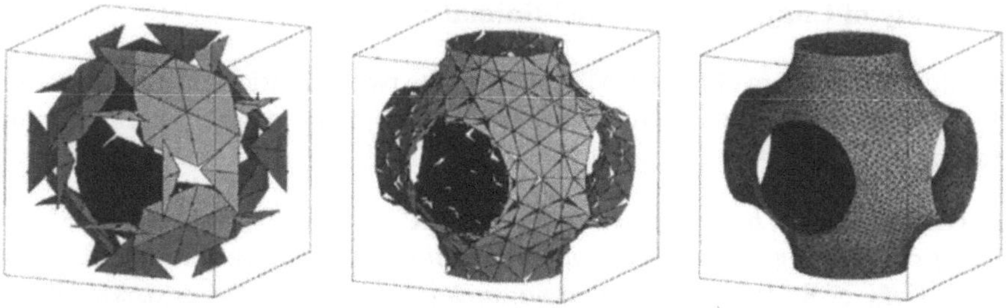

Fig. 2. Free-boundary value problem of Schwarz P-surface in a cube solved via the discrete conjugate surface construction of [6]. The solution of that method is a non-conforming discrete minimal surface which is shown here at different discretizations. (cf. color plate 2, Page 667)

For non-conforming surfaces we have the following representation of the area functional which was derived in [7]:

Lemma 3.6 *A non-conforming surface M_h^* is discrete minimal if it solves a system of equations such that at each edge midpoint m_i we have*

$$\begin{aligned}\frac{d}{dm_i}\text{area } M_h^* &= 2(\cot\alpha_{i_{-2}}\cdot(m_i - m_{i_{-1}}) + \cot\alpha_{i_{-1}}\cdot(m_i - m_{i_{-2}}) \\ &\quad + \cot\alpha_{i_1}\cdot(m_i - m_{i_2}) + \cot\alpha_{i_2}\cdot(m_i - m_{i_1})) \\ &= 0.\end{aligned} \tag{4}$$

where $\{i_{-2}, i_{-1}, i_1, i_2,\}$ denote subindices of adjacent edge midpoints m_i as shown in Figure 3 (assuming that v is the identity map in the figure, i.e. angles are measure in M_h^*).

Note, this theorem is true for the larger class of non-conforming Laplace-Beltrami harmonic maps v on a simplicial surface where the area gradient generalizes to the gradient of the Dirichlet energy, see [7].

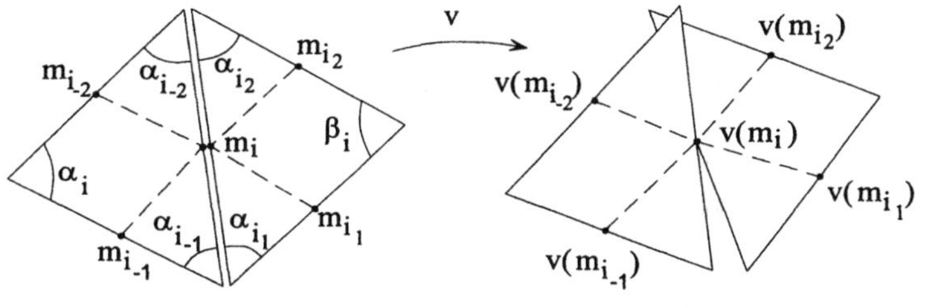

Fig. 3. A non-conforming map is given by its value on edge midpoints.

3.1 Duality of Conjugate Discrete Minimal Surfaces

The following theorem obtained in [7] justifies the two definitions of discrete minimal surfaces. It states the general relation between conjugate pairs of discrete minimal surfaces.

Theorem 3.7 *Let N be a simplicial surface in $\mathbb{R}^m$ and let S_h respectively S_h^* be the space of conforming respectively non-conforming immersions of N into $\mathbb{R}^d$. Then*

1. Let $M_h \subset \mathbb{R}^d$ be a discrete minimal surface in S_h. Then there exists a conjugate surface $M_h^* \subset \mathbb{R}^d$ in S_h^* which is critical for the area functional in S_h^*.

2. Let $M_h \subset \mathbb{R}^d$ be a discrete minimal surface in S_h^*. Then there exists a conjugate surface $M_h^* \subset \mathbb{R}^d$ in S_h which is critical for the area functional in S_h.

3. M_h^* is uniquely determined by M_h up to translation.

4. M_h and M_h^* are isometric and have the same Gauss map in the sense that corresponding triangles are congruent and parallel.

5. Applying the conjugation twice leads to

$$M_h^{**} = -M_h$$

for a suitably chosen origin.

Summarizing, the theorem shows that a pair of conjugate discrete minimal surfaces generically does not exist in the space of piecewise linear conforming elements S_h but naturally leads to the space of piecewise linear non-conforming Crouzeix-Raviart elements S_h^* [2]. S_h alone is too rigid to contain the conjugate of a minimal surface.

At first sight, this theorem seems to suggest a strategy for solving free-boundary problems for conforming meshes, namely, to simulate the *conjugate surface construction* which worked so well for smooth minimal surfaces [3][4]. That means, assume a given free boundary value problem for a conforming mesh, then solve the conjugate Dirichlet problem with a non-conforming mesh. The conjugate of the non-conforming minimal surface is then a conforming solution of the original problem.

The problem with this strategy is that the area functional for non-conforming meshes is not stable. Consider two adjacent triangles which are coplanar and whose corresponding edges enclose a positive angle. Then Figure 4 indicates that the joint area of both triangles may be reduced by moving the common edge midpoint towards the intersection of the lines through the other edge midpoints. This movement continuously reduces the common area to zero while not affecting any other parts of the surfaces. In practice this degeneration of pairwise triangles is the default behaviour when minimizing the area functional of non-conforming meshes.

4 The Alignment Energy

The stability of the area functional of conforming simplicial surfaces is often spoiled by tangential motion of vertices which leads to degenerate triangles. As noted in

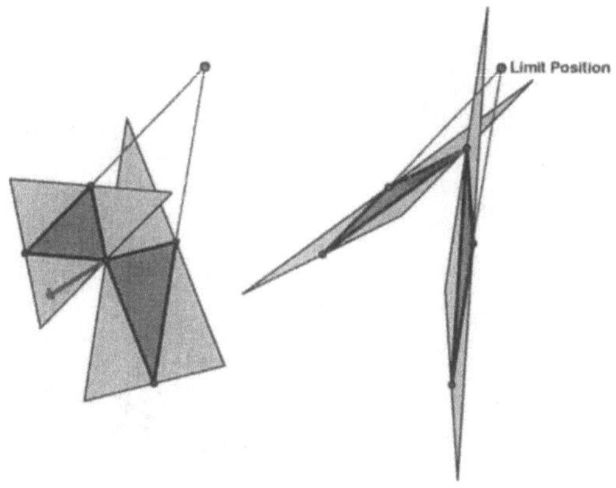

Fig. 4. This example shows the instability of the area functional for non-conforming meshes. Two adjacent coplanar triangles with non-parallel edge degenerate to two triangles with zero area.

[6] degenerate triangles may often be removed in practical implementations but in general there is no strategy available which tells existence of a discrete solution of a boundary value problem when assuming a given underlying mesh connectivity.

The situation with non-conforming meshes is even worse since basically there do not exist stable minimizers, i.e. discrete minimizers which are stable under small pertubations of their vertices.

Therefore, we will experiment with the following new energy for non-conforming surfaces which promises to be much more stable. This energy is designed to minimize the gradient of the non-conforming area functional instead of the functional itself.

Note, minimizing the gradient instead of the functional is a common practice, for example, if the numerical accuracy is to be improved.

We begin with the following definition of a rotation operator on a simplicial surface.

Definition 4.8 *Assume that the boundary of each triangle of a conforming or non-conforming surface is oriented. Two adjacent triangles may have different orientations, which means, we include non-orientable surfaces. Then the* rotation operator *J is defined as a rotation by $\frac{\pi}{2}$ within the plane of each triangle such that for each edge c the vector Jc is orthogonal to c and points into the interior of the triangle.*

Consider a pair of adjacent triangles T_1 and T_2 whose edges c_1 and c_2 are connected at their midpoint.

Definition 4.9 *The* alignment energy *of a non-conforming surface M_h^* is given by*

$$E(M_h^*) := \sum_{\text{edge } c \in \mathring{M}} |J_1 c_1 + J_2 c_2|^2 + \frac{1}{2} \sum_{\text{edge } c \in \partial M} |Jc + Rot_c Jc|^2 \qquad (5)$$

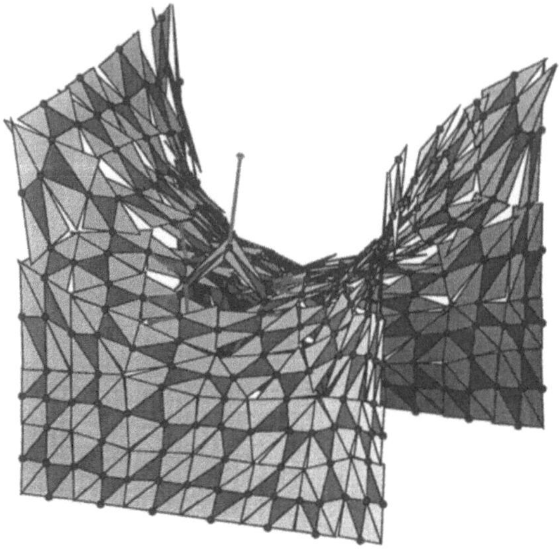

Fig. 5. Non-conforming mesh with a stencil of a vertex for the alignment energy and the induced gradients. The marked vertex has been moved away from its equilibrium position. (cf. color plate 4, Page 668)

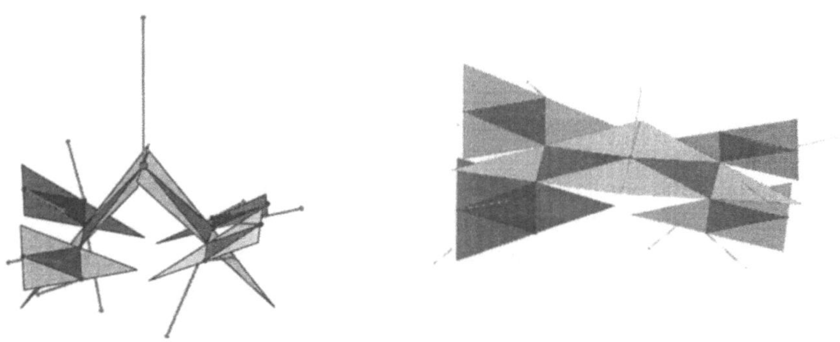

Fig. 6. Stencil of the alignment energy associated to an edge midpoint. Each vertex contributes to the alignment energy of the six triangles of its 2-star.

where c_1 *and* c_2 *represent the edge* c *in each of the two adjacent triangles* T_1 *and* T_2 *of* c*. At a boundary edge* c *the comparison vector* $Rot_c Jc$ *is produced by rotating* Jc *around* c *by* π

The alignment energy measures two entities simultaneously: first, the angle between the two corresponding midperpendiculars at an edge, and second, the difference in length of c_1 and c_2. That means, the alignment energy $|J_1 c_1 + J_2 c_2|^2$ of an edge is zero if c_1 and c_2 have equal lengths and if the two corresponding midperpendiculars are parallel, i.e. both triangles are *aligned* at the common edge midpoint m.

For computational purposes we calculate the gradient of the alignment energy. The edge midpoint m of c contributes to all edges of T_1 and T_2.

$$\frac{d}{dm}E(M) = \frac{d}{dm}\begin{pmatrix} |J_1c_1 + J_2c_2|^2 \\ +|J_{1,1}a_{1,1} + J_1a_1|^2 + |J_{1,2}b_{1,2} + J_1b_1|^2 \\ +|J_{2,1}a_{2,1} + J_2a_2|^2 + |J_{2,2}b_{2,2} + J_2b_2|^2 \end{pmatrix} \tag{6}$$

In a non-conforming triangle (m, q, r), where m, q, r are midpoints of the edges $c = -2c_m, a = -2a_m, b = -2b_m$ and where a_m, b_m, c_m are the edges of the mid-quarter triangle, we have

$$\begin{aligned} \nabla_m Jc &= (\cot\alpha + \cot\beta)NN^t \\ \nabla_m Ja &= +J\circ\pi - \cot\beta NN^t \\ \nabla_m Jb &= -J\circ\pi - \cot\alpha NN^t. \end{aligned} \tag{7}$$

The contribution of each triangle T_1 to the gradient of the alignment energy at m is

$$\begin{aligned} \frac{d}{dm}E(M)^t = 2(&(\cot\alpha_1 + \cot\beta_1)\, N_1N_1^t \cdot J_2c_2 \\ &+ (\cot\alpha_2 + \cot\beta_2)\, N_2N_2^t \cdot J_1c_1 \\ &+a_1 + \left(-J_1 - \cot\beta_1 N_1N_1^t\right) \cdot J_{1,1}a_{1,1} \\ &-b_1 + \left(+J_1 - \cot\alpha_1 N_1N_1^t\right) \cdot J_{1,2}b_{1,2}). \end{aligned} \tag{8}$$

The importance of the alignment energy is a consequence of the following observation.

Theorem 4.10 *Let M_h^* be a non-conforming surface. Then M_h^* is a discrete minimal surface if and only if M_h^* has vanishing alignment energy.*

Proof. Let T be a triangle with oriented edges $a = b + c$, and angles α and β opposite to a and b. Then we have

$$-Jc = \cot\alpha \cdot a + \cot\beta \cdot b.$$

The assumption is now an immediate consequence of theorem 4 and equation 4 since we can write the area gradient of M_h^* as

$$\frac{d}{dm_i}\text{area}\, M_h^* = J_1c_1 + J_2c_2.$$

Since the alignment energy vanishes exactly if it vanishes at every edge, a minimizer of the alignment energy is a non-conforming discrete minimal surface, and vice versa.

As a corollary of the preceeding theorem we obtain from theorem 3.7 that the discrete conjugation provides the following 1-1 pairing:

Corollary 4.11 *There exists a duality between non-conforming minimizers of the alignment energy and conforming discrete minimal surface.*

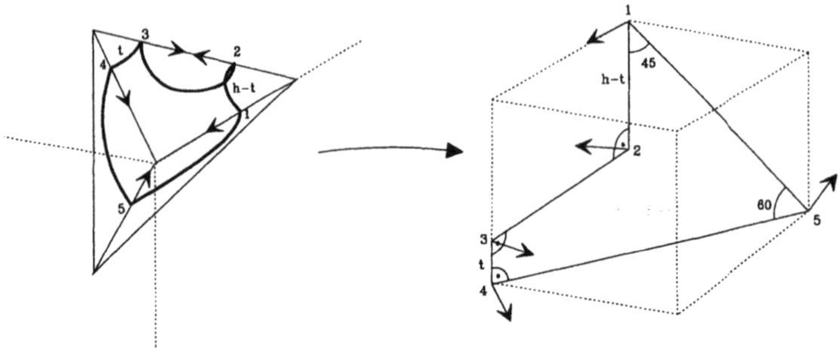

Fig. 7. Transformation of a free-boundary value problem into a family of Dirichlet boundary value problems.

4.1 Algorithm for Free-Boundary Value Problems

Corollary 4.11 suggests the following algorithm to compute solutions of free-boundary value problems for minimal surfaces. The method is similar to the algorithms in [6] for solving free-boundary value problems for minimal surfaces and in [5] for computing discrete constant mean curvature surfaces in $\mathbb{R}^3$ by minimizing surface area in $\mathbb{S}^3$. In contrast to our method, both previous methods obtained non-conforming solutions of the free-boundary value problem while we obtain conforming solutions. Following the remark on the stability of the area functional of non-conforming surfaces, conforming solutions are much better suited, for example, to compute the index of discrete minimal surfaces.

The following algorithm assumes a background knowledge of the conjugate surface construction of smooth minimal surfaces to obtain the conjugate contour Γ consisting of straight lines from a given free-boundary contour consisting of a collection of planes [4]. Note, in general the lengths of $n-4$ edges of Γ are unknown parameters (when ignoring scaling), which means, one must consider a family of contours Γ if there are more than 4 boundary segments.

Alignment Algorithm. Given a free-boundary value problem for a minimal surface, and assuming the conjugate contour Γ is known.

1. Solve the Dirichlet problem in Γ for the alignment energy among non-conforming surfaces. Denote the minimizer M_h^*.
2. Compute the conjugate conforming surface from M_h^*. The conjugate M_h is a conforming discrete minimal surface.

The first step produces a non-conforming discrete minimal surface. The second step of the algorithm is the same as the conjugation operation of discrete harmonic maps introduced in [6] except that it is applied here to non-conforming meshes. For discrete minimal surfaces the conjugation operation amounts to rotating each triangle by $\frac{\pi}{2}$ in the plane of each triangle and then assembling the triangles at corresponding edges. Note, theorem 3.7 ensures that this operation leads to a conforming discrete minimal surface.

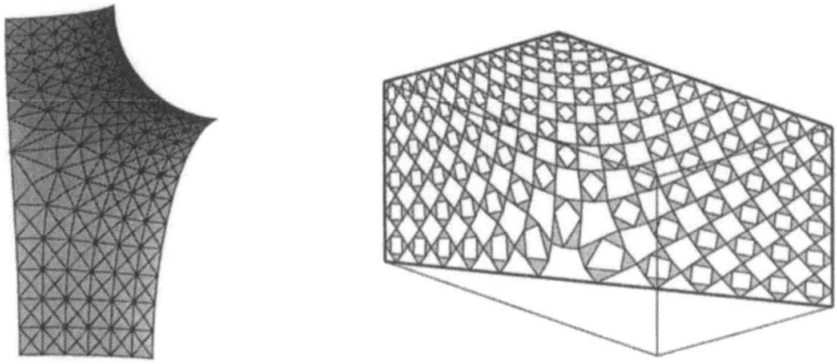

Fig. 8. Pair of conjugate fundamental patches of the H-T surface. The non-conforming mesh on the right was obtained by minimizing the alignment energy, and then conjugated to the conforming surface on the left (the right figure shows the mid-quarter of each triangle).

4.2 Comparison of Stability Properties

We consider a contour consisting of segments of straight lines, as in Figure 9, which typically appears in boundary value problems for periodic minimal surfaces. In practice we assume that the boundary vertices may freely move along their boundary segment in order to allow extension of a solution by reflection at the boundary segments, see Definition 3.4. For a given initial triangulation one observes that minimizing the area functional often leads to degenerate situations where triangles shrink to an edge. This instability of the discrete area functional is well-known and is related to tangential motions of the vertices. The same effect occurs when iteratively minimizing the Dirichlet energy, although each step is more stable. See [6] which justifies the removal of degenerate triangles.

When considering the degeneracy phenomenon from a combinatorial point of view then one can say that an immersion as a discrete minimal surface does not exist for every given combinatorial mesh. An example for the non-existence of immersions is easily constructed as shown in Figure 10 where it is obvious that the two free boundary points must move to a degenerate situation. The set of combinatorial meshes, which may be realized as discrete minimal surfaces in a given boundary contour, cannot be estimated yet but it seems that the number of realizable combinatorial meshes for a given boundary value problem may be quite restricted.

The situation in the non-conforming case is even worse. Basically, every non-conforming discrete minimal surface is unstable for the discrete area functional. A small pertubation of two adjacent triangles of a non-conforming discrete minimal surface may lead to a degenerate situation as indicated in Figure 4 for two coplanar triangles. If the common edge is not parallel then the common midpoint of this edge may be moved to the intersection of the two lines spanned by the other two edge midpoints in each triangle. It is easily seen that the area of the two quarter triangles, which is always a quarter of the full triangle, decreases to zero.

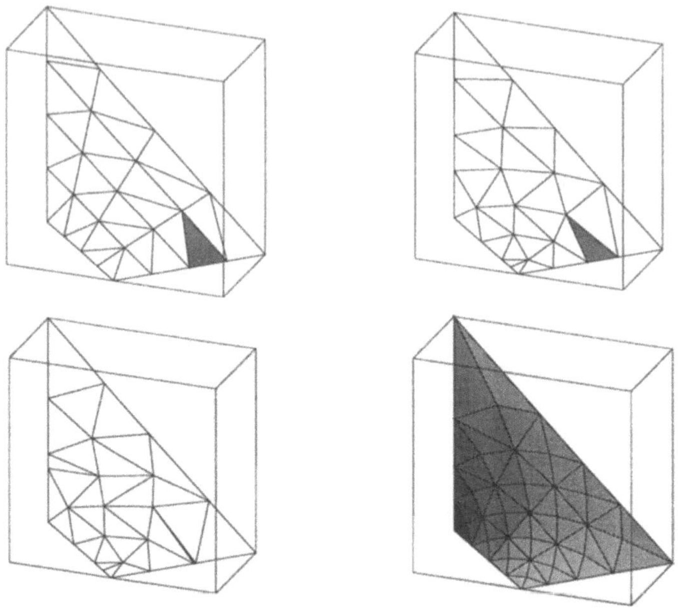

Fig. 9. Minimizing the area functional may lead to degenerate triangulations even for simple boundary contours. Although the Dirichlet energy is more stable its iterative minimization leads to the same degeneracies. In the sequence of three figures a triangle shrinks to an edge. The fourth figure at the bottom right uses an apropriate triangulation to avoid degeneration.

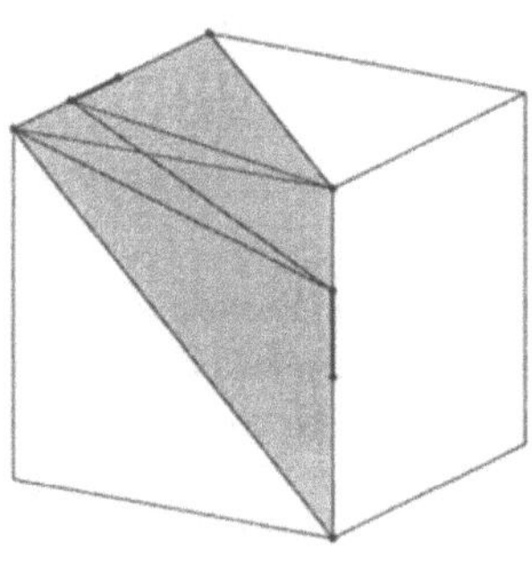

Fig. 10. Non-existence of a minimal immersion of a combinatorial mesh in a given boundary value problem. The two points will finally move to vertices of the contour as indicated by their area gradients.

The described instability makes the area functional useless for computing non-conforming minimal surfaces. This is the place where the alignment energy appears. As demonstrated in this paper the alignment energy seems to be reasonably stable and an efficient tool for computing unstable discrete minimal surfaces.

5 Examples of Discrete Triply Periodic Minimal Surfaces

We now report on some experiments we performed with triply periodic minimal surfaces. A detailed background on the minimal surfaces are given in [3][4]. Different previous approaches for the computation of these surfaces have been done. The first known computation by Anderson used the fact that the fundamental patch for the symmetry group of certain surfaces solves a free-boundary value in a tetrahedron and that the patch is a graph in tetrahedral coordinates. Pinkall and Polthier [6] introduced the conjugation of discrete conforming minimal surfaces which leads to non-conforming meshes as shown in Figure 2. This method was later extended to compute constant mean curvature surfaces by Oberknapp and Polthier [5] through discrete minimal surfaces in $\mathbb{S}^3$. Both methods used an adhoc approach to produce a conforming mesh from the obtained non-conforming surface since non-conforming meshes were not considered as valid mathematical objects during those days. Note, the averaging step can be justified to produce good results as indicated in Figure 2. Brakke has successfully computed several triply periodic minimal surfaces by specifying volume constraints [1] and by minimizing a discrete Willmore energy which is the integral of mean curvature squared.

The following experiments were produced with the algorithm 4.1 and aim at directly producing conforming discrete minimal surfaces.

The eigenvalue computations follow the discrete setup of Polthier and Rossman [9]. Note, for the experiments described in that paper more complicated unstable discrete minimal surfaces were not available yet. The discrete surfaces produced via minimizing the alignment energy turn out to have a sufficient numerical accuracy.

5.1 Index of F-Rd

A building block of the F-Rd surface is bounded by a cube as shown in Figure 12. The discrete minimal surface has a very small area gradient. As a consequence we obtain basically no tangential eigenfunctions, that means, the numerically computed index 9 is rather accurate. We plan to submit the datasets of this surface and the following I-Wp surface to the EG-Models server at www.eg-models.de to enable comparison of our results with other numerical algorithms.

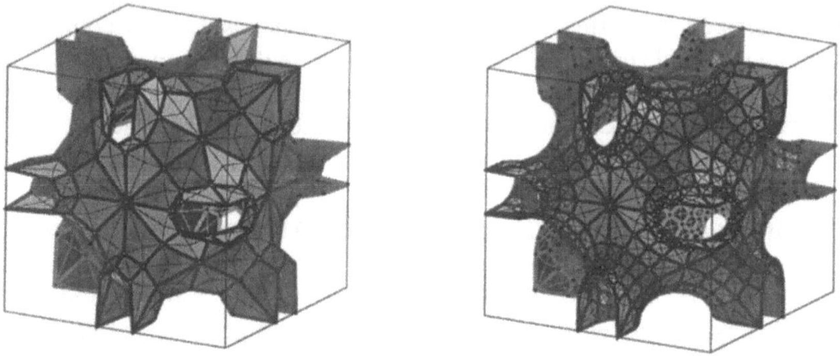

Fig. 11. Discrete versions of Neovius surface in a cube. Note the observation that in both minimizers each triangle has three coplanar adjacent triangles which leads to a tessellation with planar quadrilaterals. (cf. color plate 3, Page 667)

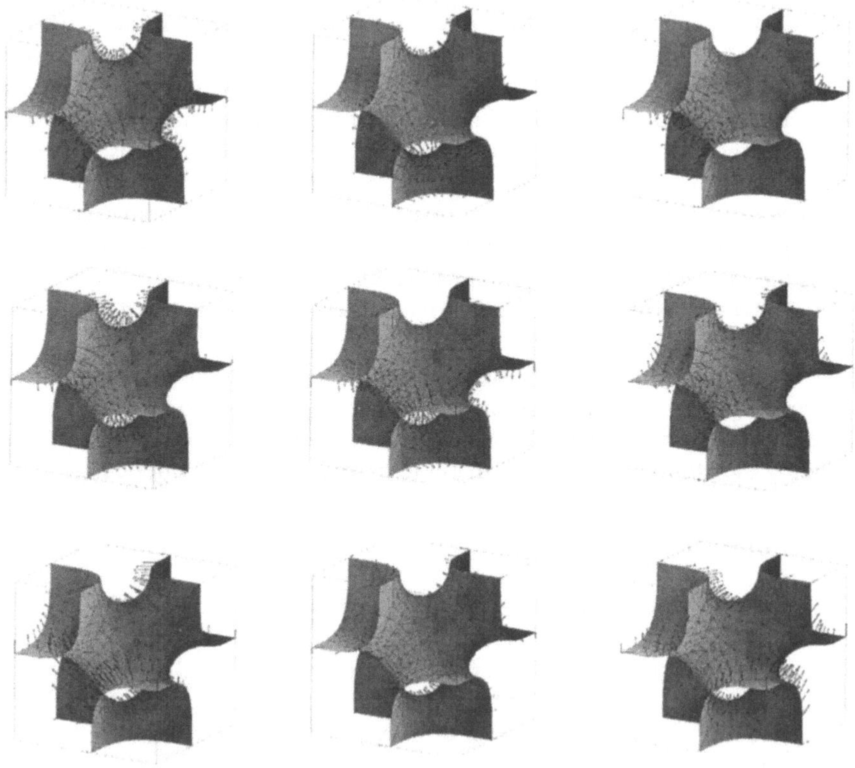

Fig. 12. Numerically found 9 eigenfunctions of the discrete F-Rd surface with negative eigenvalues.

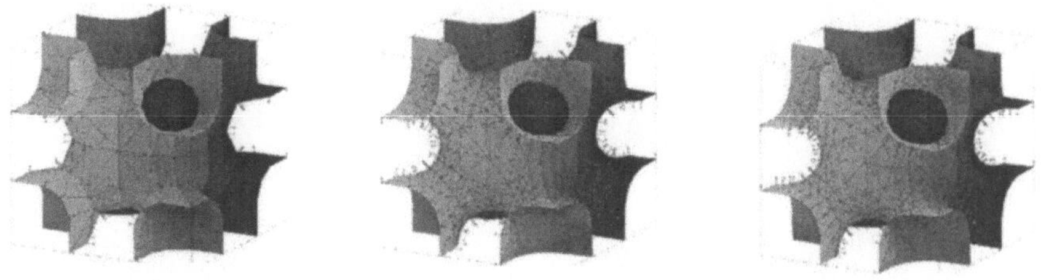

Fig. 13. I-Wp surface at different resolutions with first eigenvector.

Vertices/Elements 430/768
Area/Gradient 15.006517086292167/0.000000000053007

9 negative and first zero eigenvalues of the discrete F-Rd surface.

eValue[0] = -0.1504559679908802	eValue[5] = -0.03865424556938688
eValue[1] = -0.09904128162525966	eValue[6] = -0.004015601274444393
eValue[2] = -0.09904128162524234	eValue[7] = -0.004015601274391426
eValue[3] = -0.09904128162524005	eValue[8] = -0.004015601274380579
eValue[4] = -0.03865424556948896	eValue[9] = -2.7290174465353138E-28

5.2 Index of I-Wp

We computed the index of the I-Wp surface at different solutions. It turned out that already the very coarsest resolution gave the same answer 19 for the number of negative eigenvalues. The finer resolutions confirmed this result but did not have the same small gradient but only of the order of 1.E-6. This larger numerical error led to a few tangential eigenfunctions with small negative eigenvalues but which were clearly distinguishable from the normal eigenfunctions.

Vertices/Elements 234/384
Area/Gradient 40.187264207966614/0.000000000006771

19 negative and first zero eigenvalues of the discrete I-Wp surface.

eValue[0] = -0.632087803765472	eValue[10] = -0.13011897222983074
eValue[1] = -0.5175825952322985	eValue[11] = -0.13011897222982016
eValue[2] = -0.5175825952322736	eValue[12] = -0.1301189722297763
eValue[3] = -0.5175825952322616	eValue[13] = -0.05465367417861271
eValue[4] = -0.37793138360297224	eValue[14] = -0.054653674157824385
eValue[5] = -0.3779313836029666	eValue[15] = -0.054653674137100865
eValue[6] = -0.37793138360293166	eValue[16] = -0.04551152809511838
eValue[7] = -0.2598562604031981	eValue[17] = -0.04551152809511528
eValue[8] = -0.2249510919234355	eValue[18] = -0.045511528095097585
eValue[9] = -0.22495109192343474	eValue[19] = -8.039329159488626E-28

Fig. 14. First six (0,1,4,7,8,10) of the numerically found 19 eigenfunctions of the discrete I-Wp surface with negative eigenvalues.

References

1. K. A. Brakke. The surface evolver. *Exp. Math.*, 1(2):141–165, 1992.
2. S. C. Brenner and L. R. Scott. *The Mathematical Theory of Finite Element Methods*. Springer Verlag, 1994.
3. H. Karcher. The triply periodic minimal surfaces of A. Schoen and their constant mean curvature compagnions. *Man. Math.*, 64:291–357, 1989.
4. H. Karcher and K. Polthier. Construction of triply periodic minimal surfaces. In J. Klinowski and A. L. Mackay, editors, *Curved Surfaces in Chemical Structures*, volume 354 (1715) of *A*, pages 2077–2104. The Royal Society, London, Great Britain, phil. trans. r. soc. lond. edition, September 1996.
5. B. Oberknapp and K. Polthier. An algorithm for discrete constant mean curvature surfaces. In H.-C. Hege and K. Polthier, editors, *Visualization and Mathematics*, pages 141–161. Springer Verlag, Heidelberg, 1997.
6. U. Pinkall and K. Polthier. Computing discrete minimal surfaces and their conjugates. *Experim. Math.*, 2(1):15–36, 1993.
7. K. Polthier. Conjugate harmonic maps and minimal surfaces. Report 446, SFB 288, TU-Berlin, 2000.
8. K. Polthier. Computational aspects and discrete minimal surfaces. In D. Hoffman, editor, *Proc. of the Clay Summerschool on Minimal Surfaces*. AMS, 2002.
9. K. Polthier and W. Rossman. Discrete constant mean curvature surfaces and their index. *J. reine angew. Math*, 549(47–77), 2002.

An Adaptive Finite Element Method for Minimal Surfaces

W. Dörfler[1] and K.G. Siebert[2]

[1] Institut für Angewandte Mathematik II, Universität Karlsruhe, Kaiserstraße 12, 76128 Karlsruhe, Germany, doerfler@math.uni-karlsruhe.de
[2] Institut für Angewandte Mathematik, Universität Freiburg, Hermann-Herder-Straße 10, 79104 Freiburg, Germany, kgs@mathematik.uni-freiburg.de

Summary. Our aim is to compute surfaces of the topological type of the disc that are stationary points of the area functional. Since such computations may be very time consuming, we derive an a posteriori controlled adaptive algorithm based on a recently developed and analyzed finite element method [11] [12] [13]. Numerical results are presented for two examples.

1 Introduction

The *minimal surface* spanned by a given curve $\Gamma \subset \mathbb{R}^d$ is a stationary point of the area functional among all surfaces bounded by Γ. Such a stationary point will fulfill a nonlinear partial differential equation. If, for example, the surface can be represented as a graph of a function $u : \Omega \subset \mathbb{R}^d \to \mathbb{R}$, then it satisfies a Dirichlet problem for *the minimal surface equation*

$$\nabla \cdot \Big(\frac{\nabla u}{\sqrt{1 + |\nabla u|^2}} \Big) = 0.$$

The more general case of parametric surfaces (with topology of the unit disc) can be formulated in many ways, one of which is as follows: let $\mathcal{C}(\Gamma)$ be the set of all continuous mappings from the unit disc $D \subset \mathbb{R}^2$ to $\mathbb{R}^d$ with $u(\partial D) \subset \Gamma$ that are monotone and with prescribed values at three distinct points on ∂D (three point condition). Then $u \in \mathcal{C}(\Gamma)$ is sought that solves

$$\begin{aligned} \Delta u &= 0 \quad \text{in } D, \\ |\partial_x u| = |\partial_y u|, &\quad \partial_x u \cdot \partial_y u = 0 \quad \text{in } D. \end{aligned} \tag{1}$$

This equation shows the non-linear character of the problem, that now appears in the conformity conditions (1). The (equivalent) formulation in Definition 2.2 that is used here, relies on finding a stationary point of the Dirichlet integral. A classical result of [10] and [18] is that the minimization problem has at least one solution in $\mathcal{C}(\Gamma)$. As reference for the theory we cite [6] [7] [22], and with special emphasize, [14].

Based on the continuous variational formulation we can define a corresponding discrete problem which is, however, not simply the restriction of the continuous one to the discrete space. Moreover, the three point condition is replaced by a weaker one (see (2)) that is more suitable to this approach. For such a setting it has been proven in a sequence of papers by Dziuk and Hutchinson [11] [12] [13] that a non-degenerate minimal surface can be approximated by this method to optimal order. These references are also recommended concerning historical developments and alternative numerical approaches. The main reason for considering this method is the numerical computation of unstable minimal surfaces, that is, those stationary points that are no minimizers.

The purpose of the present work is the construction of an adaptive version for this method. Since the computational effort is very large one is interested in using locally refined meshes (to save degrees of freedoms) as well as criteria for the control of the Poisson problems that have to be solved in each step.

The main results are a posteriori error estimates of residual type, Theorem 4.9 and Theorem 4.11. In fact, the right hand side of this estimate is not completely computable, so that we have to derive practical criteria by additional heuristics, see Sect. 4.5. Based on these theoretical arguments, we propose an algorithm in Sect. 5 that differs in some aspects from the one in [12]. This algorithm is applied to two examples in Sect. 6.

Notations

For a domain $\Omega \subset \mathbb{R}^d$ we let $L^p(\Omega) = L^p(\Omega, \mathbb{R})$ and $W^{m,p}(\Omega) = W^{m,p}(\Omega, \mathbb{R})$, for $m \in \mathbb{N}_0$ and $p \in [1, \infty]$, denote the usual Lebesgue- and Sobolev spaces with norms denoted by $\|\,.\,\|_{L^p(\Omega)}$, $\|\,.\,\|_{W^{m,p}(\Omega)}$. Especially, $L^p(\Omega) = W^{0,p}(\Omega)$ and $H^m(\Omega) = W^{m,2}(\Omega)$. The notation $[\![\,.\,]\!]$ is used for the semi-norms, e.g. $[\![\,.\,]\!]_{W^{m,p}(\Omega)} = \|\nabla\,.\,\|_{L^p(\Omega)}$. The classical spaces of continuous and continuously differentiable functions are denoted by $C^m(\overline{\Omega})$, $C^{m,\alpha}(\overline{\Omega})$ for $m \in \mathbb{N}_0$ and $\alpha \in (0,1]$. Correspondingly, we define spaces $W^{m,p}(\Omega, \mathbb{R}^l)$ for $l > 1$ componentwise and use the vector norm in the function space norm.

Our parameter domain will be $D := B_1(0) \subset \mathbb{R}^2$ with its boundary $\partial D = S^1$. Since $\phi \mapsto e^{i\phi}$ is a diffeomorphism from $\mathbb{R}/2\pi$ to ∂D we will write for convenience ϕ for $e^{i\phi}$, $g(\phi)$ for $g(e^{i\phi})$ and $|\phi - \overline{\phi}|$ for $|e^{i\phi} - e^{i\overline{\phi}}|$. Thus for $g : \partial D \to \mathbb{R}$ the tangential derivatives are $g^{(m)}(\phi) := (d/d\phi)^m g(\phi)$ for $m \in \mathbb{N}$. The Sobolev spaces $W^{m,p}(\partial D)$ are then defined using

$$\|g^{(m)}\|^p_{L^p(\partial D)} := \int_{\partial D} |g^{(m)}(\phi)|^p \, d\phi,$$

$$\|g^{(m)}\|_{L^\infty(\partial D)} := \text{ess sup}\{|g^{(m)}(\phi)| : \phi \in \partial D\}.$$

Likewise the classical spaces $C^m(\partial D)$ and $C^{m,\alpha}(\partial D)$ are defined. The fractional power space $H^{1/2}(\partial D)$ is equipped with the norm $\|g\|_{H^{1/2}(\partial D)} := [\![g]\!]_{H^{1/2}(\partial D)} + \|g\|_{L^2(\partial D)}$, where

$$[\![g]\!]^2_{H^{1/2}(\partial D)} := \int_{\partial D} \int_{\partial D} \frac{|g(\phi) - g(\overline{\phi})|^2}{|\phi - \overline{\phi}|^2} \, d\overline{\phi} \, d\phi.$$

The definition of $H^{m/2}(\partial D)$ for $m = 3, 5, \dots$ is likewise. An equivalent norm for $H^m(\partial D)$ can be defined using the Fourier transformation [1, Ch. 7.48]. The definition of function spaces $W^{m,p}(\partial D, \mathbb{R}^l)$ is again componentwise.

In the following we work with a conformal parametrization of $\mathcal{C}(\Gamma)$. We fix a smooth parametrization of Γ denoted by $\gamma : S^1 \to \Gamma \subset \mathbb{R}^3$ with $\|1/|\gamma'|\,\|_{L^\infty(S^1)} < \infty$. The problem is then to find $s : \partial D \to S^1$ such that the harmonic extension of $\gamma \circ s : \partial D \to \mathbb{R}^3$ into $H^1(D, \mathbb{R}^3)$ is a stationary point of the Dirichlet integral (observe that, although identical, we use ∂D for the domain of definition of s and S^1 for its range). Using the identification $\phi \leftrightarrow e^{i\phi}$ we will interpret s as a mapping $\mathbb{R}/2\pi \to \mathbb{R}$ which gives to this set of admissible functions a linear structure. Moreover, s is subjected to additional constraints: (a) s is assumed to be *monotone*: s is positively oriented and $s^{-1}(q)$ is connected for all $q \in S^1$, (b) the minimal surface is of the topological type of the disc, (c) s fulfills the *three-point-condition*: $s(\phi_i) = s_i$, $i = 1, 2, 3$, for distinct $\phi_i \in [0, 2\pi)$, $s_i \in S^1$.
These requirements lead to the conclusion that s has to be in $C^0(\partial D) \cap H^{1/2}(\partial D)$ and it is of the form $s(\phi) = \phi + \xi(\phi)$ with ξ satisfying $\xi(\phi_i) = 0$ (without loss of generality one may take $\phi_i = s_i$). For later purposes, however, we will need a condition that can be formulated also in $H^{1/2}(\partial D)$ where point evaluations are not well-defined. So the three-point-condition that we actually use here is given by

$$\int_0^{2\pi} \xi(\phi)\, d\phi = 0, \quad \int_0^{2\pi} \xi(\phi)\, \sin(\phi)\, d\phi = 0, \quad \int_0^{2\pi} \xi(\phi)\, \cos(\phi)\, d\phi = 0. \quad (2)$$

For a more detailed description of this approach see [22] [14].

2 Theoretical Problem

2.1 Harmonic Extension

Let $f \in H^{1/2}(\partial D, \mathbb{R}^3)$ be given. Then the *harmonic extension* $\Phi(f) \in X := H^1(D, \mathbb{R}^3)$ is defined by the unique solution of the boundary value problem

$$\begin{aligned} \Delta\Phi(f) &= 0 \quad \text{on } D, \\ \Phi(f) &= f \quad \text{on } \partial D. \end{aligned} \quad (3)$$

Φ is a linear continuous mapping $H^{1/2}(\partial D, \mathbb{R}^3) \to X$ and there is a number C (independent of f) such that

$$\|\nabla\Phi(f)\|_{L^2(D)} \le C\,[\![f]\!]_{H^{1/2}(\partial D)} \quad (4)$$

and, for $f \in H^1(\partial D, \mathbb{R}^3)$,

$$\|\partial_n\Phi(f)\|_{L^2(\partial D)} = [\![f]\!]_{H^1(\partial D)} \quad (5)$$

[13, p. 530]. Note that in (4) holds equality with $C := 1$ if we define $H^s(\partial D)$-norms via Fourier transformations. This estimate holds correspondingly if we replace D by any open and bounded domain Ω whose boundary is Lipschitz-continuous. In this case C depends on Ω.

2.2 Function Spaces

We introduce the Banach space

$$H := \{\xi : \partial D \to \mathbb{R} : \|\xi\|_{H^{1/2}(\partial D)} < \infty \text{ and } \xi \text{ satisfies (2)}\} \tag{6}$$

and define the affine space $\mathcal{H}$ by

$$\mathcal{H} := Id + H = \{s = Id + \xi : \partial D \to \mathbb{R} : \xi \in H\}. \tag{7}$$

Due to (2) we have a Poincaré-type inequality

$$\|\xi\|_{L^2(\partial D)} \leq C\, [\![\xi]\!]_{H^{1/2}(\partial D)} \quad \forall \xi \in H \tag{8}$$

and therefore we can let $\|\xi\|_H := [\![\xi]\!]_{H^{1/2}(\partial D)}$. Furthermore, we define the Banach space $T := H \cap C^0(\partial D, \mathbb{R})$ with norm $\|\eta\|_T := \|\eta\|_H + \|\eta\|_{C^0(\partial D)}$ and the affine space $\mathcal{T} := Id + T$.

2.3 Energy and Its Derivatives

Let $\gamma \in C^{2,1}(S^1, \mathbb{R}^3)$. For $s \in \mathcal{H}$ let $u(s) := \Phi(\gamma \circ s) \in H^1(\partial D, \mathbb{R}^3)$ (see Proposition 2.4). Then the *energy* of s is defined by

$$E : \mathcal{H} \to \mathbb{R}_+, \quad E(s) := \frac{1}{2} \int_D |\nabla u(s)|^2.$$

For $s \in \mathcal{T}$ and $\xi \in T$ let $v(s, \xi) := \Phi(\xi\, \gamma' \circ s) \in H^1(D, \mathbb{R}^3)$ (see Propositions 2.3 and 2.4). Then E has a Fréchet-derivative given by

$$E'(s) : T \to \mathbb{R}, \quad E'(s)[\xi] = \int_D \nabla u(s) : \nabla v(s, \xi).$$

For $s \in \mathcal{T}$ and $\xi^1, \xi^2 \in T$ let $w(s, \xi^1, \xi^2) := \Phi(\xi^1 \xi^2\, \gamma'' \circ s) \in H^1(D, \mathbb{R}^3)$ (see Propositions 2.3 and 2.4). Then the second Fréchet-derivative of E is

$$E''(s) : T \times T \to \mathbb{R},$$
$$E''(s)[\xi^1, \xi^2] = \int_D \{\nabla u(s) : \nabla w(s, \xi^1, \xi^2) + \nabla v(s, \xi^1) : \nabla v(s, \xi^2)\}.$$

This shows that E is a mapping defined on $\mathcal{H}$, but it is differentiable as a mapping restricted to $\mathcal{T}$ only. However, by more regularity requirements on γ or s, the linearizations can be extended to weaker arguments.

Proposition 2.1 *([12, Prop. 3.8, 3.9]) E has the following regularity properties.*
1) If $\gamma \in C^{2,1}(S^1, \mathbb{R}^3)$, then for $s \in \mathcal{H}$: $E'(s) \in \mathbb{L}(T, \mathbb{R})$ and $E''(s) \in \mathbb{L}(T \times T, \mathbb{R})$.
2) If $\gamma \in C^{1,1}(S^1, \mathbb{R}^3)$, then for $s \in C^{0,1}(\partial D)$: $E'(s) \in \mathbb{L}(H, \mathbb{R})$, $E''(s) \in \mathbb{L}(H \times H, \mathbb{R})$.

Proof. See [12, Prop. 3.8, 3.9]. In 2) the requirement $s \in C^1(\partial D)$ was stated, however, it is easily seen by the method of proof that $s \in C^{0,1}(\partial D)$ suffices. □

Definition 2.2 *The harmonic function* $u := \Phi(\gamma \circ s) \in X$ *is a solution of the Plateau problem for* Γ, *if and only if* $s \in \mathcal{T}$ *is monotone and satisfies*

$$E'(s)[\xi] = 0 \qquad \forall \xi \in T.$$

Assume $s \in C^{0,1}(\partial D)$. *Then the solution* u *(or* s*) is called* non-degenerate *if for some* $\lambda > 0$

$$\inf_{\xi^1 \in H : \|\xi^1\|_H = 1} \ \sup_{\xi^2 \in H : \|\xi^2\|_H = 1} E''(s)[\xi^1, \xi^2] \geq \lambda.$$

2.4 Some Useful Estimates

Here we collect some (slightly modified) results from [13] for later reference. Note that $C^{k,1} \equiv W^{k+1,\infty}$. The constants in the following propositions are simple numbers.

Proposition 2.3 *([13, Prop. 3.1]) For* $f, g : \partial D \to \mathbb{R}$ *and* $k > 1/2$

$$[\![fg]\!]_{H^{1/2}(\partial D)} \leq \|f\|_{L^\infty(\partial D)} [\![g]\!]_{H^{1/2}(\partial D)} + [\![f]\!]_{H^{1/2}(\partial D)} \|g\|_{L^\infty(\partial D)},$$
$$[\![fg]\!]_{H^{1/2}(\partial D)} \leq C \|f\|_{H^k(\partial D)} \|g\|_{H^{1/2}(\partial D)} \leq C \|f\|_{W^{1,\infty}(\partial D)} \|g\|_{H^{1/2}(\partial D)}.$$

Proof. The first assertion is from [13, Prop. 3.1]. For the second part we take smooth f, g and write for simplicity $f \equiv \Phi(f)$, $g \equiv \Phi(g)$. With the trace theorem we have for all $p > 1$ and dual exponent $p' := p/(p-1)$

$$\begin{aligned} [\![fg]\!]_{H^{1/2}(\partial D)} &\leq C [\![fg]\!]_{H^1(D)} \\ &\leq C\big(\|\nabla f\|_{L^{2p'}(D)} \|g\|_{L^{2p}(D)} + \|f\|_{L^\infty(D)} \|\nabla g\|_{L^2(D)}\big). \end{aligned}$$

By elliptic regularity $[\![f]\!]_{H^s(D)} \leq C [\![f]\!]_{H^{s-1/2}(\partial D)}$ for $s \in (0,2)$ [15, Ch. 2.7] and the Sobolev embeddings $H^{1+1/p}(D) \hookrightarrow W^{1,2p'}(D)$, $H^{1/2+1/p}(\partial D) \hookrightarrow L^\infty(\partial D)$, and $H^{1/2}(\partial D) \hookrightarrow L^{2p}(\partial D)$ are known to be continuous for all $p < \infty$. Hence $[\![fg]\!]_{H^{1/2}(\partial D)} \leq C \|f\|_{H^{1/2+1/p}(\partial D)} \|g\|_{H^{1/2}(\partial D)}$. □

Proposition 2.4 *[13, Prop. 3.2] For* $g : S^1 \to \mathbb{R}$, $s : \partial D \to S^1$, *and* $k \in \{1/2, 1\}$

$$[\![g \circ s]\!]_{H^k(\partial D)} \leq \|g'\|_{L^\infty(S^1)} [\![s]\!]_{H^k(\partial D)}.$$

Proposition 2.5 *[13, Prop. 3.3] For* $g : S^1 \to \mathbb{R}$, $s_1, s_2 : \partial D \to S^1$ *and* $\sigma := s_1 - s_2$

$$\|g \circ s_1 - g \circ s_2\|_{L^2(\partial D)} \leq \|g'\|_{L^\infty(S^1)} \|\sigma\|_{L^2(\partial D)},$$
$$\begin{aligned} [\![g \circ s_1 - g \circ s_2]\!]_{H^{1/2}(\partial D)} &\leq C \|g'\|_{W^{1,\infty}(S^1)} \\ &\quad \times \big(1 + \|s_1'\|_{L^\infty(\partial D)} + \|\sigma\|_{L^\infty(\partial D)}\big) \|\sigma\|_{H^{1/2}(\partial D)}, \end{aligned}$$
$$[\![g \circ s_1 - g \circ s_2]\!]_{H^1(\partial D)} \leq C \|g'\|_{W^{1,\infty}(S^1)} \big(1 + \|s_1'\|_{L^\infty(\partial D)}\big) \|\sigma\|_{H^1(\partial D)}.$$

3 The Discrete Problem

3.1 Discretization

To discretize D we start from the decomposition of D into triangles shown in Fig. 1.(a). Finer discretizations will be obtained by further refinement using techniques such as newest node bisection (with special care to improved boundary approximation) [16]. All discretizations obtained in this way satisfy the usual requirements [3, Ch. 2.2.2] and, moreover, there is a lower positive bound for the angles in every triangle that is independent of the number of performed refinement steps (*shape regularity*). For a given discretization let $\mathcal{G}_h$ denote the set of all closed triangles, $\mathcal{N}_h$ the set of all vertices, $\mathcal{F}_h$ the set of all edges. It is convenient to distinguish between objects in the interior and at the boundary. So we use notation $\mathcal{G}_h^b, \mathcal{N}_h^b, \mathcal{F}_h^b$ for the set of triangles with at least one boundary vertex, the set of boundary vertices, and the set of edges with two boundary points, respectively. The complementary sets are denoted by $\mathcal{G}_h^i, \mathcal{N}_h^i, \mathcal{F}_h^i$. The computational domain D_h is given by $\overline{D}_h := \cup_{G\in\mathcal{G}_h} G \subsetneq \overline{D}$. ∂D intersects ∂D_h precisely at the boundary vertices $\mathcal{N}_h^b$, i.e. $\partial D \cap \partial D_h = \mathcal{N}_h^b$. The domain build up by the boundary triangles is denoted by $\overline{D}_h^b$, i.e. $\overline{D}_h^b := \cup_{G\in\mathcal{G}_h^b} G$, and D_h^i is its complement in D.

It is also convenient to enlarge the notion of discretization in the following way. Let $G \in \mathcal{G}_h^b$ with one boundary edge, say $F \in \mathcal{F}_h^b$ (the case of two boundary edges can be excluded by performing an additional refinement step if necessary). To F corresponds a segment $\tilde{F}$ of ∂D with the same end-points. Let $\mathfrak{p}$ be a diffeomorphism $\mathfrak{p} : F \to \tilde{F}$. We assign to G the larger set $\tilde{G}$ by replacing the boundary segment F by the arc $\tilde{F}$. If we let $G = \tilde{G}$ otherwise, then $\tilde{\mathcal{G}}_h := \{\tilde{G} : G \in \mathcal{G}_h\}$ is a decomposition of $\overline{D}$, while $\tilde{\mathcal{F}}_h^b := \{\tilde{F} : \tilde{F} = \mathfrak{p}(F) \text{ for } F \in \mathcal{F}_h^b\}$ is a decomposition of ∂D into curved segments. As an approximation to X we use linear finite elements, i.e. globally continuous functions on $\overline{D}$ that are linear on each $G \in \mathcal{G}_h$. For $G \in \mathcal{G}_h^b$ we can uniquely extend such a function to a linear function on $\tilde{G}$. In this sense we let

$$X_h := \left\{v \in C^0(\overline{D},\mathbb{R}^3) : v\lfloor_{\tilde{G}} \in \mathbb{P}^1(\tilde{G},\mathbb{R}^3) \text{ for all } \tilde{G} \in \tilde{\mathcal{G}}_h\right\}.$$

Note that each function in X_h is uniquely determined by its values on $\mathcal{N}_h$ and that $X_h \subset X$. The restriction of X_h to ∂D_h (continuous piecewise linear functions on ∂D_h) will be denoted by X_h^b.

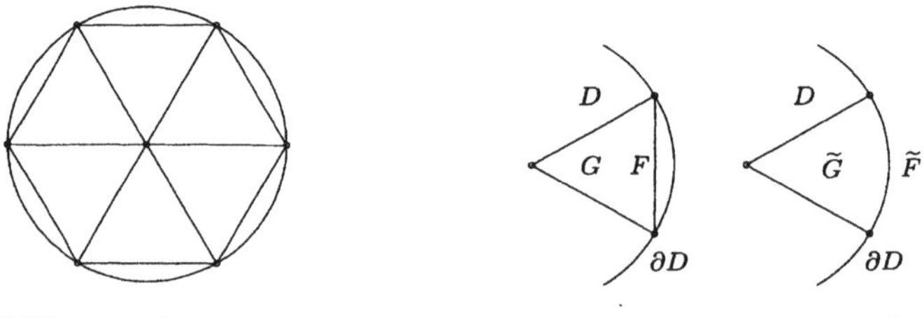

Fig. 1. **(a)** Hexagonal macro discretization (left) **(b)** Triangle at the boundary (right)

For approximating H we use the decomposition of ∂D that is defined by the boundary vertices $\mathcal{N}_h^b$ and consider continuous functions that are piecewise linear with respect to the arclength on ∂D, that is,

$$S_h := \{\eta \in C^0(\partial D) : \eta \text{ is piecewise linear with respect to arclength on } \tilde{F} \text{ for all } \tilde{F} \in \tilde{\mathcal{F}}_h^b\}.$$

Note that $S_h \neq X_h\lfloor_{\partial D}$. To incorporate the necessary conditions for fixing the conformal group we let

$$H_h := \{\eta \in S_h : \eta \text{ satisfies (2)}\}$$

and define the affine space by

$$\mathcal{H}_h := \{s \in C^0(\partial D, \mathbb{R}) : s = Id + \eta \text{ for } \eta \in H_h\}.$$

Note that $H_h \subset T \subset H$ and $\mathcal{H}_h \subset \mathcal{T} \subset \mathcal{H}$. The boundary vertices on ∂D induce a discretization of $\mathbb{R}/2\pi$ and S_h is then the usual linear finite element space on this mesh.

For $G \in \mathcal{G}_h$ let $\Omega_G \subset D$ be the union of all $G' \in \mathcal{G}_h$ that are adjacent to G (i.e. $G' \cap G \neq \emptyset$). Correspondingly, for $\tilde{F} \in \tilde{\mathcal{F}}_h^b$ we define $\omega_{\tilde{F}} \subset \partial D$ to be the arc consisting of $\tilde{F}$ and its two neighboring arcs in $\tilde{\mathcal{F}}_h^b$. For $p \in \mathcal{N}_h^b$ the set $\omega_p \subset \partial D$ consists of the two arcs in $\tilde{\mathcal{F}}_h^b$ that meet in p.

For any bounded set $N \subset \mathbb{R}^d$, for $d \in \{1,2\}$, we let $h_N := \operatorname{diam}(N)$. By construction, the quotients $h_G/h_{G'}$ are uniformly bounded for $G \in \mathcal{G}_h$ and $G' \subset \Omega_G$ independently of the choice of $\mathcal{G}_h$ in the family of meshes that have been obtained as described earlier. We express this by writing $h_G \sim h_{G'}$ for all adjacent $G, G' \in \mathcal{G}_h$. In this sense we also have that $h_F \sim h_{\tilde{F}}$ and $h_{\tilde{F}} \sim h_{\tilde{F}'}$ for all $F \in \mathcal{F}_h^b$ and all $\tilde{F}' \in \omega_{\tilde{F}}$, $h_G \sim h_F$ and $\operatorname{meas}_2(G) \sim h_F^2$ for $G \in \mathcal{G}_h$ and $F \in \mathcal{F}_h : F \subset \partial G$, and $\operatorname{meas}_2(\tilde{G} \setminus G) \sim h_G^3$ for $G \in \mathcal{G}_h^b$.

We introduce a meshsize function h on D by defining $h(x) := h_G$ for $x \in G$, $G \in \mathcal{G}_h$. We use the same symbol h also to define a meshsize function on ∂D by $h(x) := h_{\tilde{F}}$ for $x \in \tilde{F}$, $\tilde{F} \in \tilde{\mathcal{F}}_h^b$. The usage of h is always clear from the context. h is well-defined except on a set of relative measure zero in both cases.

Our main concern is to introduce discretizations with strongly varying meshsize. In our results we will need the following (in our case not too restrictive) condition on the meshsize in ∂D:

$$h_{\max}^{2-\kappa} \leq h_F \qquad \forall F \in \mathcal{F}_h^b \qquad (h_{\max} := \|h\|_{L^\infty(\partial D)}) \tag{9}$$

for some small $\kappa > 0$. Note that $h_{\max} < 1$.

3.2 Interpolation Operators

Nodal interpolation onto S_h. For $r \in C^0(\partial D)$ we define a piecewise linear function $I_h r \in S_h$ by $I_h r(p) := r(p)$ for $p \in \mathcal{N}_h^b$. It is immediate that for all $\tilde{F} \in \tilde{\mathcal{F}}_h^b$, $l \in \{0,1\}$, $m \in \{1,2\}$, $p \in [1,\infty]$

$$\|r - I_h r\|_{W^{l,p}(\tilde{F})} \le C\, h_{\tilde{F}}^{m-l}\, [\![r]\!]_{W^{m,p}(\tilde{F})}. \tag{10}$$

Quasi-interpolation onto S_h. To obtain a posteriori error estimates it is necessary to have interpolation operators on $H^{1/2}(\partial D)$, that is, operators with less regularity requirements as above. For examples we refer to [4] [21] [23]. Here, we use the idea of [21] with some modification. To each basis function $\psi_p \in S_h$ for $p \in \mathcal{N}_h^b$ we uniquely define the *dual basis function* ψ_p^*, piecewise linear with respect to arclength and with $\omega_p = \mathrm{supp}(\psi_p) = \mathrm{supp}(\psi_p^*)$, by $\int_{\partial D} \psi_p^* \psi_q = 1/2\, \mathrm{meas}_1(\omega_p)\delta_{pq}$ for all $q \in \mathcal{N}_h^b$. Then, for $r \in L^2(\partial D)$, let $P_h r := \sum_{p\in\mathcal{N}_h^b} \big(2 ⨍_{\omega_p} \psi_p^* r\big)\, \psi_p \in S_h$. It is easily seen that P_h is bounded as a mapping $H^s(\partial D) \to H^s(\partial D)$, $s \in [0,1]$, that $P_h r_h = r_h$ for all $r_h \in S_h$ and if r is linear on $\omega_{\tilde{F}}$, then $P_h r = r$ on $\tilde{F} \in \tilde{\mathcal{F}}_h^b$. With standard techniques we can show that P_h satisfies for $s \in [0,2]$ and $k \in [0,1]$ the estimates

$$\begin{aligned} &\|r - P_h r\|_{L^2(\tilde{F})} \le C\, h_{\tilde{F}}^s\, [\![r]\!]_{H^s(\omega_{\tilde{F}})}, \quad [\![P_h r]\!]_{H^k(\tilde{F})} \le C\, [\![r]\!]_{H^k(\omega_{\tilde{F}})}, \\ &|P_h r(p)| \le C\, h_{\omega_p}^{-1/2}\, \|r\|_{L^2(\omega_p)} \end{aligned} \tag{11}$$

for all $p \in \mathcal{N}_h^b$, $\tilde{F} \in \tilde{\mathcal{F}}_h^b$, and $r \in H^{1/2}(\partial D)$. The adjoint of P_h is given by $P_h^* r = \sum_{p\in\mathcal{N}_h^b} (2 ⨍_{\omega_p} \psi_p r)\, \psi_p^*$. It is a bounded operator $L^2(\partial D) \to L^2(\partial D)$ that locally preserves constants: if r equals a constant r_0 on $\omega_{\tilde{F}}$, then $P_h r = r_0$ on $\tilde{F} \in \tilde{\mathcal{F}}_h^b$. Indeed, if $\tilde{F} = \omega_p \cap \omega_q$ and thus $\omega_{\tilde{F}} = \omega_p \cup \omega_q$, then $P_h^* r\lfloor_{\tilde{F}} = r_0\big((2⨍_{\omega_p} \psi_p)\, \psi_p^* + (2⨍_{\omega_q} \psi_q)\, \psi_q^*\big) = r_0\big(\psi_p^* + \psi_q^*\big) = r_0$ (on $[0,1]$: $\psi_0^*(t) = 2-3t$, $\psi_1^*(t) = 2-3(1-t)$). Since P_h^* is a bounded mapping from $L^2(\partial D)$ into itself, we get with standard arguments

$$\|r - P_h^* r\|_{L^2(\tilde{F})} \le C h_{\omega_{\tilde{F}}} \|r\|_{H^1(\omega_{\tilde{F}})}. \tag{12}$$

L^2-Projection onto H_h. Note that P_h from above does not map H onto H_h, but S_h only. Such a property has the L^2-projection O_h defined by $\int_{\partial D} O_h \eta\, \nu_h = \int_{\partial D} \eta\, \nu_h$ for all $\nu_h \in H_h$. However, O_h is not suitable for our later purposes since it is non-local (unlike P_h). However, we show in the following that P_h "almost" maps H onto H_h, i.e.

$$\|(Id - O_h) P_h r\|_{H^k(\partial D)} \le C h_{\max}^{3/2} \|r\|_{H^{1/2}(\partial D)} \tag{13}$$

for $r \in H$ and $k \in [0,1]$.

Let $\tau^0 := Id$, $\tau^1 := \sin$, $\tau^2 := \cos$. It has been shown in [11] that $S_h = H_h \oplus \mathrm{span}\{\tau_h^0, \tau_h^1, \tau_h^2\}$, where $\{\tau_h^0, \tau_h^1, \tau_h^2\}$ is an L^2-orthonormal set and $\|\tau_h^j - \tau^j\|_{L^2(\partial D)} \le C h_{\max}^2$. Hence, for given $r \in H$, we have the representation

$$(Id - O_h) P_h r = \sum_{j=0}^{2} g_j(r) \tau_h^j,$$

with $g_j(r) := \int_{\partial D} P_h r\, \tau_h^j$. For arbitrary $r \in H$ and $r_h \in S_h$,

$$
\begin{aligned}
g_j(r) &= \int_{\partial D} \{P_h r\, \tau^j + P_h r\,(\tau_h^j - \tau^j)\} = \int_{\partial D} \{(P_h r - r)\,\tau^j + P_h r\,(\tau_h^j - \tau^j)\} \\
&= \int_{\partial D} (P_h(r - r_h) - (r - r_h))\,\tau^j + \int_{\partial D} P_h r\,(\tau_h^j - \tau^j) \\
&= -\int_{\partial D} (r - r_h)\,(Id - P_h^*)\tau^j + \int_{\partial D} P_h r\,(\tau_h^j - \tau^j).
\end{aligned}
$$

Since $\tau_h^j \in \operatorname{span}\{P_h \tau^{j'} : j' = 0,1,2\}$ we have by (11) the bound $\|\tau_h^j\|_{H^k(\partial D)} \leq C$ for $k \in [0,1]$ and with (12) and $r_h := P_h r$ we thus obtain

$$
\begin{aligned}
\|(Id - O_h) P_h r\|_{H^k(\partial D)} &\leq C \sum_{j=0}^{2} |g_j(r)| \leq C\Big(h_{\max}^{3/2} [\![r]\!]_{H^{1/2}(\partial D)} + h_{\max}^2 \|r\|_{L^2(\partial D)}\Big) \\
&\leq C\, h_{\max}^{3/2}\, \|r\|_{H^{1/2}(\partial D)}.
\end{aligned}
$$

Local mean values. For $r \in L^2(\partial D)$ and $\omega \subset \partial D$ we let $m(r,\omega) := \int_\omega r/h_\omega$ be the mean value of r over ω. It satisfies the bounds

$$
|m(r,\omega)| \leq h_\omega^{-1/2}\, \|r\|_{L^2(\omega)}, \quad \|r - m(r,\omega)\|_{L^2(\omega)} \leq C\, h_\omega^{1/2}\, [\![r]\!]_{H^{1/2}(\omega)} \tag{14}
$$

for $r \in H^{1/2}(\omega)$.

Quasi-interpolation onto X_h. We define a projection $Q_h : X \to X_h$ on a given triangulation of D. For $p \in \mathcal{N}_h^i$ we choose an edge $F_p \in \mathcal{F}_h^i$ with no boundary point and define $Q_h v(p)$ to be the mean value of v over F_p, i.e. $Q_h v(p) := m(v, F_p)$. Note that such an integral is well-defined for $v \in X$. For $p \in \mathcal{N}_h^b$ we define $Q_h v(p) := m(v, \omega_p)$. Since this definition preserves locally constant functions, we can show with similar arguments as in [21] that the following estimates do hold

$$
\begin{aligned}
&\|v - Q_h v\|_{L^2(G)} \leq C\, \|h \nabla v\|_{L^2(\Omega_G)}, \quad \|\nabla Q_h v\|_{L^2(G)} \leq C\, \|\nabla v\|_{L^2(\Omega_G)}, \\
&\|v - Q_h v\|_{L^2(\partial G)} \leq C\, \|\sqrt{h} \nabla v\|_{L^2(\Omega_G)}.
\end{aligned} \tag{15}
$$

(The construction in [21] preserves even linear functions, for our purposes preservation of constants will be sufficient. The essential point is the definition at the boundary points).

Discrete Harmonic Extension

3.3 Discrete Energy

For a given $s_h \in \mathcal{H}_h$ we define a discrete harmonic extension $u_h(s_h) := \Phi_h I_h(\gamma \circ s_h)$ (I_h from Sect. 3.2) by solving the discrete Dirichlet problem on the given discretization $\mathcal{G}_h$ of D with prescribed nodal values $\gamma(s_h(p))$ for $p \in \mathcal{N}_h^b$. The discrete energy is then defined to be

$$E_h : \mathcal{H}_h \to \mathbb{R}_+, \quad E_h(s_h) := \frac{1}{2}\int_{D_h} |\nabla u_h(s_h)|^2.$$

However, we want to extend the discrete energy to $\mathcal{H}$. Since I_h requests nodal values we have to use additionally a quasi-interpolation operator. Taking P_h from Sect. 3.2 we define an extension to E_h that coincides with the previous definition on $\mathcal{H}_h$ as follows: for $s \in \mathcal{H}$ let $\tilde{s}_h := P_h s \in S_h$ (not necessarily in $\mathcal{H}_h$) and $u_h(s) := \Phi_h I_h(\gamma \circ \tilde{s}_h)$, then

$$E_h : \mathcal{H} \to \mathbb{R}_+, \quad E_h(s) := \frac{1}{2}\int_{D_h} |\nabla u_h(s)|^2.$$

E_h is two times differentiable (in the same sense as E is) and we obtain easily the following formulas. For $s \in \mathcal{H}$ and $\xi \in H$, $\tilde{\xi}_h := P_h\xi$, let $v_h(s,\xi) := \Phi_h I_h(\tilde{\xi}_h\,\gamma' \circ \tilde{s}_h)$. Then

$$E_h'(s) : H \to \mathbb{R}, \quad E_h'(s)[\xi] = \int_{D_h} \nabla u_h(s) : \nabla v_h(s,\xi).$$

For $s \in \mathcal{H}$ and $\xi^l \in H$, $l \in \{1,2\}$, let $w_h(s,\xi^1,\xi^2) := \Phi_h I_h(\tilde{\xi}_h^1\tilde{\xi}_h^2\,\gamma'' \circ \tilde{s}_h)$. Then

$$E_h''(s) : H \times H \to \mathbb{R},$$

$$E_h''(s)[\xi^1,\xi^2] = \int_{D_h} \{\nabla u_h(s) : \nabla w_h(s,\xi^1,\xi^2) + \nabla v_h(s,\xi^1) : \nabla v_h(s,\xi^2)\}.$$

We can now state the definition of a discrete solution of the Plateau problem. Note that no monotonicity condition is required for s_h.

Definition 3.6 *The discrete harmonic function $u_h := \Phi_h I_h(\gamma \circ s_h) \in X_h$ is a solution of the discrete Plateau problem for Γ, if and only if $s_h \in \mathcal{H}_h$ satisfies*

$$E_h'(s_h)[\xi_h] = 0 \qquad \forall \xi_h \in H_h.$$

u_h (or s_h) is called non-degenerate if for some $\lambda_h > 0$

$$\inf_{\xi_h^1 \in H : \|\xi_h^1\|_H = 1} \; \sup_{\xi_h^2 \in H : \|\xi_h^2\|_H = 1} E_h''(s_h)[\xi_h^1, \xi_h^2] \geq \lambda_h.$$

4 A Posteriori Error Estimates

4.1 Error Representation

Let $u := \Phi(\gamma \circ s)$ for $s \in C^2(\partial D)$ and $u_h := \Phi_h I_h(\gamma \circ s_h)$ for $s_h \in \mathcal{H}_h$ be solutions of the Plateau-problem (Definition 2.2) and its discrete version (Definition 3.6), respectively. Let $\xi \in H$ and $\xi_h \in H_h$ be arbitrary. Then (recall Proposition 2.1)

$$\begin{aligned}
E''(s_h)\big[s-s_h,\xi\big] &= \int_0^1 E''(s_h+\tau(s-s_h))\big[s-s_h,\xi\big]\,d\tau \\
&\quad - \int_0^1 \big(E''(s_h+\tau(s-s_h)) - E''(s_h)\big)\big[s-s_h,\xi\big]\,d\tau \\
&\equiv \big(E'(s)-E'(s_h)\big)\big[\xi\big] - \mathrm{II}_1\big[s-s_h,\xi\big] \\
&= -E'(s_h)\big[\xi\big] - \mathrm{II}_1\big[s-s_h,\xi\big] \\
&\equiv -\mathrm{II}_0\big[\xi\big] - \mathrm{II}_1\big[s-s_h,\xi\big]. \qquad (16)
\end{aligned}$$

In the following we will estimate the right hand side by a term of the form $\big(A(u_h,s_h)+B(u_h,s_h,s)\big)\ \|\xi\|_H$. A depends on computable quantities only and vanishes of first order with the stepsize function h. B does also depend on the unknown function s, but vanishes of higher order with h. On the left hand side we need the non-degeneracy of the continuous problem (Definition 2.2) to guarantee existence of a constant $\lambda>0$ such that

$$\lambda\,\|s-s_h\|_H \le \sup_{\xi\in H\,:\,\|\xi\|_H=1} E''(s)\big[s-s_h,\xi\big]. \qquad (17)$$

Taking the upper bound into account this will result in the estimate

$$\|s-s_h\|_H \le \frac{1}{\lambda}\big(A(u_h,s_h)+B(u_h,s_h,s)\big)$$

(Theorem 4.9). The value of λ, however, is not known. We will therefore modify the previous argument using the non-degeneracy of the discrete problem (Definition 3.6) that gives

$$\lambda_h\,\|O_hP_h(s-s_h)\|_H \le \sup_{\xi_h\in H_h\,:\,\|\xi_h\|_H=1} E_h''(s_h)\big[O_hP_h(s-s_h),\xi_h\big] \qquad (18)$$

and obtain in case of $\lambda_h>0$ an estimate

$$\|s-s_h\|_H \le \frac{1}{\lambda_h}\big(\tilde A(u_h,s_h,s)+\tilde B(u_h,s_h,s)\big)$$

(Theorem 4.11), where the leading order term $\tilde A$ depends now on s also. These bounds for $\|s-s_h\|_H$ and the estimate (19) below will be the base of our adaptive algorithm in Sect. 5.

4.2 A Posteriori Error Control for Harmonic Extensions

The computations of harmonic extensions will be a special task in computing E_h and its derivatives. Assume that we want to compute an approximation to $w:=\Phi(g)\in X$ by $w_h:=\Phi_h(g_h)\in X_h$. Here, $g\in H^{1/2}(\partial D,\mathbb{R}^3)$ and $g_h\in X_h^b$ are given functions that have extensions in X and X_h (denoted by the same symbols), respectively. It has been shown in [9, Thm. 1] that one has in this situation an estimate

$$\|\nabla(w - w_h)\|_{L^2(D)} \leq \Theta_{\mathcal{G}_h}(w_h) + 2\,\|\nabla(g - g_h)\|_{L^2(D)} \tag{19}$$

which bounds the error in terms of known quantities. The first term is the *residual error estimator* [24, Sect. 1.2] defined by

$$\Theta_{\mathcal{G}_h}(w_h)^2 := \sum_{G\in\mathcal{G}_h} \Theta_G(w_h)^2 \equiv \sum_{F\subset\partial G:F\in\mathcal{F}_h^i} \hat{c}(G)\,h_F\,\big\|\big[\partial_n w_h\big]_F\big\|^2_{L^2(F)}, \tag{20}$$

where $\big[\partial_n w_h\big]_F$ denotes the "jump" of the normal derivative of w_h on interior edges F. The constant $\hat{c}(G)$ depends only on the shape of G and has to be chosen appropriately. The second term requires an explicitly known extension g or a higher order approximation to g, while we take $g_h = u_h$ for $g_h \in X_h^b$. Note that this error estimate takes into account also the error of using D_h instead of D. Besides the upper bound (19) for the error in terms of the computable quantity on the right there is also an estimate of the reverse type showing that this estimate is realistic up to a constant factor of moderate size [9, Thm. 2]. With this information it is possible to formulate an algorithm that computes an approximation to the harmonic extension on a sequence of regularly refined meshes within any prescribed tolerance.

4.3 A Posteriori Error Bounds

Here, we derive the estimates for the quantities II_0 and II_1 defined in (16).

Estimate for $\mathrm{II}_0\,[\xi]$

Recall that

$$\mathrm{II}_0\,[\xi] = E'(s_h)\,[\xi] = \int_D \nabla u : \nabla v$$

with $u = \Phi(\gamma \circ s_h)$, $v = \Phi(\xi\,\gamma' \circ s_h)$. Now take $u_h := \Phi_h I_h(\gamma \circ s_h)$, $v_h := \Phi_h I_h(\tilde{\xi}_h\,\gamma' \circ s_h)$, and $z_h := \Phi_h I_h(O_h\tilde{\xi}_h\,\gamma' \circ s_h)$. P_h and O_h are the operators from Sect. 3.2. Further, take any $\overline{v}_h \in X_h$ with the same boundary values as v_h and let $\hat{v}_h := Q_h v \in X_h$ as in Sect. 3.2. By definition of s_h and since $O_h\tilde{\xi}_h \in H_h$

$$\begin{aligned}
\mathrm{II}_0\,[\xi] &= \int_D \nabla u : \nabla v - \int_{D_h} \nabla u_h : \nabla \overline{v}_h + \int_{D_h} \nabla u_h : \nabla(\overline{v}_h - z_h)\\
&= \int_D \nabla(u - u_h) : \nabla v + \int_D \nabla u_h : \nabla(v - \hat{v}_h) + \int_{D_h} \nabla u_h : \nabla(\hat{v}_h - \overline{v}_h)\\
&\quad + \int_{D\setminus D_h} \nabla u_h : \nabla \hat{v}_h + \int_{D_h} \nabla u_h : \nabla(\overline{v}_h - z_h)\\
&=: J_1 + J_2 + J_3 + J_4 + J_5.
\end{aligned} \tag{21}$$

Estimating J_1. By definition of u we can write J_1 as

$$\int_D \nabla(\overline{u} - u_h) : \nabla v$$

for some $\overline{u} \in X$ with the same boundary values as u. $\overline{u}$ will be defined as follows: let $\overline{u} := u_h$ on interior edges and $\overline{u} := \gamma \circ s_h$ on ∂D. Now define $\overline{u}$ on $\tilde{G} \in \tilde{\mathcal{G}}_h$ to be the harmonic extension of the now prescribed values on $\partial\tilde{G}$. By this definition we have $\overline{u} = u_h$ on all $G \in \mathcal{G}_h$ that have no edge on the boundary. If $\tilde{G} \in \tilde{\mathcal{G}}_h$ with boundary arc $\tilde{F} \in \tilde{\mathcal{F}}_h^b$, then with (4)

$$\begin{aligned}\|\nabla(\overline{u} - u_h)\|_{L^2(\tilde{G})} &\leq C \, [\![\overline{u} - u_h]\!]_{H^{1/2}(\partial\tilde{G})} \leq C\, h_{\tilde{G}} \, [\![\overline{u} - u_h]\!]_{C^{0,1}(\partial\tilde{G})} \\ &\leq C\, h_{\tilde{G}} \, [\![\overline{u} - u_h]\!]_{C^{0,1}(\tilde{F})}.\end{aligned}$$

C depends on $\tilde{G}$, but not on $h_{\tilde{G}}$. By the assumed shape regularity of the discretizations in Sect. 3.1, this bound can in fact be chosen independently of $\tilde{G}$. Since $\overline{u}$ equals u_h at the end-points of $\tilde{F}$, we have with (10)

$$[\![\overline{u} - u_h]\!]_{C^{0,1}(\tilde{F})} \leq C h_{\tilde{F}} \, [\![\gamma \circ s_h]\!]_{C^{1,1}(\tilde{F})}.$$

With (4) and Proposition 2.3 we observe that

$$\|\nabla v\|_{L^2(D)} \leq C \, [\![\xi \, \gamma' \circ s_h]\!]_{H^{1/2}(\partial D)} \leq C \, \|\gamma' \circ s_h\|_{C^{0,1}(\partial D)} \, \|\xi\|_H \tag{22}$$

so that we finally achieve the bound

$$|J_1| \leq C \, \|\gamma' \circ s_h\|_{C^{0,1}(\partial D)} \Big(\sum_{F \in \mathcal{F}_h^b} h_F^4 \, [\![\gamma \circ s_h]\!]^2_{C^{1,1}(\tilde{F})} \Big)^{1/2} \|\xi\|_H.$$

Estimating J_2. Integrating by parts we get the decomposition

$$\begin{aligned} J_2 &= \int_D \nabla u_h : \nabla(v - \hat{v}_h) \\ &= \sum_{F \in \mathcal{F}_h^i} \int_F [\partial_n u_h]_F \cdot (v - \hat{v}_h) + \sum_{F \in \mathcal{F}_h^b} \int_{\tilde{F}} \partial_{\tilde{n}} u_h \cdot (v - \hat{v}_h). \end{aligned}$$

Here and in the following we will use the notation n for the constant normal vector on F and $\tilde{n}$ for the non-constant normal vector on $\tilde{F}$. The first term can be estimated in the usual way by $\Theta_{\mathcal{G}_h}(u_h) \, \|\nabla v\|_{L^2(D)}$ (using the techniques of [24, Sect. 1.2]) since $\hat{v}_h$ is a quasi-interpolant that satisfies (15). It remains to consider the second term. Fix $F \in \mathcal{F}_h^b$, let $\tilde{F} = \mathfrak{p}(F)$ and G be the corresponding boundary triangle. Now we insert the function $\pm \hat{v}_h \circ \mathfrak{p}^{-1}$. With

$$\big|(\hat{v}_h \circ \mathfrak{p}^{-1} - \hat{v}_h)(x)\big| \leq C\, h_F^2 \, \big|\nabla \hat{v}_h\lfloor_G\big| \quad \text{for } x \in \tilde{F}$$

we obtain

$$\left|\int_{\tilde F} \partial_{\tilde n} u_h \cdot \left(\hat v_h \circ \mathfrak{p}^{-1} - \hat v_h\right)\right| \le C\, \|\partial_{\tilde n} u_h\|_{L^2(\tilde F)}\, h_F^{5/2}\, |\nabla \hat v_h\lfloor_G|$$
$$\le C\, h_F^{3/2}\, \|\partial_{\tilde n} u_h\|_{L^2(\tilde F)}\, \|\nabla \hat v_h\|_{L^2(G)}.$$

Let $p, q \in \mathcal{N}_h^b$ be the end-points of F. The linear basis functions ψ_p, ψ_q on F will be mapped onto $\tilde\psi_p := \psi_p \circ \mathfrak{p}^{-1}$, $\tilde\psi_q := \psi_q \circ \mathfrak{p}^{-1}$ on $\tilde F$. Note that $0 \le \tilde\psi_p, \tilde\psi_q \le 1$ and $\tilde\psi_p + \tilde\psi_q = 1$ on $\tilde F$. Using mean values we get by definition of $\hat v_h$ in Sect. 3.2

$$\int_{\tilde F} \partial_{\tilde n} u_h \cdot (v - \hat v_h \circ \mathfrak{p}^{-1}) = \int_{\tilde F} \partial_{\tilde n} u_h \cdot \Big((v - m(v;\omega_p))\, \tilde\psi_p + (v - m(v;\omega_q))\, \tilde\psi_q \Big).$$

We expand the first integral on the right

$$\int_{\tilde F} \partial_{\tilde n} u_h \cdot (v - m(v;\omega_p))\, \tilde\psi_p = \int_{\tilde F} \partial_{\tilde n} u_h \cdot (\xi\gamma' \circ s_h - m(\xi\gamma' \circ s_h;\omega_p))\, \tilde\psi_p$$
$$= \int_{\tilde F} \partial_{\tilde n} u_h \cdot \Big(m(\xi;\tilde F)\, (\gamma' \circ s_h - m(\gamma' \circ s_h;\omega_p)) + (\xi - m(\xi;\tilde F))\, \gamma' \circ s_h \Big)\, \tilde\psi_p$$
$$- \int_{\tilde F} \partial_{\tilde n} u_h \cdot m\big((\xi - m(\xi;\tilde F))\, \gamma' \circ s_h;\omega_p\big)\, \tilde\psi_p$$

and get here, using (14), for the first integral the bound

$$C\, h_F\, [\![\gamma' \circ s_h]\!]_{C^{0,1}(\omega_p)}\, \|\partial_{\tilde n} u_h\|_{L^2(\tilde F)}\, \|\xi\|_{L^2(\omega_p)}$$
$$+ C\, h_F^{1/2}\, \|\partial_{\tilde n} u_h \cdot \gamma' \circ s_h\|_{L^2(\tilde F)}\, [\![\xi]\!]_{H^{1/2}(\omega_p)}.$$

In the second integral we write $\partial_{\tilde n} u_h(\phi) \cdot \gamma'(s_h(\phi_0)) = \partial_{\tilde n} u_h(\phi) \cdot \big(\gamma'(s_h(\phi)) + \gamma'(s_h(\phi_0)) - \gamma'(s_h(\phi))\big)$ for $\phi \in \tilde F$, $\phi_0 \in \omega_p$. Then this term is estimated by the same expression as the first one. Hence, with (22), (15), and (8)

$$|J_2| \le C\Bigg(\|\gamma' \circ s_h\|_{C^{0,1}(\partial D)}\, \Theta_{\mathcal{G}_h}(u_h) + \|\gamma' \circ s_h\|_{C^{0,1}(\partial D)} \Big(\sum_{F \in \mathcal{F}_h^b} h_F^3\, \|\partial_{\tilde n} u_h\|^2_{L^2(\tilde F)} \Big)^{1/2}$$
$$+ \Big(\sum_{F \in \mathcal{F}_h^b} \{ h_F^2\, [\![\gamma' \circ s_h]\!]^2_{C^{0,1}(\omega_{\tilde F})}\, \|\partial_{\tilde n} u_h\|^2_{L^2(\tilde F)} + h_F\, \|\partial_{\tilde n} u_h \cdot \gamma' \circ s_h\|^2_{L^2(\tilde F)} \} \Big)^{1/2} \Bigg)\, \|\xi\|_H.$$

Estimating J_3. We take $\overline v_h$ such that it equals $\hat v_h$ on interior nodes. Since then $\overline v_h$ equals $\hat v_h$ on interior triangles we get

$$J_3 = \int_{D_h} \nabla u_h : \nabla(\hat v_h - \overline v_h) = \int_{D_h^b} \nabla u_h : \nabla(\hat v_h - \overline v_h)$$
$$= \sum_{F \in \mathcal{F}_h^i} \int_F [\partial_n u_h]_F \cdot (\hat v_h - \overline v_h) + \sum_{F \in \mathcal{F}_h^b} \int_F \partial_n u_h \cdot (\hat v_h - \overline v_h).$$

For $p \in \mathcal{N}_h^b$ consider $F_p \in \mathcal{F}_h^i$ with $p \in F_p$, then

$$\int_{F_p} \left[\partial_n u_h\right]_{F_p} \cdot (\hat{v}_h - \overline{v}_h) = h_{F_p} h_{\omega_p}^{-1} \left[\partial_n u_h\right]_{F_p} \cdot \int_{\omega_p} \{\xi \gamma' \circ s_h - \tilde{\xi}_h(p) \gamma'(s_h(p))\}.$$

The integral is written as

$$\int_{\omega_p} \Big\{ \big(\xi - \tilde{\xi}_h(p)\big) \gamma' \circ s_h + \tilde{\xi}_h(p) \big(\gamma' \circ s_h - \gamma'(s_h(p))\big) \Big\}$$

so that with $\tilde{\xi}_h = P_h \xi$ and (11)

$$\Big| \int_{F_p} \left[\partial_n u_h\right]_{F_p} \cdot (\hat{v}_h - \overline{v}_h) \Big| \le C \left(h_{F_p}^{1/2} \|\gamma' \circ s_h\|_{L^\infty(\omega_p)} + h_{F_p} [\![\gamma' \circ s_h]\!]_{C^{0,1}(\omega_p)} \right) \times \left\| \left[\partial_n u_h\right]_{F_p} \right\|_{L^2(F_p)} \|\xi\|_{H^{1/2}(\omega_{\omega_p})}.$$

For $F \in \mathcal{F}_h^b$ with end-points $p, q \in \mathcal{N}_h^b$

$$\int_F \partial_n u_h \cdot (\hat{v}_h - \overline{v}_h) = \frac{1}{2} |F| \, \partial_n u_h \cdot \big((\hat{v}_h - \overline{v}_h)(p) + (\hat{v}_h - \overline{v}_h)(q) \big)$$

and for $(\hat{v}_h - \overline{v}_h)(p)$ we have the same expression as above. Using arguments as for $|J_2|$ we obtain

$$|J_3| \le C \Bigg(\sum_{p \in \mathcal{N}_h^b} \sum_{F \in \mathcal{F}_h^i : p \in F} \left(h_F^2 [\![\gamma' \circ s_h]\!]_{C^{0,1}(\omega_p)}^2 + h_F \|\gamma' \circ s_h\|_{L^\infty(\omega_p)}^2 \right) \left\| \left[\partial_n u_h\right]_F \right\|_{L^2(F)}^2 + \sum_{F \in \mathcal{F}_h^b} \left\{ h_F^2 [\![\gamma' \circ s_h]\!]_{C^{0,1}(\omega_{\tilde{F}})}^2 \|\partial_n u_h\|_{L^2(F)}^2 + h_F \|\partial_n u_h|_F \cdot \gamma' \circ s_h\|_{L^2(\omega_{\tilde{F}})}^2 \right\} \Bigg)^{1/2} \|\xi\|_H .$$

Estimating J_4. Using that ∇u_h and $\nabla \hat{v}_h$ are piecewise constant and that $\operatorname{meas}_2(\tilde{G} \setminus G) \le C\, h_G \operatorname{meas}_2(G)$ for each $G \in \mathcal{G}_h^b$ we obtain

$$\int_{D \setminus D_h} |\nabla u_h : \nabla \hat{v}_h| \le C \sum_{G \in \mathcal{G}_h^b} h_G \|\nabla u_h\|_{L^2(G)} \|\nabla \hat{v}_h\|_{L^2(G)} \le C \Big(\sum_{G \in \mathcal{G}_h^b} h_G^2 \|\nabla u_h\|_{L^2(G)}^2 \Big)^{1/2} \|\nabla \hat{v}_h\|_{L^2(D)} .$$

With (15) and (22) we finally get

$$|J_4| \le C \|\gamma' \circ s_h\|_{C^{0,1}(\partial D)} \Big(\sum_{G \in \mathcal{G}_h^b} h_G^2 \|\nabla u_h\|_{L^2(G)}^2 \Big)^{1/2} \|\xi\|_H .$$

Estimating J_5. Again, we can replace z_h by $\overline{z}_h$ which is redefined to be equal to $\overline{v}_h$ on interior nodes. We write

$$J_5 = \sum_{F\in\mathcal{F}_h^i} \int_F [\partial_n u_h]_F \cdot (\overline{v}_h - \overline{z}_h) + \sum_{F\in\mathcal{F}_h^b} \int_F \partial_n u_h \cdot (\overline{v}_h - \overline{z}_h)$$

$$= \sum_{p\in\mathcal{N}_h^b} h_{F_p} [\partial_n u_h]_{F_p} \cdot (\tilde{\xi}_h - O_h\tilde{\xi}_h)(p)\, \gamma'(s_h(p))$$

$$+ \sum_{F\in\mathcal{F}_h^b} \int_F \partial_n u_h \cdot I_h\big((\tilde{\xi}_h - O_h\tilde{\xi}_h)\,\gamma'\circ s_h\big)$$

with notation as for J_3. Writing

$$I_h\big((\tilde{\xi}_h - O_h\tilde{\xi}_h)\,\gamma'\circ s_h\big) = m(\gamma'\circ s_h,\tilde{F})\, I_h(\tilde{\xi}_h - O_h\tilde{\xi}_h) + I_h\big((\tilde{\xi}_h - O_h\tilde{\xi}_h)\,(\gamma'\circ s_h - m(\gamma'\circ s_h,\tilde{F}))\big)$$

in the second term we obtain

$$|J_5| \le C\Big(\sum_{p\in\mathcal{N}_h^b} |\gamma'(s_h(p))|\, h_{F_p}^{1/2}\, \big\|[\partial_n u_h]_{F_p}\big\|_{L^2(F_p)}\, |(\tilde{\xi}_h - O_h\tilde{\xi}_h)(p)|$$
$$+ \sum_{F\in\mathcal{F}_h^b} h_F^{1/2}\big(\|\partial_n u_h\cdot\gamma'\circ s_h\|_{L^2(\tilde{F})}$$
$$+ h_F [\![\gamma'\circ s_h]\!]_{C^{0,1}(\tilde{F})} \|\partial_n u_h\|_{L^2(F)}\big)\, \|\tilde{\xi}_h - O_h\tilde{\xi}_h\|_{L^\infty(\tilde{F})}\Big)$$

and using inverse inequalities like in (11) and (13)

$$|J_5| \le C\Big(\|\gamma'\circ s_h\|^2_{L^\infty(\partial D)} \sum_{p\in\mathcal{N}_h^b} \big\|[\partial_n u_h]_{F_p}\big\|^2_{L^2(F_p)}$$
$$+ \sum_{F\in\mathcal{F}_h^b} \big\{\|\partial_n u_h\cdot\gamma'\circ s_h\|^2_{L^2(\tilde{F})} + h_F^2 [\![\gamma'\circ s_h]\!]^2_{C^{0,1}(\tilde{F})} \|\partial_n u_h\|^2_{L^2(F)}\big\}\Big)^{1/2} h_{\max}^{3/2}\, \|\xi\|_H.$$

Lemma 4.7 *For $\gamma \in W^{2,\infty}(S^1,\mathbb{R}^3)$ let $u_h = \Phi_h I_h(\gamma\circ s_h) \in X_h$ for $s_h \in \mathcal{H}_h$ be a solution of the discrete Plateau-problem (Definition 3.6) on a mesh as described in Sect. 3.1, especially with the property* (9). *Then for all $\xi \in H$*

$$\big|\mathrm{II}_0[\xi]\big| \le C\Big(\|\gamma'\circ s_h\|_{W^{1,\infty}(\partial D)}\, \Theta_{\mathcal{G}_h}(u_h) + \Big(\sum_{F\in\mathcal{F}_h^b} \big\{h_F\, \|\partial_n u_h\cdot\gamma'\circ s_h\|^2_{L^2(\tilde{F})}$$
$$+ h_F^2\, \|\gamma''\circ s_h s_h'\|^2_{L^\infty(\omega_{\tilde{F}})}\, \|\partial_n u_h\|^2_{L^2(F)}\big\}\Big)^{1/2} + \|\gamma'\circ s_h\|_{W^{1,\infty}(\partial D)}$$
$$\times \Big(\sum_{G\in\mathcal{G}_h^b} h_G^2\, \|\nabla u_h\|^2_{L^2(G)} + \sum_{F\in\mathcal{F}_h^b} h_F^4\, \|\gamma''\circ s_h (s_h')^2\|^2_{L^\infty(\tilde{F})}\Big)^{1/2}\Big)\, \|\xi\|_H$$
$$\equiv C\, R(u_h,\gamma')\, \|\xi\|_H.$$

The constant C depends on properties of the mesh only.

Proof. We collect the estimates for $J_1, \dots, J_5$. Recalling $[\![\,.\,]\!]_{C^{k,1}} = [\![\,.\,]\!]_{W^{k+1,\infty}}$ we get $[\![\gamma' \circ s_h]\!]_{C^{0,1}} \leq \|\gamma'' \circ s_h s_h'\|_{L^\infty}$ and $[\![\gamma \circ s_h]\!]_{C^{1,1}} \leq \|\gamma'' \circ s_h (s_h')^2\|_{L^\infty}$. Using $\|n - \tilde{n}\|_{L^\infty(\tilde{F})} \leq C h_F$ we can replace $\partial_{\tilde{n}}$ by ∂_n in the last two terms of the estimate for J_2 with additional terms estimated as

$$\sum_{F \in \mathcal{F}_h^b} h_F^3 \, \|\partial_{\tilde{n}} u_h\|_{L^2(\tilde{F})}^2 \leq \sum_{G \in \mathcal{G}_h^b} h_G^2 \, \|\nabla u_h\|_{L^2(G)}^2.$$

The first term in the bound of J_3 is estimated in terms of $\Theta_{\mathcal{G}_h}(u_h)$. In the bound for J_5 we estimate $h_{\max}^3$ inside the sum by the corresponding h_F using our assumption (9). □

Estimate for $\mathrm{II}_1\left[s - s_h, \xi\right]$

Let $\sigma := s - s_h$. Expanding integrals yields

$$\begin{aligned}
\mathrm{II}_1\left[\sigma,\xi\right] &= \left(E''(s_h + \tau\sigma) - E''(s_h)\right)\left[\sigma,\xi\right] \\
&= \int_D \left\{\nabla u_\tau : \nabla w_\tau + \nabla v_\tau^1 : \nabla v_\tau^2\right\} - \int_D \left\{\nabla u : \nabla w + \nabla v^1 : \nabla v^2\right\} \\
&= \int_D \left\{\nabla u_\tau : \nabla w_\tau - \nabla u : \nabla w\right\} + \int_D \left\{\nabla v_\tau^1 : \nabla v_\tau^2 - \nabla v^1 : \nabla v^2\right\} \\
&= \int_D \left\{\nabla (u_\tau - u) : \nabla w_\tau + \nabla u : \nabla (w_\tau - w)\right\} \\
&\quad + \int_D \left\{\nabla (v_\tau^1 - v^1) : \nabla v_\tau^2 + \nabla v^1 : \nabla (v_\tau^2 - v^2)\right\} =: K_1 + K_2 + K_3 + K_4,
\end{aligned}$$

with notations $u = u(s_h)$, $v^1 = v(s_h, \sigma)$, $v^2 = v(s_h, \xi)$, $w = w(s_h, \sigma, \xi)$ and corresponding expressions with subscript τ have arguments $s_h + \tau\sigma$ instead of s_h.

Lemma 4.8 *For $\gamma \in W^{3,\infty}(S^1, \mathbb{R}^3)$ let $u = \Phi(\gamma \circ s) \in X$ for $s \in \mathcal{H}$ be a solution of the Plateau-problem (Definition 2.2) and $u_h = \Phi_h I_h(\gamma \circ s_h) \in X_h$ for $s_h \in \mathcal{H}_h$ be a solution of the discrete Plateau-problem (Definition 3.6). With $\sigma := s - s_h$ we have for all $\xi \in H$ and $k \in (1/2, 1]$*

$$\begin{aligned}
\left|\mathrm{II}_1\left[\sigma,\xi\right]\right| &\leq C \,\|\gamma'\|_{W^{1,\infty}(S^1)} \,\|\gamma''\|_{W^{1,\infty}(S^1)} \,\|s_h\|_{W^{1,\infty}(\partial D)} \,\|\sigma\|_{H^k(\partial D)}^2 \,\|\xi\|_H \\
&\equiv C\, \widetilde{R}(s_h, \gamma') \,\|\sigma\|_{H^k(\partial D)}^2 \,\|\xi\|_H.
\end{aligned}$$

The constant C depends on k only.

Proof. Since we now deal with exact solutions we can use (4) and (5). With Propositions 2.3–2.5 and $\|s_h\|_{W^{1,\infty}(\partial D)} = 1 + \|s_h'\|_{L^\infty(\partial D)}$ we have for $k \in (1/2, 1]$

$$\begin{aligned}
|K_1| &\le \|\nabla(u_\tau - u)\|_{L^2(D)} \|\nabla w_\tau\|_{L^2(D)} \\
&\le C \,[\![\gamma \circ (s_h + \tau\sigma) - \gamma \circ s_h]\!]_{H^{1/2}(\partial D)} \|\sigma\xi\,\gamma'' \circ (s_h + \tau\sigma)\|_{H^{1/2}(\partial D)} \\
&\le C \|\gamma'\|_{W^{1,\infty}(S^1)} \|\gamma''\|_{W^{1,\infty}(S^1)} \|s_h\|_{H^{1,\infty}(\partial D)} \|\sigma\|_{H^k(\partial D)} \|\sigma\|_{H^{1/2}(\partial D)} \|\xi\|_H, \\
|K_2| &= \Big| \int_{\partial D} \partial_n u\,(w_\tau - w) \Big| \\
&\le C \,[\![\gamma \circ s_h]\!]_{H^1(\partial D)} \|\sigma\xi\,\gamma'' \circ (s_h + \tau\sigma) - \gamma'' \circ s_h\|_{L^2(\partial D)} \\
&\le C \|\gamma'\|_{W^{1,\infty}(S^1)} \|\gamma''\|_{W^{1,\infty}(S^1)} \|s_h'\|_{L^\infty(\partial D)} \|\sigma\|^2_{L^\infty(\partial D)} \|\xi\|_H, \\
|K_3| &\le \|\nabla(v_\tau^1 - v^1)\|_{L^2(D)} \|\nabla v_\tau^2\|_{L^2(D)} \\
&\le C \|\sigma\,(\gamma' \circ (s_h + \tau\sigma) - \gamma' \circ s_h)\|_{H^{1/2}(\partial D)} \|\xi\,\gamma' \circ (s_h + \tau\sigma)\|_{H^{1/2}(\partial D)} \\
&\le C \|\gamma'\|_{W^{1,\infty}(S^1)} \|\gamma''\|_{W^{1,\infty}(S^1)} \\
&\qquad \times \big(\|s_h\|_{W^{1,\infty}(\partial D)} + \|\sigma\|_{L^\infty(\partial D)}\big) \|\sigma\|_{H^k(\partial D)} \|\sigma\|_{H^{1/2}(\partial D)} \|\xi\|_H, \\
|K_4| &= \Big| \int_{\partial D} \partial_n v^1\,(v_\tau^2 - v^2) \Big| \\
&\le C \,[\![\sigma\,\gamma' \circ s_h]\!]_{H^1(\partial D)} \|\xi\,(\gamma' \circ (s_h + \tau\sigma) - \gamma' \circ s_h)\|_{L^2(\partial D)} \\
&\le C \|\gamma'\|^2_{W^{1,\infty}(S^1)} \|s_h\|_{W^{1,\infty}(\partial D)} \big([\![\sigma]\!]_{H^1(\partial D)} + \|\sigma\|_{L^2(\partial D)}\big) \|\sigma\|_{L^2(\partial D)} \|\xi\|_H.
\end{aligned}$$

Hence we get, using $\|\sigma\|_{L^\infty(D)} \le C$ and the embedding $H^k(\partial D) \hookrightarrow L^\infty(\partial D)$,

$$\Big| \mathrm{II}_1 [\sigma, \xi] \Big| \le C \|\gamma'\|_{W^{1,\infty}(S^1)} \|\gamma''\|_{W^{1,\infty}(S^1)} \|s_h\|_{W^{1,\infty}(\partial D)} \|\sigma\|^2_{H^k(\partial D)} \|\xi\|_H.$$

□

Theorem 4.9. *For $\gamma \in W^{3,\infty}(S^1, \mathbb{R}^3)$ let $u = \Phi(\gamma \circ s) \in X$ for $s \in \mathcal{H}$ be a non-degenerate solution of the Plateau-problem (Definition 2.2) and $u_h = \Phi_h I_h(\gamma \circ s_h) \in X_h$ for $s_h \in \mathcal{H}_h$ be a solution of the discrete Plateau-problem (Definition 3.6) on a mesh as described in Sect. 3.1, especially with the property* (9). *With R as in Lemma 4.7 and $\widetilde{R}$ as in Lemma 4.8, we have for all $\xi \in H$ and $k \in (1/2, 1]$ the error estimate*

$$\lambda \|s - s_h\|_H \le C\Big(R(u_h, \gamma') + \widetilde{R}(s_h, \gamma') \|s - s_h\|^2_{H^k(\partial D)}\Big) \tag{23}$$

is valid. λ is the constant from (17), *the constant C depends on k and properties of the mesh only.*

Proof. Using (17) we get

$$\begin{aligned}
\lambda \|s - s_h\|_H &\le \sup_{\xi \in H\,:\,\|\xi\|_H = 1} E''(s)\big[s - s_h, \xi\big] \\
&\le \sup_{\xi \in H\,:\,\|\xi\|_H = 1} E''(s_h)\big[s - s_h, \xi\big] + \sup_{\xi \in H\,:\,\|\xi\|_H = 1} \big(E''(s) - E''(s_h)\big)\big[s - s_h, \xi\big]
\end{aligned}$$

and the result follows from Lemma 4.7 and 4.8. □

4.4 An Error Bound Using Discrete Non-degeneracy

In order to exploit the estimate (18) we have to estimate the remainder term (again $\sigma := s - s_h$)

$$\begin{aligned} \mathrm{II}_2[\sigma,\xi] &:= \left(E''(s_h) - E_h''(s_h)\right)[\sigma,\xi] \\ &= \int_D \left\{\nabla u : \nabla w + \nabla v^1 : \nabla v^2\right\} - \int_{D_h} \left\{\nabla u_h : \nabla w_h + \nabla v_h^1 : \nabla v_h^2\right\} \\ &= \int_D \nabla u : \nabla w - \int_{D_h} \nabla u_h : \nabla w_h + \int_D \nabla v^1 : \nabla v^2 - \int_{D_h} \nabla v_h^1 : \nabla v_h^2. \end{aligned} \quad (24)$$

Here we have $u = u(s_h)$, $u_h = u_h(s_h)$, $v^1 = v(s_h,\sigma)$, $v_h^1 = v_h(s_h,\sigma)$, $v^2 = v(s_h,\xi), v_h^2 = v_h(s_h,\xi), w = w(s_h,\sigma,\xi), w_h = w_h(s_h,\sigma,\xi)$ as defined in Sects. 2.3 and 3.3. This is the sum of two expressions that look like the first two in (21) each.

Lemma 4.10 *For $\gamma \in W^{3,\infty}(S^1,\mathbb{R}^3)$ let $u = \Phi(\gamma \circ s) \in X$ for $s \in \mathcal{H}$ be a solution of the Plateau-problem (Definition 2.2) and $u_h = \Phi_h I_h(\gamma \circ s_h) \in X_h$ for $s_h \in \mathcal{H}_h$ be a solution of the discrete Plateau-problem (Definition 3.6) on a mesh as described in Sect. 3.1, especially with the property* (9). *With $\sigma := s - s_h$, $v_h^1 := \Phi_h I_h(\tilde{\sigma}\gamma' \circ s_h) \in X_h$ and R as in Lemma 4.7 we have for all $\xi \in H$ and $k \in (1/2, 1]$*

$$\left|\mathrm{II}_2[\sigma,\xi]\right| \le C\Big(R(u_h,\gamma'')\,\|\sigma\|_{H^k(\partial D)} + R(v_h^1,\gamma')\Big)\,\|\xi\|_H.$$

Proof. We consider the first two integrals in (24) and write these in the form $J_1 + \cdots + J_4$ as for II_0 (21) (a term like J_5 does not appear). Then we can use the same arguments to estimate J_1, J_2, J_4 as before, but with γ' replaced by γ'' and $\|\xi\|_H$ by $\|\tilde{\sigma}\xi\|_H$. It follows with Proposition 2.3 and (11) that this part is bounded by $R(u_h,\gamma'')\,\|\sigma\|_{H^k(\partial D)}\,\|\xi\|_H$ for $k \in (1/2, 1]$. Assuming that v_h^1 is known, the last two integrals will give the same result as in Lemma 4.7, but with u_h replaced by v_h^1. Hence the bound is $R(v_h^1,\gamma')\,\|\xi\|_H$. □

Theorem 4.11. *For $\gamma \in W^{3,\infty}(S^1,\mathbb{R}^3)$ let $u = \Phi(\gamma \circ s) \in X$ for $s \in \mathcal{H}$ be a solution of the Plateau-problem (Definition 2.2) and $u_h = \Phi_h I_h(\gamma \circ s_h) \in X_h$ for $s_h \in \mathcal{H}_h$ be a solution of the discrete Plateau-problem (Definition 3.6) on a mesh as described in Sect. 3.1, especially with the property* (9). *With $v_h^1 := \Phi_h I_h(P_h(s - s_h)\gamma' \circ s_h) \in X_h$, R as in Lemma 4.7, and $\widetilde{R}$ as in Lemma 4.8, we have that for all $\xi \in H$, $k \in (1/2, 1]$, and some $\kappa > 0$ the error estimate*

$$\begin{aligned} \lambda_h\,\|s - s_h\|_H \le C\Big(& R(u_h,\gamma') + R(v_h^1,\gamma') + R(u_h,\gamma'')\,\|s - s_h\|_{H^k(\partial D)} \\ &+ \widetilde{R}(s_h,\gamma')\|s - s_h\|_{H^k(\partial D)}^2\Big) + \mathcal{O}\big(h_{\max}^{1-\kappa}\|s - s_h\|_H + \lambda_h\,\|s - P_h s\|_{L^2(\partial D)}\big) \end{aligned} \quad (25)$$

is valid. λ_h is the constant from (18), *the constant C depends on k and properties of the mesh only.*

Proof. We have with (18) and (13)

$$\begin{aligned}\lambda_h \|s - s_h\|_H &\leq \lambda_h\big(\|O_h P_h(s - s_h)\|_H + \|(Id - O_h)P_h(s - s_h)\|_H + \|s - P_h s\|_H\big)\\ &\leq \sup_{\xi\in H\,:\,\|\xi\|_H=1} \big\{E_h''(s_h)\big[O_h P_h(s - s_h), \xi\big]\big\}\\ &\qquad + C\lambda_h\big(h_{\max}^{3/2}\|s - s_h\|_H + \|s - P_h s\|_{L^2(\partial D)}\big).\end{aligned}$$

Writing

$$\begin{aligned}E_h''(s_h)\big[O_h P_h(s - s_h), \xi\big] &= E''(s_h)\big[s - s_h, \xi\big] + \big(E_h''(s_h) - E''(s_h)\big)\big[s - s_h, \xi\big]\\ &\qquad - E_h''(s_h)\big[(Id - O_h)P_h(s - s_h), \xi\big]\end{aligned}$$

we can estimate the first two terms on the right hand side by Theorem 4.9 and Lemma 4.10 and obtain the expressions in the bracket in (25). For the remaining term we recall the result stated in [13, Prop. 3.4] that

$$\|\nabla\Phi_h I_h f\|_{L^2(\partial D)} \leq \|f\|_{H^{1/2}(\partial D)} + C\,h_{\max}^{1/2}\|f\|_{H^1(\partial D)}$$

for all $f \in H^1(\partial D)$. Since f is in the following argument a discrete function, we can restrict ourselves to estimate $\|f\|_{H^{1/2}(\partial D)}$ only by using the inverse estimate $h_{\min}^{1/2}\|f\|_{H^1(\partial D)} \leq C\|f\|_{H^{1/2}(\partial D)}$ to get with (9) $\|\nabla\Phi_h I_h f\|_{L^2(\partial D)} \leq C\,h_{\max}^{-1/2-\kappa}\|f\|_{H^{1/2}(\partial D)}$ for some $\kappa > 0$. The decisive term is $\|(Id - O_h)P_h(s - s_h)\,\tilde{\xi}\gamma'' \circ s_h\|_{H^{1/2}(\partial D)}$ which is estimated by $\|\gamma'''\|_{L^\infty(\partial D)}\|s_h\|_{W^{1,\infty}(\partial D)}\|(Id - O_h)P_h(s - s_h)\|_{H^k(\partial D)}\|\xi\|_H$ for some $k \in (1/2, 1]$. For the third factor we can again use (13). □

Remark 4. Non-degeneracy of the discrete problem can be assumed on sufficiently refined meshes if the continuous problem is non-degenerate [13, Thm. 5.4].

4.5 Practical A Posteriori Error Control

In fact, the estimate (25) is not an a posteriori error estimate since some terms still depend on the unknown solution. However, we will demonstrate that such terms are of "higher order". Of course, on coarse meshes this might not be true and so we will present some heuristic precautions to work on a sufficiently refined mesh. Our arguments are based on the fact that for strictly monotone $\gamma \in C^4(S^1, \mathbb{R}^3)$ one has (at least) $s \in W^{3,\infty}(\partial D) \hookrightarrow H^{5/2}(\partial D)$ and $u \in H^2(D)$ [13, Prop. 2.1]. In the same paper the optimal error bound $\|s - s_h\|_H \leq C\,h_{\max}$ for non-degenerate solutions s was proved [Thm. 5.4]. Additionly we claim that

$$\begin{aligned}\|s - s_h\|_{H^s(\partial D)} &\leq C\,h_{\max}^{3/2-s}, \qquad \text{for } s \in [0, 3/2],\\ \|s - s_h\|_{L^\infty(\partial D)} &\leq C\,h_{\max}\log(h_{\max})\end{aligned} \tag{26}$$

(assuming that (9) holds).

Essential error terms. The first term on the right hand side in the estimate (23) (defined in Lemma 4.7) is of a posteriori type. It is known that $\Theta_{\mathcal{G}_h}(u_h)$ is controlled by $h_{\max}$ for $u(s_h) \in H^2(D)$. For

$$\eta^2(u_h) := \sum_{F \in \mathcal{F}_h^b} \eta_F(u_h)^2$$
$$\equiv \sum_{F \in \mathcal{F}_h^b} \left\{ h_F \, \|\partial_n u_h \cdot \gamma' \circ s_h\|^2_{L^2(\tilde{F})} + h_F^2 \, \|\gamma'' \circ s_h s_h'\|^2_{L^\infty(\omega_{\tilde{F}})} \, \|\partial_n u_h\|^2_{L^2(F)} \right\} \tag{27}$$

we expect heuristically

$$\sum_{F \in \mathcal{F}_h^b} h_F^2 \, \|\partial_n u_h\|^2_{L^2(F)} \preceq \sum_{G \in \mathcal{G}_h^b} h_F \|\nabla u_h\|^2_{L^2(G)} \preceq h_{\max} \|\nabla u_h\|^2_{L^2(D_h^b)}$$
$$\preceq h_{\max}^2 \|\nabla u_h\|^2_{L^2(D)} \preceq h_{\max}^2 \|\nabla u\|^2_{L^2(D)}.$$

The conformity relation (1) for the exact solution $u(s)$ will under our smoothness assumptions hold up to the boundary. We easily derive $\partial_n u \cdot \gamma' \circ s \, s' = 0$ and therefore, since $s' \neq 0$, $\partial_n u \cdot \gamma' \circ s = 0$. From this we obtain $\|\partial_n u \cdot \gamma' \circ s_h\|_{L^2(\partial D)} = O(h_{\max}^2)$ and further

$$\sum_{F \in \mathcal{F}_h^b} h_F \, \|\partial_n u_h \cdot \gamma' \circ s_h\|^2_{L^2(\tilde{F})} \preceq \sum_{F \in \mathcal{F}_h^b} h_F \, \|\partial_n (u_h - u) \cdot \gamma' \circ s_h\|^2_{L^2(\tilde{F})} + O(h_{\max}^4)$$

with

$$\sum_{F \in \mathcal{F}_h^b} h_F \, \|\partial_n (u_h - u)\|^2_{L^2(\tilde{F})} \preceq h_{\max} \, \|\partial_n (u_h - u)\|^2_{L^2(\partial D)}$$
$$\preceq h_{\max} \, \|s - s_h\|^2_{H^1(\partial D)} \preceq h_{\max}^2 \, \|s\|^2_{H^{3/2}(\partial D)}.$$

The size of $R(v_h^1, \gamma')$ is essentially determined by $\|\sigma\|_H$. Hence $R(u_h, \gamma')$ and $R(v_h^1, \gamma')$ will become small like $h_{\max}$.

Higher order terms. With the assumption (26) we obtain that the second term on the right in (23) is almost of the order $h_{\max}^2$. $R(u_h, \gamma'')$ in (25) (defined in Lemma 4.10) can by the arguments of the previous paragraph assumed to be of order $h_{\max}^{1/2}$ only, since $\partial_n u \cdot \gamma'' \circ s$ will not vanish. The additional factor $\|\sigma\|_{H^k(\partial D)}$ makes the whole term of order almost $h_{\max}^{3/2}$ in total. Using however (9) will turn it into an a posteriori expression with leading term already controlled in $R(u_h, \gamma')$. The final term in (25) will be estimated with (26) and (11) by $h_{\max}^{2-\kappa} \|s\|^2_{H^{3/2}(\partial D)} + \lambda_h \|h^2 s''\|_{L^2(\partial D)}$ and may thus be neglected. The structure of the higher order terms suggests to formulate a saturation measure of the form

$$\eta_{\text{sat}} := \max \left\{ h_F^2 \, \|\gamma'' \circ s_h\|_{L^\infty(\tilde{F})} \, \|\gamma''' \circ s_h\|_{L^\infty(\tilde{F})} \; : \; F \in \mathcal{F}_h^b \right\}. \tag{28}$$

5 Implementation

The adaptive solver for the minimal surface problem is implemented within the adaptive finite element toolbox ALBERT [20, 19]. Our aim is to find a discretization $\mathcal{G}_h$ and $s_h \in \mathcal{H}_h$ (approximation to a non-degenerate solution $s \in \mathcal{H}$) such that

$$E_h'(s_h)[\xi_h] = 0 \quad \forall \xi_h \in H_h$$

and $\|s - s_h\|_H \leq \varepsilon$ for any prescribed tolerance ε. In Theorem 4.9 and 4.11 we have established two bounds for the error $\|s - s_h\|_H$ and, based on heuristic arguments, extracted 4 quantities for adaptive error control:

- the residual $\eta(u_h)$ (27) controlling the deviation from the conformity relation (1),
- the error control $\Theta_{\mathcal{G}_h}$ (20) for the harmonic extensions,
- $\eta(v_h^1)$ (from $R(v_h^1, \gamma')$) to control sufficient closeness of discrete and exact second functional derivative of the energy E,
- the saturation measure η_{sat} (28) to control higher order effects.

5.1 The Adaptive Algorithm

Let a tolerance ε and a macro discretization as in Fig. 1.(a) be given and set $s_h^0 := Id$ and pre-refine the mesh until $\eta_{\text{sat}} \leq 1/10$.

Adaptive algorithm
for $k > 0$ do
- Solve for $s_h^k \in \mathcal{H}_h$ and compute $u_h^k = \Phi_h I_h(\gamma \circ s_h^k) \in X_h$.
- Compute $\eta^k := \Theta_{\mathcal{G}_h}(u_h^k) + \eta(u_h^k) + \eta(\tilde{v}_h^k)$ and η_{sat}.
- if $\eta^k < \varepsilon$ and $\eta_{\text{sat}} < 1/10\ \eta^k$ stop.
- Refine according to the local contributions to η^k and η_{sat}.

Since v_h^1 and thus $\eta(v_h^1)$ still involves the unknown solution s, we additionally compute the harmonic extension $\tilde{v}_h^k$ of $(s_h^k - s_h^{k-1})\gamma' \circ s_h^k$ in order to get an approximation to $\eta(v_h^1)$.

The construction of the harmonic extensions with a posteriori error control as a converging method is a well documented operation [8] [9] [17]. Here we use the *fixed energy fraction* marking strategy and two newest node bisections for each marked triangle (see also [19]). An analogous procedure is set up for refining the boundary arcs according to the numbers $\eta_F(u_h^k)$. Note that refinements of D and ∂D influence each other.

Linear and nonlinear solvers. The non-linear equation for s_h^k will be solved by Newton's method which reads

Newton's method
- Compute $E_h'(s_h)$ (initialized with s_h^{k-1});
- Solve for $\delta_h \in H_h$: $E_h''(s_h)[\delta_h, \xi_h] = -E_h'(s_h)[\xi_h] \quad \forall \xi_h \in H_h$;

- Update $s_h \leftarrow s_h + \delta_h$;
- Stop if $E'_h(s_h)$ or δ_h is small enough (then $s_h^k := s_h$), otherwise repeat the loop;

The computation of $E'_h(s_h)$ requires the harmonic extension $u_h(s_h)$, the operation $E''_h(s_h)[\delta_h, \,.\,]$ the discrete harmonic extension $\Phi_h I_h(\delta_h \gamma' \circ s_h)$ (see details below). The linear systems (for δ_h *and* for the harmonic extensions) are solved by conjugated gradients, preconditioned with hierarchical basis [25]. In fact, we used Newton's method with variable stepsize [5] and the stopping criteria were $1 \cdot 10^{-5}$ for $|E'_h(s_h)|$ in the nonlinear method, $1 \cdot 10^{-7}$ in the inner linear solver, and $1 \cdot 10^{-9}$ in the harmonic extensions for the residuals.

The linear algebra structure. For a better understanding of what actually is done we switch to matrix notations. On a given mesh with M nodes on ∂D and $N > M$ nodes of the discretization of $\overline{D}$ functions $r_h \in X_h^b$ and $v_h \in X_h$ are represented as $r_h = \sum_{i=1}^M r_i \psi_i$ and $v_h = \sum_{i=1}^N v_i \phi_i$ with $r_i \in \mathbb{R}$, $v_i \in \mathbb{R}^3$. These are in turn identified with vectors $\boldsymbol{r} = [r_1, \ldots, r_M] \in \mathbb{R}^M$ and $\boldsymbol{v} = [v_1, \ldots, v_N] \in \mathbb{R}^{3N}$, respectively. The enumeration of nodes in D is chosen such that $v_1, \ldots, v_M$ denote the boundary values of v_h in the corresponding enumeration at the boundary. Let us denote by $\boldsymbol{H} : \mathbb{R}^{3M} \to \mathbb{R}^{3N}$ the discrete harmonic extension of the given boundary values and $\boldsymbol{R} : \mathbb{R}^{3N} \to \mathbb{R}^{3M}$ the nodal restriction to the boundary. Then for the discrete solution $u_h = u_h(s_h)$ we have the identities

$$u_h = \sum_{i=1}^N u_i \phi_i = \sum_{i=1}^M \gamma(s_i)\phi_i + \sum_{i=M+1}^N u_i \phi_i,$$
$$\boldsymbol{u} = \boldsymbol{u}(\boldsymbol{s}) = \boldsymbol{H}\boldsymbol{\gamma}(\boldsymbol{s}), \quad \boldsymbol{\gamma}(\boldsymbol{s}) = \boldsymbol{R}\boldsymbol{u}(\boldsymbol{s}),$$

where $\boldsymbol{\gamma}(\boldsymbol{s}) := [\gamma(s_1), \ldots, \gamma(s_M)] \in \mathbb{R}^{3M}$. Note that $\boldsymbol{RH} = \boldsymbol{Id}_{\mathbb{R}^{3M}}$.

Let $\boldsymbol{S}$ be the stiffness matrix for the Laplacian on $\mathcal{G}_h$. Observe that $\boldsymbol{v} \cdot \boldsymbol{S}\boldsymbol{u} = [\boldsymbol{R}\boldsymbol{v}, 0, \ldots, 0] \cdot \boldsymbol{S}\boldsymbol{u}$ since $\boldsymbol{u}$ is discrete harmonic. The energy function is now given by ($\mathcal{E}(\boldsymbol{s}) \equiv E_h(s_h)$)

$$\begin{aligned}\mathcal{E}(\boldsymbol{s}) &= \frac{1}{2}\,\boldsymbol{u}(\boldsymbol{s}) \cdot \boldsymbol{S}\boldsymbol{u}(\boldsymbol{s}) = \frac{1}{2}\,\boldsymbol{H}\boldsymbol{\gamma}(\boldsymbol{s}) \cdot \boldsymbol{S}\boldsymbol{u}(\boldsymbol{s}) = \frac{1}{2}\,\boldsymbol{RH}\boldsymbol{\gamma}(\boldsymbol{s}) \cdot \boldsymbol{S}\boldsymbol{u}(\boldsymbol{s}) \\ &= \frac{1}{2}\,\boldsymbol{\gamma}(\boldsymbol{s}) \cdot \boldsymbol{RS}\boldsymbol{u}(\boldsymbol{s}) = \frac{1}{2}\,\boldsymbol{\gamma}(\boldsymbol{s}) \cdot \boldsymbol{RSH}\boldsymbol{\gamma}(\boldsymbol{s}).\end{aligned}$$

For vectors $\boldsymbol{r} \in \mathbb{R}^M$ and $\boldsymbol{z} \in \mathbb{R}^{3M}$ we let $\boldsymbol{r} \star \boldsymbol{z} := [r_i z_i]_i \in \mathbb{R}^{3M}$ (componentwise scalar times 3-vector) and for vectors $\boldsymbol{z} \in \mathbb{R}^M$ and $\boldsymbol{z}^1, \boldsymbol{z}^2 \in \mathbb{R}^{3M}$ we let $\boldsymbol{z}^1 \odot \boldsymbol{z}^2 := [z_i^1 \cdot z_i^2]_i \in \mathbb{R}^M$ (componentwise scalar product between 3-vectors). Assume that $\boldsymbol{r} \star \boldsymbol{z}^1 \odot \boldsymbol{z}^2 := (\boldsymbol{r} \star \boldsymbol{z}^1) \odot \boldsymbol{z}^2$. Then we can write $\boldsymbol{D_s}\boldsymbol{\gamma}(\boldsymbol{s})\boldsymbol{r} = \boldsymbol{r} \star \boldsymbol{\gamma}'(\boldsymbol{s})$ and we get for the derivatives

$$\begin{aligned}\boldsymbol{D_s}\mathcal{E}(\boldsymbol{s}) &= \boldsymbol{\gamma}'(\boldsymbol{s}) \odot \boldsymbol{RS}\boldsymbol{u}(\boldsymbol{s}), \\ \boldsymbol{D_s^2}\mathcal{E}(\boldsymbol{s})\boldsymbol{r} &= \boldsymbol{r} \star \boldsymbol{\gamma}''(\boldsymbol{s}) \odot \boldsymbol{RS}\boldsymbol{u}(\boldsymbol{s}) + \boldsymbol{\gamma}'(\boldsymbol{s}) \odot \boldsymbol{RSH}(\boldsymbol{r} \star \boldsymbol{\gamma}'(\boldsymbol{s}))\end{aligned}$$

for all $\boldsymbol{r} \in \mathbb{R}^M$. In each step of Newton's method, we have to evaluate $E_h'(s_h)$, i.e. to construct the harmonic extension $\boldsymbol{u}(\boldsymbol{s})$. For each evaluation of $E_h''(s_h)\big[\delta_h, \,.\,\big]$, we additionally have to compute the harmonic extension of $\boldsymbol{\delta} \star \gamma'(\boldsymbol{s})$.

Note, that $\mathcal{E}$ is defined on $\mathcal{S}_h$ and not on $\mathcal{H}_h$. In order to construct a sequence $s_h^k \in \mathcal{S}_h$ we start with s_h^0 in $\mathcal{S}_h$ and use an increment δ_h in H_h using the ansatz $\delta_h = O_h r_h$ for some $r_h \in \mathcal{S}_h$. Using matrix notation $\boldsymbol{\delta} = \boldsymbol{O}\boldsymbol{r}$ as well as $\boldsymbol{O}^\dagger = \boldsymbol{O}$, $\boldsymbol{s} = \boldsymbol{O}\boldsymbol{s}$ we obtain

$$\begin{aligned}
\boldsymbol{D_s}\mathcal{E}(\boldsymbol{s}) &= \boldsymbol{O}\big(\gamma'(\boldsymbol{s})\odot \boldsymbol{RSu}(\boldsymbol{s})\big),\\
\boldsymbol{D_s^2}\mathcal{E}(\boldsymbol{s})\boldsymbol{r} &= \boldsymbol{O}\big(\boldsymbol{Or}\star\gamma''(\boldsymbol{s})\odot \boldsymbol{RSu}(\boldsymbol{s}) + \gamma'(\boldsymbol{s})\odot \boldsymbol{RSH}(\boldsymbol{Or}\star\gamma'(\boldsymbol{s}))\big).
\end{aligned}$$

Thus we have to compute one projection to get $\boldsymbol{D_s}\mathcal{E}(\boldsymbol{s})$ and two projections to apply $\boldsymbol{D_s^2}\mathcal{E}(\boldsymbol{s})$ to some vector $\boldsymbol{r}$. For the projection, we are computing the L^2-projections $\cos_h$ and $\sin_h$ of cos and sin into S_h as an initialization step to the Newton solver. These functions are then used for the projection of r_h.

6 Numerical Results

In this section we present two numerical experiments: the Enneper surface and a curve on the torus. In the first example we compare the derived estimator to the true error and in the second example we demonstrate the benefits from the adaptive method. In all examples we used the hexagonal mesh shown in Fig. 1.(a) as macro discretization.

6.1 Example 1: The Enneper Surface

As a first example we choose the classical Enneper surface that has also been considered in [12]. The boundary curve is given by

$$\begin{aligned}
&\gamma : [0, 2\pi) \to \mathbb{R}^3,\\
&\gamma(\phi) \mapsto R\left[\cos(\phi) - \frac{1}{3}R^2\cos(3\phi), \sin(\phi) + \frac{1}{3}R^2\sin(3\phi), R\cos(2\phi)\right].
\end{aligned}$$

The harmonic extension $u = \Phi(\gamma \circ Id)$ is the solution of Plateau's problem. For $R \in (0, 1]$ it is an area minimizing solution while there are 3 distinct solutions for $R \in (1, \sqrt{3})$, one of which is unstable (at least for R close to $\sqrt{3}$).

We compare exact and estimated errors on a sequence of uniformly and adaptively refined meshes for $R = 0.9$ (Tables 1 and 2) and for $R = 1.1$ (Tables 3 and 4). N is the number of unknowns of the discretization of D (i.e. the number of elements in $\mathcal{N}_h^i$) and $\eta = \Theta_{\mathcal{G}_h}(u_h) + \eta(u_h) + \eta(\tilde{v}_h)$. As constants for the different estimators we used 0.5 for computing $\Theta_{\mathcal{G}_h}(u_h)$ and 0.1 for $\eta(u_h)$ and $\eta(\tilde{v}_h)$. In [13, Theorem 5.5] it is shown that the error estimate for $\|s - s_h\|_H$ gives an error estimate $[\![u - u_h]\!]_{H^1(\Omega)}$ for the harmonic extensions of the same order. Since we cannot compute $\|s - s_h\|_H$ we compare the estimator η with the error for the harmonic extension $[\![u - u_h]\!]_{H^1(\Omega)}$.

Table 1. Enneper surface for $R = 0.9$, uniform refinement

N	η	EOC	$[\![u - u_h]\!]_{H^1(\Omega)}$	EOC	$E_h(s_h)$
7	2.9125		1.3344		3.832
		0.90		0.28	
19	1.8540		1.1580		5.157
		0.98		1.02	
61	1.0469		0.6406		5.159
		1.00		1.12	
217	0.5549		0.3153		5.161
		0.99		1.11	
817	0.2883		0.1514		5.162
		0.99		1.07	
3169	0.1474		0.0735		5.162
		0.99		1.03	
12481	0.0746		0.0363		5.162

Table 2. Enneper surface for $R = 0.9$, adaptive refinement

N	η	EOC	$[\![u - u_h]\!]_{H^1(\Omega)}$	EOC	$E_h(s_h)$
7	2.9125		1.3344		3.832
		0.90		0.28	
19	1.8540		1.1580		5.157
		0.92		0.91	
69	1.0249		0.6450		5.152
		0.93		1.22	
203	0.6218		0.3335		5.170
		1.05		1.08	
613	0.3490		0.1836		5.166
		1.01		1.11	
2303	0.1793		0.0880		5.163
		0.91		0.94	
5361	0.1220		0.0591		5.163
		1.14		1.14	
9323	0.0890		0.0431		5.163

Table 3. Enneper surface for $R = 0.9$, adaptive refinement

N	η	EOC	$[\![u - u_h]\!]_{H^1(\Omega)}$	EOC	$E_h(s_h)$
7	4.7876		2.1642		7.336
		0.60		−0.29	
19	3.5445		2.4999		10.517
		1.09		1.34	
61	1.8720		1.1434		10.287
		0.95		0.26	
217	1.0256		0.9713		10.260
		1.06		1.84	
817	0.5078		0.2860		10.258
		1.02		1.24	
3169	0.2552		0.1232		10.256
		0.99		1.04	
12481	0.1291		0.0602		10.256

Table 4. Enneper surface for $R = 1.1$, adaptive refinement

N	η	EOC	$[\![u - u_h]\!]_{H^1(\Omega)}$	EOC	$E_h(s_h)$
19	3.5445		2.4999		10.517
		1.11		1.48	
55	1.9658		1.1408		10.417
		0.99		1.18	
175	1.1055		0.5757		10.308
		0.91		0.96	
1043	0.4928		0.2435		10.268
		1.23		1.14	
1821	0.3498		0.1772		10.264
		0.99		1.02	
4179	0.2320		0.1162		10.261
		1.03		1.08	
10103	0.1473		0.0720		10.258

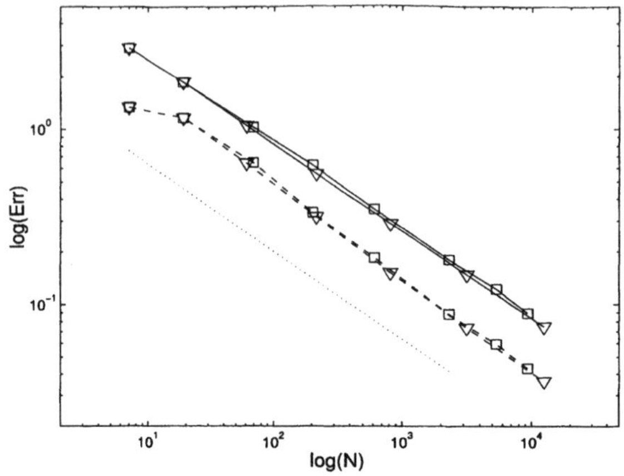

Fig. 2. Estimated (*solid*) and exact (*dashed*) errors from Tables 1 (▽) and 2 (□). The dotted line has slope $-1/2$ (corresponding to EOC$=1$)

We took 2 subsequent values for errors and corresponding N to fit κ in the assumed law 'error$= CN^\kappa + D$'. Then we let EOC $:= -2\kappa$ (since $h \approx N^{-1/2}$). The convergence history can be seen in Fig. 2. The correlation between numbers of unknowns and error fits to the results in [12, p. 20] (note that our bisection algorithm produces different meshes). Exact and estimate errors are in good correspondence for both values of R (we have an overestimation of a factor of 2) although we have only chosen $\lambda_h = 1$. Although this example is suitable to validate the error estimation, it is not well suited to show the benefit of adaptive mesh refinement (with linear elements) since the error is quickly equally-distributed and further refinement is uniform.

6.2 Example 2: A Curve on the Torus

The boundary curve γ in the second example is given by a curve on the torus that winds m-times around the smaller radius r ($r < 1$). This example has been obtained from [2]. Its parameterization is chosen to be

$$\begin{aligned} &\gamma : [0, 2\pi) \to \mathbb{R}^3, \\ &\gamma(\phi) \mapsto \big[\cos(\phi)\big(1 + r\cos(m\phi)\big), \sin(\phi)\big(1 + r\sin(m\phi)\big), r\sin(m\phi)\big]. \end{aligned} \tag{29}$$

For the computation we have chosen $m = 12$ and $r = 0.15$, $r = 0.25$. In Fig. 3 and 4 we show the final minimal surfaces for $r = 0.15$ and $r = 0.25$ on meshes with 76453, 99395 unknowns. The estimated error is about 0.1136, 0.6504 and the discrete energies $E_h(s_h)$ are 3.6830, 4.8266 resp.

In Fig. 5 we show the triangulations after the pre-refinement (with η_{sat}) and after the third refinement step for $r = 0.15$. In this example one has a varying step-size on ∂D that ranges from $h_{\min} \approx 2 \cdot 10^{-4}$ to $h_{\max} \approx 1 \cdot 10^{-3}$ showing that (9) is satisfied.

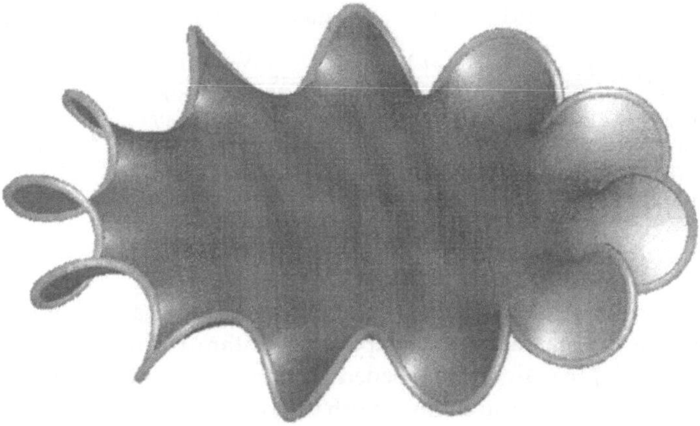

Fig. 3. Minimal surface for γ given by (29) with $m = 12$ and $r = 0.15$ (cf. color plate 12, Page 672)

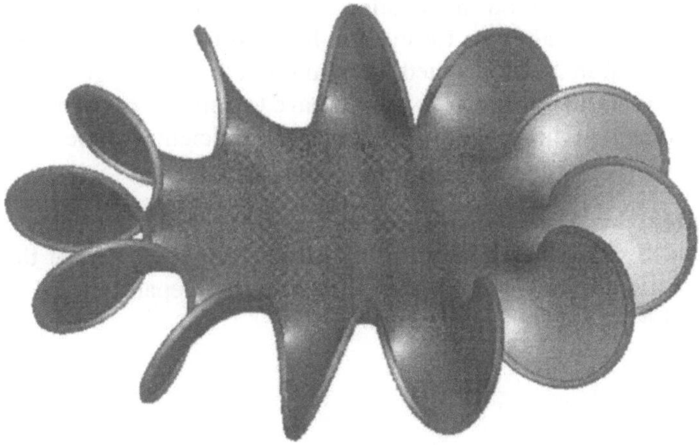

Fig. 4. Minimal surface for γ given by (29) with $m = 12$ and $r = 0.25$

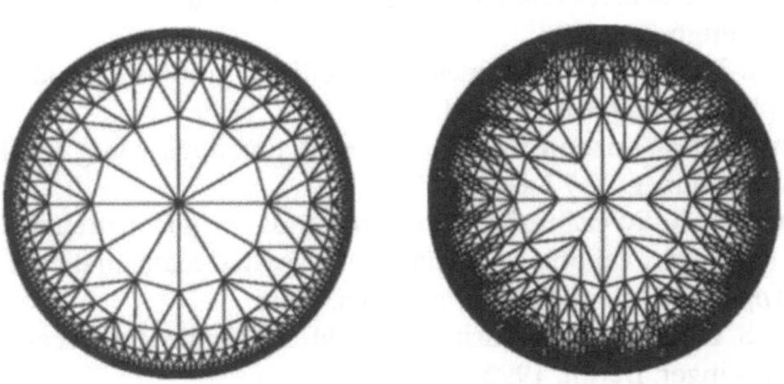

Fig. 5. Triangulations before the first and after third refinement step in case of Fig. 3

Table 5. Example 2 for $r = 0.15$ and $m = 12$, adaptive (left) and uniform (right) refinement

N	η	EOC	$E_h(s_h)$
9196	0.6779		3.726
10228	0.4844	6.32	3.707
15628	0.3179	1.99	3.692
33612	0.2053	1.14	3.686
76453	0.1136	1.44	3.683

N	η	EOC	$E_h(s_k)$
12481	0.7800		3.716
49537	0.4847	0.69	3.693
197377	0.2777	0.81	3.685

Table 5 shows the decay of the estimator during adaptive and uniform refinement for $r = 0.15$. In this example, the adaptive algorithm is much superior to global refinement. The adaptive algorithm needed 20 minutes on a PC with a 1GHz Intel-Pentium III processor for constructing a mesh with estimate $\eta = 0.2053$. Using global refinement we needed 100 minutes for reaching the mesh with estimate $\eta = 0.2777$. The calculation of the discrete solution on the finest level needed approximately 60 minutes. But without using the solution from the previous iteration as an initial guess for Newton's method the solution could not be calculated within 5 hours. Using s_h^{k-1} as an initial guess for the non-linear solver on the next mesh, reduces the initial residual in Newton's method from order 1 to $\approx 1 \cdot 10^{-2}$ or even less on refined meshes. As an outcome, we usually have to perform only one to two Newton iterations on the fine meshes. This underlines the efficiency of the derived adaptive method.

Acknowledgments. The authors thank G. Dziuk (Freiburg) for many valuable discussions and S. Hildebrandt (Bonn) for pointing out [14]. Part of this work was developed while the second author was visiting the Department of Mathematics, University of Maryland at College Park.

References

1. R. A. Adams: Sobolev spaces. Academic press, New York, 1975
2. R. Axthelm: Ein adaptives Finite Elemente Verfahren für das Plateau-problem. Diploma thesis, Univ. Freiburg., 1998
3. P. G. Ciarlet: The finite element method for elliptic problems. Studies in Mathematics and its Applications. North-Holland, Amsterdam, 1979
4. P. Clément: Approximation by finite element functions using local regularizations. RAIRO Anal. Numér., 2, 77–84, 1975
5. J. E. Dennis, R. B. Schnabel: Numerical methods for unconstrained optimization and nonlinear equations. Volume 16 of Classics in applied mathematics. Society for Industrial and Applied Mathematics, Philadelphia, 1996
6. U. Dierkes, S. Hildebrandt, A. Küster, O. Wohlrab: Minimal surfaces I. Boundary value problems. Springer, Berlin, 1992
7. U. Dierkes, S. Hildebrandt, A. Küster, O. Wohlrab: Minimal surfaces II. Boundary regularity. Springer, Berlin, 1992
8. W. Dörfler: A convergent adaptive algorithm for Poisson's equation. SIAM J. Numer. Anal., 33, 1106–1124, 1996

9. W. Dörfler, M. Rumpf: An adaptive strategy for elliptic problems including a posteriori controlled boundary approximation. Math. Comput., 67, 1361–1382, 1998
10. J. Douglas: Solution of the problem of Plateau. Trans. Am. Math. Soc., 33, 263–321, 1931
11. G. Dziuk, J. E. Hutchinson: On the approximation of unstable parametric minimal surfaces. Calc. Var., 4, 27–58, 1996
12. G. Dziuk, J. E. Hutchinson: The discrete Plateau problem: Algorithm and numerics. Math. Comput., 68, 1–23, 1999
13. G. Dziuk, J. E. Hutchinson: The discrete Plateau problem: Convergence results. Math. Comput., 68, 519–546, 1999
14. C. Imbusch, M. Struwe: Variational principles for minimal surfaces. In: J. Escher, G. Simonett (eds) Topics in nonlinear analysis, 477–498. Birkhäuser, Basel, 1999
15. J. L. Lions, E. Magenes: Problemes aux limites non homogenes et applications 1. Dunod, Paris, 1968
16. W. F. Mitchell: A comparison of adaptive refinement techniques for elliptic problems. ACM Trans. Math. Software, 15:327–346, 1989
17. P. Morin, R. H. Nochetto, K. G. Siebert:Data oscillation and convergence of adaptive FEM. SIAM J. Numer. Anal., 38, 466–488, 2000
18. T. Rado: On Plateau's problem. Ann. Math., 2, 457–469, 1930
19. A. Schmidt, K. G. Siebert: ALBERT: An adaptive hierarchical finite element toolbox. Manual. Preprint 06-2000, Univ. Freiburg., 2000
20. A. Schmidt, K. G. Siebert: ALBERT – Software for scientific computations and applications. Acta Mathematica Universitatis Comenianae, 70, 199–229, 2001
21. L. R. Scott, S. Zhang: Finite element interpolation of non-smooth functions satisfying boundary conditions. Math. Comput., 54, 483–493, 1990
22. M. Struwe: Plateau's Problem and the Calculus of Variations. Princeton University Press, Princeton, 1988
23. R. Verfürth: Error estimates for some quasi-interpolation operators. Math. Model. Numer. Anal., 33, 695–713, 1999
24. R. Verfürth: A Review of A Posteriori Error Estimation and Adaptive Mesh–Refinement Techniques. Wiley-Teubner, Chichester, 1996
25. H. Yserentant: On the multi-level splitting of finite element spaces. Numer. Math., 49, 379–412, 1986

Singular Minimal Surfaces

Ulrich Dierkes

Institut für Mathematik, Universität Duisburg, D-47048 Duisburg
dierkes@math.uni-duisburg.de

1 Introduction

In this survey I would like to report on some recent results concerning the *singular minimal surface equation*

$$\operatorname{div}\left(\frac{Du}{\sqrt{1+|Du|^2}}\right) = \frac{\alpha}{u\sqrt{1+|Du|^2}} \tag{$*$}$$

for functions $u : \Omega \to \mathbb{R}$, $\Omega \subset \mathbb{R}^n$ a domain, or $\Omega = \mathbb{R}^n$. Here α denotes some real number.

Clearly, $(*)$ is an equation of mean curvature type with (singular) mean curvature given by

$$H(u, Du) = \frac{\alpha}{u\sqrt{1+|Du|^2}}$$

and a classical solution $u \in C^2(\Omega)$ of $(*)$ is either strictly positive or strictly negative. Notice also that $(*)$ is the Euler equation of the variational integral

$$E_\alpha(u) = \int u^\alpha \sqrt{1+|Du|^2}dx \ ,$$

which in case $\alpha = 1$ might be interpreted as the last coordinate of the center of gravity of the surface

$$\mathcal{M} := \operatorname{graph}(u) = \{(x, u(x)) \in \mathbb{R}^n \times \mathbb{R}; x \in \Omega\} \ ,$$

where $\mathcal{M}$ is supposed to have constant mass density. Equation $(*)$ has already been derived by Lagrange [37], pp. 158-162, as a model equation for *"heavy surfaces in vertical gravitational fields"*, see also Cisa de Gresy [10], pp. 274-276 and Jellett [35], pp. 349-354. Somewhat later, about 1812, a more general and different approach to describing the equilibrium of a flexible, inextensible surface in a force field was given by Poisson [49], see in particular pp. 173-187. Poisson used direct arguments from mechanics and introduces two independent tensions T and T' which correspond to the intrinsic forces of the surface. If the external force field is given by $\mathcal{F} = (X, Y, Z)$, then the equilibrium condition for the surface $u = u(x,y)$, $(x,y) \in \Omega \subset \mathbb{R}^2$ can be described by the following system for the unknown functions T, T' and $u = u(x,y)$,

$$\text{(I)} \quad Xk + \frac{\partial}{\partial x}\left[\frac{T(1+q^2)}{k}\right] - \frac{\partial}{\partial y}\left[\frac{T' pq}{k}\right] = 0$$

$$\text{(II)} \quad Yk - \frac{\partial}{\partial x}\left[\frac{T pq}{k}\right] + \frac{\partial}{\partial y}\left[\frac{T'(1+p^2)}{k}\right] = 0$$

$$\text{(III)} \quad Zk + \frac{\partial}{\partial x}\left[\frac{Tp}{k}\right] + \frac{\partial}{\partial y}\left[\frac{T'q}{k}\right] = 0$$

where we have put $p := \frac{\partial u}{\partial x}$, $q := \frac{\partial u}{\partial y}$ and $k := \sqrt{1+p^2+q^2}$.

Of particular interest is the case $T = T'$, that is the tensions are assumed to be equal. In this case (I), (II) & (III) immediately imply the relation

$$\text{(IV)} \quad Z - pX - qY + \frac{T}{k^2}\left[(1+q^2)\frac{\partial^2 u}{\partial x^2} - 2pq\frac{\partial^2 u}{\partial x \partial y} + (1+p^2)\frac{\partial^2 u}{\partial y^2}\right] = 0$$

as well as the condition

$$\text{(V)} \quad X dx + Y dy + Z dz + dT = 0$$

i.e. the external force must have a potential U and $T = U + \text{const}$ (see Poisson [49], p. 179, 180).

From equations (IV) & (V) Poisson infers (cp. [49], p. 183-186):

i) The *minimal surface equation*, by taking zero external forces $\mathcal{F} = (X,Y,Z) = 0$ and $T = c$.

ii) The *capillary surface equation* by taking the external force field

$$X = -\frac{pN}{k}, \quad Y = -\frac{qN}{k}, \quad Z = \frac{N}{k},$$

where $N = a + bz$ and hence $T = \text{const}$. In this case (IV) reduces to

$$a + bz + \frac{T}{k^3}\left[(1+q^2) - 2pq\frac{\partial^2 u}{\partial x \partial y} + (1+p^2)\frac{\partial^2 u}{\partial y^2}\right] = 0,$$

which is the familiar equation for *capillary surfaces*.

iii) The equation of a *heavy surface in a gravitational field*, by taking $X = Y = 0$, $Z = g\epsilon$, $g =$ gravitational constant and $\epsilon =$ density of the surface. The tension is then given by $T = c - g\epsilon z$ and hence (IV) yields the equation

$$g\epsilon + \frac{c - g\epsilon z}{k^2}\left[(1+q^2)\frac{\partial^2 u}{\partial x^2} - 2pq\frac{\partial^2 u}{\partial x \partial y} + (1+p^2)\frac{\partial^2 u}{\partial y^2}\right] = 0,$$

which is equivalent to the *singular minimal surface equation* $(*)$ (for $n = 2$, $\alpha = 1$), upon putting $g\epsilon = 1$ and performing a translation in the x_3-direction.

Furthermore we observe that equation $(*)$ or at least its "translation invariant" version (iii) describes the shape of a "hanging roof", which, according to the architect F. Otto [48] is of importance for the construction of perfect domes. In fact, turning a hanging solution u of (iii) upside down, gives the optimal shape of a cupola. This is the two-dimensional version of a construction principle which was already known to Galilei and Hooke: Ut pendet continuum flexile, sic stabit contiguum rigidum inversum. We refer to the monograph of the architect F. Otto [48] for a modern derivation of equation $(*)$ or (iii) respectively (in particular [48, p. 290]). Architects have coined the name "Gleichgespannte Membranen unter Eigenlast (membranes of equal tension under their own weight)" or *"heavy surfaces of minimal surface type"* for solutions of (iii).

A parametric approach to equation $(*)$ in the two-dimensional case has been given by Böhme, Hildebrandt, Tausch [7]. Assuming that the boundary of a "parametric" solution (which is constructed in [7]) projects simply onto the boundary of a convex domain in $\mathbb{R}^2$, this parametric solution is in fact of "non-parametric" type, i.e. a solution of the singular minimal surface equation $(*)$, see Dierkes [14]. Of particular interest are also solutions of $(*)$ which, besides prescribed Dirichlet boundary conditions, additionally fulfill appropriate area constraints. We refer to Nitsche [46] and Dierkes, Huisken [25] for non-existence and existence results respectively for this problem.

There is still another geometric interpretation of the equation $(*)$ and the integral E_α. To this end consider the one-dimensional integral for $\alpha = 1$, i.e. $E_1 = \int_a^b u \cdot \sqrt{1+u'^2}\,dx$. As is well known, this integral is equal (up to a constant factor) to the area of the rotational symmetric graph $\mathcal{M}_{\text{rot}}$ of u, where

$$\mathcal{M}_{\text{rot}} := \{(x, u(x)\cos\varphi,\, u(x)\sin\varphi) \in \mathbb{R}^3;\, x \in [a,b],\, \varphi \in [0,2\pi]\} \subset \mathbb{R}^3 .$$

Consequently one might expect that for $\alpha = m \in \mathbb{N}$ and arbitrary $n \geq 2$ the integral $E_\alpha(u)$ corresponds to the area of the symmetric graph

$$\mathcal{M}_{\text{rot}} = \{(x, u(x)\cdot\omega) \in \mathbb{R}^n \times \mathbb{R}^{m+1};\, x \in \Omega \subset \mathbb{R}^n \text{ and } \omega \in S^m \subset \mathbb{R}^{m+1}\} .$$

(Indeed, this fact seems to be well known, see e.g. Barbosa, Do Carmo [3], Do Carmo, Dajczer [26].)

To see this let

$$\begin{aligned}\Phi : \Omega \times [0,2\pi] \times [0,\pi)^{m-1} &\to \mathbb{R}^n \times \mathbb{R}^{m+1}\\ (x,\varphi,\theta) &\mapsto \Phi(x,\varphi,\theta)\end{aligned}$$

be given by $\Phi(x,\varphi,\theta) := (x, u(x)\cdot\omega(\varphi,\theta))$, where

$$\omega : [0,2\pi] \times [0,\pi]^{m-1} \to S^m \subset \mathbb{R}^{m+1},\; S^m := \{x \in \mathbb{R}^{m+1};\, |x| = 1\} ,$$

denote polar coordinates given by $\omega = (\omega^1, \ldots, \omega^{m+1})$ and

$$\begin{array}{rlcr}
\omega^1 = & \cos\varphi\sin\theta_1\sin\theta_2 & \dots & \sin\theta_{m-1} \\
\omega^2 = & \sin\varphi\sin\theta_1\sin\theta_2 & \dots & \sin\theta_{m-1} \\
\omega^3 = & \cos\theta_1\sin\theta_2 & \dots & \sin\theta_{m-1} \\
\vdots & & \ddots & \vdots \\
\omega^m = & & & \cos\theta_{m-2}\sin\theta_{m-1} \\
\omega^{m+1} = & & & \cos\theta_{m-1}
\end{array}$$

The Jacobian $D\Phi^* \circ D\Phi : \Omega \times [0, 2\pi] \times [0, \pi]^{m-1} \to \mathbb{R}^{n+m}$ is given by the $(n+m) \times (n+m)$ matrix

$$D\Phi^* \circ D\Phi = \begin{bmatrix} \left(\delta_{ij} + u_i u_j\right)_{i,j=1,\dots,n} & 0 \\ 0 & \left(u^2 \omega_{\theta_i} \omega_{\theta_j}\right)_{i,j=0,\dots,m-1} \end{bmatrix}$$

where we have put $\theta_0 = \varphi$, $\omega_{\theta_i}\omega_{\theta_j} = \sum_{k=1}^{m+1} \omega_{\theta_i}^k \omega_{\theta_j}^k$.

On the other hand we have for polar coordinates

$$\begin{aligned} \Omega : \mathbb{R}^+ \times [0, 2\pi] \times [0, \pi]^{m-1} &\to \mathbb{R}^{m+1} \\ (r, \varphi, \theta_1, \dots, \theta_{m-1}) &\mapsto \Omega(r, \varphi, \theta) \end{aligned}$$

where $\Omega(r, \varphi, \theta) = r \cdot \omega(\varphi, \theta)$ the identity

$$\begin{aligned} \left(\frac{\partial\Omega}{\partial(r,\varphi,\theta)}\right)^* \left(\frac{\partial\Omega}{\partial(r,\varphi,\theta)}\right) &= \begin{pmatrix} \Omega_r\Omega_r & \Omega_r\Omega_\varphi \dots & \Omega_r\Omega_{\theta_{m-1}} \\ \vdots & & \vdots \\ \Omega_{\theta_{m-1}}\Omega_r & \dots \quad \dots & \Omega_{\theta_{m-1}}\Omega_{\theta_{m-1}} \end{pmatrix} \\ &= \begin{pmatrix} 1 & 0 \\ 0 & (r^2\omega_{\theta_i}\omega_{\theta_j})_{i,j=0,\dots,m-1} \end{pmatrix} . \end{aligned}$$

Since $\det\left(\frac{\partial\Omega}{\partial(r,\varphi,\theta)}\right) = \pm r^m \sin\theta_1 \sin^2\theta_2 \dots \sin^{m-1}\theta_{m-1}$ we find for the Gram determinant

$$\begin{aligned} \det(D\Phi^* \circ D\Phi) = \det\left((\delta_{ij} + u_i u_j)_{i,j=1,\dots,n}\right) \\ \cdot u^{2m} \sin^2\theta_1 \sin^4\theta_2 \dots \sin^{2m-2}\theta_{m-1} . \end{aligned}$$

Therefore the area of the rotational symmetric graph $\mathcal{M}_{\text{rot}}$ is given by

$$\begin{aligned} \mathcal{A}(\mathcal{M}_{\text{rot}}) &= \int_\Omega \int_0^{2\pi} \int_{[0,\pi]^{m-1}} \sqrt{\det(D\Phi^* \circ D\Phi)} dx\, d\varphi\, d\theta_1 \dots d\theta_{m-1} \\ &= \int_\Omega \int_0^{2\pi} \int_{[0,\pi]^{m-1}} \sqrt{1 + |Du|^2}\, u^m \sin\theta_1 \sin^2\theta_2 \dots \\ &\qquad \dots \sin^{m-1}\theta_{m-1} dx\, d\varphi \dots d\theta_{m-1} \\ &= \text{const} \cdot \int_\Omega u^m(x) \sqrt{1 + |Du(x)|^2}\, dx . \end{aligned}$$

Concluding we observe that every *positive* solution of $(*)$ above with $\alpha = m \in \mathbb{N}$ has a symmetric graph which has zero mean curvature in $\mathbb{R}^{n+m+1}$, while every *non-negative* solution yields a minimal submanifold containing a singular set, namely the set $\{u = 0\}$ (which of course is empty if $u > 0$). Hence in particular the singular solutions of $(*)$ lead to interesting new examples of co-dimension one minimal surfaces with a (possibly) controlled behaviour near singular points, and with singular sets having gaps etc., see L. Simon [54] for results in this direction.

Before proceeding further let us remark that this fact — namely that for $\alpha = m \in \mathbb{N}$ the rotational graph is a minimal submanifold — can very easily be deduced directly, cf. also Barbosa, Do Carmo [3]. To see this let $\varphi : \mathbb{R}^n \times \mathbb{R}^m \times \mathbb{R} \to \mathbb{R}$ be given by

$$\varphi(x_1, \ldots, x_{n+m+1}) = u^2(x_1, \ldots, x_n) - x_{n+1}^2 - \cdots - x_{n+m+1}^2$$

so that $\mathcal{M}_{\text{rot}} \subset \mathbb{R}^{n+m+1}$ is described by $\varphi = 0$. The mean curvature of $\mathcal{M}_{\text{rot}}$ is given by the formula

$$\begin{aligned} H &= \sum_{i=1}^{n+m+1} D_i \left(\frac{\nabla \varphi}{|\nabla \varphi|} \right) \\ &= \sum_{i=1}^{n} D_i \left(\frac{D_i u}{\sqrt{1 + |Du|^2}} \right) - \frac{1}{\sqrt{1 + |Du|^2}} \sum_{i=n+1}^{n+m+1} D_i \left(\frac{x_i}{R} \right) \end{aligned}$$

where $R := \left\{ x_{n+1}^2 + \cdots + x_{n+m+1}^2 \right\}^{1/2}$ ($= u$ on $\{\varphi = 0\}$). Hence we find for the mean curvature of $\mathcal{M}_{\text{rot}}$ the identity

$$H = \sum_{i=1}^{n} D_i \left(\frac{D_i u}{\sqrt{1 + |Du|^2}} \right) - \frac{\alpha}{u\sqrt{1 + |Du|^2}} ,$$

so that $H = 0$ iff $(*)$ holds. Note however once again that $(*)$ & E_α are meaningful for every $\alpha \in \mathbb{R}$ and obviously for $\alpha = 0$ correspond to the classical Euclidean minimal surface situation, while $\alpha = -n$ leads to minimal surfaces in the hyperbolic n-space, modeled by the upper half space $\mathbb{R}^n \times \mathbb{R}^+$ with the hyperbolic metric $g_{ij} = x_{n+1}^{-2} \delta_{ij}$ (see Lin [40] for a further discussion of this interesting case).

Similarly, if $M \subset \mathbb{R}^n \times \mathbb{R}^+$ is a hypersurface of $\mathbb{R}^{n+1}$, not necessarily a graph, locally given by a parametrization $\psi : W \mapsto \mathbb{R}^{n+1}$, $W \subset \mathbb{R}^n$, open, $\psi_{n+1} \geq 0$, we may define for $\alpha = m \in \mathbb{N}$ the rotationally symmetric surface $\mathcal{M}_{\text{rot}}$ parametrized (locally) by

$$\begin{aligned} \Phi : \ & W \times [0, 2\pi] \times [0, \pi]^{m-1} \to \mathbb{R}^{n+m+1} \\ & (x, \varphi, \theta) \mapsto \Phi(x, \varphi, \theta) = \left(\psi_1(x), \ldots, \psi_n(x), \psi_{n+1}(x) \omega(\varphi, \theta) \right) . \end{aligned}$$

It then follows that $\int_M x_{n+1}^m dH_n = \text{const} \int_{\mathcal{M}_{\text{rot}}} dH_{n+m}$, where H_k denotes k-dimensional Hausdorff measure. Moreover, we find that the mean curvature of the symmetric hypersurface $\mathcal{M}_{\text{rot}}$ is given by $H_{\mathcal{M}_{\text{rot}}}(x_1, \ldots, x_{n+m+1}) =$

$H_M(x_1, \ldots, x_{n+1}) - \frac{m\nu_{n+1}}{x_{n+1}}$ where $H_{\mathcal{M}_{\mathrm{rot}}}$, H_M denote the mean curvatures of $\mathcal{M}_{\mathrm{rot}}$ and M respectively, while ν_{n+1} stands for the last coordinate of the unit normal $\nu = (\nu_1, \ldots, \nu_{n+1})$ of the surface $M \subset \mathbb{R}^{n+1}$. Hence $\mathcal{M}_{\mathrm{rot}}$ is minimal in $\mathbb{R}^{n+m+1}$ if $\alpha = m \in \mathbb{N}$ and the mean curvature of M at $x \in \mathbb{R}^{n+1}$ satisfies the parametric version of $(*)$, namely

$$H(x) = \alpha \frac{\nu_{n+1}}{x_{n+1}} \quad \text{for all } x \in M \subset \mathbb{R}^{n+1} . \tag{1}$$

Again, $(*)$, E_α and (1) are meaningful for every $\alpha \in \mathbb{R}$, while the above interpretation (i.e. that $\mathcal{M}_{\mathrm{rot}}$ is a minimal submanifold) is only reasonable for $\alpha \in \mathbb{N}$.

2 The n-Dimensional Analogue of the Goldschmidt Solution

Let us consider the one-dimensional variational integral

$$I(u) = \int_a^b u(x)\sqrt{1+(u')^2}\, dx ,$$

where $u : [a,b] \to \mathbb{R}$ with $u \geq 0$ and its parametric counterpart $F(X) = \int_0^1 y(t)\sqrt{\dot{x}^2 + \dot{y}^2}\, dt$, with $X = (x,y) : [0,1] \to \mathbb{R} \times \mathbb{R}^+$ and both w and X are piecewise differentiable (or Lipschitz) functions. Observe that both integrals equal (up to a constant factor) the area of the corresponding rotational graph, which is hence bounded by two coaxial parallel circles in $\mathbb{R}^3$. It is well known (and easily calculated) that all extremals for I are catenaries of the form

$$u = u(x) = a \cos h \left(\frac{x - x_0}{a}\right) ,$$

which furnish non-parametric $\mathcal{F}$-extremals

$$(x(t), y(t)) = \left(t,\, a \cos h \left(\frac{t - t_0}{a}\right)\right) .$$

However, there are also parametric $\mathcal{F}$-extremals, which do not possess a non-parametric representation (in the fixed coordinate system), namely the curves

$$(x(t), y(t)) = (x_0, t) \quad , \quad t \in \mathbb{R}^+ \text{ , (i.e. vertical line segments)}$$

(and hence also every reparametrization hereof). These segments can be used for the construction of the so-called Goldschmidt solution (see [33]): Given two points $P_1 = (x_1, y_1)$ & $P_2 = (x_2, y_2)$ with $x_1 < x_2$ in the upper half plane then the Goldschmidt solution is a D^1 (i.e. piecewise C^1) parametrization of the three segments $\overline{P_1P_1'}$, $\overline{P_1'P_2'}$ and $\overline{P_2'P_2}$ where $P_1' = (x_1, 0)$, $P_2' = (x_2, 0)$ denote the projections of P_1, P_2 onto the x-axis. Note that the Goldschmidt solution is not an extremal for the parametric integral $\mathcal{F}$ (i.e. a solution of the Euler equation). However, it turns out that the Goldschmidt curve always represents a relative minimizer for $\mathcal{F}$ in the class $\mathcal{C} = \{X = (x(t), y(t)) \in \mathrm{Lip}([0,1], \mathbb{R}^2);\ X(0) = P_1,\ X(1) = P_2,\ y(t) \geq 0\ \forall t \in [0,1]\}$, i.e. a solution of an obstacle problem. Furthermore, the Goldschmidt polygon furnishes an absolute minimum of $\mathcal{F}$ in $\mathcal{C}$, if the points P_1 & P_2 are "far" apart (and this can be made absolutely precise). Clearly, the Goldschmidt curve corresponds to the two discs spanned by the boundary circles defined by revolving the points P_1 and P_2 about the x-axis, and a one-dimensional segment of zero area connecting the center of the discs, while the catenaries correspond to catenoids bounded by the two coaxial circles. Concluding we find that the two discs furnish a relative minimum for the area functional when compared to surfaces of revolution which are bounded by the two circles, and, in addition, this minimum is an absolute one if the distance of the two discs exceeds a certain critical value (which can be explicitly calculated). A further argument then shows that the discs are not only minimal with respect to rotationally symmetric perturbations but even have (locally) smaller area when compared to arbitrary surfaces (integer multiplicity 2-currents say) bounded by the circles (cp. Lawson [39]). Summarizing we note the following result; for further information see Giaquinta-Hildebrandt [30], chapt. 8.4.3

Theorem 2.1 *([33], [41], [56].) The singular problem,*

$$\mathcal{F}(X) = \int_0^1 y(t)\sqrt{\dot{x}^2 + \dot{y}^2}\, dt \to \min$$

in the class $\mathcal{C} = \{X \in \mathrm{Lip}([0,1], \mathbb{R}^2);\ X(0) = P_1,\ X(1) = P_2,\ y(t) \geq 0\}$ *has a solution. This solution is either furnished by the catenary or the Goldschmidt curve, or both of them (if the integrals are equal). Except for reparametrization there are no further solutions.*

A similar reasoning extends Theorem 2.1 to a broader class of singular, one-dimensional variational integrals which lead to corresponding results for minimal submanifolds in $\mathbb{R}^{n+2}$ (rather than $\mathbb{R}^3$). To this end consider for any $\alpha > 0$, and $X \in \mathcal{C}$ the integral

$$\mathcal{F}_\alpha(X) = \int_0^1 y^\alpha(t)\sqrt{\dot{x}^2 + \dot{y}^2}\, dt$$

and the corresponding (non-parametric) extremals $x = x(y)$ given by the integral

$$x = \lambda \int_{\sqrt[\alpha]{\lambda}}^{y} \frac{d\xi}{\sqrt{\xi^{2\alpha} - \lambda^2}} + c\,,\ y \geq \sqrt[\alpha]{\lambda}\,, \tag{2}$$

where $\lambda > 0$ is arbitrary (cp. Lawson [39], Dierkes [18]). These curves (which for $\alpha = 1$ simply are the usual catenaries) might be called "α-catenaries" and, in case $\alpha = n \in \mathbb{N}$, lead to rotationally symmetric minimal submanifolds of dimension $(n+1)$ in $\mathbb{R}^{n+2}$, the "$(n+1)$-catenoids". Again, as in the case $\alpha = 1$, there is a Goldschmidt solution, since the line segments $x = \text{const}$ are (parametric) extremals. For $\alpha > 1$ it also follows directly that the Goldschmidt solution consisting of the three line segments connecting two points P_1, P_2, is an extremal of $\mathcal{F}_\alpha$ (a so-called "discontinuous solution" in classical terminology). This curve always defines a relative strong minimizer for $\mathcal{F}_\alpha$, that — in addition — furnishes an absolute minimum if P_1, P_2 are far apart. It turns out that Theorem 2.1 also holds for the variational integral $\mathcal{F}_\alpha(\ )$ in the class $\mathcal{C}$ ([12]). In particular we observe that the $(n+1)$-catenoids or the two $(n+1)$-dimensional discs minimize area in $\mathbb{R}^{n+2}$ with respect to their boundaries.

From the analysis of the one-dimensional problem one might expect a non-existence result for connected minimal submanifolds bounded by two (or more) boundary components Γ_1, Γ_2 which are too far apart. In fact, for $\alpha = n \in \mathbb{N}$ the one-parameter family of n-catenaries (2), with $c = 0$, may be shown to have a cone K_{τ_0} of a considerable large angle of aperture as an envelope. Moreover, this cone $K_{\tau_0} := \{x \in \mathbb{R}^{n+2};\ \tau_0^2 x_{n+2}^2 > (x_1^2 + \cdots + x_{n+1}^2)\} \cup \{0\}$ has the following non-existence property.

Theorem 2.2 *([44], [45], [47], [34], [18].) Let $C = C^+ \cup \{P_0\} \cup C^-$ be a solid cone with vertex $P_0 \in \mathbb{R}^{n+2}$ which is congruent to K_{τ_0}. Then there is no connected, compact $(n+1)$-dimensional minimal submanifold $M \subset \mathbb{R}^{n+1}$ which is of class C^2 with $\partial M \subset \overline{C}$ such that both $\partial M \cap C^+$ and $\partial M \cap C^-$ are non-empty.*

Remark 2.3

i) By construction, the cone K_{τ_0} is the "largest" cone with this "non-existence" property.

ii) Theorem 2.2 extends to minimizing currents with two or more boundary components.

iii) One may prove similar non-existence results for surfaces with prescribed mean curvature (Hildebrandt [34], Dierkes [18]). These methods extend to mean curvature flow problems and prove development of singularities in finite time for suitable initial manifolds, the so-called "neck-pinching" phenomenon (Ecker [27], Winklmann [58]).

iv) The approximate value of τ_0 and the angle of aperture θ are listed in the following table:

$n+1$	2	3	4	5	6	7	8
τ_0	1.51	2.37	3.15	3.89	4.63	5.44	6.02
θ	56.46	67.15	72.40	75.60	77.81	79.59	80.58

Now we consider the n-dimensional singular variational problem

$$E_\alpha(u) = \int_\Omega u^\alpha \sqrt{1 + |Du|^2} \cdot dx \to \min$$

in a suitable class of comparison functions, defined by boundary conditions. Clearly, the most prominent example appears when $\alpha = 0$, i.e. when $E_0(u) = \mathcal{A}(u)$, the classical area integral. It is well known that the minimum problem for $\mathcal{A}$ can be solved using the theory of Caccioppoli sets, i.e. sets with finite perimeter (see the monographs Federer [29], Giusti [32], Simon [52], Massari-Miranda

[42]). In this approach $\mathcal{A}(u)$ is extended to $BV(\Omega)$, the space of functions with bounded variation: $u \in BV(\Omega) \underset{\text{def}}{\Longleftrightarrow} u \in L_1(\Omega)$ and the distributional derivatives are Radon measures. The Radon measure $\sqrt{1+|Du|^2}$ ist then given by $\int_U \sqrt{1+|Du|^2} = \sup \int_U \left\{ g_{n+1} + u \sum_{i=1}^{n} D_i g_i \right\} dx$ where the supremum is to be taken over all $g = (g_1, \dots, g_{n+1}) \in C_c^1(\Omega, \mathbb{R}^{n+1})$ such that $\sum_{i=1}^{n+1} g_i^2(x) \le 1$ and $U \subset \Omega$ is any open set. For more general integrands (like the one in E_α) such an extension may become rather complicated, because Du no longer is a function (but merely a measure). Giaquinta-Modica-Souček [31] studied variational integrands of the form $f(x,p) + g(x,u)$ and Dal Maso [11] and Anzellotti [2] dealt with the more general case $f(x,u,p)$. The idea is to define $\int_\Omega f(x,u,Du)$ on $BV(\Omega)$ through the elliptic parametric integrand that is associated to $f(x,u,p)$. However, even for integrands f that unlike E_α are not singular, this approach does not lead to regularity of minima, at least in the case where f explicitly depends on the function u, since no suitable gradient estimates are available in this case. In case the integrand $f = f(x,p)$ only depends on x and Du, existence and regularity of solutions have been established in Giaquinta-Modica-Souček [31].

Concluding we observe that although $E_\alpha(u)$ may be defined for $u \in BV(\Omega)$ it is not immediately possible to minimize the corresponding integral because of lack of coerciveness. The idea to overcome this difficulty is as old as it is simple and can be traced back at least as far as to the early papers of Bernstein and even Euler [28], namely to consider a function $v = \psi(u)$ instead of u itself, cf. [6], Giaquinta-Hildebrandt [30], chapt. 4, sect. 2.3, and also in this particular context Bemelmans [4], Bemelmans-Dierkes [5], Dierkes [17]. To obtain a suitable transformation assume first that $u \in C^1(\Omega)$, $u \ge 0$, and observe that

$$u^\alpha \sqrt{1+|Du|^2} = \sqrt{u^{2\alpha} + u^{2\alpha}|Du|^2} = \sqrt{u^{2\alpha} + \frac{1}{(1+\alpha)^2}|Du^{1+\alpha}|^2}$$

i.e. $u^\alpha \sqrt{1+|Du|^2}$ can be considered to be a function on u^α and $|Du^{1+\alpha}|$. In other words we have to require $u^{1+\alpha} \in L_1(\Omega)$ or $u \in L_{1+\alpha}(\Omega)$ and $v := u^{1+\alpha} \in BV(\Omega)$. This motivates the following

Definition 2.4 *For any $\alpha \ge 0$ the class $BV^+_{1+\alpha}(\Omega)$ is given by*

$$BV^+_{1+\alpha}(\Omega) := \left\{ u \in L_{1+\alpha}(\Omega);\ u \ge 0 \text{ and } u^{1+\alpha} \in BV(\Omega) \right\} .$$

If $u \in BV^+_{1+\alpha}(\Omega)$ then $E_\alpha(u)$ is well defined as the Radon measure

$$\int_\Omega u^\alpha \sqrt{1+|Du|^2} := \sup \left\{ \int_\Omega \left\{ u^\alpha \cdot g_{n+1} + \frac{1}{1+\alpha} u^{1+\alpha} \sum_{i=1}^{n} D_i g_i \right\} dx; \right. \\ \left. g \in C_c^1(\Omega, \mathbb{R}^{n+1}),\ |g(x)| \le 1 \right\} . \tag{3}$$

Note that for $u \in H_1^1(\Omega) \cap L_\infty(\Omega)$ it easily follows that

$$\int_\Omega u^\alpha \sqrt{1+|Du|^2} = \int_\Omega u^\alpha \sqrt{1+|Du|^2}\, dx \ ,$$

where the r.h.s. is to be interpreted as a classical Lebesgue integral. A further remark is that the "nonparametric" and "parametric" integrals are equal, i.e.

$$\int_\Omega u^\alpha \sqrt{1+|Du|^2} = \int_\Omega x_{n+1}^\alpha |D\varphi_U| \ , \tag{4}$$

where $U := \{(x,t) \in \Omega \times \mathbb{R}^+;\ 0 \le t < u(x)\}$ denotes the (non-negative) *"subgraph"* of $u \in BV_{1+\alpha}^+(\Omega)$, with characteristic function φ_U. In other words our definitions of $E_\alpha(u)$ coincides with the one given by [31] and Anzellotti [2], Dal Maso [11] at least if $u \in BV_{1+\alpha}^+$. Moreover from the very definition (3) of the integral E_α one sees that the problem is coercive in $BV_{1+\alpha}^+$ (while this is not the case in $BV(\Omega)$). Using the trace formula for BV-functions one is then lead to the following minimization problem

$$\mathcal{P}_\alpha : E_\alpha^*(u) := \int_\Omega u^\alpha \sqrt{1+|Du|^2} + \frac{1}{1+\alpha} \int_{\partial\Omega} \left| u^{1+\alpha} - \varphi^{1+\alpha} \right| dH_{n-1}$$
$$\longrightarrow \text{ minimum in } BV_{1+\alpha}^+(\Omega)$$

where we assume $\partial\Omega$ to be Lipschitz and $0 \le \varphi \in L_{1+\alpha}(\partial\Omega)$ to be given. In light of the above remarks it is natural to expect some kind of existence-regularity result, although the singular behaviour of the integrand when u tends to zero is still inherent in this formulation. However, there is the useful relation (4) which connects the non-parametric with the parametric problem and which in particular implies that any minimizer $u \in BV_{1+\alpha}^+$ of E_α^* gives rise to a (local) minimum $U = \{(x,t) \in \Omega \times \mathbb{R}^+;\ 0 \le t < u(x)\}$ for the parametric integral

$$\mathcal{E}_\alpha(V) = \int_{\Omega\times\mathbb{R}^+} x_{n+1}^\alpha |D\varphi_V| \ .$$

Concluding we observe that the regularity theory for co-dimension one area minimizing currents (for a good account see Simon [52]), or — alternatively — regularity results for *"almost minimal sets"* (see Almgren [1] and in particular for the present context Tamanini [57]) are amenable to the present situation. We have

Theorem 2.5 *([5], [17].) Let $\partial\Omega$ be Lipschitz and $\varphi \in L_{1+\alpha}(\partial\Omega)$ be non-negative, $\alpha \ge 0$. Then the problem $\mathcal{P}_\alpha$ possesses a solution $u \in BV_{1+\alpha}^+(\Omega)$. Moreover, if $n \le 6$, any minimum u is continuous in Ω and analytic in the (open) set $\Omega \cap \{x;\ u(x) > 0\}$. In general the "coincidence set" $\{x \in \Omega;\ u(x) = 0\}$ is non-empty. More precisely, we have for any $\alpha > 0$ the estimate $\alpha\big|\{u=0\}\big| \ge \alpha|\Omega| - \sup\limits_{\partial\Omega} \varphi \cdot |\partial\Omega|$, so that $\{u = 0\}$ has positive measure if the boundary data φ are "small" in terms of the "isoperimetric condition" $\sup\limits_{\partial\Omega}\varphi < \frac{\alpha|\Omega|}{|\partial\Omega|}$. Here $|\Omega|$ denotes n-dimensional Lebesgue measure, while $|\partial\Omega|$ stands for $(n-1)$-dimensional Hausdorff measure of Ω and $\partial\Omega$, respectively.*

Theorem 2.5 leaves open the following questions:

i) What is the optimal interior regularity of any minimizer $u \in BV^+_{1+\alpha}(\Omega)$?
i) Under what conditions can it be shown that the coincidence set $\{x \in \Omega;\ u(x) = 0\}$ is empty?

Ad i). Here it is conceivable that, due to the strong singular behaviour of the integral E_α, solutions to $\mathcal{P}_\alpha$ are not necessarily regular. In fact it is possible to construct merely Hölder continuous minimizers.

Theorem 2.6 *([13], [15].) Let $n \geq 2$, $\alpha > 0$ and $R > 0$ be given. There exist a number $\delta > 0$ and a function $w \in BV^+_{1+\alpha}(B_R(0)) \cap C^{0,1/2}(B_R(0))$ (depending on n, α and R of course) which minimizes*

$$E^*_\alpha(u) = \int_{B_R} u^\alpha \sqrt{1+|Du|^2} + \frac{1}{1+\alpha}\int_{\partial B_R} |u^{1+\alpha} - \delta^{1+\alpha}| dH_{n-1}$$

in the class $BV^+_{1+\alpha}(B_R)$. Furthermore, w assumes the (constant) boundary values δ and the regularity of w cannot be improved, i.e. $w \notin C^{0,1/2+\epsilon}(B_R)$.

This solution $w \in C^{0,1/2}$ might be considered as the n-dimensional analogue of the Goldschmidt discontinuous solution. However, note that the n-dimensional analogue possesses a non-parametric representation, while this does not hold in the classical case. The solution $w \in C^{0,1/2}(B_R)$ is identical to zero on a certain ball $B_r(0), r < R$, lifts off vertically-asymptotically like $\sqrt{|x|-r}$ and then assumes the constant boundary values on $|x| = R$ in a very smooth way. The proof of Theorem 2.6 uses the construction of a field of rotationally symmetric extremals from a careful analysis of a similar ordinary differential system which appears in the context of area minimizing cones over products of spheres, see Bombieri-De Giorgi-Giusti [8] and Simoes [51]. The argument might hence be viewed as the modern adaptation to a BV-setting of the classical *"field theory"* due to Weierstraß and H.A. Schwarz. (In fact, Theorem 3 of [15] only contains a somewhat weaker statement as formulated here, however, it turns out that the argument of the proofs of Theorems 2 & 3 in that paper can be easily modified to obtain the stated version in Theorem 2.6 here.)

Concluding we remark that $\frac{1}{2}$-Hölder continuity is (in general) the best possible regularity of a minimizer u of E_α that can be expected.

Ad ii) Clearly, the situation changes drastically if we can prove the coincidence set $\{u = 0\}$ of a minimizer u to be empty. Here we have

Theorem 2.7 *([22].) Suppose $\Omega \subset \mathbb{R}^n$ is a Lipschitz domain, $\alpha > 0$ and $\varphi \in L_\infty(\partial\Omega)$ satisfies the isoperimetric condition*

$$\inf_{\partial\Omega} \varphi > (1+\alpha)n \cdot c(n)\Big(1+\frac{1}{n}\Big)^{n+1} |\Omega|^{\frac{1}{n}},$$

where $c(n) := n^{-1}\omega_n^{-\frac{1}{n}}$. Then every solution $u \in BV^+_{1+\alpha}(\Omega)$ of the Dirichlet problem $\mathcal{P}_\alpha$ is strictly positive and we have $u \in C^\omega(\Omega)$.

Remark 2.8

- *If Ω is mean convex, φ is continuous at $x_0 \in \partial\Omega$ then also u is continuous at x_0 and we have*

$$\lim_{x \to x_0} u(x) = \varphi(x_0) \quad \textit{(see [16])} .$$

- *Obviously, any solution u of Theorem 2.7 satisfies the singular minimal surface equation*

$$\operatorname{div}\left(\frac{Du}{\sqrt{1+|Du|^2}}\right) = \frac{\alpha}{u\sqrt{1+|Du|^2}} \quad \textit{in } \Omega \tag{$*$}$$

and $u = \varphi$ on $\partial\Omega$ if φ is continuous and Ω is mean convex.
- *For any $\alpha > 0$ the non-parametric cones*

$$c_n^\alpha(x) := \sqrt{\frac{\alpha}{n-1}}(x_1^2 + \cdots + x_n^2)^{\frac{1}{2}}$$

are Lipschitz solutions of $(*)$, which obviously are not C^1. The minimizing and stability properties of these cones have been completely analyzed, cp. [15], [19]. Note that for $\alpha = m \in \mathbb{N}$ these cones correspond to cones over products of spheres, namely to the cones over $S^m \times S^{n-1} \subset \mathbb{R}^{m+1} \times \mathbb{R}^n$ given by $\frac{1}{m}(x_{n+1}^2 + \ldots x_{n+m+1}^2) = \frac{1}{(n-1)}(x_1^2 + \cdots + x_n^2)$. These cones have been classified by Simons [55], Bombieri-de Giorgi-Giusti [8], Lawson [39], Simoes [51] and also Lawlor [38].
- *Using Schauder theory, Dierkes and Huisken [24] proved existence results for regular solutions of the singular minimal surface equation $(*)$ under similar conditions.*
- *Of particular interest are also* "singular solutions" *of $(*)$, i.e. Lipschitz solutions like e.g. the cones c_n^α mentioned above and also more generally weak solutions which are non-negative and vanish on a certain lower dimensional set. For $\alpha = m \in \mathbb{N}$ the rotationally symmetric graph of any such singular solutions defines a codimension one minimal submanifold in $\mathbb{R}^{n+m+1}$ with a singular set, namely the set $\{(x,y,z) \in \mathbb{R}^n \times \mathbb{R}^m \times \mathbb{R};\ x \in \Omega,\ u(x) = 0 = y = z\}$. It follows from Theorems 2.5 & 2.7 that for "large" boundary values φ there is a strictly positive solution of $(*)$, while no such solution can be expected if we have $\sup_{\partial\Omega} \varphi < \alpha\frac{|\Omega|}{|\partial\Omega|}$. Hence, if a family of boundary data is considered, parametrized by $0 \le t \le 1$ say, and such that there is a positive solution for $t = 1$, while there is no solution with the given data for $t = 0$, one of course expects some singular solution somewhere in between. However, solutions of $(*)$ are in general non-unique, and it is this phenomenon which causes difficulties in this approach. Using Leray-Schauder theory L. Simon [54] was able to carry out the above program and hence constructed singular Lipschitz solutions of $(*)$ which vanish in isolated points.*

– *Since* $c(n) = n^{-1}\omega_n^{-\frac{1}{n}}$ *is the isoperimetric constant we have (by the isoperimetric inequality)*

$$c(n)|\Omega|^{\frac{1}{n}} \geq \frac{|\Omega|}{|\partial\Omega|} \, .$$

It is therefore an interesting question, whether our regularity result Theorem 2.7 remains true under an inequality of the form

$$\inf_{\partial\Omega} \varphi \geq c \frac{|\Omega|}{|\partial\Omega|}$$

for some suitable constant $c \geq 1$. *This condition would be a counterpart to the smallness condition in Theorem 2.5.*

3 Entire Solutions

Now we turn to the global behaviour of solutions to the singular minimal surface equation

$$\operatorname{div}\left(\frac{Du}{\sqrt{1+|Du|^2}}\right) = \frac{\alpha}{u\sqrt{1+|Du|^2}} \quad \text{in } \mathbb{R}^n \, . \tag{$*$}$$

There exist entire (i.e. defined on all of $\mathbb{R}^n$) solutions of $(*)$ e.g. the weak Lipschitz-solutions

$$c_n^\alpha(x) = \sqrt{\frac{\alpha}{n-1}}\,(x_1^2 + \cdots + x_n^2)^{\frac{1}{2}}$$

for any $\alpha > 0$ and $n \geq 2$. From the field construction (see [13], [15]) it also follows directly that for any $\alpha > 0$ and $n \geq 2$ there are smooth (analytic) entire solutions of the equation $(*)$, see also Keiper [36]. The stability analysis in [36] also indicates that for the case $\alpha = 1$ the above mentioned entire solutions are unstable if $n \leq 5$, while they are stable in dimensions $n \geq 6$. Hence one cannot expect a *"Bernstein property"* of equation $(*)$ without imposing any further assumptions on the solutions. Note that in dimension 2 many equations do possess this Bernstein property, i.e.

"global existence $\longrightarrow$ triviality of the solution in a specified sense (e.g. linearity)".

In fact for some classes of equations this property can be proved even for higher dimensions and we refer to L. Simon [53] and the literature cited therein for a recent account of these examples. Here we only mention two results concerning the equation $(*)$. To formulate the first one we need the notion of stability: Let $M \subset \mathbb{R}^n \times \mathbb{R}^+$, $\mathbb{R}^+ = \{t > 0\}$ be a smooth hypersurface, for example $M = \operatorname{graph} u = \{(x, u(x));\ x \in \mathbb{R}^n\}$ the graph of a smooth, positive function $u : \mathbb{R}^n \to \mathbb{R}$. Then M is called *(globally) stable*, iff the second variation of the integral $\mathcal{E}_\alpha = \int x_{n+1}^\alpha dH_n$ is non-negative. Analytically this is easily seen to be equivalent to the following inequality (see [19])

$$\int_M x_{n+1}^\alpha(|A|^2 + \frac{1}{\alpha}H^2)\eta^2 dH_n \le \int_M x_{n+1}^\alpha |\nabla\eta|^2 dH_n \tag{5}$$

for all $\eta \in C_c^1(M \cap B_r(\xi))$ and all $B_r(\xi) \subset \mathbb{R}^{n+1}$ where $\xi \in M$. Observe that the stability inequality in case $\alpha = m \in \mathbb{N}$ corresponds to the stability inequality for the area of the rotationally symmetric graph of u in $\mathbb{R}^{n+m+1}$. However, note that stability for $\mathcal{E}_\alpha$ is weaker than stability for the area of the symmetric graph, since any test function η which is admissible in (5), when considered as a a function on the rotated graph, leads to a testfunction with rotationally symmetric support on the symmetric graph, while on the other hand, in the condition for stability of the symmetric graph any test function with compact support is admissible.

We have the following non-existence result:

Theorem 3.9 *([20].) Suppose $\alpha > 0$, $\alpha + n < 4 + 2\sqrt{\frac{2}{n+\alpha}}$ (i.e. $\alpha + n < 5.23\ldots$). Then there is no entire, stable solution $u \in C^2(\mathbb{R}^n)$ of the singular minimal surface equation $(*)$.*

Remark 3.10

- *The proof of Theorem 3.9 can be found in Dierkes [20]. The main ingredient in [20] is a new "Simons inequality" for the Laplacian of the second fundamental form $|A|$ and the mean curvature H together with suitable integral curvature estimates. If $\alpha = m \in \mathbb{N}$ is an integer then, using the observation made in the introduction, namely that the rotational graph $\mathcal{M}_{\text{rot}}$ of u is a minimal submanifold in $\mathbb{R}^{n+m+1}$, the conclusion of Theorem 3.9 can also be deduced from Schoen-Simon-Yau [50]. However, one then has to deal with the analogue of the weaker stability condition (5) and — in addition — appropriate area-growth estimates have to be proved for stationary (not necessarily minimizing) rotational minimal surfaces, see [20]. In the fundamental paper [50] curvature estimates (integral and pointwise) are proved for (stable or minimizing) minimal hypersurfaces which are immersed in a Riemannian manifold.*
- *There is also a parametric version of Theorem 3.9, see [20].*
- *Curvature estimates can also be proved for solutions of $(*)$, see [21]. These estimates imply Theorem 3.9.*

The next non-existence theorem for entire solutions of $(*)$ requires controlled growth of the solutions at infinity. First recall that for the classical minimal surface equation we have the following result, due to Bombieri, De Giorgi, Miranda and Moser.

Theorem 3.11 *([9], [43].) Let $u \in C^{0,1}(\mathbb{R}^n)$ be an entire solution of the minimal surface equation* $\operatorname{div}\left(\frac{Du}{\sqrt{1+|Du|^2}}\right) = 0$ *in $\mathbb{R}^n$. Assume further that u is bounded on one side by a cone, i.e.*

$$u(x) \le c + c|x| \tag{6}$$

for all $x \in \mathbb{R}^n$ and some constant $c > 0$. Then u is an affine function.

The proof uses the a-priori gradient estimate of Bombieri-De Giorgi-Miranda [9] together with the upper bound (6) to show that any entire solution satisfying (6) has a globally bounded gradient. The assertion then follows from Moser's Bernstein theorem [43]. Recall also that — by elliptic regularity theory — any Lipschitz solution of the minimal surface equation is in fact analytic.

The situation for the singular minimal surface equation is of course different, since we have the singular minimal cones which are global weak solutions of $(*)$. Hence the following result is in this respect optimal.

Theorem 3.12 *([23].) There is no entire non-negative solution* $u \in C^{0,1}(\mathbb{R}^n)$ *of the singular minimal surface equation* $(*)$ *which has sublinear growth i.e. for which* $u(x) = o(|x|)$ *as* $|x| \to \infty$.

Remark 3.13

- *Note that by elliptic regularity we have that* $u \in C^\omega(\{u > 0\})$. *Also the zero level set* $\{u = 0\}$ *must have measure zero.*
- *Theorem 3.12 follows from a lower area estimate of the type* $\int_{M \cap B_\rho(\xi)} x_{n+1}^\alpha dH_n \geq c\rho^{n+\alpha}$, $\rho \gg 1$, *where* M *denotes the graph of* u. *However, this estimate can be shown to contradict the sublinear growth of* u. *We refer to [23] for details.*

Acknowledgement. I would like to express my gratitude to the Sonderforschungsbereich 256 in Bonn for stimulating a large part of the research which permitted me to contribute to the developments described in this survey.

References

1. F.J. Almgren: Existence and regularity almost everywhere of solutions of elliptic variational problems with constraints. Mem. A.M.S., 4, No. 165, 1976
2. G. Anzellotti: Dirichlet problem and removable singularities for functionals with linear growth. Boll. U.M.I., 18, 141–159, 1981
3. J.L. Barbosa, M. do Carmo: Helicoids, catenoids and minimal hypersurfaces of $\mathbb{R}^n$ invariant by an ℓ-parameter group of motions. An. Acad. Brasil. Cienc., 53, 403–408, 1981
4. J. Bemelmans: Gleichgewichtsfiguren zäher Flüssigkeiten mit Oberflächenspannung. Analysis, 1, 241–282, 1981
5. J. Bemelmans, U. Dierkes: On a singular variational integral with linear growth. Arch. Rat. Mech. Anal., 100, 83–103, 1987
6. S. Bernstein: Sur la généralisation du problème de Dirichlet I & II. Math. Ann., 62, 253–271, 1906 & Math. Ann., 69, 82–136, 1910
7. R. Böhme, S. Hildebrandt, E. Tausch: The two-dimensional analogue of the catenary. Pac. J. Math., 88, 247–278, 1980
8. E. Bombieri, E. De Giorgi, E. Giusti: Minimal cones and the Bernstein problem. Invent. Math., 88, 243–268, 1969
9. E. Bombieri, E. De Giorgi, M. Miranda: Una maggiorazione a priori relativa alle ipersuperfici minimali non parametriche. Arch. Rat. Mech. Anal., 32, 255-267, 1969

10. Cisa de Gresy: Considération sur l'équilibre des surfaces flexibles et inextensibles. Mem. Reale Accad. Sci. Torino 21(1), 259–294, 1818
11. G. Dal Maso: Integral representation on $BV(\Omega)$ of Γ-limits of variational integrals. Man. Math., 30, 387–416, 1980
12. U. Dierkes: Singuläre Variationsprobleme und Hindernisprobleme. Bonner Math. Schriften 155, 1984
13. U. Dierkes: Minimal hypercones and $C^{0,\frac{1}{2}}$ minimizers for a singular variational problem. Indiana Univ. Math. J., 73(4), 841–863, 1988
14. U. Dierkes: A geometric maximum principle, Plateau's problem for surfaces of prescribed mean curvature and the two-dimensional analogue of the catenary. Partial Differential Equations and Calculus of Variations, Springer Lecture Notes in Mathematics, 1357, 111–141, 1988
15. U. Dierkes: A classification of minimal cones in $\mathbb{R}^n \times \mathbb{R}^+$ and a counterexample to interior regularity of energy minimizing functions. Man. Math., 63, 173–192, 1989
16. U. Dierkes: Boundary regularity for solutions of a singular variational problem with linear growth. Arch. Rat. Mech. Anal., 105, 285–298, 1989
17. U. Dierkes: Über singuläre Lösungen gewisser mehrdimensionaler Variationsprobleme. Ann. Univ. Sarav., 3(2), 38–108, 1990
18. U. Dierkes: Maximum principles and nonexistence results for minimal submanifolds. Man. Math., 69, 203–218, 1990
19. U. Dierkes: On the non-existence of enery stable minimal cones. Ann. Inst. Henri Poincaré, Anal. Non Linéaire, 7(6), 589–601, 1990
20. U. Dierkes: A Bernstein result for energy minimizing hypersurfaces. Calc. Var. and Part. Diff. Eq., 1, 37-54, 1993
21. U. Dierkes: Curvature estimates for minimal hypersurfaces in singular spaces. Invent. math., 122, 453–473, 1995
22. U. Dierkes: On the regularity of solutions for a singular variational problem. Math. Z., 225, 657–670, 1997
23. U. Dierkes: Non-existence for entire singular minimal graphs of controlled growth. To appear.
24. U. Dierkes, G. Huisken: The N-dimensional analogue of the catenary: existence and non-existence. Pac. J. Math. (I), 141, 47–54, 1990
25. U. Dierkes, G. Huisken: The N-dimensional analogue of the catenary: Prescribed area. In: J. Jost (ed) Calculus of Variations and Geometric Analysis. International Press, 1–13, 1996
26. M. do Carmo, M. Dajczer: Rotation hypersurfaces in spaces of constant curvature. Trans. Am. Math. Soc. 277, 685–709, 1983
27. K. Ecker: Local techniques for mean curvature flow. ANU Canberra, Proc. CMA vol. 26, 107–120, 1991
28. L. Euler: Methodus inveniendi lineas curvas maximi minimive proprietate gaudentes, sive solutio problematis isoperimetrici lattissimo sensu accepti. Bousquet, Lausannae et Genevae 1744. EG5A.O.O. Ser I, vol. 24
29. H. Federer: Geometric measure theory. Springer Grundlehren 153, 1969
30. M. Giaquinta, S. Hildebrandt:Calculus of Variations I and II. Springer Grundlehren 310 & 311, 1996
31. M. Giaquinta, G. Modica, J. Souček: Functionals with linear growth in the calculus of variations I, II. Comm. Math. Univ. Carol., 20, 143–156 & 157–172, 1979
32. E. Giusti: Minimal surfaces and functions of bounded variation. Birkhäuser, Basel, 1984
33. B. Goldschmidt: Determinatio superfici minimae rotatione curva data duo puncta jungentis circa datum axem artae. Dissertation, Göttingen 1831

34. S. Hildebrandt: Maximum principles for minimal surfaces and for surfaces of continuous mean curvature. Math. Z., 128, 253–269, 1972
35. F.H. Jellett: Die Grundlehren der Variationsrechnung. Braunschweig, Verlag der Hofbuchhandlung v. Leibrock, 1860
36. J.B. Keiper: The axially symmetric n-tectum. Preprint, 1980
37. J.L. Lagrange: Mécanique analytic. Oeuvre tome onzième.
38. G.R. Lawlor: A sufficient criteria for a cone to be aera-minimizing. Diss. Stanford Univ. 1988
39. H.B. Lawson: The equivariant Plateau problem and interior regularity. Trans. Am. Math. Soc., 173, 231–249, 1972
40. F.H. Lin: On the Dirichlet problem for minimal graphs in hyperbolic space. Invent. Math., 96, 593–612, 1989
41. H. Mac Neish: Concerning the discontinuous solution in the problem of the minimum surface of revolution. Ann. Math., (2), 7, 72–80, 1905
42. U. Massari, M. Miranda: Minimal surfaces of codimension one. North Holland, Amsterdam-New York-Oxford, 1984
43. J. Moser: A new proof of De Giorgi's theorem concerning the regularity problem for elliptic differential equations. Comm. Pure Appl. Math., 13, 457–468, 1960
44. J.C.C. Nitsche: A necessary criterion for the existence of certain minimal surfaces. J. Math. Mech., 13, 659–666, 1964
45. J.C.C. Nitsche: A supplement to the condition of J. Douglas. Rend. Circ. Mat. Palermo, 13, 192–198, 1964
46. J.C.C. Nitsche: A nonexistence theorem for the two-dimensional analogue of the catenary. Analysis, 6, 143–156, 1986
47. R. Osserman, M. Schiffer: Doubly connected minimal surfaces. Arch. Rat. Mech. Anal., 58, 285–306, 1974/75
48. F. Otto: Zugbeanspruchte Konstruktionen. Bd. I, II. Berlin, Frankfurt/M., Wien: Ullstein 1962, 1966
49. S.D. Poisson: Sur les surfaces élastique. Mem. Cl. Sci. Math. Phys. Inst. France, deux, 167–225, 1812
50. R. Schoen, L. Simon, S.T. Yau: Curvature estimates for minimal hypersurfaces. Acta Math., 134, 275–288, 1975
51. P. Simoes: On a class of minimal cones in $\mathbb{R}^n$. AMS Bulletin, 80, 488–489, 1974
52. L. Simon: Lectures on geometric measure theory. ANU Canberra, Proc. CMA vol. 3, 1983
53. L. Simon: The minimal surface equation. Encylopaedia of Math. Sciences, 90, 239–266. Ed. R. Osserman, Geometry V, Springer 1997
54. L. Simon: Examples for singular minimal surfaces. To appear.
55. J. Simons: Minimal variaties in Riemannian manifolds. Ann. Math., 88, 62–105, 1968
56. M. Sinclair: The absolute minimum in the problem of the surface of revolution of minimum area. Ann. Math., 9(2), 151–155, 1908
57. I. Tamanini: Boundaries of Caccioppoli sets with Hölder-continuous normal vector. J. reine angew. Math., 334, 27–39, 1982
58. S. Winklmann: Enclosure theorems for generalized mean curvature flows. To appear in Calc. Var. and Part. Diff. Eq. 2002

Note on the Isoperimetric Profile of a Convex Body

Ernst Kuwert

Mathematisches Institut der Albert-Ludwigs-Universität Freiburg, Eckerstraße 1, D-79104 Freiburg, Germany, kuwert@mathematik.uni-freiburg.de

Summary. A solution to the relative isoperimetric problem (or partitioning problem) is a subset E of a given set $\Omega \subset \mathbb{R}^n$ (the container) with prescribed volume $|E| = V$ and minimal area $\int_\Omega |D\chi_E| = A(V)$ of the interface. Improving a result due to Sternberg & Zumbrun [8], we obtain that if Ω is convex then $A(V)^{\frac{n}{n-1}}$ is a concave function of V. As consequence we deduce that the isoperimetric ratio of E is no worse than that of the half-ball contained in $\mathbb{H} = \mathbb{R}^{n-1} \times (0,\infty)$, and that the mean curvature of the minimizer is bounded a priori.

In this note we improve an argument due to Sternberg & Zumbrun [8]. Let $\Omega \subset \mathbb{R}^n$, $n \geq 2$, be open, bounded and convex, with boundary $\partial\Omega \in C^2$ of class C^2. We put $K = \overline{\Omega}$. For $V \in [0, \mathcal{L}^n(\Omega)]$ we consider the functions

$$A(V) = \inf\{\int_\Omega |D\chi_E| : E \subset \Omega, |E| = V\} \tag{1}$$

$$\sigma(V) = \frac{A(V)^{\frac{n}{n-1}}}{\min\{V, \mathcal{L}^n(\Omega) - V\}}. \tag{2}$$

The function σ will be called the isoperimetric profile of Ω. It is shown in [8, Thm. 2.8] that the function $A(V)$ is concave. Here we show that this can in fact be improved, namely we have:

Theorem 1.1 *The function $g(V) = A(V)^{\frac{n}{n-1}}$ is concave.*

Let $\sigma_n = \omega_{n-1}^{\frac{n}{n-1}}/\alpha_n$ denote the standard isoperimetric constant of $\mathbb{R}^n$. The Theorem 1.1 implies

Corollary 1.2 *For any $V \in (0, \mathcal{L}^n(\Omega))$ we have $\sigma(V) \leq \sigma_{\mathbb{H}}$, where $\sigma_{\mathbb{H}} = 2^{-\frac{n}{n-1}}\sigma_n$ is the isoperimetric constant of the halfspace $\mathbb{H} = \mathbb{R}^{n-1} \times (0,\infty)$.*

By results of Gonzalez, Massari and Tamanini [3] and the boundary regularity due to Grüter [4], we have for any $V \in (0, \mathcal{L}^n(\Omega))$ a minimizing set $E \subset \Omega$, i. e.

$$\mathcal{L}^n(E) = V, \quad \int_\Omega |D\chi_E| = A(V), \tag{3}$$

such that for $2 \leq n \leq 7$ the set $\Sigma = \partial E \cap \Omega$ is of class $C^{1,\alpha}$ up to the boundary and smooth in the interior. In higher dimensions one has a corresponding *partial* regularity result. We refer to the work of Sternberg & Zumbrun [7, 8] for the discussion of minimizers with singularities and merely assume here that the solution is sufficiently smooth. Using the first variation formula one infers that Σ has constant mean curvature H and meets $\partial\Omega$ orthogonally along its boundary. Here we define H as arithmetic mean of the principal curvatures, such that $H = +1$ for ∂B^n with respect to the outward unit normal. In general the minimizer is not unique and in fact there can be minimizers with different H, see [5]. However, the set of minimizers for fixed V is compact in the $C^{1,\alpha}$ topology.

Now let $\varphi, \psi : \Sigma \longrightarrow \mathbb{R}$ be given and let $Z : \Sigma \longrightarrow \mathbb{R}^n$ be a given tangential vector field on Σ. Consider a variation $f_\varepsilon : \Sigma \longrightarrow \Sigma_\varepsilon \subset \mathbb{R}^n$ satisfying

$$\frac{\partial}{\partial\varepsilon} f_\varepsilon|_{\varepsilon=0} = \varphi\,\nu \quad (\nu = \text{exterior unit normal}) \tag{4}$$

$$\frac{\partial^2}{\partial\varepsilon^2} f_\varepsilon|_{\varepsilon=0} = \psi\,\nu + Z. \tag{5}$$

Let $p \in \partial K$ and write $\partial K = \{x : u(x) = 0\}$ locally around p, where w. l. o. g. $Du(p) = \nu_{\partial K}(p)$ and $\nu_{\partial K}$ is the exterior unit normal along ∂K. Then if $f_\varepsilon(p) \in \partial K$, we get

$$\begin{aligned} 0 &= \frac{\partial^2}{\partial\varepsilon^2} u(f_\varepsilon(p))|_{\varepsilon=0} \\ &= D^2u(p)\,(\varphi\,\nu, \varphi\,\nu) + Du(p)\cdot(\psi\nu + Z)(p). \end{aligned}$$

Denoting by $B_{\partial K}$ the second fundamental form of ∂K with respect to $\nu_{\partial K}$, we conclude that the condition

$$\varphi^2 B_{\partial K}(\nu,\nu) + \langle \nu_{\partial K}, Z\rangle = 0 \tag{6}$$

means that f_ε preserves ∂K in second order. For a variation with (4), (5) and (6) the first and second variational formulae yield, see for example [8],

$$\frac{d}{d\varepsilon}\mathcal{H}^{n-1}(\Sigma_\varepsilon)|_{\varepsilon=0} = (n-1)\,H \int_\Sigma \varphi\, d\mathcal{H}^{n-1}, \tag{7}$$

$$\frac{d}{d\varepsilon}\mathcal{L}^n(E_\varepsilon)|_{\varepsilon=0} = \int_\Sigma \varphi\, d\mathcal{H}^{n-1}, \tag{8}$$

$$\begin{aligned} \frac{d^2}{d\varepsilon^2}\mathcal{H}^{n-1}(\Sigma_\varepsilon)|_{\varepsilon=0} &= (n-1)\,H \int_\Sigma \psi\, d\mathcal{H}^{n-1} \\ &\quad + \int_\Sigma \{|\nabla\varphi|^2 + ((n-1)^2 H^2 - |h|^2)\,\varphi^2\}\, d\mathcal{H}^{n-1} \\ &\quad - \int_{\partial\Sigma} B_{\partial K}(\nu,\nu)\,\varphi^2\, d\mathcal{H}^{n-2}, \end{aligned} \tag{9}$$

where h is the second fundamental form of Σ with respect to ν. Finally

$$\frac{d^2}{d\varepsilon^2}\mathcal{L}^n(E_\varepsilon)|_{\varepsilon=0} = \int_\Sigma \psi\, d\mathcal{H}^{n-1} + (n-1)\,H \int_\Sigma \varphi^2\, d\mathcal{H}^{n-1}. \tag{10}$$

Multiplying (10) by $(n-1)H$ and subtracting from (9) yields

$$\begin{aligned}&\frac{d^2}{d\varepsilon^2}\mathcal{H}^{n-1}(\Sigma_\varepsilon)|_{\varepsilon=0}-(n-1)H\frac{d^2}{d\varepsilon^2}\mathcal{L}^n(E_\varepsilon)|_{\varepsilon=0}\\&=\int_\Sigma(|\nabla\varphi|^2-|h|^2\varphi^2)\,d\mathcal{H}^{n-1}-\int_{\partial\Sigma}B_{\partial K}(\nu,\nu)\varphi^2\,d\mathcal{H}^{n-2}.\end{aligned}\tag{11}$$

Using the flow of a suitably constructed vector field, one shows that for given φ,ψ,Z satisfying (6) there exists a variation f_ε with (4) and (5) such that $f_\varepsilon(E)=E_\varepsilon\subset\Omega$, $f_\varepsilon(\partial\Sigma)\subset\partial K$. Let us now assume

$$\frac{d}{d\varepsilon}\mathcal{L}^n(E_\varepsilon)|_{\varepsilon=0}=\int_\Sigma\varphi d\mathcal{H}^{n-1}\neq 0.\tag{12}$$

It follows that the function $\varepsilon\mapsto\mathcal{L}^n(E_\varepsilon)$ has a local inverse

$$\varepsilon:(V_0-\delta,V_0+\delta)\longrightarrow(-\varepsilon_0,\varepsilon_0),\tag{13}$$

where $V_0=\mathcal{L}^n(E)$. Define the function

$$a:(V_0-\delta,V_0+\delta)\to\mathbb{R},\quad a(V)=\mathcal{H}^{n-1}(\Sigma_\varepsilon)|_{\varepsilon=\varepsilon(V)}.\tag{14}$$

Using the chain rule we compute

$$\varepsilon'(V_0)=\left(\int_\Sigma\varphi\,d\mathcal{H}^{n-1}\right)^{-1}\tag{15}$$

$$a'(V_0)=\varepsilon'(V_0)\frac{d}{d\varepsilon}\mathcal{H}^{n-1}(\Sigma_\varepsilon)|_{\varepsilon=0}=(n-1)H.\tag{16}$$

Furthermore

$$\begin{aligned}\varepsilon''(V_0)&=-\varepsilon'(V_0)^3\,\frac{d^2}{d\varepsilon^2}\mathcal{L}^n(E_\varepsilon)|_{\varepsilon=0}\\a''(V_0)&=\varepsilon''(V_0)\,\frac{d}{d\varepsilon}\mathcal{H}^{n-1}(\Sigma_\varepsilon)|_{\varepsilon=0}+\varepsilon'(V_0)^2\,\frac{d^2}{d\varepsilon^2}\mathcal{H}^{n-1}(\Sigma_\varepsilon)|_{\varepsilon=0}\\&=\varepsilon'(V_0)^2\left(\frac{d^2}{d\varepsilon^2}\mathcal{H}^{n-1}(\Sigma_\varepsilon)|_{\varepsilon=0}-\varepsilon'(V_0)\frac{d}{d\varepsilon}\mathcal{H}^{n-1}(\Sigma_\varepsilon)|_{\varepsilon=0}\,\frac{d^2}{d\varepsilon^2}\mathcal{L}^n(E_\varepsilon)|_{\varepsilon=0}\right).\end{aligned}$$

Using (15), (7) and (11) we get

$$a''(V_0)=\frac{\int_\Sigma(|\nabla\varphi|^2-|h|^2\varphi^2)d\mathcal{H}^{n-1}-\int_{\partial\Sigma}B_{\partial K}(\nu,\nu)\varphi^2\,d\mathcal{H}^{n-2}}{(\int_\Sigma\varphi d\mathcal{H}^{n-1})^2}.\tag{17}$$

We now take as test function $\varphi\equiv 1$. By convexity of ∂K we have

$$\int_{\partial\Sigma}B_{\partial K}(\nu,\nu)\,\varphi^2\,d\mathcal{H}^{n-2}\geq 0.$$

The following step differs slightly from [8]. We split $h = h^0 + Hg$, where g is the metric on Σ and h^0 is tracefree. Thus we have $|h|^2 \geq (n-1)\, H^2$. Inserting this and the previous inequality into (17) yields

$$a''(V_0) \leq -\frac{(n-1)\, H^2}{a(V_0)} = -\frac{1}{n-1}\frac{a'(V_0)^2}{a(V_0)}.$$

Putting $g_+(V) = a(V)^{\frac{n}{n-1}}$ we get

$$\begin{aligned} g_+''(V_0) &= \frac{n}{n-1}\left(a(V)^{\frac{1}{n-1}}\, a'(V)\right)'(V_0) \\ &= \frac{n}{n-1}\, a(V_0)^{\frac{1}{n-1}}\left(a''(V_0) + \frac{1}{n-1}\frac{a'(V_0)^2}{a(V_0)}\right) \leq 0. \end{aligned}$$

Thus we have shown that the function $g(V) = A(V)^{\frac{n}{n-1}}$ has the following property: for any $V_0 \in \big(0, \mathcal{L}^n(\Omega)\big)$ there exists a locally defined C^2-function g_+ such that

$$\begin{cases} g(V) \leq g_+(V) & \text{for } V \in (V_0 - \delta, V_0 + \delta) \\ g(V_0) = g_+(V_0) \\ g_+''(V_0) \leq 0 \end{cases} \tag{18}$$

A standard reasoning shows that g is concave. In fact, consider $g_\alpha(V) = g(V) - \alpha\, V^2$ for $\alpha > 0$. It is sufficient to show that if $g_\alpha(V_1) = g_\alpha(V_2) = s$, then $g_\alpha(V) \geq s$ for $V_1 \leq V \leq V_2$, because this proves concavity of g_α (by subtracting linear functions) and then one may let $\alpha \searrow 0$. But if g_α attained a negative minimum on $[V_1, V_2]$, then we easily obtain a contradiction using the upper function $g_+(V) - \alpha\, V^2$. This proves the Theorem 1.1.

To prove Corollary 1.2, we note that for small V we easily get by explicit construction an upper estimate of the type

$$A(V)^{\frac{n}{n-1}} \leq \sigma_{\mathbb{H}} V + \text{ higher order terms}, \tag{19}$$

as $V \to 0$. Thus at $V = 0$ we have, since $g(0) = 0$,

$$\limsup_{V \searrow 0} \frac{g(V) - g(0)}{V} \leq \sigma_{\mathbb{H}},$$

and the concavity of g yields $g(V) \leq \sigma_{\mathbb{H}}\, V$. As $A(V)$ is symmetric with respect to $\frac{1}{2}\mathcal{L}^n(\Omega)$, the corollary follows.

Remark 1.3 *The corresponding problem for sets E contained in the complement of a convex body was recently studied by J. Choe in [2].*

The Theorem 1.1 has some further applications. As already shown in [8, Thm. 2.6], see also [6], one has

Corollary 1.4 *Assume that K is strictly convex, in the sense that the second fundamental form $B_{\partial K}$ is strictly positive. Then the surface Σ must be connected.*

Proof. If we had two different components Σ_i, $i = 1, 2$, we could find $c_i \in \mathbb{R}$ $(i = 1, 2)$, such that the variation given by $\varphi = c_i$ on Σ_i preserves volume, i. e.

$$c_1 \mathcal{H}^{n-1}(\Sigma_1) + c_2 \mathcal{H}^{n-1}(\Sigma_2) = 0,$$

and $c_i \neq 0$ for $i = 1, 2$. Choosing ψ, Z appropriately we could then arrange that $\mathcal{L}^n(E_\varepsilon) = \mathcal{L}^n(E)$ in second order, and conclude

$$\frac{d^2}{d\varepsilon^2} \mathcal{H}^{n-1}(\Sigma_\varepsilon)|_{\varepsilon=0} = -\sum_{i=1}^{2} \left(\int_{\Sigma_i} |h|^2 \, c_i^2 \, d\mathcal{H}^{n-1} + \int_{\Sigma_i} B_{\partial K}(\nu, \nu) \, c_i^2 \, d\mathcal{H}^{n-2} \right) < 0,$$

a contradiction. □

Another consequence, deduced by Sternberg and Zumbrun in [8, Cor. 2.9], is the following

Corollary 1.5 *Let $V \in (0, \frac{1}{2}\mathcal{L}^n(\Omega))$. Then the mean curvature H of the corresponding minimizer with respect to the outward unit normal is nonnegative: $H \geq 0$.*

Proof. For $0 < V_0 < \frac{1}{2}\mathcal{L}^n(\Omega)$, we have from our above calculation

$$\limsup_{h \searrow 0} \frac{A(V_0 + h) - A(V_0)}{h} \leq \lim_{h \searrow 0} \frac{a(V_0 + h) - a(V_0)}{h} = (n-1)H.$$

Thus if we had $H < 0$ for the minimizer corresponding to V_0, then by concavity we see that $A(V) \leq A(V_0) + H\,(V - V_0) < A(V_0)$ for $V > V_0$. This contradicts $A(V_0) = A(\mathcal{L}^n(\Omega) - V_0)$. □

Remark 1.6 *In Corollary 1.5 one can show the strict inequality $H > 0$, provided that K is strictly convex in the sense that $B_{\partial K}$ is strictly positive.*

We finally observe

Corollary 1.7 *Let H be the mean curvature corresponding to a minimizer with volume V, w. l. o. g. $0 < V \leq \frac{1}{2}\mathcal{L}^n(\Omega)$. Then we have the inequalities*

$$H \leq \frac{1}{n} \frac{A(V)}{V} \leq \frac{1}{n} \sigma_{\mathbb{H}}^{\frac{n-1}{n}} V^{-\frac{1}{n}}. \tag{20}$$

Proof. By concavity we have for any $V \in [0, \mathcal{L}^n(\Omega)]$

$$A(V)^{\frac{n}{n-1}} \leq A(V_0)^{\frac{n}{n-1}} + \frac{n}{n-1} A(V_0)^{\frac{1}{n-1}} (n-1)\, H(V - V_0).$$

Evaluating at $V = 0$ shows

$$H \leq \frac{1}{n} \frac{A(V_0)}{V_0} \leq \frac{1}{n} \sigma_{\mathbb{H}}^{\frac{n-1}{n}} V_0^{-\frac{1}{n}}.$$

□

Surprisingly, the last estimate is completely independent of the body. Note that for $0 < V < \frac{1}{2}\mathcal{L}^n(\Omega)$ we also have $H \geq 0$ due to Corollary 1.5. For $V = \frac{1}{2}\mathcal{L}^n(\Omega)$ we get the inequality (20) with H replaced by $|H|$.

References

1. Bürger, W., (2001): Das parametrische Partitionsproblem. Dissertation thesis, Universität Freiburg, submitted
2. Choe, J. (2001): Relative isoperimetric inequality for domains outside a convex set. Preprint, to appear in J. Inequalities Appl.
3. Gonzales, E., Massari, U., Tamanini, I. (1983): On the regularity of boundaries of sets minimizing perimeter with a volume constraint. Indiana Univ. Math. J., **32**, 25–37
4. Grüter, M.(1987): Boundary regularity for solutions of a partitioning problem. Arch. Rational Mech. Analysis, **97**, 261–270
5. Hildebrandt, S. (1987): Remarks on some isoperimetric problem. In: Cesari, L. (ed) Contributions to modern calculus of variations (Bologna 1985). Pitman Res. Notes Harlow, 108–122
6. Ros, A., Vergasta, E. (1995): Stability for hypersurfaces of constant mean curvature with free boundary. Geom. Dedicata, **56**, 19–33
7. Sternberg, P., Zumbrun, K. (1998): A Poincaré inequality with applications to volume-constrained surfaces. J. Reine Angew. Math., **503**, 63–85
8. Sternberg, P., Zumbrun, K. (1999): On the connectivity of boundaries of sets minimizing perimeter subject to a volume constraint. Comm. Analysis Geom., **7**, 199–220

Geometric Conditions on Free Boundaries

Reiner Schätzle

Mathematisches Institut der Rheinischen Friedrich-Wilhelms-Universität Bonn,
Beringstraße 6, D-53115 Bonn, Germany
schaetz@math.uni-bonn.de

Summary. On free boundaries arising in models for melting and solidification of materials, conditions are often prescribed using geometrical data of the free boundary. In smooth models approximating these free boundary problems, the geometric conditions are replaced by elliptic and parabolic equations. We describe the approximation of mean curvature flow by the Allen-Cahn equation, also with coupling, and of the Stefan problem with Gibbs-Thomson law by the quasi-stationary phase field equations.

1 Introduction

Free boundaries which separate different phases appear for example in models describing the melting and solidification of materials as the Stefan problem with Gibbs-Thomson law which reads

$$\begin{aligned} \partial_t(u+\varphi) - \Delta u &= f, \\ H_{\partial[\varphi=1]} &= u. \end{aligned} \tag{1}$$

Here u denotes the temperature of the material, and φ describes the phase where $\varphi = -1$ corresponds to the solid phase and $\varphi = 1$ to the liquid phase of the material. The solid and the liquid phase meet at the free boundary $\partial[\varphi = 1]$, and the free boundary is often called an interface. The first equation is the heat-diffusion equation taking into account the latent heat, which is released and absorbed when solidification and melting takes place, by the time derivative of the phase parameter. The second equation is the Gibbs-Thomson law which determines the melting temperature by a geometrical quantity of the free boundary namely by its mean curvature. In particular, the melting temperature is smaller for a solid ball than for a flat free boundary due to the fact that fewer solid particles are present in a neighbourhood of the free boundary which is curved towards the solid phase. Depending on the curvature, this enables supercooled liquids and superheated solids which is observed in experiments. Existence of solutions for the Stefan problem with Gibbs-Thomson law was proved in [21].

In dynamical processes, other geometric conditions can be imposed on the free boundary. For example, if the Gibbs-Thomson law holds only with a delay, we get the evolution law of the free boundary as the so-called mean curvature flow

$$V_N = -H_{\partial[\varphi=1]} + u. \tag{2}$$

Here the normal velocity V_N of the free boundary is given by the difference of its mean curvature and the forcing term u.

The numerical computation of evolutions of free boundaries may be difficult, in particular when topological changes of the free boundary occur. Therefore these problems are approximated by elliptic or parabolic regularisations as for example the Allen-Cahn-equation

$$\varepsilon \partial_t \varphi - \varepsilon \Delta \varphi + \frac{1}{\varepsilon} W'(\varphi) = c_W u, \tag{3}$$

where $W(t) := (t^2 - 1)^2$ is a double-well potential and c_W an appropriate positive constant. Here φ is an order parameter which describes the material as being liquid or solid if $\varphi \approx +1$ or $\varphi \approx -1$, respectively. The mean curvature flow (2) has a sharp free boundary evolving, whereas the Allen-Cahn-equation (3) has a transition region where φ is between ± 1. Rigorous proofs for the approximation of mean curvature flow by the Allen-Cahn-equation were given in several articles; we mention here only [1], [12] and [19]. When the underlying medium is anisotrop, this was proved in the articles [9], [10] and [11]. We give a brief sketch in §2.

The Allen-Cahn-equation appears also coupled with a parabolic equation

$$\begin{aligned} \varepsilon \partial_t \varphi - \varepsilon \Delta \varphi + \frac{1}{\varepsilon} W'(\varphi) + c_W u &= 0, \\ \partial_t u - \Delta u &= \varphi, \end{aligned} \tag{4}$$

or with an elliptic equation

$$\begin{aligned} \varepsilon \partial_t \varphi - \varepsilon \Delta \varphi + \frac{1}{\varepsilon} W'(\varphi) + c_W u &= 0, \\ -\Delta u &= \varphi. \end{aligned} \tag{5}$$

(5) is a model for thin magnetic films. These equations approximate a mean curvature flow with coupling, see [14] and [29] for (4), [3], [17] and [20] for (5), and §3 for a brief description.

Returning to the Stefan problem, it was shown in [4] by formal asymptotics that the phase field equations

$$\begin{aligned} \partial_t (u + \varphi) - \Delta u &= f, \\ \alpha(\varepsilon) \varepsilon \partial_t \varphi - \varepsilon \Delta \varphi + \frac{1}{\varepsilon} W'(\varphi) &= c_W u, \end{aligned} \tag{6}$$

approximate the Stefan problem with Gibbs-Thomson law (1), when $\alpha(\varepsilon) \to 0$ for $\varepsilon \to 0$. Rigorous proofs for the phase field equations are done only in special cases so far. The difficulty is that area of the free boundary can be lost throughout the limit procedure. In §4, we consider instead the quasi-stationary phase field equations

$$\begin{aligned} \partial_t (u + \varphi) - \Delta u &= f, \\ -\varepsilon \Delta \varphi + \frac{1}{\varepsilon} W'(\varphi) &= c_W u, \end{aligned} \tag{7}$$

for which area can be preserved in the limit procedure $\varepsilon \to 0$. Existence of solutions for the quasi-stationary phase field equations was proved in [23] und [26]. Further, the convergence of these solutions for $\varepsilon \to 0$ to solutions of the Stefan problem with Gibbs-Thomson law was established there.

2 The Allen-Cahn Equation

The evolution of a free boundary according to mean curvature flow reads

$$V_N = -H. \tag{8}$$

Although this is a parabolic equation, singularities and topological changes of the free boundary can occur in finite time which creates considerable difficulties for the numerical computation of this evolution. Therefore one considers parabolic regularisations such as the Allen-Cahn-equation

$$\varepsilon \partial_t \varphi_\varepsilon - \varepsilon \Delta \varphi_\varepsilon + \frac{1}{\varepsilon} W'(\varphi_\varepsilon) = 0, \tag{9}$$

where $W(t) := (t^2 - 1)^2$ is a double-well potential. W has exactly two minima at ± 1 which are both stable. (9) is a semi-linear parabolic equation, and, as the semi-linear term

$$\frac{1}{\varepsilon} W'$$

is monotonically increasing outside a compact interval, it has a unique solution global in time.

(9) has variational structure in the sense that it is the gradient flow

$$\varepsilon \partial_t \varphi_\varepsilon = -\delta F_\varepsilon(\varphi_\varepsilon)$$

of the functional

$$F_\varepsilon(\varphi) := \int \Big(\frac{\varepsilon}{2}|\nabla \varphi|^2 + \frac{1}{\varepsilon} W(\varphi)\Big).$$

As the minima of W are stable, the semi-linear term in (9) forces the solution to get close to ± 1. The solution is smooth and takes all intermediate values, but only on a small intermediate layer of width $O(\varepsilon)$. This intermediate layer substitutes the sharp free boundary we are going to approximate.

We do a heuristic calculation: Let (Γ_t) be a free boundary evolving smoothly and consider ist signed distance function $d(., \Gamma_t)$ to be positive in the interior of Γ_t and negative outside. We consider an approximate solution of (9)

$$\varphi_\varepsilon \approx U(\frac{d}{\varepsilon}) + \varepsilon v,$$

where $U : \mathbf{R} \to]-1,1[$ is the profile of the solution around the free boundary which has to be determined. We calculate

$$\begin{aligned}
\varepsilon\partial_t\varphi_\varepsilon &= U'(\frac{d}{\varepsilon})\partial_t d + O(\varepsilon^2),\\
\nabla\varphi_\varepsilon &= U'(\frac{d}{\varepsilon})\frac{\nabla d}{\varepsilon} + O(\varepsilon),\\
\varepsilon\Delta\varphi_\varepsilon &= U''(\frac{d}{\varepsilon})\frac{|\nabla d|^2}{\varepsilon} + U'(\frac{d}{\varepsilon})\Delta d + O(\varepsilon).
\end{aligned}$$

Plugging into (9), we obtain

$$\begin{aligned}
0 &= \varepsilon\partial_t\varphi_\varepsilon - \varepsilon\Delta\varphi_\varepsilon + \frac{1}{\varepsilon}W'(\varphi_\varepsilon) =\\
&= \frac{1}{\varepsilon}\Big(- U''(\frac{d}{\varepsilon})|\nabla d|^2 + W'(U(\frac{d}{\varepsilon}))\Big) + U'(\frac{d}{\varepsilon})\Big(\partial_t d - \Delta d\Big) + O(\varepsilon).
\end{aligned}$$

Since $|\nabla d| = 1$, we formally get

$$U'' = W'(U), \tag{10}$$

and

$$\partial_t d - \Delta d = 0 \quad \text{on } \Gamma_t. \tag{11}$$

Multiplying (10) by U', we get

$$|U'|^2 = 2W(U) + const,$$

and assuming the conditions

$$U(\pm\infty) = \pm 1, \quad U'(\pm) = 0, \quad U(0) = 0, \quad U' > 0,$$

we get

$$U' = \sqrt{2}(1 - U^2),$$

hence

$$U(t) = \tanh(\sqrt{2}t).$$

This is the profile how the intermediate layer approaches the free boundary.

To clarify (11), we observe that the normal at Γ_t is given by

$$\nu_{\Gamma_t} = \nabla d,$$

since d is the signed distance function of Γ_t. Now the mean curvature vector satisfies

$$\mathbf{H}_{\Gamma_t} = -div(\nu_{\Gamma_t})\,\nu_{\Gamma_t} = -\Delta d\,\nabla d.$$

The normal velocity of Γ_t is determined as

$$V_{\nu_t} = \mathbf{V}_{\nu_t} \nabla d = -\partial_t d,$$

when considering $x(t) \in \Gamma_t$, that is $d(x(t),t) = 0$. Therefore (11) describes the evolution of (Γ_t) as the mean curvature flow.

This procedure was rigorously carried out in [7], [12], [19] for the Brakke flow, [1], [2], and in the anisotropic case in [9], [10] and [11].

In the rigorous procedure, one defines

$$\varphi_\varepsilon^\pm = U(\frac{d \pm \dots}{\varepsilon}) + \varepsilon v_\varepsilon^\pm$$

with suitable modifications and shows that

$$\varepsilon \partial_t \varphi_\varepsilon^\pm - \varepsilon \Delta \varphi_\varepsilon^\pm + \frac{1}{\varepsilon} W'(\varphi_\varepsilon^\pm) \geq (\leq) 0$$

are sub- and supersoltuons. Then the comparison principle yields

$$\varphi_\varepsilon^- \leq \varphi_\varepsilon \leq \varphi_\varepsilon^+.$$

We see that

$$\varphi_\varepsilon^\pm(x,t) \to \begin{cases} +1 & \text{if } d(x,t) > 0\,, \\ -1 & \text{if } d(x,t) < 0\,, \end{cases}$$

hence

$$\varphi_\varepsilon(x,t) \to \begin{cases} +1 & \text{if } d(x,t) > 0\,, \\ -1 & \text{if } d(x,t) < 0\,, \end{cases} \tag{12}$$

and φ_ε approximate Γ_t.

3 Coupled Allen-Cahn Equation

In this section, we consider the Allen-Cahn equation when the forcing term is given by an elliptic equation

$$\begin{aligned} \varepsilon \partial_t \varphi_\varepsilon - \varepsilon \Delta \varphi_\varepsilon + \frac{1}{\varepsilon} W'(\varphi_\varepsilon) + c_W u_\varepsilon = 0, \\ -\Delta u_\varepsilon = \varphi_\varepsilon. \end{aligned} \tag{13}$$

This system is a model for thin magnetic films.

As the Allen-Cahn equation approximates mean curvature flow in the sense of (12), the limit problem for $\varepsilon \to 0$ is obtained as

$$\begin{aligned} V_N &= -H - u, \\ -\Delta u &= \begin{cases} 1 & \text{inside the free boundary,} \\ -1 & \text{outside the free boundary.} \end{cases} \end{aligned} \tag{14}$$

Indeed in [3], short-time existence of a smooth solution was proved for the free boundary problem (14), and in [20], convergence of (13) to (14) for $\varepsilon \to 0$ was shown in case (14) has a smooth solution.

Here we want to outline this limit procedure beyond the existence of smooth solutions for (14), thereby showing that (14) always has weak solutions global in time.

With parabolic coupling for

$$\begin{aligned} \varepsilon\partial_t\varphi_\varepsilon - \varepsilon\Delta\varphi_\varepsilon + \tfrac{1}{\varepsilon}W'(\varphi_\varepsilon) + c_W u_\varepsilon &= 0, \\ \partial_t u_\varepsilon - \Delta u_\varepsilon &= \varphi_\varepsilon, \end{aligned} \tag{15}$$

this was done in [14] and [29]. The difficulty of (13) when compared to (15) is proving compactness for u_ε as the time derivative $\partial_t u_\varepsilon$ is missing. Moreover, (13) admits no comparison principle. Instead,

$$E(\varphi, u) := \int \Big(\frac{\varepsilon}{2}|\nabla\varphi|^2 + \frac{1}{\varepsilon}W(\varphi) + |\nabla u|^2\Big)$$

is a Ljapunov-functional for (13). Putting

$$\psi_\varepsilon := H(\varphi_\varepsilon) := \frac{3}{2}\int\limits_0^{\varphi_\varepsilon} \sqrt{W},$$

in particular $H(\pm 1) = \pm 1$, the following estimates are well-known from literature

$$\begin{aligned} &|\varphi_\varepsilon| \le 1 + C\varepsilon, \\ &\varepsilon\int\limits_0^T\int |\partial_t\varphi_\varepsilon|^2, \int |\nabla\psi_\varepsilon| \le C, \\ &\int\limits_{t_1}^{t_2}\int |\partial_t\psi_\varepsilon| \le C|t_2 - t_1|^{1/2}. \end{aligned} \tag{16}$$

As

$$\| \varphi_\varepsilon - \psi_\varepsilon \|_{L^\infty(L^1)} \le \omega(\varepsilon) \tag{17}$$

with $\omega(\varepsilon) \to 0$ for $\varepsilon \to 0$, we get

$$\varphi_\varepsilon, \psi_\varepsilon \to \varphi \quad \text{pointwise almost everywhere} \tag{18}$$

where φ takes only the values ± 1 .

To pass to the limit in the Allen-Cahn equation, one has to establish uniform convergence for u_ε . As the elliptic equation is regularizing, regularity in space is immediate

$$\| u_\varepsilon \|_{L^\infty(W^{2,p})} \leq C. \tag{19}$$

The crucial point is to obtain estimates for the time differences. From (16) and (17), we see

$$\| \psi_\varepsilon(t) - \psi_\varepsilon(s) \|_{L^1} \leq \int_s^t \int |\partial_t \psi_\varepsilon| \leq C|t-s|^{1/2},$$

and

$$\| \varphi_\varepsilon(t) - \varphi_\varepsilon(s) \|_{L^1} \leq C|t-s|^{1/2} + \omega(\varepsilon). \tag{20}$$

Further we get

$$\| \varphi_\varepsilon(t) - \varphi_\varepsilon(s) \|_{L^\infty(L^1)} \leq \int_s^t \int |\partial_t \varphi_\varepsilon| \leq$$

$$\leq C|t-s|^{1/2} \Big(\int_s^t \int |\partial_t \varphi_\varepsilon|^2 \Big)^{1/2} \leq C\varepsilon^{-1/2}|t-s|^{1/2}.$$

Combined with (20), this yields

$$\| \varphi_\varepsilon(t) - \varphi_\varepsilon(s) \|_{L^1} \leq \omega(|t-s|),$$

hence

$$\| u_\varepsilon(t) - u_\varepsilon(s) \|_{C^{1,\alpha}} \leq \omega(|t-s|), \tag{21}$$

and $\{u_\varepsilon\}_\varepsilon$ is compact in C^0 in time and space. Therefore we can pass to the limit and get

$$\begin{aligned} u_\varepsilon &\to u \quad \text{uniformly}, \\ -\Delta u &= \varphi \end{aligned} \tag{22}$$

and

$$u \in C^0, \quad |\nabla u| \leq C. \tag{23}$$

Now, we use $u + h, |h| < \delta$ as a driving force in the Allen-Cahn equation

$$\varepsilon \partial_t \varphi_\varepsilon^h - \varepsilon \Delta \varphi_\varepsilon^h + \frac{1}{\varepsilon} W'(\varphi_\varepsilon^h) + c_W (u - h) = 0.$$

Clearly by uniform convergence

$$u - \delta \leq u_\varepsilon \leq u + \delta \quad \text{for } \varepsilon \leq \varepsilon(\delta),$$

hence

$$\varphi_\varepsilon^{-\delta} \leq \varphi_\varepsilon \leq \varphi_\varepsilon^\delta. \tag{24}$$

By the previous section (12), we see

$$\varphi_\varepsilon^h(x,t) \to \begin{cases} +1 & \text{if } d^h(x,t) > 0\,, \\ -1 & \text{if } d^h(x,t) < 0\,, \end{cases}$$

where d^h denots the signed distance function for the free boundary evolving by

$$V_N = -H - u + h.$$

As $d^h \to d := d^0$, we see form (24)

$$\varphi_\varepsilon(x,t) \to \begin{cases} +1 & \text{if } d(x,t) > 0\,, \\ -1 & \text{if } d(x,t) < 0\,, \end{cases}$$

and (14) is obtained from (18) and (22).

4 The Quasi-Stationary Phase Field Equations

The Stefan problem is a mathematical model that describes the melting and solidification of materials. It is a free boundary problem for parabolic equations. The temperature u has to satisfy after normalizing constants to one and the melting temperature to 0 , a heat-diffusion equation with a source term

$$\partial_t u - \Delta u = f \tag{25}$$

in the domain where $u > 0$ and $u < 0$ separately, that is in the domain where the material is liquid or solid. The amount of material which melts or solidifies is proportional to the heat flux through the free boundary where the liquid and solid phase meet. More precisely, the normal velocity is determined by

$$V_\nu = \frac{1}{2} \Big[\partial_\nu u \Big] \tag{26}$$

where ν denotes the unit normal at the free boundary and $[\, \partial_\nu u]$ the jump of the normal derivatives of u across the free boundary. Using the phase parameter φ

with $\varphi = 1$ in the liquid phase and $\varphi = -1$ in the solid phase, we can summarize (25) and (26) in one equation

$$\partial_t(u+\varphi) - \Delta u = f. \tag{27}$$

Since the free boundary $\partial[\varphi = 1]$ is determined throughout the evolution, we have to add a boundary condition on the free boundary.

As the free boundary separates the liquid and the solid phase, classically one requires the temperature on the free boundary to be the melting temperature, which we set equal to 0 for simplicity, that is

$$u = 0 \quad \text{on } \partial[\varphi = 1]. \tag{28}$$

We get the following relations for the phase and the temperature

$$\begin{aligned} &[u > 0] \subseteq [\varphi = 1] \subseteq [u \geq 0], \\ &[u < 0] \subseteq [\varphi = -1] \subseteq [u \leq 0]. \end{aligned} \tag{29}$$

The system (27), (28) is called the classical Stefan problem. Existence and uniqueness was for example proved by Friedman in [13]. We add that in the classical Stefan problem the phase parameter can also take values between ± 1. This is called mushy region.

In the classical Stefan problem, the material is in the liquid or solid phase if the temperature is above or below the melting temperature, respectively. This seems to be reasonable. But in physical experiments, one can observe a liquid with a temperature below the melting point. This phenomenon is called supercooling; the analogous situation for a solid is superheating. Superheating and supercooling appear in some mathematical models which include surface tension. For example, let u be a given temperature distribution. Now consider the functional

$$F(\varphi) := \int \Big(|\nabla \varphi| - c_W u \varphi \Big) \tag{30}$$

for φ taking only the values ± 1. $F(\varphi)$ represents the free energy of the liquid-solid configuration corresponding to φ with the given temperature distribution u. The physically acceptable stationary liquid-solid configurations for the given temperature distribution u correspond to the absolute minima of F. This penalizes the free boundary having large area. Extremals of F satisfy under suitable regularity conditions the Gibbs-Thomson law on the free boundary

$$u = H_{\partial[\varphi=1]} \tag{31}$$

where $H_{\partial[\varphi=1]} = \mathbf{H}_{\partial[\varphi=1]} \nu_{\partial[\varphi=1]}$ and $\nu_{\partial[\varphi=1]}$ is the normal at $\partial[\varphi = 1]$ pointing to the inside of $[\varphi = 1]$. We see that the melting temperature is smaller for a solid ball than for a flat free boundary. This is justified by the fact that fewer solid particles are present in a neighbourhood of the free boundary when its is curved towards the solid phase. Existence of weak solutions for the Stefan problem with Gibbs-Thomson law (27), (31) was proved by Luckhaus in [21].

Again as it may be difficult to calculate the free boundary in the Stefan problem numerically, one is looking for parabolic regularisations. As mentioned in the introduction, Caginalp showed in [4] using formal asymptotics that the phase field equations (6) approximate the Stefan problem with Gibbs-Thomson law, similarly for the quasi-stationary phase field equations

$$\begin{aligned} \partial_t(u+\varphi) - \Delta u &= f, \\ -\varepsilon\Delta\varphi + \tfrac{1}{\varepsilon}W'(\varphi) &= c_W u. \end{aligned} \tag{32}$$

Actually, the convergence procedure to the Stefan problem with Gibbs-Thomson law for $\varepsilon \to 0$ can easily be deduced from [22] when no energy is lost in the limit procedure which can be satisfied for (32).

The difficult part for the quasi-stationary phase field equations is proving existence of solutions. For $n \leq 3$, this was done for Dirichlet data by Plotnikov and Starovoitov in [23] and for Neumann data by the present author in [26]. The difficulty is that since the time derivative $\partial_t\varphi$ is missing in the second equation, compactness of φ is not immediate when (32) is approximated.

In the following, we sketch the proofs of [23] and [26]. Abbreviating $v := u+\varphi$, we rewrite (32) in the form

$$\begin{aligned} \partial_t v - \Delta u &= f, \\ -\varepsilon\Delta\varphi + \frac{1}{\varepsilon}W'(\varphi) + c_W\varphi &= c_W v. \end{aligned} \tag{33}$$

with boundary conditions

$$\begin{aligned} u = u_D \quad \text{or} \quad \partial_\nu u &= 0, \\ \partial_\nu\varphi &= 0. \end{aligned} \tag{34}$$

The second equation in (33) is the Euler-Lagrange differential equation of the functional

$$G_\varepsilon(\varphi) := \int \frac{\varepsilon}{2}|\nabla\varphi|^2 + \frac{1}{\varepsilon}W(\varphi) + \frac{c_W}{2}|\varphi|^2$$

and can be written in the form

$$G'(\varphi) = c_W v. \tag{35}$$

Now we set up a time-discrete approximation which satisfies

$$\begin{aligned} \partial_t v_h - \Delta u_h &= f, \\ G'(\varphi_h) &= c_W v_h(jh) \quad \text{on }](j-1)h, jh[, \\ \varphi_h(t) &= \varphi_{h,j} \quad\quad\;\; \text{on }](j-1)h, jh[\end{aligned} \tag{36}$$

and boundary conditions

$$u_h = u_D \quad \text{or} \quad \partial_\nu u_h = 0,$$
$$\partial_\nu \varphi_h = 0.$$

One easily obtains the following estimates

$$\| u_h \|_{L^\infty(L^2)\cap L^2(W^{1,2})}, \| v_h \|_{L^\infty(L^2)\cap L^2(W^{1,2})},$$
$$\| \varphi_h \|_{L^\infty(W^{2,2})}, \| \partial_t v_h \|_{L^2((W_0^{1,2})^*)} \leq C.$$

With this we can pass to the limit

$$u_h, \varphi_h \to u, \varphi \quad \text{weakly in } L^2(W^{1,2}),$$
$$v_h \to v \quad \text{weakly in } L^2(W^{1,2}) \text{ and strongly in } L^2(L^2).$$

From (36)

$$\partial_t v - \Delta u = f,$$
$$G'(\varphi_h) = c_W v + \tilde{v}_h$$

with

$$\| \tilde{v}_h \|_{L^\infty((W_0^{1,2})^*)} \to 0. \tag{37}$$

First we consider the case of Dirichlet data for u. We fix t such that

$$v_h(t) \to v(t) \quad \text{strongly in } L^2,$$
$$\| \varphi_h(t) \|_{W^{2,2}} \leq C.$$

Therefore $\{\varphi_h(t)\}_h$ is compact in $W^{1,2}$ and any sequence $h_k \to 0$ has a converging subsequence. We consider

$$\varphi_{h_k^i}(t) \to \varphi_i \quad \text{strongly in } W^{1,2} \text{ for } i = 1, 2.$$

Now for u, we have only the bound

$$\| u_h \|_{L^2(W^{1,2})} \leq C.$$

To simplify the procedure, we assume

$$\| u_h(t) \|_{W^{1,2}} \leq C. \tag{38}$$

Then for a further subsequence

$$u_{h_k^i}(t) \to u_i \quad \text{weakly in } W^{1,2}$$

and

$$u_2 - u_1 \in W_0^{1,2}$$

as we are considering Dirichlet data for u . Next

$$\begin{aligned}\varphi_2 - \varphi_1 &= \lim_k (\varphi_{h_k^2}(t) - \varphi_{h_k^1}(t)) = \\ &= \lim_k (v_{h_k^2}(t) - v_{h_k^1}(t)) - \lim_k (u_{h_k^2}(t) - u_{h_k^1}(t)) = \\ &= -(u_2 - u_1) \in W_0^{1,2}.\end{aligned}$$

From (37)

$$-\varepsilon \Delta \varphi_i + \frac{1}{\varepsilon} W'(\varphi_i) + c_W \varphi_i = c_W v(t),$$
$$\partial_\nu \varphi_i = 0 \quad \text{on the boundary},$$

hence

$$-\varepsilon \Delta(\varphi_2 - \varphi_1) + b(x)(\varphi_2 - \varphi_1) = 0,$$
$$\partial_\nu(\varphi_2 - \varphi_1) = 0, \varphi_2 - \varphi_1 = 0 \quad \text{on the boundary},$$

which implies

$$\varphi_2 \equiv \varphi_1$$

and

$$\varphi_h(t) \text{ converges pointwise almost everywhere in } W^{1,2}.$$

By (37)

$$G'(\varphi) = c_W v,$$

and (u, φ, v) is a solution of (33) with Dirichlet data in (34). To circumvent the simplified assumption in (38), see [23].

For Neumann boundary data for u , we proceed differently. We investigate the equation

$$G'(\varphi) = c_W v. \tag{39}$$

or likewise

$$-\varepsilon \Delta \varphi + \frac{1}{\varepsilon} W'(\varphi) + c_W \varphi = c_W v.$$

Since W' is not monotone, the solution of this equation is not unique. Actually, this equation can have infinitely many solutions, see [6], [8] and [15].

On the other hand, the number of global minimizers of

$$\varphi \mapsto G(\varphi) - \int c_W v \varphi$$

is finite, see [5]. We outline this briefly. For a global minimizer φ_0, we have

$$G''(\varphi_0) \geq 0,$$

where $G''(\varphi_0) \in L(W^{1,2} \times W^{1,2}, \mathbf{R})$ or likewise $G''(\varphi_0) : W^{1,2} \to (W^{1,2})^* \cong W^{1,2}$. If

$$ker\ G''(\varphi_0) = \{0\},$$

then φ_0 is an isolated solution of (39).

If $ker\ G''(\varphi_0) \neq \{0\}$, zero is the smallest eigenvalue of $G''(\varphi_0)$, as $G''(\varphi_0) \geq 0$, hence by Harnack inequality, it is simple and

$$ker\ G''(\varphi_0) = span\{\psi_0\},$$
$$\psi_0 > 0,$$

assuming that the domain under consideration is connected. Let

$$W^{1,2} = span\{\psi_0\} \oplus Y$$

be the orthogonal decomposition and

$$P_Y : W^{1,2} \to Y$$

be the orthogonal projection. Then writing $\varphi = \varphi_0 + t\psi_0 + y$, (39) is equivalent to

$$\begin{aligned} P_Y(G'(\varphi_0 + t\psi_0 + y) - c_W v) &= 0, \\ \langle G'(\varphi_0 + t\psi_0 + y) - c_W v, \psi_0 \rangle &= 0. \end{aligned} \tag{40}$$

Since $P_Y G''(\varphi_0)|_Y : Y \to Y$ is an isomorphism, we can resolve the first equation in (40) locally

$$P_Y(G'(\varphi_0 + t\psi_0 + y(t)) - c_W v) = 0,$$
$$y'(0) = 0,$$

and put $\gamma :]-\delta, \delta[\to \mathbf{R}$

$$\gamma(t) := \langle G'(\varphi_0 + t\psi_0 + y(t)) - c_W v, \psi_0 \rangle.$$

Therefore the solutions of (39) close to φ_0 correspond bijectively to the zeros of γ, in particular

$$\gamma(0) = 0.$$

Since W is a polynomial, all functions involved, in particular γ are real-analytic. Then either

$$\gamma(t) \neq 0 \text{ for } t \neq 0,$$

or

$$\gamma \equiv 0$$

in a neighbourhood of 0 .

This yields that φ_0 is either an isolated solution of (39) or belongs to a 1–dimensional manifold without boundary which consists only of minimizers as can easily be shown.

Since the solution set of (39) is compact, the set of global minimizers consists of finitely many isolated points and finitely many spheres S^1 . The finiteness result is proved, when we exclude these spheres. We assume on contrary that there is one such sphere

$$C \subseteq \{ \text{ global minimizers } \}.$$

Then we evaluate these minimizers at a point x_0 that is

$$\Gamma : C \to \mathbf{R}, \quad \Gamma(\varphi) := \varphi(x_0).$$

This map is continuous, hence takes a maximum on the compact set C , say

$$\Gamma(\varphi_0) \geq \Gamma.$$

Considering the structure of C close to φ_0 , we see as shown above that

$$\varphi_0 + t\psi_0 + y(t) \in C \quad \text{for } |t| < \delta.$$

Then

$$\frac{d}{dt}\Gamma(\varphi_0 + t\psi_0 + y(t))|_{t=0} = \psi_0(x_0) > 0,$$

as $y'(0) = 0$, contradicting that φ_0 is a maximum for Γ .

We see that there are only finitely many global minimizers, and locally under perturbation of v , the solutions of (39) lie on a one parameter family.

Now small perturbations of v can induce jumps of φ . On the other hand, if one chooses φ_h in the time-discrete approximation in such a way that

$$\varphi \mapsto G(\varphi) - \int c_W v \varphi$$

decreases, the jumps of φ are in the same directions along this one parameter family, to the left or the right, and these jumps add up. Only a larger perturbation of v_h can change the direction of the jumps. Since $\partial_t v_h$ is bounded in $L^2((W_0^{1,2})^*)$, these larger perturbations of v_h are controlled.

Therefore we can estimate the time differences of φ_h , get

$$\varphi_h \to \varphi \quad \text{strongly in } L^2(W^{1,2}),$$

and (u, φ, v) is a solution of (33) with Neumann data in (34).

The whole argument can be found in [26].

References

1. Barles, G., Soner, H.M., Souganidis, P.E. (1993): Front propagation and phase field theory. SIAM J. Control Optim., **31**, 439–469
2. Barles, G., Souganidis, P.E. (1998): A New Approach to Front Propagation Problems: Theory and Applications. Archive for Rational Mechanics and Analysis, **141**, 237–296
3. Bonami, A., Hilhorst, D., Logak, E. (2000): Modified motion by mean curvature: local existence and uniqueness and qualitative properties. Differential and Integral Equations, **13**, no. 10-12, 1371–1392
4. Caginalp, G. (1986): An analysis of a phase-field model of a free boundary. Archive for Rational Mechanics and Analysis, **92**, 205–245
5. Canarius, T., Schätzle, R. (1998): Finiteness and positivity results for global minimizers of a semilinear elliptic problem. Journal of Differential Equations, **148**, 212–229
6. Canarius, T., Schätzle, R. (1999): Multiple solutions for a semilinear elliptic problem. To appear in Nonlinear Analysis TMA
7. Chen, X. (1992): Generation and Propagation of interface in reaction-diffusion equations. Journal of Differential Equations, **96**, 116–141
8. Dancer, E.N. (1983): Breaking of Symmetries for Forced Equations. Mathematische Annalen, **262**, 473–486
9. Elliott, C.M., Paolini, M., Schätzle, R.: (1996): Interface estimates for the fully anisotropic Allen-Cahn equation and anisotropic mean-curvature flow. Math. Models Methods Appl. Sci., **6**, No. 8, 1103–1118
10. Elliott, C.M., Schätzle, R. (1996): The limit of the anisotropic double-obstacle Allen-Cahn equation. Proceedings of the Royal Society of Edinburgh, **126A**, 1217–1234
11. Elliott, C.M., Schätzle, R. (1997): The limit of the fully anisotropic double-obstacle Allen-Cahn equation in the non-smooth case. SIAM Journal on Mathematical Analysis, **Vol. 28**, No. 2, 274–303
12. Evans, L.C., Soner, H.M., Souganidis, P.E. (1992): Phase transitions and generalized motion by mean curvature. Communications on Pure and Applied Mathematics, **Vol. XLV**, 1097–1123
13. Friedman, A. (1968): The Stefan problem in several space variables. Transactions of the American Mathematical Society, **133**, 51–87
14. Giga, Y., Goto, S., Ishii, H. (1992): Global existence of weak solutions for interface equations coupled with diffusion equations. SIAM Journal on Mathematical Analysis, **Vol. 23**, No. 4, 821–835
15. Gurtin, M., Matano, H. (1988): On the structure of equilibrium phase transitions within the gradient theory of fluids. Quarterly of Applied Mathematics, **Vol. XLVI**, No. 2, 301–317

16. Henry, M., Hilhorst, D., Schätzle, R. (2000): Convergence to a viscosity solution for an advection-reaction-diffusion equation arising from a chemotaxis-growth model. Hiroshima Mathematical Journal
17. Hilhorst, D., Logak, E., Schätzle, R. (2000): Global existence for a nonlocal mean curvature flow as a limit of a parabolic-elliptic phase transition model. Interfaces and Free Boundaries, **2**, 267–282
18. Hilhorst, D., Peletier, L.A., Schätzle, R. (2000): $\Gamma-$limit for the Extended Fisher-Kolmogorov equation. Preprint Universität Leiden No. MI 2000-16 May, to appear in Proceedings of the Royal Society of Edinburgh
19. Ilmanen, T. (1993): Convergence of the Allen-Cahn equation to Brakke's motion by mean curvature. Journal of Differential Geometry, **38**, No. 2, 417–461
20. Logak, E. (1996): Singular limit of reaction-diffusion systems and modified motion by mean curvature. preprint
21. Luckhaus, S. (1991): The Stefan Problem with Gibbs Thomson law. Sezione di Analisi Matematica e Probabilitita, Universita di Pisa, **2.75 (591)**
22. Luckhaus, S., Modica, L. (1989): The Gibbs Thompson Relation within the Gradient Theory of Phase Transitions. Archive for Rational Mechanics and Analysis, **107**, 71–83
23. Plotnikov, P.I., Starovoitov, V.N. (1993): Stefan Problem with Surface Tension as a Limit of the Phase Field Model. Differential Equations, **29**, No. 3, 395–404
24. Reshetnyak, Y.G. (1968): Weak convergence of completely additive vector functions on a set. Siberian Mathematical Journal, **9**, 1039–1045
25. Schätzle, R. (1997): A counterexample for an approximation of the Gibbs-Thomson law. Advances in Mathematical Sciences and Applications, **Vol. 7**, No. 1, 25–36
26. Schätzle, R. (2000): The quasi-stationary phase field equations with Neumann boundary conditions. Journal of Differential Equations, **162**, No. 2, 473–503
27. Schätzle, R. (2001): Hypersurfaces with mean curvature given an ambient Sobolev function. Preprint Universität Freiburg Nr. 10/1999, habilitation thesis. To appear in Journal of Differential Geometry
28. Schätzle, R. (2002): A geometric regularity estimate via fully nonlinear ellitptic equations. International Mathematical Series, Vol. 1, Nonlinear Problems of Mathematical Physics and Related Topics, Volumes to honour Professor O.A.Ladyzenskaja to her 80th birthday.
29. Souganidis, P.E., Soravia, P. (1996): Phase field theory for FitzHugh-Nagumo type systems. SIAM Journal on Mathematical Analysis, **Vol. 27**, 1341–1359

On Generalized Mean Curvature Flow in Surface Processing

Ulrich Clarenz[1], Gerhard Dziuk[2], and Martin Rumpf[1]

[1] Institut für Mathematik, Gerhard-Mercator-Universität Duisburg, Lotharstraße 65, 47048 Duisburg, {clarenz, rumpf}@math.uni-duisburg.de

[2] Institut für angewandte Mathematik, Albert Ludwigs Universität Freiburg, Hermann-Herder-Straße 10, 79104 Freiburg, gerd@mathematik.uni-freiburg.de

1 Introduction

Geometric evolution problems for curves and surfaces and especially curvature flow problems are an exciting and already classical mathematical research field. They lead to interesting systems of nonlinear partial differential equations and allow the appropriate mathematical modeling of physical processes such as material interface propagation, fluid free boundary motion, crystal growth.

On the other hand, curves and surfaces are essential objects in computer aided geometric design and computer graphics. Here, issues are fairing, modeling, deformation, and motion. Constructive and more explicit approaches based for instance on splines are nowadays already classical tools. More recently geometric evolution problems and variational approaches have entered this research field as well and have turned out to be powerful tools. Their strength relies on the possibility to mathematically model problems on a continuous level, initially not worrying about the discretization. Furthermore, the resulting models come along with natural Galerkin discretizations which lead to consistent and simple algorithms mostly based on widespread simplicial grids approximating the continuous curves and surfaces [12, 26, 27, 36].

Not very surprisingly the resulting models mostly lead to similar geometric evolution problems as known from the above physical applications. Again systems of nonlinear partial differential equations control transport and diffusion phenomena in curve and surface processing applications.

Here, we will report on recent results concerning generalized mean curvature motion and its application in curve and surface fairing.

The processing of detailed triangulated surfaces is a fundamental topic in computer aided geometric design and in computer graphics. Nowadays, various such surfaces are delivered from different measurement techniques [9] or derived from two- or three dimensional data sets [28]. Recent laser scanning technology for example enables very fine triangulations of real world surfaces and sculptures. Also from medical image generation methods, such as CT and MRI devices or 3D ultrasound, certain surfaces of interest can be extracted - frequently in triangulated form - at a

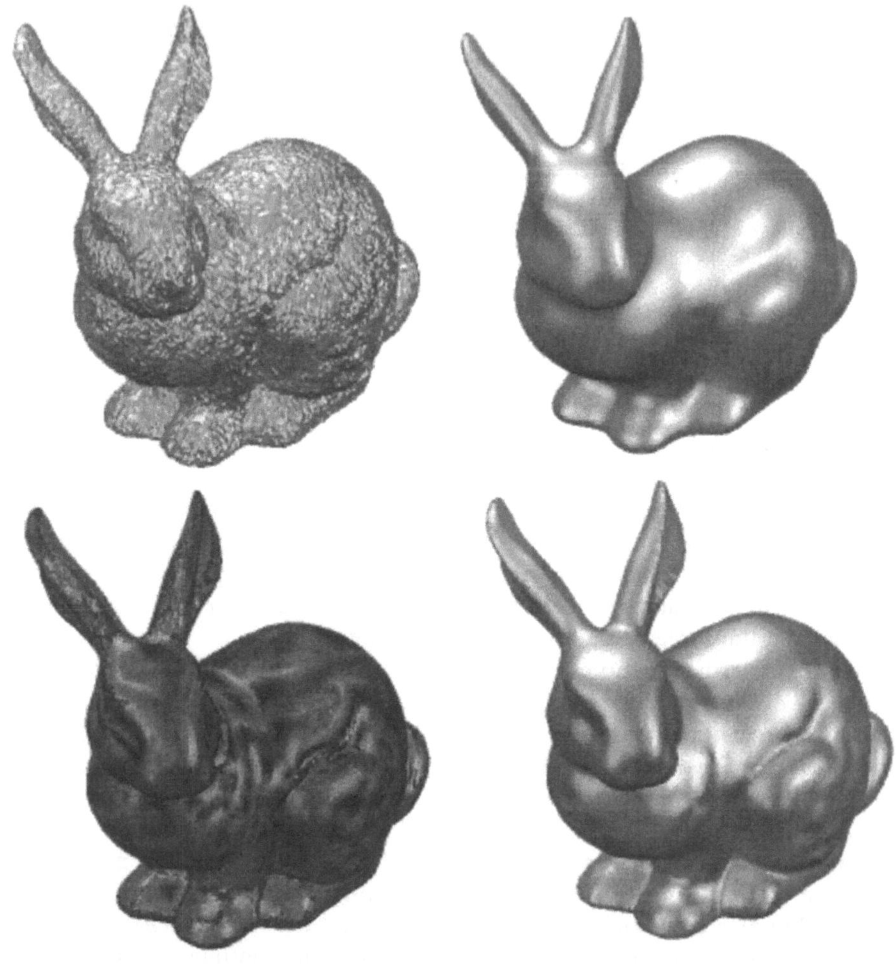

Fig. 1. *A noisy initial surface (top left) is evolved by discrete mean curvature flow (top right) and by the new anisotropic diffusion method (bottom right). Furthermore for the latter surface the dominant principal curvature - on which the diffusion tensor depends - is color coded (bottom left). The snapshots are taken at the same time-steps. (cf. color plate 8, Page 670)*

high resolution for further post processing and analysis. These surfaces are usually characterized by interesting features, such as edges and corners. On the other hand, they are typically disturbed by noise, which is often due to local measurement errors.

The aim of this paper is to discuss methods which allow the *fairing of discrete curves and surfaces*. Additionally they are able to *retain and even enhance important features such as surface edges and corners*. Figure 1 shows the performance of the basic method and compares it with a simple smoothing by mean curvature flow, the appropriate geometric "Gaussian" smoothing filter. Two alternative approaches are presented and compared:

- An anisotropic geometric diffusion problem is presented which respects edge features and in addition incorporates tangential smoothing along noisy edges.
- An anisotropic energy is formulated, where the local energy integrand reflects detected geometric features.

Both methods have in common that they are based on a local surface classification and the actual surface evolution which locally depends on the results of the classification. Furthermore they both lead to anisotropic curvature motion problems. A scale of successively smoothed representations is generated, where time is considered as the scale parameter. Such approaches have been originally developed for image processing purposes. Here we generalize them to curve and surface processing.

The methods differ with respect to the type of local classification and the model starting point. On the one hand, we model an anisotropic diffusion tensor in the geometric evolution problem. On the other hand, we define an energy density and ask for the gradient flow with respect to the corresponding energy. Furthermore, appropriate finite element algorithms based on the schemes in [15, 17] are proposed to discretize the continuous evolution problems. For convergence results in 1D we refer to [16, 17]. For 2D graphs in $\mathbb{R}^3$ an error analysis is contained in [11, 10].

The paper is organized as follows. First, in Section 2 we will discuss the background work on surface fairing by geometric smoothing and on image processing. In the following paragraph we collect some important notation. In Section 3 we introduce the different approaches to anisotropic curvature motion, i.e., the parabolic problem for an anisotropic Laplace Beltrami operator and the gradient flow for a spatially varying crystalline energy density. Their principal difference is discussed in Section 3.3. In Section 4 we derive finite element discretizations for both types of methods. Furthermore, in Section 5 the local classification types are introduced and in Section 6 the different classifications are incorporated in the geometric evolution approaches to steer the local evolution.

The volume enclosed by a surface without boundary is an important characteristic, which we should try to preserve during processing. Section 6 ends with a theorem that enables us to define generalized resp. anisotropic mean curvature motion keeping fixed the volume enclosed by the evolving surface.

2 Image and Surface Processing Background

In physics, diffusion is known as a process that equilibrates spatial variations in concentration. If we consider some initial noisy concentration or image intensity ρ_0 on a domain $\Omega \subset \mathbb{R}^2$ and seek solutions of the linear heat equation

$$\partial_t \rho - \Delta \rho = 0 \tag{1}$$

with initial data ρ_0 and natural boundary conditions on $\partial\Omega$, we obtain a scale of successively smoothed concentrations $\{\rho(t)\}_{t\in\mathbb{R}^+}$. For $\Omega = \mathbb{R}^2$ the solution of this parabolic problem coincides with the filtering of the initial data using a Gaussian filter $G_\sigma^\infty(x) = (2\pi\sigma^2)^{-1}e^{-x^2/(2\sigma^2)}$ of width or standard deviation σ, i.e., $\rho(\sigma^2/2) =$

$\overline{G_\sigma^\infty} * \rho_0$. Concerning the smoothing of disturbed surface geometries one may ask for analogues strategies. The geometrical counterpart of the Euclidian Laplacian Δ on smooth surfaces is the Laplace Beltrami operator $\Delta_{\mathcal{M}}$ [14, 5]. Thus, one obtains the geometric diffusion $\partial_t x = \Delta_{\mathcal{M}(t)} x$ for the coordinates x on the corresponding family of surfaces $\mathcal{M}(t)$.
From differential geometry [13] we know that the mean-curvature vector hn equals the Laplace Beltrami operator applied to the immersion x of a surface $\mathcal{M}$:

$$h(x)n(x) = -\Delta_{\mathcal{M}} x. \tag{2}$$

Thus geometric diffusion is equivalent to mean curvature motion (MCM)

$$\partial_t x = -h(x)n(x)\ , \tag{3}$$

where $h(x)$ is the corresponding mean curvature (here defined as the sum of the two principal curvatures), and $n(x)$ is the normal on the surface at point x. In dimensions higher than two, singularities may occur in the evolution. Generalized - so called viscosity solutions - can be defined in terms of a level set formulation

$$\partial_t \phi - |\nabla \phi| \mathrm{div}\left(\frac{\nabla \phi}{|\nabla \phi|}\right) = 0\,.$$

Existence in this context has been proved by Evans and Spruck [18]. The mean curvature h is known to be the first variation of the surface area $\int_{\mathcal{M}} dA$. We obtain for the area $\mathrm{Ar}(\omega(t))$ of a subset $\omega(t)$ of a smooth surface $\mathcal{M}$ undergoing the MCM evolution (cf. [23]) $\frac{d}{dt}\mathrm{Ar}(\omega(t)) = -\int\limits_{\omega(t)} H^2 dA$. This is one indication for the strong regularizing effect of MCM.

In the context of Finsler geometry MCM can be generalized considering a 1-homogeneous convex scalar function $\gamma(\cdot)$ and a weighted area $\int_{\mathcal{M}} \gamma(N) dA$ depending on the surface orientation. As its first variation we obtain the weighted mean curvature h_γ [38, 6]. The corresponding anisotropic curvature flow has been studied for instance by Bellettini and Paolini [2].

On triangulated surfaces as they frequently appear in geometric modeling and computer graphics applications, several authors introduced discretized MCM operators [36, 26, 27, 22, 12]

Unfortunately MCM doesn't only decrease the geometric noise due to imprecise measurement but also smoothes out geometric features such as edges and corners of the surface. Hence, we seek models which improve a simple high pass filtering.

In image processing, Perona and Malik [31] proposed a nonlinear diffusion method, which modifies the diffusion coefficient at edges. Edges are indicated by steep intensity gradients. For a given initial image ρ_0 they considered the evolution problem

$$\partial_t \rho - \mathrm{div}\left(G\left(\frac{\|\nabla \rho\|}{\lambda}\right)\nabla \rho\right) = 0 \tag{4}$$

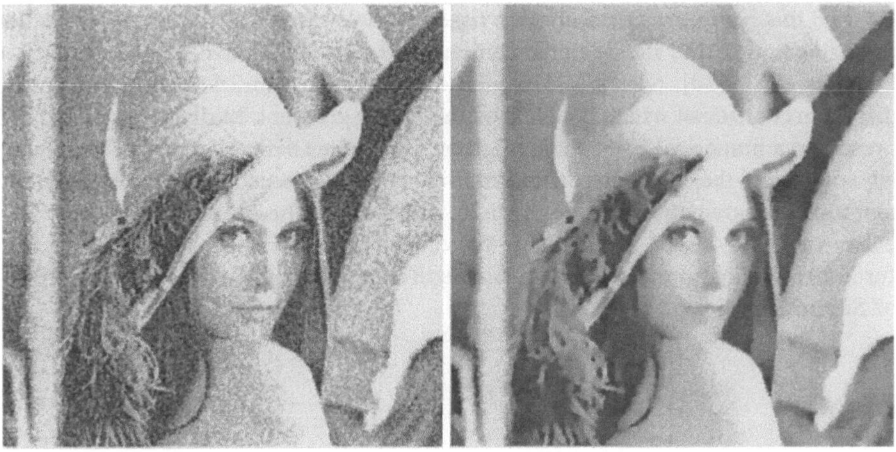

Fig. 2. *Isotropic Perona-Malik diffusion (right) is applied to a noisy initial image (left).*

for some parameter $\lambda \in \mathbb{R}^+$. For increasing time t - the scale parameter - the original image at the initial time is now successfully smoothed and image patterns are coarsened. But simultaneously edges are enhanced if one chooses a diffusion coefficient $G(.)$ which suppresses diffusion in areas of high gradients (cf. Fig. 2). A suitable choice for G is

$$G(s) = \left(1 + s^2\right)^{-1} . \tag{5}$$

Thus edges are classified by the involved parameter λ.

Kawohl and Kutev [24] gave a detailed analysis of the diffusion types in this method. Unfortunately the above original Perona and Malik model is still ill-posed because there is a true backward diffusion in areas of large gradients. Catté et al. [4] proposed a regularization method where the diffusion coefficient is no longer evaluated on the exact intensity gradient. Instead they suggested to consider the gradient evaluation on a prefiltered image.

Weickert [37] improved this method taking into account anisotropic diffusion, where the Perona Malik type diffusion is concentrated in one direction, for instance the gradient direction of a prefiltered image. This leads to an additional tangential smoothing along edges and amplifies intensity correlations along lines. Kimmel [25] generalized the scale space approach for planar images to the case of images mapped on surfaces.

Unfortunately, none of the above models is invariant under gray value transformations. In the axiomatic work by Alvarez et al. [1] general nonlinear evolution problems were derived from a set of axioms. Especially including the axiom of gray value invariance they end up with a curvature evolution model, i.e.,

$$\partial_t \phi - |\nabla \phi| \left(t \operatorname{div} \left(\frac{\nabla \phi}{|\nabla \phi|} \right) \right)^{\frac{1}{3}} = 0 .$$

Beyond this result, curvature motion based models proved as successful ingredients in segmentation and image enhancement methods. They have been considered by Pauwels et al. [30]. Sapiro [34] proposed a modification of MCM considering a diffusion coefficient which depends on the image gradient. Malladi and Sethian [29] presented a numerical level set method on 2D images called "min/max" flow which also considers the curvature evolution. In [7] and [32] anisotropic curvature motion methods on parametric surfaces (which will partially be revisited here) and on the entity of level sets of a 3D image have been considered. For an exposition and further references on geometric concepts in image processing we refer to the book of Sapiro [35].

Notation

Let us summarize notations and conventions we use in the sequel. We consider a parameter manifold $\mathcal{M}$ which essentially fixes the topological type of immersed surfaces $x : \mathcal{M} \to \mathbb{R}^{d+1}$. The parameter on $\mathcal{M}$ is denoted by ξ. For immersions x the differential Dx induces canonically a metric on $\mathcal{M}$ via the following relation which holds for all $v, w \in \mathcal{TM}$

$$g(v,w) = Dx(v) \cdot Dx(w) .$$

The scalar product in $\mathbb{R}^m$ is denoted by $\cdot$. For smooth functions on $\mathcal{M}$ we can define a gradient $\nabla_{\mathcal{M}}$ or $\text{grad}_{\mathcal{M}}$ as the representation of its differential in the metric g, i.e.,

$$df(v) =: g(\nabla_{\mathcal{M}} f, v) = g(\text{grad}_{\mathcal{M}} f, v) ,$$

for all $v \in \mathcal{TM}$. The Levi-Civita connection will be denoted by ∇. With the Levi-Civita connection at hand we define the divergence of a vector field v by

$$\text{div}_{\mathcal{M}} v = \text{tr}\,(\nabla_{\bullet} v) ,$$

which is nothing but the trace of the endomorphism $w \mapsto \nabla_w v$. This definition generalizes to the divergence of mappings $z : \mathcal{M} \to \mathbb{R}^{d+1}$ by:

$$\text{div}_{\mathcal{M}} z = \text{tr}\,\left(Dx^{-1}[Dz(\cdot)]^{tan}\right) ,$$

where "^{tan}" represents the tangential component of Dz. For tangential mappings z, i.e., $z = Dx(v)$, with $v \in \mathcal{TM}$ we have the identity $\text{div}_{\mathcal{M}} z = \text{div}_{\mathcal{M}} v$.

The manifolds $\mathcal{M}$ are assumed to be oriented and the normal mapping will be denoted by $n : \mathcal{M} \to S^d$, where d is the dimension of the manifold $\mathcal{M}$. This enables us to define the shape operator S by

$$g(S_{\mathcal{T}_x\mathcal{M}} w, v) := \partial_w N \cdot Dx(v) .$$

The trace of the shape operator is the classical mean curvature $h = \text{tr}\, S_{\mathcal{T}_x\mathcal{M}}$. We often write $S = S_{\mathcal{T}_x\mathcal{M}}$ if a misunderstanding is ruled out.

The Laplacian $\operatorname{div}_{\mathcal{M}} \nabla_{\mathcal{M}}$ is denoted by $\Delta_{\mathcal{M}}$. We will use the concept of tangential gradients. The tangential gradient for a function $u \in C^1(\mathbb{R}^{d+1})$ is defined as

$$\underline{D}u = \nabla_{\mathbb{R}^{d+1}} u - (n \cdot \nabla_{\mathbb{R}^{d+1}} u)\, n\,.$$

For the components of $\underline{D}u$ we have $\underline{D}_i u = dx_i(\nabla_{\mathcal{M}} u)$. We sometimes identify $x(\mathcal{M})$ and $\mathcal{M}$ and in this sense $\underline{D}u = \nabla_{\mathcal{M}} u$. Note the relation for the Laplacian $\Delta_{\mathcal{M}} u = \sum_i \underline{D}_i \underline{D}_i u$. For more details on tangential gradients we refer to [21, Chapter 16]. Integration over $\mathcal{M}$ leads to the L^2-scalar product of L^2-functions f, g on $\mathcal{M}$:

$$(f, g) := \int_{\mathcal{M}} f \cdot g \, dA\,.$$

Finally, let us from now on use Einstein summation convention.

3 Anisotropic Curvature Motion

3.1 Generalized Mean Curvature Motion

Let $x : \mathcal{M} \to \mathbb{R}^{d+1}$, $\mathcal{M}$ an orientable manifold of dimension d, be an immersion with normal $n : \mathcal{M} \to S^d$. In this section we consider general endomorphisms of the tangent space

$$a : T\mathcal{M} \to T\mathcal{M}$$

and the corresponding generalized mean curvature flow:

$$\partial_t x = -h_a n, \quad h_a = \operatorname{tr}(a \circ S)\,, \tag{6}$$

where S denotes the shape operator. The variational formulation of this problem is:

$$(\partial_t x, \vartheta) = -\int_{\mathcal{M}} h_a \, (n \cdot \vartheta) \, dA$$

for all $\vartheta \in C_0^1(\mathcal{M}, \mathbb{R}^{d+1})$. In case of classical mean curvature h, the relation $\Delta_{\mathcal{M}} x = -hn$ leads to a re-formulation of the above equation:

$$(\partial_t x, \vartheta) = \int_{\mathcal{M}} \Delta_{\mathcal{M}} x \cdot \vartheta \, dA = -\int_{\mathcal{M}} g(\nabla_{\mathcal{M}} x, \nabla_{\mathcal{M}} \vartheta) dA.$$

Let us already emphasize that this formulation will enable the later discretization by finite elements (cf. [15]). Now, we seek a generalization of the equation $\Delta_{\mathcal{M}} x = -hn$ that involves generalized mean curvatures. The next theorem gives such a generalization. The idea is to make use of the operator $\Delta_a \cdot = \operatorname{div}_{\mathcal{M}}(a \nabla_{\mathcal{M}} \cdot)$. We will see that the application of this operator to x leads to tangential components which are given by the divergence of the endomorphism a. For this reason we remind of the definition and basic properties of the divergence:

Definition 3.1 *The divergence of an endomorphism*

$$a : \mathcal{TM} \to \mathcal{TM}$$

is a vector field given by

$$g(\mathrm{div}_{\mathcal{M}} a, v) := \mathrm{tr}(\nabla_{\bullet} a^* v),$$

where the above equation is supposed to hold for all $v \in \mathcal{TM}$.

From this definition we obtain the representation:

$$\begin{aligned} g(\mathrm{div}_{\mathcal{M}} a, v) &= g(\nabla_{e_i} a^* v, e_i) \\ &= g(v, \nabla_{e_i} a\, e_i) \end{aligned}$$

and thus $\mathrm{div}_{\mathcal{M}} a = \nabla_{e_i} a\, e_i$ for an orthonormal basis $\{e_1, \ldots, e_d\} \subset \mathcal{T}_\xi \mathcal{M}$. Hence we obtain in coordinates

$$\mathrm{div}_{\mathcal{M}} a = g^{ij} \nabla_{\frac{\partial}{\partial \xi_i}} a \frac{\partial}{\partial \xi_i}.$$

Now we are prepared to formulate

Theorem 3.2 *Let $x : \mathcal{M} \to \mathbb{R}^{d+1}$ be an immersion of an orientable d-dimensional manifold. If $a : \mathcal{TM} \to \mathcal{TM}$ is differentiable, linear and symmetric on each tangent space, then there is a second order differential operator Θ_a with*

$$\Theta_a x = -h_a n,$$

where $h_a = \mathrm{tr}\,(a \circ S)$. The second order operator is given by

$$\Theta_a(\cdot) = \Delta_a(\cdot) - (\mathrm{div}_{\mathcal{M}} a)(\cdot)\,.$$

A proof of the above theorem can be found in [6]. It shows that we are able to express the velocity $-h_a n$ via projection of $\mathrm{div}_{\mathcal{M}}(a\, \nabla_{\mathcal{M}} x)$ onto the space spanned by the normal n. Consequently, the equation $\partial_t x = -h_a n$ can be written as follows:

$$\begin{aligned} v &= \mathrm{div}_{\mathcal{M}}(a\, \nabla_{\mathcal{M}} x) \\ \partial_t x &= (v \cdot n)\, n\,. \end{aligned} \tag{7}$$

Once more, let us emphasize that this formulation with the operator in divergence form enables us to generalize the algorithm in [15] for mean curvature motion (see sections 4 and 6.1).

3.2 Anisotropic Energies and Gradient Flow

As it is well known mean curvature evolution $\partial_t x = -hn$ can be considered as L^2-gradient flow w.r.t. the area functional. In this section we study more generally, so called anisotropic energies. Here the functional $\int_{\mathcal{M}} dA$ is generalized to the energy

$$E_\gamma[x] = \int_{\mathcal{M}} \gamma(\xi, n(\xi))\, dA(\xi)\,, \tag{8}$$

where

$$\begin{aligned} \gamma : \mathcal{M} \times \mathbb{R}^{d+1} &\longrightarrow \mathbb{R}^+ \\ (\xi, z) &\longmapsto \gamma(\xi, z) \end{aligned} \tag{9}$$

is an integrand of class $C^2(\mathcal{M} \times (\mathbb{R}^{d+1} - \{0\}))$ which is positively homogeneous of degree 1 in the z-variable.

The corresponding functional not only depends on the normal mapping of x but also on the parameter ξ. Thus favorable normal directions w.r.t. the energy E_γ may differ locally.

The derivative of E involves the generalized mean curvature $h_\gamma = \mathrm{tr}\,(a_\gamma \circ S)$ induced by γ, where

$$\begin{aligned} a_\gamma : T\mathcal{M} &\to T\mathcal{M} \\ (\xi, v) &\mapsto (\xi, Dx^{-1} \circ D_z^2\gamma(\xi, n(\xi)) \circ Dx(v)). \end{aligned}$$

The endomorphism $D_z^2\gamma(\xi, n(\xi)) : \mathbb{R}^{d+1} \to \mathbb{R}^{d+1}$ can be interpreted as an endomorphism on the orthogonal complement $n(\xi)^\perp$ of n. Indeed – due to the homogeneity of γ – we have

$$D_z^2\gamma(\xi, n)n = 0$$

and therefore, a_γ is well defined. The application of the derivative of E_γ to $\vartheta \in C_0^1(\mathcal{M}, \mathbb{R}^{d+1})$ is given by

$$\langle E'_\gamma[x], \vartheta \rangle = \int_{\mathcal{M}} h_\gamma(\vartheta \cdot n)\, dA.$$

For more details and proofs we refer to [6], [33], [38]. The gradient flow corresponding to E_γ is

$$\partial_t x = -h_\gamma n\,.$$

Wulff-Shape and Frank-Diagram

Let us restrict for a moment to the case, that γ is constant w.r.t. the ξ-variable, i.e., $\gamma(\xi, z) = \gamma(z)$. In the following we will explain that there is a bijection between convex bodies and these anisotropic integrands resp. energies. For the visualization of the energy E_γ one frequently uses the so called Wulff-shape and Frank-diagram. First let us consider a compact convex body $\mathcal{K} \subset \mathbb{R}^d$. Then we have

Proposition 3.3 *There is a 1-homogeneous and convex function $\gamma : \mathbb{R}^d \to \mathbb{R}$, called the support function of the compact convex body $\mathcal{K}$, such that*

$$\mathcal{K} = \{x \in \mathbb{R}^d \,|\, x \cdot \nu \le \gamma(\nu) \textit{ for all } \nu \in \mathbb{R}^d\}\,.$$

Conversely, the set $\mathcal{K}$ defined by the above relation is convex if γ is convex and 1-homogeneous. In case $\gamma(\nu) > 0$ for $\nu \neq 0$, the origin is contained in int $\mathcal{K}$.

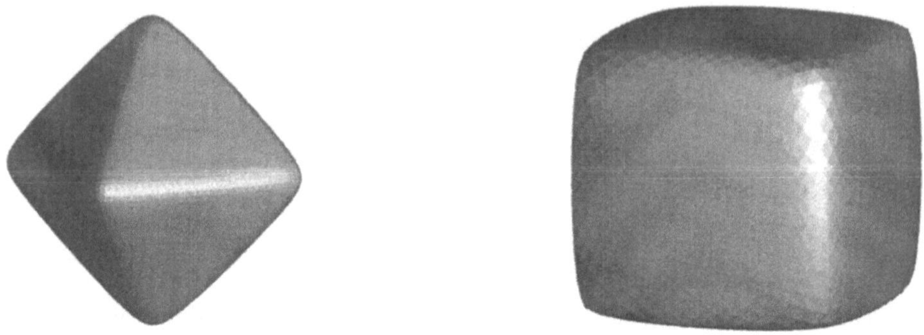

Fig. 3. A typical example of a Wulff-shape and the corresponding Frank-diagram

The above proposition shows a bijective relation between convex bodies and convex 1-homogeneous functions. A proof can be found in the classical introduction to convex analysis [3].

Now we fix an integrand γ of an energy E. The corresponding convex body $\mathcal{K}$ will be denoted by $\mathcal{W}_\gamma$ and is called Wulff-shape. (This denomination goes back to the work [39] of the cristallograph G. Wulff from 1901.) Introducing the dual γ^* of γ which is given by

$$\gamma^*(z^*) = \sup_{z \in S^{d-1}} \frac{z \cdot z^*}{\gamma(z)},$$

$\mathcal{W}_\gamma$ can be described in a different way:

$$\mathcal{W}_\gamma = \{x \in \mathbb{R}^d \,|\, \gamma^*(x) \leq 1\}.$$

The justification of the name *dual* is the content of the next

Proposition 3.4 *Let $\gamma : \mathbb{R}^d \to \mathbb{R}^+$ be a convex integrand then the duality relation $\gamma^{**} := (\gamma^*)^* = \gamma$ holds.*

For a proof of this proposition we refer to [2]. The dual shape $\mathcal{F}_\gamma$ of the Wulff-shape $\mathcal{W}_\gamma$ is according to Proposition 3.4 given by

$$\mathcal{F}_\gamma = \{z \in \mathbb{R}^d \,|\, \gamma(z) \leq 1\}$$

and is called the Frank diagram.

Example 3.5 *Consider the integrand in $\mathbb{R}^d$:*

$$\gamma_p(z) = \left(\sum_i |z_i|^p\right)^{\frac{1}{p}}, \quad p \in [1, \infty],$$

then the dual integrand is given by

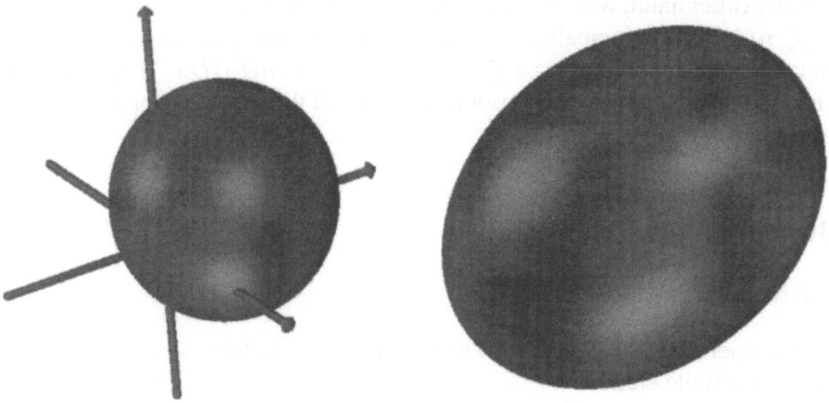

Fig. 4. Here, the Wulff shape is a sphere whose center is not the origin.

$$\gamma_p^*(z^*) = \left(\sum_i |z_i^*|^{p^*} \right)^{\frac{1}{p^*}} = \gamma_{p^*}(z^*),$$

where $\frac{1}{p} + \frac{1}{p^*} = 1$.

Remark 3.6 *For the computation of Frank diagrams we derive a simple algorithm. Given a convex body with boundary* $\partial\mathcal{W}_\gamma$ *we would like to determine the boundary* $\partial\mathcal{F}_\gamma$ *of the corresponding Frank diagram. To this aim consider a point* ν *of the surface and the normal* n *in* ν*. The support function* γ *of the convex body evaluated in* $\frac{n}{n \cdot \nu}$ *equals 1. Therefore we consider the map* $f : \partial\mathcal{W}_\gamma \to \mathbb{R}^d$

$$f(\nu) = \frac{n}{n \cdot \nu}$$

and we have $\gamma(f(\nu)) = 1$.

Wulff-shapes are known to be the solution of an isoperimetric problem. The boundary of $\mathcal{W}_\gamma$ is the minimizer of E_γ in the class of surfaces enclosing the same volume. For details we refer to [19] and [20].

3.3 Comparing the Two Approaches

The preceding section contained two methods to define generalized resp. anisotropic mean curvature motion.

The generalized mean curvature approach consists in introducing an anisotropic tensor $a : \mathcal{TM} \to \mathcal{TM}$, without specifying this tensor in detail. In this sense one might think of $a = a(\xi)$ depending on the parameter $\xi \in \mathcal{M}$.

On the other hand, we considered a gradient flow defined by the energy $E_\gamma[x] = \int_{\mathcal{M}} \gamma(\xi, n(\xi))\, dA$. Principally, this flow is nothing but a special case of a generalized mean curvature flow choosing $a(\xi) = Dx^{-1} D_z^2 \gamma(\xi, n(\xi)) Dx$. We point at the fact that for this special choice the tensor a depends on the geometry of x via the surface normal n.

4 Discretization

4.1 Generalized Mean Curvature

Using the operator Θ_a we can give a weak formulation of (6), which can numerically be solved by finite elements:

$$\int_{\mathcal{M}} \partial_t x \cdot \vartheta\, dA = \int_{\mathcal{M}} \Theta_a x \cdot \vartheta\, dA.$$

Concerning numerical realization of the Θ_a-operator we emphasis that the tangential term $\mathrm{div}_{\mathcal{M}} a$ has to be taken into consideration. Otherwise, on a triangulated surface one observes strong tangential shifts leading to irregular meshes. The operator $\mathrm{div}_{\mathcal{M}} a$ is not easy to handle numerically. Therefore we use a projection of the Δ_a-operator as follows:

$$\Theta_a x = (\Delta_a x \cdot n) n.$$

Let us identify from now on $\mathcal{M}$ and $x(\mathcal{M})$. Furthermore, for time discretization, we follow [15] and view in each time-step $x^{(k+1)}$ as a mapping from $\mathcal{M}^{(k)}$ onto $\mathcal{M}^{(k+1)}$ instead of a mapping defined on a fixed parameter domain. For a semi-implicit time discretization of (7) we are lead to:

$$\int_{\mathcal{M}^{(k)}} v^{(k+1)} \cdot \vartheta\, dA = \int_{\mathcal{M}^{(k)}} g\left(a^{(k)} \nabla_{\mathcal{M}^{(k)}} (x^{(k)} + \tau v^{(k+1)}), \nabla_{\mathcal{M}^{(k)}} \vartheta\right) dA$$

for $\vartheta \in C_0^1(\mathcal{M})$ and

$$x^{(k+1)} = x^{(k)} + \tau \left(v^{(k+1)} \cdot n^{(k)}\right) n^{(k)}.$$

This scheme defines approximations $x^{(k)}$ of $x(\tau k)$.

Next, we discuss the spatial discretization. To clarify the notation we will always denote discrete quantities with upper case letters to distinguish them from the corresponding continuous quantities in lower case letters.

We restrict our considerations to the case $d = 2$ and consider a polyhedron consisting of triangles $\mathcal{M}_h$ which is an approximation of $\mathcal{M}$. X will be the parametrization of $\mathcal{M}_h$, i.e., X is in each component an affine linear function on the triangles of $\mathcal{M}_h$. The corresponding finite element space

$$\mathcal{V}_h = \{\Phi \in C^0(\mathcal{M}_h) \mid \Phi_T \in \mathcal{P}_1, T \in \mathcal{M}_h\} \tag{10}$$

consisting of those functions being affine linear on each triangle of $\mathcal{M}_h$ has the basis $\{\Phi_j\}_{j=1}^J$, where J is the number of vertices of $\mathcal{M}_h$ and $\Phi_j(X_i) = \delta_{ij}$ for all vertices X_i. We can represent X as

$$X = \overline{X}_j \Phi_j,$$

i.e., $X \in [\mathcal{V}_h]^3$. The discretization of $a : \mathcal{TM} \to \mathcal{TM}$ will be denoted by A and will be considered as constant symmetric positive definite endomorphism of $\mathbb{R}^3$ on each triangle.

The gradient-operator on $\mathcal{M}_h$ is now $\nabla_{\mathcal{M}_h}$, the gradient on the Lipschitz surface $\mathcal{M}_h$ (see [15] for more details). For the fully discrete semi implicit scheme we introduce the discrete velocities $V^{(k)} \in [\mathcal{V}_h]^3$ and obtain for all $\Phi \in \mathcal{V}_h$:

$$\int_{\mathcal{M}_h^{(k)}} V^{(k+1)} \cdot \Phi = \int_{\mathcal{M}_h^{(k)}} \left(A^{(k)} \nabla_{\mathcal{M}_h^{(k)}} (X^{(k)} + \tau V^{(k+1)}) \cdot \nabla_{\mathcal{M}_h^{(k)}} \Phi \right),$$

$$X^{(k+1)} = X^{(k)} + \tau (V^{(k+1)} \cdot N^{(k)}) N^{(k)},$$

where $N^{(k)}$ is given on each vertex $\overline{X}_j$ by the weighted sum $N_j^{(k)} = \frac{\sum_{i=1}^{\ell_j} |T_i| N_{T_i}^{(k)}}{\left|\sum_{i=1}^{\ell_j} |T_i| N_{T_i}^{(k)}\right|}$ and $N_{T_i}^{(k)}$ is the normal on $T_i \in \mathcal{M}_h$ with $\overline{X}_j \in \overline{T}_i$. Here ℓ_j is the number of elements T_i with node $\overline{X}_j$. Thus starting with $\mathcal{M}_h^{(0)}$ we obtain a sequence of triangulated surfaces $\mathcal{M}_h^{(k)}$ approximating $\mathcal{M}(t)$ at time $t = k\tau$. The implementation of this discrete scheme is straightforward. We introduce the mass matrix $M^{(k)}$ and the stiffness matrix $L^{(k)}$:

$$M_{ij}^{(k)} = \int_{\mathcal{M}_h^{(k)}} \Phi_i \Phi_j \, dA, \quad L_{ij}^{(k)} = \int_{\mathcal{M}_h^{(k)}} \left(A^{(k)} \nabla_{\mathcal{M}_h^{(k)}} \Phi_i \right) \cdot \nabla_{\mathcal{M}_h^{(k)}} \Phi_j \, dA$$

The computation of $V^{(k+1)}$ requires the solution of the linear systems

$$(M^{(k)} - \tau L^{(k)}) \overline{V}_l^{(k+1)} = L^{(k)} \overline{X}_l^{(k)}, \tag{11}$$

where we set $\overline{V}_l^{(k+1)} = (\overline{V}_{1,l}^{(k+1)}, \dots, \overline{V}_{J,l}^{(k+1)})$ and $\overline{X}_l^{(k+1)} \in \mathbb{R}^J$ analogue. For more details we refer to [7].

The above algorithm is clearly applicable for the case $d = 1$. Let us have a look at this simple case only considering spatial discretization. The system we have to solve is:

$$\int_{\mathcal{M}_h(t)} \langle V_h, \Phi_k \rangle + \int_{\mathcal{M}_h(t)} \langle A \nabla_{\mathcal{M}_h} X_h, \nabla_{\mathcal{M}_h} \Phi_k \rangle = 0$$

$$\partial_t X_h = \langle V_h, N_h \rangle N_h$$

4.2 Anisotropic Gradient Flow

The 1D case.
In this section we want to derive a numerical scheme for energies as defined in (8) in case of curves, thus $\mathcal{M} = S^1$. The algorithm we derive here is related to [17]. Our starting point for the computation of the generalized mean curvature flow

$$\partial_t x = -h_\gamma n$$

is the weak formulation

$$\int_{S^1} x_t \vartheta \, |x_\xi| d\xi + \int_{S^1} \gamma_z(\xi, x_\xi^\perp) \vartheta_\xi^\perp d\xi = 0 \ \text{ for all } \ \vartheta \in C^1(S^1, \mathbb{R}^2) \,. \qquad (12)$$

The symbol $^\perp$ denotes a rotation by 90°. For the spatial discretization of this weak form of curvature flow consider a decomposition of $S^1 = \mathbb{R}/2\pi$ into intervals:

$$[0, 2\pi] = \bigcup_{j=1}^{J} I_j, \quad I_j = [\xi_{j-1}, \xi_j], \quad h_j = |I_j|.$$

The underlying finite element space is $[\mathcal{S}_h]^2$, where

$$\mathcal{S}_h = \{\Phi \in C^0(S^1) \,|\, \Phi_{|I_j} \in \mathcal{P}_1, \, j = 1, \dots, J\}$$

The discrete solution $X : (0, T) \to [\mathcal{S}_h]^2$ is represented as:

$$X(t, \xi) := \overline{X}_j(t) \Phi_j(\xi).$$

Now we assume γ to be a piece-wise linear function for fixed $z \in S^1$, i.e.,

$$\gamma(\xi, z) = \Phi_j(\xi) \gamma_j(z).$$

A discrete solution X fulfills

$$\int_{S^1} \partial_t X |X_\xi| \Phi d\xi + \int_{S^1} \gamma_z(\xi, X_\xi^\perp) \Phi_\xi^\perp d\xi = 0 \ \text{ for all } \ \Phi \in [\mathcal{S}_h]^2.$$

Taking $\Phi = \Phi_j \in \mathcal{S}_h$ separately for each component as test function one obtains:

$$\int_{S^1} \gamma_z^\perp(\xi, X_\xi^\perp) \Phi_{j,\xi} d\xi =$$

$$\frac{1}{2}\left[\gamma_{j-1,z}^\perp(\overline{X}_j^\perp - \overline{X}_{j-1}^\perp) + \gamma_{j,z}^\perp(\overline{X}_j^\perp - \overline{X}_{j-1}^\perp)\right]$$

$$-\frac{1}{2}\left[\gamma_{j,z}^\perp(\overline{X}_{j+1}^\perp - \overline{X}_j^\perp) + \gamma_{j+1,z}^\perp(\overline{X}_{j+1}^\perp - \overline{X}_j^\perp)\right]$$

Lumping of masses leads to:

$$\frac{1}{2}\partial_t\overline{X}_j(|\overline{X}_j - \overline{X}_{j-1}| + |\overline{X}_{j+1} - \overline{X}_j|)$$

$$-\frac{1}{2}\left[\gamma^{\perp}_{j-1,z}(\overline{X}^{\perp}_j - \overline{X}^{\perp}_{j-1}) + \gamma^{\perp}_{i,z}(\overline{X}^{\perp}_j - \overline{X}^{\perp}_{j-1})\right]$$

$$+\frac{1}{2}\left[\gamma^{\perp}_{j,z}(\overline{X}^{\perp}_{j+1} - \overline{X}^{\perp}_j) + \gamma^{\perp}_{j+1,z}(\overline{X}^{\perp}_{j+1} - \overline{X}^{\perp}_j)\right] = 0.$$

Introducing $q_j = |\overline{X}_j - \overline{X}_{j-1}|, \tau_j = (\overline{X}_j - \overline{X}_{j-1})/q_j$, $N_j = \tau_j^{\perp}$ and using the zero homogeneity of $\gamma_{j,z}$ in the z-variable we can rewrite the above equation as:

$$\begin{aligned}&\partial_t\overline{X}_j(q_j + q_{j+1}) - \gamma^{\perp}_{j-1,z}(N_j)\\ &- \gamma^{\perp}_{j,z}(N_j) + \gamma^{\perp}_{j,z}(N_{j+1}) + \gamma^{\perp}_{j+1,z}(N_{j+1}) = 0\end{aligned} \tag{13}$$

For all i, j the following relation is valid:

$$\begin{aligned}\gamma^{\perp}_{j,z}(N_i) &= (\gamma^{\perp}_{j,z}(N_i)\cdot\tau_i)\tau_i + (\gamma^{\perp}_{j,z}(N_i)\cdot N_i)N_i\\ &= -(\gamma_{j,z}(N_i)\cdot N_i)\tau_i + (\gamma_{j,z}(N_i)\cdot\tau_i)N_i\\ &= -\gamma_j(N_i)\tau_i + (\gamma_{j,z}(N_i)\cdot\tau_i)N_i.\end{aligned}$$

Equation (13) now can be formulated as:

$$\begin{aligned}&\partial_t\overline{X}_j(q_j + q_{j+1}) + \gamma_{j-1}(N_j)\tau_j + \gamma_j(N_j)\tau_j\\ &\quad - \gamma_j(N_{j+1})\tau_{j+1} - \gamma_{j+1}(N_{j+1})\tau_{j+1} =\\ &\quad (\gamma_{j-1,z}(N_j)\cdot\tau_j)N_j + (\gamma_{j,z}(N_j)\cdot\tau_j)N_j\\ &\quad - (\gamma_{j,z}(N_{j+1})\cdot\tau_{j+1})N_{j+1} - (\gamma_{j+1}(N_{j+1})\cdot\tau_{j+1})N_{j+1}\end{aligned}$$

and with $g_{i,j} = \gamma_i(N_j), g'_{i,j} = \gamma_{i,z}(N_j)\tau_j$ this leads to:

$$\partial_t \overline{X}_j (q_j + q_{j+1}) - \left(\frac{g_{j-1,j}}{q_j} + \frac{g_{j,j}}{q_j} \right) \overline{X}_{j-1}$$
$$+ \left(\frac{g_{j-1,j}}{q_j} + \frac{g_{j,j}}{q_j} + \frac{g_{j,j+1}}{q_{j+1}} + \frac{g_{j+1,j+1}}{q_{j+1}} \right) \overline{X}_j$$
$$- \left(\frac{g_{j,j+1}}{q_{j+1}} + \frac{g_{j+1,j+1}}{q_{j+1}} \right) \overline{X}_{j+1}$$
$$+ \left(\frac{g'_{j-1,j}}{q_j} + \frac{g'_{j,j}}{q_j} \right) \overline{X}^{\perp}_{j-1}$$
$$- \left(\frac{g'_{j-1,j}}{q_j} + \frac{g'_{j,j}}{q_j} + \frac{g'_{j,j+1}}{q_{j+1}} + \frac{g'_{j+1,j+1}}{q_{j+1}} \right) \overline{X}^{\perp}_j$$
$$+ \left(\frac{g'_{j,j+1}}{q_{j+1}} + \frac{g'_{j+1,j+1}}{q_{j+1}} \right) \overline{X}^{\perp}_{j+1} = 0$$

We would like emphasize that this scheme is intrinsic, in the sense that it is independent of the parametrization. For the time-discretization we use a semi-implicit scheme, where q_j, $g_{i,j}$ and $g'_{i,j}$ are treated explicitly. For more details see [17].

The 2D case.
In the following we describe the discretization of the generalized mean curvature flow for anisotropic energies (8) depending only on the surface normal. The discretization will be such that it only contains the first derivatives of the anisotropy function γ. In abstract form the problem reads

$$(\partial_t x, \vartheta) = -\langle E'_\gamma[x], \vartheta \rangle \tag{14}$$

for every test function $\vartheta \in C^1_0(\mathcal{M}, \mathbb{R}^{d+1})$. For practical reasons we use the concept of tangential gradients for a reformulation of (14). Easy calculations show, that

$$\langle E'_\gamma[x], \vartheta \rangle = \int_{\mathcal{M}} h_\gamma \, n \cdot \vartheta \, dA = \int_{\mathcal{M}} \operatorname{tr} \left(D^2_z \gamma(n) \nabla_{\mathcal{M}} n \right) \, n \cdot \vartheta \, dA$$
$$= \int_{\mathcal{M}} \sum_{k,l=1}^{d+1} \gamma_{z_k z_l}(n) \, \underline{D}_k n_l \, n \cdot \vartheta \, dA$$
$$= - \int_{\mathcal{M}} \sum_{k,l=1}^{d+1} \underline{D}_k \left(\gamma(n) \, \underline{D}_k x_l - \sum_{m=1}^{d+1} \gamma_{z_m}(n) \, n_l \, \underline{D}_k x_m \right) \vartheta_l \, dA$$
$$= \int_{\mathcal{M}} \sum_{k,l=1}^{d+1} \left(\gamma(n) \, \underline{D}_k x_l \, \underline{D}_k \vartheta_l - \sum_{m=1}^{d+1} \gamma_{z_m}(n) \, n_l \, \underline{D}_k x_m \, \underline{D}_k \vartheta_l \right) \, dA$$
$$= \int_{\mathcal{M}} \gamma(n) \, \nabla_{\mathcal{M}} x \cdot \nabla_{\mathcal{M}} \vartheta - \sum_{k,l=1}^{d+1} \gamma_{z_k}(n) \, n_l \, \nabla_{\mathcal{M}} x_k \cdot \nabla_{\mathcal{M}} \vartheta_l \, dA.$$

Thus anisotropic mean curvature flow (14) can be written as

$$\begin{aligned}\int_{\mathcal{M}} \partial_t x \cdot \vartheta + \int_{\mathcal{M}} \gamma(n) \nabla_{\mathcal{M}} x \cdot \nabla_{\mathcal{M}} \vartheta \, dA \\ = \int_{\mathcal{M}} \sum_{k,l=1}^{d+1} \gamma_{z_k}(n) n_l \nabla_{\mathcal{M}} x_k \cdot \nabla_{\mathcal{M}} \vartheta_l \, dA.\end{aligned} \tag{15}$$

for every $\vartheta \in C_0^1(\mathcal{M}, \mathbb{R}^{d+1})$. The right hand side of (15) can be simplified using the relation

$$\underline{D}_j x_k = \delta_{jk} - n_j n_k.$$

Hence

$$\int_{\mathcal{M}} \sum_{k,l=1}^{d+1} \gamma_{z_k}(n) \, n_l \, \nabla_{\mathcal{M}} x_k \cdot \nabla_{\mathcal{M}} \vartheta_l \, dA = \int_{\mathcal{M}} \sum_{k,l=1}^{d+1} \gamma_{z_k}(n) \, n_l \, \underline{D}_k \vartheta_l \, dA.$$

But as we shall see, the right hand side in (15) is numerically more convenient after time discretization. For isotropic mean curvature flow, i. e. for $\gamma(z) = |z|$, this right hand side vanishes. We also observe that the left hand side of (15) decouples with respect to the components of x.

For the discretization we restrict ourselves to $d = 2$. We use the following time discretization:

$$\begin{aligned}\int_{\mathcal{M}^{(k)}} \frac{1}{\tau} \left(x^{(k+1)} - x^{(k)} \right) \cdot \vartheta \, dA + \\ \int_{\mathcal{M}^{(k)}} \gamma(n^{(k)}) \, \nabla_{\mathcal{M}^{(k)}} x^{(k+1)} \cdot \nabla_{\mathcal{M}^{(k)}} \vartheta \, dA \\ = \sum_{l,m=1}^{d+1} \int_{\mathcal{M}^{(k)}} \gamma_{z_m}(n^{(k)}) \, n_l^{(k)} \, \nabla_{\mathcal{M}^{(k)}} x_m^{(k)} \cdot \nabla_{\mathcal{M}^{(k)}} \vartheta_l \, dA.\end{aligned} \tag{16}$$

Here τ is the time step and $g^{(k)}$ stands for the evaluation of a generic function g on the k–th time level. $x^{(0)}$ is given by the initial surface.

The discretization in space is based on piece-wise linear finite elements similar to the approach above in Section 4.1.

We compute a sequence of time step solutions $X^{(k)}$ for $k = 0, \cdots$, where the initial polyhedron $X^{(0)}$ is the interpolant of $x^{(0)}$. The fully discrete scheme then reads: for every $\Phi \in [\mathcal{V}_h]^3$

$$\int_{\mathcal{M}_h^{(k)}} \frac{1}{\tau}\left(X^{(k+1)} - X^{(k)}\right) \cdot \Phi \, dA +$$
$$\int_{\mathcal{M}_h^{(k)}} \gamma(N^{(k)}) \nabla_{\mathcal{M}_h^{(k)}} X^{(k+1)} \cdot \nabla_{\mathcal{M}_h^{(k)}} \Phi \, dA$$
$$= \sum_{l,m=1}^{3} \int_{\mathcal{M}_h^{(k)}} \gamma_{z_m}(N^{(k)}) \, N_l^{(k)} \, \nabla_{\mathcal{M}_h^{(k)}} X_m^{(k)} \cdot \nabla_{\mathcal{M}_h^{(k)}} \Phi_l \, dA. \qquad (17)$$

This linear system of equations is quite easy to implement. For given $\mathcal{M}_h^{(k)}$ we define the mass matrix $M^{(k)}$ as in Section 4.1, the stiffness matrix $L^{(k)}$ by

$$L_{ij}^{(k)} = \int_{\mathcal{M}_h^{(k)}} \gamma(N^{(k)}) \nabla_{\mathcal{M}_h^{(k)}} \Phi_i \cdot \nabla_{\mathcal{M}_h^{(k)}} \Phi_j \, dA$$

$(i, j = 1, \ldots, J)$, and the right hand sides $C_l^{(k)}$ by

$$C_{l,j}^{(k)} = \sum_{m=1}^{3} \int_{\mathcal{M}_h^{(k)}} \gamma_{z_m}(N^{(k)}) \, N_l^{(k)} \, \nabla_{\mathcal{M}_h^{(k)}} X_m^{(k)} \cdot \nabla_{\mathcal{M}_h^{(k)}} \Phi_j \, dA$$

$(l = 1, 2, 3, j = 1, \ldots, J)$. In every time step of the scheme we have to solve 3 linear systems with the same matrix for the computation of $\overline{X}_l^{(k+1)} \in \mathbb{R}^J$:

$$\left(\frac{1}{\tau} M^{(k)} + L^{(k)}\right) \overline{X}_l^{(k+1)} = \frac{1}{\tau} L^{(k)} \overline{X}_l^{(k)} + C_l^{(k)}.$$

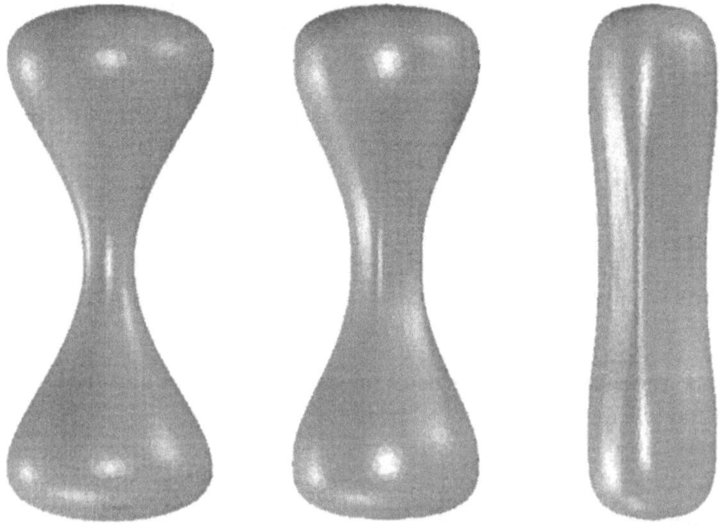

Fig. 5. From left to right the initial surface and two evolution steps of an anisotropic mean curvature flow are depicted (graphically scaled).

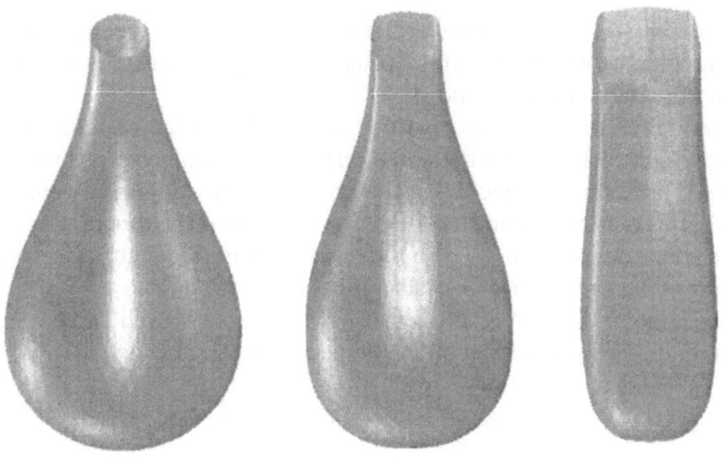

Fig. 6. The sliced surfaces of Figure 5 are shown.

The discrete normal $N^{(k)}$ is piece-wise constant and so the matrix $L^{(k)}$ and the right hand sides $C_l^{(k)}$ follow the usual setup of the stiffness matrix in a finite element code via a loop over the triangles. Only a constant factor has to be included on every triangle.

In Figure 5 and Figure 6 a result of an anisotropic evolution is shown. The anisotropy is similarly chosen as in Figure 3.

5 Curve and Surface Classification

In this section we will describe two methods for the local classification of curves and surfaces. This will enable us to distinguish smooth areas on the curve or surface from edge or corner regions. Both methods will be robust with respect to noise. Later on in Section 6 we will make use of these local classifications to process noisy curves and surfaces.

5.1 Classification via Curvature Analysis

In this approach, the quantity for the detection of edges is the curvature tensor, in case of codimension 1 represented by the symmetric shape operator $S_{\mathcal{T}_x\mathcal{M}}$ (note that we identify $x(\mathcal{M})$ and $\mathcal{M}$ and consequently x and ξ). An edge is supposed to be indicated by one sufficiently large eigenvalue of $S_{\mathcal{T}_x\mathcal{M}}$.

We define a tensor depending on $S_{\mathcal{T}_x\mathcal{M}}$, which considered as diffusion tensor (see Section 6.1) enables us to decrease diffusion significantly at edges indicated by $S_{\mathcal{T}_x\mathcal{M}}$. This tensor locally classifies the curve or surface.

Furthermore we will introduce a threshold parameter λ (see equation (19)) for the identification of edges. Roughly speaking, this parameter enables us to define what is an important characteristic.

The evaluation of the shape operator on a noisy surface might be misleading with respect to the original but unknown surface and its edges. Thus we prefilter the surface $\mathcal{M}$ before we evaluate the shape operator. The straightforward "geometric Gaussian" filter is a short time-step $\tau = \sigma^2/2$ of mean curvature motion. Hence, we compute a shape operator $S_{\mathcal{T}_x\mathcal{M}_\sigma}$ on the resulting prefiltered surface $\mathcal{M}_\sigma$. Here the parameter σ is the "geometric Gaussian" filterwidth.

For every point x on $\mathcal{M}_\sigma$ the tensor $a^\sigma_{\mathcal{T}_x\mathcal{M}_\sigma}$ is supposed to be a symmetric, positive definite, linear mapping on the tangent space $\mathcal{T}_x\mathcal{M}_\sigma$:

$$a^\sigma_{\mathcal{T}_x\mathcal{M}_\sigma}(x) : \mathcal{T}_x\mathcal{M}_\sigma \to \mathcal{T}_x\mathcal{M}_\sigma \, .$$

There is an orthonormal basis $\{w^{1,\sigma}, w^{2,\sigma}\}$ of $\mathcal{T}_x\mathcal{M}_\sigma$ such that $S_{\mathcal{T}_x\mathcal{M}_\sigma}$ is represented by

$$S_{\mathcal{T}_x\mathcal{M}_\sigma} = \begin{pmatrix} \kappa^{1,\sigma} & 0 \\ 0 & \kappa^{2,\sigma} \end{pmatrix} \tag{18}$$

because of the symmetry of the shape operator. Now we consider a tensor which is defined with respect to the above orthonormal basis as follows:

$$a^\sigma_{\mathcal{T}_x\mathcal{M}_\sigma} = \begin{pmatrix} G\left(\frac{\kappa^{1,\sigma}}{\lambda}\right) & 0 \\ 0 & G\left(\frac{\kappa^{2,\sigma}}{\lambda}\right) \end{pmatrix} \tag{19}$$

with the function G as defined in the introduction.

Hence, due to the anisotropy defined in (19), a point belongs to an edge if there is a principal direction of curvature on $\mathcal{M}_\sigma$ with large curvature compared to λ. If the second principal curvature is small w.r.t. λ, we regard the first direction as orthogonal to an edge on the surface. At corners both principal curvatures of $\mathcal{M}_\sigma$ are large.

In summary, analyzing our classification tensor leads to a surface classification as follows: Smooth parts of the surface can be characterized by $a^\sigma_{\mathcal{T}_x\mathcal{M}_\sigma} \backsim \text{diag}[1,1]$. An edge can be defined via the relation $a^\sigma_{\mathcal{T}_x\mathcal{M}_\sigma} \backsim \text{diag}[1,0]$. In this case, the direction along the edge is given by $w^{2,\sigma}$ where we assume here $|\kappa^{1,\sigma}| >> |\kappa^{2,\sigma}|$. In this setting, corners are given by the relation $a^\sigma_{\mathcal{T}_x\mathcal{M}_\sigma} \backsim \text{diag}[0,0]$. We may introduce as an edge-indicator the function $\eta_\lambda(x) = \text{tr}\; a^\sigma_{\mathcal{T}_x\mathcal{M}_\sigma}$. Depending on the parameter λ edges and corners are given by $\eta_\lambda < 1$.

5.2 Classification via Moments

The second approach to local surface classification is to use the zero surface moments. This allows to distinguish smooth regions from the vicinity of edges on the curve or surface. Hence, we compute for $x : \mathcal{M} \to \mathbb{R}^{d+1}$ the barycenter $M^0_\epsilon(x(\xi)), \xi \in \mathcal{M}$,

Fig. 7. The curvature classification scheme, where the trace of the classification operator is color coded. Red means that the trace is approximately 0, blue indicates smooth domains where the trace equals 2. (cf. color plate 10, Page 671)

of $x(\mathcal{M}) \cap B_\epsilon(x(\xi))$, where $B_\epsilon(x(\xi))$ is the Euclidian ϵ-ball. The parameter ϵ will be called the scanning-width.

It is well known, that the difference $n_\epsilon(\xi) = M_\epsilon^0(x(\xi)) - x(\xi)$ (denoted as the ϵ-normal) is independent of translations and in addition $|n_\epsilon(\xi)|$ is independent of rotations. We will show that a locally smooth surface is characterized by a quadratic scaling of n_ϵ in ϵ and close to edges we observe a linear scaling. Let us first consider a locally smooth curve or surface. For a smooth function η on a Euclidian ϵ-ball $B_\epsilon(0) \subset \mathbb{R}^d, d = 1, 2$, we have:

$$\begin{aligned}
\fint_{B_\epsilon} \eta &= \fint_{B_\epsilon} \eta(0) + \nabla\eta(0) \cdot x + \frac{1}{2}\nabla^2\eta(0)x \cdot x \, dx + o(\epsilon^2) \\
&= \eta(0) + \frac{1}{2}\fint_{B_\epsilon(0)} \nabla^2\eta(0)x \cdot x \, dx + o(\epsilon^2) \\
&= \eta(0) + \frac{1}{2}\lambda_i \fint_{B_\epsilon(0)} x_i^2 dx + o(\epsilon^2),
\end{aligned}$$

where each λ_i is an eigenvalue of $\nabla\eta^2(0)$. Therefore we have:

$$\begin{aligned}
\fint_{B_\epsilon} \eta &= \eta(0) + \frac{1}{2} \cdot \frac{1}{d} \fint_{B_\epsilon(0)} |x|^2 dx \cdot \operatorname{tr} \nabla^2\eta(0) + o(\epsilon^2) \\
&= \eta(0) + \frac{1}{2} \cdot \frac{1}{d} \cdot \frac{1}{4-d}\epsilon^2 \Delta\eta(0) + o(\epsilon^2).
\end{aligned}$$

In a first step we replace in our considerations the Euclidean ball $B_\epsilon(x(\xi)) \subset \mathbb{R}^{d+1}$ by a geodesic ball $\widetilde{B}_\epsilon(x(\xi)) \subset \mathcal{M}$ and compute the barycenter $\widetilde{M_\epsilon^0}(x(\xi))$ of $\widetilde{B}_\epsilon(x(\xi)) \cap$

$\mathcal{M}$. For the evaluation of $\int_{\widetilde{B}_\epsilon(x(\xi))} x(\xi)dA(\xi)$ we use normal coordinates, i.e.,

$$g_{ij}(0) = \delta_{ij} \quad , \partial_k g_{ij}(0) = 0 \, .$$

Then one obtains for the Laplacian on $\mathcal{M}$ at $x(\xi)$:

$$(\Delta_{\mathcal{M}} f)(x(\xi)) = (\partial_i \partial_i f)(0) = \Delta f(0) \, ,$$

where $f \in C^2(\mathcal{M})$. For the difference $\widetilde{M}^0_\epsilon(x(\xi)) - x(\xi)$ we obtain

$$\begin{aligned}
\widetilde{M}^0_\epsilon(x(\xi)) - x(\xi) &= \fint_{B_\epsilon(0)} x dA - x(\xi) \\
&= \frac{1}{2d} \cdot \frac{1}{4-d} \epsilon^2 \Delta x(0) + o(\epsilon^2) \\
&= \frac{1}{2d} \cdot \frac{1}{4-d} \epsilon^2 \Delta_{\mathcal{M}} x(\xi) + o(\epsilon^2) \\
&= -\frac{1}{2d} \cdot \frac{1}{4-d} \epsilon^2 h(\xi) n(\xi) + o(\epsilon^2) \, ,
\end{aligned}$$

where h is the mean curvature of $\mathcal{M}$. The ratio $|\widetilde{B}_\epsilon(x(\xi))| / |B_\epsilon(x(\xi)) \cap \mathcal{M}|$ converges to 1, where $\epsilon \to 0$. So far we have shown:

Theorem 5.7 *Let $x : \mathcal{M} \to \mathbb{R}^{d+1}$, $d = 2, 3$, be an immersion. For $\xi \in \mathcal{M}$ consider a ball of radius ϵ with center $x(\xi)$ and the ϵ-normal $n_\epsilon(\xi) = M^0_\epsilon(x(\xi)) - x(\xi)$. Then n_ϵ scales quadratically in ϵ and*

$$n_\epsilon(\xi) = -\epsilon^2 \frac{1}{2d(4-d)} h(\xi) n(\xi) + o(\epsilon^2) \, .$$

Now we are going to examine the scaling w.r.t. ϵ in a non-smooth situation. Here, we restrict ourselves to curves. The generalization to surfaces in higher dimension will be subject of future work.

On account of the invariance of translation and rotation of $|M^0_\epsilon(x) - x|$ one can consider a normalized situation:

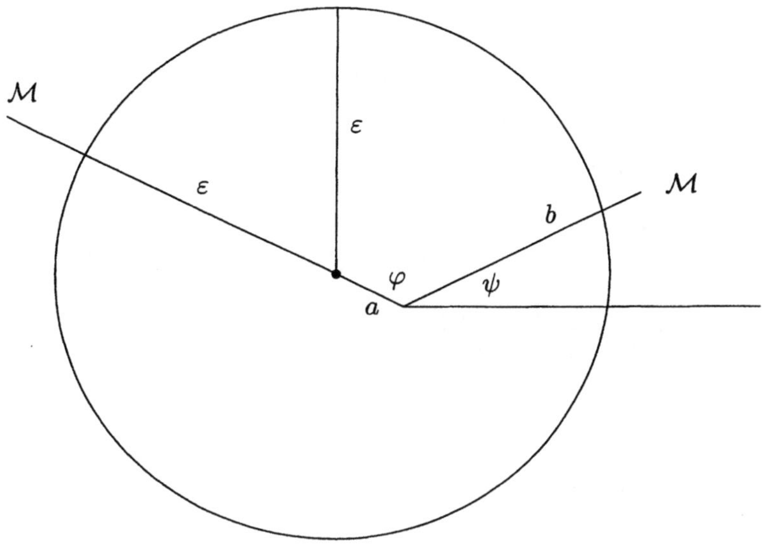

The center of the circle in the picture above is the curve-point x. For the barycenter of $\Gamma \cap B_\epsilon(x)$ we have

$$M_\epsilon^0 = \frac{(\epsilon+a)^2}{2(\epsilon+a+b)}\begin{pmatrix}-\cos\psi\\ \sin\psi\end{pmatrix} + \frac{b^2}{2(\epsilon+a+b)}\begin{pmatrix}\cos\psi\\ \sin\psi\end{pmatrix}.$$

and thus for $n_\epsilon(x)$ we obtain

$$M_\epsilon^0(x) - x = \left[\frac{(\epsilon+a)^2}{2(\epsilon+a+b)} - a\right]\begin{pmatrix}-\cos\psi\\ \sin\psi\end{pmatrix} + \frac{b^2}{2(\epsilon+a+b)}\begin{pmatrix}\cos\psi\\ \sin\psi\end{pmatrix}.$$

The length b is depending on a and φ:

$$b^2 - 2ab\cos\varphi + a^2 - \epsilon^2 = 0$$

and one computes for b:

$$b = b(\epsilon, a, \varphi) = a\cos\varphi + \sqrt{a^2\cos^2\varphi + \epsilon^2 - a^2}.$$

We now can determine $|M_\epsilon^0(x) - x|^2$ just knowing the values a, ϵ and φ:

$$\begin{aligned}
&|M_\epsilon^0(x) - x|^2 \\
&= \left|\left[\frac{(\epsilon+a)^2}{2(\epsilon+a+b(\epsilon,a,\varphi))} - a\right]\begin{pmatrix}-\cos\psi\\ \sin\psi\end{pmatrix} + \frac{b^2(\epsilon,a,\varphi)}{2(\epsilon+a+b(\epsilon,a,\varphi))}\begin{pmatrix}\cos\psi\\ \sin\psi\end{pmatrix}\right|^2 \\
&= f_1^2(\epsilon,a,\varphi) + f_2^2(\epsilon,a,\varphi) + 2f_1(\epsilon,a,\varphi)f_2(\epsilon,a,\varphi)[\sin^2\psi - \cos^2\psi] \\
&= f_1^2(\epsilon,a,\varphi) + f_2^2(\epsilon,a,\varphi) + 2f_1(\epsilon,a,\varphi)\cdot\cos\varphi,
\end{aligned}$$

where

$$f_1(\epsilon, a, \varphi) = \frac{(\epsilon + a)^2}{2(\epsilon + a + a\cos\varphi + \sqrt{a^2\cos^2\varphi + \epsilon^2 - a^2})} - a$$

$$= \epsilon \left[\frac{\left(1+\frac{a}{\epsilon}\right)^2}{2\left(1+\frac{a}{\epsilon}+\frac{a}{\epsilon}\cos\varphi + \sqrt{\left(\frac{a}{\epsilon}\right)^2\cos^2\varphi + 1 - \left(\frac{a}{\epsilon}\right)^2}\right)} - \frac{a}{\epsilon} \right]$$

$$f_2(\epsilon, a, \varphi) = \frac{[a\cos\varphi + \sqrt{a^2\cos^2\varphi + \epsilon^2 - a^2}]^2}{2(\epsilon + a + a\cos\varphi + \sqrt{a^2\cos^2\varphi + \epsilon^2 - a^2})}$$

$$= \epsilon \frac{\left[\frac{a}{\epsilon}\cos\varphi + \sqrt{\left(\frac{a}{\epsilon}\right)^2\cos^2\varphi + 1 - \left(\frac{a}{\epsilon}\right)^2}\right]^2}{2\left(1+\frac{\epsilon}{a}+\frac{\epsilon}{a}\cos\varphi + \sqrt{\left(\frac{a}{\epsilon}\right)^2\cos^2\varphi + 1 - \left(\frac{a}{\epsilon}\right)^2}\right)}$$

Using the following definitions

$$g_1(\gamma, \varphi) = \frac{(1+\gamma)^2}{2(1+\gamma+\gamma\cos\varphi + \sqrt{\gamma^2\cos^2\varphi + 1 - \gamma^2}} - \gamma$$

$$g_2(\gamma, \varphi) = \frac{[\gamma\cos\varphi + \sqrt{\gamma^2\cos^2\varphi + 1 - \gamma^2}]^2}{1+\gamma+\gamma\cos\varphi + \sqrt{\gamma^2\cos^2\varphi + 1 - \gamma^2}}$$

we arrive at:

$$\left|\frac{M_\epsilon^0(x) - x}{\epsilon}\right|^2 = g_1^2\left(\frac{a}{\epsilon}, \varphi\right) + g_2^2\left(\frac{a}{\epsilon}, \varphi\right) + 2\cos\varphi g_1\left(\frac{a}{\epsilon}, \varphi\right) g_2\left(\frac{a}{\epsilon}, \varphi\right) \quad (20)$$

and analyzing the local shape on two scales ϵ_1, ϵ_2 one is able to determine a and φ, the distance of x to the edge and the apex angle respectively. The rescaled ϵ-normal $\eta_\epsilon(x) := \epsilon^{-1}|n_\epsilon(x)|$ may serve as an edge indicator. If $\epsilon^{-1}|\eta_\epsilon(x)|$ exceeds a given threshold λ then we expect an edge in a neighborhood of x.

Let us point out that the above computation can be regarded as a first order approximation in case that in an edge two non-linear curve pieces intersect.

6 Applications in Curve and Surface Processing

We are now prepared to discuss the application of anisotropic curvature flow as a powerful multiscale method in curve and surface processing. Hence, we will consider two different approaches. One is based on the definition of an anisotropic diffusion tensor which incorporates the local surface classification based on the approximate shape operator. The other one takes into account the local moment analysis and derives a position dependent anisotropic surface energy integrand. In both cases the

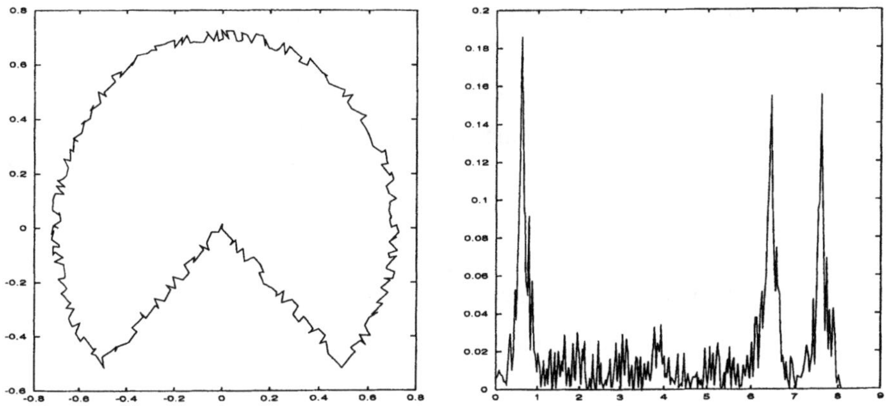

Fig. 8. Left the curve we want to analyze via moments, right the graph of $|n_\epsilon|$, where $\epsilon = 0.2$. On the abscissa of the graph the arc length is drawn; obviously three edges are clearly depicted by $|n_\epsilon|$.

solution of the resulting parabolic problem - either the diffusion problem itself or the gradient flow corresponding to the defined energy - deliver a multiscale of surface representations. For increasing time, in the evolution noise is reduced. Simultaneously features on the curves and surfaces are preserved. Let us already emphasize that especially the gradient flow approach is capable to evolve sharp edges in regions where the classification gives strong evidence for an edge feature. Thus, to our knowledge this is the first multiscale method, which not only retains features for longer times, but robustly enhances them.

We consider a noisy initial curve or surface $\mathcal{M}_0$. Both approaches share the general algorithm outline:

- A local classification is involved to figure out the expected curve or surface shape especially distinguishing between smooth areas, edges and corners (cf. Section 5).
- In time the multiscale evolution is driven by forces derived from the current local classification (cf. Section 3).

Thereby a family of surfaces $\{\mathcal{M}(t)\}_{t\in\mathbb{R}_0^+}$ is generated, where the time t serves as the scale parameter. Spatial discretization based on finite element and time-step algorithms are applied to implement the multiscale methods numerically on usual triangulated curves or surfaces (cf. Section 4). Thus, with respect to our building blocks *classification* and *evolution*, the complete algorithms cycle through the local classification and the next time step. The classification results and the curve or surface metric are always considered explicitly in every time-step.

The resulting methods lead to spatial displacement and the volume enclosed by $\mathcal{M}(t)$ is changed in the evolution. An additional force f in the evolution depending on certain integrated curvature expressions leads to volume preservation and further improves our methods.

6.1 Generalized Mean Curvature Motion in Surface Processing

We will explain how one can use the classification via curvature analysis in our general scheme. To this aim we take into account the tensor $a^{\sigma}_{\mathcal{T}_x\mathcal{M}_\sigma}$ introduced in Section 5. This tensor classifies the surface and indicates edges together with its tangential edge direction as well as corners. Now, we consider a diffusion method involving a closely related tensor $a^{\sigma}_{\mathcal{T}_x\mathcal{M}}$. We use $a^{\sigma}_{\mathcal{T}_x\mathcal{M}_\sigma}$ for its definition.

W.r.t. surface processing, our model integrates tangential smoothing along edges into the multiscale approach. To define the actual diffusion on $\mathcal{T}_x\mathcal{M}$ we decompose a vector $z \in \mathbb{R}^3$ in the orthogonal basis $\{w^{1,\sigma}, w^{2,\sigma}, N^\sigma\}$ (cf. Section 5.1), i.e.,

$$z = (z \cdot w^{1,\sigma})w^{1,\sigma} + (z \cdot w^{2,\sigma})w^{2,\sigma} + (z \cdot N^\sigma)N^\sigma$$

where $\{w^{1,\sigma}, w^{2,\sigma}\} \subset \mathbb{R}^3$ denotes the embedded tangent vectors corresponding to the basis $w^{1,\sigma}$, $w^{2,\sigma}$ (see equation (18)) and N^σ is the surface normal of $\mathcal{M}_\sigma$. Thus, we define the diffusion coefficient $a^{\sigma}_{\mathcal{T}_x\mathcal{M}}$ in a sloppy but intuitive way by

$$\begin{aligned} a^{\sigma}_{\mathcal{T}_x\mathcal{M}}\, z := \Pi_{\mathcal{T}_x\mathcal{M}}\Big(& G(\kappa^{1,\sigma})(z \cdot w^{1,\sigma})w^{1,\sigma} \\ & + G(\kappa^{2,\sigma})(z \cdot w^{2,\sigma})w^{2,\sigma} + (z \cdot N^\sigma)N^\sigma \Big), \end{aligned} \tag{21}$$

Here $\Pi_{\mathcal{T}_x\mathcal{M}}$ denotes the orthogonal projection onto the tangent space $\mathcal{T}_x\mathcal{M}$ and we identify the operator on the abstract tangent space and the endomorphism in $\mathbb{R}^3$.

Using $a^{\sigma}_{\mathcal{T}_x\mathcal{M}}$ as diffusion tensor ends up with the following type of parabolic surface evolution problem. Given an initial compact embedded manifold $\mathcal{M}_0$ in $\mathbb{R}^3$, we compute a one parameter family of manifolds $\{\mathcal{M}(t)\}_{t\in\mathbb{R}_0^+}$ with corresponding coordinate mappings $x(t)$ which solves the system of anisotropic geometric evolution equations:

$$\partial_t x - \mathrm{div}_{\mathcal{M}(t)}(a^{\sigma}_{\mathcal{T}_x\mathcal{M}}\nabla_{\mathcal{M}(t)} x) = 0 \quad \text{on} \quad \mathbb{R}^+ \times \mathcal{M}(t), \tag{22}$$

and satisfies the initial condition

$$\mathcal{M}(0) = \mathcal{M}_0.$$

Hence, due to the anisotropy defined in (19), we enforce a signal enhancement in a principal direction of curvature with curvature larger than λ. If the second principal curvature is smaller than λ we regard the first direction as orthogonal to an important edge on the surface which is going to be sharpened. Simultaneously, in the other direction - the tangent direction along the edge - we invoke smoothing. At approximate corners both principal curvatures are large, thus sharpening takes place in both directions.

To avoid tangential velocity components in the evolution we project the velocity in equation (22) in normal direction and obtain

$$\partial_t x - (\mathrm{div}_{\mathcal{M}}\left(a^{\sigma}_{\mathcal{T}_x\mathcal{M}}\nabla_{\mathcal{M}} x\right) \cdot n)n = 0\,, \tag{23}$$

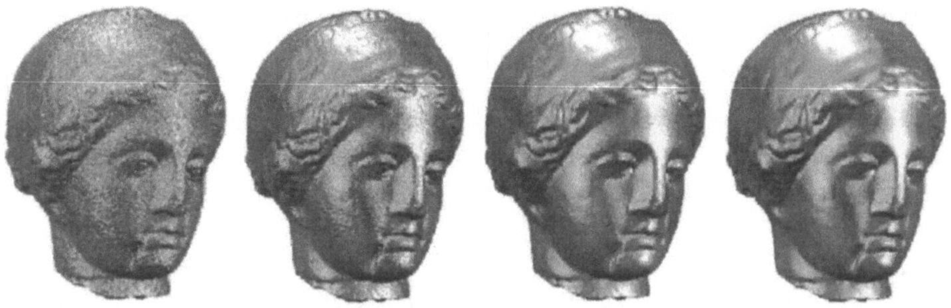

Fig. 9. *The initial surface (top left) and three timesteps from the generalized mean curvature evolution of a venus head consisting of* 268714 *triangles are shown.*

Fig. 10. *Here, for the evolution shown in Figure 9, the norm of the dominant principal curvature is color coded. (cf. color plate 9, Page 670)*

which is an example of the system (7) and indeed our diffusion method is a generalized mean curvature evolution.

In Figure 9 three time-steps of the evolution are shown. Figure 10 demonstrates that curvature is reduced significantly throughout the evolution.

6.2 A Gradient Flow Approach

We will proceed analogously as above but change the type of local classification and consider the gradient flow with respect to a surface energy.

The classification is now with respect to moments. As we saw in Section 5 we can distinguish a non-smooth situation from a smooth one by use of different scaling properties of the ϵ-normal. W.r.t. the definition of a normal velocity we want to have isotropic mean curvature evolution in smooth areas. Close to edges and corners we aim at the definition of an anisotropic mean curvature evolution respecting the local shape of the surface. This can be done by prescribing locally different Wulff-shapes.

Indeed, we consider a gradient flow for the energy (8). This implies that in $x(\xi)$ we prescribe the Wulff shape defined by $\gamma(\xi, \bullet)$.

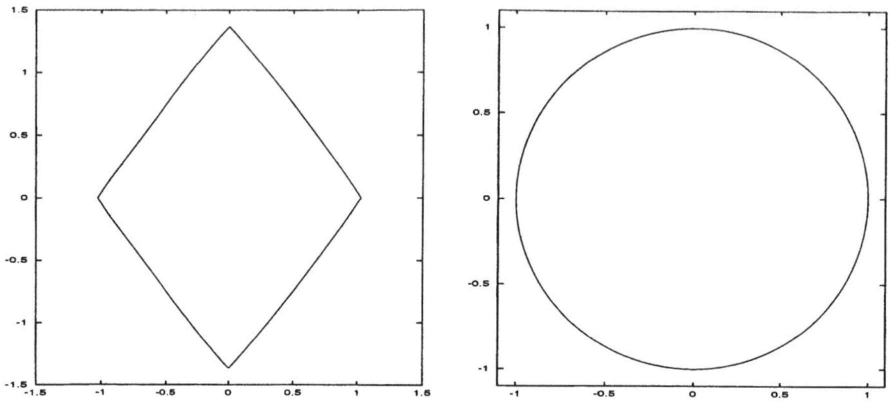

Fig. 11. Two prototypes of prescribed Wulff shapes occurring typically in the evolution steered by the gradient flow of energy (8).

If a point ξ is classified to be smooth and the distance to non-smooth domains is large enough (belonging by definition to the set $\mathcal{S}$), the local Wulff-shape is the sphere, i.e., one has the isotropic integrand $\gamma(\xi, n) = |n|$.

In the present paper we confine to the case of curves concerning the definition of γ in the remaining set $\mathcal{M} - \mathcal{S}$.

On curves, we just have to determine *edges* together with their *apex angles* and the *directions of the apex angle.* Assume that for a given scanning-width ϵ a neighborhood $\mathcal{N}_\epsilon$ of an edge is defined by

$$\mathcal{N}_\epsilon = \{x \in \mathcal{M} \mid \epsilon^{-1}\|n_\epsilon\| \geq \lambda\}$$

where λ is some threshold parameter. A point ξ_0, where n_ϵ achieves a local maximum in $\mathcal{N}_\epsilon$ is defined to be an edge. This is motivated by the observation that in (20) the left hand side is maximal for $a = 0$. The corresponding direction of the apex angle in $x(\xi_0)$ is given by $n_\epsilon(\xi_0)$. It remains to determine the apex angle. Using two scanning-widths ϵ and $r\epsilon$, where r is a positive constant (in our applications we choose $r = 2$), we are able to compute the apex angle by relation (20). Now we define for all $\xi \in \mathcal{N}_\epsilon$ the integrand $\gamma(\xi, n) = \gamma_0(n)$ and γ_0 determines a Wulff-shape whose apex angle and direction of apex angle coincides with the computed quantities of the curve on $\mathcal{N}_\epsilon$. Furthermore we choose γ_0 as an even integrand, i.e., $\gamma_0(n) = \gamma_0(-n)$. This enables us to treat convex and concave situations equivalently. By definition Wulff-shapes are convex bodies and on a first glance it seems to be impossible to enhance concave situations. But by the choice of even integrands it is energetic favourable to enhance a concave edge instead of convexifying the situation. Finally, we use interpolation in $\gamma(\cdot, n)$ which lead to a convex blending from the non-smooth Wulff-shape to the sphere a long the curve if we move from the neighborhood of an edge to a smooth region (see Figure 11).

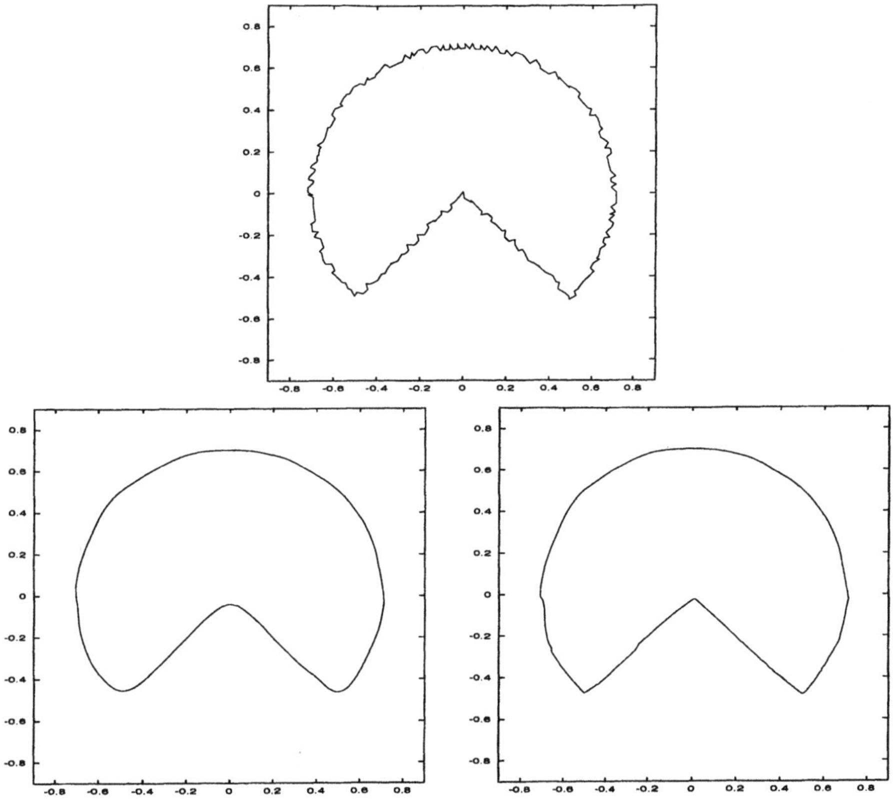

Fig. 12. On top the initial surface is depicted, bottom left a time-step of a mean curvature evolution and bottom right the same time-step is chosen for the anisotropic evolution.

Figure 12 shows a comparison between mean curvature evolution and our gradient flow approach. In Figures 13 and 14 we see results of this method.

Volume Conservation

Now we come to a theorem that allows to define an algorithm for image processing that keeps the enclosed volume fixed.

Theorem 6.8 *Let $a : \mathcal{T}_x\mathcal{M} \to \mathcal{T}_x\mathcal{M}$ be an endomorphism of the tangent-space in every point on $\mathcal{M}$. Then the enclosed volume of the surface does not change under the evolution*

$$\partial_t x - \left(div_{\mathcal{M}} \left(a^{\sigma}_{\mathcal{T}_x\mathcal{M}} \nabla_{\mathcal{M}} x\right) \cdot n\right) n = \overline{h}(t) n$$

if we choose $\overline{h}(t) := \frac{1}{\int_{\mathcal{M}(t)} dA} \int_{\mathcal{M}(t)} h_a \, dA.$

This theorem is an immediate consequence of the following

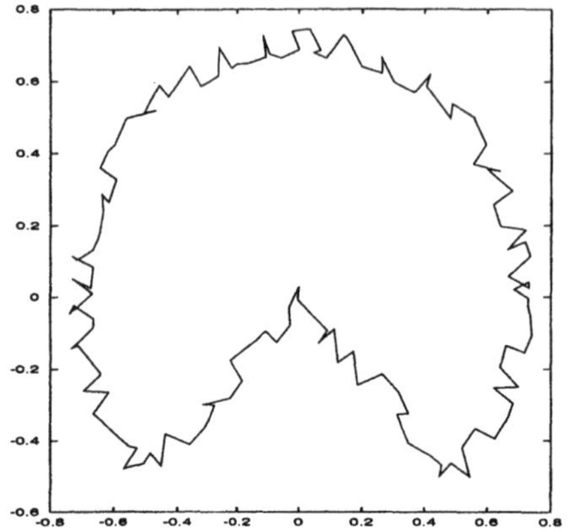

Fig. 13. The method is able to cope with strong noise. Left the initial curve is shown, right the result of the algorithm minimizing the energy (8).

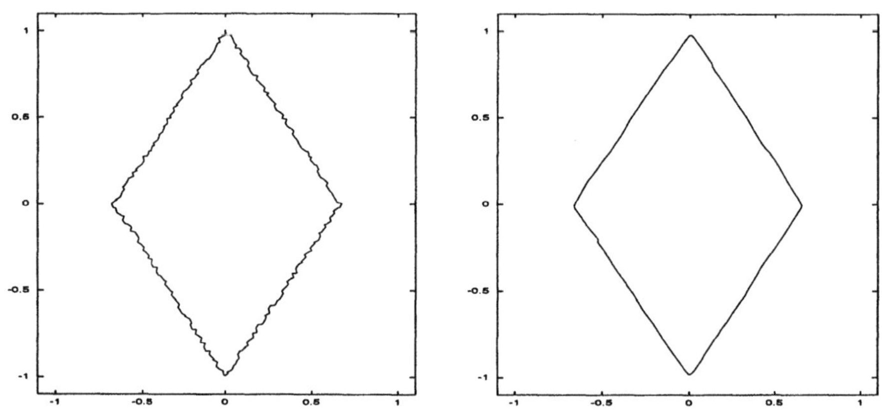

Fig. 14. This figure shows that the gradient flow method is able to handle acute and obtuse angles.

Proposition 6.9 *Consider an evolution of a surface $x(t) : \mathcal{M} \to \mathbb{R}^{d+1}$ in normal direction, i.e., $\partial_t x = \varphi\, n$. The change of the enclosed volume in time is given by*

$$\frac{d}{dt}[Vol(\mathcal{M}(t))]_{t=t_0} = (d+1) \int_{\mathcal{M}(t_0)} \varphi \, dA \,.$$

Proof. *For the time derivative of the volume we obtain*

$$\begin{aligned}
&\partial_t \int_{\mathcal{M}} (x \cdot n)\, dA \\
&= \int_{\mathcal{M}} (\partial_t x \cdot n)\, dA + \int_{\mathcal{M}} (x \cdot \partial_t n)\, dA + \int_{\mathcal{M}} (x \cdot n)\, div_{\mathcal{M}}(\varphi n)\, dA \\
&= \int_{\mathcal{M}} \varphi\, dA - \int_{\mathcal{M}} (x \cdot Dx(\nabla_{\mathcal{M}} \varphi)\, dA + \int_{\mathcal{M}} (x \cdot n)\, \varphi h\, dA
\end{aligned}$$

For the relations used in the above equalities see [6] and [8]. Integration by parts leads to:

$$\begin{aligned}
&\partial_t \int_{\mathcal{M}} (x \cdot n)\, dA \\
&= \int_{\mathcal{M}} \varphi\, dA + \int_{\mathcal{M}} div_{\mathcal{M}} x\, \varphi\, dA = (d+1) \int_{\mathcal{M}} \varphi\, dA
\end{aligned}$$

and we can finish our proof. □

Choosing φ with vanishing mean value leads to constant volume under the evolution and thus Theorem 6.8 is proved.

References

1. L. Alvarez, F. Guichard, P. L. Lions, J. M. Morel: Axioms and fundamental equations of image processing. Arch. Ration. Mech. Anal., 123, 199–257, 1993
2. G. Belletini, M. Paolini: Anisotropic motion by mean curvature in the context of finsler geometry. Hokkaido Math. J., 25, 537–566, 1996
3. T. Bonnesen, W. Fenchel: Konvexe Körper. Springer, 1934
4. F. Catté, P. L. Lions, J. M. Morel, T. Coll: Image selective smoothing and edge detection by nonlinear diffusion. SIAM J. Numer. Anal., 29, 182–193, 1992
5. I. Chavel. Eigenvalues in Riemannian Geometry. Academic Press, 1984
6. U. Clarenz. Enclosure theorems for extremals of elliptic parametric functionals. Calc. Var., online publication DOI 10.1007/s005260100128, 2001
7. U. Clarenz, U. Diewald, M. Rumpf: Nonlinear anisotropic diffusion in surface processing. Proc. Visualization 2000, 397–405, 2000
8. U. Clarenz, H. von der Mosel: Compactness theorems and an isoperimetric inequality for critical points of elliptic parametric functionals. Calc. Var., 12, 85–107, 2001
9. B. Curless, M. Levoy: A volumetric method for building complex models from range images. In: Computer Graphics (SIGGRAPH '96 Proceedings), 303–312, 1996
10. K. Deckelnick, G. Dziuk: A fully discrete numerical scheme for weighted mean curvature flow. to appear in Numer. Math.
11. K. Deckelnick, G. Dziuk: Discrete anisotropic curvature flow of graphs. Math. Modelling Numer. Anal. (RAIRO), 33, 1203–1222, 1999
12. M. Desbrun, M. Meyer, P. Schroeder, A. Barr: Implicit fairing of irregular meshes using diffusion and curvature flow. In: Computer Graphics (SIGGRAPH '99 Proceedings), 317–324, 1999
13. U. Dierkes, S. Hildebrandt, A. Küster, O. Wohlrab: Minimal Surfaces. Grundlehren der Mathematischen Wissenschaften. 295. Berlin: Springer-Verlag, 1992
14. M. P. do Carmo: Riemannian Geometry. Birkhäuser, Boston–Basel–Berlin, 1993

15. G. Dziuk: An algorithm for evolutionary surfaces. Numer. Math., 58, 603–611, 1991
16. G. Dziuk: Convergence of a semi-discrete scheme for the curve shortening flow. Mathematical Models and Methods in Applied Sciences, 4(4), 589–606, 1994
17. G. Dziuk: Discrete anisotropic curve shortening flow. SIAM J. Numer. Anal., 36, 1808–1830, 1999
18. L. Evans, J. Spruck: Motion of level sets by mean curvature. J. Diff. Geom., 33(3), 635–681, 1991
19. I. Fonseca: The wulff theorem revisited. Proc. Roy. Soc. London A, 432, 125–145, 1991
20. I. Fonseca, S. Müller: A uniqueness proof for the wulff theorem. Proc. Roy. Soc. Edinb. A, 119, 125–136, 1991
21. D. Gilbarg, N. Trudinger: Elliptic partial differential equations of second order. Grundlehren der Mathematischen Wissenschaften. 224. Berlin-Heidelberg-New York: Springer-Verlag, 1992
22. I. Guskov, W. Sweldens, P. Schroeder: Multiresolution signal processing for meshes. In: Computer Graphics (SIGGRAPH '99 Proceedings), 1999
23. G. Huisken: The volume preserving mean curvature flow. J. Reine Angew. Math., 382, 35–48, 1987
24. B. Kawohl, N. Kutev: Maximum and comparison principle for one-dimensional anisotropic diffusion. Math. Ann., 311 (1), 107–123, 1998
25. R. Kimmel: Intrinsic scale space for images on surfaces: The geodesic curvature flow. Graphical Models and Image Processing, 59(5), 365–372, 1997
26. L. Kobbelt: Discrete fairing. In: Proceedings of the 7th IMA Conference on the Mathematics of Surfaces, 101–131, 1997
27. L. Kobbelt, S. Campagna, J. Vorsatz, H.-P. Seidel: Interactive multi-resolution modeling on arbitrary meshes. In: Computer Graphics (SIGGRAPH '98 Proceedings), 105–114, 1998
28. W. Lorensen, H. Cline: Marching cubes: A high resolution 3d surface construction algorithm. Computer Graphics, 21(4), 163–169, 1987
29. R. Malladi, J. Sethian: Image processing: Flows under min/max curvature and mean curvature. Graphical Models and Image Processing, 58(2), 127–141, 1996
30. E. Pauwels, P. Fiddelaers, L. Van Gool: Enhancement of planar shape through optimization of functionals for curves. IEEE Trans. Pattern Anal. Mach. Intell., 17, 1101–1105, 1995
31. P. Perona, J. Malik: Scale space and edge detection using anisotropic diffusion. In: IEEE Computer Society Workshop on Computer Vision, 1987
32. T. Preußer, M. Rumpf: A level set method for anisotropic geometric diffusion in 3D image processing. To appear in SIAM J. Appl., 2002
33. K. Räwer: Stabile Extremalen parametrischer Doppelintegrale in $\mathbb{R}^3$. Preprint SFB 256, Univ. Bonn, 321, 1993
34. G. Sapiro: Vector (self) snakes: A geometric framework for color, texture, and multiscale image segmentation. Proc. IEEE International Conference on Image Processing, 1996
35. G. Sapiro: Geometric Partial Differential Equations and Image Processing. Cambridge University Press, 2001
36. G. Taubin: A signal processing approach to fair surface design. In: Computer Graphics (SIGGRAPH '95 Proceedings), 351–358, 1995
37. J. Weickert: Foundations and applications of nonlinear anisotropic diffusion filtering. Z. Angew. Math. Mech., 76, 283–286, 1996
38. B. White: The space of m-dimensional surfaces that are stationary for a parametric elliptic functional. Indiana Univ. Math. J., 36, 567–602, 1987
39. G. Wulff: Zur Frage der Geschwindigkeit des Wachsthums und der Auflösung der Kristallflächen. Zeitschrift der Kristallographie, 34, 449–530, 1901

A Finite Element Level Set Method for Anisotropic Mean Curvature Flow with Space Dependent Weight

Klaus Deckelnick[1] and Gerhard Dziuk[2]

[1] Centre for Mathematical Analysis and Its Applications, School of Mathematical Sciences, University of Sussex, Falmer Brighton BN1 9QH, UK
K.P.Deckelnick@sussex.ac.uk

[2] Institut für Angewandte Mathematik, Albert-Ludwigs-Universiät Freiburg, Hermann-Herder-Straße 10, 79104 Freiburg, Germany
gerd@mathematik.uni-freiburg.de

1 Introduction

The subject of this paper is the evolution of hypersurfaces $\Gamma(t) \subset \mathbb{R}^d$ according to the law

$$v = H_\gamma + f \qquad \text{on } \Gamma(t). \tag{1}$$

Here, v is the normal velocity of $\Gamma(t)$ and f is a given function. Furthermore, H_γ denotes the anisotropic mean curvature with respect to the positive weight function $\gamma = \gamma(x,p) : \Omega \times S^{d-1} \to \mathbb{R}$. Here, x represents the space variables, while $p \in S^{d-1}$ indicates the dependence of γ on the orientation of the hypersurface. We think of γ as being extended to $\Omega \times \mathbb{R}^d$ in such a way that γ is convex and 1–homogeneous in p. The weighted mean curvature H_γ then is the first variation of

$$\int_\Gamma \gamma(x, \nu(x)) dA(x), \tag{2}$$

where $\nu(x)$ is the unit normal to Γ. For more information on this aspect see §2 and [1].

There are various applications of (1) in material science, in which $\Gamma(t)$ typically represents the interface between two different phases (e.g. solid/liquid) or between crystals of different orientation. In such a situation the weight function γ stands for the interfacial energy. A well–known example is the mathematical model for dendritic growth ([16], [13]), which couples the heat equation in the solid and the liquid phase of the material to the Stefan condition and a slightly more general version of (1), which both hold on the interface. Thus, the relation (1), which in this context is referred to as the Gibbs–Thomson law, occurs as a subproblem of a moving boundary value problem and tracking the interface $\Gamma(t)$ forms part of its solution.

While in the above mentioned situation the weight function γ will in general only depend on the orientation of the hypersurface, the law (1) with an explicit dependence of γ on x has recently found applications in image and surface processing. Here, anisotropic surface diffusion is used in order to smoothen noisy images or surfaces while simultaneously enhancing edges or corners. This is achieved by introducing an anisotropic energy γ which reflects detected local geometric features, [2, 15, 4].

In order to solve the problem (1) analytically or numerically, a suitable description of the hypersurface $\Gamma(t)$ is required. The choice of description will depend on properties of the problem under consideration (such as the occurrence of topological changes) and popular choices are the parametric approach, the level set method [14] and the phase field method. The level set and phase field approaches have the advantage that they are capable of handling topological changes in the evolving hypersurfaces.

In this work we shall be concerned with a level set method for solving (1). The PDE satisfied by the level set function is nonlinear, singular and degenerate. Nevertheless, the existence of a unique viscosity solution has been obtained in [12], [3] under quite general assumptions on the data of the problem. This solution will in general be only Lipschitz–continuous in space and time. Due to the degenerate and singular nature of the PDE and the low regularity of the viscosity solution, its numerical approximation and the corresponding convergence analysis are very difficult. Our approach here is based on the fact that the viscosity solution can be approximated locally uniformly by the (smooth) solutions of a sequence of regularized problems. It is therefore natural to discretise the regularized equation, for which we shall use a finite element approach. For a fixed value of the regularization parameter and the resulting semidiscrete scheme we shall obtain optimal error estimates. The corresponding analysis involves energy arguments for geometric quantities like the normal or the normal velocity. Even though these quantities are nonlinear expressions of the level set function and its derivatives, their use leads to a natural and fairly transparent convergence analysis.

The paper is organized as follows: We will first define what we mean by weighted mean curvature flow of level sets with space dependent anisotropy and introduce the differential equation satisfied by the level set function. We then regularize the problem with the help of a suitable extension of the weight function and discretise the resulting partial differential equation using piecewise linear finite elements. The main part of this paper is devoted to the convergence proof and we conclude with computational results.

Let us mention some results which are related to our proof. In [6] the authors proved error estimates for the isotropic mean curvature flow of twodimensional graphs. A generalization to the anisotropic situation is contained in [7]. The time discretization was discussed in [11], [9] and [5]. [8] contains a proof of the fact that the discrete solution of the regularized equation in fact approximates the viscosity solution. This is done for the isotropic case.

2 Anisotropic Mean Curvature Flow

Throughout this paper we shall use weight functions of the following type. Let Ω be a subset of $\mathbb{R}^n$. An admissible anisotropy or weight function $\gamma = \gamma(x,p)$ is a positive function $\gamma : \overline{\Omega} \times \mathbb{R}^n \to \mathbb{R}^+$, which for fixed $x \in \overline{\Omega}$ satisfies $\gamma(x,\cdot) \in C^3(\mathbb{R}^n \setminus \{0\})$

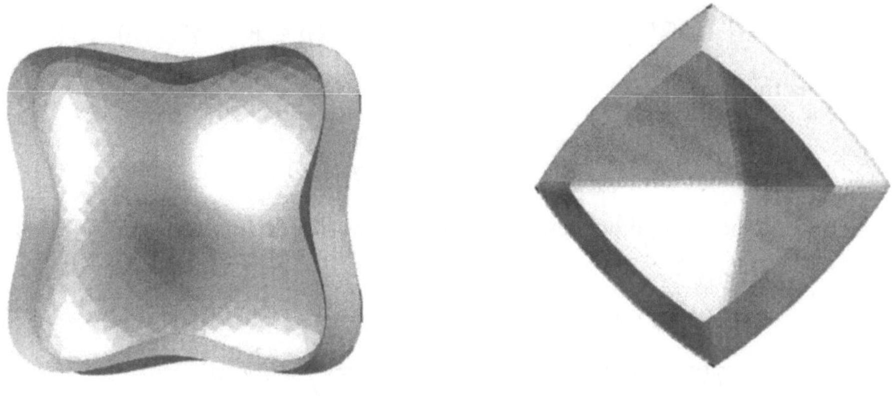

Fig. 1. Frank-diagram and the corresponding Wulff-shape (scaled) for a non-convex anisotropy.

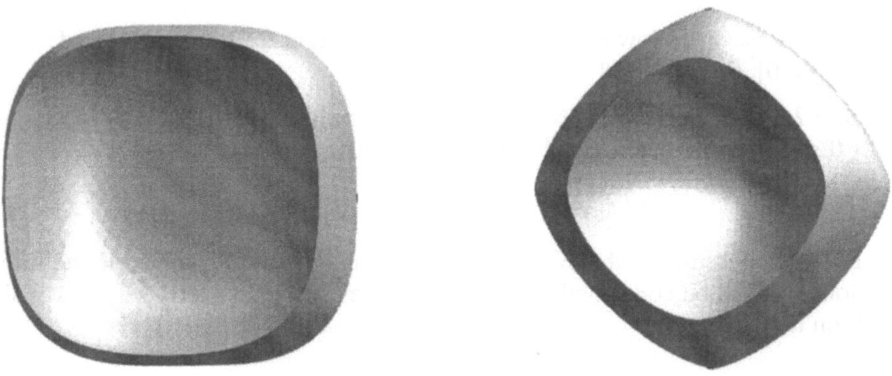

Fig. 2. Frank-diagram and the corresponding Wulff-shape (scaled) for a convex anisotropy.

and $\gamma(\cdot, p) \in L^\infty(\Omega)$ for $p \in \mathbb{R}^n$, which is positively homogeneous of degree one,

$$\gamma(x, \lambda p) = \lambda \gamma(x, p), \ \text{for}\ p \in \mathbb{R}^n, \lambda > 0, \tag{3}$$

and which is convex in the sense that there exists a constant $c_0 > 0$ such that

$$\gamma_{pp}(x, p) q \cdot q \geq c_0 |q|^2 \ \text{for all}\ p, q \in \mathbb{R}^n \ \text{with}\ p \cdot q = 0, |p| = 1 \tag{4}$$

and all $x \in \Omega$. By γ_{pp} we denote the matrix of the second derivatives of γ with respect to p. Analogously we use γ_{ppp} and γ_p. We also assume that

$$\sup_{x \in \Omega, p \in S^{n-1}} \left(|\gamma_p(x, p)| + |\gamma_{pp}(x, p)| + |\gamma_{ppp}(x, p)| \right) < \infty.$$

We emphasize that we allow the weight functions γ to depend on the space variables. Anisotropy can be visualized using the Frank diagram $\mathcal{F}$ and the Wulff shape $\mathcal{W}$. In our case these sets vary with $x \in \Omega$.

$$\mathcal{F}(x) = \{p \in \mathbb{R}^n \mid \gamma(x,p) \le 1\}, \quad \mathcal{W}(x) = \{p \in \mathbb{R}^n \mid \gamma^*(x,p) \le 1\}.$$

Here γ^* is the dual of γ,

$$\gamma^*(x,q) = \sup_{p \in \mathbb{R}^n \setminus \{0\}} \frac{p \cdot q}{\gamma(x,p)}.$$

As a consequence of homogeneity and smoothness of $\gamma(x,\cdot)$ we have for $\lambda > 0$ and all $p \in \mathbb{R}^n \setminus \{0\}$,

$$\gamma_p(x,p) \cdot p = \gamma(x,p), \; \gamma_{pp}(x,p)p = 0, \tag{5}$$

$$\gamma_p(x,\lambda p) = \gamma_p(x,p), \; \gamma_{pp}(x,\lambda p) = \frac{1}{\lambda}\gamma_{pp}(x,p), \tag{6}$$

$$\gamma_{ppp}(x,\lambda p) = \frac{1}{\lambda^2}\gamma_{ppp}(x,p).$$

We will use the following simple result about weight functions γ. A proof can be found in [10], Proposition 1.

Lemma 2.1 *Let γ be an admissible weight function. Then there exists a constant $c_1 > 0$ such that for $p, q \in S^{n-1}$ with $|p+q| > 0$ and $x \in \Omega$*

$$\gamma(x,q) - \gamma_p(x,p) \cdot q \ge c_1 |q-p|^2. \tag{7}$$

Let us now return to the evolution law (1). As already mentioned in the Introduction we shall be concerned with hypersurfaces which are level sets of a scalar function $u = u(x,t)$,

$$\Gamma(t) = \{x \in \mathbb{R}^d \,|\, u(x,t) = 0\}.$$

At points where $\nabla u \neq 0$, normal ν and normal velocity v of Γ are given by

$$\nu = \frac{\nabla u}{|\nabla u|}, \qquad v = \frac{u_t}{|\nabla u|}. \tag{8}$$

Next, for a given anisotropy function $\gamma = \gamma(x,p)$, anisotropic mean curvature H_γ can be formally introduced as the first variation of the weighted area

$$\int_\Gamma \gamma(x,\nu(x))\,dA(x).$$

In the context of the level set description of $\Gamma(t)$, this leads to the following expression for H_γ:

$$H_\gamma = \nabla \cdot \gamma_p(\cdot,\nabla u) = \sum_{j,k=1}^{d} \gamma_{p_j p_k}(\cdot,\nabla u) u_{x_j x_k} + \sum_{j=1}^{d} \gamma_{p_j x_j}(\cdot,\nabla u). \tag{9}$$

In conclusion, for $n = d$ the level set function u has to satisfy the following nonlinear, singular and degenerate PDE

$$u_t - |\nabla u| \, \nabla \cdot \gamma_p(x, \nabla u) = f(x) \, |\nabla u| \tag{10}$$

or equivalently,

$$\frac{u_t}{|\nabla u|} - \sum_{j,k=1}^{d} \gamma_{p_j p_k}(x, \nabla u) u_{x_j x_k} - \sum_{j=1}^{d} \gamma_{p_j x_j}(x, \nabla u) = f(x). \tag{11}$$

Note that for the choices $\gamma(x, p) = |p|$, $f = 0$ we recover the well–studied level set equation

$$u_t - \sum_{j,k=1}^{d} \left(\delta_{jk} - \frac{u_{x_j} u_{x_k}}{|\nabla u|^2} \right) u_{x_j x_k} = 0 \tag{12}$$

and all level sets of u formally evolve according to (isotropic) mean curvature flow.

3 Regularization

The nonlinear equation (10) is degenerate parabolic and not defined where $\nabla u = 0$, a situation, which occurs for example, when the topology of Γ changes. We regularize the equation by using an extension of the anisotropy to $d+1$ space dimensions. We assume that there exists an admissible weight function $\overline{\gamma}$ with $n = d + 1$,

$$\overline{\gamma} = \overline{\gamma}(x_1, \ldots, x_d, p_1, \ldots, p_{d+1})$$

such that

$$\overline{\gamma}(x_1, \ldots, x_d, p_1, \ldots, p_d, 0) = \gamma(x_1, \ldots, x_d, p_1, \ldots, p_d).$$

$\overline{\gamma}$ does not depend on x_{d+1}. In the following we denote this extension again by γ.

Rather than treating (10) we introduce for a (small) positive parameter ε the regularized problem

$$\frac{u_t^\varepsilon}{\sqrt{\varepsilon^2 + |\nabla u^\varepsilon|^2}} - \nabla \cdot \gamma_p(x, \nabla u^\varepsilon, -\varepsilon) = f(x) \tag{13}$$

or written explicitly,

$$\frac{u_t^\varepsilon}{\sqrt{\varepsilon^2 + |\nabla u^\varepsilon|^2}} - \sum_{j,k=1}^{d} \gamma_{p_j p_k}(x, \nabla u^\varepsilon, -\varepsilon) u^\varepsilon_{x_j x_k} - \sum_{j=1}^{d} \gamma_{p_j x_j}(x, \nabla u^\varepsilon, -\varepsilon) = f(x).$$

We consider this differential equation on $\Omega \times (0, T)$, where $\Omega \subset \mathbb{R}^d$ is a bounded smooth domain and $T > 0$ is some final time. Furthermore, we prescribe the initial and boundary conditions

$$u^\varepsilon(x, t) = u_0(x), \quad (x, t) \in \overline{\Omega} \times \{0\} \cup \partial\Omega \times [0, T]. \tag{14}$$

In what follows we shall assume that the problem (13), (14) has a unique solution u^ε, which satisfies

$$\|u_t^\varepsilon\|_{L^\infty(\Omega\times(0,T))} + \|\nabla u^\varepsilon\|_{L^\infty(\Omega\times(0,T))} + \int_0^T \|\nabla u_t^\varepsilon\|_{L^\infty(\Omega)}dt + \int_0^T \|u_t^\varepsilon\|^2_{H^2(\Omega)}dt \le M < \infty. \tag{15}$$

The above regularity properties will allow us to carry out an error analysis for our finite element method to be presented in the next section.

4 Finite Element Method

Although the differential equation (13) is not in divergence form, we can introduce the following variational form, which is consistent with the underlying law of motion (1),

$$\int_\Omega \frac{u_t^\varepsilon}{\sqrt{\varepsilon^2 + |\nabla u^\varepsilon|^2}}\varphi + \sum_{j=1}^d \int_\Omega \gamma_{p_j}(\cdot, \nabla u^\varepsilon, -\varepsilon)\varphi_{x_j} = \int_\Omega f\varphi \qquad \forall \varphi \in H_0^1(\Omega). \tag{16}$$

This is the basis for a finite element method. For this let $\mathcal{T}_h$ be a family of triangulations of Ω with maximum mesh size $h := \max_{S\in\mathcal{T}_h} \operatorname{diam}(S)$. We denote by Ω_h the corresponding discrete domain, i.e.

$$\bar{\Omega}_h = \bigcup_{S\in\mathcal{T}_h} S$$

and assume that all vertices on $\partial\Omega_h$ also lie on $\partial\Omega$. Furthermore we suppose that the triangulation is nondegenerate in the sense that

$$\max_{S\in\mathcal{T}_h} \frac{\operatorname{diam}(S)}{\rho_S} \le \kappa$$

where the constant $\kappa > 0$ is independent of h and ρ_S denotes the radius of the largest ball which is contained in S.

As a discrete function space we use the space of piecewise linear finite elements on the discrete domain Ω_h,

$$X_h := \{v_h \in C^0(\bar{\Omega}_h) \,|\, v_h \text{ is a linear polynomial on each } S \in \mathcal{T}_h\},$$

and

$$\mathring{X}_h := X_h \cap H_0^1(\Omega_h).$$

A function from $\mathring{X}_h$ is always set zero outside Ω_h.

There exists an interpolation operator $I_h : H^2(\Omega_h) \to X_h$ mapping $H^2(\Omega_h) \cap H_0^1(\Omega)$ into $\mathring{X}_h$ such that

$$\|v - I_h v\|_{L^2(\Omega_h)} + h\|\nabla(v - I_h v)\|_{L^2(\Omega_h)} \le ch^2\|v\|_{H^2(\Omega_h)} \tag{17}$$

for all $v \in H^2(\Omega_h)$.

A spatially discrete solution of (16) is a function $u_h^\varepsilon : \Omega_h \times [0, T) \to \mathbb{R}$ with $u_h^\varepsilon(\cdot, t) \in X_h$, $(t \in [0, T))$ such that

$$\int_{\Omega_h} \frac{u_{ht}^\varepsilon}{\sqrt{\varepsilon^2 + |\nabla u_h^\varepsilon|^2}} \varphi_h + \sum_{j=1}^d \int_{\Omega_h} \gamma_{p_j}(\cdot, \nabla u_h^\varepsilon, -\varepsilon)\varphi_{hx_j} = \int_{\Omega_h} f\varphi_h \tag{18}$$

for all $\varphi_h \in \mathring{X}_h, t \in (0, T)$ and

$$u_h^\varepsilon(x, t) = I_h u_0(x), \quad (x, t) \in \overline{\Omega_h} \times \{0\} \cup \partial\Omega_h \times [0, T]. \tag{19}$$

5 Convergence

From now on we will use the following abbreviations for the geometric quantities which appear in the differential equations.

$$Q(v) = \sqrt{\varepsilon^2 + |\nabla v|^2}, \quad \nu(v) = \frac{(\nabla v, -\varepsilon)}{\sqrt{\varepsilon^2 + |\nabla v|^2}} = \frac{(\nabla v, -\varepsilon)}{Q(v)},$$
$$V(v) = -\frac{v_t}{\sqrt{\varepsilon^2 + |\nabla v|^2}} = -\frac{v_t}{Q(v)}.$$

Theorem 5.2 *Suppose that u^ε is the a solution of (13), (14) with finite norms (15) and let u_h^ε be the solution of (18), (19). Then*

$$\int_0^T \int_\Omega (V(u^\varepsilon) - V(u_h^\varepsilon))^2 Q(u_h^\varepsilon) + \sup_{(0,T)} \int_\Omega |\nu(u^\varepsilon) - \nu(u_h^\varepsilon)|^2 Q(u_h^\varepsilon) \le c(\varepsilon)h^2.$$

The constant $c(\varepsilon)$ depends on the solution u^ε of (13), (14) through the norms occurring in (15).

Proof. We prove the result for the case of a convex domain Ω. The general case only adds technical difficulties which can be treated with standard methods from numerical analysis (see e. g. [5]). We may therefore assume from now on that $\Omega_h \subset \Omega$.

Because of (5) the difference between smooth (16) with $\varphi = \varphi_h$ and discrete (18) variational equation implies

$$-\int_{\Omega_h} (V(u^\varepsilon) - V(u_h^\varepsilon))\,\varphi_h + \sum_{j=1}^d \int_{\Omega_h} (\gamma_{p_j}(\cdot, \nu(u^\varepsilon)) - \gamma_{p_j}(\cdot, \nu(u_h^\varepsilon)))\,\varphi_{hx_j} = 0$$

for every $\varphi_h \in \mathring{X}_h$. Since the boundary values are independent of time, we may use

$$\varphi_h = I_h u_t^\varepsilon - u_{ht}^\varepsilon = (u_t^\varepsilon - u_{ht}^\varepsilon) - (u_t^\varepsilon - I_h u_t^\varepsilon)$$

as a test function and get

$$\begin{aligned}
&-\int_{\Omega_h} (V(u^\varepsilon) - V(u_h^\varepsilon))\,(u_t^\varepsilon - u_{ht}^\varepsilon) \qquad (20)\\
&\qquad + \sum_{j=1}^{d} \int_{\Omega_h} \left(\gamma_{p_j}(\cdot,\nu(u^\varepsilon)) - \gamma_{p_j}(\cdot,\nu(u_h^\varepsilon))\right)(u_{tx_j}^\varepsilon - u_{htx_j}^\varepsilon)\\
&= -\int_{\Omega_h} (V(u^\varepsilon) - V(u_h^\varepsilon))\,(u_t^\varepsilon - I_h u_t^\varepsilon)\\
&\qquad + \sum_{j=1}^{d} \int_{\Omega_h} \left(\gamma_{p_j}(\cdot,\nu(u^\varepsilon)) - \gamma_{p_j}(\cdot,\nu(u_h^\varepsilon))\right)(u_{tx_j}^\varepsilon - (I_h u_t^\varepsilon))_{x_j}).
\end{aligned}$$

Let us estimate the terms in this equation separately. We begin with the two main estimates for the terms on the left hand side of (20). Firstly,

$$\begin{aligned}
&-\int_{\Omega_h} (V(u^\varepsilon) - V(u_h^\varepsilon))\,(u_t^\varepsilon - u_{ht}^\varepsilon) \qquad (21)\\
&= \int_{\Omega_h} (V(u^\varepsilon) - V(u_h^\varepsilon))^2 Q(u_h^\varepsilon) + \int_{\Omega_h} V(u^\varepsilon)(Q(u^\varepsilon) - Q(u_h^\varepsilon))(V(u^\varepsilon) - V(u_h^\varepsilon))\\
&\geq \int_{\Omega_h} (V(u^\varepsilon) - V(u_h^\varepsilon))^2 Q(u_h^\varepsilon)\\
&\qquad -\frac{1}{\varepsilon}\|u_t^\varepsilon\|_{L^\infty(\Omega)} \int_{\Omega_h} |\nu(u^\varepsilon) - \nu(u_h^\varepsilon)||V(u^\varepsilon) - V(u_h^\varepsilon)|Q(u_h^\varepsilon)\\
&\geq \frac{3}{4}\int_{\Omega_h} (V(u^\varepsilon) - V(u_h^\varepsilon))^2 Q(u_h^\varepsilon) - \frac{1}{\varepsilon^2}\|u_t^\varepsilon\|_{L^\infty(\Omega)}^2 \int_{\Omega_h} |\nu(u^\varepsilon) - \nu(u_h^\varepsilon)|^2 Q(u_h^\varepsilon).
\end{aligned}$$

Here we have used Young's inequality and the fact that

$$|Q(u^\varepsilon) - Q(u_h^\varepsilon)| \leq \frac{1}{\varepsilon}|\nu(u^\varepsilon) - \nu(u_h^\varepsilon)|Q(u^\varepsilon)Q(u_h^\varepsilon). \qquad (22)$$

The main term on the left hand side of (20) is the one which contains the anisotropy. Observing that the $(d+1)$–st component of the vector

$$\nu(u^\varepsilon)Q(u^\varepsilon) - \nu(u_h^\varepsilon)Q(u_h^\varepsilon)$$

equals zero we conclude

$$\sum_{j=1}^{d} \int_{\Omega_h} \left(\gamma_{p_j}(\cdot, \nu(u^\varepsilon)) - \gamma_{p_j}(\cdot, \nu(u_h^\varepsilon))\right) \partial_t (u_{x_j}^\varepsilon - u_{hx_j}^\varepsilon) \tag{23}$$

$$= \int_{\Omega_h} (\gamma_p(\cdot, \nu(u^\varepsilon)) - \gamma_p(\cdot, \nu(u_h^\varepsilon))) \cdot \partial_t \left(\nu(u^\varepsilon)Q(u^\varepsilon) - \nu(u_h^\varepsilon)Q(u_h^\varepsilon)\right)$$

$$= -\int_{\Omega_h} (\gamma_p(\cdot, \nu(u^\varepsilon)) - \gamma_p(\cdot, \nu(u_h^\varepsilon))) \cdot \{\nu(u_h^\varepsilon)\partial_t Q(u_h^\varepsilon) + \partial_t(\nu(u_h^\varepsilon))Q(u_h^\varepsilon) \\ -\partial_t(\nu(u^\varepsilon))Q(u^\varepsilon) - \nu(u^\varepsilon)\partial_t Q(u^\varepsilon)\}$$

$$= I_1 + I_2 + I_3 + I_4. \tag{24}$$

We use the homogeneity properties (5) of γ. For the terms $I_1, \dots, I_4$ this gives

$$I_1 = -\int_{\Omega_h} (\gamma_p(\cdot, \nu(u^\varepsilon)) - \gamma_p(\cdot, \nu(u_h^\varepsilon))) \cdot \nu(u_h^\varepsilon)\partial_t Q(u_h^\varepsilon) \tag{25}$$

$$= \int_{\Omega_h} (\gamma(\cdot, \nu(u_h^\varepsilon) - \gamma_p(\cdot, \nu(u^\varepsilon)) \cdot \nu(u_h^\varepsilon))\, \partial_t Q(u_h^\varepsilon)$$

$$= \frac{d}{dt} \int_{\Omega_h} (\gamma(\cdot, \nu(u_h^\varepsilon)) - \gamma_p(\cdot, \nu(u^\varepsilon)) \cdot \nu(u_h^\varepsilon))\, Q(u_h^\varepsilon)$$

$$+ \int_{\Omega_h} \gamma_{pp}(\cdot, \nu(u^\varepsilon))\nu(u_h^\varepsilon) \cdot \partial_t \nu(u^\varepsilon)\, Q(u_h^\varepsilon)$$

$$+ \int_{\Omega_h} \gamma_p(\cdot, \nu(u^\varepsilon)) \cdot \partial_t \nu(u_h^\varepsilon)\, Q(u_h^\varepsilon) - \int_{\Omega_h} \partial_t \gamma(\cdot, \nu(u_h^\varepsilon))\, Q(u_h^\varepsilon).$$

For the second integral we have

$$I_2 = -\int_{\Omega_h} (\gamma_p(\cdot, \nu(u^\varepsilon)) - \gamma_p(\cdot, \nu(u_h^\varepsilon))) \cdot \partial_t(\nu(u_h^\varepsilon))\, Q(u_h^\varepsilon) \tag{26}$$

$$= -\int_{\Omega_h} \gamma_p(\cdot, \nu(u^\varepsilon)) \cdot \partial_t(\nu(u_h^\varepsilon))Q(u_h^\varepsilon) + \int_{\Omega_h} \partial_t \gamma(\cdot, \nu(u_h^\varepsilon))Q(u_h^\varepsilon),$$

and I_4 can be written as

$$I_4 = \int_{\Omega_h} (\gamma_p(\cdot, \nu(u^\varepsilon)) - \gamma_p(\cdot, \nu(u_h^\varepsilon)))\, \nu(u^\varepsilon)\partial_t Q(u^\varepsilon) \tag{27}$$

$$= \int_{\Omega_h} \gamma(\cdot, \nu(u^\varepsilon))\partial_t Q(u^\varepsilon) - \int_{\Omega_h} \gamma_p(\cdot, \nu(u_h^\varepsilon)) \cdot \nu(u^\varepsilon)\partial_t Q(u^\varepsilon).$$

We add (25), (26), (27) and I_3,

$$I_3 = \int_{\Omega_h} (\gamma_p(\cdot, \nu(u^\varepsilon)) - \gamma_p(\cdot, \nu(u_h^\varepsilon)))\, \partial_t \nu(u^\varepsilon)\, Q(u^\varepsilon), \tag{28}$$

and arrive at the following expression for (23):

$$\sum_{j=1}^{d}\int_{\Omega_h}(\gamma_{p_j}(\cdot,\nu(u^\varepsilon))-\gamma_{p_j}(\cdot,\nu(u_h^\varepsilon)))\,\partial_t(u_{x_j}^\varepsilon-u_{hx_j}^\varepsilon) \tag{29}$$
$$= \frac{d}{dt}\int_{\Omega_h}(\gamma(\cdot,\nu(u_h^\varepsilon))-\gamma_p(\cdot,\nu(u^\varepsilon))\cdot\nu(u_h^\varepsilon))\,Q(u_h^\varepsilon)$$
$$+\int_{\Omega_h}(\gamma(\cdot,\nu(u^\varepsilon))-\gamma_p(\cdot,\nu(u_h^\varepsilon))\cdot\nu(u^\varepsilon))\,\partial_t Q(u^\varepsilon)$$
$$+\int_{\Omega_h}(\gamma_p(\cdot,\nu(u^\varepsilon))-\gamma_p(\cdot,\nu(u_h^\varepsilon)))\,\partial_t\nu(u^\varepsilon)\,Q(u^\varepsilon)$$
$$+\int_{\Omega_h}\gamma_{pp}(\cdot,\nu(u^\varepsilon))\nu(u_h^\varepsilon)\cdot\partial_t\nu(u^\varepsilon)\,Q(u_h^\varepsilon).$$
$$= \frac{d}{dt}\int_{\Omega_h}(\gamma(\cdot,\nu(u_h^\varepsilon))-\gamma_p(\cdot,\nu(u^\varepsilon))\cdot\nu(u_h^\varepsilon))\,Q(u_h^\varepsilon)$$
$$+\int_{\Omega_h}\{\gamma(\cdot,\nu(u^\varepsilon))-[\gamma(\cdot,\nu(u_h^\varepsilon))+\gamma_p(\cdot,\nu(u_h^\varepsilon))\cdot(\nu(u^\varepsilon)-\nu(u_h^\varepsilon))]\}\partial_t Q(u^\varepsilon)$$
$$-\int_{\Omega_h}\{\gamma_p(\cdot,\nu(u_h^\varepsilon))-[\gamma_p(\cdot,\nu(u^\varepsilon))+\gamma_{pp}(\cdot,\nu(u^\varepsilon))(\nu(u_h^\varepsilon)-\nu(u^\varepsilon))]\}$$
$$\cdot\partial_t\nu(u^\varepsilon)\,Q(u^\varepsilon)$$
$$+\int_{\Omega_h}\gamma_{pp}(\cdot,\nu(u^\varepsilon))(\nu(u_h^\varepsilon)-\nu(u^\varepsilon))\cdot\partial_t\nu(u^\varepsilon)(Q(u_h^\varepsilon)-Q(u^\varepsilon)).$$

Two terms in the above formula contain Taylor expansions of the anisotropy function γ with respect to the p–variable. If we let $\nu_s = s\nu(u^\varepsilon)+(1-s)\nu(u_h^\varepsilon)$, $\tilde\nu_s = s\nu(u_h^\varepsilon)+(1-s)\nu(u^\varepsilon)$ and observe that $\nu_s,\ \tilde\nu_s\neq 0$ for all $s\in[0,1]$ we can continue

$$= \frac{d}{dt}\int_{\Omega_h}(\gamma(\cdot,\nu(u_h^\varepsilon))-\gamma_p(\cdot,\nu(u^\varepsilon))\cdot\nu(u_h^\varepsilon))\,Q(u_h^\varepsilon)$$
$$+\int_{\Omega_h}\int_0^1(1-s)\gamma_{pp}(\cdot,\nu_s)(\nu(u^\varepsilon)-\nu(u_h^\varepsilon))\cdot(\nu(u^\varepsilon)-\nu(u_h^\varepsilon))\partial_t Q(u^\varepsilon)$$
$$-\sum_{k=1}^{d+1}\int_{\Omega_h}\int_0^1(1-s)\gamma_{p_kpp}(\cdot,\tilde\nu_s)(\nu(u^\varepsilon)-\nu(u_h^\varepsilon)\cdot(\nu(u^\varepsilon)-\nu(u_h^\varepsilon))\partial_t\nu(u^\varepsilon)_k\,Q(u^\varepsilon)$$
$$+\int_{\Omega_h}\gamma_{pp}(\cdot,\nu(u^\varepsilon))(\nu(u_h^\varepsilon)-\nu(u^\varepsilon))\cdot\partial_t\nu(u^\varepsilon)(Q(u_h^\varepsilon)-Q(u^\varepsilon)).$$

In order to estimate the second and the third integral we require a lower bound on $|\nu_s|$ and $|\tilde\nu_s|$ respectively. To this purpose we calculate for $s\in[0,1]$

$$|\nu_s|^2 = s^2+(1-s)^2+2s(1-s)\frac{\nabla u^\varepsilon\cdot\nabla u_h^\varepsilon+\varepsilon^2}{Q(u^\varepsilon)Q(u_h^\varepsilon)}$$
$$\geq 1-2s(1-s)\Big(1-\frac{\nabla u^\varepsilon}{Q(u^\varepsilon)}\cdot\frac{\nabla u_h^\varepsilon}{Q(u_h^\varepsilon)}\Big)\geq 1-\frac{1}{2}\Big(1+\frac{|\nabla u^\varepsilon|}{Q(u^\varepsilon)}\Big).$$

Since (15) implies that $|\nabla u^\varepsilon| \le M$ we obtain

$$|\nu_s|^2 \ge 1 - \frac{1}{2}(1 + \frac{M}{\sqrt{\varepsilon^2 + M^2}})$$

so that (6) yields

$$|\gamma_{pp}(\cdot, \nu_s)| = |\gamma_{pp}(\cdot, \frac{\nu_s}{|\nu_s|})| \frac{1}{|\nu_s|} \le c(\varepsilon, M) \sup_{x \in \Omega, p \in S^d} |\gamma_{pp}(x, p)|$$

uniformly in $s \in [0, 1]$. Thus, we can estimate

$$\begin{aligned}
&\int_{\Omega_h} \int_0^1 (1-s)\gamma_{pp}(\cdot, \nu_s)(\nu(u^\varepsilon) - \nu(u_h^\varepsilon)) \cdot (\nu(u^\varepsilon) - \nu(u_h^\varepsilon))\partial_t Q(u^\varepsilon) \\
&\le \int_{\Omega_h} \int_0^1 (1-s)|\gamma_{pp}(\cdot, \nu_s)||\nu(u^\varepsilon) - \nu(u_h^\varepsilon)|^2 |\nabla u_t^\varepsilon| \\
&\le c(\varepsilon, M)\|\nabla u_t^\varepsilon\|_{L^\infty(\Omega)} \int_{\Omega_h} |\nu(u^\varepsilon) - \nu(u_h^\varepsilon)|^2 Q(u_h^\varepsilon).
\end{aligned}$$

If we apply a similar argument to the third integral above and also use the simple inequality $|Q(u^\varepsilon)\partial_t \nu(u^\varepsilon)| \le c|\nabla u_t^\varepsilon|$ we finally obtain from (29)

$$\begin{aligned}
&\sum_{j=1}^d \int_{\Omega_h} \left(\gamma_{p_j}(\cdot, \nu(u^\varepsilon)) - \gamma_{p_j}(\cdot, \nu(u_h^\varepsilon))\right) \partial_t(u_{x_j}^\varepsilon - u_{hx_j}^\varepsilon) \qquad (30) \\
&\ge \frac{d}{dt} \int_{\Omega_h} \left(\gamma(\cdot, \nu(u_h^\varepsilon)) - \gamma_p(\cdot, \nu(u^\varepsilon)) \cdot \nu(u_h^\varepsilon)\right) Q(u_h^\varepsilon) \\
&\quad - c(\varepsilon, M)\|\nabla u_t^\varepsilon\|_{L^\infty(\Omega)} \int_{\Omega_h} |\nu(u^\varepsilon) - \nu(u_h^\varepsilon)|^2 Q(u_h^\varepsilon).
\end{aligned}$$

Let us now estimate the right hand side of (20).

$$\begin{aligned}
&-\int_{\Omega_h} \left(V(u^\varepsilon) - V(u_h^\varepsilon)\right)(u_t^\varepsilon - I_h u_t^\varepsilon) \qquad (31) \\
&\le \left(\int_{\Omega_h} \left(V(u^\varepsilon) - V(u_h^\varepsilon)\right)^2 Q(u_h^\varepsilon)\right)^{\frac{1}{2}} \left(\int_{\Omega_h} \frac{(u_t^\varepsilon - I_h u_t^\varepsilon)^2}{Q(u_h^\varepsilon)}\right)^{\frac{1}{2}} \\
&\le \frac{1}{\sqrt{\varepsilon}} \left(\int_{\Omega_h} \left(V(u^\varepsilon) - V(u_h^\varepsilon)\right)^2 Q(u_h^\varepsilon)\right)^{\frac{1}{2}} \|u_t^\varepsilon - I_h u_t^\varepsilon\|_{L^2(\Omega_h)}
\end{aligned}$$

and with (17)

$$\begin{aligned}
&\le \frac{c}{\sqrt{\varepsilon}} h^2 \|u_t^\varepsilon\|_{H^2(\Omega)} \left(\int_{\Omega_h} \left(V(u^\varepsilon) - V(u_h^\varepsilon)\right)^2 Q(u_h^\varepsilon)\right)^{\frac{1}{2}} \\
&\le \frac{1}{4} \int_{\Omega_h} \left(V(u^\varepsilon) - V(u_h^\varepsilon)\right)^2 Q(u_h^\varepsilon) + \frac{c}{\varepsilon} h^4 \|u_t^\varepsilon\|_{H^2(\Omega)}^2.
\end{aligned}$$

For the second term on the right hand side of (20) we have

$$\begin{aligned} &\sum_{j=1}^{d}\int_{\Omega_h}\left(\gamma_{p_j}(\cdot,\nu(u^\varepsilon))-\gamma_{p_j}(\cdot,\nu(u_h^\varepsilon))\right)\partial_t(u^\varepsilon_{x_j}-(I_hu^\varepsilon))_{x_j}) \qquad (32)\\ &\le c\left(\int_{\Omega_h}|\nu(u^\varepsilon)-\nu(u_h^\varepsilon)|^2Q(u_h^\varepsilon)\right)^{\frac12}\left(\int_{\Omega_h}\frac{|\nabla(u_t^\varepsilon-I_hu_t^\varepsilon)|^2}{Q(u_h^\varepsilon)}\right)^{\frac12}\\ &\le c\int_{\Omega_h}|\nu(u^\varepsilon)-\nu(u_h^\varepsilon)|^2Q(u_h^\varepsilon)+\frac{c}{\varepsilon}h^2\|u_t^\varepsilon\|^2_{H^2(\Omega)}. \end{aligned}$$

We collect the estimates (21), (30), (31), (32) and get

$$\begin{aligned} &\frac12\int_{\Omega_h}\left(V(u^\varepsilon)-V(u_h^\varepsilon)\right)^2Q(u_h^\varepsilon) \qquad (33)\\ &\qquad+\frac{d}{dt}\int_{\Omega_h}\left(\gamma(\cdot,\nu(u_h^\varepsilon))-\gamma_p(\cdot,\nu(u^\varepsilon))\cdot\nu(u_h^\varepsilon)\right)Q(u_h^\varepsilon)\\ &\le c(\varepsilon,M)\left(\|\nabla u_t^\varepsilon\|_{L^\infty(\Omega)}+\|u_t^\varepsilon\|^2_{L^\infty(\Omega)}\right)\int_{\Omega_h}|\nu(u^\varepsilon)-\nu(u_h^\varepsilon)|^2Q(u_h^\varepsilon)\\ &\qquad+\frac{c}{\varepsilon}h^4\|u_t^\varepsilon\|^2_{H^2(\Omega)}. \end{aligned}$$

We integrate with respect to time and use Lemma 2.1 to estimate

$$\gamma(\cdot,\nu(u_h^\varepsilon))-\gamma_p(\cdot,\nu(u^\varepsilon))\cdot\nu(u_h^\varepsilon)\ge c_1|\nu(u^\varepsilon)-\nu(u_h^\varepsilon)|^2.$$

We thus finally arrive at the estimate

$$\begin{aligned} &\int_0^t\int_{\Omega_h}\left(V(u^\varepsilon)-V(u_h^\varepsilon)\right)^2Q(u_h^\varepsilon)\,dt+\int_{\Omega_h}|\nu(u^\varepsilon)-\nu(u_h^\varepsilon)|^2Q(u_h^\varepsilon) \qquad (34)\\ &\le c(\varepsilon,M)\int_0^t\left(\|\nabla u_t^\varepsilon\|_{L^\infty(\Omega)}+\|u_t^\varepsilon\|^2_{L^\infty(\Omega)}\right)\int_{\Omega_h}|\nu(u^\varepsilon)-\nu(u_h^\varepsilon)|^2Q(u_h^\varepsilon)\,dt\\ &\quad+\frac{c}{\varepsilon}h^4\int_0^t\|u_t^\varepsilon\|^2_{H^2(\Omega)}\,dt+\frac{c}{\varepsilon}h^2\|u_0\|^2_{H^2(\Omega)}. \end{aligned}$$

A Gronwall argument completes the proof of Theorem 5.2.

6 Computational Results

For the implementation we use a semi-implicit time discretization as it was studied in [7]. It is such that in every time step a linear system of equations has to be solved numerically, which is done with a conjugate gradient method. λ was chosen in accordance with the stability condition discussed in [5].

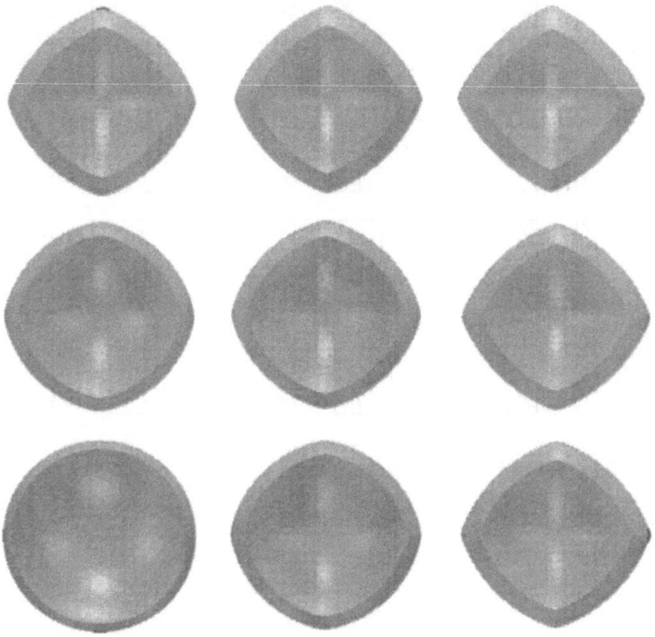

Fig. 3. Space dependent Frank-diagrams for the weight function (35) on $\Omega = (0,1) \times (0,1)$.

Algorithm 6.3 *Set* $u_h^{\varepsilon(0)} = I_h u_0$. *For given* $u_h^{\varepsilon(m)} \in X_h$, *compute* $u_h^{\varepsilon(m+1)} \in X_h$ *such that* $u_h^{\varepsilon(m+1)} - I_h u_0 \in \mathring{X}_h$ *and*

$$\frac{1}{\tau}\int_{\Omega_h} \frac{1}{Q(u_h^{\varepsilon(m)})}(u_h^{\varepsilon(m+1)} - u_h^{\varepsilon(m)})\varphi_h + \sum_{i=1}^{d}\int_{\Omega_h} \gamma_{p_i}(\cdot,\nu(u_h^{\varepsilon(m)}))\varphi_{hx_i}$$
$$+ \quad \lambda \int_{\Omega_h} \frac{\gamma(\cdot,\nu(u_h^{\varepsilon(m)}))}{Q(u_h^{\varepsilon(m)})}\nabla(u_h^{\varepsilon(m+1)} - u_h^{\varepsilon(m)}) \cdot \nabla\varphi_h = 0$$

for all $\varphi_h \in \mathring{X}_h$.

We computed the solution of (10) on $\Omega = (0,1) \times (0,1)$ with the right hand side $f = 1$ and initial and boundary data $u_0(x_1, x_2) = \sin(14x_1 - 7x_2 + 0.25)$ and the anisotropy

$$\gamma(x,p) = \frac{|x|}{\sqrt{2}}\sum_{j=1}^{3}\sqrt{\delta^2|p|^2 + p_j^2} + \left(1 - \frac{|x|}{\sqrt{2}}\right)|p|, \qquad (35)$$

with $\delta = 10^{-8}$. In Figure 3 we show the Frank diagrams for this weight function at the points $x = (\frac{i}{2}, \frac{i}{2})$, $i, j = 0, 1, 2$. The anisotropy function is such that it is nearly not admissible near some parts of $\partial\Omega$. This is why corrugated level sets appear there.

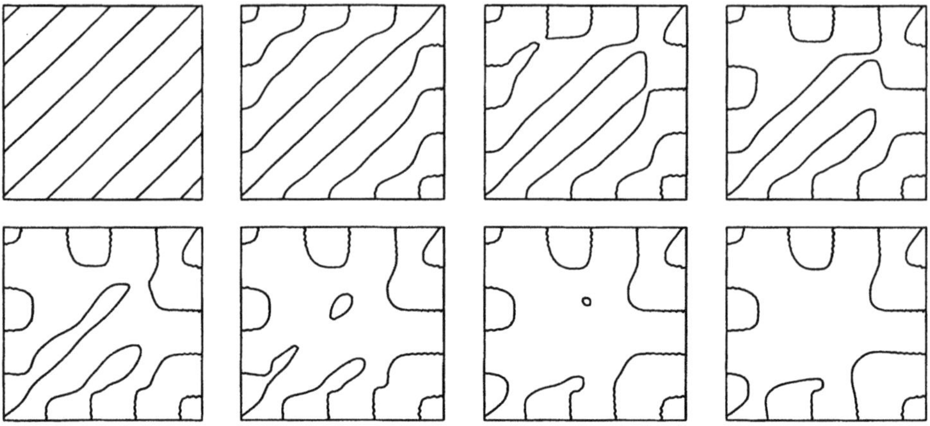

Fig. 4. Evolution of the zero level set under anisotropy (35).

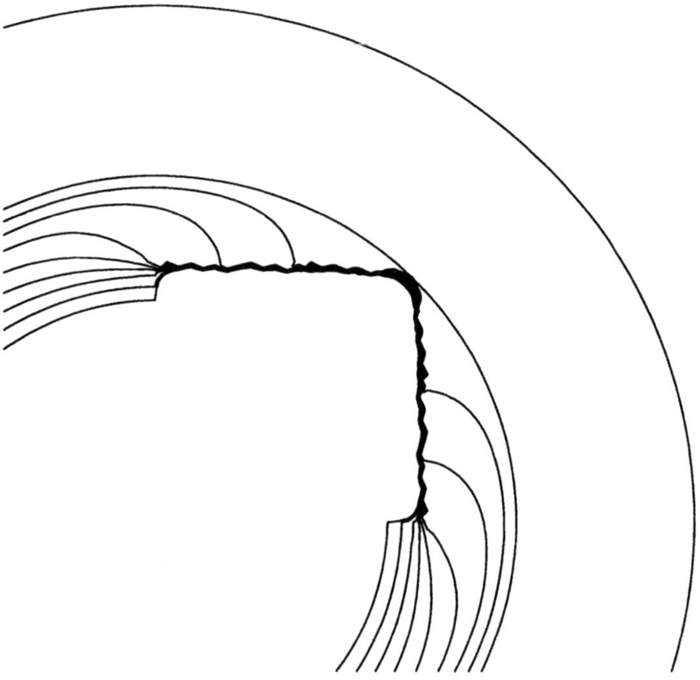

Fig. 5. The detection of an object by anisotropic mean curvature flow with weight function (36) in detail. Zero level sets from right to left with increasing time.

In Figure 4 we show the zero level set of the solution. The initial data are such that the zero level set consists of straight lines. This set then deforms into the direction of the Wulff shapes corresponding to the Frank diagrams given by (35). These Wulff shapes change with the modulus $|x|$ of the space variable from a sphere to a smoothed cube.

In Figure 5 we show some time steps of the evolution of an initial circle under anisotropic mean curvature flow with the weight function

$$\gamma(x,p) = \left(\chi_{\Omega\setminus Q}(x) + 0.01\,\chi_Q(x)\right)\,|p| \tag{36}$$

with $Q = [0,0.5] \times [0,0.5]$ and the unit disk as domain Ω. We have thus chosen a weight function which is discontinuous with respect to the variable x. This is a typical test case for object detection. The object to be detected is the square Q. The zero level set of the solution of (10) with weight function (36) starts as a circle near the boundary of the unit disk. It then moves inwards until it meets the square and wraps around it.

Acknowledgements. This work was supported by the Deutsche Forschungsgemeinschaft via Graduiertenkolleg "Nichtlineare Differentialgleichungen: Modellierung, Theorie, Numerik, Visualisierung".
The program GRAPE was used for the graphical presentation.

References

1. G. Bellettini, M. Paolini: Anisotropic motion by mean curvature in the context of Finsler geometry. Hokkaido Math. J., 25, 537–566, 1996
2. V. Caselles, R. Kimmel, G. Sapiro, C. Sbert: Minimal surfaces: a geometric three dimensional segmentation approach. Numer. Math., 77, 423–451, 1997
3. Y.-G. Chen, Y. Giga, S. Goto: Uniqueness and existence of viscosity solutions of generalized mean curvature flow equations. Diff. Geom, 33, 749–786, 1991
4. U. Clarenz, G. Dziuk, M. Rumpf: On generalized mean curvature flow in surface processing. This volume.
5. K. Deckelnick, G. Dziuk: A fully discrete numerical scheme for weighted mean curvature flow. Numer. Math., to appear
6. K. Deckelnick, G. Dziuk: Convergence of a finite element method for non–parametric mean curvature flow. Numer. Math., 72, 197–222, 1995
7. K. Deckelnick, G. Dziuk: Discrete anisotropic curvature flow of graphs. Math. Modelling Numer. Anal., 33, 1203–1222, 1999
8. K. Deckelnick, G. Dziuk: Convergence of numerical schemes for the approximation of level set solutions to mean curvature flow. Preprint Mathematische Fakultät Freiburg, 00-17, 2000
9. K. Deckelnick, G. Dziuk: Error estimates for a semi–implicit fully discrete finite element method scheme for the mean curvature flow of graphs. Interfaces and Free Boundaries, 2, 341–359, 2000
10. G. Dziuk: Discrete anisotropic curve shortening flow. SIAM J. Numer. Anal., 36, 1808–1830, 1999

11. G. Dziuk: Numerical schemes for the mean curvature flow of graphs. In: P. e. a. Argoul (ed) IUTAM Symposium on Variations of Domains and Free-Boundary Problems in Solid Mechanics., 63–70. Kluwer Academic Publishers, 1999
12. L. Evans, J. Spruck: Motion of level sets by mean curvature I. J. Diff. Geom., 33, 636–681, 1991
13. M. Fried, A. Veeser: Simulation and numerical analysis of dendritic growth. In: B. Fiedler (ed) Ergodic theory, analysis, and efficient simulation of dynamical systems. 225–252, Springer, 2001
14. S. Osher, J. A. Sethian: Fronts propagating with curvature dependent speed: Algorithms based on Hamilton-Jacobi formulations. J. Comp. Phys., 79 12–49, 1988
15. M. Preußer, T. Rumpf: A level set method for anisotropic geometric diffusion in 3d image processing. Report SFB 256 Bonn, 37, 2000
16. A. Schmidt: Computation of three dimensional dendrites with finite elements. J. Comp. Phys., 125, 293–312, 1996

Optimal Regularity Results via A-Harmonic Approximation

Frank Duzaar[1], Joseph F. Grotowski[1] and Klaus Steffen[2]

[1] Mathematisches Institut der Universität Erlangen-Nürnberg, Bismarckstr. 1 1/2, D-91054 Erlangen, Germany, {duzaar, grotow}@mi.uni-erlangen.de

[2] Mathematisches Institut der Heinrich-Heine-Universität Düsseldorf Universitätsstraße 1, D-40225 Düsseldorf, Germany, steffen@cs.uni-duesseldorf.de

Summary. We discuss a new approach to regularity theory for almost minimizers of variational integrals in geometric measure theory or in the classical calculus of variations. This method is direct, exhibiting the dependence of the regularity estimates on the structural data of the variational integrand in explicit form; it requires only weak growth and smoothness assumptions on the integrand; it allows a unified treatment of interior and boundary regularity; and it leads to new regularity results which give the best possible modulus of continuity for the derivative of the almost minimizer in a variety of situations.

1 Introduction

In the regularity theory for minimizers of elliptic variational integrals, both in the classical calculus of variations and in the setting of geometric measure theory, a crucial step is the comparison of the given minimizer to a solution of a linear homogeneous elliptic system of partial differential equations with constant coefficients. For $\mathbb{R}^N$-valued functions h of n variables $x \in \mathbb{R}^n$, such systems are determined by a symmetric bilinear form $A \in \bigodot^2 \operatorname{Hom}(\mathbb{R}^n, \mathbb{R}^N)$ which satisfies the Legendre-Hadamard condition (also called rank one ellipticity; see (2) below) and may be written, in a weak formulation on a domain $U \subset \mathbb{R}^n$,

$$\int_U A(Dh, D\varphi)\, dx = 0 \quad \text{for all } \varphi \in C^1_{\mathrm{c}}(U, \mathbb{R}^N)\,. \tag{1}$$

(The subscript "c" indicates compact support of the test functions φ in U.) The solutions h to (1) will be called *A-harmonic functions*, because in the case that A is the canonical inner product on $\operatorname{Hom}(\mathbb{R}^n, \mathbb{R}^N)$ they are simply the classical $\mathbb{R}^N$-valued harmonic functions on U.

In the context of variational integrals

$$\mathcal{F}(u) = \int_U f(x, u(x), Du(x))\, dx$$

for $\mathbb{R}^N$-valued functions u on a domain $U \subset \mathbb{R}^n$ the form A is usually obtained by computing the Hessian of the integrand $f(x, y, p)$ with respect to the variable

p at suitable points $(x_0, u(x_0), p_0)$. If u is minimizing for $\mathcal{F}$, then one can deduce estimates for u near x_0 by comparing u to suitable A-harmonic functions on balls centered at x_0.

In geometric measure theory, where one considers parametric integrals

$$\mathbf{F}(T) = \int_{M_T} F(z, \zeta_T(z))\vartheta_T(z)\, d\mathcal{H}^n z$$

of n-currents T on $\mathbb{R}^{n+N}$, described by an n-rectifiable supporting set $M_T \subset \mathbb{R}^{n+N}$, an orientation n-vector field ζ_T for M_T, and a multiplicity function ϑ_T with values in the positive integers, one first passes to the nonparametric integrand associated with F and some reference n-plane in $\mathbb{R}^{n+N}$. One then defines A as the Hessian of this nonparametric integrand $f(x, y, p)$ with respect to the variable p at a point $(x_0, y_0, 0)$, where (x_0, y_0) is in the support of T, and in the case that T is minimizing for $\mathbf{F}$ one can deduce estimates for T near (x_0, y_0) by comparing T to the graph of suitable A-harmonic functions on balls with center x_0. We may assume the reference n-plane to be $\mathbb{R}^n \times \{0\}^N$ and we will refer to directions in $\mathbb{R}^n \times \{0\}^N$ as *horizontal* and directions in $\{0\}^n \times \mathbb{R}^N$ as *vertical.*

Different methods have been utilized to define A-harmonic functions suitable for the comparison argument in the regularity proofs. One method is *blow-up* by a sequence of expanding homotheties $x \mapsto r_k^{-1}(x - x_0)$ in the horizontal directions, where $r_k \downarrow 0$, and by suitably related expanding homotheties $y \mapsto \varepsilon_k^{-1}(y - y_0)$ in the vertical directions, where also $0 < \varepsilon_k \to 0$. Under appropriate hypotheses one can show that the rescaled minimizing functions u_k then converge, in some sense, to an A-harmonic function respectively the rescaled minimizing currents T_k converge to the graph of an A-harmonic function. From this convergence and standard estimates for solutions to constant coefficient linear elliptic systems one then deduces, by contradiction, corresponding estimates for the u_k respectively T_k with k sufficiently large. For the original minimizers u respectively T these estimates then imply a decay property for the excess, a quantity that is of central importance in the regularity proofs, and this excess decay finally leads to (partial) regularity theorems. In the context of geometric measure theory the blow-up method was introduced by DeGiorgi [13] and Almgren [2] (see also [24], Chap. 5), and it has been applied to minimizing functions u of variational integrals $\mathcal{F}$ by a number of authors; as representative, we mention here [48], [33], [26], [30] and [37].

Another method of defining an A-harmonic comparison function h for the minimizer u near x_0 is the *solution of the Dirichlet problem* for (1) on a small ball $U \subset \mathbb{R}^n$ centered at x_0 with boundary condition $h = u$ on ∂U. The minimizing property must then be exploited to obtain further information about u, e.g. higher integrability $|Du| \in L^p$ for some $p > 2$ (when 2 is the natural integrability exponent for the problem) in order to be able to deduce excess decay for u at x_0 from an argument comparing $\mathcal{F}(u)$ to $\mathcal{F}(h)$. Such higher integrability can be derived by means of a lemma of Gehring type (see [27]) from a reversed Hölder inequality for u, which in turn is based on the Sobolev-Poincaré inequality and a Caccioppoli inequality (also known as reverse Poincaré inequality) obtained from a direct comparison construction. This approach was introduced into the calculus of variations by Giaquinta and

Modica [31] (see also [28], [29]) and it has been used many times since then, see e.g. [32], [8].

In the context of geometric measure theory Schoen and Simon [51] gave a short, direct proof for the (partial) regularity of $\mathbf{F}$-minimizing rectifiable n-currents T, which is based on the local approximation of T by the graph of a Lipschitz function $g : \mathbb{R}^n \to \mathbb{R}^N$ and on the comparison of T to the graph of the A-harmonic function h determined on a small ball by the boundary values of a smoothing g_ε of g.

The two methods of defining A-harmonic comparison functions described above in very vague terms have certain disadvantages. The blow-up method is indirect in nature, and hence it does not yield explicit information about the dependence of the degree of smoothness of the minimizers on the structural data of the variational problem. As a consequence, one usually cannot expect to be able to prove the optimal regularity of the minimizers which is allowed by the data by using this method. (One may be able to do this a posteriori, however, in a second step after some degree of smoothness has been established by the blow-up method.) A similar remark applies to the method based on a reversed Hölder inequality and higher integrability $|Du| \in L^p$, because the dependence of the exponent $p > 2$ on the data is not explicit. Another disadvantage is that it is far from straightforward to extend these methods to cover the boundary situation. For example, we mention the paper [38], where the modifications that need to be made to the regularity proof of [24], 5.3 in order to establish a corresponding boundary regularity theorem for minimizing n-currents are given; this paper is 70 pages long. The regularity proof in [51] uses a delicate choice of the smoothing g_ε of the Lipschitz approximation g for T and $C^{2,\alpha}$ Schauder estimates for the A-harmonic function h with $h = g_\varepsilon$ on the boundary of the ball. In the corresponding boundary situation one would have to consider a half ball which is not sufficiently smooth at the boundary to allow such estimates, hence it is not clear at all how one could adapt the proof of [51] to prove boundary regularity.

We note here that, in the context of the classical calculus of variations, boundary regularity was obtained for a particular class of minimizers in [41] using the Gehring lemma technique.

In the present article we want to give an exposition of a third method of determining A-harmonic comparison functions, which avoids the disadvantages just described. This method is based on being able to obtain a good approximation of certain functions $g \in H^{1,2}(U, \mathbb{R}^N)$ (for U a ball in $\mathbb{R}^n$) by A-harmonic functions $h \in H^{1,2}(U, \mathbb{R}^N)$, in both the L^2 topology and in the weak topology of the Sobolev space $H^{1,2}$. The requirement is that the functions to be approximated are already approximatively A-harmonic in a certain sense. The precise statement is given in Lemma 1.1 below. In the applications of this lemma one chooses for g a scaled version of the minimizing function u respectively of a Lipschitz function whose graph is a good approximation of the minimizing n-current T. The required approximate A-harmonicity of g is a consequence of the minimizing property of u respectively T, and the comparison argument involving h and u, g respectively T, g gives an estimate which, when combined with a Caccioppoli inequality, then yields the desired excess decay. Of course, this describes the method only very roughly. In Sections 2 and 3 below we will give more details and we will formulate precisely some of the new optimal regularity results that can be obtained with this method.

There are several conceptual and technical advantages of this procedure, which we call the *method of A-harmonic approximation*. The method is direct, in particular it does not use an indirect blow-up argument, and all the steps can be carried out with complete control of the dependence of the estimates on the structural data. As a consequence, with this method one is able to derive an explicit modulus of continuity for the derivative of the minimizing function respectively for the tangent plane field of the minimizing current. The method needs only natural and quite weak smoothness conditions for the variational integrands, and the resulting modulus of continuity for the derivative of the minimizers is frequently the best one that can be expected on the basis of these smoothness assumptions. In this sense the method leads to optimal regularity results. The method of A-harmonic approximation applies equally well to questions of interior regularity and to questions of boundary regularity. The method is very flexible, because strictly speaking it does not use minimality or the Euler equation; the approximate A-harmonicity can be viewed as a differential inequality in weak form. For this reason the method also applies to functions and currents which are only almost minimizing (see Sections 2, 3 for a definition), a fact that considerably widens the range of applicability of the regularity results, and it can also be used to prove new regularity results for weak solutions to elliptic nonlinear systems of partial differential equations in a variety of situations.

Harmonic approximation has been used for a long time in regularity theory, e.g. by Allard [1] in his regularity theorem for stationary varifolds, in the special case of the area integrand where one approximates the minimizers by graphs of ordinary harmonic functions (see also [49], Chapter 21 and [50], Chapter 1). However, it seems that the first application of the method of A-harmonic approximation to general elliptic parametric variational integrals was given only recently in [22], where new and optimal interior and boundary regularity theorems were proved for almost minimizing rectifiable integer multiplicity currents. Subsequently it was shown in [17], [19], [16], [18], [35], [36], [43] that the method can also be used to derive new and often optimal regularity theorems for (almost) minimizing functions of quasiconvex integrals in the classical calculus of variations and for weak solutions to elliptic nonlinear systems of partial differential equations. In Section 2 and Section 3 below we concentrate on giving an exposition of the results obtained in [22] and [17].

We conclude the introduction with a precise statement of the A-harmonic approximation lemma, and for completeness we also include the simple proof which is modelled after [49], Chapter 21 and [50], Chapter 1. To fix notation we recall that $A \in \bigodot^2 \operatorname{Hom}(\mathbb{R}^n, \mathbb{R}^N)$ satisfies the *Legendre-Hadamard condition* (or condition of *rank one ellipticity*) if for some $\lambda > 0$ (the *ellipticity constant*) we have

$$A(p,p) \geq \lambda |p|^2 \quad \text{for all } p \in \operatorname{Hom}(\mathbb{R}^n, \mathbb{R}^N) \text{ of rank one,} \tag{2}$$

where $|p|^2 = \operatorname{trace}(p^* p)$ is the square of the natural Euclidean norm on $\operatorname{Hom}(\mathbb{R}^n, \mathbb{R}^N)$. Representing A by a matrix $(a_{i\,j}^{\alpha\beta})$ and p of rank one by $(\xi^i \eta_\alpha)$, condition (2) reads

$$\sum_{i,j=1}^{N} \sum_{\alpha,\beta=1}^{n} a_{i\,j}^{\alpha\beta} \xi^i \xi^j \eta_\alpha \eta_\beta \geq \lambda |\xi|^2 |\eta|^2 \quad \text{for all } \xi \in \mathbb{R}^N \text{ and } \eta \in \mathbb{R}^n.$$

We say that $\Lambda \in (0, \infty)$ is an *upper bound* for A if

$$|A(p,q)| \leq \Lambda |p||q| \quad \text{for all } p, q \in \operatorname{Hom}(\mathbb{R}^n, \mathbb{R}^N). \tag{3}$$

By $\mathbf{U}$ we denote the open unit ball in $\mathbb{R}^n$.

Lemma 1.1 (A-harmonic approximation) *Corresponding to $\lambda \leq \Lambda$ in $(0,\infty)$ and $\varepsilon > 0$ there exists $\delta(n, N, \lambda, \Lambda, \varepsilon) > 0$ with the following property: whenever $A \in \bigodot^2 \operatorname{Hom}(\mathbb{R}^n, \mathbb{R}^N)$ is rank one elliptic with ellipticity constant $\lambda > 0$ and upper bound $\Lambda \in (0, \infty)$, $\varepsilon > 0$ is given, and $g \in H^{1,2}(\mathbf{U}, \mathbb{R}^N)$ with $\int_{\mathbf{U}} |Dg|^2\, dx \leq 1$ is approximatively A-harmonic in the sense that*

$$\left| \int_{\mathbf{U}} A(Dg, D\varphi)\, dx \right| \leq \delta(n, N, \lambda, \Lambda, \varepsilon) \sup |D\varphi|$$

holds for all $\varphi \in C_c^1(\mathbf{U}, \mathbb{R}^N)$, then there exists an A-harmonic function $h \in H^{1,2}(\mathbf{U}, \mathbb{R}^N)$ such that

$$\int_{\mathbf{U}} |h - g|^2\, dx \leq \varepsilon \quad \textit{and} \quad \int_{\mathbf{U}} |Dh|^2\, dx \leq 1\,.$$

Proof. Supposing the lemma to be false we have existence of a sequence of quadratic forms $A_k \in \bigodot^2 \operatorname{Hom}(\mathbb{R}^n, \mathbb{R}^N)$ with fixed ellipticity constant $\lambda > 0$ as in (2) and upper bound $\Lambda < \infty$ as in (3), and a sequence of $g_k \in H^{1,2}(\mathbf{U}, \mathbb{R}^N)$ with $\|Dg_k\|_{L^2} \leq 1$ and

$$\left| \int_{\mathbf{U}} A_k(Dg_k, D\varphi)\, dx \right| \leq \frac{1}{k} \sup |D\varphi| \tag{4}$$

for all $\varphi \in C_c^1(\mathbf{U}, \mathbb{R}^N)$, such that for some $\varepsilon > 0$ the inequality

$$\int_{\mathbf{U}} |g_k - h_k|^2\, dx > \epsilon \tag{5}$$

is valid for each k and each A_k-harmonic function $h_k \in H^{1,2}(\mathbf{U}, \mathbb{R}^N)$ with $\|Dh_k\|_{L^2} \leq 1$.

Clearly, we may assume $\int_{\mathbf{U}} g_k\, dx = 0$ and apply Poincaré's inequality and Rellich's theorem to obtain, after having passed to a subsequence, $g_k \to g$ strongly in $L^2(\mathbf{U}, \mathbb{R}^N)$ and weakly in $H^{1,2}(\mathbf{U}, \mathbb{R}^N)$ and $A_k \to A$ in $\bigodot^2 \operatorname{Hom}(\mathbb{R}^n, \mathbb{R}^N)$. Then $\|Dg\|_{L^2} \leq 1$, and by rewriting

$$\begin{aligned} \int_{\mathbf{U}} A(Dg, D\varphi)\, dx &= \int_{\mathbf{U}} A(Dg - Dg_k, D\varphi)\, dx \\ &\quad + \int_{\mathbf{U}} (A - A_k)(Dg_k, D\varphi)\, dx + \int_{\mathbf{U}} A_k(Dg_k, D\varphi)\, dx \end{aligned}$$

and using (4) we infer that g is A-harmonic.

Rank one ellipticity of A_k, i.e. (2), implies (by Fourier transform and Plancherel's theorem, i.e. Gårding's inequality in the case of constant coefficients) the coercivity condition

$$\int_{\mathbf{U}} A_k(D\eta, D\eta)\, dx \geq \lambda \int_{\mathbf{U}} |D\eta|^2\, dx \quad \text{for } \eta \in H_0^{1,2}(\mathbf{U}, \mathbb{R}^N). \tag{6}$$

hence we can choose h_k as the solution to the Dirichlet problem corresponding to A_k and the boundary values of g, i.e.

$$\begin{aligned} h_k - g &\in H_0^{1,2}(\mathbf{U}, \mathbb{R}^N) \\ \int_{\mathbf{U}} A_k(Dh_k, D\eta)\, dx &= 0 \quad \text{for } \eta \in H_0^{1,2}(\mathbf{U}, \mathbb{R}^N). \end{aligned} \tag{7}$$

Then

$$\begin{aligned} \lambda \int_{\mathbf{U}} &|Dh_k - Dg|^2\, dx \\ &\leq \int_{\mathbf{U}} A_k(Dh_k - Dg, Dh_k - Dg)\, dx \quad \text{by (6),} \\ &= - \int_{\mathbf{U}} A_k(Dg, Dh_k - Dg)\, dx \quad \text{by (7),} \\ &= \int_{\mathbf{U}} (A - A_k)(Dg, Dh_k - Dg)\, dx \quad \text{by } A\text{-harmonicity of } g, \\ &\leq |A_k - A| \left[\int_{\mathbf{U}} |Dh_k - Dg|^2\, dx \right]^{1/2} \text{on account of } \|Dg\|_{L^2} \leq 1, \end{aligned}$$

and in view of $A_k \to A$ we have $h_k \to g$ strongly in $H^{1,2}$. But this implies $\|h_k - g_k\|_{L^2} \leq \|h_k - g\|_{L^2} + \|g - g_k\|_{L^2} \to 0$ as $k \to \infty$ and the same assertion is true when we replace h_k by the scaled function $\min(1, \|Dh_k\|_{L^2}^{-1}) h_k$, contradicting (5). □

A boundary version of the A-harmonic approximation lemma can be proved in the same way. In the statement one simply replaces $\mathbf{U}$ by the half ball $\mathbf{U}_+ = \{x \in \mathbf{U} : x_n > 0\}$ and $H^{1,2}(\mathbf{U}, \mathbb{R}^N)$ by the subspace $H_{(0)}^{1,2}(\mathbf{U}_+, \mathbb{R}^N)$ of $H^{1,2}(\mathbf{U}_+, \mathbb{R}^N)$ which consists of functions with vanishing boundary trace on the equatorial $(n-1)$-disk $\mathbf{U} \cap \partial \mathbf{U}_+$ in the boundary of $\mathbf{U}_+$. To adapt the proof one merely has to observe that Poincaré's inequality is valid in $H_{(0)}^{1,2}(\mathbf{U}_+, \mathbb{R}^N)$, Rellich's compactness theorem holds for the embedding $H_{(0)}^{1,2}(\mathbf{U}_+, \mathbb{R}^N) \to L^2(\mathbf{U}_+, \mathbb{R}^N)$, and $H_{(0)}^{1,2}(\mathbf{U}_+, \mathbb{R}^N)$ is a closed subspace of $H^{1,2}(\mathbf{U}_+, \mathbb{R}^N)$.

We remark that the proof by contradiction given above for the lemma of A-harmonic approximation does not mean that regularity proofs using this lemma are of an indirect nature. The lemma guarantees the existence of a positive function $\delta(n, N, \lambda, \Lambda, \varepsilon)$ determined by the rank one elliptic symmetric bilinear forms on $\mathrm{Hom}(\mathbb{R}^n, \mathbb{R}^N)$, and this function enters our regularity estimates in a way that allows to derive optimal moduli of continuity for the derivative respectively for the tangent plane field despite the fact that $\delta(n, N, \lambda, \Lambda, \varepsilon)$ is not known explicitly. For the purposes of regularity theory, we therefore need not address the issue of giving explicit (positive lower) bounds for $\delta(n, N, \lambda, \Lambda, \varepsilon)$.

2 Almost Minimizers of Elliptic Parametric Integrals in Geometric Measure Theory

In this section we discuss how the method of A-harmonic approximation can be applied to prove interior and boundary regularity theorems for almost minimizers to elliptic parametric integrands in the setting of geometric measure theory. We follow [22] where new and optimal regularity results haven been obtained. We do not intend to give a full proof of these here, but we will at least give precise formulations of the results and we want to make clear how the A-harmonic approximation lemma, Lemma 1.1, is applied in the proof.

Recall that an n-dimensional *integer multiplicity rectifiable current* (or *n-current* for short) T on $\mathbb{R}^{n+N}$ is a generalized oriented surface which can be represented by integration as a linear functional on the space of smooth n-forms ϕ with compact support in $\mathbb{R}^{n+N}$ as follows:

$$\langle T, \phi \rangle = \int_{M_T} \langle \phi, \zeta_T \rangle \vartheta_T \, d\mathcal{H}^n = \int_{\mathbb{R}^{n+N}} \langle \phi, \zeta_T \rangle \, d\mu_T \,,$$

where $\mathcal{H}^n$ denotes the n-dimensional Hausdorff measure on $\mathbb{R}^{n+N}$, the *supporting set* $M_T \subset \mathbb{R}^{n+N}$ is countably $(\mathcal{H}^n, n)$ rectifiable (i.e. it can be covered by countably many n-submanifolds of class C^1 in $\mathbb{R}^{n+N}$ and a set of zero $\mathcal{H}^n$-measure), ϑ_T is a positive integer valued $\mathcal{H}^n$-summable *multiplicity function*, ζ_T is an n-measurable *orientation n-vectorfield* for M_T (i.e. $\zeta_T(z)$ equals, at $\mathcal{H}^n$ almost every point $z \in M_T$, the exterior product of some orthonormal basis in the n-dimensional approximate tangent space $\operatorname{Tan}^n(\mathcal{H}^n \llcorner M_T, z)$), and we have written μ_T for the Radon measure $\mathcal{H}^n \llcorner \vartheta_T$ associated with T, understanding that ϑ_T is extended by the value zero on $\mathbb{R}^{n+N} \setminus M_T$. We note that summability of ϑ_T is equivalent to finiteness of the *mass* (or *n-area*)

$$\mathbf{M}(T) = \mu_T(\mathbb{R}^{n+N}) = \int_{M_T} \vartheta_T \, d\mathcal{H}^n \,.$$

The *boundary operator* ∂ on n-currents is defined as the dual of the exterior derivative d, i.e. $\langle \partial T, \psi \rangle = \langle T, d\psi \rangle$ for smooth $(n-1)$-forms ψ with compact support. We refer to [24] and [49] for the basic notions and notation related to currents; notation not explained in this paper are taken from [24]. Our usage is consistent with this reference, except that what we have called an integer multiplicity rectifiable current above would be a locally rectifiable current with finite mass in the language of [24].

A *parametric integrand* of degree n on an open set $\mathcal{O}$ in $\mathbb{R}^{n+N}$ is a continuous map

$$F : \mathcal{O} \times \bigwedge_n \mathbb{R}^{n+N} \to \mathbb{R}$$

that is positively homogeneous of degree one in the second variable, i.e. $F(z, r\zeta) = rF(z,\zeta)$ for $r > 0$. The *integral* of F over an n-current T as above (with $\vartheta_T \equiv 0$ outside $\mathcal{O}$) is then

$$\mathbf{F}(T) = \int_{M_T} F(z, \zeta_T(z))\vartheta_T(z)\, d\mathcal{H}^n z = \int_{\mathcal{O}} F(z, \zeta_T(z))\, d\mu_T z\,,$$

provided the integral exists and is finite. The most important example is the *area integrand* $M(z,\zeta) = |\zeta|$. Of course, for minimizers of this special integrand much stronger regularity results and estimates on the size of the singular set are known than the ones discussed here and valid for minimizers of general elliptic parametric integrands (see [24], Chapter 5, [25], [39], [4]).

Our *structure conditions* for the integrand $F(z,\zeta)$ and its first and second order total derivatives $D_\zeta F(z,\zeta), D_\zeta^2 F(z,\zeta)$ with respect to the variable $\zeta \in \bigwedge_n \mathbb{R}^{n+N}$ are the following:

(F1) There exists a constant $\Lambda \in [1,\infty)$ such that for all $z \in \mathbb{R}^{n+N}$ and $0 \neq \zeta \in \bigwedge_n \mathbb{R}^{n+N}$ the following inequalities hold

$$|\zeta| \le F(z,\zeta) \le \Lambda|\zeta|, \qquad |D_\zeta F(z,\zeta)| \le \Lambda, \qquad |D_\zeta^2 F(z,\zeta)| \le \Lambda|\zeta|^{-1}.$$

(F2) Each of $F(z,\zeta)$, $D_\zeta F(z,\zeta)$, and $D_\zeta^2 F(z,\zeta)$ depend continuously on (z,ζ) with $\zeta \neq 0$, and we have moduli of continuity μ, ν such that

$$\begin{aligned} |F(z,\zeta) - F(\tilde{z},\zeta)| &\le \mu(|z-\tilde{z}|)|\zeta| \\ |D_\zeta F(z,\zeta) - D_\zeta F(\tilde{z},\zeta)| &\le \mu(|z-\tilde{z}|) \end{aligned}$$

for all $z, \tilde{z}, \zeta$ and

$$|D_\zeta^2 F(z,\zeta) - D_\zeta^2 F(z,\tilde{\zeta})| \le \nu(|\zeta - \tilde{\zeta}|)$$

for all z and all $\zeta, \tilde{\zeta}$ with $|\zeta| = 1 = |\tilde{\zeta}|$. (Here we assume $\mu(0+) = 0$, $\nu(0+) = 0$, and some further conditions will later be imposed on μ.)

(F3) The integrand F is a uniformly *elliptic integrand* with *ellipticity constant* $\lambda \in (0,1]$, i.e. for every $a \in \mathbb{R}^{n+N}$ the functional $\mathbf{F}_a$ corresponding to the "frozen integrand" $F_a(z,\zeta) = F(a,\zeta)$ satisfies

$$\mathbf{F}_a(T) - \mathbf{F}_a(S) \ge \lambda[\mathbf{M}(T) - \mathbf{M}(S)]\,,$$

whenever S, T are compactly supported n-currents on $\mathbb{R}^{n+N}$ with the same boundary and S is flat, i.e. represented by an $\mathcal{H}^n$-measurable subset of some n-dimensional affine subspace of $\mathbb{R}^{n+N}$ with constant orientation n-vector field and with $\mathcal{H}^n$-summable positive integer valued multiplicity (cf. [10] and [24], 5.1.2 for this variant of Almgren's original ellipticity condition [2]).

We note that the uniformity assumptions with respect to the space variable $z \in \mathbb{R}^{n+N}$ contained in (F1)-(F3) are no restriction for regularity theory, which is local in nature; further (F1),(F2) are quite weak assumptions. Since we consider only positive integrands and we may scale by a positive factor, it is no loss of generality to assume $|F(z,\zeta)| \geq |\zeta|$ in (F1). For the continuity modulus $\mu(r)$ of F and $D_\zeta F$ with respect to the variable z we assume that μ is nondecreasing on $[0,\infty)$ with $\mu(0) = 0 = \mu(0+)$ and with $r \mapsto \frac{1}{r}\mu(r)$ nonincreasing; this is no essential restriction. In the course of the regularity proof one needs some further growth conditions for $\mu(r)$ as $r \downarrow 0$. In any case, Hölder moduli $\mu(r) = cr^\beta$ with $\beta \in (0,1)$ will be allowed and also certain Dini moduli $\mu(r)$ with $r^{-\beta}\mu(r) \to \infty$ as $r \to 0+$ for all $\beta > 0$.

An n-current T is called $\mathbf{F}$-*minimizing* in an open set $\mathcal{O} \subset \mathbb{R}^{n+N}$ if the inequality $\mathbf{F}(T) \leq \mathbf{F}(T+X)$ holds for all closed n-currents X (i.e. $\partial X = 0$) with compact support in $\mathcal{O}$. To widen the range of applicability of the theory with a view to applications it is important to weaken this notion of minimality. Following Almgren [3] and Bombieri [10] we consider a relaxed minimizing condition which allows modifications $T+X$ of T with $\partial X = 0$ and $\mathbf{F}(T+X)$ smaller than $\mathbf{F}(T)$, but only by a factor $1+\omega(r)$ close to 1 if X has support in a ball of small radius r. We adopt the definition from [10]:

Definition 2.2 (almost minimizing currents) *For a given function $\omega : [0,\infty) \to [0,\infty]$ with $\omega(0) = 0 = \omega(0+)$ we say that an n-current T is* almost minimizing *for $\mathbf{F}$ and ω in $\mathcal{O}$, or $(\mathbf{F},\omega)$-minimizing for short, if*

$$\mathbf{F}(T) \leq \mathbf{F}(T+X) + \omega(r)\mathbf{M}(T \llcorner \operatorname{spt} X + X)$$

holds for all closed n-currents X with compact support $\operatorname{spt} X \subset\subset \mathcal{O}$ *and all r such that* $\operatorname{spt} X$ *is contained in a ball of radius r in $\mathbb{R}^{n+N}$.* □

We will make similar assumptions on $\sqrt{\omega}$ to those made for the modulus μ from (F2), as both quantities enter the regularity estimates in a similar way. In particular, we will assume that $\omega(r)$ is nondecreasing and $r^{-2}\omega(r)$ is nonincreasing in r. Further conditions on ω will be imposed later. Note that by scaling with homotheties of $\mathbb{R}^{n+N}$ we may always assume that ω and μ are as small as we like on $(0,1]$. All our conditions will be satisfied by $\omega(r) = cr^{2\beta}$ with $\beta \in (0,1)$ and also by certain Dini-type moduli $\omega(r)$ with $r^{-2\beta}\omega(r) \to \infty$ as $r \downarrow 0$ for all $\beta > 0$. To illustrate the significance of the notion of almost minimality we now give some examples. For more details on the preceding examples and further discussions of the significance of the $(\mathbf{F},\omega)$-minimizing property we refer to [3], [53]–[55], and [22].

Example 2.3 (every $C^{1,\beta}$ submanifold is almost minimizing)*: If T is given by an oriented uniform $C^{1,\beta}$ n-submanifold $\mathcal{T}$ in $\mathbb{R}^{n+N}$, then T is almost minimizing for the n-area $\mathbf{M}$ and for $\omega(r) = Cr^{2\beta}$ with suitable constant C. More precisely, the n-current defined by integration over $\mathcal{T}$ with multiplicity 1 has this property. This is similar to the corresponding example in [53], and is quite easily proved. The significance of this example is that it shows the optimality of the modulus of continuity for ζ_T given by our regularity theorems below in the case of a smooth integrand and* *a Hölder modulus $\sqrt{\omega}$ of almost minimality. In fact, the argument can be extended to more general moduli allowed by ($\mu\omega1$) – ($\mu\omega3$) below, again showing the optimality of the regularity results (cf. [17]).* □

Example 2.4 (solution to an obstacle problem): *If T minimizes $\mathbf{F}$ subject to an obstacle condition $\operatorname{spt} T \subset \mathbb{R}^{n+N} \setminus \mathcal{O}$, where $\mathcal{O} \subset \mathbb{R}^{n+N}$ is an open "obstacle", then T is almost minimizing (without constraint) for $\mathbf{F}$ and $\omega(r) = Cr^{2\beta}$, provided $\mathcal{O}$ is uniformly of class $C^{1,\beta}$ at the boundary. (See [3] and, in a more special situation, [45], [53]).* □

Example 2.5 (volume-constrained problems): *Suppose that the n-current T minimizes $\mathbf{F}$ subject to a volume constraint $\mathbf{V}(T) = v$. Here $\mathbf{V}$ may be any (weighted) oriented volume functional, by which we mean an additive functional on n-currents such that $|\mathbf{V}(X)| \le \gamma\mathbf{M}(Q)$ holds whenever X is a closed n-current and Q an $(n+1)$-current with $X = \partial Q$. If the first variation of $\mathbf{V}$ on T does not vanish and X is supported in a small closed ball $\mathbf{B}_r^{n+N}(z)$, then we can correct the volume $\mathbf{V}(T + X)$ by passing to a modification $\widetilde{T}$ of T such that $\widetilde{T} = T$ near $\operatorname{spt} X$, $\mathbf{V}(\widetilde{T}+X) = v$, and $|\mathbf{F}(\widetilde{T}+X) - \mathbf{F}(T+X)| \le C|\mathbf{V}(X)|$. Since T is $\mathbf{F}$ minimizing with the constraint $\mathbf{V}(T) = v$, this implies*

$$\mathbf{F}(T) \le \mathbf{F}(T+X) + C|\mathbf{V}(X)| \le \mathbf{F}(T+X) + \gamma C\mathbf{M}(Q)\,. \tag{8}$$

Now, comparing Q to the cone over X with vertex z one sees

$$\mathbf{M}(Q) \le \frac{r}{n+1}\mathbf{M}(X) \le \frac{r}{n+1}[\mathbf{M}(T\llcorner \operatorname{spt} X + X) + \mathbf{M}(T\llcorner \operatorname{spt} X)]\,, \tag{9}$$

and one deduces, taking into account the smallness of r and $\mathbf{M} \le \mathbf{F} \le \Lambda\mathbf{M}$, that T is almost minimizing for $\mathbf{F}$ and $\omega(r) = Cr$. (See [3], [15] in the case of codimension $N = 1$ and [34], [55] for a more special case.) □

Example 2.6 (functionals of mean curvature type): *Suppose that the n-current T is $\mathbf{F} + \mathbf{V}$ minimizing on $\mathbb{R}^{n+N}$ where the volume functional $\mathbf{V}$ is as in the preceding example. With $X = \partial Q$ and γ as above we then have*

$$\mathbf{F}(T) \le \mathbf{F}(T+X) + \mathbf{V}(X) \le \mathbf{F}(T+X) + \gamma\mathbf{M}(Q)\,, \tag{10}$$

similarly to (8), and hence (9) again implies that T is $(\mathbf{F}, \omega)$-minimizing with $\omega(r) = Cr$. Functionals of the type $\mathbf{F} + \mathbf{V}$ occur as energies in the variational approach to hypersurfaces with prescribed mean curvature (see [22] for references). In [20] currents with the almost minimizing property (10) for the area integrand $\mathbf{F} = \mathbf{M}$ were studied systematically in codimension $N = 1$.

We note that the regularity theory [25] for mass-minimizing currents in codimension 1 has been extended to $\mathbf{M} + \mathbf{V}_H$-minimizing currents in [20]. Here $\mathbf{V}_H(T)$ is the oriented volume enclosed by T and a fixed reference n-current T_0 with $\partial T = \partial T_0$ where the volume is computed with a given weight function H, i.e. $\mathbf{V}_H(T)$ is the integral of H over the unique top dimensional current Q (a "set with integer multiplicities") of finite mass such that $\partial Q = T - T_0$. The minimizers of $\mathbf{M} + \mathbf{V}_H$ then *are (generalized) hypersurfaces of prescribed mean curvature H in $\mathbb{R}^{n+1}$. Moreover, complete boundary regularity of minimizers for $\mathbf{M} + \mathbf{V}_H$, as proved in [39] in the minimal hypersurface case $H \equiv 0$, was also shown in the case of bounded Lipschitz functions H in [21]. As we have remarked already, these regularity results are much stronger than those that can be obtained for (almost) minimizers of elliptic parametric integrals in general.* □

Consider now an n-current T which is $(\mathbf{F}, \omega)$-minimizing in $\mathcal{O} \subset \mathbb{R}^{n+N}$, where F satisfies the general assumptions (F1)-(F3). The aim of interior *regularity theory* is to prove that, for a large set of points $z \in \mathcal{O} \cap \operatorname{spt}(T) \setminus \operatorname{spt}(\partial T)$ (the *regular set*) and for suitable rotations Γ_z of $\mathbb{R}^{n+N}$ fixing z, the rotated current $(\Gamma_z)_{\#}T$ is represented locally near z by the graph, taken with suitable orientation and constant multiplicity, of a smooth function $g \in C^1(\mathbb{R}^n, \mathbb{R}^N)$. Moreover, we want to establish the best possible modulus of continuity for the derivative Dg, i.e. for the tangent plane field of T near z, which is allowed by the data (F, ω).

To fix a setting suitable for the treatment of this problem we denote by $\mathbf{p}$ and $\mathbf{q}$ the projections of $\mathbb{R}^{n+N} = \mathbb{R}^n \times \mathbb{R}^N$ onto the first and second factor respectively and we consider integer multiplicity rectifiable n-currents T on $\mathbb{R}^{n+N}$ with zero boundary $(\partial T) \llcorner \mathbf{C} = 0$ in the open unit cylinder $\mathbf{C} = \mathbf{U} \times \mathbb{R}^N$ (recall that $\mathbf{U}$ is the open unit ball in $\mathbb{R}^n$), with the origin in the support of T, and with projection $\mathbf{p}_{\#}T$ represented by $\mathbf{U}$ with the standard orientation n-vector $\mathbf{e}$ of $\mathbb{R}^n$ and with constant positive multiplicity $m \in \mathbb{Z}_+$. Replacing T by $T \llcorner \mathbf{C}$, the current obtained from T by restricting the integration of test differential n-forms to $M_T \cap \mathbf{C}$, it is no loss of generality to assume that T vanishes outside $\mathbf{C}$ or, more precisely, $T = T \llcorner \mathbf{C}$. Finally, we require that the multiplicity ϑ_T of T is at least m, i.e. $\vartheta \geq m$ holds μ_T almost everywhere. We collect all these conditions on T by writing

$$T \in \mathcal{T}_i(n, N, m)\,,$$

where $\mathcal{T}_i(n, N, m)$ is called the class of *admissible currents*, the subscript i indicating here that the interior regularity problem is considered.

A crucial quantity in regularity theory is the *cylindrical excess* which is defined, for n-currents T on $\mathbb{R}^{n+N}$ and $r > 0$, by the scaled difference of the mass of T in a cylinder of radius r and the mass of its projection on the cross section of this cylinder, i.e.

$$\mathbf{E}(T, r) = r^{-n}[\mathbf{M}(T \llcorner \mathbf{C}_r) - \mathbf{M}(\mathbf{p}_{\#}(T \llcorner \mathbf{C}_r)] \geq 0\,, \tag{11}$$

where $\mathbf{C}_r = \mathbf{U}_r \times \mathbb{R}^N$ is the cylinder over the open ball $\mathbf{U}_r$ in $\mathbb{R}^n$ with center 0 and radius r. For the excess of T in We will write $\mathbf{E}(T, x, r)$ for the excess of T in translated cylinders $\mathbf{C}_r(x) = (x, 0) + \mathbf{C}_r = \mathbf{U}_r(x) \times \mathbb{R}^N$ with $x \in \mathbb{R}^n$. If $T \in \mathcal{T}_i(n, N, m)$, then for $r \in (0, 1]$ we have

$$\mathbf{E}(T, r) = r^{-n}\mathbf{M}(T \llcorner \mathbf{C}_r) - m\boldsymbol{\alpha}(n)\,,$$

where $\boldsymbol{\alpha}(n) = \mathcal{L}^n(\mathbf{U})$ denotes the Lebesgue measure of the unit n-ball, and we can rewrite this as

$$\mathbf{E}(T,r) = r^{-n}\int_{\mathbf{C}_r}(1-\zeta_T\cdot\mathbf{e})\,d\mu_T = \tfrac{1}{2}\int_{\mathbf{C}_r}|\zeta_T-\mathbf{e}|^2 d\mu_T\,, \tag{12}$$

showing the similarity of the excess of an admissible n-current to Dirichlet's integral of a function in $H^{1,2}(\mathbf{U}_r,\mathbb{R}^N)$. Note that $r^{-n}\mathbf{E}(T,r)$ is nondecreasing with respect to r. The main problem of regularity theory is to prove that $\mathbf{E}(T,r)$ decreases to zero with a certain rate as $r\downarrow 0$.

The regularity theorem will eventually assert that an $(\mathbf{F},\omega)$-minimizing admissible n-current T is regular near the origin provided that the excess $\mathbf{E}(T,1)$ is sufficiently small (and some additional smallness conditions are satisfied by μ and ω on $(0,1]$; these can usually be achieved by suitable initial scaling).

We now give an *outline of the regularity proof* describing the various steps. In the rest of this section we denote by $c\in(0,1]$ and $C\in[1,\infty)$ constants that may depend on the dimension n, the codimension N, the multiplicity $m\in\mathbb{Z}_+$, the bound Λ for F from (F1), and the ellipticity constant λ from (F3). Dependence on other quantities such as parameters $\varepsilon>0$ or moduli of continuity $\delta(\,\cdot\,)$ will be indicated by writing $c=\mathrm{const}(\varepsilon,\delta(\,\cdot\,),\dots)$ and $C=\mathrm{Const}(\varepsilon,\delta(\,\cdot\,),\dots)$; such constants may also depend on n,N,m,Λ,λ. Apart from the conditions on μ from (F2) and on ω that have already been mentioned we also initially assume $\mu(1)\le 1$, $\omega(1)\le 1$.

The first step of the proof is a *lower mass estimate* for T in balls $\mathbf{U}_r^{n+N}(z)\subset\mathbb{R}^{n+N}$ with $z\in\operatorname{spt}T$ and $\mathbf{U}_r^{n+N}(z)\cap\operatorname{spt}\partial T=\emptyset$:

$$\mathbf{M}(T\llcorner\mathbf{U}_r^{n+N}(z))\ge cr^n\quad\text{for } 0<r\le 1-|\mathbf{p}z|\,. \tag{13}$$

This is well known and easily proved by comparing T in $\mathbf{U}_r^{n+N}(z)$ with an n-current S of least mass satisfying $\partial S=\partial(T\llcorner\mathbf{U}_r^{n+N}(z))$. From the isoperimetric inequality for S one then gets a differential inequality for $\Phi(r)=\mathbf{M}(T\llcorner\mathbf{U}_r^{n+N}(z))$ which gives (13) by integration.

The next step is the *height bound* for T in the cylinder $\mathbf{C}_{1/2}$, i.e. a bound for the norm of the vertical components $\mathbf{q}z$ of points $z\in\mathbf{C}_{1/2}\cap\operatorname{spt}T$,

$$\sup\{|\mathbf{q}z| : z\in\mathbf{C}_{1/2}\cap\operatorname{spt}T\}\le C\mathbf{E}(T,1)^{1/2n}\,, \tag{14}$$

provided a smallness condition $\mathbf{E}(T,1)\le c$ is satisfied. This is again well known and proved as follows: for a unit vector $v\in\{0\}^n\times\mathbb{R}^N$ and the half spaces $H(s)=\{z\in\mathbb{R}^{n+N} : v\cdot z>s\}$ one considers the currents $T_s=T\llcorner H(s)$ in a range $r_0<s<r$ such that $\mathbf{M}(T_{r_0})\le\frac{1}{2}\mathbf{M}(T)$. From the condition $\vartheta_T\ge m$ for $T\in\mathcal{T}_i(n,N,m)$ one then deduces that $\mathbf{p}_\#T_s$ cannot cover more than three quarters of $\mathbf{U}$, if $E=\mathbf{E}(T,1)$ is small, and a relative isoperimetric inequality implies

$$c(\mathbf{M}(T_s)-E)^{1-1/n}\le c\mathbf{M}(\mathbf{p}_\#T_s)^{1-1/n}\le\mathbf{M}((\partial T_s)\llcorner\mathbf{C})\,. \tag{15}$$

Now, if $\mathbf{M}(T_s)\ge\sqrt{E}$ for $r_0<s<r$, then integration of (15) and standard slicing estimates ([24], 4.3.2 (2)) yield $(\frac{1}{2}\sqrt{E})^{1-1/n}(r-r_0)\le C\sqrt{E}$, i.e. $r-r_0\le CE^{1/2n}$. On the other hand, if $\mathbf{M}(T_r)\le\sqrt{E}$, then it follows directly from (13) that $T_s=0$ for $s-r>CE^{1/2n}$. Clearly, these observations imply (14) taking into account

$0 \in \operatorname{spt} T$. We refer to [24], Chapter 5, [10], Section IV, [51], Appendix, and [22], Section 2 for details.

The next step in the procedure is technically important; it consists of approximating $T \llcorner \mathbf{U}_\rho$ for a fixed $\rho > 0$ (say $\rho = \frac{1}{17}$ as in [22], Sec. 3) by the current T_g associated with a Lipschitz map $g : \mathbf{U}_\rho \to \mathbb{R}^N$ with $\operatorname{Lip} g \leq 1$ by taking the graph of g with multiplicity m and with appropriate orientation. This *Lipschitz approximation* was carried out in [51], Lemma 3 modifying [1], 8.12 (cf. [49], 20). The idea is to define the "good set" G as the set of $x \in \mathbf{U}_\rho$ such that $\mathbf{E}(T, x, \sigma) \leq \eta$ for $0 < \sigma < 1 - \rho$ where $\eta > 0$ will be suitably chosen. Considering $z = (x, y)$, $\widetilde{z} = (\widetilde{x}, \widetilde{y})$ in $(G \times \mathbb{R}^{n+N}) \cap \operatorname{spt} T$ one deduces from the height estimate (14) (after translating x to the origin and scaling by a homothety with factor $|x - \widetilde{x}|^{-1}$) the inequality

$$|y - \widetilde{y}| \leq C\eta^{1/2n}|x - \widetilde{x}| \,.$$

Choosing η such that $C\eta^{1/2n} \leq 1$ it follows that, in $G \times \mathbb{R}^N$, the support of T coincides with the graph of a function $g : \mathbf{U}_\rho \to \mathbb{R}^N$ satisfying $\operatorname{Lip} g \leq 1$. The construction allows good estimates for the difference of T and T_g in $\mathbf{C}_\rho$ and for the "bad set" $B = \mathbf{U}_\rho \setminus G$ in terms of the excess $\mathbf{E}(T, 1)$, for example

$$\mathbf{M}((T - T_g) \llcorner \mathbf{C}_\rho) + \mathcal{L}^n(B) + \int_G |Dg|^2 dx \leq C\mathbf{E}(T, 1) \,. \tag{16}$$

Moreover, denoting by B_γ for $\gamma > 0$ the union of B and the set of points $x \in \mathbf{U}_\rho$ with $|Dg(x)| \geq \gamma$ one can show

$$\mathbf{M}(T \llcorner (B_\gamma \times \mathbb{R}^N)) + \mathcal{L}^n(B_\gamma) \leq C\gamma^{-2}\mathbf{E}(T, 1) \,. \tag{17}$$

One can also achieve $g(0) = 0$, and from (14) a bound $|g| \leq CE^{1/2n}$ follows on $\mathbf{U}_\rho$. See [51], Lemma 3 and [22], Lemma 3.1 for complete proofs.

A crucial next step in our method of A-harmonic approximation is now to establish *approximate A-harmonicity* of the Lipschitz approximation g. Here A is the Hessean $D_p^2 f(0, 0, 0)$ of the nonparametric integrand $f(x, y, p)$ associated with F and the reference plane $\mathbb{R}^n \times \{0\}^N$, and the deviation of g from being exactly A-harmonic is again measured in terms of the excess $\mathbf{E}(T, 1)$. Note that A is rank one elliptic by (F3) (see [24], 5.1.10), and we may assume that the ellipticity constant λ from (F3) equals the ellipticity constant in (2). The approximate A-harmonicity of an approximating function for T was first proved by Bombieri [10], Lemmas 12 and 13. The idea is to consider the vertical deformations $h_t(z) = z + t(0, \varphi(\mathbf{p}z))$ associated with $\varphi \in C_c^1(\mathbf{U}_\rho, \mathbb{R}^N)$ and to compare $\mathbf{F}(T \llcorner \mathbf{C}_\rho)$ to $\mathbf{F}((h_{t\#}T) \llcorner \mathbf{C}_\rho)$ on the one hand, using the $(\mathbf{F}, \omega)$ minimality of T and the structure conditions (F1),(F2), and to $\mathbf{F}(T_g \llcorner \mathbf{C}_\rho)$ on the other hand, using the estimates (16), (17) for the Lipschitz approximation. The connection to A is made by a Taylor expansion of the nonparametric integrand f in

$$\mathbf{F}(T_g \llcorner \mathbf{C}_\rho) = \int_{\mathbf{U}_\rho} f(x, g(x), Dg(x)) \, dx \,,$$

also using the modulus ν of $D^2_\zeta F$ from (F2). The desired result then follows by suitably choosing t with $|t| \approx \mu(1)+\omega(1)$. For the somewhat lengthy but straightforward calculations we refer to [10] and [22], but we want to give a precise formulation of the result here, because it is essential for the later application of Lemma 1.1 and for obtaining the optimal modulus of continuity for Dg.

For this we associate with μ from (F2) the modified modulus

$$\widehat{\mu}(r) = \sup\{\mu(s) - (s/r)^2 : 0 < 2s \leq 1\} \tag{18}$$

for $r > 0$. One verifies that $\sqrt{\widehat{\mu}}$ has similar properties to μ, viz. $\widehat{\mu}(r)$ is nondecreasing with $\widehat{\mu}(0+) = 0$ and $\frac{1}{r}\sqrt{\widehat{\mu}(r)}$ is nonincreasing. Moreover, $\frac{1}{4}\mu(r)^2 \leq \widehat{\mu}(r) \leq \mu(r)$ holds for $0 < r \leq \frac{1}{2}$. For example, if $\mu(r) = cr^\beta$ with $\beta \in (0,1]$, then $\widehat{\mu}(r) = \mathrm{const}(c,\beta)r^{2\beta/(2-\beta)}$. The reason for introducing $\widehat{\mu}$ is that we obtain a better modulus of continuity $Cr^{\beta/(2-\beta)}$ in the end than $Cr^{\beta/2}$, which would be the result if we had not replaced μ by the smaller quantity $\widehat{\mu}$. The following lemma is the version of approximate A-harmonicity taken from [22], Lemma 3.2.

Lemma 2.7 (approximate A-harmonicity) *Suppose $T \in \mathcal{T}_i(n,N,m)$ is $(\mathbf{F},\omega)$-minimizing in $\mathbf{C}$ with height not exceeding 1 in $\mathbf{C}_\rho$, $\chi = \widehat{\mu}(1)+\omega(1) \leq 1$, $E = \mathbf{E}(T,1) \leq c$, and g is the Lipschitz approximation associated with T above. Then*

$$\left|\int_{\mathbf{U}_\rho} A(Dg, D\varphi)\,dx\right| \leq C\Big[\gamma^{-2}E + \sqrt{\chi}(1+E) \tag{19}$$
$$+ (\sqrt{E} + \sqrt{\chi})\nu(\gamma + \sqrt{\chi})\Big] \sup|D\varphi|$$

holds for all parameters $\gamma \in (0,1]$ and all test functions $\varphi \in C^1_c(\mathbf{U}_\rho, \mathbb{R}^N)$. □

The essential point here is that in later arguments χ/E will be small so that, with an appropriate choice of γ, the right-hand side of (19) is a small quantity times $\sqrt{E}\sup|D\varphi|$.

As is well known ([2], [3], [10], [24], 5.3, [38], [51]), the crucial step in the regularity proof is *excess improvement by tilting*, an idea introduced by Almgren. If T is $(\mathbf{F},\omega)$-minimizing with small excess $\mathbf{E}(T,1)$ in the cylinder $\mathbf{C}$ of radius 1 perpendicular to the n-plane $S_0 = \mathbb{R}^n \times \{0\}^N$, then one shows that the excess of T in a cylinder of smaller radius perpendicular to some "tilted" n-plane S close to S_0 is substantially smaller that $\mathbf{E}(T,1)$. Iterating this one obtains a decay rate of the excess in concentric cylinders of radius r perpendicular to some fixed n-plane as $r \downarrow 0$, and from this the regularity of T in a small cylinder eventually follows.

In the method we discuss here, the idea is to choose S as the graph of the linearization at the origin of the A-harmonic function h which is associated by Lemma 1.1 to the approximately A-harmonic Lipschitz function g above. More precisely, keeping the notations and hypotheses of Lemma 2.7 we consider $\widetilde{g} = c(E + \sigma^{-1}\chi)^{-1/2}g$ with arbitrary $\sigma \in (0,1]$ and with c suitably chosen to obtain

$$\int_{\mathbf{U}_\rho} |D\widetilde{g}|\,dx \leq 1$$

with the help of (16). Using Lemma 2.7 we then apply Lemma 1.1 (scaled by a homothety and recalling that $\rho = \frac{1}{17}$ is fixed) to obtain, for given $\varepsilon \in (0,1]$, an A-harmonic function $h : \mathbf{U}_\rho \to \mathbb{R}^N$ satisfying

$$\int_{\mathbf{U}_\rho} |g-h|^2 dx \le \varepsilon C(E + \sigma^{-1}\chi) \tag{20}$$

and

$$\int_{\mathbf{U}_\rho} |h|^2 dx \le C(E + \sigma^{-1}\chi), \tag{21}$$

provided the smallness condition

$$\gamma^{-2}\sqrt{E} + \sqrt{\sigma} + \nu(\gamma + \sqrt{\chi}) \le c\delta(n, N, \lambda, \Lambda, \varepsilon) \tag{22}$$

is satisfied. Defining then $L = Dh(0) \in \mathrm{Hom}(\mathbb{R}^n, \mathbb{R}^N)$, $S = \operatorname{graph} L$, and letting v be the component of $(0, h(0))$ orthogonal to S, one infers from (20), (21), by Taylor expansion of h and the standard estimate

$$\sup_{\mathbf{U}_{\rho/2}} |Dh|^2 + \sup_{\mathbf{U}_{\rho/2}} |D^2 h|^2 \le C \int_{\mathbf{U}_\rho} |Dh|^2 dx \tag{23}$$

for the A harmonic function h (cf. [11], Teorema 9.2), a bound

$$\int_{\mathbf{C}_\eta} \operatorname{dist}(z - v, S)^2 d\mu_T(z) \le C\left(\gamma^{-2}E^{1/n} + \eta^{n+4} + \varepsilon\right)(E + \sigma^{-1}\chi) \tag{24}$$

for $0 < \eta \le \frac{1}{2}\rho$ as well as bounds

$$\|L\| \le C\sqrt{E + \sigma^{-1}\chi}, \qquad \|v\| \le C(\sqrt{E + \sigma^{-1}\chi} + E^{1/2n}). \tag{25}$$

Now, (24) means that we have estimated the square integral of the height of $T \llcorner \mathbf{C}_\eta$ over the tilted n-plane S by a small quantity times $E + \sigma^{-1}\chi$ (and $\sigma^{-1}\chi$ will also be small compared to E in later arguments). To obtain the desired excess improvement we need an inequality that bounds the excess in a cylinder by the square integral of the height in some larger concentric cylinder. Recalling that the excess resembles the Dirichlet integral in the classical calculus of variations such an inequality will be called *Caccioppoli inequality* or *reverse Poincaré inequality*. It was proved by Bombieri in [10], Section 5 using a comparison of the $(\mathbf{F}, \omega)$-minimizing n-current $T \in \mathcal{T}_i(n, N, m)$ with a current $\Sigma_{r,\vartheta}$ which, for $0 < \vartheta \ll 1$, is represented inside $\mathbf{C}_{(1-\vartheta)r}$ by the disk $\mathbf{U}_{(1-\vartheta)r} \times \{0\}^N$ with multiplicity m and appropriate orientation, which coincides with T outside the closure of $\mathbf{C}_r$, and which suitably interpolates in-between. The ellipticity of the integrand F in the geometric form of condition (F3) also enters the proof in an essential way. The following version of the inequality is taken from [22]:

Lemma 2.8 (reverse Poincaré inequality) *Suppose* $T \in \mathcal{T}_i(n, N, m)$ *is* $(\mathbf{F}, \omega)$*-minimizing with height not exceeding 1 in* $\mathbf{C}$ *and* $\chi = \widehat{\mu}(1) + \omega(1)$. *Then the following inequalities hold:*

$$\mathbf{E}(T, \tfrac{1}{2}) \leq \vartheta \mathbf{E}(T, 1) + C\vartheta^{-1}\left[\chi \mathbf{M}(T) + \int_{\mathbf{C}} |\mathbf{q}z|^2 d\mu_T(z)\right] \tag{26}$$

for all parameters $\vartheta > 0$, *and*

$$\mathbf{E}(T, \tfrac{1}{5}) \leq C\left[\chi \mathbf{M}(T) + \int_{\mathbf{C}} |\mathbf{q}z|^2 d\mu_T(z)\right]. \tag{27}$$

□

The "weak Poincaré inequality" (26) is proved in [22] by a refinement of the arguments in [10], Section V. The strong form (27) then follows with a covering argument taken from [50], 2.8. Since the term χ/E is small in the later applications of Lemma 2.8, the extra summand $\chi \mathbf{M}(T)$ in the right-hand side of (27) does not matter. We note that the condition $0 \in \operatorname{spt} T$ contained implicitly in the hypothesis $T \in \mathcal{T}_i(n, N, m)$ is not needed in the proof, and also that multiplicity $m = 0$ of $\mathbf{p}_{\#}T$ can be allowed.

Using the height estimate and elementary geometric considerations it is now straightforward to combine inequality (24) with a scaled and rotated version of the reversed Poincaré inequality given above in order to derive the desired excess improvement. The result, given as Lemma 5.2 in [22], is the following:

Lemma 2.9 (excess improvement by tilting) *Under the hypotheses of Lemma 2.8 and given* $\vartheta \in (0, \frac{1}{20}\rho)$ *there exists a linear isometry* Γ *of* $\mathbb{R}^{n+N}$ *such that*

$$\mathbf{E}(\Gamma_{\#}T, \vartheta) \leq C\vartheta^2 \mathbf{E}(T, 1) + \mathrm{Const}(\vartheta)\chi \tag{28}$$

holds, provided a smallness condition

$$\mathbf{E}(T, 1) + \chi \leq \mathrm{const}(\vartheta, \nu(\,\cdot\,)) \tag{29}$$

is satisfied, where $\nu(\,\cdot\,)$ *is from (F2). Moreover,*

$$\|\Gamma - \mathrm{id}\| \leq \mathrm{Const}(\vartheta)\sqrt{\mathbf{E}(T, 1) + \chi}\,. \tag{30}$$

□

The last step to prove *excess decay* is now to iterate (28). This is standard, see [2], [3], [10], and [24], 5.3. The iteration produces a sequence of currents T_k and linear isometries Γ_k such that, with $\boldsymbol{\mu}_\vartheta(z) = \vartheta z$,

$$T_{k+1} = (\boldsymbol{\mu}_{\vartheta\#}^{-1} \Gamma_{k\#} T_k) \llcorner \mathbf{C}$$

and (28) – (30) are valid for T_k, and one must secure a limit Γ of the products $\Gamma_k \Gamma_{k-1} \cdots \Gamma_0$ as $k \to \infty$. Assuming for simplicity that $r^{-2\alpha}\chi(r) = r^{-2\alpha}(\widehat{\mu}(r) +$

$\omega(r)$) is nonincreasing for some fixed $\alpha \in (0,1)$, we can use (28) and (30) to reduce this problem to that of establishing convergence of the series $\sum_k \chi(\vartheta^k)$. By the general properties of $\chi(\cdot)$, this is equivalent to finiteness of the integral $\int_0^s \frac{1}{r}\sqrt{\chi(r)}\,dr$ for some $s > 0$. We thus have to assume for our moduli $\omega(\cdot)$ from Definition 2.2 and $\widehat{\mu}(\cdot)$ from (18) the following *conditions of Dini type*:

$$\Omega(r) = \left[\int_0^s \tfrac{1}{r}\sqrt{\omega(r)}\,dr\right]^2 < \infty\,, \quad \widehat{\mathsf{M}}(r) = \left[\int_0^s \tfrac{1}{r}\sqrt{\widehat{\mu}(r)}\,dr\right]^2 < \infty\,. \tag{31}$$

Note that on account of the inequality $\widehat{\mu}(r) \le \mu(r)$ for small r, the second condition is implied by $\int_0^s \frac{1}{r}\sqrt{\mu(r)}\,dr < \infty$ where μ is from (F2). Note also that (31) is satisfied by $\omega(r) = cr^{2\beta}$, $\mu(r) = cr^{\beta}$ with $\beta \in (0,1]$, and $\Omega(r) = Cr^{2\beta}$, $\widehat{\mathsf{M}}(r) = Cr^{2\beta/(2-\beta)}$ in this case. Condition (31) for ω already occurs in [3], [10], [53] and is, in a sense, necessary for obtaining C^1 regularity of $(\mathbf{F},\omega)$-minimizers as shown by examples in [53] for the area integrand.

With the help of assumption (31), the existence of the limit rotation Γ can now be established, and (28) for the T_k with a suitable choice of ϑ implies the following excess decay for $\Gamma_{\#}(T \llcorner \mathbf{C}_{1/2})$, see Lemma 5.4 in [22]. In this lemma we use α from above and the definition

$$\mathsf{X}(r) = \widehat{\mathsf{M}}(r) + \Omega(r)\,. \tag{32}$$

Lemma 2.10 (excess decay) *For $\beta \in [\alpha, 1)$ there exists $\varepsilon_0 = \mathrm{const}(\beta, \nu(\cdot))$ such that the following holds: whenever $T \in \mathcal{T}_i(n, N, m)$ is $(\mathbf{F},\omega)$-minimizing in $\mathbf{C}$ with*

$$\mathbf{E}(T,1) + \mathsf{X}(1) \le \varepsilon_0\,,$$

then there exists a linear isometry Γ of $\mathbb{R}^{n+N}$ such that

$$\mathbf{E}(\Gamma_{\#}(T \llcorner \mathbf{C}_{1/2}), r) \le C\left[r^{2\beta}\mathbf{E}(T,1) + \mathsf{X}(r)\right] \quad \textit{for } 0 < r \le \tfrac{1}{3} \tag{33}$$

and

$$\mathbf{E}(T,r) + \|\Gamma - \mathrm{id}\|^2 \le C[\mathbf{E}(T,1) + \mathsf{X}(1)] \quad \textit{for } 0 < r \le 1\,, \tag{34}$$

where $C = \mathrm{Const}(\beta)$. □

The last step in the proof of an interior "ε-regularity-theorem" is now well known from [2], [3], [10], and [24], 5.3. After having rotated T suitably one first infers from Lemma 2.10 by translating and scaling that $\mathbf{E}(T, x, r) + \mathsf{X}(r) \le \mathrm{Const}(\beta)\varepsilon$ is uniformly small for $x \in \mathbf{U}_{\rho/2}$ and $0 < r \le 1 - |x|$, if $\mathbf{E}(T,1) + \mathsf{X}(1) \le \varepsilon$ and ε is sufficiently small. Recalling the definition of the good set for the Lipschitz approximation g of T one deduces that the bad set is empty if ε is small enough, i.e. T is represented in $\mathbf{U}_\rho$ by the graph of g with multiplicity m and appropriate orientation. Applying Lemma 2.10 again to translated and scaled copies of T one further sees that all points $x \in \mathbf{U}_\rho$ are Lebesgue points of Dg and that the mean L^2 deviation of Dg from its Lebesgue value at x on a ball of radius r in $\mathbf{U}_\rho$ can be estimated by the quantity appearing in the right-hand side of (33). These observations then give immediately a modulus of continuity for Dg. To formulate the result precisely, we reiterate here the assumptions we have made at various stages for μ from (F2) and for ω from Definition 2.2:

- **($\mu\omega$1)** $\mu(\,\cdot\,)$ and $\omega(\,\cdot\,)$ are positive and nondecreasing on $(0,\infty)$ with $\mu(0)=0=\mu(0+)$ and $\omega(0)=0=\omega(0+)$;
- **($\mu\omega$2)** $r\mapsto r^{-\alpha}$ and $r\mapsto r^{-\alpha}\sqrt{\omega(r)}$ are nondecreasing for some $\alpha\in(0,1)$;
- **($\mu\omega$3)** $\widehat{\mathsf{M}}(r)$ and $\Omega(r)$ from (31) are finite for $0<r\ll 1$.

Theorem 2.11 (interior ε-regularity-theorem) *Given $\beta\in[\alpha,1)$ there exists $\varepsilon=$ const$(\beta,\nu(\,\cdot\,))$ such that the following is true: if $T\in\mathcal{T}_i(n,N,m)$ is $(\mathbf{F},\omega)$-minimizing in $\mathbf{C}$, where F satisfies (F1)-(F3) and $\mu(\,\cdot\,),\omega(\,\cdot\,)$ satisfy ($\mu\omega$1) – ($\mu\omega$3), and if*

$$\mathbf{E}(T,1)+\widehat{\mathsf{M}}(1)+\Omega(1)\le\varepsilon\,,$$

then $T\llcorner\mathbf{C}_{\rho/2}$ is represented by the graph of a function $g\in C^1(\mathbf{U}_{\rho/2},\mathbb{R}^N)$, taken with multiplicity m, and Dg has the following modulus of continuity

$$|Dg(x)-Dg(\widetilde{x})|\le C\left[|x-\widetilde{x}|^{2\beta}\mathbf{E}(T,1)+\widehat{\mathsf{M}}(|x-\widetilde{x}|)+\Omega(|x-\widetilde{x}|)\right]^{1/2} \tag{35}$$

for $x,\widetilde{x}\in\mathbf{U}_{\rho/2}$, where $C=$ Const(β). □

Recall that $\rho=\frac{1}{17}$ was fixed. Of course, we can admit any $\rho\in(0,2)$ in the theorem, provided we allow ε and C to depend on ρ as $\rho\uparrow 2$. All the hypotheses on μ and ω are satisfied if $\omega(r)=cr^{2\beta}$ and $\mu(r)=cr^{\beta'}$ with $\beta,\beta'\in(0,1)$, and Theorem 2.11 then yields the modulus $C[r^{\beta}+r^{\beta'/(2-\beta')}]$ for Dg. By the examples discussed previously in this section this modulus is optimal if F is of class $C^{1,1}$ or at least $\beta'\ge 2\beta/(1+\beta)$. One would expect that $r^{\beta'/(2-\beta')}$ could be replaced by $r^{\beta'}$ here, but we do not currently know how to prove this. The optimal version of the interior ε-regularity-theorem above was proved in [22], Theorem 6.1. Since one can always achieve smallness of $\widehat{\mathsf{M}}(1)+\Omega(1)$ by scaling, the theorem can be applied to an n-current T locally at every point $z\in\operatorname{spt}(T)\setminus\operatorname{spt}(\partial T)$ where ϑ_T has a local minimum and T has an oriented n-plane with positive integer multiplicity as tangent cone. This implies regularity of T, with modulus (35) for the tangent plane field, on a dense open subset of $\operatorname{spt}(T)\setminus\operatorname{spt}(\partial T)$ (see [24], 5.3). In special cases one can estimate the dimension of the singular set; see [25] for the area integrand with codimension $N=1$, [52] for general elliptic integrands with $N=1$, and Almgren's monumental work [4] for the area integrand with arbitrary dimension n and codimension N.

As we have mentioned in the introduction, one of the advantages of the method of A-harmonic approximation we have described here is that it allows a unified treatment of interior and boundary regularity, and to conclude this section we want to discuss briefly the question of *boundary regularity*. Indeed, if we impose a *flat boundary condition* on our n-currents T, i.e. $(\partial T)\llcorner\mathbf{C}$ is represented by the $(n-1)$-dimensional unit disk $\mathbf{U}^{n-1}\times\{0\}^{1+N}$ with multiplicity 1 and suitable orientation, then most of the arguments go through with obvious small modifications. As *admissible currents in the boundary situation*

$$T\in\mathcal{T}_b(n,N,m)$$

we then have to consider integer multiplicity rectifiable n-currents $T=T\llcorner\mathbf{C}$ on $\mathbb{R}^{n+N}$ with flat boundary condition such that the projection $\mathbf{p}_{\#}T$ is represented by the upper half ball $\mathbf{U}_+$ in $\mathbb{R}^n$ with positive multiplicity $m\in\mathbb{Z}_+$ and with the orientation $\mathbf{e}$ of $\mathbb{R}^n$ plus the lower half ball $\mathbf{U}_-$ with multiplicity $m-1\geq 0$ and with orientation $\mathbf{e}$. A height bound as in (14) is valid if we additionally require the density condition $\vartheta_T\geq m$ on $\mathbf{U}_+\times\mathbb{R}^N$ and $\vartheta\geq m-1$ on $\mathbf{U}_-\times\mathbb{R}^N$ almost everywhere with respect to μ_T (see [22], Remark 2.3). However, this is of no value because in the course of the regularity proof we have to tilt the n-plane $\mathbb{R}^n\times\{0\}^N$ (leaving $\mathbb{R}^{n-1}\times\{0\}^{1+N}$ fixed), and this additional density condition is not preserved a priori by such tiltings. Another problem that occurs is that the proof of Lemma 2.7 now gives approximate A-harmonicity of g as in (19) only separately on $\mathbf{U}_+$ and on $\mathbf{U}_-$ but not on $\mathbf{U}$.

For these reasons we have to restrict ourselves to the case of multiplicity $m=1$ in the boundary regularity discussion as was already done in [38], the first boundary regularity result of the type we discuss here. Then all of the steps of the proof sketched above go through, the Lipschitz approximation g vanishes on $\mathbf{U}^{n-1}\times\{0\}$, and the linear isometries used in the excess improvement by tilting leave $\mathbb{R}^{n-1}\times\{0\}^{1+N}$ pointwise fixed. Details are given in [22], and the result is that Theorem 2.11 holds in complete analogy for $T\in\mathcal{T}_b(n,N,1)$ except that now $T\llcorner\mathbf{C}_{\rho/2}$ is represented by the graph of a C^1 function on $(\overline{\mathbf{U}_+}\cap\mathbf{U}_{\rho/2})\times\mathbb{R}^N$ with multiplicity 1.

However, this is not satisfactory for the following reason: if T is $(\mathbf{F},\omega)$-minimizing for a smooth integrand F and ∂T is represented locally by an $(n-1)$-dimensional oriented submanifold $\mathcal{B}$ of $\mathbb{R}^{n+N}$ with multiplicity 1 and with continuity modulus κ for the tangent plane field of $\mathcal{B}$, then we can flatten the boundary locally by transforming $\mathcal{B}$ into $\mathbf{U}^{n-1}\times\{0\}^{1+N}$ with a diffeomorphism Ψ^{-1} of class $C^{1,\kappa(\cdot)}$ and apply the analogue of Theorem 2.11 to the transformed current $\widetilde{T}=\Psi_{\#}^{-1}T$ and to the transformed integrand $\widetilde{F}=\Psi_{\#}F$ (provided the hypotheses of the theorem are satisfied by $\widetilde{T}$ and $\widetilde{F}$). However, the continuity modulus of $\widetilde{F}(z,\zeta)$ and $D_\zeta\widetilde{F}(z,\zeta)$ with respect to the space variable z will not be better than κ, in general, hence one only gets C^1 regularity for T with the modulus $(\widehat{\mathsf{K}}+\Omega)^{1/2}$, where $\widehat{\mathsf{K}}$ is defined in terms of κ in analogy to (18) and (31). For example, if $\kappa(r)=cr^\beta$ with $0<\beta<1$ and $\omega(r)=cr^{2\beta}$ then the analogue of Theorem 2.11 will give a modulus $Cr^{\beta/(2-\beta)}$ at best for the tangent plane field of T, whereas the expected boundary regularity of T in this situation is $C^{1,\beta}$.

However, if we choose Ψ in a natural way and analyze the transformed integrand $\widetilde{F}$ carefully, then we find $\widetilde{\mu}\leq C(\mu+\kappa)$ for the modulus $\widetilde{\mu}$ associated with $\widetilde{F}$ as in (F2), but for directions orthogonal to the boundary we obtain the better modulus $C\mu$ for $\widetilde{F}$, i.e.

$$|\widetilde{F}(z,\zeta) - \widetilde{F}(\widetilde{z},\zeta)| \le C\mu(|z-\widetilde{z}|)|\zeta|,$$
$$|D_\zeta\widetilde{F}(z,\zeta) - D_\zeta\widetilde{F}(\widetilde{z},\zeta)| \le C\mu(|z-\widetilde{z}|),$$

if $z - \widetilde{z} \perp \mathbb{R}^{n-1} \times \{0\}^{1+N}$. Now, the structure of the proof based on A-harmonic approximation allows us to go through all the estimates and verify, for an integrand F with modulus μ in (F2) and with modulus μ_0 in directions perpendicular to $\mathbb{R}^{n-1} \times \{0\}^{1+N}$, that for $(\mathbf{F},\omega)$-minimizing currents in $\mathcal{T}_b(n,N,1)$ the analogue of Lemma 2.9 holds with $\chi = \widehat{\mu_0}(1) + \mu(1)^2 + \omega(1)$. Consequently, versions of Lemma 2.10 and of the ε-regularity-theorem are also valid for $(\mathbf{F},\omega)$-minimizing $T \in \mathcal{T}_b(n,N,1)$ with $\mathsf{X}(r) = \widehat{\mathsf{M}_0}(r) + \mathsf{M}(r)^2 + \Omega(r)$, where $\widehat{\mathsf{M}_0}$ and Ω are defined by (31) and (18) as before and

$$\mathsf{M}(s) = \int_0^s \tfrac{1}{r}\mu(r)\,dr\,.$$

Note that this may give a better modulus of continuity for Dg than the direct analogue of Theorem 2.11, because $\mu_0 \le \mu$ and $\mu(r)^2 \le 4\widehat{\mu}(r)$, and hence $\widehat{\mathsf{M}_0}(r) + \mathsf{M}(r)^2 \le 5\widehat{\mathsf{M}}(r)$ for $0 < r \ll 1$. Indeed, if we apply this result to the transformed integrand $\widetilde{F}$ above we obtain the optimal boundary regularity theorem [22], Theorem 6.4. To state it we introduce conditions

($\kappa\mu\omega$1)-($\kappa\mu\omega$3) analogous to $(\mu\omega1)$ – $(\mu\omega3)$ with $(\mu\omega1)$, $(\mu\omega2)$ also required for κ and with $(\mu\omega3)$ replaced by $\mathsf{K}(r)^2 + \widehat{\mathsf{M}}(r) + \Omega(r) < \infty$.

Theorem 2.12 (boundary ε-regularity-theorem) *Given $\beta \in [\alpha, 1)$ there exists $\varepsilon_1 = \mathrm{const}(\beta, \nu(\,\cdot\,))$ such that the following holds: whenever the n-current T is $(\mathbf{F},\omega)$-minimizing in $\mathbf{C}$, its boundary ∂T is represented in $\mathbf{C}$ by the graph of a function $\psi \in C^1(\mathbf{U}^{n-1}, \mathbb{R}^{1+N})$ with $\psi(0) = 0$, $D\psi(0) = 0$ and modulus of continuity κ of $D\psi$, the projection $\mathbf{p}(\mathrm{graph}\,\psi)$ divides $\mathbf{U}$ into two components D_+, D_- on which $\mathbf{p}_\# T$ has constant multiplicity 1 and 0 respectively, conditions (F1)-(F3) are valid for F and conditions $(\kappa\mu\omega1)$-$(\kappa\mu\omega3)$ for κ, μ from (F2) and ω, and the smallness condition*

$$\mathbf{E}(T,1) + \mathsf{K}(1)^2 + \widehat{\mathsf{M}}(1) + \Omega(1) \le \varepsilon_1$$

is satisfied, then $T \llcorner \mathbf{C}_{1/50}$ is represented by the graph of a C^1 function $g : \mathbf{U}_{1/50} \cap \overline{D_+} \to \mathbb{R}^N$ and Dg has the modulus of continuity

$$|Dg(x) - Dg(\widetilde{x})| \le C\left[r^{2\beta}\mathbf{E}(T,1) + \mathsf{K}(r)^2 + \widehat{\mathsf{M}}(r) + \Omega(r)\right]^{\frac{1}{2}} \tag{36}$$

for $x, \widetilde{x} \in \mathbf{U} \cap \overline{D_+}$ with $|x - \widetilde{x}| \le r$, where $C = \mathrm{Const}(\beta)$. □

For example, if $\kappa(r) = cr^\beta$, $\omega(r) = cr^{2\beta}$ with $\beta \in (0,1)$ and F is of class $C^{1,1}$ (or $\mu(r) \le Cr^{\beta'}$ with $2\beta/(1+\beta) \le \beta' < 1$), then (36) yields the modulus of continuity Cr^β for Dg up to the boundary, which is clearly optimal. Theorem 2.12 may be applied to $(\mathbf{F},\omega)$-minimizing currents on $\mathbb{R}^{n+N}$, with boundary ∂T represented by an oriented $(n-1)$-submanifold $\mathcal{B}$ of multiplicity 1 and of class $C^{1,\kappa(\cdot)}$, at each boundary point $z \in \operatorname{spt} \partial T$ where the n-dimensional density $\Theta^n(\mu_T, z) = \frac{1}{2}$, and the theorem then implies that spt T is, locally at z, an n-dimensional C^1 submanifold with boundary $\mathcal{B}$ and with modulus of continuity of its tangen n-plane field given by (36). In fact, there exists $\vartheta > 0$, depending on F, ω and κ, such that the inequality $\Theta^n_*(\mu_T, z) \leq \frac{1}{2} + \vartheta$ for the lower density is sufficient for this conclusion. This was shown in [38] (see also [22], Section 7). Unfortunately, one does not know in general whether the density condition is satisfied by some boundary points. In the case of codimension $N = 1$ it may be possible to verify it with the help of suitable barriers. For area minimizing currents and $N = 1$ one knows complete boundary regularity from Hardt and Simon [39] without restrictive assumptions on multiplicities or on the density in boundary points. The result of Hardt and Simon has been extended in [21] to energy minimizing hypersurfaces of prescribed mean curvature.

3 Almost Minimizers of Quasi-Convex Integrals in the Classical Calculus of Variations

In this section we consider applications of the technique of A-harmonic approximation in the setting of the calculus of variations. For the purposes of illustration we concentrate on the results contained in the paper [17], which is concerned with partial regularity for almost minimizers of quasi-convex integrals. This is the setting in which the parallels with the situation considered in Section 2 are most apparent. At the end of the section, we briefly discuss other results in the setting of the calculus of variations, primarily results on partial regularity for elliptic systems in divergence form.

One of the fundamental problems in the calculus of variations is the study of existence and regularity of minimizers of regular functionals subject to given boundary conditions. To fix ideas we consider a functional

$$\mathcal{F}(u) = \int_U f(x, u(x), Du(x))\, dx\,, \tag{37}$$

for $x \in U$, a domain in $\mathbb{R}^n$, u mapping U into $\mathbb{R}^N$; then $\mathcal{F}$ is called *regular* if $f(x, y, p)$ is convex in p. Appropriate conditions on f can be imposed to ensure that the Euler equation corresponding to $\mathcal{F}$ is elliptic, or at least degenerate elliptic; however even under reasonable assumptions on f, in the case of systems of equations (i.e. $N{>}1$) one cannot, in general, expect that minimizers of $\mathcal{F}$ will be classical, i.e. C^2-solutions. This was first shown by De Giorgi [14], cf. also [44]; we refer the reader to [28], Chapter II.3 for further discussion. Given this fact, it is of interest to consider questions of partial regularity. The *regular set* of a solution u is defined by

$$\operatorname{Reg} u = \{x \in U \mid u \text{ is continuous on a neighbourhood of } x\},$$

and the *singular set* by

$$\operatorname{Sing} u = U \setminus \operatorname{Reg} u.$$

Partial regularity theory involves obtaining estimates on the size of $\operatorname{Sing} u$ (i.e. showing that $\operatorname{Sing} u$ has zero n-dimensional Lebesgue measure or – better – controlling the Hausdorff dimension of $\operatorname{Sing} u$), as well as showing higher regularity on $\operatorname{Reg} u$. We refer the reader to the monographs [28], [29] and the literature contained therein for discussions of results concerning the existence and regularity of minimizers (and, more generally, of stationary points) of regular functionals.

The requirement that the integrand be convex in the gradient is quite restrictive. There are a number of interesting and fundamental problems in the calculus of variations which are not regular; in addition, weak lower semicontinuity, a canonical notion for establishing the existence of minimizers, is implied by convexity (in appropriate function spaces, e.g. Sobolev spaces), but not vice versa. This lead Morrey to introduce the notion of *quasi convexity* in the paper [46]; we postpone giving a precise definition, and simply note here that Morrey showed that, in many circumstances, quasi convexity and weak lower semicontinuity are equivalent, and refer the reader additionally to [12], [9] and [5] for discussion, literature and further references.

The first results on partial regularity for minimizers of general quasi-convex integrands were obtained by Evans [23]. He considered integrals of the form $\mathcal{F}(u) = \int_U f(Du(x))\,dx$ and showed, under the principle assumption of uniform strict quasi convexity (see (f 3) below), that a minimizer u of such a functional satisfies $\mathcal{L}^n(\operatorname{Sing} u) = 0$, and further that Du is Hölder continuous on $\operatorname{Reg} u$ for all exponents between 0 and 1; see [23, Section 2] for precise statements. These results were extended in [26], [6] and [32] to more general functionals of the form (37) under assumptions comparable to those of the paper [17], and an additional assumption concerning the Hölder continuity of the integrand $f(x,u,p)$ in x and u In particular in this latter setting, Hölder continuity of Du on the regular set for some exponent depending on the Hölder continuity of the integrand f can be shown: see [26], Section 2; [32], Theorem 1.1; and [5], Theorem II.2.

In [17] we consider a more general class of functions than minimizers, namely *almost minimizers*.

Definition 3.13 *Consider a functional $\mathcal{F}$ as in (37) defined on $H^{1,2}_{loc}(U,\mathbb{R}^N)$, and $\omega\colon [0,\infty) \to [0,\infty]$ with $\omega(0) = 0 = \omega(0+)$. A function $u \in H^{1,2}_{loc}(U,\mathbb{R}^N)$ is called* $(\mathcal{F},\omega)$-minimizing at $x_0 \in U$ *if, for all $\rho > 0$ with $\mathbf{U}_\rho(x_0) \subset\subset U$, there holds:*

$$\mathcal{F}(u;\mathbf{U}_\rho(x_0)) \le \mathcal{F}(u+\varphi;\mathbf{U}_\rho(x_0)) + \omega(\rho)\int_{\mathbf{U}_\rho(x_0)} (1+|Du|^2+|D\varphi|^2)\,dx \tag{38}$$

for all $\varphi \in H^{1,2}_0(\mathbf{U}_\rho(x_0),\mathbb{R}^N)$. A function u is $(\mathcal{F},\omega)$-minimizing *if u is $(\mathcal{F},\omega)$-minimizing at each $x_0 \in U$.*

Note that the case of ω identically vanishing corresponds to the case of $\mathcal{F}$-minimizers. We will shortly give examples that show that solutions of a number of problems in the calculus of variations (precisely, minimizers subject to certain constraints) are almost minimizers of suitable functionals, showing that this is in fact a useful notion.

As noted above and discussed in more detail in the introduction, there are close ties between the current setting and the study of elliptic parametric variational problems in geometric measure theory discussed in Section 2.

The main regularity result of the paper [17] is given in Theorem 3.14. For integrals of the form $\mathcal{F}(u) = \int_U f(Du(x))\,dx$ under reasonable conditions on f (the principal one being uniform strict quasi convexity) and the function ω, we show that $(\mathcal{F}, \omega)$-minimizers are regular away from a set of zero-measure. In addition we obtain an optimal local modulus of continuity for Du on Reg u. The structure of the proof and the nature of the definition of an almost minimizer enable the result to be extended easily to families of such integrals. As an easy corollary we then obtain partial regularity for *minimizers* of integrals of the form $\mathcal{F}(u) = \int_U f(x, Du(x))\,dx$ where f is quasiconvex, but where we only require a Dini condition (cf. [40], Section 1) on the continuity of the coefficients in x. In particular, the coefficients are not a priori required to be Hölder continuous in x, in contrast to the results of [26], [32] and [6] (of course, the results there admit u-dependency, which is not the case in [17]). Indeed even for minimizers of regular integrals of the form $\mathcal{F}(u) = \int_U f(x, Du(x))\,dx$, in the case of systems (i.e. $N > 1$) this appears to be the first time that partial regularity results have been obtained for coefficients which are not Hölder continuous. Note however that there are a number of results for scalar valued problems; we mention here specifically [40] and [42]. We also note here that a comparable but more restrictive definition of an almost minimizer was introduced by Anzellotti [7]. In that paper the author shows partial regularity for almost minimizers of the (regular) functional with integrand given by $a^{\alpha\beta}(x)D_\alpha u D_\beta u + g(x)$ for suitably regular $a^{\alpha\beta}$ and g; see [7], Theorem 1.5. Anzellotti required a sharper inequality than (38) in his definition of an almost minimizer. In addition, he required Hölder continuity for the function ω for his partial regularity result.

We next discuss the structure conditions imposed on f and ω; here we essentially conform to the notation in [17], though for compatibility with Section 2 we have made a few minor changes. We consider functionals of the form

$$\mathcal{F}(u) = \int_U f(Du(x))\,dx,$$

where U is a domain in $\mathbb{R}^n$, and $f : \operatorname{Hom}(\mathbb{R}^n, \mathbb{R}^N) \to \mathbb{R}$ satisfies the following conditions:

(f1) there exist positive constants C_1 and C_2 such that, for all $p \in \operatorname{Hom}(\mathbb{R}^n, \mathbb{R}^N)$

$$C_1^{-1}|p|^2 - C_2 \le f(p) \le C_1|p|^2 + C_2;$$

(f2) the function f is C^2 and there exists a nonnegative constant Λ such that for all $p \in \operatorname{Hom}(\mathbb{R}^n, \mathbb{R}^N)$ there holds $|D^2 f(p)| \le \Lambda$;

(f3) the function f is *(uniformly) strictly quasi-convex*, i.e. there exists positive λ such that for all $\mathbf{U}_\rho(x_0) \subset\subset U$, $p \in \operatorname{Hom}(\mathbb{R}^n, \mathbb{R}^N)$ and $\varphi \in C_0^1(\mathbf{U}_\rho(x_0), \mathbb{R}^N)$

there holds:

$$\int_{\mathbf{U}_\rho(x_0)} \Big(f(p+D\varphi)-f(p)\Big)\,dx \geq \lambda \int_{\mathbf{U}_\rho(x_0)} |D\varphi|^2\,dx.$$

The reader should compare these conditions to the conditions (F1)-(F3) appearing in Section 2. Note in particular that the upper bound in (f 1) follows from (f 2), and the lower bound is only useful for questions of existence, cf. [47], 4.4.7, [23], p. 228. We record it here largely for completeness in the examples which follow. Note also that condition (f 3) implies that $D^2 f(p)$ satisfies the Legendre-Hadamard condition (2) uniformly for $p \in \mathrm{Hom}(\mathbb{R}^n, \mathbb{R}^N)$, see [47], 4.4.3 and 4.4.1 or [24], 5.1.10. From condition (f 2) we can deduce the existence of a family of monotone nondecreasing, concave functions $\nu(M,\,\cdot\,)\colon [0,\infty) \to [0,\infty)$ for $M > 0$ satisfying $\nu(M,0) = 0$ and

$$|D^2 f(p) - D^2 f(\tilde{p})| \leq \nu(M, |p-\tilde{p}|^2)$$

for all $p, \tilde{p} \in \mathrm{Hom}(\mathbb{R}^n, \mathbb{R}^N)$ with $|p| \leq M$.

In a similar manner to Section 2, we also consider a function $\omega\colon [0,\infty) \to [0,\infty]$, and define

$$\Omega(r) = \Big(\int_0^r \frac{\sqrt{\omega(\rho)}}{\rho}\,d\rho\Big)^2 .$$

We impose the conditions:

(ω1) ω is nondecreasing on $(0,\infty)$, and $\omega(0) = 0 = \omega(0+)$;
(ω2) $r \mapsto \omega(r)/r^{2\alpha}$ is nonincreasing for some $\alpha \in (0,1)$;
(ω3) $\Omega(r)$ is finite for some $r > 0$.

Note that (ω3) shows that $\Omega(r)$ is in fact finite for all positive r.

We are now in a position to state the main result, which is taken from Theorem 2.2 and Corollary 5.3 of [17]. The mean of a function g on $\mathbf{U}_\rho(x_0)$ will be denoted by $g_{x_0,\rho}$.

Theorem 3.14 (optimal partial regularity) *On a domain $U \subseteq \mathbb{R}^n$ consider a function ω satisfying (ω1)–(ω3), and a function f which satisfies (f3) and (f2). Let $\mathcal{F}$ be the functional on $H^{1,2}(U,\mathbb{R}^N)$ given by $\mathcal{F}(u) = \int_U f(Du(x))\,dx$. Let $u \in H^{1,2}(U,\mathbb{R}^N)$ be $(\mathcal{F},\omega)$-minimizing on U. Then there exists a relatively closed subset of U, $\mathrm{Sing}\,u$, such that*

$$u \in C^1(U \setminus \mathrm{Sing}\,u)\,.$$

Further $\mathrm{Sing}\,u \subseteq \Sigma_1 \cup \Sigma_2$, *where here*

$$\Sigma_1 = \{x_0 \in U \,:\, \liminf_{\rho\to 0+} \fint_{\mathbf{U}_\rho(x_0)} |Du - (Du)_{x_0,\rho}|^2\,dx > 0\}, \quad \textit{and}$$

$$\Sigma_2 = \{x_0 \in U \,:\, \sup_{\rho>0} |(Du)_{x_0,\rho}| = \infty\}\,;$$

in particular $\mathcal{L}^n(\mathrm{Sing}\,u) = 0$.

In addition, in a neighbourhood of any $x_0 \in U \setminus \operatorname{Sing} u$ and for any β with $\alpha < \beta < 1$, Du has a modulus of continuity given by

$$\mu(r) = C\big[r^{2\beta} + \Omega(r)\big]^{1/2},$$

where C is a constant depending only on $\limsup_{\rho\to 0} |(Du)_{x_0,\rho}|$, on β, on the dimensions n and N, on the structural parameters λ, Λ and α, and on the functions $\omega(\,\cdot\,)$ and $\nu(\limsup_{\rho\to 0} |(Du)_{x_0,\rho}|\,,\,\cdot\,)$. □

With a view to applications it is also useful to be able to consider a different functional at each point, i.e. a functional of the form

$$\mathcal{F}_{x_0}(u) = \int_U f_{x_0}(Du(x))\,dx,$$

for $x_0 \in U$. Given a family of such functionals, the analogues of (f2) and (f3) are easily formulated, and a comparable regularity result to Theorem 3.14 follows in this case: we refer the reader to [17], Corollary 2.3 for the details. In particular, we note here (cf. [17], Example 1) that the corollary can be applied to u which minimizes a functional of the form

$$G(u) = \int_U g(x, Du)\,dx$$

under suitable conditions on the frozen coefficients $g_{x_0}(p) = g(x_0, p)$, namely that the analogues of (f2) and (f3) hold, and that there holds

$$|g(x,p) - g(\tilde{x},p)| \le \omega(|x-\tilde{x}|)(1+|p|^2)$$

for all $x, \tilde{x} \in U$ and all $p \in \operatorname{Hom}(\mathbb{R}^n, \mathbb{R}^N)$ for some ω satisfying (ω1)–(ω3). Setting

$$G_{x_0}(u) = \int_U g_{x_0}(Du)\,dx = \int_U g(x_0, Du)\,dx,$$

we deduce that u is (G_{x_0}, ω)–minimizing at each $x_0 \in U$.

We now give a few examples of almost minima, and applications of the partial regularity result. We note here that some of the examples are related to examples appearing in [7], and refer the reader to [17], Section 2 for more details, and additional examples. The following examples correspond to Examples 3, 2 and 4 from [17].

Example 3.15 (almost minimizers of the Dirichlet integral; optimality)*: Every function $u\colon U \to \mathbb{R}^N$ of class $C^{1,\alpha}$ is an almost minimizer of the Dirichlet integral with $\omega(\rho) = C\rho^{2\alpha}$, for some constant $C > 0$.*

Note in particular that Example 3.15 shows that the regularity theorem is optimal in the case of Hölder-continuous moduli of continuity. This is also true for an arbitrary ω satisfying conditions (ω1)-(ω3); see [17], Section 2. □

Example 3.16 (solutions of an obstacle problem)*: Consider the problem of minimizing $\int_U |Dv(x)|^2\,dx$ amongst all functions $v \in H_0^{1,2}(U, \mathbb{R}^N)$ satisfying*

$$v^i \geq \psi^i \qquad \textit{on } U, \qquad (i = 1 \dots N),$$

where the given functions $\psi^i \in C^{1,\beta}(\overline{U}, \mathbb{R}^N)$ *are nonpositive on* ∂U. *Then one can show that a minimizer* u *is an almost minimizer of the Dirichlet integral with* $\omega(\rho) = C\rho^{2\beta}$. □

Example 3.17 (volume-constrained minimizers)*: For fixed* $v_0 \in H^{1,2}(U, \mathbb{R}^N)$ *we define* $\mathcal{H}_{v_0}$ *to be the set of functions* v *in* $H^{1,2}(U, \mathbb{R}^N)$ *such that* $v = v_0$ *on* ∂U *and* $\int_U v\, dx = \int_U v_0\, dx$. *We then consider* $u \in \mathcal{H}_{v_0}$ *such that*

$$\int_U |Du|^2\, dx \leq \int_U |Dv|^2\, dx \tag{39}$$

for all $v \in \mathcal{H}_{v_0}$, *that is, the function* u *minimizes the Dirichlet integral amongst all functions satisfying a given (vector-valued, signed) volume constraint. One can show that* u *is an almost minimizer for the Dirichlet integral, for a function* $\omega(r) = Cr$. □

We next give an outline of the proof of the regularity theorem. The reader will note the analogies with the outline of the proof of the regularity result for almost minimizers of elliptic parametric integrals given in Section 2. Given that the analogies have also been discussed in Section 1, we will refrain from pointing out specific comparisons in the discussion which follows. We will also adhere to similar conventions concerning the notation for constants to those given in Section 2: $c \in (0, 1)$, $C \in [1, \infty)$ will denote constants which may a priori depend on n, N, λ and Λ, and dependencies on other quantities will be indicated in the text.

For ease of exposition we will restrict our attention to the situation that $D^2 f$ is uniformly continuous. In conjunction with (f2) this leads to the existence of a monotone nondecreasing, concave function $\nu : [0, \infty) \to [0, \infty)$ satisfying $\nu(0) = 0$ and

$$|D^2 f(p) - D^2 f(\tilde{p})| \leq \nu(|p - \tilde{p}|^2) \tag{40}$$

for all $p, \tilde{p} \in \mathrm{Hom}(\mathbb{R}^n, \mathbb{R}^N)$. We refer the reader to the proof of Corollary 5.3 in [17] for the (relatively minor) modifications necessary in the general case.

The first step is to establish a suitable Caccioppoli inequality (sometimes also referred to as an inequality of reverse Poincaré type). The following is Lemma 3.2 in [17].

Lemma 3.18 (Caccioppoli inequality) *Let* f *satisfy (f2) and (f3), and* ω *satisfy (ω1). Let* $\mathcal{F}$ *be the functional on* $H^{1,2}(U, \mathbb{R}^N)$ *given by* $\mathcal{F}(u) = \int_U f(Du(x))\, dx$. *Then there exist positive constants* $\rho_1 = \mathrm{const}(\lambda, \omega(\cdot))$ *and* C *such that for every* $\mathbf{U}_\rho(x_0) \subset\subset U$ *with* $\rho \leq \rho_1$, $p_0 \in \mathrm{Hom}(\mathbb{R}^n, \mathbb{R}^N)$ *and every* $u \in H^{1,2}(\mathbf{U}_\rho(x_0), \mathbb{R}^N)$ *which is* (F, ω)*–minimizing at* x_0 *there holds:*

$$\int_{\mathbf{U}_{\rho/2}(x_0)} |Du - p_0|^2\, dx \tag{41}$$

$$\leq C\Big[\rho^{-2} \int_{\mathbf{U}_\rho(x_0)} |u - p_0(x - x_0)|^2\, dx + \boldsymbol{\alpha}(n)\omega(\rho)\rho^n(1 + |p_0|^2)\Big].$$

The proof is close to the proof of similar (albeit somewhat less general) Caccioppoli inequalities given in [23], Lemma 5.1 and [32], Proposition 4.1: we sketch the argument very briefly here.

For $\frac{\rho}{2} \le t < s \le \rho$ we choose a cut-off function $\eta \in C_c^\infty(\mathbf{U}_\rho(x_0), [0,1])$, $\eta \equiv 1$ on $\mathbf{U}_t(x_0)$, $\eta \equiv 0$ outside $\mathbf{U}_s(x_0)$ and $|\nabla \eta| \le 2/(s-t)$. We set

$$\varphi = \eta(u - p_0(x - x_0)), \quad \psi = (1-\eta)(u - p_0(x - x_0)).$$

Given that $D\varphi + D\psi = Du - p_0$, we can use (f 3) to see

$$\lambda \int_{\mathbf{U}_s(x_0)} |D\varphi|^2 \, dx \le \int_{\mathbf{U}_s(x_0)} [f(p_0 + D\varphi) - f(p_0)] \, dx = I + II + III\,,$$

where

$$I = \int_{\mathbf{U}_s(x_0)} [f(Du - D\psi) - f(Du)] \, dx\,,$$

$$II = \int_{\mathbf{U}_s(x_0)} [f(Du) - f(Du - D\varphi)] \, dx \qquad \text{and}$$

$$III = \int_{\mathbf{U}_s(x_0)} [f(p_0 + D\psi) - f(p_0)] \, dx\,.$$

We can then estimate II via the $(\mathcal{F}, \omega)$-minimality of u, and I and III via (f 2) to obtain the inequality

$$\int_{\mathbf{U}_t(x_0)} |Du - p_0|^2 \, dx \le c \int_{\mathbf{U}_s(x_0)} |Du - p_0|^2 \, dx + \frac{C}{(s-t)^2} \int_{\mathbf{U}_s(x_0)} |v|^2 \, dx + C\alpha_n \omega(\rho) \rho^n (1 + |p_0|^2). \tag{42}$$

Since the coefficient of the first term on the right-hand side of (42) is strictly smaller than 1, we can apply the elementary technical lemma [26], Lemma 3.2 to complete the proof. □

The remainder of the proof of Theorem 3.14 revolves around gaining appropriate estimates of the *excess*

$$\Phi(x_0, \rho, p_0) = \fint_{\mathbf{U}_\rho(x_0)} |Du - p_0|^2 \, dx,$$

and in particular on excesses of the form $\Phi(x_0, \rho, (Du)_{x_0,\rho})$ (which we will denote by $\Phi(x_0, \rho)$, where no confusion is likely to arise) for ρ tending to zero.

We define a bilinear form A by setting

$$A(\xi, \eta) = D^2 f(p_0)(\xi, \eta).$$

The next result is then termed *approximate A-harmonicity*, and is given in Lemma 4.1 of [17].

Lemma 3.19 (approximate A-harmonicity) *Under the same conditions as Lemma 3.18 there exists a constant C such that for every $u \in H^{1,2}(U,\mathbb{R}^N)$ that is $(\mathcal{F},\omega)$–minimizing at x_0, every ball $\mathbf{U}_\rho(x_0) \subset\subset U$ and every $p_0 \in \mathrm{Hom}(\mathbb{R}^n,\mathbb{R}^N)$ we have:*

$$\left|\rho^{-n}\int_{\mathbf{U}_\rho(x_0)} A(Du - p_0, D\varphi)\,dx\right| \le C\Big[\omega^{1/2}(\rho)(1+\Phi+|p_0|^2)+\nu^{1/2}(\Phi)\Phi^{1/2}\Big]\sup_{\mathbf{U}_\rho(x_0)}|D\varphi|$$

for all $\varphi \in C_0^1(\mathbf{U}_\rho(x_0),\mathbb{R}^N)$. Here we have abbreviated $\Phi(x_0,\rho,p_0)$ by Φ. □

The name is obviously suggestive: smallness of the right-hand side means that the function $u - p_0(x - x_0)$ is L^2-close to being A-harmonic. In particular, we can apply Lemma 3.19 to the the rescaled function $w(x) = \frac{1}{\gamma}(u(x) - p_0(x - x_0))$, where $\gamma = C[\Phi(x_0,\rho,p_0)+\delta^{-2}\omega(\rho)(1+|p_0|^2)^2]^{1/2}$. Further imposing the *smallness condition*

$$\nu(\Phi(x_0,\rho,p_0)) + \Phi(x_0,\rho,p_0) \le \delta^2/2\,, \tag{43}$$

it is straightforward to verify that we are in a situation to apply the A-harmonic approximation lemma, Lemma 1.1, to deduce the existence of an A-harmonic $h \in H^{1,2}(\mathbf{U}_\rho(x_0),\mathbb{R}^N)$ satisfying

$$\rho^{-n}\int_{\mathbf{U}_\rho(x_0)} |Dh|^2\,dx \le 1 \qquad \text{and} \tag{44}$$

$$\rho^{-n-2}\int_{\mathbf{U}_\rho(x_0)} |w - h|^2\,dx \le \varepsilon. \tag{45}$$

The fact that h is A-harmonic means that we can estimate its derivatives of first and second orders as in (23). Using this estimate along with (44) in the Taylor expansion of h about x_0 we obtain the inequality:

$$\sup_{x\in\mathbf{U}_{2\theta\rho}(x_0)} |h(x) - h(x_0) - Dh(x_0)(x - x_0)|^2 \le C\theta^4\rho^2.$$

This yields (keeping in mind also (45)) the integral inequality

$$(2\theta\rho)^{-n-2}\int_{\mathbf{U}_{2\theta\rho}(x_0)} |u - u_{x_0,2\theta\rho} - (p_0 + \gamma Dh(x_0))(x - x_0)|^2 dx \le C\left(\theta^{-n-2}\varepsilon + \theta^2\right)\left(\Phi(x_0,\rho,p_0) + \delta^{-2}\omega(\rho)(1+|p_0|^2)^2\right).$$

This inequality can then be combined with Lemma 3.18, our Caccioppoli inequality, to obtain an estimate on $\Phi(x_0,\theta\rho)$ in terms of $\Phi(x_0,\rho)$ which can be termed *excess improvement*. The resultant estimate is formulated in Proposition 5.1 of [17].

Proposition 3.20 (Excess improvement) *Consider u satisfying the conditions of Theorem 3.14, and β fixed, $\alpha < \beta < 1$. Then we can find positive constants C, C_1 and δ, and $\theta \in (0,1)$ (with C_1, θ and δ depending possibly on β) such that the smallness conditions $\rho \leq \rho_1$,*

$$\nu(\Phi(x_0,\rho)) + \Phi(x_0,\rho) \leq \delta^2/2\,, \tag{46}$$

and

$$C\omega(\rho)(1 + |(Du)_{x_0,\rho}|^4) \leq \delta^2 \tag{47}$$

together imply the growth condition

$$\Phi(x_0,\theta\rho) \leq \theta^{2\beta}\Phi(x_0,\rho) + C\omega(\rho)(1 + |(Du)_{x_0,\rho}|^4).$$

Here ρ_1 is given in Lemma 3.18. □

The next lemma shows that we can iterate Proposition 3.20, assuming only that the quantities $\Phi(x_0,\rho)$ and ρ are sufficiently small for some $\mathbf{U}_\rho(x_0)$.

Lemma 3.21 (excess decay) *For $M_0 > 0$ there exist $\rho_0(M_0)$ and $\Phi_0(M_0)$ such that if the conditions*

$$\begin{array}{lll} (i) & |(Du)_{x_0,\rho}| \leq M_0, & \\ (ii) & \quad\rho \quad\leq \rho_0(M_0) & \text{and} \\ (iii) & \Phi(x_0,\rho) \leq \Phi_0(M_0) & \end{array}$$

are satisfied on $\mathbf{U}_\rho(x_0) \subset\subset U$, then the smallness conditions (46) and (47) are fulfilled on $\mathbf{U}_{\theta^j\rho}(x_0)$ for all $j \in \mathbb{N}$. Furthermore the limit

$$\Upsilon_{x_0} = \lim_{j\to\infty} (Du)_{x_0,\theta^j\rho}$$

exists, and there is a constant $C(\alpha,\beta,M_0)$ such that for all $r < \rho$ there holds:

$$\fint_{\mathbf{U}_r(x_0)} |Du - \Upsilon_{x_0}|^2\,dx \leq C\Big(\Big(\frac{r}{\rho}\Big)^{2\beta}\Phi(x_0,\rho) + \Omega(r)\Big).$$

□

The remainder of the proof of Theorem 3.14 is then standard. For any point x_0 belonging to neither Σ_1 nor Σ_2, we can find a ball $\mathbf{U}_\sigma(x_0)$ such that we can apply Lemma 3.21 on $\mathbf{U}_r(z)$ for any $z \in \mathbf{U}_\sigma(x_0)$ and r sufficiently small (independent of z) to deduce

$$\fint_{\mathbf{U}_r(z)} |Du - \Upsilon_z|^2\,dx \leq C\Big(\Big(\frac{r}{\rho}\Big)^{2\beta}\Phi(z,\rho) + \Omega(r)\Big). \tag{48}$$

Comparing this estimate for two points $z, \widetilde{z} \in \mathbf{U}_\sigma(x_0)$ yields the desired modulus of continuity.

The same basic procedure can be applied to establish, for example, partial regularity results for nonlinear elliptic systems in divergence form. In the paper [19] the authors give an elementary, self-contained approach to partial regularity for second order nonlinear elliptic systems of divergence type. In the paper [16] these results have been extended to allow more general coefficients, and in the paper [18] to higher order systems. Partial regularity results at the boundary have been obtained in the papers [35], [36]. Finally we note here that the results of [17] have been extended to functionals with integrands depending on derivatives of higher order in [43].

References

1. Allard, W.K.: On the first variation of a varifold. Annals of Math., **95**, 417–491, 1972
2. Almgren, F.J.: Existence and regularity almost everywhere of solutions to elliptic variational problems among surfaces of varying topological type and singularity structure. Annals of Math., **87**, 321–391, 1968
3. Almgren, F.J.: Existence and regularity almost everywhere of solutions to elliptic variational problems with constraints. Mem. Amer. Math. Soc., **165**, 1976
4. Almgren, F. J.: **Q**-valued functions minimizing Dirichlet's integral and the regularity of area minimizing rectifiable currents up to codimension two. Princeton Univ. 1984, 3745 g; see also Bull. Amer. Math. Soc., **8**, 327–328 (1983) and Almgren's big regularity paper, ed. Scheffer, V., Taylor, J. E., World Scientific Singapore 2000
5. Acerbi, E., Fusco, N.: Semicontinuity problems in the calculus of variations. Arch. Ration. Mech. Anal., **86**, 125–145, 1984
6. Acerbi, E., Fusco, N.: A regularity theorem for minimizers of quasiconvex integrals. Arch. Ration. Mech. Anal., **99**, 261–281, 1987
7. Anzellotti, G.: On the $C^{1,\alpha}$–regularity of ω–minima of quadratic functionals. Boll. Unione Mat. Ital., VI. Ser., C, Anal. Funz. Appl., **2**, 195-212, 1983
8. Anzellotti, G., Giaquinta, M.: Convex functions and partial regularity. Arch. Ration. Mech. Anal., **102**, 243–272, 1988
9. Ball, J.M.: Convexity conditions and existence theorems in nonlinear elasticity. Arch. Ration. Mech. Anal., **63**, 337–403, 1977
10. Bombieri, E.: Regularity theory for almost minimal currents. Arch. Ration. Mech. Anal., **7**, 99–130, 1982
11. Campanato, S.: Equazioni ellittiche del IIe ordine e spazi $\mathcal{L}^{2,\lambda}$. Ann. Mat. Pura Appl., **69**, 321–381, 1965
12. Dacorogna, B.: Direct methods in the calculus of variations. Springer, Berlin–Heidelberg–New York, 1989
13. De Giorgi, E.: Frontiere orientate di misura minima. Seminaro Mat. Scuola Norm. Sup. Pisa, 1–56, 1961
14. De Giorgi, E.: Un esempio di estremali discontinue per un problema variazionale di tipo ellitico. Boll. Unione Mat. Ital., IV. Ser., **1**, 135–137, 1968
15. Duzaar, F.: Boundary regularity for area minimizing currents with prescribed volume. J. Geom. Anal., **7**, 585–592, 1997
16. Duzaar, F., Gastel, A.: Nonlinear elliptic systems with Dini continuous coefficients. Archiv der Math., **78**, 58–73, 2002

17. Duzaar, F., Gastel, A., Grotowski, J.F.: Partial regularity for almost-minimizers of quasiconvex integrals. SIAM J. Math. Analysis, **32**, 665–687, 2000
18. Duzaar, F., Gastel, A., Grotowski, J.F.: Optimal partial regularity for nonlinear elliptic systems of higher order. J. Math. Sci., Tokyo, **8**, 463–499, 2001
19. Duzaar, F., Grotowski, J.F.: Partial regularity for nonlinear elliptic systems: the method of A-harmonic approximation. Manuscr. Math., **103**, 267–298, 2000
20. Duzaar, F., Steffen, K.: λ minimizing currents. Manuscr. Math., **80**, 403–447, 1993
21. Duzaar, F., Steffen, K.: Boundary regularity for minimizing currents with prescribed mean curvature. Calc. Var. Partial Diff. Equ., **1**, 355–406, 1993
22. Duzaar, F., Steffen, K.: Optimal interior and boundary regularity for almost minimal currents to elliptic integrands. J. Reine Angew. Math., **546**, 73–138, 2002
23. Evans, L.C.: Quasiconvexity and partial regularity in the calculus of variations. Arch. Ration. Mech. Anal., **95**, 227–252, 1986
24. Federer, H.: Geometric Measure Theory. Springer, Berlin–Heidelberg–New York, 1969
25. Federer, H.: The singular set of area minimizing rectifiable currents with codimension one and of area minimizing flat chains modulo two with arbitrary codimension. Bull. Amer. Math. Soc., **76**, 767–771, 1970
26. Fusco, N., Hutchinson, J.: $C^{1,\alpha}$ partial regularity of functions minimising quasiconvex integrals. Manuscr. Math., **54**, 121–143, 1985
27. Gehring, F.W.: The L^p–integrability of the partial derivatives of a quasiconformal map. Acta Math., **130**, 265–277, 1973
28. Giaquinta, M.: Multiple Integrals in the Calculus of Variations and Nonlinear Elliptic Systems. Princeton University Press, Princeton, 1983
29. Giaquinta, M.: Introduction to Regularity Theory for Nonlinear Elliptic Systems. Birkhäuser, Basel–Boston–Berlin, 1993
30. Giaquinta, M., Ivert, P.-A.: Partial regularity for minima of variational integrals. Ark. Math., **25**, 221–229, 1987
31. Giaquinta, M., Modica, G.: Regularity results for some classes of higher order nonlinear elliptic systems. J. Reine Angew. Math., **311/312**, 125–169, 1979
32. Giaquinta, M. Modica, G.: Partial regularity of minimizers of quasiconvex integrals. Ann. Inst. H. Poincaré, Analyse non linéaire, **3**, 185-208, 1986
33. Giusti, E., Miranda: Sulla regolarità delle soluzioni deboli di una classe di sistemi ellittici quasi-lineari. Arch. Ration. Mech. Anal., **31**, 173–184, 1968
34. Gonzalez, E., Massari, U., Tamanini, I.: On the regularity of boundaries of sets minimizing perimeter with a volume constraint. Indiana Univ. Math. J., **32**, 25–37, 1983
35. Grotowski, J.F.: Boundary regularity for nonlinear elliptic systems. To appear in: Calc. Var. Partial Differ. Equ.
36. Grotowski, J.F.: Boundary regularity for quasilinear elliptic systems. To appear in: Commun. Partial Differ. Equations.
37. Hamburger, C.: A new partial regularity proof for solutions of nonlinear elliptic equations. Manuscr. Math., **95**, 11–31, 1998
38. Hardt, R.: On boundary regularity for integral currents or flat chains modulo two minimizing the integral of an elliptic integrand. Commun. Partial Diff. Equations, **2**, 1163–1232, 1977
39. Hardt, R., Simon, L.: Boundary regularity and embedded solutions for the oriented Plateau problem. Annals of Math., **110**, 439–486, 1979
40. Hartman, P., Wintner, A.: On uniform Dini conditions in the theory of linear partial differential equations of elliptic type. Am. J. Math., **77**, 329–354, 1955
41. Jost, J., Meier, M.: Boundary regularity for minima of certain quadratic functionals. Math. Ann., **262**, 549–561, 1983

42. Kovats, J.: Fully nonlinear elliptic equations and the Dini condition. Commun. Partial Diff. Equations, **22**, 1911–1927, 1997
43. Kronz, M.: Partial regularity results for minimizers of quasiconvex functionals of higher order. Ann. Inst. H. Poincaré, Analyse non linéaire, **19**, 81–112, 2002
44. Maz'ya, V.G.: Examples of nonregular solutions of quasilinear elliptic equations with analytic coefficients. Funkts. Anal. Prilozh., **2**, 53–57 (1968); translated in Funct. Anal. Appl., **2**, 230–234, 1968
45. Miranda, M.: Frontiere orientate con ostacoli. Ann. Univ. Ferrara, **16**, 29–37, 1971
46. Morrey, C.B.: Quasi-convexity and the lower semicontinuity of multiple integrals. Pacific J. Math., **2**, 25–53, 1952
47. Morrey, C.B.: Multiple Integrals in the Calculus of Variations. Springer, Heidelberg–New York, 1966
48. Morrey, C.B.: Partial regularity results for non-linear elliptic systems. J. Math. Mech., **17**, 649–670, 1968
49. Simon, L.: Lectures on Geometric Measure Theory. Proc. CMA, Vol. 3, ANU Canberra, 1983
50. Simon, L.: Theorems on Regularity and Singularity of Energy Minimizing Maps. Birkhäuser, Basel–Boston–Berlin, 1996
51. Schoen, R., Simon, L.: A new proof of the regularity for rectifiable currents which minimize parametric elliptic functionals. Indiana Univ. Math. J., **31**, 413–43, 1982
52. Schoen, R., Simon, L., Almgren, F.J.: Regularity and singularity estimates on hypersurfaces minimizing parametric elliptic variational integrals. Acta Math., **139**, 217–265, 1977
53. Tamanini, I.: Regularity results for almost minimal oriented hypersurfaces in $\mathbb{R}^n$. Quad. Dipt. Mat. Uni. Lecce, **1–1984**, 1984
54. Tamanini, I.: Boundaries of Caccioppoli sets with Hölder-continuous normal vector. J. Reine Angew. Math., **334**, 27–39, 1982
55. Tamanini, I.: Variational problems of least area type with constraints. Ann. Univ. Ferrara, **34**, 183–217, 1988

Dominance Functions for Parametric Lagrangians

Stefan Hildebrandt and Heiko von der Mosel

Mathematisches Institut, Universität Bonn, Beringstraße 1, D-53115 Bonn
heiko@math.uni-bonn.de

Summary. We discuss the concept of dominance functions for parametric Lagrangians together with important examples and various applications to the existence and regularity theory for minimizers of parametric functionals, and for the construction of unstable stationary surfaces. The focus lies on the construction of a perfect dominance function based on ideas of Morrey.

1 Introduction

Parametric Lagrangians and the Plateau problem. Let $\mathcal{F}$ be a parametric double integral of the form

$$\mathcal{F}_{\Omega}(X) := \int_{\Omega} F(X, X_u \wedge X_v)\, dudv, \tag{1}$$

where F is of class $C^0(\mathbb{R}^n \times \mathbb{R}^N)$, $n \geq 2$, $N = n(n-1)/2$, satisfying the homogeneity condition

$$F(x, tz) = tF(x, z) \ \text{ for all } \ t > 0,\, x \in \mathbb{R}^n,\ z \in \mathbb{R}^N. \tag{H}$$

In addition, we assume that F is *positive definite*, that is, there are constants m_1, m_2 with $0 < m_1 \leq m_2$ such that

$$m_1|z| \leq F(x, z) \leq m_2|z| \ \text{ for all } \ (x, z) \in \mathbb{R}^n \times \mathbb{R}^N. \tag{D}$$

Then $\mathcal{F}_{\Omega}(X)$ is well-defined by (1) for any domain $\Omega \subset \mathbb{R}^2$ and any mapping $X \in H^{1,2}(\Omega, \mathbb{R}^n)$, where

$$p_1 \wedge p_2 := (p_1^j p_2^k - p_1^k p_2^j)_{j<k} \in \mathbb{R}^N$$

denotes the bivector generated by the two vectors $p_1 = (p_1^1, \dots, p_1^n)$, $p_2 = (p_2^1, \dots, p_2^n) \in \mathbb{R}^n$. Notice that the Lagrange identity

$$|p_1 \wedge p_2|^2 = |p_1|^2 |p_2|^2 - (p_1 \cdot p_2)^2 \tag{2}$$

leads to

$$|p_1 \wedge p_2| \leq \frac{1}{2}(|p_1|^2 + |p_2|^2)$$

with equality if and only if

$$|p_1|^2 = |p_2|^2 \quad \text{and} \quad p_1 \cdot p_2 = 0. \tag{3}$$

The homogeneity condition (H) implies that $\mathcal{F}_\Omega$ is parameter invariant, i.e., invariant with respect to reparametrizations by means of C^1-diffeomorphisms with positive Jacobian.

A special Lagrangian of this kind is the *area integrand* $A(z) := |z|$ satisfying (D) with $m_1 = m_2 = 1$, which leads to the *area functional*

$$\mathcal{A}_\Omega(X) := \int_\Omega A(X_u \wedge X_v)\, dudv \quad \text{for} \quad X \in H^{1,2}(\Omega, \mathbb{R}^n).$$

The parametric integrand F is said to be *semi-elliptic* on $K \times \mathbb{R}^N$, K a closed set in $\mathbb{R}^n$, if

$$F(x,z) \ \text{ is convex in } \ z \in \mathbb{R}^N \ \text{ for any } \ x \in K. \tag{C}$$

Admitting arbitrary closed sets $K \subset \mathbb{R}^n$ in condition (C) is useful for the treatment of parametric variational problems in Riemannian manifolds. In the sequel, however, we assume for simplicity that $K = \mathbb{R}^n$.

We call F *elliptic* if for every $R_0 > 0$ there is some $\lambda_F(R_0) > 0$ such that $F - \lambda_F(R_0)A$ is semi-elliptic on $\overline{B_{R_0}}(0) \times \mathbb{R}^N$. If $F \in C^2(\mathbb{R}^n \times (\mathbb{R}^N - \{0\}))$, then condition (C) is equivalent to the assumption that $F_{zz}(x,z)$ is positive semi-definite for $z \neq 0$, whereas F is elliptic if and only if for any $R_0 > 0$ there is some $\lambda_F(R_0)$ such that $F_{zz}(x,z) - \lambda_F(R_0)A_{zz}(z)$ is positive semi-definite for all $(x,z) \in \overline{B_{R_0}}(0) \times (\mathbb{R}^N - \{0\})$.

Let

$$P_z^\perp \zeta := \zeta - |z|^{-2}(z \cdot \zeta)z \quad \text{for} \quad z \neq 0 \tag{4}$$

denote the projection of $\zeta \in \mathbb{R}^N$ onto the hyperplane perpendicular to z in $\mathbb{R}^N$. Then we can express the ellipticity of F as

$$|z|\zeta \cdot F_{zz}(x,z)\zeta \geq \lambda_F(R_0)|P_z^\perp \zeta|^2 = \lambda_F(R_0)|z|^{-2}|\zeta \wedge z|^2 \tag{5}$$

for some $\lambda_F(R_0) > 0$ and any $x \in \overline{B_{R_0}}(0) \subset \mathbb{R}^n$, $z, \zeta \in \mathbb{R}^N$, $z \neq 0$, since

$$|z|\zeta \cdot A_{zz}(z)\zeta = |\zeta|^2 - |z|^{-2}(z \cdot \zeta)^2 = |P_z^\perp \zeta|^2 = |z|^{-2}|\zeta \wedge z|^2 \tag{6}$$

by virtue of (2). The identity (6) implies in particular that the area integrand A is elliptic, where we may choose $\lambda_F(R_0) = 1$ for all $R_0 > 0$.

Conformally parametrized extremals of the area functional, i.e., surfaces $X \in H^{1,2}(\Omega, \mathbb{R}^n)$ satisfying

$$|X_u|^2 = |X_v|^2 \quad \text{and} \quad X_u \cdot X_v = 0 \quad \text{a.e. on } \Omega,$$

which are stationary points of $\mathcal{A}$, are minimal surfaces. More generally, surfaces of prescribed mean curvature H in $\mathbb{R}^3$ may be regarded as conformally parametrized extremals of the parametric functional

$$\mathcal{E}_\Omega(X) := \int_\Omega E(X, X_u \wedge X_v)\, dudv$$

with the Lagrangian

$$E(x,z) := |z| + Q(x) \cdot z, \quad x, z \in \mathbb{R}^3, \tag{7}$$

where Q is a differentiable vector field on $\mathbb{R}^3$ with $2\operatorname{div} Q = H$. Again, E is elliptic satisfying (5) with $\lambda_F(R_0) = 1$ for all $R_0 > 0$. However, E is positive definite only if

$$Q_0 := \sup_{x \in \mathbb{R}^3} |Q(x)| < 1. \tag{8}$$

The parametric Lagrangian given by the *discrete l^r-norm,* $1 \le r \le \infty$,

$$I(z) := (|z^1|^r + |z^2|^r + |z^3|^r)^{1/r}, \quad z = (z^1, z^2, z^3) \in \mathbb{R}^3, \tag{9}$$

leads to an anisotropic interface energy

$$\mathcal{I}_\Omega(X) := \int_\Omega I(X_u \wedge X_v)\, dudv$$

used in mathematical image processing to detect and sharpen edges [2]. Note, however, that I is not even semi-elliptic in general; a regularized version of (9), e.g., for $r = 1$ is given by

$$I_\epsilon(z) := \sum_{i=1}^{3} \sqrt{\epsilon |z|^2 + (z^i)^2}, \tag{10}$$

which is elliptic for each $\epsilon > 0$. We also refer to the work of J. Taylor on “crystalline structures” (see [29]–[35]), where the surface $W := \{\zeta \in \mathbb{R}^3 : \zeta \cdot z \le F(z) \text{ for all } z \in \mathbb{R}^3 \text{ with } |z| = 1\}$ known as *Wulff shape* is supposed to be a polyhedron.

Various geometric boundary value problems for general parametric functionals of the type (1) have been studied. For the sake of brevity we will restrict our attention to the ***Plateau problem***. For a closed rectifiable Jordan curve $\Gamma \subset \mathbb{R}^n$ and the fixed domain $B := \{(u,v) : u^2 + v^2 < 1\}$ we consider the nonempty class $\mathcal{C}(\Gamma)$ of competing surfaces $X \in H^{1,2}(B, \mathbb{R}^n)$ whose trace $X|_{\partial B}$ on ∂B is a continuous, weakly monotonic mapping of ∂B onto Γ. Setting $\mathcal{F} := \mathcal{F}_B$, $\mathcal{A} := \mathcal{A}_B$, we have the following existence result for the minimization problem “$\mathcal{F} \to \min \quad \text{in } \mathcal{C}(\Gamma)$” proved in [14],[15]:

Theorem 1.1 *If $F \in C^0(\mathbb{R}^n \times \mathbb{R}^N)$ satisfies* (H),(D), *and* (C), *then there exists an $\mathcal{F}$-minimizing surface $X \in \mathcal{C}(\Gamma)$ of class $C^0(\overline{B}, \mathbb{R}^n) \cap C^{0,\alpha}(B, \mathbb{R}^n) \cap H^{1,q}_{\text{loc}}(B, \mathbb{R}^n)$ for $\alpha = m_1/m_2 \in (0,1]$ and some $q > 2$, which is conformally parametrized, i.e. X satisfies $\mathcal{F}(X) = \inf_{\mathcal{C}(\Gamma)} \mathcal{F}$, and*

$$|X_u|^2 = |X_v|^2, \quad X_u \cdot X_v = 0 \quad a.e. \text{ on } B. \tag{11}$$

If Γ satisfies a chord-arc condition, any minimizer X of $\mathcal{F}$ in $\mathcal{C}(\Gamma)$ is of class $C^\beta(\overline{B}, \mathbb{R}^n)$ for some $\beta \in (0,1)$.

We note that the existence of continuous minimizers for $\mathcal{F}$ in $\mathcal{C}(\Gamma)$ was proved by Sigalov, Cesari, and Danskin in the years 1951–1952 (cf. the references in [22], [23]). In 1961, Morrey [22] gave a new, greatly simplified existence proof, adding Hölder continuity of minimizers in the interior. His new approach consisted in minimizing a modified functional $\mathcal{F}^* := \int_B f^*(X, \nabla X)\, dudv$ with the Lagrangian

$$f^*(x, p_1, p_2) := \left\{F^2(x, p_1 \wedge p_2) + c[4^{-1}(|p_1|^2 + |p_2|^2)^2 + (p_1 \cdot p_2)^2]\right\}^{1/2}$$

under the subsidiary condition $\int_B |\nabla^2 X|^2\, dudv \le R$, and then letting $R \to \infty$. In all of this work it was not shown that there exists an a.e. conformally parametrized minimizer. Later on, in his treatise [23], Morrey presented a modification of his method replacing f^* by a very interesting dominance function f^{**} thereby obtaining a conformally parametrized minimizer of $\mathcal{F}$ in $\mathcal{C}(\Gamma)$. However, there seems to be a gap in his proof since f^{**} is only shown to be rank-one convex, and so the corresponding functional $\mathcal{F}^{**}(X) := \int_B f^{**}(X, X_u, X_v)\, dudv$ need not be weakly lower semicontinuous in $H^{1,2}(B, \mathbb{R}^n)$ since this property would require quasiconvexity of f^{**}.

Another existence proof is due to Y.G. Reshetnyak [24] who in 1962 obtained the same result as Morrey [22] but with a very different method. On p. 767 the author states without proof that "the parameters u and v of the function yielding the minimum of the functional $I(x, F, Q)$ and obtained by the above proof are isothermal".

The result stated in Theorem 1.1, in particular the existence of conformally parametrized minimizers, was proved by the present authors, first in a slightly weaker form (cf. [14]), and then in the final form of [15] using an idea provided by S. Müller. The basic idea of our approach was to minimize first the new functional $\mathcal{F}^\epsilon := \mathcal{F} + \epsilon\mathcal{D}$ penalized by Dirichlet's integral $\mathcal{D}$ thus obtaining "approximate" minimizers X^ϵ of the original problem which are conformally parametrized. Passing to the limit $\epsilon \to +0$, one obtains the desired minimizer of $\mathcal{F}$ in conformal parameters. It seems to us that this is crucial for proving higher regularity results because an *arbitrary* continuous minimizer of $\mathcal{F}$ cannot even be expected to be of class C^1. First steps of proving higher regularity were carried out in [15] and [16], using a modification of Morrey's perfect dominance function. For details we refer to the discussion below; cf. Theorems 1.3–1.7.

An existence result for a free boundary value problem for the functional (1) can be found in [20]; the semi-free boundary problem with a given boundary curve and a support surface was solved in [16], and the Douglas problem, where one seeks multiply connected $\mathcal{F}$-minimizing surfaces bounded by several Jordan curves, was treated in [18] and, more generally, in [19].

Dominance functions. In the sequel we call $f(x,p) := F(x, p_1 \wedge p_2)$ for $p = (p_1, p_2) \in \mathbb{R}^n \times \mathbb{R}^n \cong \mathbb{R}^{2n}$ the *associated Lagrangian* of the parametric integrand F.

Definition 1.2 (i) *Let $F(x, z)$ be a parametric Lagrangian with the associated Lagrangian $f(x,p)$. Then a function $G : \mathbb{R}^n \times \mathbb{R}^{2n} \to \mathbb{R}$ is said to be a* dominance function for F *if it is continuous and satisfies the following two conditions:*

$$f(x,p) \leq G(x,p) \ \textit{for any} \ (x,p) \in \mathbb{R}^n \times \mathbb{R}^{2n}, \tag{D1}$$

$$f(x,p) = G(x,p) \quad \textit{if and only if} \quad |p_1|^2 = |p_2|^2, \ p_1 \cdot p_2 = 0. \tag{D2}$$

(ii) *A dominance function G of the parametric Lagrangian F is called* quadratic *if*

$$G(x, tp) = t^2 G(x,p) \ \textit{for all} \ t > 0, (x,p) \in \mathbb{R}^n \times \mathbb{R}^{2n}, \tag{D3}$$

and it is said to be positive definite *if there are two numbers μ_1, μ_2 with $0 < \mu_1 \leq \mu_2$, such that*

$$\mu_1 |p|^2 \leq G(x,p) \leq \mu_2 |p|^2 \ \textit{for any} \ (x,p) \in \mathbb{R}^n \times \mathbb{R}^{2n}. \tag{D4}$$

(iii) *A function $G \in C^0(\mathbb{R}^n \times \mathbb{R}^{2n}) \cap C^2(\mathbb{R}^n \times (\mathbb{R}^{2n} - \{0\}))$ is called a* perfect dominance function for the parametric Lagrangian F *if it satisfies* (D1)–(D4), *and if for any $R_0 > 0$ there is a constant $\lambda_G(R_0) > 0$ such that*

$$\xi \cdot G_{pp}(x,p)\xi \geq \lambda_G(R_0)|\xi|^2 \ \textit{for} \ |x| \leq R_0 \ \textit{and} \ p, \xi \in \mathbb{R}^{2n}, p \neq 0. \tag{E}$$

Condition (E) *means that*

$$G_{p^j_\alpha p^k_\beta}(x,p)\xi^j_\alpha \xi^k_\beta \geq \lambda_G(R_0)\xi^j_\alpha \xi^j_\alpha, \quad \lambda_G(R_0) > 0.$$

Here and in the sequel we use the following *convention:* Greek indices $\alpha, \beta, \dots$ run from 1 to 2, and Latin indices $j, k, \dots$ from 1 to n. Correspondingly, repeated indices $\alpha, \beta, \dots, j, k, \dots$, are to be summed from 1 to 2, or n, respectively.

Let us consider some *examples.*

[1] The integrand

$$D(p) := \frac{1}{2}|p|^2 = \frac{1}{2}|p_1|^2 + \frac{1}{2}|p_2|^2, \quad p = (p_1, p_2)$$

of the Dirichlet energy $\mathcal{D}$ is a perfect dominance function for the area integrand $A(z) = |z|$ with the associated Lagrangian

$$a(p) := |p_1 \wedge p_2| = \sqrt{|p_1|^2 |p_2|^2 - (p_1 \cdot p_2)^2}.$$

[2] The integrand E defined in (7) with $n = 3$ possesses the dominance function

$$G(x,p) := \frac{1}{2}|p|^2 + Q(x) \cdot (p_1 \wedge p_2), \tag{12}$$

satisfying (D1)–(D3). If (8) holds, and if $Q \in C^2(\mathbb{R}^n)$, then $G(x,p)$ is a perfect dominance function where one can choose $\lambda_G(R_0) = 1 - Q_0 > 0$ in the ellipticity condition (E) for all $R_0 > 0$.

[3] The preceeding example can be generalized from $n = 3$ to all $n \geq 2$ replacing (7) by

$$E(x,z) := |z| + \sum_{i<k} \omega_{ik}(x) z^{ik}, \tag{13}$$

where $z = (z^{ik})_{i<k}$, and both z^{ik} and ω_{ik} are thought to be skew symmetric (no summation convention here). The associated Lagrangian is

$$e(x,p) := |p_1 \wedge p_2| + \sum_{i<k} \omega_{ik}(x)(p_1^i p_2^k - p_2^i p_1^k), \tag{14}$$

and

$$G(x,p) := D(p) + V(x,p) \quad \text{with} \quad V(x,p) := \sum_{i<k} \omega_{ik}(x)(p_1^i p_2^k - p_2^i p_1^k) \tag{15}$$

satisfies all properties of a perfect dominance function for E with the possible exception of (D4) which will be satisfied for $(x,p) \in \mathbb{R}^n \times \mathbb{R}^{2n}$ if

$$\sup_{\mathbb{R}^n} \sum_{i<k} |\omega_{ik}|^2 < 1.$$

The Euler equations corresponding to G are

$$\Delta X^l = \sum_{\substack{i<k \\ i \neq l}} q_{ikl}(x)(X_u^i X_v^k - X_v^i X_u^k), \tag{16}$$

where $q_{ikl} := \omega_{ik,l} + \omega_{kl,i} + \omega_{li,k}$, and the sum is to be extended over all pairs (i,k) with $i < k$ and $i, k \neq l$. For $n = 3$ these equations reduce to

$$\Delta X = 2 \operatorname{div} Q(x) X_u \wedge X_v \tag{17}$$

if we set $Q_1 = \omega_{23}$, $Q_2 = \omega_{31}$, $Q_3 = \omega_{12}$, because in this case the admissible triples $i\,k\,l$ are permutations of $1\,2\,3$.

The whole setting can be carried over to a Riemannian metric $ds^2 = g_{ik}(x)dx^i dx^k$. For instance, $e(x,p)$ and $G(x,p)$ in the above example become

$$e(x,p) := (\|p_1\|^2 \|p_2\|^2 - \langle p_1, p_2 \rangle)^{1/2} + V(x,p),$$
$$G(x,p) := \frac{1}{2}(\|p_1\|^2 + \|p_2\|^2) + V(x,p),$$

where $\langle p_\alpha, p_\beta\rangle := g_{ik}(x)p^i_\alpha p^k_\beta$, $\|p_1\| := \langle p_1, p_1\rangle^{1/2}$, $\|p_2\| := \langle p_2, p_2\rangle^{1/2}$, and the conformality relations (11) are to be replaced by

$$\|X_u\|^2 = \|X_v\|^2, \quad \langle X_u, X_v\rangle = 0.$$

Here we want to stay in the Euclidean case, but it should be mentioned that one is automatically forced to pass to the Riemannian case if one wants to study the boundary behaviour of minimizers.

[4] For general parametric integrands $F(x,z)$ with the properties (H),(D) and with the associated Lagrangians $f(x,p)$, Morrey introduced in [22, pp. 571–572] the dominance function

$$G(x,p) := \tag{18}$$
$$\left\{ f^2(x,p) + \frac{1}{4}(m_1+m_2)^2 \left[\frac{1}{4}(|p_1|^2 - |p_2|^2)^2 + (p_1\cdot p_2)^2\right]\right\}^{1/2},$$

satisfying (D1)–(D4) with $\mu_1 := m_1/2$, $\mu_2 := m_2/2$. A simpler choice is

$$G(x,p) := g(x,p)D(p), \tag{19}$$

where

$$g(x,p) := \begin{cases} m_1 & \text{for } z = 0 \\ F(x, z/|z|) & \text{for } z \neq 0 \end{cases}, \qquad z := p_1 \wedge p_2.$$

Notice, however, that in general these dominance functions are far from being *perfect*, since their singular set

$$\Pi := \{(p_1, p_2) \in \mathbb{R}^{2n} : p_1 \wedge p_2 = 0\},$$

coincides with the singular set of the associated Lagrangian $f(x,p)$, for which we cannot expect C^2-smoothness across Π due to the homogeneity condition (H) on the corresponding parametric integrand $F(x,z)$. Only when F is a *weighted area integrand*, i.e., $F(x,z) = \gamma(x)|z|$ with $\gamma \in C^2(\mathbb{R}^n)$, $m_1 \le \gamma(x) \le m_2$ for all $x \in \mathbb{R}^n$, one obtains a perfect dominance function of the form (19) with $g(x,p) := \gamma(x)$.

In Section 2 we will give the detailed construction of a rank-one convex dominance function $G \in C^{2,\alpha}(\mathbb{R}^n \times (\mathbb{R}^N - \{0\}))$ for an elliptic positive definite parametric integrand F of class $C^2(\mathbb{R}^n \times (\mathbb{R}^N - \{0\}))$ following the ideas of Morrey sketched in [23, pp. 391–394]. This in turn will enable us to prove the existence of a perfect dominance function for a large class of parametric integrands. In fact, we have

Theorem 1.3 *Let* $F^* \in C^0(\mathbb{R}^n \times \mathbb{R}^N) \cap C^2(\mathbb{R}^n \times (\mathbb{R}^N - \{0\}))$ *satisfy* (H),(D) with constants m_1^*, m_2^*, and (5) *with*

$$\lambda^* := \inf_{R_0 \in (0,\infty]} \lambda_{F^*}(R_0) > 0. \tag{20}$$

Then for

$$k > k_0 := 2\,[m_2^* - \min\{\lambda^*, m_1^*/2\}]$$

the parametric Lagrangian F defined by

$$F(x,z) := kA(z) + F^*(x,z) \tag{21}$$

possesses a perfect dominance function. In particular, any conformally parametrized minimizer of

$$\mathcal{F}(X) = \int_B F(X, X_u \wedge X_v)\, dudv,$$

where F is of the form (21), *is of class $H^{2,2}_{\text{loc}}(B,\mathbb{R}^n) \cap C^{1,\sigma}(B,\mathbb{R}^n)$.*

Remark 1.4 *In the applications regarding higher regularity of conformally parametrized $\mathcal{F}$-minimizers one generally has control over the size of the minimizer X, that is, an L^∞-estimate of the form*

$$\sup_{w\in\overline{B}} |X(w)| \le R < \infty$$

is known. Then one can relax condition (20) to

$$\lambda^* := \inf_{R_0\in(0,R]} \lambda_{F^*}(R_0) > 0.$$

Applications. *Existence problems.* Dominance functions were used to prove the existence of minimal surfaces and, more generally, of surfaces with prescribed mean curvature in Euclidean space and in Riemannian manifolds satisfying geometric boundary conditions. We refer to the more recent papers by K. Steffen [27] and F. Duzaar & K. Steffen [6], where this topic is surveyed from a general point of view. The gist of the method is to minimize the functional

$$\mathcal{G}(X) := \int_B G(X, \nabla X)\, dudv \tag{22}$$

instead of $\mathcal{F}$ where G is a perfect dominance function for F. This approach leads to conformally parametrized minimizers which then can be proved to be regular, whereas minimization of the original functional $\mathcal{F}$ does not necessarily yield regular extremals. The use of $\mathcal{G}$ has a smoothing effect as it suppresses the badly parametrized extremals. However, it is not a priori clear that $\mathcal{F}$ and $\mathcal{G}$ have the same infima. To prove this one has to apply fairly general results on conformal or quasiconformal mappings as was pointed out by Courant [4, pp. 116–118], and Morrey [21]. Using this "mapping idea" one verifies that

$$\inf_{\mathcal{C}} \mathcal{F} = \inf_{\mathcal{C}} \mathcal{G} \tag{23}$$

provided that the class $\mathcal{C}$ of admissible surfaces X is invariant with respect to reparametrizations, which is the case for "geometric" boundary conditions. In principle it is clear how this should be done, but it will be quite tedious to carry out the details, and for every class $\mathcal{C}$ the discussion is to be repeated in a varied form. However, as soon as one has established relation (23), one can pass from the minimum problem "$\mathcal{F} \to \min \quad \text{in } \mathcal{C}$" to the equivalent problem "$\mathcal{G} \to \min \quad \text{in } \mathcal{C}$" to establish the existence of a conformally parametrized minimizer of $\mathcal{F}$ in $\mathcal{C}$. The new problem can be solved if its Lagrangian G is quasiconvex in p and satisfies (D4) because these two properties imply (sequential) weak lower semicontinuity of $\mathcal{G}$ in $H^{1,2}(B, \mathbb{R}^n)$. Yet it is not clear whether the semi-ellipticity condition (C) on F guarantees that F possesses a dominance function $G(x, p)$ which is quasiconvex in p. For instance, we do not know if the two dominance functions (18) and (19) are polyconvex and therefore quasiconvex. These remarks explain why it is useful to work with the penalized Lagrangian $F^\epsilon := F + \epsilon D$ as we have done in our existence proofs (cf. [14]–[16] and also [18], [19]). First, (C) implies that F is quasiconvex in p, and adding the convex function ϵD does not destroy this property. Thus we can minimize the corresponding functional $\mathcal{F}^\epsilon$, and letting $\epsilon \to +0$ we obtain a minimizer of $\mathcal{F}$ which is seen to be conformally parametrized. As we do not use (23) in this procedure, we obtain this important relation as a by-product.

In the examples $\boxed{1}$ – $\boxed{3}$, however, the passage from "$\mathcal{F} \to \min \quad \text{in } \mathcal{C}$" to "$\mathcal{G} \to \min \quad \text{in } \mathcal{C}$" works very well since the dominance functions (12) and (15) are even convex if $\sup_{\mathbb{R}^3} |Q| < 1$, or $\sup_{\mathbb{R}^n} \sum_{i<k} \omega_{ik}^2 < 1$, respectively.

Higher regularity of conformally parametrized minimizers. The classical approach to Plateau's problem where one minimizes the Dirichlet integral $\mathcal{D}$ instead of the area functional $\mathcal{A}$ does not only lead to a minimal surface as a solution but also provides the simple Euler equation $\Delta X = 0$ from which all regularity properties of minimal surfaces can be deduced.

In a similar way Morrey proceeded in his solution of Plateau's problem in a Riemannian manifold, and also the work on surfaces with prescribed mean curvature uses the idea to derive regularity from the dominating functional $\mathcal{G}$ and not from $\mathcal{F}$.

The key to proving higher regularity of conformally parametrized $\mathcal{F}$-minimizers $X \in \mathcal{C}(\Gamma)$ is the analogous observation that any such surface is also the solution of the minimum problem "$\mathcal{G} \to \min \quad \text{in } \mathcal{C}(\Gamma)$," where $\mathcal{G}$ is the functional belonging to any dominance function $G(x, p)$ for $F(x, z)$. To be more precise, we have proved in [14, pp. 265–266] and [15, Thm. 1.8]

Theorem 1.5 *Suppose that $F \in C^0(\mathbb{R}^n \times \mathbb{R}^N)$ satisfies* (H),(D),(C), *and let G be an arbitrary dominance function of F with the corresponding variational integral*

$$\mathcal{G}(X) := \mathcal{G}_B(X) = \int_B G(X, \nabla X)\, dudv.$$

Then we have:

(i) $\inf_{\mathcal{C}(\Gamma)} \mathcal{F} = \inf_{\mathcal{C}(\Gamma)} \mathcal{G}$.
(ii) *Any minimizer of $\mathcal{G}$ in $\mathcal{C}(\Gamma)$ is a conformally parametrized minimizer of $\mathcal{F}$ in $\mathcal{C}(\Gamma)$.*
(iii) *Conversely, any conformally parametrized minimizer of $\mathcal{F}$ in $\mathcal{C}(\Gamma)$ is a minimizer of $\mathcal{G}$ in $\mathcal{C}(\Gamma)$.*

Notice that Theorem 1.5 implies the by no means trivial result

$$\inf_{\mathcal{C}(\Gamma)} \mathcal{A} = \inf_{\mathcal{C}(\Gamma)} \mathcal{D},$$

which follows here, in contrast to the classical approach, without any kind of sophisticated conformal mapping theorem, as was already pointed out before.

Using Part (iii) of Theorem 1.5 we can exploit the weak Euler equation of the functional $\mathcal{G}$,

$$\delta\mathcal{G}(X,\phi) = 0 \quad \text{for all} \quad \phi \in \mathring{H}^{1,2}(B,\mathbb{R}^n) \cap L^\infty(B,\mathbb{R}^n) \tag{24}$$

with

$$\delta\mathcal{G}(X,\phi) := \int_B [G_p(X,\nabla X)\cdot\nabla\phi + G_x(X,\nabla X)\cdot\phi]\, dudv \tag{25}$$

to prove

Theorem 1.6 ([15, Thm. 1.9]) *Suppose that F satisfies* (H),(D),(C), *and that F possesses a perfect dominance function G. Then any conformally parametrized minimizer X of $\mathcal{F}$ in $\mathcal{C}(\Gamma)$ is of class $H^{2,2}_{\mathrm{loc}}(B,\mathbb{R}^n) \cap C^{1,\sigma}(B,\mathbb{R}^n)$ for some $\sigma > 0$.*

Notice that the Euler equation (24) is preferable to the corresponding Euler equation

$$\delta\mathcal{F}(X,\phi) = 0 \quad \text{for all} \quad \phi \in \mathring{H}^{1,2}(B,\mathbb{R}^n) \cap L^\infty(B,\mathbb{R}^n) \tag{26}$$

of the original parametric functional $\mathcal{F}$, where

$$\delta\mathcal{F}(X,\phi) := \int_B [f_p(X,\nabla X)\cdot\nabla\phi + f_x(X,\nabla X)\cdot\phi]\, dudv. \tag{27}$$

Indeed, working with (24) one can use the much better regularity and convexity properties of $G(x,p)$ to prove Theorem 1.6 in case $G(x,p)$ is a perfect dominance function for $F(x,z)$.

One should note also that a priori it is not obvious at all whether the terms in (27) make sense, since $f_p(x,p)$ does not exist if $p \in \Pi$. But with a perfect[1] dominance function $G(x,p)$ at hand, the expressions in (25) are well-defined and one can show that the respective gradients f_x, G_x, and f_p, G_p coincide for $x \in \mathbb{R}^n$ and $p \in \mathbb{R}^{2n}$ satisfying (3). Then one can use the conformality relations (11) proved in Theorem 1.1 to show that (27) is well-defined and that

$$\delta\mathcal{F}(X,\phi) = \delta\mathcal{G}(X,\phi) \quad \text{for all} \quad \phi \in \mathring{H}^{1,2}(B,\mathbb{R}^n) \cap L^\infty(B,\mathbb{R}^n). \tag{28}$$

This is carried out in the present paper in Corollary 2.10 in Section 2.

If the boundary contour Γ is of class C^3 the regularity result Theorem 1.6 can be extended[2] up to the boundary, see [17] for the details:

[1] Actually, any dominance function of class $C^1(\mathbb{R}^n \times (\mathbb{R}^N - \{0\}))$ satisfying (D3) could be used to prove (28).

[2] Since regularity considerations are of local nature, Theorems 1.6 and 1.7 carry over to solutions of the Douglas problem treated in [18],[19], and also to the partially free boundary problem as carried out in [16, Thm. 1.12].

Theorem 1.7 *Let $\Gamma \subset \mathbb{R}^n$ be of class C^3. Suppose that F satisfies* (H),(D), *and* (C), *and that F possesses a perfect dominance function G. Then any conformally parametrized minimizer X of $\mathcal{F}$ in $\mathcal{C}(\Gamma)$ is of class $H^{2,2}(B,\mathbb{R}^n) \cap C^{1,\sigma}(\overline{B},\mathbb{R}^n)$ for some $\sigma > 0$.*

According to Theorem 1.3 these regularity results are valid for parametric functionals with integrands of the type (21) which indeed possess a perfect dominance function. It is an open problem at present whether or not *any* elliptic, positive definite parametric Lagrangian of class $C^2(\mathbb{R}^n \times (\mathbb{R}^N - \{0\})$ possesses a perfect dominance function. It also remains open if a conformally parametrized minimizer of $\mathcal{F}$ has only *isolated* branch points in B or even in $\overline{B}$. U. Clarenz and the second author, however, use the Hopf maximum principle, to show in [3] how to *exclude* boundary branch points for $\mathcal{F}$-stationary surfaces X that have no interior branch points under the hypothesis that Γ lies on the boundary of a convex body. For such a boundary configuration and even elliptic parametric integrands B. White [37] has shown the existence of smooth embedded minimizing disks. On the other hand, our regularity results Theorems 1.6 and 1.7 might be close to optimal, as J.E. Taylor [31] has proved the following theorem:

If $F \in C^0(\mathbb{R}^3) \cap C^3(\mathbb{R}^3 - \{0\})$ is an essentially noneven elliptic parametric Lagrangian independent of the spatial variable, then there exists an oriented closed analytic Jordan curve Γ on the sphere S^2 and a Lipschitz immersion X of the oriented disk (which is not an embedding) having Γ as boundary such that the value $\mathcal{F}(X)$ of the corresponding parametric functional $\mathcal{F}$ is less than the value $\mathcal{F}(Z)$ of any Lipschitz embedding Z of the oriented disk having Γ as boundary.

Here, a parametric Lagrangian $F(x,z)$ is called *essentially noneven*, if it cannot be written in the form $F(x,z) = c\tilde{F}(x,z) + Q(x) \cdot z$, where $c > 0$, $\tilde{F}$ is even in z, i.e., $\tilde{F}(x,z) = \tilde{F}(x,-z)$, and where $\operatorname{div} Q = 0$. Because of this result one cannot necessarily expect that minimizers in the class of immersions, and even more so minimizers in the more general class considered in Theorems 1.6 and 1.7, are as well-behaved as minimizers in the class of embeddings, even if they are of the type of the disk. F.J. Almgren, R. Schoen and L. Simon [1], [25] showed that the $\mathcal{F}$-minimizing integral current obtained by methods of geometric measure theory is a smooth embedded surface away from its boundary, and R. Hardt [8] showed that if the boundary curve Γ lies on the boundary of a convex body then it is also smooth at the boundary. However, these theorems say nothing about minimizing $\mathcal{F}$ among surfaces of specified topological type in contrast to our results and the result of White [37] mentioned above.

Unstable stationary surfaces. E. Heinz used the dominance function G defined in (12) with $Q(x) := 2H \cdot x/3$ for $H \in \mathbb{R}$ with $|H| < 1/2$, to construct unstable surfaces of constant mean curvature H contained in the unit ball around the origin of $\mathbb{R}^3$, and bounded by a closed Jordan curve , see [10],[11]. Starting with polygonal boundaries he used Courant's approach to work with a differentiable function of M variables whose critical points are in one-to-one correspondence with conformally parametrized solutions of the mean curvature system (17), which map $M + 3$

prescribed points on ∂B onto the $M+3$ vertices of the given polygonal boundary contour. Essential ingredients for this method are certain convexity and minimum properties of the functional $\mathcal{G}$ defined in (22), which lead to the existence of *unique* $\mathcal{G}$-minimizers subject to Dirichlet boundary data. Moreover, the regularity properties of $\mathcal{G}$ are crucial for proving that, for a given convergent sequence of boundary data, the corresponding $\mathcal{G}$-minimizers converge in the C^1-norm.

There is work in progress by R. Jakob adapting Heinz's method to parametric integrands F of the type considered in Theorem 1.3. The regularity properties of G might provide an alternative existence proof for unstable $\mathcal{F}$-stationary surfaces compared to the contribution of Shiffman [26]. Shiffman considered parametric integrands of the form $\tilde{F}(z) := kA(z) + F(z)$ with k sufficiently large, where F is a general parametric integrand of class C^1 which does not depend on x. To obtain the necessary convexity and continuity properties Shiffman replaced $\tilde{F}$ by the dominance function $\tilde{G}(p) := kD(p) + f(p)$, $f(p) := F(p_1 \wedge p_2)$. But, unfortunately, his arguments regarding the actual construction of the unstable stationary surface towards the end of his work do not seem to be complete, in contrast to Heinz's detailed proof for H-surfaces in [9],[10], or the contributions regarding more general variational problems by G. Strömer, see, e.g., [28].

2 Construction of a Perfect Dominance Function

The following result due to Morrey [23, pp. 391–394] guarantees the existence of a quadratic, positive definite dominance function with the remarkable property that the singular set Π of $f(x,.)$ in $\mathbb{R}^{2n}$ reduces to the singleton $\{0\}$ in $\mathbb{R}^{2n}$ for $G(x,.)$.

Theorem 2.8 *Let $F(x,z) \in C^0(\mathbb{R}^n \times \mathbb{R}^N) \cap C^{k,\mu}_{\mathrm{loc}}(\mathbb{R}^n \times (\mathbb{R}^N - \{0\}))$ be a parametric Lagrangian satisfying* (H) *and* (D). *Then there exists a function $G(x,p)$ for $F(x,z)$ satisfying* (D1)–(D4) *with $\mu_1 = m_1/4$, $\mu_2 = m_2/2$. In addition,*

$$G \in C^0(\mathbb{R}^n \times \mathbb{R}^N) \cap C^{k,\nu}_{\mathrm{loc}}(\mathbb{R}^n \times (\mathbb{R}^N - \{0\})). \tag{29}$$

Before proving this result we list various properties of this dominance function G, which follow from the homogeneity condition (D3) on $F(x,z)$ and from the improved regularity of $G(x,p)$ reflected in (29), see [15, Lemmas 3.1–3.3] for details. Part (iii) below in particular justifies the weak Euler equation for $\mathcal{G}$, (24),(25) mentioned in the introduction.

Corollary 2.9 (i) *$G_x(x,p)$, $G_{xx}(x,p)$ are positively homogeneous in p of degree two, $G_p(x,p)$, $G_{px}(x,p)$ are positively homogeneous in p of degree one, and $G_{pp}(x,p)$ is positively homogeneous in $p \neq 0$ of degree zero. Consequently, G, G_x, G_{xx}, G_p, G_{px}, are continuous in $\mathbb{R}^n \times \mathbb{R}^{2n}$ with*

$$\begin{gathered} G(x,0)=0, \quad G_x(x,0)=0, \quad G_{xx}(x,0)=0, \\ G_p(x,0)=0, \quad G_{px}(x,0)=0 \end{gathered} \tag{30}$$

for all $x \in \mathbb{R}^n$, and G_{pp} is bounded and continuous on $\mathbb{R}^n \times (\mathbb{R}^{2n} - \{0\})$.

(ii) *There are constants* $c_0(R_0), c_1(R_0), c_2(R_0) > 0$, *such that for* $|x| \le R_0$ *and for* $p \in \mathbb{R}^n \times \mathbb{R}^{2n}$ *we have*

$$|G(x,p)| + |G_x(x,p)| + |G_{xx}(x,p)| \le c_0(R_0)|p|^2,$$
$$|G_p(x,p)| + |G_{px}(x,p)| \le c_1(R_0)|p|,$$

and if $p \neq 0$, *then*

$$|G_{pp}(x,p)| \le c_2(R_0).$$

(iii) *For any surface* $X \in H^{1,2}(B,\mathbb{R}^n) \cap L^\infty(B,\mathbb{R}^n)$ *and any* $\phi \in \mathring{H}^{1,2}(B,\mathbb{R}^n) \cap L^\infty(B,\mathbb{R}^n)$ *we have*

$$\lim_{\epsilon \to 0} \epsilon^{-1}[\mathcal{G}(X+\epsilon\phi) - \mathcal{G}(X)] = \delta\mathcal{G}(X,\phi),$$

where $\delta\mathcal{G}(X,\phi)$ *is defined by*

$$\delta\mathcal{G}(X,\phi) := \int_B [G_p(X,\nabla X)\cdot\nabla\phi + G_x(X,\nabla X)\cdot\phi]\, dudv.$$

If, in addition, X *is a minimizer of* $\mathcal{G}$ *in* $\mathcal{C}(\Gamma)$, *then* $\delta\mathcal{G}(X,\phi) = 0$.

Using these properties as well as (D1) and (D2) we can also justify the validity of the Euler equation (26),(27) for $\mathcal{F}$. For notational convenience we introduce the algebraic surface

$$\Pi_0 := \{p = (p_1,p_2) \in \mathbb{R}^{2n} : |p_1|^2 = |p_2|^2,\ p_1\cdot p_2 = 0\}.$$

Lemma 2.10 *Suppose that* $G \in C^2(\mathbb{R}^n \times (\mathbb{R}^{2n} - \Pi))$ *is a dominance function of a parametric Lagrangian* F *of class* $C^0(\mathbb{R}^n \times \mathbb{R}^N) \cap C^2(\mathbb{R}^n \times (\mathbb{R}^N - \{0\}))$ *with the associated Lagrangian* $f \in C^2(\mathbb{R}^n \times (\mathbb{R}^{2n} - \Pi))$. *Then we have*

$$G_x(x,p) = f_x(x,p), \quad G_p(x,p) = f_p(x,p), \quad G_{px}(x,p) = f_{px}(x,p) \tag{31}$$

for $(x,p) \in \mathbb{R}^n \times \Pi_0$, *and*

$$\xi \cdot G_{pp}(x,p)\xi \ge \xi \cdot f_{pp}(x,p)\xi \ \text{ for } \ (x,p) \in \mathbb{R}^n \times (\Pi_0 - \{0\}),\ \xi \in \mathbb{R}^{2n}.$$

In particular, if $G \in C^2(\mathbb{R}^n \times (\mathbb{R}^{2n} - \{0\}))$ *and if* $X \in H^{1,2}(B,\mathbb{R}^n) \cap L^\infty(B,\mathbb{R}^n)$ *is conformally parametrized then*

$$\delta\mathcal{F}(X,\phi) = \delta\mathcal{G}(X,\phi) \quad \text{for all} \quad \phi \in \mathring{H}^{1,2}(B,\mathbb{R}^n) \cap L^\infty(B,\mathbb{R}^n),$$

where $\delta\mathcal{G}(X,\phi)$ *and* $\delta\mathcal{F}(X,\phi)$ *are given by* (25) *and* (27), *respectively.*

Proof. Since G is a dominance function of F we have

$$G(x,p) - f(x,p) \ge 0 \ \text{ on } \ \mathbb{R}^n \times \mathbb{R}^{2n}, \ \text{ and} \tag{32}$$
$$G(x,p) - f(x,p) = 0 \ \text{ on } \ \mathbb{R}^n \times \Pi_0. \tag{33}$$

For for fixed $x \in \mathbb{R}^n$ and $p = (p_1, p_2) \in \Pi_0 - \{0\}$ one uses (H) to calculate

$$\begin{aligned} f_x(x,p) &= F_x(x, p_1 \wedge p_2) = F_x(x, \frac{p_1 \wedge p_2}{|p_1 \wedge p_2|})|p_1 \wedge p_2|, \\ f_p(x,p) &= F_z(x, p_1 \wedge p_2)(p_1 \wedge p_2)_p = F_z(x, p_1 \wedge p_2)(p_1 \wedge p_2)_p, \\ f_{px}(x,p) &= F_{zx}(x, p_1 \wedge p_2)(p_1 \wedge p_2)_p = F_{zx}(x, p_1 \wedge p_2)(p_1 \wedge p_2)_p, \end{aligned}$$

hence

$$\begin{aligned} |f_x(x,p)| &\le \sup_{|z|=1} |F_x(x,z)||p|^2, \\ |f_p(x,p)| &\le \sqrt{n} \sup_{|z|=1} |F_z(x,z)||p|, \\ |f_{px}(x,p)| &\le \sqrt{n} \sup_{|z|=1} |F_{zx}(x,z)||p|, \end{aligned}$$

which allows us to extend f_x, f_p, and f_{px} continuously to $\mathbb{R}^n \times [(\mathbb{R}^{2n} - \Pi) \cup \{0\}]$ by setting

$$f_x(x,0) = 0, \quad f_p(x,0) = 0, \quad f_{px}(x,0) = 0.$$

This and property (33) together with (30) in Lemma 2.9 and $\Pi \cap \Pi_0 = \{0\}$ now imply

$$G_x(x,p) = f_x(x,p) \text{ on } \mathbb{R}^n \times \Pi_0,$$

and (32) together with (33) yield $G_p(x,p) - f_p(x,p) = 0$ for all $(x,p) \in \mathbb{R}^n \times \Pi_0$, which leads to $G_{px}(x,p) - f_{px}(x,p) = 0$ for all $(x,p) \in \mathbb{R}^n \times \Pi_0$. Finally we get $\xi \cdot (G_{pp}(x,p) - f_{pp}(x,p))\xi \ge 0$ for $(x,p) \in \mathbb{R}^n \times (\Pi_0 - \{0\})$, again by (32) and (33), since $\Pi \cap \Pi_0 = \{0\}$. The last assertion follows from (31). □

Proof of Theorem 2.8. We introduce two functions $\tau : \mathbb{R}^{2n} \to \mathbb{R}$ and $\sigma : \mathbb{R}^n \times \mathbb{R}^{2n} \to \mathbb{R}$ by

$$\tau(p) := \frac{a(p)}{D(p)} \text{ for } p \neq 0, \quad \tau(0) := 1, \tag{34}$$

and

$$\sigma(x,p) := \frac{f(x,p)}{a(p)} \text{ for } p \notin \Pi, \quad \sigma(x,p) := m_1 \text{ for } p \in \Pi. \tag{35}$$

Then we have

$$0 \le \tau(p) \le 1 \text{ and } m_1 \le \sigma(x,p) \le m_2. \tag{36}$$

Set

$$G(x,p) := D(p)[(m_1/2) + \omega(\tau(p), \sigma(x,p))], \tag{37}$$

where $\omega : [0,1] \times [m_1, \infty) \to \mathbb{R}$ is a function to be chosen later on. From (34) and (35) we infer that

$$\tau(\gamma p) = \tau(p) \text{ and } \sigma(x, \gamma p) = \sigma(x, p) \text{ for any } \gamma > 0$$

which implies property (D3). Now we set

$$u(t, s) := \frac{(s - (m_1/2))t}{(m_1/2) + (s - m_1)t} \quad \text{for } t \in [0,1], s \in [m_1, \infty), \tag{38}$$

and notice first that $u(t, m_1) = t$ for all $t \in [0, 1]$. Then we compute the first and second derivative of $u(t, s)$ with respect to t:

$$\begin{aligned} u_t(t,s) &= \frac{(m_1/2)(s-(m_1/2))}{((m_1/2)+(s-m_1)t)^2} \geq \frac{(m_1/2)(s-(m_1/2))}{(s-(m_1/2))^2} \\ &= \frac{m_1}{2(s-(m_1/2))} > 0 \end{aligned} \tag{39}$$

and

$$u_{tt}(t, s) = -2\frac{(m_1/2)(s - (m_1/2))(s - m_1)}{((m_1/2) + (s - m_1)t)^3} \leq 0.$$

Hence

$$u_t(t, s) > 0 \text{ for } s \geq m_1 \text{ and } u_{tt}(t, s) < 0 \text{ for } s > m_1. \tag{40}$$

Therefore $u(., s)$ is increasing and strictly concave for $s > m_1$. Moreover we have

$$u(0, s) = 0 \text{ and } u(1, s) = 1 \text{ for } s \geq m_1 \tag{41}$$

as well as

$$\lim_{s \to \infty} u(t, s) = \begin{cases} 0 & \text{for } t = 0 \\ 1 & \text{for } 0 < t \leq 1. \end{cases}$$

Note also that

$$u(t, s_1) < u(t, s_2) \text{ for } 0 < t < 1, \, m_1 \leq s_1 < s)2, \tag{42}$$

since the derivative of $u(t, s)$ with respect to s may be calculated as

$$u_s = \frac{m_1(t - t^2)}{2[(m_1/2) + (s - m_1)t]^2}, \tag{43}$$

which is positive for $t \in (0, 1)$. For two values u_0 and u_1 with $0 < u_0 < u_1 < 1$ we choose a smooth cut-off function $\eta : [0, 1] \to \mathbb{R}$ such that

$$\begin{aligned} &\eta \in C_0^\infty((u_0, 1]), \quad 0 < \eta(r) < 1 \text{ for } u_0 < r < 1, \quad \eta(1) = 1, \\ &\frac{1}{1 - u_0} < \eta'(1) \leq \frac{1}{1 - u_1}, \quad \eta''(r) > 0 \text{ for } r \in (u_0, 1]. \end{aligned} \tag{44}$$

Using the functions $u(s, t)$ and $\eta(r)$ we now define the function $\omega(t, s)$ appearing in (37) as

$$\omega(t,s) := [(m_1/2) + (s - m_1)t]\eta(u(t,s)) \text{ for } (t,s) \in [0,1] \times [m_1, \infty). \quad (45)$$

This leads to

$$G(x,p) = D(p)\left\{\frac{m_1}{2} + [(m_1/2) + (\sigma - m_1)\tau]\eta(u(\tau,\sigma))\right\}, \quad (46)$$

where $\tau := \tau(p)$, $\sigma := \sigma(x,p)$. By (36) and (38) we have $0 \le u(\tau,\sigma) \le 1$. Thus $\eta(u(\tau,\sigma))$ is well-defined, and $0 \le \eta(u(\tau,\sigma)) \le 1$. We finally obtain

$$\begin{aligned} \frac{m_1}{2} &\le \frac{m_1}{2} + [(m_1/2) + (\sigma - m_1)\tau]\eta(u(\tau,\sigma)) \\ &\le \frac{m_1}{2} + \frac{m_1}{2} + (\sigma - m_1) \underset{(36)}{\le} m_2, \end{aligned}$$

whence

$$(m_1/2)D(p) \le G(x,p) \le m_2 D(p),$$

that is, property (D4) holds with $\mu_1 = (m_1/4)$ and $\mu_2 = (m_2/2)$.

To prove (29) we note that G is of class $C^{k,\nu}$ in a sufficiently small neighbourhood of any point $(x,p_0) \in \mathbb{R}^n \times (\mathbb{R}^{2n} - \Pi)$. If $p_0 \in \Pi - \{0\}$ on the other hand, then $0 \le \tau(p) \ll 1$ for $|p - p_0| \ll 1$, whence

$$0 \le u(\tau(p), m_2) < u_0 \text{ for } |p - p_0| \ll 1.$$

On account of (36) and (42) we also have

$$0 \le u(\tau(p), \sigma(x,p)) \le u(\tau(p), m_2).$$

Therefore, $\eta(u(\tau(p), \sigma(x,p))) = 0$ if $|p - p_0| \ll 1$, and consequently,

$$G(x,p) = (m_1/2)D(p) \text{ for } |p - p_0| \ll 1,\ p_0 \in \Pi - \{0\} \text{ and } x \in \mathbb{R}^n. \quad (47)$$

This shows that G is of class C^∞ in a neighbourhood of any point p_0 in $\Pi - \{0\}$.

It remains to be shown that $G(x,p)$ satisfies (D1) and (D2).

Case 1. $p = 0$. Then $p \in \Pi$ and $G(x,p) = 0 = f(x,p)$, by (34),(35),(44) and (46).

Case 2. $p \in \Pi - \{0\}$. Then $f(x,p) = F(x,0) = 0$ and $G(x,p) \ge \mu_1 |p|^2 > 0$ by (D4) which was shown above, whence $f(x,p) < G(x,p)$. Since $\Pi \cap \Pi_0 = \{0\}$ we also have $p \notin \Pi_0$.

Case 3. $p \notin \Pi$. Then $p_1 \wedge p_2 \ne 0$, and so $a(p)$, $D(p)$ and $f(x,p)$ are all positive. Thus we can write

$$f(x,p) = \frac{f(x,p)}{a(p)} \frac{a(p)}{D(p)} D(p) = \sigma(x,p)\tau(p)D(p).$$

If we compare this with (37) we obtain:

$G(x,p)$ is a dominance function for $F(x,z)$ if the following property holds true:

($\star$) *For any $(t,s) \in [0,1] \times [m_1, m_2]$ we have $ts \le (m_1/2) + \omega(t,s)$ with equality if and only if $t = 1$.*

To verify ($\star$) we investigate the function $\omega(t, s)$, which is given by (45). Differentiating ω with respect to t we get

$$\begin{aligned}\omega_t(t,s) &\underset{(45)}{=} (s-m_1)\eta(u) + [(m_1/2)+(s-m_1)t]\eta'(u)u_t(t,s)\\ &\underset{(39)}{=} (s-m_1)\eta(u) + \eta'(u)\frac{[(m_1/2)+(s-m_1)t](m_1/2)(s-(m_1/2))}{[(m_1/2)+(s-m_1)t]^2}\\ &= (s-m_1)\eta(u) + \eta'(u)\frac{m_1(s-(m_1/2))/2}{(m_1/2)+(s-m_1)t},\end{aligned}$$

where we have set $u = u(t, s)$. Notice that by (40) $u = 0$ if and only if $t = 0$ in which case one finds by the properties (44) of the cut-off function η that $\omega_t(0, s) = 0$.

We infer from (38) that

$$((m_1/2)+(s-m_1)t)\,u = (s-(m_1/2))t,$$

which leads to

$$\frac{u}{t} = \frac{2}{m_1}[s-(m_1/2)-(s-m_1)u] \tag{48}$$

for $t \neq 0$ (or equivalently, for $u \neq 0$), and

$$t = \frac{m_1 u}{2[s-(m_1/2)-(s-m_1)u]} \tag{49}$$

for $t \neq 0$ (i.e., $u = 0$). Consequently we obtain

$$\begin{aligned}\omega_t(t,s) &= (s-m_1)\eta(u) + \eta'(u)\frac{m_1 u}{2t}\\ &\underset{(48)}{=} (s-m_1)\eta(u) + \eta'(u)[s-(m_1/2)-(s-m_1)u],\end{aligned} \tag{50}$$

which now also holds for $t = 0$ (i.e. $u = 0$). Thus, fixing $s \in [m_1, m_2]$, we can express ω_t as a function of u, since the mapping $t \mapsto u(t, s)$ is invertible. By differentiating ω_t with respect to u we obtain

$$\begin{aligned}\frac{\partial}{\partial u}\left[\omega_t\big|_{s\text{ fixed}}\right] &= [s-(m_1/2)-(s-m_1)u]\eta''(u)\\ &\geq [s-(m_1/2)-(s-m_1)]\eta''(u) = (m_1/2)\eta''(u) \geq 0,\end{aligned} \tag{51}$$

if we take $0 \leq u \leq 1$, $s \geq m_1$ and $\eta''(r) \geq 0$ for $r \in [0, 1]$ into account. Hence, for fixed s, the function $\omega_t(t(r, s), s)$ is nondecreasing with respect to $r \in [0, 1]$, and so it assumes its maximum at $r = 1$. By (50) and $\eta(1) = 1$ we obtain

$$\begin{aligned}\omega_t(t(1,s),s) &= (s-m_1)\eta(1) + [s-(m_1/2)-(s-m_1)]\eta'(1)\\ &= s-m_1+(m_1/2)\eta'(1) \leq s-m_1+(m_1/2)(1-u_1)^{-1}\end{aligned} \tag{52}$$

for all $r \in [0, 1]$ by (44). Because of

$$u_1 \in (0, 1/2) \tag{53}$$

we have $(m_1/2)(1-u_1)^{-1} < m_1$, and we arrive at $\omega_t(t(r,s),s) < s$ for all r in $[0,1]$, i.e., $\omega_t(t,s) < s$ for $0 \le t \le 1$, $m_1 \le s \le m_2$. Then the function $\chi : [0,1] \times [m_1, m_2] \to \mathbb{R}$ defined by $\chi(t,s) := (m_1/2) + \omega(t,s) - ts$ satisfies $\chi_t(t,s) = \omega_t(t,s) - s < 0$, hence $\chi(.,s)$ is a strictly decreasing function on $[0,1]$ for any fixed value of $s \in [m_1, m_2]$. We have $\chi(1,s) = (m_1/2) + \omega(1,s) - s$, and from (45),(41) and (44) $\omega(1,s) = s - (m_1/2)$, i.e., $\chi(1,s) = 0$ and $\chi(t,s) > 0$ for $0 \le t < 1$. This implies

$$\frac{m_1}{2} + \omega(t,s) > ts \text{ for } 0 \le t < 1, \quad \text{and } \frac{m_1}{2} + \omega(1,s) = s,$$

which establishes the proof of ($\star$), and thus of Theorem 2.8. □

Before we show how to refine the construction in Theorem 2.8 to obtain a perfect dominance function satisfying the ellipticity condition (E) we derive a couple of simple estimates for the gradient F_z of a parametric integrand $F(x,z)$, and for the Hessian of the associated Lagrangian $f(x,p)$.

Lemma 2.11 *Let $F \in C^0(\mathbb{R}^n \times \mathbb{R}^N) \cap C^1(\mathbb{R}^n \times (\mathbb{R}^N - \{0\}))$ be a parametric Lagrangian satisfying* (H),(D) *and* (C). *Then we have*

(i)

$$F(x,z) - z \cdot F_z(x,z_0) \ge 0 \textit{ for all } x \in \mathbb{R}^n,\ z, z_0 \in \mathbb{R}^N,\ z_0 \ne 0. \tag{54}$$

In particular,

$$|F_z(x,z_0)| \le m_2 \textit{ for all } (x,z_0) \in \mathbb{R}^n \times (\mathbb{R}^N - \{0\}). \tag{55}$$

(ii) *If, in addition, $F \in C^2(\mathbb{R}^n \times (\mathbb{R}^N - \{0\}))$ satisfies* (5), *then for $f(x,p) = F(x, p_1 \wedge p_2)$ one has*

$$\xi \cdot f_{pp}(x,p)\,\xi \ge \frac{\lambda_F(R_0)}{a(p)} \left|P^{\perp}_{p_1 \wedge p_2}[(\zeta \wedge p_2) - (\eta \wedge p_1)]\right|^2 \tag{56}$$
$$+2F_z(x, p_1 \wedge p_2) \cdot (\zeta \wedge \eta)$$

for all $\xi = (\zeta, \eta) \in \mathbb{R}^n \times \mathbb{R}^n \cong \mathbb{R}^{2n}$, $(x,p) \in \overline{B_{R_0}}(0) \times (\mathbb{R}^{2n} - \Pi)$, where $\lambda_F(R_0)$ is the constant appearing in (5). *In particular, we have*

$$\xi \cdot a_{pp}(x,p)\,\xi = \frac{1}{a(p)} \left|P^{\perp}_{p_1 \wedge p_2}[(\zeta \wedge p_2) - (\eta \wedge p_1)]\right|^2 \tag{57}$$
$$+ \frac{p_1 \wedge p_2}{|p_1 \wedge p_2|} \cdot (\zeta \wedge \eta).$$

Remark 2.12 *In the rank-one case, i.e., if $\xi = \zeta \otimes \Lambda$ for some $\zeta \in \mathbb{R}^n$, $\Lambda = (\lambda_1, \lambda_2) \in \mathbb{R}^2$, the estimate (56) reduces to*

$$\xi \cdot f_{pp}(x,p)\,\xi \geq \frac{\lambda_F(R_0)}{a(p)} \left|P^{\perp}_{p_1\wedge p_2}[\zeta \wedge (\lambda_1 p_2 - \lambda_2 p_1)]\right|^2$$

which implies that rank $f_{pp}(x,p) = n-2$ *for all* $\Lambda \neq 0$*, since* $p \in \mathbb{R}^{2n} - \Pi$ *has linearly independent column vectors* p_1, p_2*. In the language of Morrey this means:* f *is the integrand of a parametric regular problem, or* f *satisfies the parametric Legendre-condition, see [23, p. 392].*

Proof of Lemma 2.11 . (i) By (H) and (C) we know that for $z_0 \neq 0$

$$F(x,z) - z\cdot F_z(x,z_0) = F(x,z) - F(x,z_0) - (z-z_0)F_z(x,z_0) \geq 0,$$

which proves (54). If $F_z(x,z_0) = 0$ then (55) is trivially satisfied. For $F_z(x,z_0) \neq 0$ we may choose $z := F_z(x,z_0)$ in (54) to get

$$F(x, F_z(x,z_0)) - |F_z(x,z_0)|^2 \geq 0,$$

hence by (D) and (H),

$$|F_z(x,z_0)|^2 \leq F(x, \frac{F_z(x,z_0)}{|F_z(x,z_0)|})|F_z(x,z_0)| \leq m_2|F_z(x,z_0)| \quad \text{for all} \quad z_0 \neq 0.$$

(ii) We omit the arguments $(x,p) \in \overline{B_{R_0}}(0) \times (\mathbb{R}^{2n} - \Pi)$ of f and $(x, p_1\wedge p_2) \in \overline{B_{R_0}}(0) \times (\mathbb{R}^N - \{0\})$ of F and write for $\xi = (\zeta,\eta) \in \mathbb{R}^n \times \mathbb{R}^n \cong \mathbb{R}^{2n}$:

$$\xi \cdot f_{pp}(x,p)\,\xi = \zeta \cdot f_{p_1p_1}\zeta + \zeta\cdot f_{p_1p_2}\eta + \eta \cdot f_{p_2p_1}\zeta + \eta\cdot f_{p_2p_2}\eta.$$

In order to calculate the derivatives of f we use the identity

$$(p_1 \wedge p_2)^{(j,k)} = \varepsilon^{(j,k)}_{st} p_1^s p_2^t$$

for the (j,k)-th component of the bivector $p_1 \wedge p_2 \in \mathbb{R}^N$, where (j,k) denotes a double index with components $j,k \in \{1,\dots,n\}, j<k$, and where ε is the permutation tensor defined as

$$\varepsilon^{(j,k)}_{st} := \begin{cases} 1 & \text{if } j=s \text{ and } k=t \text{ and } j<k, \\ -1 & \text{if } j=t \text{ and } k=s \text{ and } j<k, \\ 0 & \text{otherwise.} \end{cases}$$

We obtain

$$\begin{aligned} f_{p_1^i} &= F_{z^{(j,k)}} \cdot (p_1\wedge p_2)^{(j,k)}_{p_1^i} = F_{z^{(j,k)}}\varepsilon^{(j,k)}_{it} p_2^t, \\ f_{p_2^i} &= F_{z^{(j,k)}} \cdot (p_1\wedge p_2)^{(j,k)}_{p_2^i} = F_{z^{(j,k)}}\varepsilon^{(j,k)}_{si} p_1^s, \end{aligned}$$

where we used the convention to sum over all repeated ordered double indices, and for the second derivatives we arrive at

$$
\begin{aligned}
f_{p_1^i p_1^k} &= F_{z^{(j,k)} z^{(j',k')}} \varepsilon_{it}^{(j,k)} p_2^t \varepsilon_{k\tau}^{(j',k')} p_2^\tau, \\
f_{p_1^i p_2^k} &= F_{z^{(j,k)} z^{(j',k')}} \varepsilon_{it}^{(j,k)} p_2^t \varepsilon_{\sigma k}^{(j',k')} p_1^\sigma + F_{z^{(j,k)}} \varepsilon_{ik}^{(j,k)}, \\
f_{p_2^i p_1^k} &= F_{z^{(j,k)} z^{(j',k')}} \varepsilon_{si}^{(j,k)} p_1^s \varepsilon_{k\tau}^{(j',k')} p_2^\tau + F_{z^{(j,k)}} \varepsilon_{ik}^{(j,k)}, \\
f_{p_2^i p_2^k} &= F_{z^{(j,k)} z^{(j',k')}} \varepsilon_{si}^{(j,k)} p_1^s \varepsilon_{\sigma k}^{(j',k')} p_1^\sigma.
\end{aligned}
$$

Thus we obtain

$$
\begin{aligned}
\xi \cdot f_{pp} \xi = {} & F_{z^{(j,k)} z^{(j',k')}} \Big[\varepsilon_{it}^{(j,k)} \zeta^i p_2^t \varepsilon_{k\tau}^{(j',k')} \zeta^k p_2^\tau + \varepsilon_{it}^{(j,k)} \zeta^i p_2^t \varepsilon_{\sigma k}^{(j',k')} \eta^k p_1^\sigma \\
& + \varepsilon_{si}^{(j,k)} \eta^i p_1^s \varepsilon_{k\tau}^{(j',k')} \zeta^k p_2^\tau + \varepsilon_{si}^{(j,k)} \eta^i p_1^s \varepsilon_{\sigma k}^{(j',k')} \eta^k p_1^\sigma \Big] \\
& + F_{z^{(j,k)}} [\varepsilon_{ik}^{(j,k)} \zeta^i \eta^k + \varepsilon_{ki}^{(j,k)} \zeta^k \eta^i] \\
= {} & (\zeta \wedge p_2) \cdot F_{zz}(\zeta \wedge p_2) + (\zeta \wedge p_2) \cdot F_{zz}(p_1 \wedge \eta) \\
& + (p_1 \wedge \eta) \cdot F_{zz}(\zeta \wedge p_2) + (p_1 \wedge \eta) \cdot F_{zz}(p_1 \wedge \eta) \\
& + 2F_z \cdot (\zeta \wedge \eta) \\
= {} & [(\zeta \wedge p_2) - (\eta \wedge p_1)] \cdot F_{zz}[(\zeta \wedge p_2) - (\eta \wedge p_1)] \\
& + 2F_z \cdot (\zeta \wedge \eta) \\
\underset{(5)}{\geq} {} & \frac{\lambda_F(R_0)}{a(p)} \left| P^\perp_{p_1 \wedge p_2} [(\zeta \wedge p_2) - (\eta \wedge p_1)] \right|^2 \\
& + 2F_z(x, p_1 \wedge p_2) \cdot (\zeta \wedge \eta).
\end{aligned}
$$

□

The quadratic form of the Hessian matrix of the dominance function G constructed in Theorem 2.8 can be estimated from below as follows:

Theorem 2.13 *Under the assumptions of Theorem 2.8 with the additional hypothesis that the parametric Lagrangian F is elliptic, satisfying* (5), *the dominance function G given in Theorem 2.8 satisfies*

$$
\begin{aligned}
\xi \cdot G_{pp}(x,p) \xi \geq {} & ((m_1/2) + \omega - \tau \omega_t) |\xi|^2 \\
& + (\omega_s \lambda_F(R_0) + \tau \omega_t - \sigma \omega_s) \left| P^\perp_{p_1 \wedge p_2} [(\zeta \wedge p_2) - (\eta \wedge p_1)] \right|^2 / \tau a \\
& + 2 \left[\omega_s F_z(x, p_1 \wedge p_2) + (\tau \omega_t - \sigma \omega_s) \frac{p_1 \wedge p_2}{|p_1 \wedge p_2|} \right] \cdot (\zeta \wedge \eta)/\tau \qquad (58) \\
& + D \begin{bmatrix} \tau_p \cdot \xi \\ \sigma_p \cdot \xi \end{bmatrix} \cdot \begin{pmatrix} \omega_{tt} & \omega_{ts} - (\omega_s/\tau) \\ \omega_{st} - (\omega_s/\tau) & \omega_{ss} \end{pmatrix} \begin{bmatrix} \tau_p \cdot \xi \\ \sigma_p \cdot \xi \end{bmatrix}
\end{aligned}
$$

for all $(x,p) \in \overline{B_{R_0}}(0) \times (\mathbb{R}^{2n} - \Pi)$, $\xi = (\zeta, \eta) \in \mathbb{R}^{2n}$, where $\omega = \omega(\tau, \sigma)$, τ, and σ are defined in (45),(34),(35) *in the proof of Theorem 2.8. For $(x,p) \in \mathbb{R}^n \times (\Pi - \{0\})$ we have*

$$\xi \cdot G_{pp}(x,p)\,\xi = m_1|\xi|^2/2 \;\; for\; all \;\; \xi \in \mathbb{R}^{2n}. \tag{59}$$

Proof. The identity (59) follows immediately from (47) derived in the proof of Theorem 2.8.

In the following computations we will frequently omit the argument $(x,p) \in \overline{B_{R_0}(0)} \times (\mathbb{R}^{2n} - \{0\})$ for simplicity. We use the notation

$$d \otimes e = (d^k_\alpha e^i_\beta), \quad 1 \le k, i \le n, \quad 1 \le \alpha, \beta \le 2,$$

for the dyadic product of two vectors $d, e \in \mathbb{R}^{2n}$. Differentiating (37) with respect to p one obtains

$$G_p = D_p[(m_1/2) + \omega] + D[\omega_t \tau_p + \omega_s \sigma_p], \quad D_p = p,$$

and

$$\begin{aligned} \xi \cdot G_{pp}\,\xi &= ((m_1/2) + \omega)|\xi|^2 + D[\omega_t \xi \cdot \tau_{pp}\,\xi + \omega_s \xi \cdot \sigma_{pp}\,\xi] \\ &\quad + \xi \cdot [\omega_t D_p \otimes \tau_p + \omega_s D_p \otimes \sigma_p + \omega_t \tau_p \otimes D_p + \omega_s \sigma_p \otimes D_p]\,\xi \\ &\quad + \xi \cdot D[\omega_{tt} \tau_p \otimes \tau_p + \omega_{ts} \tau_p \otimes \sigma_p + \omega_{st} \sigma_p \otimes \tau_p + \omega_{ss} \sigma_p \otimes \sigma_p]\,\xi. \end{aligned} \tag{60}$$

By (34) one has

$$\begin{aligned} \tau &= \frac{a}{D}, \quad \tau_p = \frac{a_p}{D} - \frac{aD_p}{D^2} = \frac{a_p}{D} - \tau\frac{D_p}{D}, \\ \tau_{pp} &= \frac{a_{pp}}{D} - \frac{aD_{pp}}{D^2} - \frac{a_p \otimes D_p}{D^2} - \frac{D_p \otimes a_p}{D^2} - \frac{2aD_p \otimes D_p}{D^3}, \end{aligned} \tag{61}$$

and we infer

$$\begin{aligned} \xi \cdot \tau_{pp}\,\xi &= \frac{\xi \cdot a_{pp}\,\xi}{D} - \frac{a\xi \cdot D_{pp}\,\xi}{D^2} - \xi \cdot [a_p \otimes D_p + D_p \otimes a_p - 2aD_p \otimes D_p/D]\,\xi/D^2 \\ &= \frac{\xi \cdot a_{pp}\,\xi}{D} - \frac{a|\xi|^2}{D^2} - \xi \cdot [a_p \otimes D_p + D_p \otimes a_p - 2\tau D_p \otimes D_p]\,\xi/D^2 \\ &\underset{(2)}{=} \frac{\xi \cdot a_{pp}\,\xi}{D} - \frac{a|\xi|^2}{D^2} \\ &\quad - \xi \cdot [(D\tau_p + \tau D_p) \otimes D_p + D_p \otimes (D\tau_p + \tau D_p) - 2\tau D_p \otimes D_p]\,\xi/D^2 \\ &= \frac{\xi \cdot a_{pp}\,\xi}{D} - \frac{a|\xi|^2}{D^2} - \xi \cdot [D\tau_p \otimes D_p + DD_p \otimes \tau_p]\,\xi/D^2. \end{aligned}$$

Furthermore, with $f = a\sigma$ one obtains

$$f_p = a\sigma_p + \sigma a_p \quad \text{and} \quad f_{pp} = a\sigma_{pp} + \sigma a_{pp} + a_p \otimes \sigma_p + \sigma_p \otimes a_p,$$

whence

$$\begin{aligned} \xi \cdot f_{pp}\,\xi &= a\xi \cdot \sigma_{pp}\,\xi + \sigma\xi \cdot a_{pp}\,\xi + \xi \cdot [a_p \otimes \sigma_p + \sigma_p \otimes a_p)\,\xi \\ &= a\xi \cdot \sigma_{pp}\,\xi + \sigma\xi \cdot a_{pp}\,\xi \\ &\quad + \xi \cdot [D\tau_p \otimes \sigma_p + \tau D_p \otimes \sigma_p + D\sigma_p \otimes \tau_p + \tau\sigma_p \otimes D_p]\,\xi, \end{aligned}$$

since $a_p = D\tau_p + \tau D_p$. Hence

$$\xi \cdot f_{pp}\,\xi - \sigma\xi \cdot a_{pp}\,\xi = a\xi \cdot \sigma_{pp}\,\xi \\ +\xi \cdot [D\tau_p \otimes \sigma_p + \tau D_p \otimes \sigma_p + D\sigma_p \otimes \tau_p + \tau\sigma_p \otimes D_p]\,\xi.$$

Inserting this information into (60) we arrive at

$$\begin{aligned}\xi \cdot G_{pp}\,\xi = {} & ((m_1/2) + \omega - \tau\omega_t)|\xi|^2 + \omega_s(\xi \cdot f_{pp}\,\xi - \sigma\xi \cdot a_{pp}\,\xi)/\tau \\ & -\omega_s\xi \cdot [D\tau_p \otimes \sigma_p + \tau D_p \otimes \sigma_p + D\sigma_p \otimes \tau_p + \tau\sigma_p \otimes D_p]\,\xi/\tau \\ & +\omega_t\xi \cdot a_{pp}\,\xi - \omega_t\xi \cdot [\tau_p \otimes D_p + D_p \otimes \tau_p]\,\xi \\ & +\omega_t\xi \cdot [D_p \otimes \tau_p + \tau_p \otimes D_p]\,\xi + \omega_s\xi \cdot [D_p \otimes \sigma_p + \sigma_p \otimes D_p]\,\xi \\ & +D\xi \cdot [\omega_{tt}\tau_p \otimes \tau_p + \omega_{ts}\tau_p \otimes \sigma_p + \omega_{st}\sigma_p \otimes \tau_p + \omega_{ss}\sigma_p \otimes \sigma_p]\,\xi \\[1ex] = {} & ((m_1/2) + \omega - \tau\omega_t)|\xi|^2 + \omega_s\xi \cdot f_{pp}\,\xi/\tau + (\omega_t - (\sigma\omega_s/\tau))\xi \cdot a_{pp}\,\xi \\ & -\omega_s D\xi \cdot [\tau_p \otimes \sigma_p + \sigma_p \otimes \tau_p)\xi/\tau \\ & +D\xi \cdot [\omega_{tt}\tau_p \otimes \tau_p + \omega_{ts}\tau_p \otimes \sigma_p + \omega_{st}\sigma_p \otimes \tau_p + \omega_{ss}\sigma_p \otimes \sigma_p]\,\xi \\[1ex] = {} & ((m_1/2) + \omega - \tau\omega_t)|\xi|^2 + \omega_s\xi \cdot f_{pp}\,\xi/\tau + (\omega_t - (\sigma\omega_s/\tau))\xi \cdot a_{pp}\,\xi \\ & +D[\omega_{tt}(\tau_p \cdot \xi)^2 + 2(\omega_{ts} - (\omega_s/\tau))(\tau_p \cdot \xi)(\sigma_p \cdot \xi) + \omega_{ss}(\sigma_p \cdot \xi)^2].\end{aligned}$$

By Part (ii) of Lemma 2.11 it follows that

$$\begin{aligned}\xi \cdot G_{pp}\,\xi \geq {} & ((m_1/2) + \omega - \tau\omega_t)|\xi|^2 \\ & +\omega_s\Big\{\lambda_F(R_0)\left|P^{\perp}_{p_1\wedge p_2}[(\zeta \wedge p_2) - (\eta \wedge p_1)]\right|^2/a + 2F_z(x, p_1 \wedge p_2) \cdot (\zeta \wedge \eta)\Big\}/\tau \\ & +(\omega_t - (\sigma\omega_s/\tau))\left\{\frac{1}{a}\left|P^{\perp}_{p_1\wedge p_2}[(\zeta \wedge p_2) - (\eta \wedge p_1)]\right|^2 + 2\frac{p_1 \wedge p_2}{|p_1 \wedge p_2|} \cdot (\zeta \wedge \eta)\right\} \\ & +D\begin{bmatrix}\tau_p \cdot \xi \\ \sigma_p \cdot \xi\end{bmatrix} \cdot \begin{pmatrix}\omega_{tt} & \omega_{ts} - (\omega_s/\tau) \\ \omega_{st} - (\omega_s/\tau) & \omega_{ss}\end{pmatrix}\begin{bmatrix}\tau_p \cdot \xi \\ \sigma_p \cdot \xi\end{bmatrix} \\[1ex] = {} & ((m_1/2) + \omega - \tau\omega_t)|\xi|^2 \\ & +(\omega_s\lambda_F(R_0) + \tau\omega_t - \sigma\omega_s)\left|P^{\perp}_{p_1\wedge p_2}[(\zeta \wedge p_2) - (\eta \wedge p_1)]\right|^2/\tau a \\ & +2\left[\omega_s F_z(x, p_1 \wedge p_2) + (\tau\omega_t - \sigma\omega_s)\frac{p_1 \wedge p_2}{|p_1 \wedge p_2|}\right] \cdot (\zeta \wedge \eta)/\tau \\ & +D\begin{bmatrix}\tau_p \cdot \xi \\ \sigma_p \cdot \xi\end{bmatrix} \cdot \begin{pmatrix}\omega_{tt} & \omega_{ts} - (\omega_s/\tau) \\ \omega_{st} - (\omega_s/\tau) & \omega_{ss}\end{pmatrix}\begin{bmatrix}\tau_p \cdot \xi \\ \sigma_p \cdot \xi\end{bmatrix}.\end{aligned}$$

□

Theorem 2.14 *Let $F(x,z) \in C^0(\mathbb{R}^n \times \mathbb{R}^N) \cap C^{k,\nu}(\mathbb{R}^n \times (\mathbb{R}^N - \{0\}))$ for some $k \in \mathbb{N}$, $k \geq 2$, $\nu \in [0,1]$ be a parametric Lagrangian satisfying* (H),(D) *and* (5) *with*

$$\lambda := \inf_{R_0 \in (0,\infty]} \lambda_F(R_0) > 0. \tag{62}$$

Then for any $\epsilon > 0$ there exists a dominance function $G^\epsilon(x,p)$ of F satisfying (D1)–(D4) *with $\mu_1 = (m_1/4)$, $\mu_2 = (m_2/2)$. In addition $G^\epsilon(x,p)$ is of class $C^1(\mathbb{R}^n \times \mathbb{R}^N) \cap C^{k,\nu}(\mathbb{R}^n \times (\mathbb{R}^N - \{0\}))$ and satisfies*

$$\xi \cdot G^\epsilon_{pp}(x,p)\,\xi \geq \frac{2}{\tau}\Big[\omega_s P^\perp_{p_1\wedge p_2}(F_z(x,p_1\wedge p_2)) + (\tau/a)\omega_t(p_1\wedge p_2)\Big]\cdot(\zeta\wedge\eta) \\ +[\min\{\lambda, m_1/2\} - \epsilon]\,|\xi|^2 \tag{63}$$

for all $\xi = (\zeta,\eta) \in \mathbb{R}^{2n}$, $(x,p) \in \mathbb{R}^n \times (\mathbb{R}^{2n} - \Pi)$, and

$$(\bar\zeta \otimes \Lambda)\cdot G^\epsilon_{pp}(x,p)(\bar\zeta\otimes\Lambda) \geq [\min\{\lambda, m_1/2\} - \epsilon]\,|\bar\zeta|^2|\Lambda|^2, \tag{64}$$

for all $\bar\zeta \in \mathbb{R}^n$, $\Lambda \in \mathbb{R}^2$, $(x,p) \in \mathbb{R}^n \times (\mathbb{R}^{2n} - \Pi)$, where ω, τ, σ are defined as in the proof of Theorem 2.8. Moreover, for $(x,p) \in \mathbb{R}^n \times (\Pi - \{0\})$ we have

$$\xi \cdot G^\epsilon_{pp}(x,p)\,\xi = m_1|\xi|^2/2 \ \textit{for all}\ \xi \in \mathbb{R}^{2n}. \tag{65}$$

If $\lambda \geq m_1/2$ then the full statement (including the estimates (63)–(65)*) holds true for $\epsilon = 0$ as well.*

Remark 2.15 *The second inequality (64) is known as the strict rank-one-condition, or strict Legendre-Hadamard-condition, see [7, vol. I, p.231].*

Proof of Theorem 2.14. We will choose a particular cut-off function $\eta = \eta^\epsilon$ which is admissible in the proof of Theorem 2.13 to prove the estimate (63). Then the particular choice $\xi := \bar\zeta \otimes \Lambda$ in (63) leads to (64). Equation (65) follows directly from (59) in Theorem 2.13, since η^ϵ is admissible for the construction in the proof of that theorem. All the other statements have been proven in Theorem 2.8.

Hence it suffices to prove (63). For that we recall (50) and calculate suppressing the argument $(t,s) \in [0,1] \times [m_1, m_2]$

$$\begin{aligned} \omega_{tt} &= (s-m_1)\eta'(u)u_t + \eta''(u)u_t[s-(m_1/2)-(s-m_1)u] \\ &\quad -\eta'(u)(s-m_1)u_t \\ &= \eta''(u)u_t[s-(m_1/2)-(s-m_1)u] \geq \eta''(u)u_t[m_1/2] \geq 0, \end{aligned} \tag{66}$$

by (36), $0 \leq u \leq 1$, and since $\eta''(u) \geq 0$ and $u_t \geq 0$, by (44) and (39). Differentiating ω with respect to s we get by (45)

$$\omega_s(t,s) = t\eta(u(t,s)) + [(m_1/2)+(s-m_1)t]\eta'(u(t,s))u_s(t,s)$$

and therefore we obtain from (43) at $(t,s) \in [0,1] \times [m_1, m_2]$

$$\begin{aligned} \omega_s &= t\eta(u) + [(m_1/2)+(s-m_1)t]\eta'(u)\frac{m_1(t-t^2)}{2[(m_1/2)+(s-m_1)t]^2} \\ &= t\eta(u) + \eta'(u)t(1-u), \end{aligned} \tag{67}$$

because

$$\frac{m_1(t-t^2)}{2[(m_1/2)+(s-m_1)t]}=t(1-u), \tag{68}$$

where we used (38) to deduce

$$1-u=\frac{[(m_1/2)+(s-m_1)t]-(s-(m_1/2))t}{(m_1/2)+(s-m_1)t}=\frac{m_1(1-t)}{2[(m_1/2)+(s-m_1)t]}\,.$$

Differentiating (67) we obtain

$$\begin{aligned}\omega_{ss}&=t\eta'(u)u_s+\eta''(u)u_st(1-u)-\eta'(u)tu_s\\&=\eta''(u)u_st(1-u)\geq 0,\end{aligned} \tag{69}$$

again since $\eta''(u)\geq 0$, $u_s\geq 0$, by (44) and (43) for $0\leq u,t\leq 1$. Observe that by (49) $t=m_1u\{2[s-(m_1/2)+(m_1-s)u]\}^{-1}$, and

$$\begin{aligned}u_s(1-u)u&\underset{(43)}{=}\frac{m_1(t-t^2)}{2[(m_1/2)+(s-m_1)t]^2}(1-u)u\\&\underset{(68)}{=}\frac{t(1-u)^2u}{(m_1/2)+(s-m_1)t}\\&\underset{(38)}{=}\frac{t^2(1-u)^2(s-(m_1/2))}{[(m_1/2)+(s-m_1)t]^2}\\&\underset{(39)}{=}2t^2(1-u)^2u_t/m_1.\end{aligned} \tag{70}$$

Using (70) one obtains from (69)

$$\begin{aligned}\omega_{ss}&=2\eta''(u)t^2(1-u)^2u_tt(m_1u)^{-1}\\&\underset{(49)}{=}\frac{\eta''(u)t^2(1-u)^2u_t}{s-(m_1/2)+(m_1-s)u},\end{aligned}$$

hence

$$\omega_{tt}\omega_{ss}=(\eta''(u))^2u_t^2t^2(1-u)^2.$$

Moreover, from (67) we infer

$$\begin{aligned}\omega_{ts}=\omega_{st}&=\eta(u)+t\eta'(u)u_t+\eta''(u)u_tt(1-u)+\eta'(u)(1-u)-\eta'(u)tu_t\\&=\eta(u)+\eta'(u)(1-u)+t(1-u)\eta''(u)u_t\,.\end{aligned}$$

Consequently, again by (67),

$$\begin{aligned}[\omega_{ts}-(\omega_s/t)]^2&=[\eta(u)+\eta'(u)(1-u)+t(1-u)\eta''(u)u_t\\&\qquad-\eta(u)-\eta'(u)(1-u)]^2\\&=t^2(1-u)^2(\eta''(u))^2u_t^2=\omega_{tt}\omega_{ss}.\end{aligned} \tag{71}$$

According to (66),(69) and (71) evaluated at the point $(t,s):=(\tau,\sigma)\in[0,1]\times[m_1,m_2]$ the matrix

$$\begin{pmatrix} \omega_{tt} & \omega_{ts} - (\omega_s/\tau) \\ \omega_{ts} - (\omega_s/\tau) & \omega_{ss} \end{pmatrix}$$

in (58) is positive semi-definite, hence the last term in (58) is nonnegative.

The third expression on the right-hand side of (58) coincides with the first on the right-hand side of (63), because one can use (4) to write

$$\begin{aligned} &P^{\perp}_{p_1 \wedge p_2}(F_z(x, p_1 \wedge p_2)) \\ &\quad = F_z(x, p_1 \wedge p_2) - (F_z(x, p_1 \wedge p_2) \cdot (p_1 \wedge p_2)) \frac{p_1 \wedge p_2}{|p_1 \wedge p_2|^2} \\ &\quad \underset{(H)}{=} F_z(x, p_1 \wedge p_2) - \frac{F(x, p_1 \wedge p_2)}{|p_1 \wedge p_2|} \frac{p_1 \wedge p_2}{|p_1 \wedge p_2|} \\ &\quad \underset{(35)}{=} F_z(x, p_1 \wedge p_2) - \sigma(x, p) \frac{p_1 \wedge p_2}{|p_1 \wedge p_2|} . \end{aligned}$$

Now we look at the coefficient of the second term on the right-hand side of (58) in Theorem 2.13:

$$\begin{aligned} &\tau \omega_t - (\sigma - \lambda_F(R_0))\omega_s \\ &\underset{(50),(67)}{=} \tau\{(\sigma - m_1)\eta(u) + \eta'(u)[\sigma - (m_1/2) - (\sigma - m_1)u]\} \\ &\qquad -(\sigma - \lambda_F(R_0))[\tau\eta(u) + \eta'(u)\tau(1-u)] \\ &= \tau\{[\sigma - (m_1/2) - (\sigma - m_1)u - (\sigma - \lambda_F(R_0))(1-u)]\eta'(u) \\ &\qquad +[\sigma - m_1 - (\sigma - \lambda_F(R_0))]\eta(u)\} \\ &= \tau\{[\lambda_F(R_0) - (m_1/2) + (m_1 - \lambda_F(R_0))u]\eta'(u) - (m_1 - \lambda_F(R_0))\eta(u)\} \\ &=: \tau\vartheta(u). \end{aligned}$$

By (39) τ is a nonnegative monotonically increasing function of u, and so is ϑ if we choose η appropriately. In fact, we have

$$\begin{aligned} \vartheta_u &= (m_1 - \lambda_F(R_0))\eta'(u) + [\lambda_F(R_0) - (m_1/2) + (m_1 - \lambda_F(R_0))u]\eta''(u) \\ &\quad -(m_1 - \lambda_F(R_0))\eta'(u) = \alpha(u), \end{aligned}$$

where $\alpha(u) := \lambda_F(R_0) - (m_1/2) + (m_1 - \lambda_F(R_0))u$ satisfies $\alpha_u(u) = m_1 - \lambda_F(R_0)$, which means that for $\lambda_F(R_0) \geq m_1$ the function $\alpha(u)$ is decreasing in u, and for $\lambda_F(R_0) \in [0, m_1]$ it is nondecreasing in u. This implies for $u \in [0,1]$

$$\alpha(u) \geq \begin{cases} \alpha(1) = m_1/2 & \text{if } \lambda_F(R_0) \in [m_1, \infty], \\ \alpha(0) = \lambda_F(R_0) - m_1/2 & \text{if } \lambda_F(R_0) \in [0, m_1]. \end{cases}$$

In other words,

$$\vartheta_u(u) \geq 0 \quad \text{for all} \quad u \in [0,1] \quad \text{if} \quad \lambda_F(R_0) \geq m_1/2. \tag{72}$$

If, on the other hand, $\lambda_F(R_0) < m_1/2$, we set

$$u_0 := \frac{(m_1/2) - \lambda}{m_1 - \lambda}, \tag{73}$$

where $\lambda > 0$ is the quantity defined in our assumption (62), and where u_0 is the constant appearing in the definition of the cut-off function η in (44). One easily checks with (62) that $0 < u_0 < 1/2$ so that our previous choice of $u_1 \in (0, 1/2)$ (see (53)) can be modified by taking u_1 sufficiently close to $1/2$ to guarantee the condition $0 < u_0 < u_1$ assumed in (44). Moreover we have $\vartheta_u(u) \geq 0$ for all $u \in [0,1]$ by (73) and (44), since, for $\lambda_F(R_0) \in [\lambda, m_1/2)$ and for

$$0 \leq u < \frac{(m_1/2) - \lambda_F(R_0)}{m_1 - \lambda_F(R_0)} \leq u_0,$$

one has $\eta''(u) = 0$. Hence $0 \leq \vartheta(u) \leq \vartheta(1)$ for all $u \in [0,1]$, $\tau(1) = 1$, and therefore

$$\begin{aligned} 0 &= \tau(0)\vartheta(0) \leq \tau(u)\vartheta(u) \leq \tau(1)\vartheta(1) \\ &= [\lambda_F(R_0) - (m_1/2) + m_1 - \lambda_F(R_0)]\eta'(1) - (m_1 - \lambda_F(R_0))\eta(1) \\ &\underset{(44)}{=} m_1\eta'(1)/2 - m_1 + \lambda_F(R_0), \end{aligned}$$

which proves that the coefficient of the second term on the right-hand side of (58) is nonnegative.

Recall from (44) and (53) that $u_1 \in (u_0, 1/2)$ is the only restriction on the parameter u_1 so far, and if $\lambda \geq m_1/2$, the lower bound u_0 is no restriction at all, since then $u_0 > 0$ can be chosen arbitrarily small.

Finally we are going to investigate the coefficient of the first term in (58),

$$\varphi(t) := (m_1/2) + \omega(t,s) - t\omega_t,$$

regarded as a function of t for fixed $s \in [m_1, m_2]$ with

$$\varphi_t(t) = \omega_t - \omega_t - t\omega_{tt} \leq 0,$$

by (66). That means, for $\tau \in [0,1]$, $\sigma \in [m_1, m_2]$,

$$\begin{aligned} \varphi(0) &\geq \varphi(\tau) \geq \varphi(1) \underset{(45),(41),(44)}{=} (m_1/2) + \sigma - (m_1/2) - \omega_t(1,\sigma) \\ &\underset{(41),(52)}{=} \sigma - (\sigma - m_1) - \eta'(1)m_1/2 \\ &= m_1 - \eta'(1)m_1/2 = m_1(1 - (\eta'(1)/2)). \end{aligned}$$

Without loss of generality we may assume that

$$0 < \epsilon < \min\{\lambda, m_1/2\}. \tag{74}$$

Recall (44), in particular

$$\frac{1}{1-u_0} < \eta'(1) \leq \frac{1}{1-u_1}. \tag{75}$$

If $\lambda \geq m_1/2$ the parameter $u_0 > 0$ may be chosen arbitrarily small as mentioned above, hence we can choose

$$u_0^\epsilon := \frac{\epsilon}{m_1 + 2\epsilon} \quad \text{and} \quad u_1^\epsilon := \frac{2\epsilon}{m_1 + 2\epsilon} \underset{(74)}{\in} (u_0, 1/2),$$

which, according to (75), leads to

$$\eta'(1) \leq \frac{m_1 + 2\epsilon}{m_1}, \tag{76}$$

and thus,

$$\begin{aligned}(m_1/2) + \omega - \tau\omega_t = \varphi(\tau) \geq \varphi(1) &= m_1(1 - \eta'(1)/2) \\ &\geq m_1\left(1 - \frac{1}{2(1-u_1^\epsilon)}\right) \\ &= m_1\left(1 - \frac{m_1 + 2\epsilon}{2m_1}\right) = \frac{m_1}{2} - \epsilon > 0. \end{aligned} \tag{77}$$

If $\lambda < m_1/2$ we choose

$$u_1^\epsilon := \frac{(m_1/2) - \lambda + \epsilon}{m_1 - \lambda + \epsilon} \in (u_0, 1/2) \text{ for } \epsilon > 0,$$

where now u_0 is given by (73), so that by (75)

$$\eta'(1) \leq 2m_1^{-1}(m_1 - \lambda + \epsilon). \tag{78}$$

Thus we get

$$(m_1/2) + \omega - \tau\omega_t \geq \lambda - \epsilon > 0. \tag{79}$$

With these choices the cut-off function $\eta = \eta^\epsilon$ depends on $\epsilon > 0$, and hence also $G = G^\epsilon$, and the construction is finished. The estimates (77) and (79) show that

$$(m_1/2) + \omega - \tau\omega_t \geq \min\{\lambda, m_1/2\} - \epsilon > 0,$$

which concludes the proof of Theorem 2.14. □

As a consequence we obtain

Corollary 2.16 *The dominance function $G^\epsilon(x,p)$ of Theorem 2.14 satisfies*

$$\xi \cdot G^\epsilon_{pp}(x,p)\,\xi \geq 2\left[\min\{\lambda, m_1/2\} - m_2 - \epsilon\right]|\xi|^2$$

for all $\xi \in \mathbb{R}^{2n}$, $(x,p) \in \mathbb{R}^n \times (\mathbb{R}^{2n} - \{0\})$.

Proof. Recalling (49)–(52) we find

$$0 \underset{(50)}{=} \omega_t(t(0,s),s) \underset{(51)}{\leq} \omega_t(t(r,s),s) \underset{(51)}{\leq} \omega_t(t(1,s),s) \underset{(52)}{=} s-m_1+(m_1/2)\eta'(1),$$

which implies by (76) in the case $\lambda \geq m_1/2$, and by (78) in the case $\lambda < m_1/2$, and by (36),

$$|\omega_t| \leq m_2 + \epsilon - \min\{\lambda, m_1/2\}.$$

Consequently,

$$|2a^{-1}\omega_t(p_1 \wedge p_2) \cdot (\zeta \wedge \eta)| \leq [m_2 + \epsilon - \min\{\lambda, m_1/2\}]\,|\xi|^2. \tag{80}$$

On the other hand, by (55) of Lemma 2.11, we have $|P^{\perp}_{p_1 \wedge p_2}(F_z(x, p_1 \wedge p_2))| \leq m_2$, and by (67), $\omega_s/\tau = \eta(u) + \eta'(u)(1-u)$, hence $(\omega_s/\tau)_u = \eta''(u)(1-u) \geq 0$ according to (44). This implies

$$0 = \eta(0) + \eta'(0) = (\omega_s/\tau)|_{u=0} \leq (\omega_s/\tau) \leq (\omega_s/\tau)|_{u=1} = \eta(1) = 1.$$

Therefore,

$$\left|\frac{2\omega_s}{\tau} P^{\perp}_{p_1 \wedge p_2}(F_z(x, p_1 \wedge p_2))(\zeta \wedge \eta)\right| \leq m_2|\xi|^2,$$

which together with (80) and (63) implies our assertion. □

Proof of Theorem 1.3. Take $\epsilon = \epsilon(k) := (k - k_0)/4$ and the dominance function $G^{*\epsilon}(x,p)$ for $F := F^*$ constructed in Theorem 2.14. Then according to Corollary 2.16

$$\begin{aligned} \xi \cdot G^{*\epsilon}_{pp}(x,p)\,\xi &\geq 2[\min\{\lambda^*, m_1^*/2\} - m_2^* - \epsilon]\,|\xi|^2 \\ &= -[k_0 + 2\epsilon]\,|\xi|^2 \end{aligned} \tag{81}$$

for all $\xi \in \mathbb{R}^{2n}$, $(x,p) \in \mathbb{R}^n \times (\mathbb{R}^{2n} - \{0\})$. Set $G(x,p) := kD(p) + G^{*\epsilon}(x,p)$. Then G is of class $C^2(\mathbb{R}^n \times (\mathbb{R}^{2n} - \{0\}))$ and satisfies (D1)–(D4). Furthermore, for $p \neq 0$ we have by (81)

$$\begin{aligned} \xi \cdot G_{pp}(x,p)\,\xi &= k|\xi|^2 + \xi \cdot G^{*\epsilon}_{pp}(x,p)\xi \\ &\geq [k - k_0 - 2\epsilon]|\xi|^2 = 2^{-1}(k - k_0)|\xi|^2, \end{aligned}$$

so that also the ellipticity condition (E) holds for $G(x,p)$ with a uniform constant $\lambda_G(R_0) := (k - k_0)/2$ for all $R_0 > 0$. □

References

1. F.J. Almgren, R. Schoen, L. Simon: Regularity and singularity estimates on hypersurfaces minimizing parametric elliptic variational integrals. Acta Math., **139**, (1977), 217–265

2. U. Clarenz, G. Dziuk, M. Rumpf: On generalized mean curvature flow. To appear in this volume
3. U. Clarenz, H. von der Mosel; On surfaces of prescribed F-mean curvature. To appear
4. R. Courant: Dirichlet's Principle, Conformal Mapping, and Minimal Surfaces. Interscience Publishers, New York 1950
5. U. Dierkes, S. Hildebrandt, A. Küster, O. Wohlrab: Minimal Surfaces, vols I & II. Grundlehren der math. Wissenschaften, **295** & **296**, Springer, Berlin 1992
6. F. Duzaar, K. Steffen: The Plateau problem for parametric surfaces with prescribed mean curvature. In: J. Jost (ed) Geometric analysis and the calculus of variations. For Stefan Hildebrandt. 13–70, International Press, Cambridge 1996
7. M. Giaquinta, S. Hildebrandt: Calculus of Variations, vols I & II. Grundlehren der math. Wissenschaften, **310** & **311**, Springer, Berlin Heidelberg New York 1996
8. R. Hardt: On boundary regularity for integral currents or flat chains modulo two minimizing the integral of an elliptic integrand. Comm. P.D.E., **2**, (1977), 1163–1232
9. E. Heinz: Über die Existenz einer Fläche konstanter mittlerer Krümmung bei vorgegebener Berandung. Math. Ann., **127**, (1954), 258–287
10. E. Heinz: On surfaces of constant mean curvature with polygonal boundaries. Arch. Rat. Mech., **36**, (1970), 335–347
11. E. Heinz: Unstable surfaces of constant mean curvature. Arch. Rat. Mech., **38**, (1970), 257–267
12. S. Hildebrandt: Randwertprobleme für Flächen mit vorgeschriebener mittlerer Krümmung und Anwendungen auf die Kapillaritätstheorie. Math. Z., **112**, (1969), 205–213
13. S. Hildebrandt: On the Plateau problem for surfaces of constant mean curvature. Comm. Pure Appl. Math., **23**, (1970), 97–114
14. S. Hildebrandt, H. von der Mosel: On two-dimensional parametric variational problems. Calc. Var., **9**, (1999), 249–267
15. S. Hildebrandt, H. von der Mosel: Plateau's problem for parametric double integrals: Part I. Existence and regularity in the interior. Preprint, **745**, (2001) SFB 256, University of Bonn, Preprint, **88**, (2001) MPI Mathematics Leipzig. To appear in Comm. Pure Appl. Math.
16. S. Hildebrandt, H. von der Mosel: The partially free boundary problem for parametric double integrals. Preprint, **1**, (2002) SFB 611 University of Bonn, To appear in Nonlinear Problems in Mathematical Physics and Related Topics I, International Mathematical Series Vol. I, Kluwer/Plenum, London 2002
17. S. Hildebrandt, H. von der Mosel: Plateau's problem for parametric double integrals, II. Boundary regularity. To appear
18. M. Kurzke: Geometrische Variationsprobleme auf mehrfach zusammenhängenden ebenen Gebieten. Bonner Mathematische Schriften, **341**, (2001)
19. M. Kurzke, H. von der Mosel: The Douglas problem for parametric double integrals. To appear
20. L.J. Lipkin: A free boundary problem for parametric integrals of the calculus of variations. Rend. Circ. Mat. Palermo (2), **17**, (1968), 33–67
21. C.B. Morrey: The problem of Plateau on a Riemannian manifold. Ann. Math., **49**, (1948), 807–851
22. C.B. Morrey: The parametric variational problem for double integrals. Comm. Pure Appl. Math., **14**, (1961), 569–575
23. C.B. Morrey: Multiple integrals in the calculus of variations. Grundlehren der math. Wissenschaften, **130**, Springer, Berlin 1966

24. Y.G. Reshetnyak: New proof of the theorem on existence of an absolute minimum for two-dimensional variational problems in parametric form (in Russian). Sibirsk. Matemat. Zhurnal, **3**, (1962), 744–768
25. R. Schoen, L. Simon: A new proof of the regularity theorem for rectifiable currents which minimize parametric elliptic functionals. Ind. Univ. Math. J., **31**, (1982), 415–434
26. M. Shiffman: Instability for double integral problems in the calculus of variations. Ann. Math., (2) **45**, (1944), 543–576
27. K. Steffen: Parametric surfaces of prescribed mean curvature. In: Calculus of variations and geometric evolution problems (Cetraro 1996), 211–265, Lecture Notes in Math., **1713**, Springer, Berlin 1999
28. G. Strömer: Instabile Lösungen der Eulerschen Gleichungen gewisser Variationsprobleme. Arch. Rat. Mech. Anal., **79**, (1982), 219–239
29. J.E. Taylor: Existence and structure of solutions to a class of nonelliptic variational problems. In: Symposia Mathematica, Vol. XIV (Convegno di Teoria Geometrica dell'Integrazione e Varietà Minimali, INDAM, Roma, Maggio 1973), 499–508, Academic Press, London, 1974
30. J.E. Taylor: Crystalline variational problems. Bull. Amer. Math. Soc., **84**, (1978), 568–588
31. J.E. Taylor: Nonexistence of F-minimizing embedded disks. Pac. J. Math., **88**, (1980), 279–283
32. J.E. Taylor: Constructing crystalline minimal surfaces. In: E. Bombieri (ed) Seminar on minimal submanifolds, 275–292, Ann. of Math. Stud., **103**, Princeton Univ. Press, Princeton, NJ, 1983
33. J.E. Taylor: Complete catalog of minimizing embedded crystalline cones. In: Geometric measure theory and the calculus of variations (Arcata, Calif., 1984), 379–403, Proc. Symp. Pure Math., **44**, Amer. Math. Soc., Providence, RI, 1986
34. J.E. Taylor, F.J. Almgren: Optimal crystal shapes. In: Variational methods for free surface interfaces (Menlo Park, Calif., 1985), 1–11, Springer, New York, 1987
35. J.E. Taylor: On the global structure of crystalline surfaces. Discrete Comput. Geom., **6**, (1991), 225–262
36. H. Wente: An existence theorem for surfaces of constant mean curvature. J. Math. Anal. Appl., **26**, (1969), 318–344
37. B. White: Existence and regularity of smooth embedded surfaces of prescribed genus that minimize parametric even elliptic functionals on 3-manifolds. J. Diff. Geom., **33**, (1991), 413–443

Convex Variational Problems with Linear Growth

Michael Bildhauer[1] and Martin Fuchs[2]

[1] F.R. 6.1 Mathematik, Universität des Saarlandes, Postfach 151150, D-66041 Saarbrücken
bibi@math.uni-sb.de
[2] F.R. 6.1 Mathematik, Universität des Saarlandes, Postfach 151150, D-66041 Saarbrücken
fuchs@math.uni-sb.de

1 Introduction

For a bounded Lipschitz domain $\Omega \subset \mathbb{R}^n$, $n \geq 2$, and a function $u_0 \in W_1^1(\Omega;\mathbb{R}^N)$ we consider the variational problem

$$J[w] = \int_\Omega f(\nabla w)\,\mathrm{d}x \to \min \quad \text{in } u_0 + \overset{\circ}{W}{}_1^1(\Omega;\mathbb{R}^N)\,, \tag{1}$$

where $f\colon \mathbb{R}^{nN} \to [0,\infty)$ is a strictly convex integrand of linear growth, i.e.

$$a|Z| - b \leq f(Z) \leq A|Z| + B \quad \text{for all } Z \in \mathbb{R}^{nN} \tag{2}$$

holds with suitable constants a, $A > 0$, b, $B \in \mathbb{R}$. Clearly (1) fails to have solutions in general, therefore we introduce the set

$$\mathcal{M} = \Big\{ u \in \mathrm{BV}(\Omega;\mathbb{R}^N) : u \text{ is the } L^1\text{-limit of some } J\text{-minimizing sequence } \{u_k\} \subset u_0 + \overset{\circ}{W}{}_1^1(\Omega;\mathbb{R}^N)\Big\}$$

of generalized minimizers of problem (1). It is well known that the elements u of $\mathcal{M}$ naturally occur as minimizers of suitable relaxed versions of problem (1), precisely (see [11] or [7]) we have: let

$$K[w] = \int_\Omega f(\nabla^a w)\,\mathrm{d}x + \int_\Omega f_\infty\Big(\frac{\nabla^s w}{|\nabla^s w|}\Big)\,\mathrm{d}|\nabla^s w| + \int_{\partial\Omega} f_\infty\big((u_0 - w)\otimes\nu\big)\,\mathrm{d}\mathcal{H}^{n-1}\,, \quad w \in \mathrm{BV}(\Omega;\mathbb{R}^N)\,,$$

where ν is the outward unit normal to $\partial\Omega$, f_∞ is the recession function of f, and $\nabla^a w$ and $\nabla^s w$ denote the regular and singular part of ∇w w.r.t. the Lebesgue measure, respectively. Then

- $K[w] \to \min$ in $BV(\Omega;\mathbb{R}^N)$ has at least one solution;
- $$\inf_{u_0 + \overset{\circ}{W}{}_1^1(\Omega;\mathbb{R}^N)} J = \inf_{\mathrm{BV}(\Omega;\mathbb{R}^N)} K;$$

- u is K-minimizing $\Leftrightarrow u \in \mathcal{M}$.

It should be noted that there exists a formally different approach to relaxation based on the notion of a suitable Lagrangian (see [20] and [17]): let

$$L(w,\kappa) = \int_\Omega \operatorname{div} \kappa \cdot (u_0 - w)\,\mathrm{d}x - \int_\Omega f^*(\kappa)\,\mathrm{d}x + \int_\Omega \kappa : \nabla u_0\,\mathrm{d}x\,,$$
$$w \in \mathrm{BV}(\Omega;\mathbb{R}^N),\quad \kappa \in \mathcal{U} := \left\{\sigma \in L^\infty(\Omega;\mathbb{R}^{nN}) :\ \operatorname{div}\sigma \in L^n(\Omega;\mathbb{R}^N)\right\}\,,$$

where f^* is the conjugate function of f. We introduce

$$\tilde{J}[w] = \sup_{\sigma\in\mathcal{U}} L(w,\sigma)$$

as relaxation of J to the space $BV(\Omega;\mathbb{R}^N)$, but in [7] we showed that K and $\tilde{J}$ coincide on $\mathrm{BV}(\Omega;\mathbb{R}^N)$.

In our note we like to investigate the regularity properties of generalized minimizers $u \in \mathcal{M}$. To this purpose we assume in addition to (2) that f is of class $C^2(\mathbb{R}^{nN})$. Then, according to [2], we know $\nabla u \in C^{0,\alpha}(\Omega_0;\mathbb{R}^{nN})$ for any $0 < \alpha < 1$, where Ω_0 is an open subset of Ω with full Lebesgue measure, provided we have $D^2 f(Z)(Y,Y) > 0$ for any $Z, Y \in \mathbb{R}^{nN}$, $Y \neq 0$. Another regularity result concerning the scalar case $N = 1$ is established in [10]: generalized minimizers $u \in \mathcal{M}$ are smooth in the interior of Ω if f satisfies a minimal surface type ellipticity condition. In order to get rid of this quite restrictive assumption we introduce the class of μ-elliptic, linear growth integrands.

Definition 1.1 *Let $f \in C^2(\mathbb{R}^{nN})$ satisfy (2) together with*

$$|\nabla f(Z)| \le M < \infty \quad \textit{for any} \quad Z \in \mathbb{R}^{nN}\,. \tag{3}$$

Then, we say that f is μ-elliptic for some number $\mu > 1$ if and only if

$$\lambda(1+|Z|^2)^{-\frac{\mu}{2}}|Y|^2 \le D^2 f(Z)(Y,Y) \le \Lambda(1+|Z|^2)^{-\frac{1}{2}}|Y|^2 \tag{4}$$

holds with positive constants λ, Λ and for all Y, $Z \in \mathbb{R}^{nN}$.

Remark 1.2 *It is easy to show that (3) and (4) imply (2).*

In Sect. 2 we will give a list of examples satisfying (4) for any given number $\mu > 1$.

From now on we restrict ourselves to the scalar case $N = 1$, the reader will find further comments concerning vectorial problems in [3].

Let us first recall some recent results on the regularity of generalized minimizers.

Theorem 1.3 *([6]) Consider an integrand f as in Definition 1.1 and assume $\mu < 1 + 2/n$. Then we have:*

i) $\mathcal{M} \subset C^{1,\alpha}(\Omega)$ for any $0 < \alpha < 1$.
ii) If u, $v \in \mathcal{M}$, then there is a real number c such that $u = v + c$.

So the strong assumption $\mu < 1 + 2/n$ implies regularity together with uniqueness up to a constant. Unfortunately the condition on μ becomes more and more restrictive if $n \to \infty$. But we can compensate this effect by considering good boundary data.

Theorem 1.4 *([4]) Suppose that $u_0 \in L^\infty(\Omega)$ and let the assumptions of Theorem 1.3 hold for some $\mu < 3$. Then i) and ii) of Theorem 1.3 continue to hold.*

The next result concerns the limit case $\mu = 3$:

Theorem 1.5 *([4]) Suppose that we are in the situation of Theorem 1.4 with $\mu = 3$. Then there is a generalized minimizer $u^* \in \mathcal{M}$ such that*

i) $\nabla^s u^* = 0$.

ii) For any $\Omega' \Subset \Omega$ we have

$$\int_{\Omega'} |\nabla u^*| \ln^2(1 + |\nabla u^*|^2)\, \mathrm{d}x < \infty \,.$$

iii) u^* is (up to a constant) the unique solution of the problem

$$\int_\Omega f(\nabla w)\, \mathrm{d}x + \int_{\partial\Omega} f_\infty\big((u_0 - w)\nu\big)\, \mathrm{d}\mathcal{H}^{n-1} \to \min \quad in \quad W_1^1(\Omega)\,.$$

Note that from iii) of Theorem 1.5 we deduce that the relaxed problem

$$K[w] \to \min \quad \text{in} \quad \mathrm{BV}(\Omega)$$

admits a solution in $W_1^1(\Omega)$.

The main concern of our paper is to emphasize the role of the limit exponent $\mu = 3$. To this purpose we first extend Theorem 1.5 to integrands with an additional smooth x-dependence (Sect. 3), in a second step we construct an example which shows that we can not go beyond the limit case $\mu = 3$ in Theorem 3.8 below. This strongly indicates that Theorem 1.5 is sharp in the sense that also in the x-independent situation studied in Theorem 1.5 there exist generalized minimizers which are not of class W_1^1 if we replace the exponent $\mu = 3$ by some slightly larger number.

2 Examples of μ-Elliptic Integrands

i) Let us start with the most prominent example: the minimal surface integrand $f(Z) = \sqrt{1 + |Z|^2}$ satisfies Definition 1.1 with the limit exponent $\mu = 3$. However, there is much better information on account of the geometric structure of this example, in particular we have

$$\frac{c_1}{\sqrt{1+|Z|^2}}\left[|Y|^2 - \frac{(Y \cdot Z)^2}{1+|Z|^2}\right] \le D^2 f(Z)(Y,Y)$$
$$\le \frac{c_2}{\sqrt{1+|Z|^2}}\left[|Y|^2 - \frac{(Y \cdot Z)^2}{1+|Z|^2}\right] \qquad (5)$$

for all $Z, Y \in \mathbb{R}^n$ with some real numbers c_1, c_2.

Given an integrand satisfying this condition, Ladyzhenskaya/Ural'tseva ([13]) and Giaquinta/Modica/Souček ([10]) then use Sobolev's inequality for functions defined on minimal hypersurfaces (compare [15] and [9]) as an essential tool for proving their regularity results.

ii) We fix $\mu > 1$, let

$$\varphi(r) = \int_0^r \int_0^s (1+t^2)^{-\frac{\mu}{2}} \, dt \, ds \, , \quad r \in \mathbb{R}_0^+ \, ,$$

and consider $\Phi_\mu(Z) = \varphi(|Z|)$. By direct calculations it is easy to see that Φ_μ is an μ-elliptic integrand of linear growth in the sense of Definition 1.1, in particular (4) is satisfied with the optimal exponent μ.

If we choose $\mu = 1$, then we obtain an integrand of nearly linear growth which, at least for large $|Z|$, behaves like $|Z| \ln(1 + |Z|)$.

If $\mu = 2$, then the explicit representation reads as

$$\Phi_2(Z) = |Z| \arctan |Z| - \frac{1}{2} \ln(1 + |Z|^2) \, .$$

In the limit case $\mu = 3$, it is easy to perform the integrations and the minimal surface example is exactly recovered.

iii) On account of the above observation we need to give some examples with limit ellipticity $\mu = 3$, with linear growth and which are not of minimal surface structure in the sense of (5).

A somehow technical example can be constructed by considering the function Φ_μ for some fixed $1 < \mu < 3$, where we now "destroy" ellipticity by inserting "linear pieces" (compare the construction of integrands with (s, μ, q)-growth given in [8]). Then, on one hand, the inequality on the right-hand side of (5) is no longer valid. On the other hand, we have degenerate ellipticity and by adding some minimal surface part we get an "$\mu = 3$"-elliptic energy density which is not of minimal surface type.

We did this piecewise construction for the following reason: the Ansatz $f(Z) = g(|Z|)$ automatically leads (more or less) to the minimal surface structure. In particular, if $Z \perp Y \in \mathbb{R}^n$ and if f is of linear growth, hence $g'(t) \to c$ as $t \to +\infty$, then

$$D^2 f(Z)(Y,Y) = \frac{g'(|Z|)}{|Z|} |Y|^2 \approx (1 + |Z|^2)^{-\frac{1}{2}} |Y|^2$$

if $|Z|$ is sufficiently large.

Nevertheless, there exists a different natural class of examples where the above structure is lost: the idea is to replace $|Z|$ by the distance to a convex set by the way obtaining a variety of interesting energy densities. Let us just sketch a very easy example in the case $n = 2$, $Z = (z_1, z_2)$. Denote by C the upper unit half disc, i.e. $C = \{Z : |Z| < 1, \, z_2 > 0\}$ (for the sake of simplicity we neglect a smoothing procedure at the edges). Note that the distance function $\rho(Z) := \mathrm{dist}(Z, C)$ coincides (up to the constant 1) in the upper half plane (for $|Z| > 1$) with $|Z|$. Now let

$$f(Z) = \sqrt{1 + \rho^2(Z)} \, .$$

We are mainly interested in the points $Z = (0, z_2), z_2 < 0, |z_2| \gg 1$: it is immediately verified that in this case

$$D^2 f\big((0,z_2)\big)(e_1,e_1) = 0 ,$$
$$D^2 f\big((0,z_2)\big)(e_2,e_2) = \big(1 + |\rho^2\big((0,z_2)\big)|\big)^{-\frac{3}{2}} = (1 + |Z|^2)^{-\frac{3}{2}} ,$$

where $e_i, i = 1, 2$, denotes the ith unit coordinate vector. In particular we observe that the minimal surface structure is completely destroyed on account of the degeneracy of C. This of course induces degeneracy of f as well. The first way of obtaining μ-elliptic integrands with linear growth evidently is to change the geometry in a suitable way. We prefer a simple and more anisotropic idea: let (for $|Z| > 1$)

$$\tau\big((z_1,z_2)\big) = \frac{1}{2} z_1 + \rho\big((z_1,z_2)\big) ,$$
$$\tilde{f}\big((z_1,z_2)\big) = \sqrt{1 + \tau^2\big((z_1,z_2)\big)} .$$

Then there is a positive constant c such that

$$c^{-1}|Z| \le \tau\big((z_1,z_2)\big) \le c|Z|$$

for all $|Z|$ sufficiently large, and if $z_2 < 0, |z_2| \gg 1$, then we obtain in both coordinate directions e_i, $i = 1, 2$, with suitable constants c_i

$$D^2 \tilde{f}\big((0,z_2)\big)(e_i,e_i) = c_i(1 + |\tau\big((0,z_2)\big)|^2)^{-\frac{3}{2}} .$$

Summarizing the properties of $\tilde{f}$ we see that this function is of linear growth and satisfies the μ-ellipticity condition with limit exponent $\mu = 3$. Moreover, $\tilde{f}$ does not satisfy the minimal surface ellipticity condition (5), and there is no chance to get something analogous: given the points $(0, z_2)$ as above, both eigenvalues of $D^2\tilde{f}\big((0,z_2)\big)$ grow like $(1 + |Z|^2)^{-3/2}$.

3 Smooth x-Dependence

Now we are going to prove that Theorem 1.5 remains valid if we admit an additional smooth x-dependence of the energy density f. This will enable us in the next section to discuss the sharpness of our results. To be precise, let us suppose that we now have

Assumption 3.6. *There are constants $c_1, \ldots, c_7$ such that for all $x \in \overline{\Omega}$, for all P, U, $V \in \mathbb{R}^n$ and for any $\gamma = 1, \ldots, n$*

i) the variational integrand $f = f(x,P)$ is of linear growth in P, uniformly w.r.t. x, i.e.

$$a|P| - b \le f(x,P) \le A|P| + B$$

holds with constants which are not depending on x;

ii) $f(x,P)$ *is of class* $C^2(\overline{\Omega}\times\mathbb{R}^n)$ *and any of the derivatives occurring below exist;*

iii) $\left|\nabla_P f(x,P)\right| \le c_1$*;*

iv) $c_2(1+|P|^2)^{-\frac{3}{2}}|U|^2 \le D_P^2 f(x,P)(U,U) \le c_3(1+|P|^2)^{-\frac{1}{2}}|U|^2$*;*

v) $\left|\partial_\gamma \nabla_P f(x,P)\right| \le c_4$*;*

vi) $\left|\partial_\gamma \partial_\gamma \nabla_P f(x,P)\right| \le c_5$*;*

vii) $\left|\partial_\gamma D_P^2 f(x,P)(U,V)\right| \le c_6\left|D_P^2 f(x,P)(U,V)\right| + \dfrac{c_7}{1+|P|^2}|U||V|$.

Remark 3.7 *Maybe, assumption 3.6vii) needs some brief comment: if we want to include integrands of the type* $f(x,P) = g(\alpha(x)P)$ *with some scalar function* α *in our considerations, then we can not expect that* $\partial_\gamma D_P^2 f$ *and* $D_P^2 f$ *define equivalent bilinear forms on* $\mathbb{R}^n$*. However, the admissible perturbation on the right-hand side of vii) in particular is weak enough to be verified for the counterexample of the next section.*

The x-dependent variant of Theorem 1.5 then reads as

Theorem 3.8 *Theorem 1.5 remains valid also for energy densities satisfying Assumption 3.6.*

Proof. We follow the lines of [4] (see also [3]) and start by letting

$$J_\delta[w] := \frac{\delta}{2}\int_\Omega |\nabla w|^2\,\mathrm{d}x + \int_\Omega f(x,\nabla w)\,\mathrm{d}x\ , \quad w \in u_0 + \overset{\circ}{W}{}_2^1(\Omega)\ , \quad \delta \in (0,1)\ .$$

Here and in the following we may assume in addition that $u_0 \in L^\infty \cap W_2^1(\Omega)$, the well known approximation procedure needed to handle the case $u_0 \in L^\infty \cap W_1^1(\Omega)$ is outlined, for instance, in [5]. Next, let u_δ denote the unique solution of the variational problem

$$J_\delta[w] \to \min \quad \text{in } u_0 + \overset{\circ}{W}{}_2^1(\Omega)$$

and abbreviate $f_\delta(x,P) = \frac{\delta}{2}|P|^2 + f(x,P)$. Then the main properties of the regularization $\{u_\delta\}$ are summarized in

Lemma 3.9 *i) The regularizing sequence* $\{u_\delta\}$ *is a* J*-minimizing sequence from* $u_0 + \overset{\circ}{W}{}_1^1(\Omega)$*;*

ii) there is a real number c, independent of δ*, such that*

$$\delta\int_\Omega |\nabla u_\delta|^2\,\mathrm{d}x \le c\ , \qquad \int_\Omega |\nabla u_\delta|\,\mathrm{d}x \le c\ ;$$

iii) u_δ *is of class* $W^2_{2,loc} \cap W^1_{\infty,loc}(\Omega)$*;*

iv)

$$\int_\Omega \nabla_P f_\delta(x, \nabla u_\delta) \cdot \nabla\varphi \, dx = 0 \quad \textit{for all} \;\; \varphi \in C_0^\infty(\Omega) \; ;$$

v) for all $\varphi \in C_0^\infty(\Omega)$*,* $\gamma = 1, \dots, n$*, we also have*

$$\int_\Omega D_P^2 f_\delta(x, \nabla u_\delta)(\partial_\gamma \nabla u_\delta, \nabla\varphi) \, dx + \int_\Omega (\partial_\gamma \nabla_P f_\delta)(x, \nabla u_\delta) \cdot \nabla\varphi \, dx = 0 \, ,$$

where iii)–v) are valid for any $\delta \in (0,1)$*.*

Proof. For i) we observe that the additional x-dependence does not affect the corresponding arguments of [5] (compare [18] for the case of integrands depending on the modulus of the gradient). Alternatively, we can follow the reasoning outlined in [10] which is based on Reshetnyak's lower semicontinuity theorem (see [16]). Claim ii) is immediate by $J_\delta[u_\delta] \le J_\delta[u_0] \le J_1[u_0]$ and the linear growth of f, iv) is the Euler equation for u_δ which implies iii) by Theorem 5.2, Chapter 4, of [12]. With the higher integrability and differentiability given in iii), we finally may differentiate iv) to obtain v). □

Next we state the main

Lemma 3.10 *Suppose that the hypotheses of Theorem 3.8 are valid and let* $\{u_\delta\}$ *be given as above. Then, for any domain* $\Omega' \Subset \Omega$*, there is a real number* $c(\Omega')$ *– independent of* δ *– such that*

$$\int_{\Omega'} |\nabla u_\delta| \ln^2(1 + |\nabla u_\delta|^2) \, dx \le c(\Omega') < \infty \, .$$

Proof. Let us abbreviate $\Gamma_\delta = 1 + |\nabla u_\delta|^2$, $\omega_\delta = \ln(\Gamma_\delta)$ and fix some ball $B_{2r}(x_0) \Subset \Omega$. Given $\eta \in C_0^\infty(B_{2r}(x_0))$, $0 \le \eta \le 1$, $\eta \equiv 1$ on $B_r(x_0)$, we conclude from Lemma 3.9, iii), and from standard density arguments that $\varphi = u_\delta \omega_\delta^2 \eta^2$ is an admissible choice in Lemma 3.9, iv), hence

$$\begin{aligned}
&\int_{B_{2r}(x_0)} \nabla_P f(x, \nabla u_\delta) \cdot \nabla u_\delta \omega_\delta^2 \eta^2 \, dx + \delta \int_{B_{2r}(x_0)} |\nabla u_\delta|^2 \omega_\delta^2 \eta^2 \, dx \\
&= - \int_{B_{2r}(x_0)} \nabla_P f(x, \nabla u_\delta) \cdot u_\delta \big[\nabla \omega_\delta^2 \eta^2 + \nabla \eta^2 \omega_\delta^2\big] \, dx \\
&\quad - \delta \int_{B_{2r}(x_0)} \nabla u_\delta \cdot u_\delta \big[\nabla \omega_\delta^2 \eta^2 + \nabla \eta^2 \omega_\delta^2\big] \, dx \, . \qquad (6)
\end{aligned}$$

Note that the equations iv) and v) of Lemma 3.9 remain unchanged if we replace $f(x, P)$ by $f_{x_0} := f(x, P) - \nabla_P f(x_0, 0) \cdot P$. Moreover, without changing the constants c_2–c_7, the inequalities iv)–vii) of Assumption 3.6 are valid with f replaced by

f_{x_0}. Finally, we find a positive number c_1 such that iii) of Assumption 3.6 holds uniformly w.r.t. x_0 for any f_{x_0} as above. As a consequence, we may assume w.l.o.g. that $\nabla_P f(x_0, 0) = 0$. This implies by Assumption 3.6, iv),

$$\nabla_P f(x_0, P) \cdot P = \int_0^1 \frac{\mathrm{d}}{\mathrm{d}\theta} \nabla_P f(x_0, \theta P) \cdot P \,\mathrm{d}\theta \geq c_2 |P| \int_0^{|P|} (1 + \rho^2)^{-\frac{3}{2}} \,\mathrm{d}\rho \,,$$

thus, recalling Assumption 3.6, v), we may choose r sufficiently small such that for all $x \in B_{2r}(x_0)$

$$\nabla_P f(x, P) \cdot P \geq c'(1 + |P|^2)^{\frac{1}{2}} - c'' \tag{7}$$

holds for some real numbers c', c'' which are not depending on x. Inequality (7) implies that the left-hand side of (6) is greater than or equal to

$$\int_{B_{2r}(x_0)} \left[c' \Gamma_\delta^{\frac{1}{2}} \omega_\delta^2 \eta^2 - c'' \omega_\delta^2 \eta^2 \right] \mathrm{d}x + \delta \int_{B_{2r}(x_0)} |\nabla u_\delta|^2 \omega_\delta^2 \eta^2 \,\mathrm{d}x \,. \tag{8}$$

Since $|\nabla_P f|$ and $|u_\delta|$ are bounded, we find the following estimate for the right-hand side of (6) (using Young's inequality with $\varepsilon > 0$ fixed)

$$\begin{aligned} \text{r.h.s} \leq\ & c \int_{B_{2r}(x_0)} \eta^2 \left[\varepsilon \Gamma_\delta^{\frac{1}{2}} \omega_\delta^2 + \varepsilon^{-1} \Gamma_\delta^{-\frac{1}{2}} |\nabla \omega_\delta|^2 \right] \mathrm{d}x \\ & + c(r) \int_{B_{2r}(x_0)} \omega_\delta^2 \,\mathrm{d}x \\ & + c\delta \int_{B_{2r}(x_0)} \eta^2 \left[\varepsilon |\nabla u_\delta|^2 \omega_\delta^2 + \varepsilon^{-1} |\nabla \omega_\delta|^2 \right] \mathrm{d}x \\ & + c(r)\delta \int_{B_{2r}(x_0)} |\nabla u_\delta| \omega_\delta^2 \,\mathrm{d}x \,. \end{aligned} \tag{9}$$

Clearly $\int_{B_{2r}(x_0)} \omega_\delta^2 \,\mathrm{d}x$ and $\delta \int_{B_{2r}(x_0)} |\nabla u_\delta| \omega_\delta^2 \,\mathrm{d}x$ are uniformly bounded with respect to δ (compare Lemma 3.9, ii)). Hence (6), (8) and (9) imply after absorbing terms (for ε sufficiently small)

$$\begin{aligned} \int_{B_r(x_0)} \Gamma_\delta^{\frac{1}{2}} \omega_\delta^2 \,\mathrm{d}x \leq c \Bigg[1 & + \int_{B_{2r}(x_0)} \Gamma_\delta^{-\frac{1}{2}} |\nabla \omega_\delta|^2 \eta^2 \,\mathrm{d}x \\ & + \delta \int_{B_{2r}(x_0)} |\nabla \omega_\delta|^2 \eta^2 \,\mathrm{d}x \Bigg] \,. \end{aligned} \tag{10}$$

Given (10), we observe that a.e.

$$|\nabla \omega_\delta|^2 \leq c \frac{1}{1 + |\nabla u_\delta|^2} |\nabla^2 u_\delta|^2 \,,$$

thus we may use Assumption 3.6, *iv*), with the result

$$\int_{B_r(x_0)} \Gamma_\delta^{\frac{1}{2}} \omega_\delta^2 \, dx \le c\Big[1 + c\int_{B_{2r}(x_0)} \Big(\Gamma_\delta^{-\frac{1}{2}} + \delta\Big)\Gamma_\delta^{-1}|\nabla^2 u_\delta|^2 \eta^2 \, dx\Big] \tag{11}$$
$$\le c\Big[1 + c\int_{B_{2r}(x_0)} D^2 f_\delta(\nabla u_\delta)(\partial_\gamma \nabla u_\delta, \partial_\gamma \nabla u_\delta)\eta^2 \, dx\Big] .$$

Here and in the following we always take the sum w.r.t. repeated Greek indices $\gamma = 1, \dots, n$. Now, Lemma 3.10 is proved once we have found a uniform bound for the right-hand side of (11). To this purpose we observe that the starting integrability of u_δ is good enough (recall Lemma 3.9, iii)) to take $\varphi = \eta^2 \partial_\gamma u_\delta$, η as above, as an admissible test function in the differentiated Euler equation v) of Lemma 3.9 (of course we again need some standard density argument). As a result we obtain

$$\begin{aligned}
&\int_{B_{2r}(x_0)} D_P^2 f_\delta(x, \nabla u_\delta)(\partial_\gamma \nabla u_\delta, \partial_\gamma \nabla u_\delta)\eta^2 \, dx \\
&= -2\int_{B_{2r}(x_0)} D_P^2 f_\delta(x, \nabla u_\delta)(\partial_\gamma \nabla u_\delta, \nabla \eta)\eta \partial_\gamma u_\delta \, dx \\
&\quad -2\int_{B_{2r}(x_0)} (\partial_\gamma \nabla_P f_\delta)(x, \nabla u_\delta) \cdot \partial_\gamma u_\delta \nabla \eta \eta \, dx \\
&\quad -\int_{B_{2r}(x_0)} (\partial_\gamma \nabla_P f_\delta)(x, \nabla u_\delta) \cdot \partial_\gamma \nabla u_\delta \eta^2 \, dx \\
&=: I + II + III \, .
\end{aligned} \tag{12}$$

By Assumption 3.6, v), we have

$$|II| \le c(\eta)\int_{B_{2r}(x_0)} |\nabla u_\delta| \, dx \le c \, ,$$

whereas the first integral on the right-hand side of (12) is handled with Young's inequality for $\varepsilon > 0$ sufficiently small

$$\begin{aligned}
|I| \le \varepsilon &\int_{B_{2r}(x_0)} D_P^2 f_\delta(x, \nabla u_\delta)(\partial_\gamma \nabla u_\delta, \partial_\gamma \nabla u_\delta)\eta^2 \, dx \\
&+ c\varepsilon^{-1}\int_{B_{2r}(x_0)} |D_P^2 f_\delta(x, \nabla u_\delta)||\nabla \eta|^2 |\nabla u_\delta|^2 \, dx \, .
\end{aligned}$$

Here the second integral on the right-hand side is uniformly bounded, the first one can be absorbed on the left-hand side of (12), hence it remains to find an upper bound for III. We perform a partial integration to obtain

$$\begin{aligned}
III &= \int_{B_{2r}(x_0)} (\partial_\gamma \partial_\gamma \nabla_P f_\delta)(x, \nabla u_\delta) \cdot \nabla u_\delta \eta^2 \, dx \\
&\quad + \int_{B_{2r}(x_0)} (\partial_\gamma D_P^2 f_\delta)(x, \nabla u_\delta)(\partial_\gamma \nabla u_\delta, \nabla u_\delta)\eta^2 \, dx \\
&\quad + \int_{B_{2r}(x_0)} (\partial_\gamma \nabla_P f_\delta)(x, \nabla u_\delta) \cdot \nabla u_\delta \partial_\gamma \eta^2 \, dx \\
&=: III_1 + III_2 + III_3 \, .
\end{aligned}$$

Assumption 3.6, vi), shows that $|III_1|$ is bounded independent of δ, the uniform estimate for $|III_3|$ again follows from v) of Assumption 3.6. Finally, for the consideration of $|III_2|$ we make use of Assumption 3.6, vii), which, together with Young's inequality, gives for $\varepsilon > 0$ (note that the γ-derivative of the δ-part vanishes)

$$\begin{aligned} |III_2| \le c \int_{B_{2r}(x_0)} & |D_P^2 f(x, \nabla u_\delta)(\partial_\gamma \nabla u_\delta, \nabla u_\delta)|\eta^2 \, dx \\ +c \int_{B_{2r}(x_0)} & (1 + |\nabla u_\delta|^2)^{-1} |\nabla^2 u_\delta||\nabla u_\delta|\eta^2 \, dx \\ \le c\varepsilon \int_{B_{2r}(x_0)} & D_P^2 f(x, \nabla u_\delta)(\partial_\gamma \nabla u_\delta, \partial_\gamma \nabla u_\delta)\eta^2 \, dx \\ +c\varepsilon^{-1} \int_{B_{2r}(x_0)} & D_P^2 f(x, \nabla u_\delta)(\nabla u_\delta, \nabla u_\delta)\eta^2 \, dx \\ +c\varepsilon \int_{B_{2r}(x_0)} & (1 + |\nabla u_\delta|^2)^{-\frac{3}{2}} |\nabla^2 u_\delta|^2 \eta^2 \, dx \\ +c\varepsilon^{-1} \int_{B_{2r}(x_0)} & (1 + |\nabla u_\delta|^2)^{-\frac{1}{2}} |\nabla u_\delta|^2 \eta^2 \, dx \, . \end{aligned} \tag{13}$$

Note that, on account of the ellipticity Assumption 3.6, iv), the third integral on the right-hand side is estimated by the first one which in turn is absorbed on the left-hand side of (12). The remaining two integrals on the right-hand side of (13) are handled with the linear growth of f and Lemma 3.10 is proved. □

Since the regularization $\{u_\delta\}$ is a J-minimizing sequence (recall Lemma 3.9, i)), each L^1-cluster point u^* is a generalized minimizer which satisfies on account of Lemma 3.10 the higher integrability claimed in Theorem 3.8. Once higher integrability of the gradient is shown, the last assertion (concerning uniqueness) can be taken from [7] and the proof of Theorem 3.8 is complete. □

4 A Counterexample to Theorem 3.8 in Case $\mu > 3$

We proceed with an example on the sharpness of Theorem 3.8. The idea originates from [10], Example 3.2, where the authors restrict themselves to the one-dimensional situation. We follow the proposal of Giaquinta, Modica and Souček and give a rigorous proof that the arguments extend to higher-dimensional annuli Ω. What is more, the example given in [10] is degenerated which is not the case in the modification outlined below. As a consequence, we precisely can verify the assumptions of Sect. 3 with the exception that we now have $\mu > 3$.

The general setting is the following: let $n = 2$ and $|x| = \sqrt{x_1^2 + x_2^2} = r$. We fix some positive numbers $0 < \rho_1 < \rho_2$, $\rho := (\rho_1 + \rho_2)/2$ and choose

$$\Omega := \left\{ x \in \mathbb{R}^2 : \rho_1 < r < \rho_2 \right\} .$$

Moreover, $\alpha\colon \Omega \to \mathbb{R}$ is defined by

$$\alpha(r) := 1 + \gamma|r - \rho|^2 ,$$

where the positive parameter γ is chosen later on (see (18) and (29)). If $k > 2$ is fixed, then the energy density under consideration reads as

$$f(x,P) = f(r,P) = \begin{cases} (1+\alpha(r)|P|^k)^{\frac{1}{k}} & \text{if } |P| > \varepsilon , \\ h(r,P) & \text{if } |P| \le \varepsilon . \end{cases}$$

Here $h(r,P)$ is chosen such that $f(x,P)$ is strictly convex, non-degenerate in P and such that $f(x,P)$ of class $C^2(\overline{\Omega}\times\mathbb{R}^2)$. For an explicit construction we may consider

$$\tilde{f}(P) = \begin{cases} (1+|P|^k)^{\frac{1}{k}} & \text{if } |P| > \varepsilon , \\ \tilde{h}(P) & \text{if } |P| \le \varepsilon , \end{cases}$$

together with the Ansatz

$$\tilde{h}(P) = a + b(1+|P|^l)^{\frac{1}{l}} + c|P|^2 ,$$

where a, b and c are suitable constants and $l > k$. The requirement that $\tilde{f}$ is of class $C^2(\mathbb{R}^2)$ in particular implies

$$b = (1+|\varepsilon|^k)^{\frac{1}{k}-2}(1+|\varepsilon|^l)^{2-\frac{1}{l}}|\varepsilon|^{k-l}\frac{k-2-|\varepsilon|^k}{l-2-|\varepsilon|^l} > 0 ,$$

$$c = \frac{1}{2}|\varepsilon|^{k-2}(1+|\varepsilon|^k)^{\frac{1}{k}-2}\left[(1+|\varepsilon|^k) - (1+|\varepsilon|^l)\frac{k-2-|\varepsilon|^k}{k-2-|\varepsilon|^l}\right] > 0 .$$

We then let $h(r,P) := \tilde{h}([\alpha(r)]^{1/k}P)$. Finally, the choice of the second parameter $0 < \varepsilon < 1$ will be made in inequality (26) below.

Theorem 4.11 *With the above notation, the variational problem*

$$J[w] := \int_\Omega f(x,\nabla w)\, dx \to \min \quad \text{in } u_0 + \overset{\circ}{W}{}_1^1(\Omega)$$

does not admit a generalized minimizer $v \in \mathcal{M}$ of class $W_1^1(\Omega)$ if u_0 is supposed to satisfy $u_0(\rho_1) = -a$ and $u_0(\rho_2) = a$ for a constant $a > 0$ sufficiently large (see (30)). Here and in the following – with a slight abuse of notation – we write $u(x_1,x_2) = u(r)$ whenever u is merely depending on $|x|$.

Remark 4.12 *i) Note that the ellipticity exponent of f is given by $\mu = k+1 > 3$, hence we really obtain an example on the sharpness of our results.*

ii) Moreover, it should be emphasized that the boundary values u_0 may be chosen as a function of class $C^\infty(\overline{\Omega})$.

Proof. Assume by contradiction that $v \in W_1^1(\Omega)$ is a generalized minimizer. Then the proof of Theorem 4.11 splits into three steps.

Step 1. First of all we note that by the symmetry of the problem and with the obvious meaning of notation (after introducing polar coordinates) we have

$$v(r, \varphi) = v(r) \ . \tag{14}$$

In fact, consider the regularization $\{u_\delta\}$ of Sect. 3 which clearly satisfies $u_\delta(r, \varphi) = u_\delta(r)$ since for any real number φ_0 the function $u_\delta(r, \varphi + \varphi_0)$ is J_δ-minimizing with respect to the boundary values u_0 as well. Hence, uniqueness of minimizers proves the claim for u_δ. Next, consider a L^1-cluster point $u^* \in \mathcal{M}$ of the sequence $\{u_\delta\}$, in particular we have $u^* = u^*(r)$. It was already mentioned above that there is an open set Ω_0 of full measure such that $u^* \in C^{1,\alpha}(\Omega_0)$. As an immediate consequence (see [5]) we obtain

$$\sigma = \nabla f(\nabla u^*) \quad \text{in } \Omega_0 \ .$$

Moreover, following an idea of [19], a minimax inequality is proved in [7] (compare [3] for some additional details) which implies

$$\sigma = \nabla f(\nabla w) \quad \text{in } \Omega_0$$

for any generalized minimizer $w \in \mathcal{M}$, hence

$$\nabla u^* = \nabla^a u^* = \nabla w = \nabla^a w \quad \text{in } \Omega_0 \ .$$

This, together with $u^* = u^*(r)$ and $v \in W_1^1(\Omega)$, immediately gives assertion (14).

Step 2. We next claim that v takes the boundary data u_0 in the sense that the trace of v on $\partial\Omega$ is just u_0, i.e.

$$v(\rho_1) = -a \quad \text{and} \quad v(\rho_2) = a \ . \tag{15}$$

In order to prove (15) we consider the comparison function

$$w(r) = \begin{cases} v(r) - v(\rho_1) - a & \rho_1 < r < \rho \ , \\ v(r) - v(\rho_2) + a & \rho \leq r < \rho_2 \ , \end{cases}$$

and assume by contradiction that (15) fails to be true. If we observe that

$$f_\infty(r, P) = \alpha^{\frac{1}{k}}(r)|P| \ , \quad \rho_1 < r < \rho_2 \ , \quad P \in \mathbb{R}^2 \ ,$$

then we obtain

$$K[w] = \int_\Omega f(r, \nabla^a w)\, \mathrm{d}x + \int_{\partial B_\rho(0)} \mathrm{d}|\nabla^s w| \ . \tag{16}$$

Here we used the fact that $\nabla^s w$ is supported on $\partial B_\rho(0)$ and that w takes its boundary data u_0 on $\partial\Omega$. From [1], Theorem 3.77, p. 171, one gets

$$\nabla^s w \llcorner \partial B_\rho(0) = \big(v(\rho_1) - v(\rho_2) + 2a\big)\frac{x}{|x|}\, d\mathcal{H}^1 .$$

Thus, (16) may be rewritten as

$$\begin{aligned} K[w] &= \int_\Omega f(r, \nabla^a w)\, \mathrm{d}x + 2\pi\rho\big|v(\rho_1) - v(\rho_2) + 2a\big| \\ &\le \int_\Omega f(r, \nabla v)\, \mathrm{d}x + 2\pi\rho\Big(\big|v(\rho_1) + a\big| + \big|v(\rho_2) - a\big|\Big) . \end{aligned} \tag{17}$$

Now choose γ sufficiently large such that for $i = 1, 2$

$$\frac{\rho}{\rho_i \alpha^{\frac{1}{k}}(\rho_i)} < 1 . \tag{18}$$

Then we obtain

$$K[w] < \int_\Omega f(r, \nabla v)\, \mathrm{d}x + \sum_{i=1}^{2} \alpha^{\frac{1}{k}}(\rho_i) \int_{\partial B_{\rho_i}(0)} |u_0 - v(\rho_i)|\, d\mathcal{H}^1 = K[v] ,$$

hence the desired contradiction since the characterization of K as stated in the introduction remains valid with an additional smooth x-dependence.

Step 3. Now we make use of the Euler equation for the generalized minimizer v which takes the standard form since v is assumed to be of class $W_1^1(\Omega)$, i.e. we have

$$\int_\Omega \nabla_P f(r, \nabla v) \cdot \nabla\psi\, \mathrm{d}x = 0 \quad \text{for all } \psi \in C_0^1(\Omega) . \tag{19}$$

In particular, this is true for test functions $\psi = \psi(r) \in C_0^1((\rho_1, \rho_2))$. In the following the derivative w.r.t. r is denoted by "$\dot{}$". Then, again using polar coordinates, $\nabla v = (\cos\varphi \dot{v}, \sin\varphi \dot{v})$ and with the notation $\nabla_P f(r, \nabla v) = g(r, |\dot{v}|)\nabla v$ we obtain from (19)

$$\int_{\rho_1}^{\rho_2} \int_0^{2\pi} g(r, |\dot{v}|)\dot{v}\dot{\psi} r\, \mathrm{d}r\, \mathrm{d}\varphi = 0 \quad \text{for all } \psi = \psi(r) \in C_0^1((\rho_1, \rho_2)) .$$

As a consequence, there is a real number $\lambda \in \mathbb{R}$ such that for all $r \in (\rho_1, \rho_2)$

$$g(r, |\dot{v}|)\dot{v} r = \lambda . \tag{20}$$

With the representation

$$g(r, |\dot{v}|) = \begin{cases} \big(1 + \alpha(r)|\dot{v}|^k\big)^{\frac{1}{k}-1}\alpha(r)|\dot{v}|^{k-2} & \text{if } |\dot{v}| > \varepsilon , \\ b\big(1 + \alpha(r)|\dot{v}|^l\big)^{\frac{1}{l}-1}\alpha(r)|\dot{v}|^{l-2} + 2c[\alpha(r)]^{\frac{2}{k}} & \text{if } |\dot{v}| < \varepsilon \end{cases}$$

we have to distinguish two cases.

Case 1. If $|\dot{v}| < \varepsilon$, then using the formulas for b and c we immediately see that $g(r, |\dot{v}|) \le c|\varepsilon|^{k-2}$, in particular

$$\varepsilon > |\dot{v}| = \frac{|\lambda|}{g(r,|\dot{v}|)r} \geq c|\lambda| \tag{21}$$

for some positive constant c.

Case 2. If $|\dot{v}| > \varepsilon$, then (20) implies by elementary calculations (note that $|\lambda| > 0$ in the case at hand)

$$|\dot{v}|^{-k}\alpha^{-\frac{k}{k-1}} + \alpha^{-\frac{1}{k-1}} = \big(|\lambda|/r\big)^{-\frac{k}{k-1}} . \tag{22}$$

Observe that, as a consequence of (22),

$$\big(|\lambda|/r\big)^{k} \leq \alpha(r) . \tag{23}$$

Now, again with some simple computations, (22) gives

$$|\dot{v}| = \frac{\big(|\lambda|/r\big)^{\frac{1}{k-1}}}{\alpha^{\frac{1}{k}}\Big(\alpha^{\frac{1}{k-1}} - \big(|\lambda|/r\big)^{\frac{k}{k-1}}\Big)^{\frac{1}{k}}} . \tag{24}$$

Summarizing both cases we have the formulas (21) and (24), respectively, for $|\dot{v}|$.

We then choose $\lambda_0 > 0$ sufficiently small such that (24) (which is independent of the parameter $\varepsilon > 0$) implies $|\dot{v}| \leq 1$ and assume that $|\lambda| < \lambda_0$. Then, by (21) and (24) we see that $|\dot{v}| \leq 1$ for all $r \in (\rho_1, \rho_2)$. On the other hand, v takes its boundary data and $v(r)$ is of class $W_1^1((\rho_1,\rho_2))$, hence $v(r)$ is an absolutely continuous function and we may write

$$a = \frac{1}{2}|v(\rho_1) - v(\rho_2)| \leq \frac{1}{2}\int_{\rho_1}^{\rho_2} |\dot{v}(r)|\, \mathrm{d}r \leq \frac{\rho_2 - \rho_1}{2} . \tag{25}$$

This gives a contradiction if a is sufficiently large and we may assume $|\lambda| \geq \lambda_0$ which was chosen independent of ε. Hence, if Case 1 holds true, then (21) yields

$$\varepsilon \geq c|\lambda_0| , \tag{26}$$

and we choose ε sufficiently small such that this is not possible. Once it is established that Case 2 holds for all $r \in (\rho_1, \rho_2)$, we obtain from (23)

$$|\lambda| \leq \inf_{r \in (\rho_1,\rho_2)} \alpha^{\frac{1}{k}}(r) r . \tag{27}$$

Moreover, (24) gives the right representation and using (23), (27), $\alpha \geq 1$ and $k > 2$ we estimate

$$\begin{aligned} |\dot{v}| &\leq \frac{\alpha^{\frac{1}{k(k-1)}}}{\alpha^{\frac{1}{k}}\Big(\alpha^{\frac{1}{k-1}} - \big(|\lambda|/r\big)^{\frac{k}{k-1}}\Big)^{\frac{1}{k}}} \leq \frac{\alpha^{\frac{2-k}{k(k-1)}}}{\Big(\alpha^{\frac{1}{k-1}} - \big(|\lambda|/r\big)^{\frac{k}{k-1}}\Big)^{\frac{1}{k}}} \\ &\leq \left[\alpha^{\frac{1}{k-1}} - \Big(r^{-1}\inf_{r \in (\rho_1,\rho_2)} [\alpha^{1/k}(r) r]\Big)^{\frac{k}{k-1}}\right]^{-\frac{1}{k}} =: \big[h(r)\big]^{-\frac{1}{k}} . \end{aligned} \tag{28}$$

Here we first note that $h(r)$ is independent of λ, in particular $h(r)$ does not depend on the boundary values u_0 given in terms of a. Moreover, $h(r) \geq 0$ is evident by definition. Finally, the zeros of $h(r)$ are of finite number and simultaneously (by (18) interior) minima of $h(r)$. With Taylor's formula we obtain the expansion

$$h(r) \approx c(r - r_0)^2 \quad \text{near the zeros } r_0 \text{ of } h(r) ,$$

since $h''(r_0)$ is not vanishing. In fact, assume by contradiction that

$$h(r_0) = 0 , \quad h'(r_0) = 0 , \quad h''(r_0) = 0 .$$

This leads (by elementary calculations) to

$$r_0^{2+k} \leq \frac{k(1+k)\rho^k}{2\gamma} , \tag{29}$$

hence if $\gamma = \gamma(k, \rho_1, \rho_2)$ is chosen sufficiently large, then no such zeros are possible for the radii under consideration. With the above expansion we may choose $a < \infty$ such that

$$\int_{\rho_1}^{\rho_2} \left[h(r)\right]^{-\frac{1}{k}} \, \mathrm{d}r < a . \tag{30}$$

This proves the theorem since (28) and (30) contradict (25). □

Remark 4.13 *Let us again concentrate on the regularizing sequence $\{u_\delta\}$ with L^1-cluster point u^* as studied in Sect. 3. Then it is not difficult to locate* $\mathrm{spt}\nabla^s u^*$ *in the situation at hand. To this purpose denote by $\rho_{0,i}$, $i = 1, \dots, M$, the minima (of finite number, lying in the interior of (ρ_1, ρ_2) by (18)) of the function $\alpha^{\frac{1}{k}}(r)r$ on (ρ_1, ρ_2). We then have*

$$\mathrm{spt}\, \nabla^s u^* \subset \bigcup_{i=1}^{M} \partial B_{\rho_{0,i}}(0) .$$

In fact, the sequence of radially symmetric functions $\{u_\delta\} = \{u_\delta(r)\}$ yields a minimizing sequence of the one-dimensional energy ($\Omega = I = (\rho_1, \rho_2)$)

$$J_I[w] := \int_I f(r, |\dot{w}(r)|) r \, \mathrm{d}r$$

with respect to $W_1^1(I)$-comparison functions $w(\rho_1) = -a$, $w(\rho_2) = a$. In this sense, u^* provides a generalized J_I-minimizer. Now we again extend the ideas of [10] and let $\dot{u}_a^*$, $\dot{u}_s^*$ denote the Lebesgue-decomposition of $\dot{u}^*$ in absolutely continuous and singular part, respectively. Moreover, Corollary 3.33, [1], p. 140, on the decomposition of functions of bounded variation defined on intervals allows us to choose $\tilde{v}(r) \in W_1^1(I)$ such that for almost all $r \in I$

$$\dot{\tilde{v}}(r) = \dot{u}_a^*(r) \quad \textit{and} \quad \tilde{v}(\rho_1) = u^*(\rho_1) .$$

Next let $v(r)$ differ from $\tilde{v}(r)$ just by additional jumps at the points $\rho_{0,i}$ such that

$$\dot{v}_s(r) = \frac{1}{M}\sum_{i=1}^{M} \delta_{\rho_{0,i}} \int_I \dot{u}_s^* \, ,$$

where $\delta_{\rho_{0,i}}$ denotes the Dirac-measure centered at $\rho_{0,i}$, $i = 1, \dots, M$. Note that

$$\int_I \dot{u}_a^*(t)\,\mathrm{d}t + \int_I \dot{u}_s^* = u^*(\rho_2) - u^*(\rho_1) \, ,$$

$$\int_I \dot{v}_a(t)\,\mathrm{d}t + \left(\int_I \dot{u}_s^*\right) \frac{1}{M}\sum_{i=1}^{M} \int_I \delta_{\rho_{0,i}} = v(\rho_2) - v(\rho_1)$$

also implies $v(\rho_2) = u^*(\rho_2)$. Thus we obtain

$$\begin{aligned}
&\int_I f(r, |\dot{u}_a^*(r)|) r\,\mathrm{d}r + \int_I \alpha^{\frac{1}{k}}(r) r\,\mathrm{d}|\dot{u}_s^*|(r) \\
&\leq \int_I f(r, |\dot{v}_a(r)|) r\,\mathrm{d}r + \int_I \alpha^{\frac{1}{k}}(r) r\,\mathrm{d}|\dot{v}_s|(r) \\
&\leq \int_I f(r, |\dot{u}_a^*(r)|) r\,\mathrm{d}r + \min_{r \in (\rho_1, \rho_2)} \alpha^{\frac{1}{k}}(r) r \int_I |\dot{u}_s^*| \leq K[u^*]
\end{aligned}$$

and our claim is proved. □

Remark 4.14 *Although u^* as discussed above is not of class $W_1^1(\Omega)$ and although we do not know whether u^* is of class $C^{2,\alpha}$ on the complement of* $\mathrm{spt}\nabla^s u^*$, *we might conjecture that there exist analogous examples in the case $\mu = 3$ providing W_1^1-minimizers of*

$$\int_\Omega f(\nabla w)\,\mathrm{d}x + \int_{\partial\Omega} f_\infty\big((u_0 - w)\nu\big)\,\mathrm{d}\mathcal{H}^{n-1} \to \min \, ,$$

which are smooth on the complement of a finite number of interior spheres. However, if solutions of this kind exist, then they are caused by the non-convexity of Ω. In fact, consider a smooth convex domain Ω, assume that $n \geq 2$, $N = 1$, and suppose that there is a $W_1^1(\Omega)$-solution which is of class $C^{2,\alpha}$ near the boundary $\partial\Omega$. Then, on account of the uniqueness of solutions (up to a constant), we apply Hilbert-Haar arguments (compare [14]) to see that the singular set is empty. In this sense, as the typical behaviour, singularities must concentrate near the boundary.

Remark 4.15 *In order to show rigorously that our regularity theory breaks down if $\mu > 3$, we have to ensure that the energy density f studied in Theorem 4.11 satisfies Assumption 3.6 (of course now with ellipticity exponent $\mu = k+1$). Here it is clearly sufficient to consider ($P \in \mathbb{R}^n$)*

$$\tilde{f}(P) = (1 + |P|^k)^{\frac{1}{k}} \, , \quad k > 2 \, ,$$

and to study Assumption 3.6 w.r.t.

$$f(x,P) = \tilde{f}(\alpha(x)P)\,, \quad \alpha(x) = (1+|x|^2)^{\frac{1}{k}}\,,$$

whenever $|P| > 1$ *and* $x \in B_1(0) \subset \mathbb{R}^n$. *To this purpose we first observe that direct calculations yield in the case* $|P| > 1$

$$D^2\tilde{f}(P)\cdot P = (1+|P|^k)^{\frac{1}{k}-2}(k-1)|P|^{k-2}P \tag{31}$$

and, as a direct consequence,

$$\big|D^3\tilde{f}(P)(P,U,V)\big| \le c\big|D^2\tilde{f}(P)(U,V)\big| + c(1+|P|^2)^{-\frac{1+k}{2}}|U||V| \tag{32}$$

for all $U,\ V \in \mathbb{R}^n$. *For the discussion of* f *we just have to verify v), vi) and vii) of Assumption 3.6, where v) immediately follows from (31). Now note that for* $1 \le \gamma \le n$ *and* $|P| > 1$

$$\begin{aligned}\partial_\gamma\partial_\gamma\nabla_P f(x,P) = {} & \partial_\gamma\partial_\gamma\alpha(x)\nabla\tilde{f}(\alpha(x)P)\\ & +2\partial_\gamma\alpha(x)D^2\tilde{f}(\alpha(x)P)\cdot P\partial_\gamma\alpha(x)\\ & +\alpha(x)D^3\tilde{f}(\alpha(x)P)(P,P)\big[\partial_\gamma\alpha(x)\big]^2\\ & +\alpha(x)D^2\tilde{f}(\alpha(x)P)\cdot P\partial_\gamma\partial_\gamma\alpha(x) =: \sum_{i=1}^{4} I_i\,.\end{aligned}$$

Clearly I_1 *is uniformly bounded and the same follows for* I_2 *and* I_4 *from (31).* I_3 *is estimated with the help of (32)*

$$\begin{aligned}\big|D^3\tilde{f}(\alpha(x)P)(P,P)\big| & \le \frac{1}{\alpha(x)}\big|D^3\tilde{f}(\alpha(x)P)(\alpha(x)P,P)\big|\\ & \le c\big|D^2\tilde{f}(\alpha(x)P)\big||P| + c(1+|P|^2)^{-\frac{k}{2}} \le c\,,\end{aligned}$$

hence we have vi). Finally vii) is established by observing

$$\begin{aligned}\partial_\gamma D_P^2 f(x,P)(U,V) = {} & 2\alpha(x)\partial_\gamma\alpha(x)D^2\tilde{f}(\alpha(x)P)(U,V)\\ & +\big[\alpha(x)\big]^2 D^3\tilde{f}(\alpha(x)P)(P,U,V)\partial_\gamma\alpha(x)\end{aligned}$$

if we once more recall (32). □

References

1. Ambrosio, L., Fusco, N., Pallara, D. (2000): Functions of Bounded Variation and Free Discontinuity Problems. Oxford Science Publications, Clarendon Press, Oxford
2. Anzellotti, G., Giaquinta, M. (1988): Convex functionals and partial regularity. Arch. Rat. Mech. Anal., **102**, 243–272

3. Bildhauer, M. (2001): Convex Variational Problems with Linear, Nearly Linear and/or Anisotropic Growth Conditions. Habilitationsschrift (submitted), Saarland University, Saarbrücken
4. Bildhauer, M. (2002): A priori gradient estimates for bounded generalized solutions of a class of variational problems with linear growth. To appear in J. Convex Anal.
5. Bildhauer, M., Fuchs, M. (1999): Regularity for dual solutions and for weak cluster points of minimizing sequences of variational problems with linear growth. Zap. Nauchn. Sem. St.-Petersburg Odtel. Math. Inst. Steklov (POMI), **259**, 46–66
6. Bildhauer, M., Fuchs, M. (2000): On a class of variational integrals with linear growth satisfying the condition of μ-ellipticity. Preprint Bonn University/SFB 256 no. 681
7. Bildhauer, M., Fuchs, M. (2002): Relaxation of convex variational problems with linear growth defined on classes of vector-valued functions. To appear in Algebra and Analiz **14**
8. Bildhauer, M., Fuchs, M., Mingione, G. (2001): A priori gradient bounds and local $C^{1,\alpha}$-estimates for (double) obstacle problems under nonstandard growth conditions. Z. Anal. Anw., **20**, 959–985
9. Bombieri, E., DeGiorgi, E., Miranda, M. (1969): Una maggiorazione a priori relativa alle ipersuperfici minimali non parametriche. Arch. Rat. Mech. Anal., **32**, 255–267
10. Giaquinta, M., Modica, G., Souček, J. (1979): Functionals with linear growth in the calculus of variations. Comm. Math. Univ. Carolinae, **20**, 143–172
11. Giusti, E. (1984): Minimal Surfaces and Functions of Bounded Variation. Monographs in Mathematics 80, Birkhäuser, Boston-Basel-Stuttgart
12. Ladyzhenskaya, O.A., Ural'tseva, N.N. (1964): Linear and Quasilinear Elliptic Equations. Nauka, Moskow (in Russian)
English translation (1968): Academic Press, New York
13. Ladyzhenskaya, O.A., Ural'tseva, N.N. (1970): Local estimates for gradients of solutions of non-uniformly elliptic and parabolic equations. Comm. Pure Appl. Math., **23**, 667–703
14. Massari, U., Miranda, M. (1984): Minimal Surfaces of Codimension One. North-Holland Mathematics Studies 91, North-Holland, Amsterdam
15. Miranda, M. (1967): Disequaglianze di Sobolev sulle ipersuperfici minimali. Rend. Sem. Mat. Univ. Padova, **38**, 69–79
16. Reshetnyak, Y. (1968): Weak convergence of completely additive vector functions on a set. Sibirsk. Maz. Ž., **9**, 1386–1394 (in Russian)
English translation (1968): Sib. Math. J., **9**, 1039–1045
17. Seregin, G. (1985): Variational-difference scheme for problems in the mechanics of ideally elastoplastic media. Zh. Vychisl. Mat. Fiz., **25**, 237–352 (in Russian)
English translation (1985): U.S.S.R Comp. Math. and Math. Phys., **25**, 153–165
18. Seregin, G.(1990): Differential properties of solutions of variational problems for functionals with linear growth. Problemy Matematicheskogo Analiza, Vypusk **11**, Isazadel'stvo LGU, 51–79 (in Russian)
English translation (1993): J. Soviet Math., **64**, 1256–1277
19. Seregin, G. (1996): Twodimensional variational problems in plasticity theory. Izv. Russian Academy of Sciences, **60**, 175–210 (in Russian) English translation (1996): Izvestiya Mathematics, **60**, 179–216
20. Strang, G., Temam, R. (1980): Duality and relaxations in the theory of plasticity. J. Méchanique, **19**, 1–35

II. Nonlinear Partial Differential Equations

Studying Nonlinear pde by Geometry in Matrix Space

Bernd Kirchheim[1], Stefan Müller[1] and Vladimír Šverák[2]

[1] Max Planck Institute for Mathematics in the Sciences, Inselstr. 22–26, D–04103 Leipzig, Germany, {k, sm}@mis.mpg.de
[2] Department of Mathematics, University of Minnesota, Minneapolis, MN 55455, USA sverak@math.umn.edu

Summary. We outline an approach to study the properties of nonlinear partial differential equations through the geometric properties of a set in the space of $m \times n$ matrices which is naturally associated to the equation. In particular, different notions of convex hulls play a crucial role. This work draws heavily on Tartar's work on oscillations in nonlinear pde and compensated compactness and on Gromov's work on partial differential relations and convex integration. We point out some recent successes of this approach and outline a number of open problems, most of which seem to require a better geometric understanding of the different convexity notions.

1 Introduction

The purpose of this paper is to outline a connection between nonlinear partial differential equations and simply stated but largely unexplored questions about certain convex hulls, such as the rank-one convex hull or the separately convex hull, in the space of $m \times n$ matrices. While many of the underlying ideas are old and go back to the pioneering work of Tartar [66, 67], Gromov [27, 28, 56] and DiPerna [21] there have been a number of recent new successes of this approach including the construction of elliptic and parabolic 2×2 systems with nowhere C^1 solutions [44, 45, 46], an analysis of Lipschitz maps with finitely many gradients [32], the existence of solutions in mathematical models of martensitic phase transitions [6, 11, 41] and a large number of other applications of Gromov's method of convex integration and its variants and extensions, see e.g. [14, 15, 33, 64, 47] for further discussion and references. At the same time a theory of the relevant convexity notions in matrix space is beginning to emerge [40, 33, 34] even though many basic questions remain open.

In a nutshell, the situation can be described as follows. Many nonlinear systems of pdes for a map $u : \Omega \subset \mathbb{R}^n \to \mathbb{R}^d$ can be naturally expressed as a combination of a linear systems of pdes

$$A(Dv) := \sum_{i=1}^{n} A_i \partial_i v = 0 \qquad (1)$$

and a pointwise nonlinear constraint

$$v(x) \in K \subset \mathbb{R}^d \quad \text{a.e.} \tag{2}$$

Then one considers the cone Λ related to one dimensional solutions $v(x) = h(\langle x, \xi \rangle)$ which is defined by

$$\Lambda := \left\{ \xi \in \mathbb{R}^n : \exists a \in \mathbb{R}^d \sum_{i=1}^{n} \xi_i A_i a = 0 \right\}. \tag{3}$$

Equivalently Λ characterizes the directions of one dimensional high frequency oscillations compatible with (1). Given a cone Λ we say that K is lamination convex (with respect to Λ) if for any two points $A, B \in K$ with $B - A \in \Lambda$ the whole segment $[A, B]$ belongs to K. The lamination convex hull $K^{lc,\Lambda}$ is the smallest lamination convex set containing K (Gromov [28], who works in the more general setting of jet bundles, calls this the P-convex hull).

The key point in Gromov's method of convex integration (which is a far reaching generalization of the work of Nash [49] and Kuiper [38] on isometric immersions) is that (1) and (2) admit many interesting solutions provided that $K^{lc,\Lambda}$ is sufficiently large. In applications to elliptic and parabolic systems we always have $K^{lc,\Lambda} = K$ so that Gromov's approach does not directly apply. It turns out, however, that for the construction of Lipschitz (rather than C^1) solutions one can work with the Λ-convex hull K^Λ, defined by duality. More precisely for a compact set K a point does not belong to K^Λ if and only if there exists a Λ-convex function which separates it from K. A crucial fact is that K^Λ can be much larger than the hull $K^{lc,\Lambda}$. This difference already arises for a set consisting of four matrices which form a so-called T_4-configuration (see Section 3.2 below). This surprising fact was observed independently in different contexts [53, 3, 50, 9, 69], we learned it from Tartar. In connection with suitable approximations and general position arguments it leads to surprising consequences. We illustrate this by three examples.

Theorem 1.1 ([45]) *(elliptic systems with nowhere C^1 solutions) Let Ω be the unit ball in $\mathbb{R}^2$. There exists a smooth function $\phi : \mathbb{R}^{2\times 2} \to \mathbb{R}$ which is strongly quasiconvex and satisfies $|D^2\phi| \le C$, and a Lipschitz map $w : \Omega \to \mathbb{R}^2$, which is a weak solution of the elliptic system*

$$-\operatorname{div} D\phi(\nabla w) = 0, \tag{4}$$

such that w is not C^1 in any open subset of Ω. Moreover the system (4) *admits solutions with compact support.*

We remark in passing that this counterexample is quite different from the classical counterexamples [7, 19, 26] and their more recent extensions [30, 62] which are all based on singularities at a point or more generally a set of lower dimension. Scheffer [53] used T_4-configurations as a basis of counterexamples to regularity. He proved a weaker version of Theorem 1.1 with w in the Sobolev space $W^{1,1}$ and ϕ satisfying the Legendre-Hadamard condition (7). Unfortunately, the work [53] has not appeared in a journal and the original ideas there remained largely unknown and had to be re-discovered by various authors.

Note that (4) is the Euler-Lagrange equation of the functional

$$I(w) = \int_{\Omega} \phi(\nabla w)\, dx. \tag{5}$$

By a classical result of Evans [22] *minimizers* of I are smooth outside a closed set of measure zero if ϕ satisfies the assumptions stated in Theorem 1.1 (the same is true for local minimizers, see [36]). Thus general stationary points of I can behave much worse than minimizers.

We recall that a (continuous) integrand $\phi : \mathbb{R}^{m\times n} \to \mathbb{R}$ is called *strongly quasiconvex* if

$$\int_{T^n} \phi(X + \nabla\eta) - \phi(X)\, dx \geq c \int_{T^n} |\nabla\eta|^2\, dx \tag{6}$$

for some $c > 0$, all $X \in \mathbb{R}^{m\times n}$ and all periodic Lipschitz maps $\eta : T^n \to \mathbb{R}^m$ (equivalently one can consider test functions η on bounded domains with zero boundary conditions). If (6) holds with $c = 0$ we say that ϕ is quasiconvex. Using small amplitude test functions $\eta(x)$, Taylor expansion and Fourier transform one easily sees that (6) implies that ϕ is uniformly rank-one convex and one obtains the Legendre-Hadamard (or strong ellipticity) condition

$$D^2\phi(X)(a \otimes b, a \otimes b) \geq c|a|^2|b|^2, \tag{7}$$

so that (4) is indeed an elliptic system. Recently L. Székelyhidi has shown that the conclusion of Theorem 1.1 also holds for a suitable strictly polyconvex integrand ϕ, i.e. a strictly convex function of F and $\det F$.

The failure of regularity can be extended to parabolic systems with smooth initial data and a small and Hölder continuous right hand side.

Theorem 1.2 ([46]) *(parabolic systems with nowhere C^1 solutions) Let Ω be the unit ball in $\mathbb{R}^2$. Let $\eta > 0, T > 0, \alpha \in (0,1)$. Then there exists a function $\phi : \mathbb{R}^{2\times 2} \to \mathbb{R}$ such that ϕ is strongly quasiconvex, smooth and $|D^2\phi| \leq C$, a function $f \in C^\alpha(\Omega \times [0,T]; \mathbb{R}^2)$ with $||f||_{C^\alpha} < \eta$ and a Lipschitz solution $w : \Omega \times [0,T] \to \mathbb{R}$ of the parabolic system*

$$\partial_t w - \operatorname{div} D\phi(\nabla w) = f \quad in \;\; \Omega \times (0,T) \tag{8}$$

and

$$w(\cdot, 0) \equiv 0, \quad w(t,x) = 0 \;\; for\; x \in \partial\Omega \tag{9}$$

such that w is nowhere C^1 in $\Omega \times (0,T)$.

This system exhibits some other unusual features, such as failure of uniqueness and of the energy inequality.

Our last example concerns Lipschitz maps whose gradient takes only finitely many values (except on a set of measure zero).

Theorem 1.3 ([32]) *(maps with finitely many gradients). Let Ω be the unit ball in $\mathbb{R}^2$. There exist five matrices $A_1, \dots, A_5 \in \mathbb{R}^{2\times 2}$ with*

$$\mathrm{rk}(A_i - A_j) \neq 1 \tag{10}$$

and a Lipschitz map $u : \Omega \to \mathbb{R}^2$ which satisfies

$$\nabla u \in \{A_1, \dots, A_5\} \quad a.e. \tag{11}$$

and $\nabla u \not\equiv A_i$.

Interestingly, the corresponding statement for four matrices turns out to be false [12]. The condition (10) rules out trivial maps which depend only on one direction and whose gradient takes two values. It also implies that the sets $\Omega_i = \{x : \nabla(x) = A_i\}$ must be very complicated. Indeed if Ω_i and Ω_j meet at a smooth (or rectifiable) boundary then a straightforward blow-up argument shows that $A_i - A_j$ must have rank one and the common boundary of Ω_i and Ω_j is flat with normal b, where $A_i - A_j = a \otimes b$. In this context (10) can be seen as an ellipticity condition for the partial differential relation (11). Nonetheless, as in Theorem 1.1, ellipticity is not strong enough to rule out large scale oscillations of ∇u. These are, in are certain sense, encoded in the T_4-configurations alluded to above (see Section 3.2 below).

The rest of this paper is organized as follows. We specialize to the situation that $v \in \mathbb{R}^{m\times n}$ and that the differential constraint (1) is simply $\operatorname{curl} v = 0$ (where the curl is taken along rows). Then the combination of (1) and (2) leads to the first order partial differential relation $\nabla u \in K$ and Λ is the cone of rank-one matrices. In Section 2 we review the general results on convex integration and reduce the existence of (highly oscillatory) solutions to the computation of the rank-one convex hull of K. To illustrate this idea we outline in Section 3 the main ideas of the proof of Theorem 1.1. One first finds one T_4-configuration in the relevant set K and then uses a dimension counting argument to show that the abstract conditions reviewed in Section 2 are satisfied.

The constructions related to Theorems 1.1–1.3 all use the simplest set K which has the property that $K^{lc,\Lambda} = K$, but K^{Λ} is much bigger, the T_4-configuration. We hope that a better understanding of the geometry of rank-one convexity will lead to new applications and to insights how to formulate structure conditions on ϕ which guarantee regularity results for elliptic systems (so far nothing is known beyond monotonicity or quasimonotonicity [25, 70, 29] of $D\phi$).

To this end we first recall in Section 4 some general tools to study rank-one convex hulls and then consider in Sections 5–7 a number of case studies. In Section 5 we study the set K related to the simplest polyconvex integrand $\phi(F) = (\det F)^2$ and we show that its rank-one convex hull is trivial. Interestingly, it is not known whether the same holds for the set related to the integrand $\varepsilon|F|^2 + (\det F)^2$ for small positive ε. Already the restriction to diagonal matrices leads to interesting questions about separately convex functions which we discuss in Section 6. In Section 7 we discuss an example related to compactness for hyperbolic conservation laws. It leads to a set K which is degenerate in the sense that while K contains no rank-one connections its tangent spaces do. The study of such sets was initiated in a pioneering paper by DiPerna [21], following the program outlined by Tartar [66, 67]. Finally in Section 8 we give a brief outlook.

2 Convex Integration and Rank-One Convex Hulls

For simplicity we will mostly consider the situation $v \in \mathbb{R}^{m\times n}$ and the simplest linear differential constraint $\operatorname{curl} v = 0$ (where the curl is taken by rows). If we restrict attention to simply connected domains Ω in $\mathbb{R}^n$ then the combination of (1) and (2) reduces to the first order partial differential relation

$$\nabla u \in K \quad \text{a.e. in } \Omega, \tag{12}$$

where $K \subset \mathbb{R}^{m\times n}$ is given and where we seek a map $u : \Omega \subset \mathbb{R}^n \to \mathbb{R}^m$.

In this case the cone Λ defined by (3) is simply the cone of rank-one matrices and for brevity we use the notation $K^{lc} := K^{lc,\Lambda}$ and $K^{rc} := K^{\Lambda}$ for the lamination convex hull and the rank-one convex hull. A function $f : \mathbb{R}^{m\times n} \to \mathbb{R}$ is rank-one convex if it is convex along each rank-one line $A + ta \otimes b$. For a compact set K the rank-one convex hull K^{rc} consists of all points which cannot be separated by rank-one convex functions, i.e.

$$K^{rc} := \{A \in \mathbb{R}^{m\times n} : f(A) \le \sup_K f, \ \text{ for all rank-one convex } f\}. \tag{13}$$

For a general set E we set

$$E^{rc} := \bigcup_{K\subset E \text{ compact}} K^{rc}. \tag{14}$$

If $m = 1$ or $n = 1$ then K^{rc} and K^{lc} agree with the convex hull.

2.1 Convex Integration for Open Sets

The key result in the theory of convex integration is that the partial differential relation (12) admits many solutions if K^{rc} is large. Here we just recall the relevant results and refer to [45] for the proofs and further discussion. We first consider the case that K is open. In the following we say that a map $u : \Omega \to \mathbb{R}^m$ is *piecewise affine* if it is Lipschitz and there exists finite or countably many open sets Ω_i such that u is affine on Ω_i and the union of the Ω_i has full measure.

Theorem 2.4 ([45], Thm. 3.1) *Let $K \subset \mathbb{R}^{m\times n}$ be open and let $L \subset K^{rc}$ be compact. Let $u_0 : \Omega \to \mathbb{R}^m$ be a piecewise affine map with $\nabla u_0 \in L$ a.e. Then there exists a piecewise affine map $u : \Omega \to \mathbb{R}^m$ such that*

$$\nabla u \in K \quad \text{a.e. in } \Omega, \tag{15}$$

$$u = u_0 \quad \text{on } \partial\Omega. \tag{16}$$

Remark 2.5 *In fact there exist many such solutions u. One can show that u_0 admits a fine approximation by solutions of* (15)*, i.e. for each continuous function η with $\eta > 0$ in Ω there exists a solution u with $|u - u_0|(x) < \eta(x)$.*

2.2 Closed Sets and In-Approximations

One crucial step in convex integration is the passage from open sets K to general sets (which may have high codimension). It is now understood that at least for Lipschitz solutions this can be done in different ways, e.g. by the Baire category theorem [14, 15] (for earlier applications to odes see e.g. [10, 17, 18]), a refinement of it using Baire-1 functions or the Banach-Mazur game [33] or by direct construction [64, 47]. As in [41] we follow here Gromov's original approach based on an in-approximation. Whatever approach one uses the basic theme is the same as in the construction of continuous, nowhere differentiable functions: at each step of the construction one adds a highly oscillatory correction whose frequency is much larger and whose amplitude is much smaller than those of the previous corrections. This leads to strong convergence of the gradient in L^1 but typically to a very irregular limiting Lipschitz map.

For simplicity we consider only compact sets K (for the application to elliptic systems discussed in Section 3 below it suffices to intersect the set in (24) with a large ball).

Definition 2.6 *We say that a sequence of open sets $\{U_i\}$ is an in-approximation of a compact set K if $U_i \subset U_{i+1}^{rc}$ and $\sup_{X \in U_i} \operatorname{dist}(X, K) \to 0$ as $i \to \infty$.*

Theorem 2.7 ([45], Thm. 3.2.) *Suppose that the compact set K admits an in-approximation by open sets $\{U_i\}$. Let $u_0 : \Omega \to \mathbb{R}^m$ be a C^1 map which satisfies*

$$\nabla u_0 \in U_1 \quad \text{in } \Omega. \tag{17}$$

Then there exists a Lipschitz map $u : \Omega \to \mathbb{R}^m$ such that

$$\nabla u \in K \quad \text{a.e. in } \Omega, \tag{18}$$

$$u = u_0 \quad \text{on } \partial\Omega. \tag{19}$$

Remark 2.8 *Again each such u_0 admits in fact a fine approximation by solutions u of* (18).

An illustrative example is given by equidimensional isometric (Lipschitz) immersions. In this case $K = O(n) = \{F \in \mathbb{R}^{n\times n} : F^T F = Id\}$. We have $K^{rc} = K^{lc} = \{F : F^T F \leq Id\}$ and an in-approximation of K is given by $U_i = \{F : \lambda_i Id < F^T F < Id\}$, where $0 < \lambda_1 < \lambda_2 < \ldots 1$, $\lambda_i \to 1$. Let $||F|| = \sup\{|Fx| : |x| \leq 1\}$ denote the operator norm of F with respect to the Euclidean norm in $\mathbb{R}^n$. Then any C^1 map u_0 with $\sup ||\nabla u_0|| < 1$ can be approximated in C^0 by Lipschitz maps u with $\nabla u \in O(n)$ which satisfy the same boundary conditions. For other examples, including applications to models of solid-solid phase transitions in crystals, see [44, 15].

3 Elliptic Systems with Nowhere Smooth Solutions

In this section we sketch the proof of Theorem 1.1, following partly the exposition in [46].

3.1 Reduction to First Order Systems

For definiteness let Ω be the unit ball in $\mathbb{R}^2$. We seek solutions $w : \Omega \to \mathbb{R}^2$ of the system

$$-\operatorname{div} D\phi(\nabla w) = 0, \tag{20}$$

where

$$\phi \text{ strongly quasiconvex, smooth, } |D^2\phi| \leq C. \tag{21}$$

In particular ϕ satisfies the Legendre-Hadamard condition (7). Now the condition (20) is equivalent to the existence of a potential W such that $D\phi(\nabla w)J = \nabla W$ where J is the 90° rotation. If we introduce

$$u = \begin{pmatrix} w \\ W \end{pmatrix}, \quad u : \Omega \subset \mathbb{R}^2 \to \mathbb{R}^4 \tag{22}$$

then (20) is equivalent to

$$\nabla u \in K \subset \mathbb{R}^{4\times 2}, \tag{23}$$

where

$$K = \left\{ \begin{pmatrix} X \\ Y \end{pmatrix} : Y = D\phi(X)J,\ X \in \mathbb{R}^{2\times 2} \right\}, \quad J = \begin{pmatrix} 0 & -1 \\ 1 & 0 \end{pmatrix}. \tag{24}$$

3.2 T_k-Configurations

To construct 'wild' solutions of (23) and hence (20) we have to show that K^{rc} is sufficiently large so that in-approximation of K can be constructed. A first attempt might be to show that K^{lc} is large. This, however, is doomed since the Legendre-Hadamard condition (7) implies that $K^{lc} = K$. Indeed if $A = \binom{X}{Y}$ and $A + \binom{a\otimes n}{b\otimes n}$ belong to K, then $D\phi(X + ta \otimes n) - D\phi(X) = (b \otimes n)J^T = b \otimes Jn$. Hence $\langle D\phi(X + ta \otimes n) - D\phi(X), a \otimes n\rangle = 0$ and this contradicts the strict convexity of the map $t \to \phi(X + ta \otimes n)$, see also [4].

A crucial observation is that there are simple sets which have a nontrivial rank-one convex hull K^{rc} even though K^{lc} is trivial.

Definition 3.9 *Let $k \geq 4$ and consider k-tuples $\mathbf{M} = (M_1, \ldots, M_k)$ of matrices $M_j \in \mathbb{R}^{m\times n}$. We say that $\mathbf{M}$ is a T_k-configuration if there exist rank-one matrices $C_1, \ldots, C_k$ with $\sum_{j=1}^k C_j = 0$, scalars $\kappa_1, \ldots, \kappa_k$ with $\kappa_j > 0$ and matrices $P_j \in \mathbb{R}^{m\times n}$ such that the relations*

$$P_{j+1} - P_j = C_j, \quad M_j - P_{j+1} = \kappa_j C_j$$

hold, where the index j is counted modulo k (see Fig. 1).

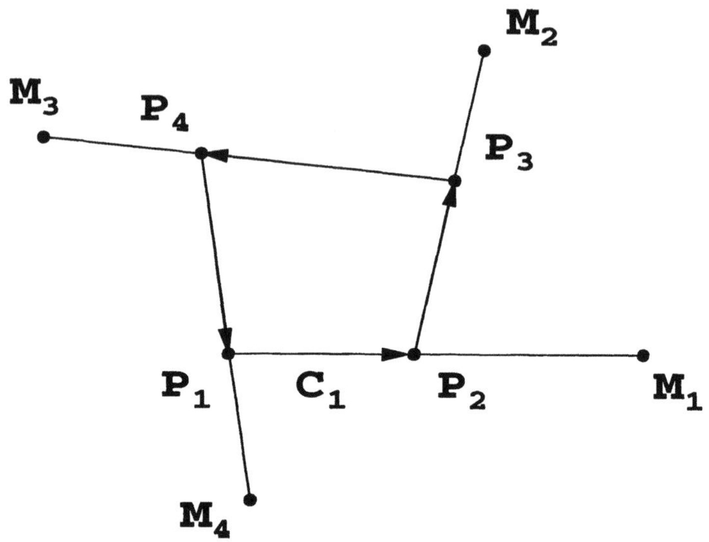

Fig. 1. T_4-configuration with $P_1 = P$, $P_2 = P + C_1$, $P_3 = P + C_1 + C_2$, $P_4 = P + C_1 + C_2 + C_3$. The lines indicate rank-1 connections. Note that the figure need not be planar

Let $\mathbf{M}$ be a T_k-configuration and let $K = \{M_1, \ldots, M_k\}$. One easily sees that every rank-one convex function which vanishes on K must also be nonpositive at all the P_j (see Fig. 1). Hence the P_j belong to K^{rc}. On the other hand there may be no rank-one connection in K. The simplest example arises already for $k = 4$ in diagonal 2×2 matrices. One may take

$$M_1 = -M_3 = \begin{pmatrix} 3 & 0 \\ 0 & -1 \end{pmatrix}, M_2 = -M_4 = \begin{pmatrix} 1 & 0 \\ 0 & 3 \end{pmatrix}.$$

We emphasize that in general a T_4-configuration need not lie in a plane. To construct 'wild' solutions of (23) we will show in the next subsections that K contains sufficiently many T_4-configurations.

3.3 Embedding a T_4-Configuration

The first observation is that there exists a strongly quasiconvex and smooth function $\phi : \mathbb{R}^{2\times 2} \to \mathbb{R}$ (with $|D^2\phi| \le C$) such that the set

$$K = \left\{ \begin{pmatrix} X \\ D\phi(X)J \end{pmatrix} : X \in \mathbb{R}^{2\times 2} \right\} \subset \mathbb{R}^{4\times 2}, J = \begin{pmatrix} 0 & -1 \\ 1 & 0 \end{pmatrix}$$

admits a T_4-configuration $\mathbf{M}^0$ with $M_i^0 \in K$, see [45], Lemma 4.3. One may take

$$M_1^0 = \begin{pmatrix} 3 & 0 \\ 0 & -1 \\ 0 & -1 \\ 3 & 0 \end{pmatrix}, M_2^0 = \begin{pmatrix} 1 & 0 \\ 0 & 3 \\ 0 & 3 \\ 1 & 0 \end{pmatrix}, M_3^0 = \begin{pmatrix} -3 & 0 \\ 0 & 1 \\ 0 & 1 \\ -3 & 0 \end{pmatrix}, M_4^0 = \begin{pmatrix} -1 & 0 \\ 0 & -3 \\ 0 & -3 \\ -1 & 0 \end{pmatrix}. \tag{25}$$

To illustrate the geometric ideas behind the construction of such an integrand we first sketch the construction of a ϕ which satisfies at least the Legendre-Hadamard condition (7). In this case it essentially suffices to construct ϕ on diagonal 2×2 matrices. Then one can use general extension arguments in the spirit of Proposition 4.15 below to conclude. Let

$$f\begin{pmatrix} s \\ t \end{pmatrix} = \phi\begin{pmatrix} s & 0 \\ 0 & t \end{pmatrix}, \quad \begin{pmatrix} \sigma \\ \tau \end{pmatrix} = \nabla f.$$

We look for T_4-configurations in the set

$$\begin{pmatrix} s & 0 \\ 0 & t \\ 0 & -\sigma(s,t) \\ \tau(s,t) & 0 \end{pmatrix}.$$

The Legendre-Hadamard condition reduces to

$$\frac{\partial \sigma}{\partial s} \geq c, \quad \frac{\partial \tau}{\partial t} \geq c, \quad c > 0. \tag{26}$$

If we drop the constraint that (σ, τ) is a gradient we can easily embed a T_4-configuration in the plane $\{\sigma = \tau = 0\}$. It suffices to make the ansatz

$$\sigma(s,t) = s - g(t), \quad \tau(s,t) = t - h(s).$$

Then the Legendre-Hadamard condition imposes no constraint on g and h and hence we can embed in the plane $\{\sigma = \tau = 0\}$ any set which is a graph both over s and over t. For definiteness we chose the T_4-configuration $(-3,1), (1,3), (3,-1), (-1,-3)$.

One can check, however, that the plane $\{\sigma = \tau = 0\}$ cannot contain a T_4-configuration if (σ, τ) is a gradient (it suffices to integrate around the inner square of the T_4-configuration and to use (26) to obtain a contradiction). To overcome this difficulty we tilt the T_4-configuration, i.e. we look for a linear map $\mathbb{R}^2 \to \mathbb{R}^{4\times 2}$ which preserves rank-one connections. A natural choice is

$$(s,t) \mapsto \begin{pmatrix} s & 0 \\ 0 & t \\ 0 & \mu t \\ \mu s & 0 \end{pmatrix},$$

where μ is a constant. This leads to the conditions $\sigma = -\mu t, \tau = \mu s$ or equivalently

$$\nabla f\begin{pmatrix} s \\ t \end{pmatrix} = \mu J\begin{pmatrix} s \\ t \end{pmatrix}, \quad \text{for } \begin{pmatrix} s \\ t \end{pmatrix} = J^k\begin{pmatrix} 3 \\ -1 \end{pmatrix}, \quad k \in \{0,1,2,3\}, \tag{27}$$

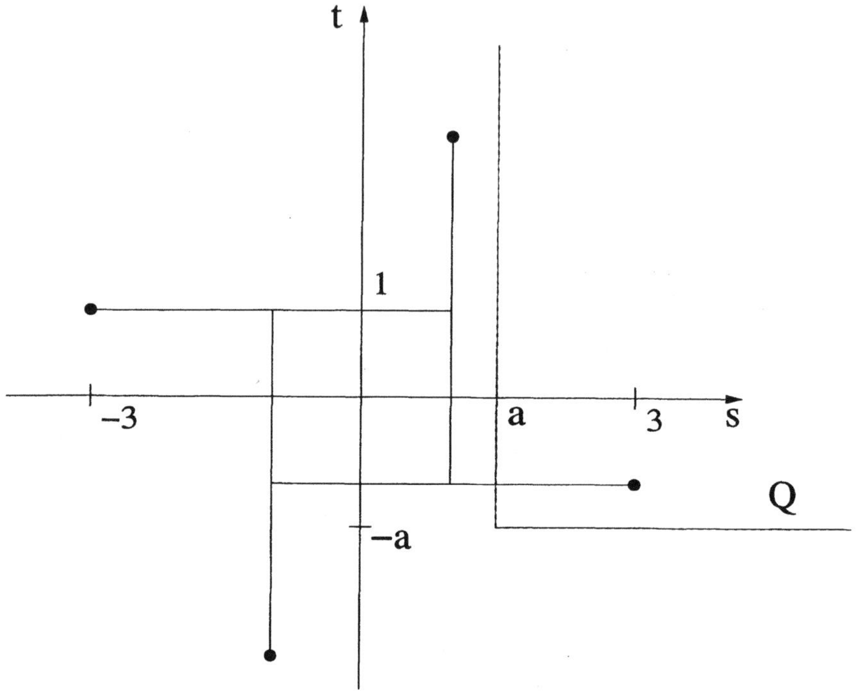

Fig. 2. Construction of an integrand for which the set K contains a T_4-configuration. The function f_0 is bilinear in the quadrant $Q = \{(s,t) : s > a, t > -a\}$ and vanishes outside Q. It vanishes in particular on the rotated quadrants $J^k Q$, where $k = 1,2,3$ and where J is the $90°$ rotation

where J is the 90° rotation. We make an ansatz for f which reflects the 90° rotation symmetry of the planar T_4-configuration:

$$f\begin{pmatrix} s \\ t \end{pmatrix} = \sum_{k=0}^{3} f_0\left(J^k \begin{pmatrix} s \\ t \end{pmatrix}\right) + \lambda(s^2 + t^2), \quad \text{where } \lambda > 0.$$

Then it suffices to verify (27) for $k = 0$ and this is equivalent to

$$\langle \nabla f(x), x \rangle = 0, \quad \text{with } x = \begin{pmatrix} 3 \\ -1 \end{pmatrix}. \tag{28}$$

We now make the choice (see Fig. 2)

$$f_0\begin{pmatrix} s \\ t \end{pmatrix} = (s-a)_+\,(t-a)_+, \quad \text{where } 1 < a < 3$$

and where $s_+ = \max(s, 0)$. In the quadrant $\{s > a, t > -a\}$ we have $f(s,t) = f_0(s,t) + \lambda(s^2 + t^2)$ and one easily checks that (28) is satisfied if, for example,

$a = 5/4$, $\lambda = 1/20$. Multiplying of f by a suitable factor we thus see that the T_4-configuration (25) lies in K.

To construct a quasiconvex ϕ for which K contains (25) we first use that the rank-one convex function $s_+ t_+$ on diagonal 2×2 matrices can be extended to a quasiconvex function on symmetric 2×2 matrices, see [59]. Then ϕ can be extended to all 2×2 matrices by adding a high quadratic penalty for the skew-symmetric part, see [45] for the details.

3.4 Families of T_4-Configurations and Dimension Counting

In order to construct an in-approximation of K, which is needed for the application of Theorem 2.7, we first show that

$$\mathcal{K} := K \times K \times K \times K \subset (\mathbb{R}^{4\times 2})^4 \tag{29}$$

contains not only the special T_4-configuration $\mathbf{M}^0$ but an eight-dimensional family of T_4-configurations. Then we will show that the corresponding corner points P_j cover an open set in the eight dimensional space $\mathbb{R}^{4\times 2}$ and use this fact to construct the in-approximation.

The dimension of the set of T_4-configurations can be guessed by a simple parameter count. First note that the rank-one cone in $\mathbb{R}^{4\times 2}$ is a five dimensional manifold (away from its vertex). In view of Definition 3.9 the set

$$\mathcal{M} = \{\mathbf{M} \in (\mathbb{R}^{4\times 2})^4 : \mathbf{M} \text{ is a } T_4 - \text{configuration}\} \subset \mathbb{R}^{32}. \tag{30}$$

involves $4 \times 5 + 4 \times 1 + 8 = 32$ parameters (namely the C_j, the κ_j and P_1) which are subject to 8 constraints (namely $\sum C_j = 0$). Hence we expect $\mathcal{M}$ to be a 24 dimensional manifold in the neighbourhood of $\mathbf{M}^0$. Since $\mathcal{K}$ is 16 dimensional, we expect the intersection to be have dimension 8, as desired.

To verify this and to actually construct the in-approximation we consider the maps

$$\begin{aligned} \pi_j : \mathcal{M} \cap \mathcal{K} &\longrightarrow \mathbb{R}^{4\times 2} \\ (M_1, M_2, M_3, M_4) &\longmapsto P_j \\ \mu_j : \mathcal{M} \cap \mathcal{K} &\longrightarrow \mathbb{R}^{4\times 2} \\ (M_1, M_2, M_3, M_4) &\longmapsto M_j \end{aligned}$$

Let $T_{M_j^0} K$ be the tangent space of K at M_j^0, let $Q_j^\perp$ denote the projection onto its orthogonal complement and define the map

$$\begin{aligned} \psi_j : \mathcal{M} \cap \mathcal{K} &\longrightarrow \mathbb{R}^{4\times 2} \\ (M_1, M_2, M_3, M_4) &\longmapsto (M_j, Q_j^\perp (P_j - P_j^0)) \end{aligned}$$

Proposition 3.10 *([45]) There exists a choice of ϕ such that $\mathrm{M}^0 \in \mathcal{K}$ and*

(i) in a neighbourhood of M^0 the sets $\mathcal{M}$ and $\mathcal{K}$ intersect transversely in an eight dimensional manifold,
(ii) π_j and ψ_j are local diffeomorphisms from a neighbourhood of M^0 in $\mathcal{M} \cap \mathcal{K}$ to open sets in $\mathbb{R}^{4\times 2}$.

With this result at hand one can construct an in-approximation as follows (see Fig. 3). Let $\mathcal{O}_1 \subset \mathcal{O}_2 \subset \ldots$ be (small) open neighbourhoods of M^0 in $\mathcal{M} \cap \mathcal{K}$ which are diffeomorphic to eight dimensional balls. For $1/2 < \lambda_2 < \lambda_3 < \ldots < 1$ consider the maps $\lambda_i \mu_j + (1-\lambda_i)\pi_j$ and define for $i \geq 2$ the sets $\mathcal{U}_i$ by

$$\mathcal{U}_i = \cup_{j=1}^4 \mathcal{U}_i^j, \quad \mathcal{U}_i^j = (\lambda_i \mu_j + (1-\lambda_i)\pi_j)(\mathcal{O}_i).$$

Using the non-degeneracy of ψ_j one can show that the $\mathcal{U}_i$ are open (if λ_2 is chosen sufficiently close to 1), see [45]. Let $\mathcal{U}_1 = \mathcal{U}_2^{rc}$. Then $\{\mathcal{U}_i\}_{i\geq 1}$ is an in-approximation of K. In fact it is an in-approximation of the set K intersected with a small neighbourhood of the set $\{M_1^0, M_2^0, M_0^3, M_0^4\}$. Moreover $\mathcal{U}_1$ contains the points $P_1^0, P_2^0, P_3^0, P_4^0$. Hence it also contains 0 .

This shows that (23) admits a non-trivial solution with zero boundary conditions whose gradient is always close to the set $\{M_1^0, M_2^0, M_0^3, M_0^4\}$. It is easy to see that one can achieve in addition $0 \in K$. Hence extension of the solution by zero yields a solution of (4) with compact support.

To see that this solution is nowhere C^1 one has to trace back the general construction used in the proof of Theorem 2.7 a bit more carefully, see [45] for the details. The main point is that at each step of the construction a (locally) affine map is replaced by a piecewise affine map whose gradient takes values near all of the four points M_i^0. This leads to a limit map whose gradient has an oscillation of order 1 in every open set.

3.5 Polyconvex Examples and Obstructions to T_4

Can one carry out the construction outlined above also for an integrand ϕ which is uniformly polyconvex (i.e. $\phi(X) = g(X, \det X)$ where g is uniformly convex)? It turns out that T_4-configurations are no longer sufficient.

Proposition 3.11 *Suppose that $\phi : \mathbb{R}^{2\times 2}$ is strictly polyconvex. Then the set K given by* (24) *does not contain any T_4-configuration.*

Nonetheless Székelyhidi [65] has shown that there exists a uniformly polyconvex ϕ such that the elliptic system (4) admits a Lipschitz, nowhere C^1 solution. His proof uses the fact that one can embed sufficiently many T_5-configurations in K.

The statement of the proposition above is in fact the corollary of a slightly more general result. It gives some insight into the consequences of the monotonicity condition on the gradient of the convex function g representing our polyconvex integrand. It states that K_ϕ can not support a discrete laminate whose second $\mathbb{R}^{2\times 2}$-coordinate

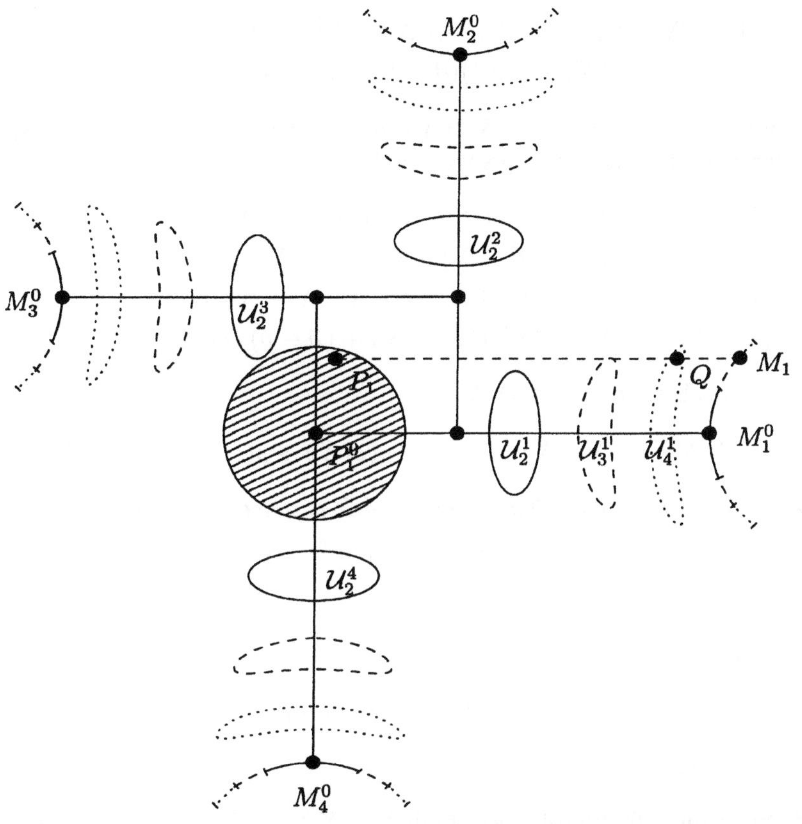

Fig. 3. Schematic illustration of the sets $\mathcal{U}_i^j \subset \mathbb{R}^{4\times 2}$. The solid (resp. dashed, or dotted) lines through the point M_1^0 are the sets $\mu_1(\mathcal{O}_2)$, $\mu_1(\mathcal{O}_3)$, $\mu_1(\mathcal{O}_4)$, respectively, i.e. the projections of the sets $\mathcal{O}_i \subset (\mathbb{R}^{4\times 2})^4$ to the first component. The sets $\mu_1(\mathcal{O}_i)$ are not open in $\mathbb{R}^{4\times 2}$ since they are contained in K. The shaded set is $\pi_1(\mathcal{O}_4)$ and is open in $\mathbb{R}^{4\times 2}$ and the sets $\mathcal{U}_i^1 = [(1-\lambda_i)\pi_1 - \lambda_i\mu_1](\mathcal{O}_i)$ are also open. A typical point Q in $\mathcal{U}_4^1$ is given by $(1-\lambda_4)\pi_1(\mathbf{M}) + \lambda_4 M_1$, where $\mathbf{M} = (M_1, M_2, M_3, M_4) \in \mathcal{O}_4$. In particular Q lies on the rank-one segment $[P_1, M_1]$ and hence in the rank-one convex hull of $\{M_1, M_2, M_3, M_4\}$. It also lies in the rank-one convex hull of four points in $\mathcal{U}_5$ which are close the M_i

is just a linear image of the first $\mathbb{R}^{2\times 2}$-coordinate, see Proposition 3.12 below. As we will see this in particular rules out the existence of a T_4 configuration in K_ϕ.

We begin with a little algebra. Let

$$\phi(X) = g(X, \det X) \text{ for } X \in \mathbb{R}^{2\times 2},$$

where $g : \mathbb{R}^5 \to \mathbb{R}$ is strictly convex. Using that

$$\nabla\phi(X) = \nabla_X G(X, \det X) + G_{,5}(X, \det X) \operatorname{cof} X,$$

and writing $\rho(X) = G_{,5}(X, \det X)$ we obtain from the strict monotonicity of ∇G that for any two different $X, \tilde{X} \in \mathbb{R}^{2\times 2}$

$$0 < \langle \nabla\phi(\tilde{X}) - \nabla\phi(X) - \rho(\tilde{X}) \operatorname{cof} \tilde{X} + \rho(X) \operatorname{cof} X, \tilde{X} - X \rangle \\ + (\rho(\tilde{X}) - \rho(X))(\det \tilde{X} - \det X).$$

We abbreviate $\rho = \rho(X)$, $Y = \nabla\phi(X)$, $\tilde{\rho} = \rho(\tilde{X})$ and $\tilde{Y} = \nabla\phi(\tilde{X})$. Expanding the difference of the determinants we conclude

$$\begin{aligned} 0 &< \langle \tilde{Y} - Y, \tilde{X} - X \rangle + \langle \rho \operatorname{cof} X - \tilde{\rho} \operatorname{cof} \tilde{X}, \tilde{X} - X \rangle \\ &\quad + (\tilde{\rho} - \rho)(\langle \operatorname{cof} X, \tilde{X} - X \rangle + \det(\tilde{X} - X)) \\ &= \langle \tilde{Y} - Y, \tilde{X} - X \rangle - \langle \tilde{\rho}(\operatorname{cof} \tilde{X} - \operatorname{cof} X), \tilde{X} - X \rangle + (\tilde{\rho} - \rho) \det(\tilde{X} - X) \\ &= \langle \tilde{Y} - Y, \tilde{X} - X \rangle - 2\tilde{\rho} \det(\tilde{X} - X) + (\tilde{\rho} - \rho) \det(\tilde{X} - X). \end{aligned}$$

Therefore, we have for any $X \neq \tilde{X}$

$$0 < \langle \nabla\phi(\tilde{X}) - \nabla\phi(X), \tilde{X} - X \rangle - (\rho(X) + \rho(\tilde{X})) \det(X - \tilde{X}). \tag{31}$$

Proposition 3.12 *Suppose that $\phi : \mathbb{R}^{2\times 2} \to \mathbb{R}$ is strictly polyconvex, and that for*

$$K = \left\{ \begin{pmatrix} X_i \\ Y_i \end{pmatrix} \; ; \; i = 1, \ldots, n \right\} \subset K_\phi$$

there is a matrix $A \in \mathbb{R}^{2\times 2}$ such that

$$(Y_i - Y_j) = A(X_i - X_j) \textit{ for all } i, j \leq n.$$

Then $K^{rc} = K$.

Proof. From the definition of K we obtain $\nabla\phi(X_i) = -Y_i J$. Since for all $X \in \mathbb{R}^{2\times 2}$ the equation $XJ = J \operatorname{cof} X$ holds, we infer from (31) that for any $i \neq j$

$$\begin{aligned} 0 &< \langle -A(X_i - X_j)J, X_i - X_j \rangle - (\rho(X_i) + \rho(X_j)) \det(X_i - X_j) \\ &= \langle -AJ \operatorname{cof}(X_i - X_j), X_i - X_j \rangle - (\rho(X_i) + \rho(X_j)) \det(X_i - X_j) \\ &= \operatorname{tr}((-AJ) \operatorname{cof}(X_i - X_j)(X_i - X_j)^T) - (\rho(X_i) + \rho(X_j)) \det(X_i - X_j) \\ &= \det(X_i - X_j)(\operatorname{tr}(-AJ) - \rho(X_i) - \rho(X_j)). \end{aligned}$$

We denote

$$\sigma(X) = -(\frac{1}{2} \operatorname{tr}(AJ) + \rho(X))$$

and have therefore

$$\det(X_i - X_j)(\sigma(X_i) + \sigma(X_j)) > 0 \text{ for } i \neq j.$$

Lemma 3.13 below ensures now that in case $K^{rc} \neq K$ we find a closed cycle $i_0, i_1, \ldots, i_l, i_{l+1} \in \{1, \ldots, n\}$ with $i_0 = i_l$, $i_1 = i_{l+1}$ such that

$$\det(X_{i_k} - X_{i_{k-1}}) \det(X_{i_{k+1}} - X_{i_k}) < 0 \text{ for all } k = 1, \ldots, l.$$

Up to a shift of indices we can therefore suppose that

$$(-1)^k(\sigma(X_{i_k}) + \sigma(X_{i_{k+1}})) > 0 \text{ if } k = 1, \ldots, l$$

and that l is even. Summing over this cycle we get

$$-\sigma(X_{i_1}) + \sigma(X_{i_2}) - \sigma(X_{i_2}) \cdots + \sigma(X_{i_{l+1}}) > 0.$$

In other words, $0 > 0$ - this contradiction finishes our proof.

Lemma 3.13 *Assume a laminate* $\mu \in \mathcal{M}^{rc}(\mathbb{R}^{2\times 2})$ *is supported in a finite set* $\{X_1, \ldots, X_n\}$ *satisfying* $\det(X_i - X_j) \neq 0$ *if* $i \neq j$. *(See Section 4.2 for the definition of* $\mathcal{M}^{rc}$.*) Then there is a closed cycle*

$$i_0, i_1, \ldots, i_l, i_{l+1} \in \{1, \ldots, n\} \textit{ with } i_0 = i_l, i_1 = i_{l+1}$$

satisfying

$$\det(X_{i_k} - X_{i_{k-1}}) \det(X_{i_{k+1}} - X_{i_k}) < 0 \textit{ for } k = 1, \ldots, l.$$

Proof. We claim that

$$\text{for each } i \leq n \text{ there are } j, k : \det(X_j - X_i) \det(X_k - X_i) < 0. \tag{32}$$

Indeed, if $\det X_j - X_i$, has a fixed sign for all j different from i then one can show that the laminate μ is either a Dirac mass at X_i or does not charge X_i (see Proposition 4.17 below). This contradicts the hypothesis that the support of μ is exactly $\{X_1, \ldots, X_n\}$.

A little combinatorial argument will now show that (32) implies the existence of the required cycle. Indeed, (32) certainly enforces the existence of

$$i_1, \ldots, i_l, i_{l+1} \text{ with } i_1 = i_{l+1}, (-1)^k \det(X_{i_{k+1}} - X_{i_k}) < 0 \text{ if } 1 \leq k \leq l.$$

Consequently, the only problem that might occur is that l is odd and so $\det(X_{i_2} - X_{i_1}), \det(X_{i_l} - X_{i_1}) > 0$. But, we observe that then necessarily

$$\det(X_{i_k} - X_{i_1}) > 0 \text{ also for all } k \in \{3, \ldots, l-1\}.$$

Indeed, if this inequality fails then we extend the starting negative connection i_1, i_k into the adjacent positive connection and keep then running in this direction inside the already built cycle back to i_1. In other words, for even k we can take $i_k, i_1, i_2, \ldots, i_k, i_1$ as the desired cycle and if k is odd then $i_l, i_1, i_k, i_{k+1}, \ldots i_l, i_1$ does the job.

However, by (32) also X_{i_1} has to have a negative connection, so we find

$$X_{i_0} \text{ with } \det(X_{i_0} - X_{i_1}) < 0 \text{ and } X_{i_0} \notin \{X_{i_1}, \ldots, X_{i_l}\}.$$

The same reasoning as before now shows that

$$\det(X_{i_0} - X_{i_k}) < 0 \text{ for all } k = 2, \ldots, l.$$

Next, we find $X_{i_{-1}}$ with $\det(X_{i_{-1}} - X_{i_k}) > 0$ if $k \geq 0$, and so on. Obviously, repeating this argument finally leads to a contradiction, because our set $\{X_1, \ldots, X_n\}$ is finite.

Proof of Proposition 3.11. In light of Proposition 3.12 we only need to show that for any T_4-configuration $M_i = (X_i, Y_i) \in \mathbb{R}^{2\times 2} \times \mathbb{R}^{2\times 2}$ there exists an $A \in \mathbb{R}^{2\times 2}$ such that $Y_i - Y_j = A(X_i - X_j)$. According to Definition 3.9 we find $n_i, x_i, y_i \in \mathbb{R}^2$ and $\kappa_i > 0$ for $i = 1, \ldots, 4$ such that

$$\sum_{i=1}^{4} \begin{pmatrix} x_i \\ y_i \end{pmatrix} \otimes n_i = 0 \text{ and } M_{i+1} - M_i = (\kappa_{i+1} + 1) \begin{pmatrix} x_{i+1} \\ y_{i+1} \end{pmatrix} \otimes n_{i+1} - \kappa_i \begin{pmatrix} x_i \\ y_i \end{pmatrix} \otimes n_i.$$

So we are done, if we find $A \in \mathbb{R}^{2\times 2}$ with $y_i = Ax_i$ for all i. For this purpose we notice that any consecutive n_i, n_{i+1} are linearly independent and that x_1, x_2, x_3, x_4 span the whole $\mathbb{R}^2$. Indeed, the first statement is obvious from $\operatorname{rk}(M_i - M_{i+1}) > 1$. So suppose the second fails, then all matrices $X_1, \ldots, X_4$ one contained in a single rank-one plane $S \subset \mathbb{R}^{2\times 2}$. Of course, ϕ must be strictly convex on S which gives the monotonicity condition

$$0 < \langle \nabla\phi(X_i) - \nabla\phi(X_j), X_i - X_j \rangle = -\langle (Y_i - Y_j)J, X_i - X_j \rangle$$

for all $i \neq j$. Because $\langle YJ, X \rangle = (X_{11}Y_{12} - X_{12}Y_{11}) + (X_{21}Y_{22} - X_{22}Y_{21})$ we see that a certain sum of quadratic minors is negative on all differences in the set $\{M_1, \ldots, M_4\}$. By [60] this even implies that all polyconvex measures on $\{M_1, \ldots, M_4\}$ have to be Dirac masses. So it is clear that after reshuffling the indices if necessary, we can assume that $\{n_1, n_2\}$ and $\{x_3, x_4\}$ are both bases for $\mathbb{R}^2$.

Next, we observe that

$$u_1 \otimes v_1 + u_2 \otimes v_2 = (u_1 \; u_2) \begin{pmatrix} v_1 \\ v_2 \end{pmatrix},$$

if the u_i's are column and the v_i's are row vectors from $\mathbb{R}^2$. So we have

$$-(x_1 \; x_2) \begin{pmatrix} n_1 \\ n_2 \end{pmatrix} = (x_3 \; x_4) \begin{pmatrix} n_3 \\ n_4 \end{pmatrix}.$$

In other words our zero sum assumption implies

$$(x_1 \; x_2) = (x_3 \; x_4)B \text{ and } (y_1 \; y_2) = (y_3 \; y_4)B, \text{ for } B = -\begin{pmatrix} n_3 \\ n_4 \end{pmatrix} \begin{pmatrix} n_1 \\ n_2 \end{pmatrix}^{-1}.$$

Thus, if we define A by $y_j = Ax_j$ for $j = 3, 4$ then

$$(y_1 \; y_2) = (y_3 \; y_4)B = [A(x_3 \; x_4)]B = A[(x_3 \; x_4)B] = A(x_1 \; x_2)$$

as required. □

3.6 Beyond T_k

In the above examples it is enough to embed a T_4-configuration (or a T_5-configuration) to show that K^{rc} is sufficiently rich. The same strategy essentially works for Theorems 1.2 and 1.3. To understand more general examples and to find structure conditions which would exclude 'wild' solutions one would like to compute (or at least estimate) K^{rc} rather than just trying to embed T_k-configurations (we will see below examples that K^{rc} can be nontrivial even if K contains no T_k-configuration). Ultimately this will require a deeper understanding of the geometry of rank-one convexity. Due to the high dimensions of the rank-one cone and the surrounding space this seems rather difficult at the moment. To build up some intuition we discuss below some examples where by different means one can reduce the dimensionality of the problem and thus gain some geometric insight.

One interesting example is set K related to the simplest polyconvex integrand

$$\varepsilon|F|^2 + (\det F)^2, \quad \varepsilon \geq 0.$$

For $\varepsilon = 0$ we show in Section 5 that $K^{rc} = K$. Interestingly, it is not known whether the same holds for the seemingly better case $\varepsilon > 0$, see Section 6 for some partial results. In both cases no partial regularity results for weak solutions are known, except for the case of local minimizers.

In Section 7 we raise the question what one can say about the local rank-one convex hull. One issue is whether one can formulate higher order conditions which give triviality of the hull when the tangent space of K contains rank-one lines. We first review some general tools to study rank-one convex hulls.

4 Tools to Study Rank-One Convex Hulls

In this section we collect some definitions and tools which are useful for the study of rank-one convex hulls and related hulls. For more detailed accounts see e.g. [13, 52, 42].

4.1 Rank-One Convex, Quasiconvex, and Polyconvex Hulls

In the following we always assume that $m, n \geq 2$ since otherwise all the convexity notions introduced below agree with ordinary convexity. A function $f : \mathbb{R}^{m\times n} \to \mathbb{R}$ is rank-one convex if is convex on each rank-one line, it is polyconvex if it can be expressed as a convex functions of minors (subdeterminants) and it is quasiconvex if

$$\int_{T^n} f(A + \nabla\eta) - f(A)\,dx \geq 0 \tag{33}$$

for all $A \in \mathbb{R}^{m\times n}$ and for all periodic Lipschitz maps $\eta : T^n \to \mathbb{R}^m$. We have the following implications:

$$f \text{ polyconvex } \Rightarrow f \text{ quasiconvex } \Rightarrow f \text{ rank-one convex.} \tag{34}$$

For a compact set we define the rank-one convex, quasiconvex and polyconvex hull as the set of those points which can not be separated by the corresponding class of functions.

$$K^* := \{A \in \mathbb{R}^{m\times n} : f(A) \le \sup_K f : f \text{ is } *\}, \quad * = \{\text{rc,qc,pc}\}. \tag{35}$$

For a general set E we set

$$E^* := \bigcup_{K \subset E \text{ compact}} K^*, \quad * = \{\text{rc,qc,pc}\}. \tag{36}$$

In view of (34)

$$K^{pc} \supset K^{qc} \supset K^{rc}. \tag{37}$$

As we have seen in Sections 2 and 3 above a large rank-one convex hull allows one to construct many solutions of $\nabla u \in K$. The quasiconvex convex hull is related to the stability of the partial differential relation under weak convergence. More specifically if

$$\nabla u^{(j)} \overset{*}{\rightharpoonup} \nabla u \quad \text{and} \quad \operatorname{dist}(\nabla u^{(j)}, K) \to 0 \text{ in } L^1 \tag{38}$$

then

$$\nabla u \in \bar{K}^{qc}, \tag{39}$$

and this property characterizes K^{qc} for compact K. The polyconvex hull provides an upper bound for both K^{qc} and K^{rc}.

4.2 The Dual Objects: Laminates and Gradient Young Measures

Let $\mathcal{P}(K)$ denote the set of probability measures supported on K. For $\mu \in \mathcal{P}(K)$ we denote by $\bar{\mu} = \int A \, d\mu(A)$ its barycentre. We consider the following subsets of $\mathcal{P}(K)$ which satisfy a Jensen's inequality with respect to the above convexity notions.

$$\begin{aligned}
\mathcal{M}^{rc}(K) &= \{\mu \in \mathcal{P}(K) : \int f(A)\, d\mu(A) \ge f(\bar{\mu}),\ \forall f \text{ rc}\} \\
\mathcal{M}^{qc}(K) &= \{\mu \in \mathcal{P}(K) : \int f(A)\, d\mu(A) \ge f(\bar{\mu}),\ \forall f \text{ qc}\} \\
\mathcal{M}^{pc}(K) &= \{\mu \in \mathcal{P}(K) : \int f(A)\, d\mu(A) \ge f(\bar{\mu}),\ \forall f \text{ pc}\} \\
&= \{\mu \in \mathcal{P}(K) : \int M(A)\, d\mu(A) = M(\bar{\mu}), \text{ for all minors } M\}.
\end{aligned}$$

For a compact set K we have

$$K^* = \{\bar{\mu} : \mu \in \mathcal{M}^*(K)\}, \quad * \in \{\text{rc,qc,pc}\}.$$

The elements of $\mathcal{M}^{qc}(\mathbb{R}^{m\times n})$ are called homogeneous gradient Young measures since they arise as distribution functions of the gradients of periodic functions. More precisely let $\mu \in \mathcal{M}^{qc}(\mathbb{R}^{m\times n})$ be a measure with compact support and let $\bar{A} = \bar{\mu}$

be its barycentre. Then there exists periodic maps $u^{(j)} : T^n \to \mathbb{R}^m$ with uniform Lipschitz constant such that the measures μ^j defined by $\int_{\mathbb{R}^{m\times n}} \eta(A)d\mu^j(A) := \int_{T^n} \eta(\bar{A} + \nabla u^{(j)})\, dx$ converge weak* to μ, see [31, 63]. For a compact set K the set $\mathcal{M}^{qc}(K)$ is trivial (i.e. contains only Dirac masses) if and only if all sequences $\nabla u^{(j)}$ satisfying (38) converge strongly in L^1.

The elements of $\mathcal{M}^{rc}(K)$ are called laminates and they can be obtained as weak* limits of so called laminates of finite order (with uniformly bounded support), see [51, 45]. The class $\mathcal{L}$ of laminates of finite order is defined by inductive splitting along rank-one segments as follows. First, each Dirac mass δ_A belongs to $\mathcal{L}$. Second, suppose that $\lambda_i \geq 0$, $\sum_{i=1}^m \lambda_i = 1$, that $\nu = \sum \lambda_i \delta_{A_i} \in \mathcal{L}$ and that $A_m = (1-s)B_1 + sB_2$ is a rank-one convex combination of B_1 and B_2. Then $\mu = \sum_{i=1}^{m-1} \lambda_i \delta_{A_i} + (1-s)\lambda_m \delta_{B_1} + s\lambda_m \delta_{B_2}$ also belongs to $\mathcal{L}$. Thus laminates are the distribution functions of gradients which arise from essentially one-dimensional constructions.

4.3 Localization and Extension

Rank-one convexity is a local property (while qc and pc are not, at least for $m \geq 3$, see [35]) and this greatly simplifies the construction of separating functions. We first recall that it suffices to separate locally.

Proposition 4.14 ([45], Lemma 2.3) *Let K be compact and let U be an open neighbourhood of K^{rc}. Suppose that $f : U \to \mathbb{R}$ is rank-one convex in U, i.e. convex on each rank-one segment entirely contained in U. Then there exists a rank-one convex function $F : \mathbb{R}^{m\times n} \to \mathbb{R}$ which agrees with f in a neighbourhood of K^{rc}.*

One can also extend from lower dimensional sets, see [58] for the details.

Proposition 4.15 *Let L be a subspace $\mathbb{R}^{m\times n}$ and suppose that $f : L \to \mathbb{R}$ is C^2 and is rank-one convex on L. Let $\delta > 0$ and let $E \subset L$ be compact. Then there exists a rank-one convex function $F : \mathbb{R}^{m\times n} \to \mathbb{R}$ which satisfies $\sup_E |F - f| \leq \delta$.*

If K^{rc} has several components then the rank-one convex hull can be computed for each piece separately.

Proposition 4.16 ([33], Thm. 4.7) *Let K be a compact set.*

(i) Let B be bounded. Then

$$K^{rc} \cap B = [(K \cap B) \cup (K^{rc} \cap \partial B)]^{rc} \cap B. \tag{40}$$

(ii) Let $C_1, \ldots C_k$ be disjoint compact sets and suppose $K^{rc} \subset \cup_i C_i$. Then $K^{rc} = \cup_i (K \cap C_i)^{rc}$.

Part (ii) appears already in [51, 40] (see [39] for a detailed proof) and it can easily be deduced from part (i) by taking B as a sufficiently small neighbourhood U_i of C_i which does not intersect the other C_j. The following application of the proposition will be useful later.

Proposition 4.17 *Let K be a compact subset of $\{X \in \mathbb{R}^{2\times 2} : \det X > 0\}$. Then $(K \cup \{0\})^{rc} = K^{rc} \cup \{0\}$ and every laminate supported on $K \cup \{0\}$ is either supported on K or is a Dirac mass at zero. In particular, if*

$$E = \{X \in \mathbb{R}^{2\times 2} : \det X = 1\} \cup \{0\}, \quad \textit{then } E^{rc} = E.$$

Remark 4.18 *Astala and Faraco [2] have shown that the same assertion holds for the quasiconvex hull and measures in $\mathcal{M}^{qc}(K \cup \{0\})$, i.e., gradient Young measures. Their proof uses ideas from the theory of quasiregular maps, in particular a careful analysis of the Beltrami equation.*

Proof. By compactness there exist $\varepsilon > 0, R > 0$ such that $\det X \geq \varepsilon$ and $|X| \leq R$ for all $X \in K$. The polyconvex function $f(X) = \varepsilon|X| - R\det X$ is ≤ 0 in $K \cup \{0\}$, but is positive for $0 < |X| < 2\varepsilon/R$. Thus we can apply Proposition 4.16 with $C_1 = \{X \in \mathbb{R}^{2\times 2} : |X| \leq \varepsilon/R\}$ and $C_2 = \{X \in \mathbb{R}^{2\times 2} : |X| \geq 2\varepsilon/R\}$ and we obtain $(K \cup \{0\})^{rc} = K^{rc} \cup \{0\}$.

Let μ be a laminate supported on K, with barycentre $\bar{\mu}$. Suppose first $\bar{\mu} \in K^{rc}$. Let U_i be small neighbourhoods of C_i and define $g : U = U_1 \cup U_2 \to \mathbb{R}$ by $g = -1$ on U_1 and $g = 1$ on U_2. Since g is constant on each component of U it is trivially rank-one convex. By Proposition 4.14 there exists a rank-one convex function $\tilde{g} : \mathbb{R}^{2\times 2} \to \mathbb{R}$ which agrees with g on K. Hence $1 = \tilde{g}(\bar{\mu}) \leq \int \tilde{g}\, d\mu = \mu(K) - \mu(\{0\})$. Thus μ must be supported on K. If $\bar{\mu} = 0$ one concludes similarly that $\mu = \delta_0$ by starting from the function $-g$. □

5 The Simplest Polyconvex Integrand

Here we consider the simplest polyconvex integrand

$$\phi(X) = \frac{1}{2}(\det X)^2.$$

The Euler-Lagrange equation $\operatorname{div} D\phi(\nabla w) = 0$ is equivalent to the first order partial differential relation

$$\nabla u \in K := \left\{ \begin{pmatrix} X \\ Y \end{pmatrix} \in \mathbb{R}^{4\times 2} : Y = \det X (\operatorname{cof} X) J,\ X \in \mathbb{R}^{2\times 2} \right\}, \tag{41}$$

where J is the $90°$ rotation.

Theorem 5.19 $K^{rc} = K$.

To prove this result it is convenient to make the change of variables $\binom{X}{Y} \to \binom{X}{-JY}$. This bijection maps rank-one lines onto rank-one lines and therefore does not affect the computation of rank-one convex hulls. Since $-J \operatorname{cof} X J = X$ it thus suffices to consider the set

$$\tilde{K} = \left\{ \begin{pmatrix} X \\ Y \end{pmatrix} \in \mathbb{R}^{4\times 2} : Y = (\det X)X,\ X \in \mathbb{R}^{2\times 2} \right\}. \tag{42}$$

The set $\tilde{K}$ is contained in the cone

$$\Sigma = \left\{ \begin{pmatrix} \alpha_1 & \alpha_2 \\ \beta_1 & \beta_2 \\ \gamma_1 & \gamma_2 \\ \delta_1 & \delta_2 \end{pmatrix} : \alpha \wedge \gamma = \beta \wedge \delta = 0,\ \alpha \wedge \delta = \beta \wedge \gamma \right\}, \tag{43}$$

where $\alpha \wedge \gamma = \alpha_1\gamma_2 - \alpha_2\gamma_1$ etc. Since Σ is defined by minors it is polyconvex and thus $\tilde{K}^{rc} \subset \Sigma$. To separate points in $\Sigma \setminus \tilde{K}$ we first use separating rank-one convex functions defined only on Σ. Then we extend these function to $\mathbb{R}^{4\times 2}$. Since Σ is not smooth this requires some care.

5.1 A Rank-One Foliation of Σ

The construction of rank-one convex functions in Σ is largely simplified by the fact that most of Σ can be decomposed into simpler sets which contain all the rank-one connections.

Proposition 5.20 *We have*

$$\Sigma = \bigcup_{\lambda \in \mathbb{R}} L_\lambda \cup L_\infty \cup N, \tag{44}$$

where

$$L_\lambda = \left\{ \begin{pmatrix} X \\ \lambda X \end{pmatrix} : X \in \mathbb{R}^{2\times 2} \right\}, \quad L_\infty = \left\{ \begin{pmatrix} 0 \\ Y \end{pmatrix} : Y \in \mathbb{R}^{2\times 2} \right\},$$

$$N = \left\{ a \otimes b : a \in \mathbb{R}^4, b \in \mathbb{R}^2 \right\}.$$

Moreover if $A, B \in \Sigma$ satisfy $\operatorname{rk}(A - B) = 1$ *then the rank-one line trough A and B lies entirely in N or entirely in one of the spaces L_λ, where $\lambda \in \mathbb{R} \cup \{\infty\}$.*

Proof. Clearly N, L_λ and L_∞ are contained in Σ. To see the converse suppose suppose first that $A = (\alpha, \beta, \gamma, \delta)^T$ satisfies $\alpha \wedge \beta \neq 0$ or $\alpha \wedge \delta = \beta \wedge \gamma \neq 0$. Then $\alpha \neq 0$ and $\beta \neq 0$ and the first two conditions in the definition of Σ give $\gamma = \lambda\alpha$, $\delta = \mu\beta$ while the last condition gives $\mu = \lambda$. Hence $A \in L_\lambda$. Now assume $\alpha \wedge \beta = \alpha \wedge \delta = \beta \wedge \gamma = 0$. If $\gamma \wedge \delta \neq 0$ then $\alpha = \lambda\gamma$, $\beta = \mu\delta$ and this yields $\alpha = \beta = 0$, so that $A \in L_\infty$. Finally, if also $\gamma \wedge \delta = 0$ then all 2×2 minors of A are zero. Hence $A \in N$.

Since Σ is given by minors any rank-one line

$$A(t) = \begin{pmatrix} X(t) \\ Y(t) \end{pmatrix} = \begin{pmatrix} X + tx \otimes n \\ Y + ty \otimes n \end{pmatrix}$$

lies entirely in Σ if two points lie in Σ. If $X(t) \equiv 0$ then $A(t)$ lies in L_∞. Next assume that $X(t) \not\equiv 0$ but $\det X(t) \equiv 0$. We claim that $A(t) \in N$. Indeed $X(t) \neq 0$ for all but one value of t. If $A(t)$ was not in N then $Y(t) = \lambda(t)X(t)$. Since $\det X(t) = 0$ this yields again $A(t) \in N$ for all but one t and hence for all t. Finally assume that $\det X(t) \not\equiv 0$. Then $\det X(t) \neq 0$ for all but one t. In particular $Xn^\perp \neq 0$ where $n^\perp = Jn$ is perpendicular to n. Moreover $Y(t) = \lambda(t)X(t)$. Applying this identity to $n^\perp$ we deduce that $\lambda(t)$ must be constant, so that the rank-one line lies in one L_λ. □

5.2 Separating Functions on Σ

The foliation of Σ almost allows one to compute the rank-one convex hull separately on all the spaces L_λ and on N. More precisely we make the following ansatz for the separating function

$$f(X,Y) = h(X,Y) - \begin{cases} \varphi(\lambda) \det Y & \text{if } \binom{X}{Y} \in L_\lambda, \\ 0 & \text{if } \binom{X}{Y} \in N \cup L_\infty. \end{cases} \tag{45}$$

Here $h : \Sigma \to \mathbb{R}$ is rank-one convex and $\varphi \in C_0^\infty(\mathbb{R})$. Since each rank-one line stays either in L_λ or in N the function f is automatically rank-one convex. On $\tilde{K}$ the expression for f simplifies since $\tilde{K} \cap (N \cup L_\infty) = \{0\}$ and in $\tilde{K} \cap L_\lambda$ we have $\lambda = \det X$. We thus have

$$f(X,Y) = h(X, X \det X) - \varphi(\det X)(\det X)^3 \tag{46}$$

Now we specify h and ϕ in dependence of the point $\bar{A} = \binom{\bar{X}}{\bar{Y}} \in \Sigma \setminus \tilde{K}$ we seek to separate.

Case 1: $\bar{A} \in L_{\bar{\lambda}}, \bar{\lambda} \notin \{0, \infty\}$. The point $A = \binom{X}{Y}$ belongs to $\tilde{K} \cap L_{\bar{\lambda}}$ if and only if

$$Y \in E := \{Z \in \mathbb{R}^{2\times 2} : \det Z = \bar{\lambda}^3\} \cup \{0\}.$$

By Proposition 4.17 the rank-one convex hull of E is trivial. Hence there exists a rank-one convex function $\bar{h} : \mathbb{R}^{2\times 2} \to \mathbb{R}$ such that $\bar{h}(\bar{Y}) = 1$ and $\bar{h} \le 0$ on E. Set $h(X,Y) = \bar{h}(Y)$. In view of (46) it suffices to show that (for each R) there exists φ such that

$$\varphi(\bar{\lambda}) = 0,$$

$$\bar{h}(X \det X) - \varphi(\det X)(\det X)^3 \le \frac{1}{2}, \quad \forall X \text{ with } |X| \le R$$

To see that this is possible first note that if $\det X = \bar{\lambda}$ then $\bar{h}(X \det X) \le 0$ by construction. Hence $\bar{h}(X \det X) \le 1/2$ for $|\det X - \bar{\lambda}| \le \delta_0(R)$. Similarly for $\det X = 0$ we have $\bar{h}(X \det X) = 0$. Hence $\bar{h}(X \det X) \le 1/2$ for $|\det X| \le$

$\delta_0(R)$. For the remaining values of $\det X$ the desired inequality can be achieved by a suitable choice of φ (note that $\det X \leq R^2$ so that there is no obstruction to choosing φ with compact support).

Case 2: $\bar{A} = \binom{\bar{X}}{0} \in L_0 \setminus N$. We have $\det \bar{X} \neq 0$ and we take

$$h(X,Y) = \det X \operatorname{sgn}(\det \bar{X}).$$

Then we can choose φ such that

$$|\lambda| - \varphi(\lambda)\lambda^3 \leq \frac{|\det \bar{X}|}{2}, \quad \text{for } |\lambda| \leq R^2. \tag{47}$$

Case 3: $\bar{A} \in N \cup L_\infty$. Let

$$h(X,Y) = |Y|^2, \quad \varphi = R^2 \hat{\varphi},$$

where $\lambda^2 - \hat{\varphi}(\lambda)\lambda^3 \leq R^{-2}|\bar{Y}|^2/2$. On the set $\tilde{K}$ the function f is bounded by

$$|X \det X|^2 - \varphi(\det X)(\det X)^3 \tag{48}$$

$$\leq R^2[(\det X)^2 - \hat{\varphi}(\det X)(\det X)^3] \leq \frac{|\bar{A}|^2}{2}. \tag{49}$$

5.3 Extension

We show that the function $f : \Sigma \to \mathbb{R}$ constructed above can be approximated (uniformly of compact subsets of Σ) by functions which are rank-1 convex in a neighbourhood of Σ. In view of Proposition 4.14 this will finish the proof of Theorem 5.19.

The main point is to define an analogue of the parameter λ in (45) for a general matrix $A \in \mathbb{R}^{4\times 2}$. Consider the set

$$\tilde{\Sigma} = \left\{ \begin{pmatrix} X \cos\alpha \\ X \sin\alpha \end{pmatrix} : X \in \mathbb{R}^{2\times 2}, \alpha \in (-\pi/2, \pi/2] \right\}.$$

A point $A = \binom{X}{Y} \in R^{4\times 2}$ has a unique best approximation $\pi_{\tilde{\Sigma}}(A)$ in $\tilde{\Sigma}$ if and only if

$$P(A) := (|X|^2 - |Y|^2)^2 + 4\langle X, Y\rangle^2 \neq 0.$$

In this case the optimal angle $\bar{\alpha}(A)$ is determined by the condition

$$\begin{pmatrix} \cos 2\bar{\alpha} \\ \sin 2\bar{\alpha} \end{pmatrix} = \frac{1}{P(A)^{1/2}} \begin{pmatrix} |X|^2 - |Y|^2 \\ 2\langle X, Y\rangle \end{pmatrix}.$$

On the set $P \neq 0$ the closest point projection $\pi_{\tilde{\Sigma}}$ is smooth. We define $\hat{\lambda}(A) = \tan \bar{\alpha}(A)$ and replace the ansatz (45) by

$$f_\delta(A) = h(A) - \varphi(\hat{\lambda}(A))\eta(\frac{P(A)}{\delta^4}) \det Y, \tag{50}$$

where $\varphi \in C_0^\infty(\mathbb{R})$, where $1-\eta \in C_0^\infty(-1,1)$ with $\eta_{|(-1/2,1/2)} = 1$ and where $\delta > 0$ is a small parameter. Then f_δ is well-defined on $\mathbb{R}^{4\times 2}$ and smooth. We first claim that on Σ the function f_δ is close to f and nearly rank-one convex.

Clearly $f_\delta = f$ on N since $\det Y = 0$ on N. On each L_λ (including $\lambda = \infty$) we have $P(A) = |A|^4$. Hence $f_\delta = f$ on $\Sigma \setminus B_\delta$ and thus $\sup_\Sigma |f - f_\delta| \le C\delta^2$. Using again that $P(A) = |A|^4$ on L_λ and the homogeneity of $\hat{\lambda}$ and det it is easy to verify that for each rank-one line $A + ta\otimes b$ in Σ we have $D^2 f_\delta(A)(a\otimes b, a\otimes b) \ge -C|a|^2|b|^2$. Thus the proof is concluded by the following extension result.

For $A = (\alpha, \beta, \gamma, \delta)^T \in \mathbb{R}^{4\times 2}$ consider the minors

$$U_1(A) = \alpha \wedge \gamma, \quad U_2(A) = \beta \wedge \delta, \quad U_3(A) = \alpha\wedge\delta - \beta\wedge\gamma$$

and recall that Σ was defined as the set where all three minors vanish.

Proposition 5.21 *Suppose that $f \in C^2(\mathbb{R}^{4\times 2})$. Assume that for all rank-one lines $A + ta\otimes b$ contained in Σ one has*

$$\begin{aligned} D^2 f_\delta(A)(a\otimes b, a\otimes b) &\ge 0 && \textit{if } |A| \ge \delta, \\ D^2 f_\delta(A)(a\otimes b, a\otimes b) &\ge -C_0|a|^2|b|^2 && \textit{if } |A| < \delta. \end{aligned}$$

Then there exists a smooth convex function $g : \mathbb{R}^{4\times 2} \to \mathbb{R}$ with

$$0 \le g(A) \le CC_0\delta|A|$$

and for each $\varepsilon > 0$ there exists $\mu \ge 0$ such that the function

$$F := f_\delta + g + \mu\sum_{i=1}^{3} U_i^2 + \varepsilon|A|^2$$

is rank-one convex in a neighbourhood of Σ.

Proof. First note that there exists a smooth convex function g with the properties stated in the proposition which satisfies $D^2 g(A) \ge C_0 \,\mathrm{Id}$ for $|A| < \delta$. Fix $\varepsilon > 0$ and suppose the assertion of the proposition was false. Then there exist points $A_k \to A$ with $A \in \Sigma$, rank-one directions $B_k = a_k \otimes b_k$ with $|B_k| = 1$ and $B_k \to B$, and $\mu_k \to \infty$ such that

$$D^2 f_\delta(A_k)(B_k, B_k) + D^2 g(A_k)(B_k, B_k) + 2\mu_k \sum_{i=1}^{3}\langle DU_i(A_k), B_k\rangle^2 \le -2\varepsilon.$$

Taking the limit $k \to \infty$ we see that

$$D^2 f_\delta(A)(B,B) + D^2 g(A)(B,B) \le -2\varepsilon \tag{51}$$

and $\langle DU_i(A), B\rangle = 0$. Hence $A + tB$ is a rank-one line on Σ and thus (51) leads to a contradiction with the hypotheses on $D^2 f$ and the choice of g. □

Remark 5.22 *The main idea behind formula (50) can be understood as follows. Let $\Sigma_{reg} = \bigcup_{\lambda\in\mathbb{R}} L_\lambda \setminus \{0\}$. We note that λ can be thought of as a smooth function on Σ_{reg} defined by $A \in L_{\lambda(A)}$. For $\varphi \in C_0^\infty(\mathbb{R})$ the function $A \to \varphi(\lambda)$ can be smoothly extended from $\Sigma_{reg} \cap \mathbb{S}^{(4\times 2)-1}$ to all of this unit sphere. Such a function can now be extended to a smooth zero homogeneous function $\tilde{\varphi}$ on $\mathbb{R}^{4\times 2} \setminus \{0\}$ which still agrees with the original function on all of Σ_{reg}. Moreover, the function*

$$A = \begin{pmatrix} X \\ Y \end{pmatrix} \to \tilde{\varphi}(A)\det(Y)$$

is 2-homogeneous on $\mathbb{R}^{4\times 2}$, smooth away from 0 and rank-one convex on Σ. The role of η and Proposition 5.21 is to handle the singularity at 0. Since the second derivatives of $\tilde{\varphi}(A)\det(Y)$ are bounded, this does not present a problem.

5.4 More General Polyconvex Integrands

It seems natural to expect that Theorem 5.19 can be extended to all the strictly polyconvex integrands

$$\phi(F) = \frac{1}{2}\varepsilon|F|^2 + \frac{1}{2}(\det F)^2, \quad \varepsilon > 0, \tag{52}$$

but whether this is true is not known. If we restrict X and Y to diagonal matrices then the corresponding set $\tilde{K}$ becomes

$$\tilde{K}_{\text{diag}} = \left\{ \begin{pmatrix} s & 0 \\ 0 & t \\ (st)s + \varepsilon t & 0 \\ 0 & (st)t + \varepsilon s \end{pmatrix} : s, t \in \mathbb{R} \right\}.$$

This set is contained in a four-dimensional subspace and on this subspace rank-one convexity in $\mathbb{R}^{4\times 2}$ reduces to separate convexity in $\mathbb{R}^2 \oplus \mathbb{R}^2$. Even in this simplified setting it is not known whether $\tilde{K}_{\text{diag}}^{rc} = \tilde{K}_{\text{diag}}$. In the next section we will establish this result at least for finite sets (this in particular implies that no T_k-configuration can be embedded in $\tilde{K}_{\text{diag}}$). To do so we study separate convexity in more detail.

6 Separate Convexity

6.1 Separate Convexity in $\mathbb{R}^2$

This corresponds to the cone $\Lambda = \mathbb{R}\times\{0\}\cup\{0\}\times\mathbb{R} \subset \mathbb{R}^2$ which arises if we restrict the rank-one convex cone in 2×2 matrices to diagonal matrices. This situation is relatively well understood. In particular every nontrivial configuration must contain a T_4-configuration.

Proposition 6.23 ([69], Remark 10; [40], Proposition 5.3) *Let K be a compact set in diagonal 2×2 matrices.*

(i) Every point $A \in K^{rc}$ is contained in the rank-one convex hull of a subset of K consisting of at most five points.
(ii) If K contains no rank-one connections but $K^{rc} \neq K$ then K must contain a T_4-configuration.

There also exists an efficient algorithm for the computation of the rank-one convex hull [40]. Moreover on diagonal 2×2 matrices rank-one convexity and quasiconvexity agree, in the sense that the spaces $\mathcal{M}^{rc}$ of laminates and $\mathcal{M}^{qc}$ of gradient Young measures agree [43].

Separate convexity in $\mathbb{R}^n = \mathbb{R} \oplus \ldots \oplus \mathbb{R}$ which arises by restricting rank-one convexity in $\mathbb{R}^{n\times n}$ to diagonal matrices is already more subtle, see [40]. Here we are more interested in separate convexity in $\mathbb{R}^2 \oplus \mathbb{R}^2$ and as an intermediate step we consider $\mathbb{R}^2 \oplus \mathbb{R}$.

One key tool is the following separation argument for sets which are supported in two opposite quadrants [69]. In order to use the same notions as in Section 4 we formally view separate convexity in $\mathbb{R}^m \oplus \mathbb{R}^n$ as a special case of rank-one convexity (see (54)).

Lemma 6.24 *For $x_0, a \in \mathbb{R}^m$ and y_0, b in $\mathbb{R}^n$ consider the generalized quadrants*

$$Q_{\pm,\pm} = \{(x,y) \in \mathbb{R}^m \oplus \mathbb{R}^n : \pm\langle x - x_0, a\rangle > 0,\ \pm\langle y - y_0, b\rangle > 0\}$$

and the set

$$Q_{00} = \{(x,y) \in \mathbb{R}^m \oplus \mathbb{R}^n : \langle x - x_0, a\rangle = \langle y - y_0, b\rangle = 0\}.$$

Let the set $K \subset Q_{++} \cup Q_{--} \cup Q_{00}$ be compact and let μ be laminate supported on K, with barycentre $\bar{\mu}$. Then one of the following three assertions holds

(i) $\bar{\mu} \in Q_{00}$ and $\operatorname{supp} \mu \subset Q_{00}$,
(ii) $\bar{\mu} \in Q_{++}$ and $\operatorname{supp} \mu \subset Q_{++} \cup Q_{00}$,
(iii) $\bar{\mu} \in Q_{--}$ and $\operatorname{supp} \mu \subset Q_{--} \cup Q_{00}$.

If, in addition, $K \cap Q_{++}$ and $K \cap Q_{--}$ are compact (e.g., if K is finite) then in (ii) and (iii) one has $\operatorname{supp} \mu \subset Q_{++}$ and $\operatorname{supp} \mu \subset Q_{--}$, respectively.

Proof. We may suppose $x_0 = y_0 = 0$. Consider the separately affine function

$$f(x,y) = \langle x, a\rangle\langle y, b\rangle.$$

Then $f(\bar{\mu}) = \int_K f\, d\mu \geq 0$. If $f(\bar{\mu}) = 0$ then $\operatorname{supp} \mu \subset Q_{00}$ and hence $\bar{\mu} \in Q_{00}$. If $f(\bar{\mu}) > 0$ then $\bar{\mu} \in Q_{++} \cup Q_{--}$. Suppose $\bar{\mu} \in Q_{++}$ and consider the separately convex function

$$g(x,y) = \langle x, a\rangle_+ \langle y, b\rangle_+ - \langle x, a\rangle\langle y, b\rangle.$$

where $a_+ = \max(a, 0)$. Then $g \leq 0$ on K and $g < 0$ on $K \cap Q_{--}$. Since $0 = g(\bar{\mu}) \leq \int g \, d\mu$ it follows that $\operatorname{supp} \mu \subset Q_{++} \cup Q_{00}$. If $K \cap Q_{++}$ is compact we can find $\varepsilon > 0$ such that $\min(\langle x, a \rangle, \langle y, b \rangle) \geq \varepsilon$ for all $(x, y) \in K \cap Q_{++}$. In addition we may assume that $\langle \bar{\mu}_1, a \rangle \geq \varepsilon$ and $\langle \bar{\mu}_2, b \rangle \geq \varepsilon$, where $\bar{\mu} = (\bar{\mu}_1, \bar{\mu}_2)$. Then we can replace $\langle x, a \rangle$ with $\langle x, a \rangle - \varepsilon$ and $\langle y, b \rangle$ with $\langle y, b \rangle - \varepsilon$ in the definition of g and we conclude easily. The case $\bar{\mu} \in Q_{--}$ is analogous. □

6.2 Separate Convexity in $\mathbb{R}^2 \oplus \mathbb{R}$

Here we consider the cone

$$\Lambda = \mathbb{R}^2 \times \{0\} \cup \{0\} \times \mathbb{R} \subset \mathbb{R}^3. \tag{53}$$

We can view Λ as a subset of the rank-one cone in 3×2 matrices if we identify $\mathbb{R}^3$ with the space

$$L = \left\{ \begin{pmatrix} x & 0 \\ y & 0 \\ 0 & z \end{pmatrix} : x, y, z \in \mathbb{R} \right\}. \tag{54}$$

With this identification in mind we continue to write K^{rc} instead of K^Λ for the rank-one convex hull. We wish to understand how the more complicated geometry of separate convexity in $\mathbb{R}^2 \oplus \mathbb{R}$ can be used to construct examples which do not exist in $\mathbb{R} \oplus \mathbb{R}$. We first summarize the results and then turn to the proofs.

Proposition 6.25 *Let L be given by* (54). *Suppose that $K \subset L$ consists of at most five points and contains no rank-one connection. If $K^{rc} \neq K$ then K contains a T_4-configuration.*

Proposition 6.26 *Let L be given by* (54). *There exists a set $K \subset L$ which consists of six points, contains no T_4-configuration and has a non-trivial rank-one convex hull $K^{rc} \neq K$.*

For the construction see Fig. 6. The short proof that the example contains no T_k configuration is given below. An interesting class of sets without rank-one connections are (monotone) graphs over curves without self-intersections. For curves which are 'spiral-like' we can show that K^{rc} is trivial (see Fig. 4).

Proposition 6.27 *Identify L given by* (54) *with $\mathbb{R}^3$. Let $\gamma : [0, T] \to \mathbb{R}^2$ have a regular C^1-image (i.e. if the curve is parametrized by arclength, the derivative exists everywhere and varies continuously) which satisfies for all $t \in [0, T)$ that $\gamma(t)$ is not in the convex hull of $\{\gamma(s) : s > t\}$. Set $K = \{(\gamma(t), t) : t \in [0, T]\}$. Then $K^{rc} = K$.*

Without the hypotheses on the convex hull the result may fail, since one can embed a T_4 configuration (see Fig. 4). Interestingly, even with that hypotheses the result can fail if we allow Lipschitz curves rather than C^1 curves, see Example 6.28 and Fig. 7 below.

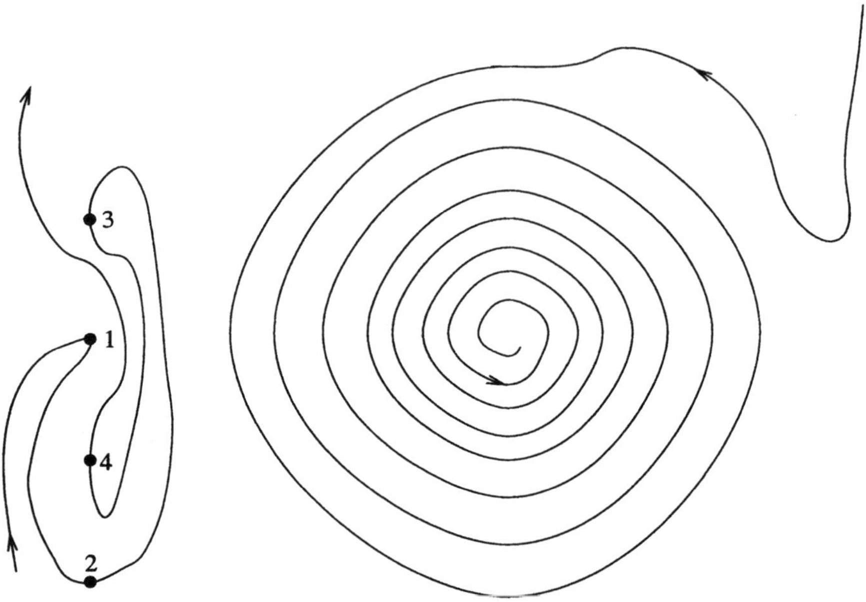

Fig. 4. The figure on the left shows the T_4 configuration embedded on a non selfintersecting curve. The highly nonconvex spiral on the right, however, leads to a trivial rank-one convex hull

Proof of Proposition 6.25. The proof shows that the simple geometric separation argument, Lemma 6.24, becomes quite powerful when combined with the localization formula (40) in Proposition 4.16 (i). We first eliminate the simpler cases when K contains four or less points.

If K contains three or less points, then it is well-known that the absence of rank-one connections implies $K^{rc} = K$. For two points one can use a suitable 2×2 minor to show this. We give a proof for three points for our example since the same argument can be used for four points. Let $K = \{P_1, P_2, P_3\}$ with $P_i = (p_i, z_i) = (x_i, y_i, z_i)$. Since there are no rank-one connections we can assume (after a permutation of indices if necessary) that $z_1 < z_2 < z_3$. We claim that $p_1 \in \{p_2, p_3\}^{co}$. If this fails then there exists $a \in \mathbb{R}^2$ such that $\langle p_i - p_1, a \rangle > 0$, for $i = 2, 3$. Thus the points P_2 and P_3 lie in the generalized quadrant

$$Q_{++} = \{(p, z) : \langle p - p_1, a \rangle > 0,\ z - z_1 > 0\},$$

and $K \subset Q_{++} \cup Q_{00}$. It follows from Lemma 6.24 that a nontrivial laminate supported on K must be supported on $\{P_2, P_3\}$, but this possibility has already been ruled out. Therefore $p_1 \in \{p_2, p_3\}^{co}$. Similarly we conclude $p_3 \in \{p_1, p_2\}^{co}$. Thus $p_1 = p_2 = p_3$ and K lies in a rank-one line, in contradiction with our assumption.

If $K = \{P_1, P_2, P_3, P_4\}$ we can again assume $z_1 < z_2 < z_3 < z_4$ and the separation argument given above implies that $p_1 \in \{p_2, p_3, p_4\}^{co}$ and $p_4 \in \{p_1, p_2, p_3\}^{co}$. Thus the convex set $\{p_1, p_2, p_3, p_4\}^{co}$ has only the two extreme points p_2, p_3 and

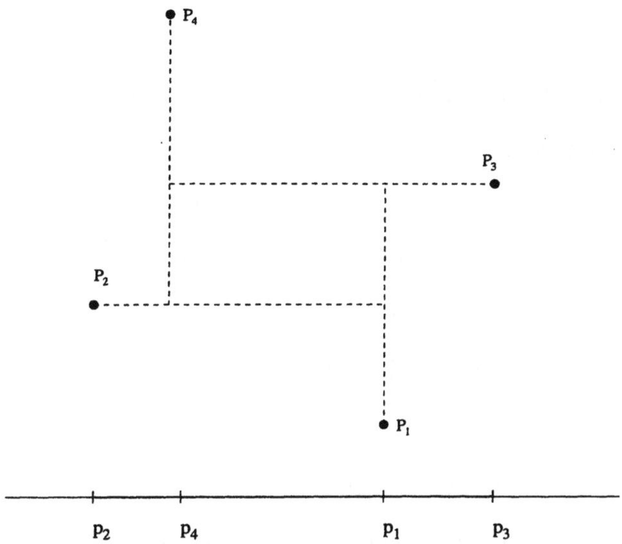

Fig. 5. Finding the T_4-configuration in the plane spanned by $[p_2, p_3]$ and the z-axis

hence $\{p_1, p_2, p_3, p_4\} \subset \{p_2, p_3\}^{co}$. If $\{p_1, p_2\}^{co} \cap \{p_3, p_4\}^{co} = \emptyset$ we can again use the separation argument to obtain a contradiction. Hence $\{p_1, p_2\}^{co} \cap \{p_3, p_4\}^{co} \neq \emptyset$ and drawing the picture of K in the planar strip $[p_2, p_3] \times \mathbb{R} \subset \mathbb{R}^2 \times \mathbb{R}$ we see the usual T_4-configuration (see Fig. 5).

Finally, if K contains five points, we suppose again $z_1 < z_2 < \ldots < z_5$. As before, we infer $p_1, p_5 \in \{p_2, p_3, p_4\}^{co}$. If p_2, p_3 and p_4 lie on a line l then K is contained in $l \times \mathbb{R}$ and hence K must contain a T_4-configuration by Proposition 6.23 (ii) for separate convexity in $\mathbb{R}^2$ (alternatively this can be checked directly arguing as above and distinguishing a few cases).

Using separately convex functions of the type $(p, z) \to (\langle p, a\rangle - b)_\pm (z - c)_\pm$ that are chosen to vanish on K, we see that $(p, z) \notin K^{rc}$ if

$$\left.\begin{array}{l} \bullet\ \ z < z_1 \text{ or } z > z_5, \text{ or} \\ \bullet\ \ (z \in [z_1, z_2) \text{ but } p \neq p_1) \text{ or } (z \in (z_4, z_5] \text{ but } p \neq p_5) \text{ or ,} \\ \bullet\ \ (z \in [z_2, z_3) \text{ but } p \notin [p_1, p_2]) \text{ or } (z \in (z_3, z_4] \text{ but } p \notin [p_4, p_5]). \end{array}\right\} \quad (55)$$

Now we will use the localization formula for the rank-one convex hull (see Proposition 4.16(i))

$$K^{rc} \cap U = [(K \cap U) \cup (K^{rc} \cap \partial U)]^{rc} \cap U, \quad (56)$$

which is valid for any compact K and any bounded U. Let B_R be an open ball which contains K^{co} and let

$$U = \{(p, z) :\ z < z_3,\ \operatorname{dist}(p, [p_1, p_2]) < z_3 - z\} \cap B_R.$$

Then

$$\partial U \subset ([p_1, p_2] \times \{z_3\}) \cup \{(p, z) : z < z_3, \mathrm{dist}(p, [p_1, p_2]) = z - z_3\} \cup \partial B_R.$$

Thus (56) and (55) yield

$$\begin{aligned} K^{rc} \cap \{(p, z) : z < z_3\} &= K^{rc} \cap U \\ &\subset U \cap (\{P_1, P_2\} \cup ([p_1, p_2] \times \{z_3\} \cap K^{rc}))^{rc}. \end{aligned} \tag{57}$$

Analogously, above the z_3-level we see

$$K^{rc} \cap \{(p, z) : z > z_3\} \subset (\{P_4, P_5\} \cup ([p_4, p_5] \times \{z_3\} \cap K^{rc}))^{rc}. \tag{58}$$

We claim that

$$C := K^{rc} \cap \{(p, z) : z = z_3\} = \{P_3\}.$$

Once this is shown were are done. Indeed (57) and (58) then imply that $K^{rc} \setminus \{(p, z) : z = z_3\} = \{P_1, P_2, P_4, P_5\}$ since the rank-one convex hull of three points without rank-one connections is trivial.

The compact convex set C is the convex hull of its extreme points. Suppose that $C \neq \{P_3\}$ and let $P_0 = (p_0, z_3) \in C \setminus \{P_3\}$ be an extreme point. We claim that $p_0 \in [p_1, p_2] \cap [p_4, p_5]$. To see this suppose first $p_0 \notin [p_1, p_2]$. Choosing a ball $B_\varepsilon(P_0)$ around P_0 which is so small that $\bar{B}_\varepsilon(P_0) \cap (K \cup ([p_1, p_2] \times \mathbb{R})) = \emptyset$ we see from the localization formula that

$$P_0 \in (K^{rc} \cap \partial B_\varepsilon(P_0))^{rc} \subset [(C \cap \partial B_\varepsilon(P_0)) \cup (K^{rc} \cap \{(p, z) : z > z_3\})]^{co}.$$

Since in the computation of the convex hull points above $\{z = z_3\}$ can not be compensated for by points below $\{z = z_3\}$ we get $P_0 \in (C \cap \partial B_\varepsilon(p_0))^{co}$. This contradicts the extremality of P_0 in C. Similarly one shows $p_0 \in [p_4, p_5]$. For future reference we note that $p_0 \neq p_2$. Otherwise $p_2 \in [p_4, p_5]$ and thus $p_5 = p_2$, since p_5 is in the convex hull of p_2, p_3 and p_4 and these three points do not lie on a line. This contradicts the assumption that P_2 and P_5 are not rank-one connected. Similarly $p_0 \neq p_4$.

We next claim that the segments $[p_1, p_2]$ and $[p_4, p_5]$ intersect transversely in a single point, which we denote again by p_0. Indeed, otherwise $p_1, p_5 \in [p_4, p_2]$ (since $p_1, p_5 \in \{p_2, p_3, p_4\}^{co}$) and thus p_4, p_1, p_5, p_2 form a T_4-configuration in the plane generated by the line through p_4 and p_2 and the z-axis. Thus the intersection must be transversal, p_0 and P_0 are uniquely determined and $C \subset [P_0, P_3]$.

Moreover, the segment $[p_0, p_3]$ intersects both $[p_1, p_2]$ and $[p_4, p_5]$ in $\{p_0\}$ only, since else we would find a T_4-configuration over one of the edges $[p_2, p_3]$ or $[p_4, p_3]$ (note that $p_0 \in [p_2, p_3]$ implies $p_5 = p_0$ since $p_4 \notin [p_2, p_3]$). Because $C \subset [P_0, P_3]$ the inclusion (57) yields

$$K^{rc} \cap \{(p, z) : z < z_3\} \subset \{P_1, P_2, P_0\}^{rc} = \{P_1, P_2, P_0\},$$

This figure presents the six point configuration with a nontrivial separately convex hull in $\mathbb{R}^2 \oplus \mathbb{R}$. The set K consists of $\{1, 2, 3, 4, 5, 6\}$ - each point at the corresponding height.
The picture below shows the projection into the base plane - at each point it is indicated for which height about this given point we are in the separately convex hull of the set.
The six smaller pictures on the right present the intersections of the hull with a plane of given height $1, \dots, 6$. Here $\bullet$ denotes a point from the original set, and $\times$ and $\circ$ denote points in the hull, but not in K - the $\times$-points are extreme on their vertical line, while the $\circ$-points are not.

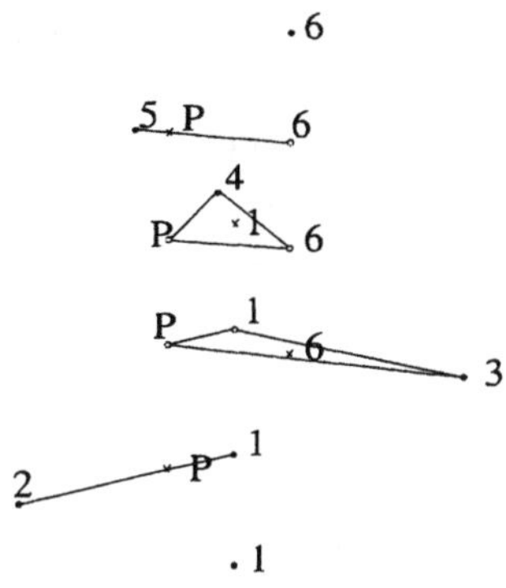

To prove that a separately convex function which vanishes on K can not be positive in any of the $\times$ or $\circ$-points it is enough to check that each $\times$-point is not convex extreme in its corresponding plane. A special feature is the occurence of the auxiliar point P - this makes it difficult to find simple grid-based algorithms to compute the hull.

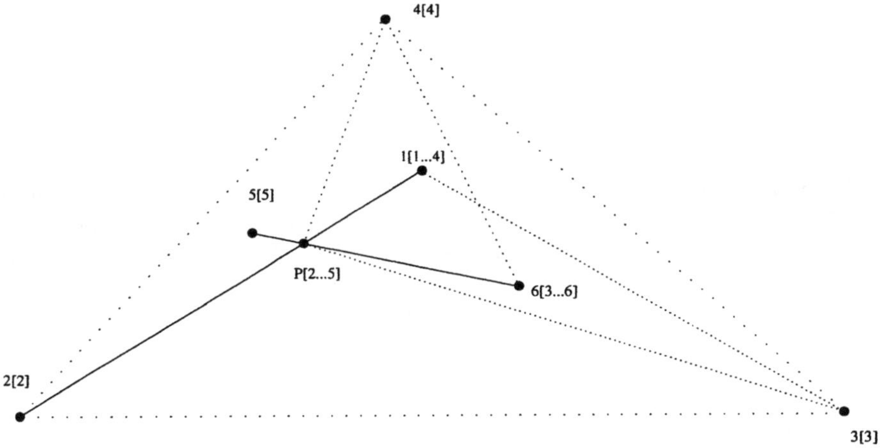

Fig. 6. The new C_6-configuration

except when $p_0 = p_1$. Suppose first $p_0 \neq p_1$. Then $K^{rc} \cap \{(p, z) : z \in (z_2, z_3)\} = \emptyset$. As before this contradicts the fact P_0 is not in K (i.e. different from P_3) but extreme in C. If $p_0 = p_1$, we have $p_0 \neq p_5$ and we can use (58) to find

$$K^{rc} \cap \{(p, z) : z > z_3\} \subset \{P_4, P_5, P_0\}^{rc} = \{P_4, P_5, P_0\}$$

and to get the same contradiction. In conclusion we must have $C := K^{rc} \cap \{(p, z) : z = z_3\} = \{P_3\}$ and the proof is finished. $\square$

Proof of Proposition 6.26. For the construction of the set K see Fig. 6. Now suppose that the set K contained a T_k configuration $(M_1, \dots, M_k)$. Then $M_{j+1} - M_j = (1+\kappa_{j+1})C_{j+1} - \kappa_j C_j$, where the C_j are rank-one matrices. Since there are no rank-one connections in K, the C_j must alternate between vertical and horizontal vectors. Thus if C_j is horizontal, both C_{j-1} and C_{j+1} are vertical and thus the projections of M_{j-1}, M_j and M_{j+1} to the base plane lie on a line. Fig. 6 shows that no three points in K have this property. □

Proof of Proposition 6.27. If the conclusion fails then $T > 0$ and, perhaps after cutting away unused parts of the curve, we can certainly find a laminate μ which is not concentrated on a proper closed subinterval of $[0, T]$. In other words

$$(\gamma(0), 0), (\gamma(T), T) \in \operatorname{supp}(\mu) \subset K. \tag{59}$$

The separation argument which is a central theme of this section now tells that for any $t \in (0, T)$ the sets $\gamma([0, t))$ and $\gamma((t, T])$ can not strictly lie on two different sides of some line. Indeed, else there is an $a \in \mathbb{R}^2$ such that

$$\langle \gamma(r), a\rangle < \langle \gamma(t), a\rangle < \langle \gamma(s), a\rangle \text{ for } r < t < s. \tag{60}$$

Then $\{(\gamma(s), s) : s \in [0, T]\} \subset Q_{++} \cup Q_{--} \cup Q_{00}$, where

$$Q_{\pm\pm} = \{\pm\langle \gamma(s) - \gamma(t), a\rangle > 0,\ \pm(s - t) > 0\}.$$

Thus Lemma 6.24 implies that $\operatorname{supp}\mu$ is contained in $\mathbb{R}^2 \times [t, T]$, or in $\mathbb{R}^2 \times [0, t]$ or in $\mathbb{R}^2 \times \{t\}$. This contradicts (59).

This shows that (60) can not occur, and we will finish our proof by verifying that the assumption on γ anyhow enforces (60) to hold for some t. The assumptions of Proposition 6.27 and (60) involve γ rather than K and remain unchanged if we (monotonously) reparametrize γ and apply some affine map of the plane to it. Therefore, due to the hypothesis $\gamma(0) \notin (\gamma((0, T])^{co}$ we can suppose in the sequel that

- $\gamma' : [0, T] \to \mathbb{S}^1$ is continuous,
- $\gamma(0) = (0, 0)$
- $\gamma_2(t) \geq 0$ for all t and hence $\gamma_2'(0) \geq 0$.

First, we assume $\gamma_2(t) > 0$ for all $t \in (0, T]$. Then a simple compactness argument gives that for $\delta > 0$ but sufficiently small

$$\langle \gamma(t), (\delta\gamma_1'(0), 1)\rangle > 0 \text{ if } t \in (0, T].$$

Because also

$$\langle \gamma'(t), (\delta\gamma_1'(0), 1)\rangle > 0 \text{ for all } t \text{ positive but small enough,}$$

the mean value theorem says that for k sufficiently large

$$\langle \gamma(t), (\delta\gamma_1'(0), 1)\rangle < \langle \gamma(\tfrac{1}{k}), (\delta\gamma_1'(0), 1)\rangle < \langle \gamma(s), (\delta\gamma_1'(0), 1)\rangle \text{ if } t < \frac{1}{k} < s.$$

So we arrived at the impossible relation (60) and are done with the first case.

Thus, we can in addition, perhaps after a reparametrisation and a reflection at the x_2 axis, suppose that

$$\gamma(T_0) = (-1, 0) \text{ for some } T_0 \in (0, T].$$

Hence

$$\gamma(t) \notin (0, \infty) \times \{0\} \text{ for } t > 0$$

since otherwise $\gamma(0) \in [\gamma(t), \gamma(T_0)]$. If $\gamma'(0) \neq (1, 0)$ then as before we find $\delta > 0$ such that

$$\langle \gamma'(0), (-\delta, 1) \rangle > 0 \text{ and } \langle \gamma(t), (-\delta, 1) \rangle > 0 \text{ for all } t > 0.$$

Again we conclude that (60) holds for $a = (-\delta, 1)$ and $t > 0$ but sufficiently small. So we are left with the most difficult case when

- $\gamma'(0) = (1, 0)$.

Now we can moreover require that T_0 was chosen such that $\gamma(T_0)$ is the point on $\gamma((0, T]) \cap (\mathbb{R} \times \{0\})$ which is closest to 0. Then

- $\gamma(t) \notin (-1, \infty) \times \{0\}$ if $t \in (0, T]$.

We use again compactness arguments to choose a few more constants. First fix

- $\delta_0 > 0$ with $\gamma_1' > 1/2$ on $[0, 2\delta_0]$, then
- $\eta > 0$ such that for any $t \in [\delta_0, T]$ we have $\gamma_2(t) > 2\eta$ or $\gamma_1(t) \leq -9/10$. Finally, pick
- $\delta_1 \in (0, \delta_0)$ with $10(\operatorname{diam}(K) + 10)\gamma_2(t) < \eta$ for all $t \in [0, \delta_1]$.

Because $\frac{d}{dt} \log(\gamma_2)$ has to be unbounded from above in any neighbourhood of the root of γ_2 at 0, we find $t_0 \in (0, \delta_1)$ with

$$\frac{\gamma_2'(t_0)}{\gamma_1'(t_0)} > \frac{\gamma_2(t_0)}{\gamma_1(t_0) + \frac{1}{2}}. \tag{61}$$

Finally, we select t_1 to maximize

$$c_t = \frac{\gamma_2(t) - \gamma_2(t_0)}{\gamma_1(t) - \gamma_1(t_0)} \text{ under the constraint } \gamma_1(t) \leq -\frac{9}{10}.$$

The key observation is now that there is an $\varepsilon > 0$ such that

(i) $c_t > c_{t_1} + 2\varepsilon$ if $t \in [0, t_0)$ or $\gamma_1(t) > \gamma(t_0)$, and
(ii) $\gamma_2(t) \geq \gamma_2(t_0) + c_{t_1}(\gamma_1(t) - \gamma_1(t_0))$ if $t \in (t_0, T]$ and $\gamma_1(t) \leq \gamma_1(t_0)$.

It is easy to check that these two conditions give for $t = t_0$ and $a = (-c_{t_1} - \varepsilon, 1)$ the forbidden separation (60).

Now for $\gamma_1(t) \leq -9/10$ assertion (ii) is just a reformulation of the maximality property of c_{t_1}. If $\gamma_1(t) \in (-9/10, \gamma_1(t_0)]$ then $t > t_0$ implies $t > \delta_0$. Thus $\gamma_2(t) > \eta > \gamma_2(t_0)$. Since $c_{t_1} \geq c_{T_0} > 0$ this proves assertion (ii).

It remains to verify (i). First we note that $c_t - c_{t_1}$ is larger then some positive constant for t near t_0 – this is just a consequence of (61), which gives

$$\frac{\gamma_2'(t_0)}{\gamma_1'(t_0)} > \frac{\gamma_2(t_0)}{\gamma_1(t_0)+\frac{1}{2}} > \frac{\gamma_2(t_0)-\gamma_2(t_1)}{\gamma_1(t_0)-\gamma_1(t_1)} = c_{t_1}.$$

If $c_t = c_{t_1}$ and $t \in [0,t_0)$ then $\gamma(t) \in \{\gamma(t_1),\gamma(t_0)\}^{co}$. Combining this with the above estimate for t near t_1 we conclude that $\inf_{t\in[0,t_0)} c_t - c_{t_1} > 0$. Similarly $\inf_{t\in(t_0,\delta_0]} c_t - c_{t_1} > 0$ follows from $\gamma(t_0) \notin \{\gamma(t_1),\gamma(t)\}^{co}$ for $t > t_0$. The last situation to deal with is $\gamma_1(t) > \gamma_1(t_0)$ but $t > \delta_0$. Then the bound is a consequence of

$$\gamma_2(t_0) + c_{t_1}(\gamma_1(t)-\gamma_1(t_0)) \le \gamma_2(t_0) + \frac{\gamma_2(t_0)}{(\frac{1}{2})}\text{diam(K)} < \eta < \gamma_2(t) - \eta.$$

□

Example 6.28 *We construct an example which shows that the conclusion of Proposition 6.27 need not to hold for Lipschitz curves. We fix in the plane the two rays*

$$g_+ = \{(x, \frac{1}{2}x) : x \ge 0\} \textit{ and } g_- = \{(x, -\frac{1}{2}x) : x \ge 0\},$$

and the point $P_0 = (1,0)$. For $\lambda \in (0,1)$ we define $P_1 = \lambda P_0$ and the two segments $s_i^+ = [P_i P_i^+]$, where $P_i^+ \in g_+$ satisfies $(P_i^+ - P_i)\|(1,2)$ for $i = 0,1$. Together with the connecting segment $c_0 = [P_1^+ P_0^+]$ they form $M_0 = s_0^+ \cup c_0 \cup s_1^+$. We use the linear map $A_\lambda = \mathrm{diag}(\lambda,-\lambda)$ to build the selfsimilar set

$$M = \bigcup_{i=0}^{\infty} M_i \cup \{0\} \textit{ where } M_{i+1} = A_\lambda(M_i),$$

consisting of the segments $s_{i+1}^- = A_\lambda(s_i^+)$, $s_{i+1}^+ = A_\lambda(s_i^-)$ and $c_{i+1} = A_\lambda(c_i)$ and containing the points $P_{i+1} = A_\lambda(P_i)$, $P_{i+1}^- = A_\lambda(P_i^+)$ and $P_{i+1}^+ = A_\lambda(P_i^-)$, see Figure 7.

Let $X_1 = (P_1 + P_1^-)/2$ and denote by $X_0 = X_0(\lambda)$ the intersection of the line through X_1 and $P_2^+(\lambda)$ with the line containing s_0^+. It is clear that for $\lambda \to 1_-$ the point $X_0(\lambda)$ tends to P_0^+, but if λ gets smaller, it will leave s_0^+ trough its endpoint P_0. Hence, we can find $\lambda_0 \in (0,1)$ such that $X_0(\lambda_0) = (P_0 + P_0^+)/2 = A_\lambda^{-1}(X_1)$.

Now we can consider any injective continuous curve $\gamma : [0,T] :\to \mathbb{R}^2$ with $\gamma([0,T]) = M = M(\lambda_0)$ and $\gamma(0) = 0$. Because $s_{i+1}^+ \cup c_{i+1} \cup s_{i+1}^-$ intersects the convex hull of $(s_i^+ \cup c_i \cup s_i^-)$ only in an extreme point and since

$$\gamma \textit{ first runs through } s_{i+1}^+ \cup c_{i+1} \cup s_{i+1}^- \textit{ and then through } s_i^+ \cup c_i \cup s_i^-, \tag{62}$$

it is easy to verify that $\gamma(t)$ never belongs to the convex hull of $\gamma((t,T])$. On the other hand, $K = \{(\gamma(t),t) : t \in [0,T]\}$ has a nontrivial rank-one convex hull. To verify this note that $X_0 \in [P_0^+, X_1]$ and thus

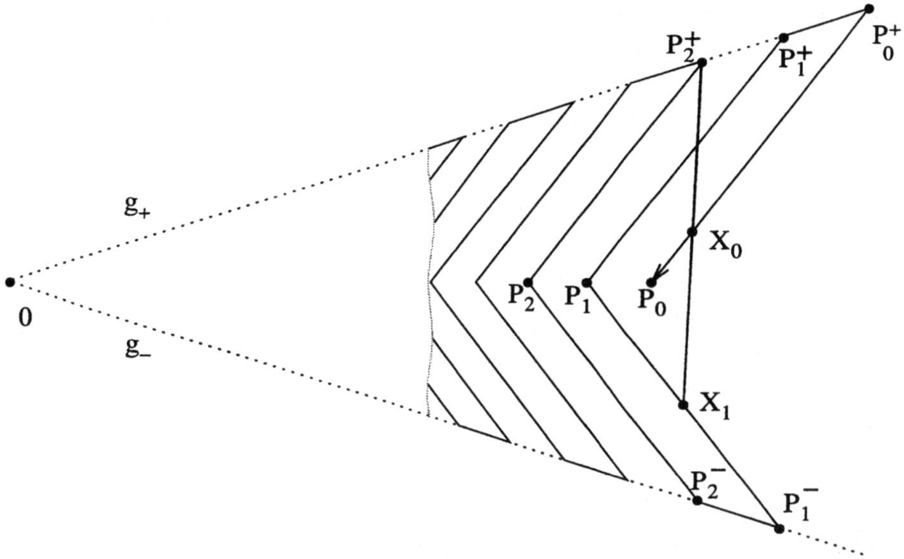

Fig. 7. A continuous curve with small future but large hull

$$(X_0, \gamma^{-1}(P_2^+)) \in [(P_2^+, \gamma^{-1}(P_2^+)), (X_1, \gamma^{-1}(P_2^+))].$$

Now

$$(X_1, \gamma^{-1}(P_2^+)) \in [(X_1, \gamma^{-1}(P_3^-)), (X_1, \gamma^{-1}(X_1))]$$

which implies that

$$(X_0, \gamma^{-1}(P_2^+)) \in (K \cup \{(X_1, \gamma^{-1}(P_3^-))\})^{rc}.$$

Using the (A_λ)-selfsimilarity of M to iterate this we have

$$(X_0, \gamma^{-1}(P_2^+)) \in (K \cup \{(X_i, \gamma^{-1}(P_{i+2}^{(-1)^i}))\})^{rc} \text{ for all } i.$$

Since $\lim_{i\to\infty}(X_i, \gamma^{-1}(P_{i+2}^{(-1)^i})) = (0,0) \in K$ *and any separately convex function is continuous, we get as required*

$$(X_0, \gamma^{-1}(P_2^+)) \in K^{rc} \setminus K.$$

□

6.3 Separate Convexity in $\mathbb{R}^2 \oplus \mathbb{R}^2$

We consider the integrand

$$\phi(X) = \frac{1}{2}|X|^2 + h(\det X), \quad X \in \mathbb{R}^{2\times 2}, \tag{63}$$

the corresponding Euler-Lagrange equation div $D\phi(\nabla w) = 0$ and its reformulation as a first order partial differential relation

$$\nabla u \in K_\phi = \left\{ \begin{pmatrix} X \\ D\phi(X)J \end{pmatrix} : X \in \mathbb{R}^{2\times 2} \right\}. \tag{64}$$

Theorem 6.29 *Suppose $h : \mathbb{R} \to \mathbb{R}$ is differentiable and strictly convex. Suppose further that E is a finite subset of K and that the elements of E are of the form $\binom{X}{Y}$, where X is a diagonal 2×2 matrix. Then $K^{rc} = K$ and the laminates $\mathcal{M}^{rc}(K)$ supported on K are Dirac masses.*

Proof. Geometrically the heart of the matter is again a good foliation as in Section 5. After some transformation we will see that E lies in a certain two-dimensional set Γ in some $\mathbb{R}^2 \oplus \mathbb{R}^2$, where the $\mathbb{R}^2$'s correspond to the rank-one directions. The point is that Γ can be foliated by a one-parameter family of curves whose projection to either $\mathbb{R}^2$ is a foliation (of one quadrant) by strictly convex curves (see Step 4 below; the different curves are distinguished by $d = st$). Then the separation result, Lemma 6.24, in connection with the finiteness of E can be used once more to finish the argument. First we need to make some normalizations and to show that E lies in one quadrant.

Step 1: Reduction to $h'(0) = 0$. Adding a linear term to h does not affect the Euler-Lagrange equation since det is a null Lagrangian, i.e. for all function w we have $\operatorname{div}(D\det)\nabla w = \operatorname{div}\operatorname{cof}\nabla w = 0$. On the level of the partial differential relation we have

$$\begin{aligned} K_{\phi+\lambda\det} &= \left\{ \begin{pmatrix} X \\ D\phi(X)J + \lambda \operatorname{cof} X\, J \end{pmatrix} : X \in \mathbb{R}^{2\times 2} \right\} \\ &= \left\{ \begin{pmatrix} X \\ Y + \lambda JX \end{pmatrix} : \begin{pmatrix} X \\ Y \end{pmatrix} \in K_\phi \right\}. \end{aligned}$$

Since $\binom{X}{Y} \mapsto \binom{X}{Y+\lambda JX}$ is a linear isomorphism which preserves rank-one lines we may suppose $h'(0) = 0$.

Step 2: Special separately convex functions. We now write out the elements $\binom{X}{Y}$ of K_ϕ for which X is a diagonal matrix more explicitly

$$\begin{pmatrix} X \\ D\phi(X)J \end{pmatrix} = \begin{pmatrix} X \\ X + h'(\det X)\operatorname{cof} X\, J \end{pmatrix} = \begin{pmatrix} s & 0 \\ 0 & t \\ 0 & -s - h'(st)t \\ t + h'(st)s & 0 \end{pmatrix}.$$

After exchanging rows and multiplying one row by -1 (both of which correspond to making a linear change of the dependent variable u in (64)) we may suppose that

$$E \subset \left\{ \begin{pmatrix} s & 0 \\ \sigma & 0 \\ 0 & t \\ 0 & \tau \end{pmatrix} : s, t \in \mathbb{R}, \sigma = t + h'(st)s, \tau = s + h'(st)t \right\}. \tag{65}$$

On the linear subspace

$$L = \{A \in \mathbb{R}^{4\times 2} : A_{12} = A_{22} = A_{31} = A_{41} = 0\},$$

which contains E, rank-one convexity agrees with separate convexity in the variables (A_{11}, A_{21}) and (A_{32}, A_{42}).

In particular, for each $a, b \in \mathbb{R}^2$ we have the separately convex functions

$$(A_{11}, A_{21}, A_{32}, A_{42}) \mapsto \langle (A_{11}, A_{21}), a\rangle_{\pm} \langle (A_{32}, A_{42}), b\rangle_{\pm}$$

at our disposal.

Step 3: Restriction to generalized quadrants. We argue by contradiction. If the claim fails then there is a minimal set $E_0 \subset E$ which supports a nontrivial laminate μ_0. Our goal is to contradict minimality by a separation argument similar to [59, 69]. We denote by $P_0 = (s_0, \sigma_0, t_0, \tau_0)$ the centre of mass of μ_0.

Recall that strict convexity and the normalization condition $h'(0) = 0$ imply that $h'(x)x > 0$ for $x \neq 0$. Thus all $(s, \sigma, t, \tau) \in E$ satisfy

$$|\tau| \geq |s|, \ \tau s \geq 0 \quad \text{and } s = 0 \Leftrightarrow \tau = 0. \tag{66}$$

Thus we can apply Lemma 6.24 to the generalized quadrants

$$Q_{\pm\pm} = \{(s, \sigma, t, \tau) : \pm s > 0, \pm\tau > 0\}.$$

Since E_0 is finite this shows that either $s_0 > 0, \tau_0 > 0$ and $s > 0, \tau > 0$ in E_0, or $s_0 < 0, \tau_0 < 0$ and $s < 0, \tau < 0$ on E_0, or $s_0 = \tau_0 = 0$ and $s = \tau = 0$ on E_0. In the last case we have $\sigma = t$ on E_0 and therefore the quadratic minor $\sigma t \hat{=} A_{21}A_{32} - A_{22}A_{31}$ has a (strictly) positive value at all non vanishing differences of elements in E_0. This implies that even all polyconvex measures supported on E_0 must be Dirac masses [60] and thus yields a contradiction. Hence we have shown that there exists a $c_s \in \{-1, 1\}$ such that

$$c_s s_0, c_s \tau_0 > 0, \quad c_s s, c_s \tau > 0 \text{ on } E_0.$$

Since the problem is invariant under the exchange of variables $(s, \tau) \leftrightarrow (t, \sigma)$ we also find $c_t \in \{-1, 1\}$ such that

$$c_t t_0, c_t \sigma_0 > 0 \quad c_t t, c_t \sigma > 0 \text{ on } E_0.$$

Multiplication of all variables by -1 leaves the right hand side of (65) invariant. Hence we may assume $c_s = 1$. If we multiply only t and σ by -1 and replace $h(x)$ by its reflection $h(-x)$ then (65) remains invariant. Hence we may suppose $c_s = c_t = 1$, i.e.

$$s_0, \sigma_0, t_0, \tau_0 > 0, \quad s, \sigma, t, \tau > 0 \text{ on } E_0. \tag{67}$$

Step 4: Separation. Let $d = \min\{st : (s, \sigma, t, \tau) \in E_0\}$. Since E_0 is finite we have $d > 0$. Pick any point $P_1 \in E_0$ with $s_1 t_1 = d$. Projecting E_0 into the (s, σ) and (t, τ) planes we see that the set

$$E_0^s := \{(s,\sigma) : (s,\sigma,t,\tau) \in E_0\}$$

is contained in the (closed) epigraph of the strictly convex function

$$s \mapsto \frac{d}{s} + h'(d)s,$$

because for fixed $s > 0$ the expression on the right hand side is increasing in $d > 0$. Since (s_1, σ_1) lies on the graph of this strictly convex function it is an extreme point of E_0^s. Thus there exists $a_s \in \mathbb{R}^2$ with

$$\langle (s_1,\sigma_1), a_s\rangle < \langle (s,\sigma), a_s\rangle \quad \text{for } (s,\sigma,t,\tau) \in E_0 \setminus \{P_1\}. \tag{68}$$

Here we used that fact that the projection from E_0 to E_0^s is injective since K_ϕ contains no rank-one connections (this in turn follows from the strict convexity of h). In view of the invariance under the exchange of variables $(s,\sigma) \leftrightarrow (t,\tau)$ the same reasoning yields $a_t \in \mathbb{R}^2$ with

$$\langle (t_1,\tau_1), a_t\rangle < \langle (t,\tau), a_t\rangle \quad \text{for } (s,\sigma,t,\tau) \in E_0 \setminus \{P_1\}. \tag{69}$$

Invoking once more Lemma 6.24 we deduce that μ_0 is either supported on $\{P_1\}$ or on $E_0 \setminus \{P_1\}$. This contradicts the minimality of μ_0 and the proof is finished. □

6.4 A Laminate without Discrete Approximations

One might hope to extend Theorem 6.29 to more general sets through an approximation by finite sets. The following example shows, however, that the rank-one convex hull can shrink drastically, when we pass from continua to discrete sets. It was motivated by J.M. Ball's construction of sets of gradients with no rank-one connections ([5]) and is based on separate convexity in $\mathbb{R} \oplus \mathbb{R} \oplus \mathbb{R}$, which arises by restricting rank-one convexity in $\mathbb{R}^{3\times 3}$ to diagonal matrices.

Proposition 6.30 *We consider separate convexity in $\mathbb{R} \oplus \mathbb{R} \oplus \mathbb{R}$ and define for $t \in [0,1]$*

$$p(t) = \begin{pmatrix} t \\ t \\ t \end{pmatrix}, \quad p_i(t) = p(t) + te_i \ \textit{for}\ i = 1,2,3.$$

Let $S_i = p_i([0,1])$. Then $p(t) \in (S_1 \cup S_2 \cup S_3)^{rc}$ for $t \in [0,1]$ and there is only one laminate supported in $S_1 \cup S_2 \cup S_3$ which generates this point. Moreover, this laminate does not charge points. More precisely,

$$\{\mu \in \mathcal{M}^{rc}(S_1 \cup S_2 \cup S_3) : \bar{\mu} = p(t)\} = \{\mu_t\},$$

where $\mu_0 = \delta_0$ and where for $t > 0$

$$\mu_t(A) = \frac{1}{t^3}\int_0^t \sum_{i=1}^{3} \chi_A(p_i(\tau))\tau^2\, d\tau \ \textit{if}\ A \subset \mathbb{R} \oplus \mathbb{R} \oplus \mathbb{R}.$$

In particular, if K is a compact subset of $(S_1 \cup S_2 \cup S_3)$ but $p_i(t_0) \notin K$, then $p(t) \notin K^{rc}$ for $t \in (t_0, 1]$.

Proof of Proposition 6.30. In the sequel we identify $\mathbb{R}\oplus\mathbb{R}\oplus\mathbb{R}$ with $\mathbb{R}^3$ and suppose the measures μ_t are defined as above. First we show that for all $t_0 \in [0,1]$ and all $f : \mathbb{R}^3 \to \mathbb{R}$ separately convex the Jensen-type inequality

$$f(p(t_0)) \leq \int f \, d\mu_{t_0} \tag{70}$$

holds. Choosing for f constant and linear functions we can than conclude that μ_{t_0} is a probability measure, $\bar{\mu}_{t_0} = p(t_0)$ and $\mu_{t_0} \in \mathcal{M}^{rc}(S_1 \cup S_2 \cup S_3)$.

To establish (70) we notice that for any $f : \mathbb{R}^3 \to \mathbb{R}$ separately convex, for $x \in \mathbb{R}^3$ and $t_1, t_2, t_3 > 0$ we have

$$\limsup_{t\to 0_+} \frac{f(x+p(t)) - f(x)}{t} \leq \sum_{i=1}^{3} s_i, \text{ where } s_i = \frac{f(x+t_i e_i) - f(x)}{t_i}. \tag{71}$$

Indeed, after addition of an affine function and translation we may suppose that $x = 0 = (s_1, s_2, s_3)$. Suppose now that the upper limit of these difference quotients was positive. Since separately convex functions are locally Lipschitz, a suitable subsequence of the rescaled functions $f_\varepsilon(y) = \varepsilon^{-1}(f(\varepsilon y) - f(0))$ converges (uniformly on compact subsets) to a limit f_0 with $f_0(p(1)) > 0$. Moreover f_0 is globally Lipschitz, separately convex and satisfies $f_0(te_i) \leq 0 = f_0(0)$ for $t > 0$ and $i = 1, 2, 3$. This implies that for any $y \in \mathbb{R}^3$ and i the function $t \to f_0(y + te_i)$ is nonincreasing. Otherwise $t \mapsto f_0(y + te_i)$ is bounded from below by a function with positive slope and $f_0(y + te_i) - f_0(te_i)$ tends to infinity as $t \to +\infty$, in contradiction with the Lipschitz property of f_0. Using that $t \mapsto f_0(y + te_i)$ is nonincreasing we obtain

$$0 < f_0(p(1)) \leq f_0(e_1 + e_2) \leq f_0(e_1) \leq f_0(0) = 0,$$

a contradiction proving (71).

Since $t \mapsto f(p(t))$ is Lipschitz it is differentiable for almost every $t \in [0,1]$ and (71) implies that

$$\frac{d}{dt} f(p(t)) \leq \frac{1}{t}(f(p_1(t)) + f(p_2(t)) + f(p_3(t)) - 3f(p(t))).$$

This is equivalent to

$$3f(p(t)) + t\frac{d}{dt} f(p(t)) \leq f(p_1(t)) + f(p_2(t)) + f(p_3(t)),$$

and therefore

$$t_0^3 f(p(t_0)) \leq \int_0^{t_0} \sum_{i=1}^{3} f(p_i(t)) t^2 \, dt = \int f \, d(t_0^3 \mu_{t_0})$$

follows and ensures (70).

The uniqueness of μ_{t_0} requires a bit more effort, so assume there is an other $\mu' \in \mathcal{M}^{rc}(S_1 \cup S_2 \cup S_3)$ with $\bar{\mu'} = p(t_0)$. Since these two measures differ, we find some

$$f : S_1 \cup S_2 \cup S_3 \to \mathbb{R} \text{ such that } \int f \, d\mu_{t_0} > \int f \, d\mu'$$

and, after adding a constant and making a suitable approximation if necessary, we also may assume that for $i = 1, 2, 3$

$$t \to f(p_i(t)) \text{ is } C^\infty(\mathbb{R}) \text{ with support in } (\eta_0, 2] \text{ for some } \eta_0 > 0.$$

Now, for $\varepsilon > 0$ let $g_\varepsilon : \mathbb{R} \to \mathbb{R}$ be the solution of

$$g_\varepsilon'(t) = \begin{cases} \frac{1}{t}(f(p_1(t)) + f(p_2(t)) + f(p_3(t)) - 3g_\varepsilon(t)) - \varepsilon & \text{if } t > \eta_0, \\ -\frac{\varepsilon}{4}t & \text{if } t \le \eta_0, \end{cases}$$

with $g_\varepsilon(0) = 0$. It is easily checked that g_ε is also C^∞. Since $\lim_{\varepsilon \to 0_+} g_\varepsilon(t) = \int f \, d\mu_t$ for $t \in [0,1]$ we can choose $\varepsilon = \varepsilon_0 > 0$ such that

$$g = g_{\varepsilon_0} \text{ satisfies } \int f \, d\mu' < g(t_0).$$

For convenience we also introduce the smooth functions

$$s_i(t) = \begin{cases} \frac{1}{t}(f(p_i(t)) - g(t)) & \text{if } t > \eta_0, \\ \frac{\varepsilon}{4} & \text{if } t \le \eta_0 \end{cases}, i = 1, 2, 3,$$

and fix a finite constant c_0 such that

$$10 \sum_{i=1}^{3} |s_i'(t)| + |s_i''(t)| < c_0 \text{ if } t \in [-4, 4].$$

Finally, we define the auxiliary function

$$f_1(x) = \prod_{j \in \{1,2,3\} \setminus \{i\}} (x_j - x_i) \text{ if } x \in \mathbb{R}^3 \text{ and } x_i = \min_j x_j,$$

and the function

$$F(x) = g(t) + \langle x - p(t), \begin{pmatrix} s_1(t) \\ s_2(t) \\ s_3(t) \end{pmatrix} \rangle + c_0 f_1(x) \text{ for } x \in \mathbb{R}^3 \text{ and } t = \min_j x_j.$$

Suppose we had already shown that F is separately convex in the open cube $(-4, 4)^3$ of $\mathbb{R}^3$. Then Proposition 4.14 combined with Proposition 4.15 imply that F satisfies the Jensen inequality for all laminates supported in $(-2, 2)^3$. Because $F(p_i(t)) = g(t) + ts_i(t) = f(p_i(t))$ and $F(p(t)) = g(t)$ for $t \in [0, 1]$, we obtain the contradiction

$$\int F \, d\mu' = \int f \, d\mu' < g(t_0) = F(p(t_0)) = F(\bar{\mu'})$$

finishing our proof.

Therefore, it remains to study separate convexity of F near an arbitrary point $x^0 \in (-4,4)^3$. Essentially only $x^0 \in (0,4)^3$ is interesting but the general case is not more complicated. So, let $t^0 = \min_j x_j^0$ and we consider

$$h(r) = f(x^0 + re_i) \text{ for } i \in \{1,2,3\} \text{ fixed and } r \text{ near } 0.$$

First, suppose $t^0 < x_i^0$. Hence $\min_j(x^0 + re_j) \equiv t^0$, and

$$h(r) = (x_i^0 + r - t^0)s_i(t^0) + c_0 f_1(x^0 + re_i) + const$$

is obviously convex in r.

Next, let $t^0 = x_i^0 < \min_{j\neq i} x_j^0$. Then $\min_j(x^0 + re_i) = t^0 + r$ and

$$h(r) = g(t^0 + r) + \sum_{j\neq i}(x_j^0 - t - r)s_j(t^0 + r) + c_0 \prod_{j\neq i}(x_j^0 - t^0 - r).$$

Since $g'(t) = s_1(t) + s_2(t) + s_3(t) - \varepsilon_0$, we have

$$h'(r) = s_i(t^0 + r) - \varepsilon_0 + \sum_{j\neq i}(x_j^0 - t^0 - r)s_j'(t^0 + r) + c_0(2r + 2t^0 - \sum_{j\neq i} x_j^0)$$

and

$$h''(0) = s_i'(t^0) + \sum_{j\neq i} -s_j'(t) + (x_j^0 - t^0)s_j''(t^0) + 2c_0 \geq 2c_0 - c_0 > 0.$$

So we are left with the last case when $t^0 = x_i^0 = x_j^0 \leq x_k^0$ where $\{i,j,k\} = \{1,2,3\}$. For $r < 0$ we compute h and h' as before and obtain

$$\lim_{r\to 0_-} h'(r) = s_i(t^0) - \varepsilon_0 + (x_k^0 - t^0)(s_i'(t^0 + r) - c_0).$$

For $r > 0$ we have

$$h(r) = g(t^0) + \langle (x^0 + re_i) - p(t^0), s(t^0)\rangle + r(x_k^0 - t^0)$$

and thus

$$\lim_{r\to 0_+} h'(r) = s_i(t^0) + (x_k^0 - t^0) \geq \lim_{r\to 0_-} h'(r) + \varepsilon_0$$

and again h has a (local) subdifferential at zero. This finishes the proof. □

7 Local Hulls, Degenerate Sets, and Hyperbolic Conservation Laws

In this section we formulate some 'local' problems for the various hulls K^* and the classes of measures $\mathcal{M}^*$ associated with a given set $K \subset \mathbb{R}^{m\times n}$. As we sketch below, such problems are relevant for example in connection with compactness properties of L^∞ entropy solutions of $l \times 2$ systems of hyperbolic conservation laws.

Let us consider a smooth submanifold $K \subset \mathbb{R}^{m\times n}$. We introduce the following properties of K.

(P1) Each point $A \in K$ has a neighbourhood $U \subset \mathbb{R}^{m\times n}$ such that $(K \cap \bar{U})^{rc} = K \cap \bar{U}$.
(P2) Each point $A \in K$ has a neighbourhood $U \subset \mathbb{R}^{m\times n}$ such that $\mathcal{M}^{rc}(K \cap \bar{U})$ is trivial.
(P3) Each point $A \in K$ has a neighbourhood $U \subset \mathbb{R}^{m\times n}$ such that $(K \cap \bar{U})^{pc} = K \cap \bar{U}$.
(P4) Each point $A \in K$ has a neighbourhood $U \subset \mathbb{R}^{m\times n}$ such that $\mathcal{M}^{pc}(K \cap \bar{U})$ is trivial.

A sufficient condition for (P1) and (P2) to be satisfied is that the tangent spaces $T_A K$ do not contain any rank-one connections, see e.g. [67] (this condition in fact implies the stronger assertion obtained by replacing 'rc' with 'qc'). The same condition is also sufficient for (P3) and (P4) when $n = 2$. This follows from the fact that a rank-one convex quadratic form on $\mathbb{R}^{m\times n}$ is polyconvex, see [54].

If $T_A K$ does contain rank-one connections, the situation is more complicated. For simplicity, assume that for some $A_0 \in K$ the rank-one cone Λ and $T_{A_0} K$ intersect transversely (away from the origin). Then $T_A K \cap \Lambda$ behaves 'well' for A close to A_0, and one might speculate that there is some natural 'higher order' condition which would imply (P1)–(P4), or at least some of these properties, in a neighbourhood of A_0. This situation arises in connection with compactness properties of L^∞ entropy solutions of hyperbolic conservation laws, as was pointed out already in DiPerna's work [21], where some very interesting examples are considered and shown to have property (P4).

It seems that the problem of determining whether (P1)–(P4) are satisfied remains open even in some simple and very natural situations. Consider for example the following problem taken from the above paper by DiPerna. We look at the one dimensional wave equation

$$\varphi_{tt} - a(\varphi_x)_x = 0, \tag{72}$$

where a is strictly convex, increasing and satisfies $a(0) = 0$. There is a natural energy associated with (72):

$$\eta(\varphi_t, \varphi_x) = \frac{1}{2}\varphi_t^2 + F(\varphi_x), \quad \text{where } F(\xi) = \int_0^\xi a(s)\, ds.$$

One has $\eta_t - q_x = 0$ for each regular solution of (72), where $q = q(\varphi_t, \varphi_x) = \varphi_t a(\varphi_x)$. Letting $u = \varphi_t$, $v = \varphi_x$ we write (72) as a first order system

$$\begin{aligned} u_t - a(v)_x &= 0, \\ v_t - u_x &= 0. \end{aligned} \tag{73}$$

L^∞ entropy solutions of (73) can be defined as L^∞ functions (u, v) satisfying

$$\begin{aligned} u_t - a(v)_x &= 0, \\ v_t - u_x &= 0, \\ \eta_t - q_x &\le 0 \end{aligned} \tag{74}$$

in the sense of distributions. One can also add further entropies to (74), to get a more restrictive class of solutions, see [21]. In view of Murat's lemma [48] it is reasonable - at least in a first approximation - to replace the inequality in (74) by an equality when studying compactness properties. This enables us to introduce stream functions as follows.

$$\begin{aligned} (u, -a(v)) &= (\psi^1_x, -\psi^1_t), \\ (v, -u) &= (\psi^2_x, -\psi^2_t), \\ (\eta, -q) &= (\psi^3_x, -\psi^3_t). \end{aligned}$$

This can be rewritten as

$$\nabla \psi \in K \subset \mathbb{R}^{3\times 2},$$

with

$$K = \left\{ \begin{pmatrix} u & a(v) \\ v & u \\ \eta(u,v) & q(u,v) \end{pmatrix} : u, v \in \mathbb{R} \right\}.$$

As far as we know it is an open problem to determine for which functions a the set $K \subset \mathbb{R}^{3\times 2}$ given above satisfies any of the conditions (P1)–(P4). We refer the reader to the paper of DiPerna [21] for a study of a situation with additional entropies, in which one can determine that (P4) holds.

8 Outlook

Here we formulate some further questions and briefly mention a few related results.

Question 1. Is it possible to characterize (or at least give some non-trivial examples of) smooth uniformly rank-one convex functions $\phi : \mathbb{R}^{2\times 2} \to \mathbb{R}$ for which $K_\phi^{pc} = K_\phi$?

Perhaps we can expect that weak solutions of the Euler-Lagrange equation $\operatorname{div} D\phi(\nabla u) = 0$ for ϕ with $K_\phi^{pc} = K_\phi$ have some extra regularity properties.

Characterizing ϕ with $K_\phi^{qc} = K_\phi$ would be even more interesting but that looks out of reach.

Question 2. Is it possible to identify some geometric properties of smooth "elliptic" sets $K \subset \mathbb{R}^{m\times 2}$ which imply a-priori estimates (for example C^α-estimates) for smooth solutions of $\nabla u \in K$?

By a smooth elliptic set we mean a set set K which is a smooth submanifold of $\mathbb{R}^{m\times 2}$ such that the tangent spaces $T_A K$ contain no rank-one connections.

While rank-one connections, T_4-configurations or, in general, non-trivial measures in $\mathcal{M}^{rc}$ are obstructions to regularity of Lipschitz solutions of $\nabla u \in K$, their rôle in connection with a-priori estimates for smooth solutions is less clear. Of course, they do become relevant even in this setting if they occur "infinitesimally", but that is not possible for "elliptic" sets.

Question 3. Is it possible to develop some general methods which would work for the local problems in Section 7?

The construction of examples by convex integration rests on the subtle combination of one-dimensional constructions (and hence is closely related to rank-one convexity).

Question 4. Can one extend convex integration using genuinely multi-dimensional constructions as building blocks?

This would allow one to obtain interesting examples under the weaker (and more natural) condition that the quasiconvex hull is sufficiently trivial. The main difficulty is that it is much harder to 'glue' multidimensional constructions. One interesting test case is an example by Šverák (see, e.g., [42], Section 4.7, equation (4.25)) of a set $K \subset \mathbb{R}^{6\times 2}$ for which $\nabla u \in K$ admits periodic solutions. It is not known whether solutions with compact support exist. For possible extensions of convex integration see also [28], Section 2.4.12.

Question 5. Is there an effective algorithm to decide whether a given probability measure supported on a finite subset of $\mathbb{R}^{2\times 2}$ is a laminate?

Such an algorithm would enable us to effectively test (at least in a probability sense) whether rank-one convexity implies quasi-convexity on 2×2 matrices. For example, one would generate (by some random or pseudo-random process) on a computer piecewise affine maps $u : T^2 \to \mathbb{R}^2$ and check whether the corresponding distribution μ of their gradients given by

$$\langle f, \mu \rangle = \frac{1}{|T^2|} \int_{T^2} f(\nabla u(x))\, dx$$

are laminates. At present no effective algorithm for this is known. For the related question to determine numerically the rank-one convex hull of a function, see [Do 02] and the references therein.

Many simple geometric questions about rank-one convexity are open, even in low dimensions. In view of applications the case of 2×2 matrices is particular interesting.

Question 6. Does each T_k-configuration in $\mathbb{R}^{2\times 2}$ contain some T_4-configuration? More generally, does every compact set $K \subset \mathbb{R}^{2\times 2}$ with nontrivial rank-one convex hull contain a T_4-configuration?

This is even open for symmetric 2×2 matrices. The answer to both questions is yes for diagonal 2×2 matrices (see Proposition 6.23). The answer to the second question is no for 3×2 matrices (see Proposition 6.26). Székelyhidi [65] constructs a T_5-configuration in $\mathbb{R}^{4\times 2}$ which does not contain a T_4-configuration.

Many examples arise from nonlinear elasticity and thus have at least an $SO(n)$ symmetry.

Question 7. How can exploit symmetries efficiently?

A nice example is the computation of the rank-one convex and polyconvex hull for an energy function which describes nematic elastomers [20, 55]. The role of discrete symmetries which arise in crystalline materials is also important [24], but largely unexplored.

Question 8. Is there an efficient algorithm to decide if $K^{rc} = K$? Can one at least efficiently check whether K contains a T_k configuration?

A typical ingredient in efficient algorithms is the use of extreme points in connection with the Krein-Milman theorem. With a sufficiently abstract (dual) definition of extreme points the Krein-Milman theory holds in our setting [1, 37, 71]. The question is whether one can obtain a more geometric characterization of extreme points. For a nearly optimal result for 2×2 matrices see [33], Thm. 4.20.

Question 9. Can one combine Theorems 1.1 and 1.3? In other words, is there an elliptic system (with a quasiconvex or polyconvex energy function) which admits a solution whose gradient takes only finitely many values?

This might be easier if one goes beyond 2×2 systems. In this case $v = \binom{X}{Y}$, with $X, Y \in \mathbb{R}^{n \times n}$ and $A(Dv) = \binom{\text{curl}\, X}{\text{div}\, Y}$.

Question 10. Is the quasiconvex hull of the set K related to $\phi(X) = (\det X)^2$ trivial?

This is related to the compactness and regularity properties of solutions to the corresponding (degenerate) Euler-Lagrange equation

$$\text{div}[(\det \nabla w) \operatorname{cof} \nabla w] = 0.$$

Since $\text{div} \operatorname{cof} \nabla w = 0$ one can argue formally that all solutions must satisfy $\det \nabla w = \text{const}$ and this argument is correct for C^1 solutions. If one applies the formal argument to the 'linearized' problem

$$\text{div}[f(x) \operatorname{cof} \nabla w] = 0$$

one would similarly conclude that $f = \text{const}$, at least if $\det \nabla w \neq 0$. There exist, however, a Lipschitz map $w : \mathbb{R}^2 \to \mathbb{R}^2$ with $\det \nabla w \in \{-1, 1\}$ a.e. and a non constant f with values in $\{-1, 1\}$ which satisfy the above equation.

In this paper we have mainly studied sets, but the study of rank-one convex functions is equally interesting. Many seemingly simple questions are open.

Question 11. Can one characterize the traces of separately convex functions $f : \mathbb{R} \oplus \mathbb{R} \to \mathbb{R}$ on the diagonal?

Tartar showed that any smooth function can arise as a trace, see [69], Remark 11 there. Motivated by results due to Šverák and Preiss mentioned in [69], it was shown that C^1 is a necessary and $C^{1,\alpha}$ a sufficient condition for being a trace. The lack of a precise characterization seems related to our partial understanding of rank-one convexification procedures. The situation is even less clear if we go into 3 or more dimensions.

Finally, throughout this work we have focused on oscillations effects only and restricted attention to bounded sets K. If one drops this assumptions one also needs to study the possible interaction of oscillation and concentration effects and new tools are required, see e.g. [68]. More on the technical side one can ask to which differential operators A one can extend the general theory (for questions of compactness, A gradient Young measures, and relaxation see [23, 8]).

Acknowledgements. We thank Laszlo Székelyhidi for many interesting discussions, in particular on T_5 configurations and on the laminate without discrete approximations. S.M. and V.S. wish to thank the members SFB 256 and the DFG for their longstanding support and encouragement and for providing an atmosphere of open and inspiring exchange. Special thanks go to W. Ballmann, U. Hamenstädt and J. Lohkamp who organised a seminar on convex integration during V.S.'s stay in Bonn in 1992/93. S.M and V.S. were also supported by a Max Planck Research award, and V.S. was supported by the NSF grant DMS 9877055. B.K and S.M. would like to thank the IMA and the Department of Mathematics at the University of Minnesota for their hospitality during numerous visits.

References

1. E.M. Alfsen, *Compact convex sets and boundary integrals*, Springer, 1971.
2. K. Astala and D. Faraco, Quasiregular mappings and Gradient Young measures, to appear in *Proc. Roy. Soc. Edinburgh.*
3. R. Aumann and S. Hart, Bi-convexity and bi-martingales, *Israel J. Math.* **54** (1986), 159–180.
4. J.M. Ball, Strict convexity, strong ellipticity and regularity in the calculus of variations, *Math. Proc. Cambridge Phil. Soc.* **87** (1980), 501–513.
5. J.M. Ball, Sets of gradients with no rank-one connections, *J. Math. Pures Appl.* **69** (1990), 241–259.
6. J.M. Ball and R.D. James, Fine phase mixtures as minimizers of energy, *Arch. Rat. Mech. Anal.* **100** (1987), 13–52.
7. E. Bombieri, E. De Giorgi and E. Giusti, Minimal cones and the Bernstein theorem, *Invent. Math.* **7** (1969), 243–269.
8. A. Braides, I. Fonseca and G. Leoni, A-quasiconvexity: relaxation and homogenization, *ESAIM Control Optim. Calc. Var.* **5** (2000), 539–577.
9. E. Casadio-Tarabusi, An algebraic characterization of quasi-convex functions, *Ricerche Mat.* **42** (1993), 11–24.
10. A. Cellina, On the differential inclusion $x \in [-1, 1]$, *Atti Accad. Naz. Lincei Rend. Sci. Fis. Mat. Nat.* **69**, 1–6.
11. M. Chipot and D. Kinderlehrer, Equilibrium configurations of crystals, *Arch. Rat. Mech. Anal.* **103** (1988), 237–277.
12. M. Chlebik and B. Kirchheim, Rigidity for the four gradient problem, Preprint MPI-MIS 35/2000.
13. B. Dacorogna, *Direct methods of the calculus of variations*, Springer, 1989.

14. B. Dacorogna and P. Marcellini, General existence theorems for Hamilton-Jacobi equations in the scalar and vectorial cases, *Acta Math.* **178** (1997), 1–37.
15. B. Dacorogna and P. Marcellini, *Implicit partial differential equations*, Birkhäuser, 1999.
16. G. Dolzmann, *Variational Methods for Crystalline Microstructure - Analysis and Computation*, to appear as Lecture Notes in Mathematics, Springer.
17. F.S. De Blasi and G. Pianigiani, Baire category approach to the existence of solutions of multivalued differential equations in Banach spaces, *Funkcial. Ekvac.* **25** (1982), 153–162.
18. F.S. De Blasi and G. Pianigiani, Non convex valued differential inclusions in Banach spaces, *J. Math. Anal. Appl.* **157** (1991), 469–494.
19. E. De Giorgi, Un esempio di estremali discontinue per un problema variazionale di tipo ellittico, *Boll. U.M.I.* **4** (1968), 135–137.
20. A. DeSimone and G. Dolzmann, Macroscopic response of nematic elastomers via relaxation of a class of $SO(3)$ invariant energies, *Arch. Rat. Mech. Anal.* **161** (2002), 181–204.
21. R.J. DiPerna, Compensated compactness and general systems of conservation laws, Trans. Amer. Math. Soc. **292** (1985), 383–420.
22. L.C. Evans, Quasiconvexity and partial regularity in the calculus of variations, *Arch. Rat. Mech. Anal.* **95** (1986), 227–252.
23. I. Fonseca and S. Müller, A-quasiconvexity, lower semicontinuity, and Young measures, *SIAM J. Math. Anal.* **30** (1999), 1355–1390.
24. G. Friesecke, personal communication.
25. M. Fuchs, Regularity theorems for nonlinear systems of partial differential equations under natural ellipticity conditions, *Analysis* **7** (1987), 83–93.
26. E. Giusti and M. Miranda, Un esempio di soluzioni discontinue per un problema di minimo relativo ad un integrale regolare del calcolo delle variazioni, *Boll. U.M.I* **2** (1968), 1–8.
27. M. Gromov, Convex integration of differential relations, Izw. Akad. Nauk S.S.S.R **37** (1973), 329–343.
28. M. Gromov, *Partial differential relations*, Springer, 1986.
29. C. Hamburger, Quasimonotonicity, regularity and duality for nonlinear systems of partial differential equations, *Ann. Mat. Pure Appl.* Ser. IV **169** (1995), 321–354.
30. W. Hao, S. Leonardi, J. Nečas, An example of irregular solution to a nonlinear Euler-Lagrange elliptic system with real analytic coefficients, *Ann. SNS Pisa Ser. IV* **23** (1996), 57–67.
31. D. Kinderlehrer and P. Pedregal, Characterizations of Young measures generated by gradients, *Arch. Rat. Mech. Anal.* **115** (1991), 329–365.
32. B. Kirchheim, Deformations with finitely many gradients and stability of quasiconvex hulls, *C. R. Acad. Sci. Paris Sér. I Math.* **332** (2001), 289–294.
33. B. Kirchheim, Habilitation thesis, University Leipzig, 2001.
34. J. Kolář, Non-compact lamination convex hulls, Preprint MPI-MIS 17/2001.
35. J. Kristensen, On the non-locality of quasiconvexity, *Ann. Inst. H. Poincaré Anal. Non Linéaire* **16** (1999), 1–13.
36. J. Kristensen and A. Taheri, Partial regularity of strong local minimizers in the multidimensional calculus of variations, Preprint MPI-MIS 59/2001.
37. M. Kružík, Bauer's maximum principle and hulls of sets, *Calc. Var.* **11** (2000), 321–332.
38. N.H. Kuiper, On C^1 isometric embeddings, I., *Nederl. Akad. Wetensch. Proc.* **A 58** (1955), 545–556.

39. J. Matoušek, On directional convexity, *Discrete Comput. Geom.* **25** (2001), 389–403.
40. J. Matoušek and P. Plecháč, On functional separately convex hulls, *Discrete Comput. Geom.* **19** (1998), 105–130.
41. S. Müller and V. Šverák, Attainment results for the two-well problem by convex integration, *Geometric analysis and the calculus of variations*, (J. Jost, ed.), International Press, 1996, 239–251.
42. S. Müller, *Variational models for microstructure and phase transitions*, in: Calculus of Variations and Geometric Evolution Problems, Cetaro 1996 (F. Bethuel, G. Huisken, S. Müller, K. Steffen, S. Hildebrandt, and M. Struwe, eds.), Springer, Berlin, 1999.
43. S. Müller, Rank-one convexity implies quasiconvexity on diagonal matrices, *Int. Math. Research Not.* 1999, 1087–1095.
44. S. Müller and V. Šverák, Unexpected solutions of first and second order partial differential equations, *Documenta Math.*, Special volume Proc. ICM 1998, Vol. II, 691–702.
45. S. Müller and V. Šverák, Convex integration for Lipschitz mappings and counterexamples to regularity, Preprint MPI-MIS 26/1999.
46. S. Müller, M.O. Rieger and V. Šverák, Parabolic systems with nowhere smooth solutions, Preprint.
47. S. Müller and M. Sychev, Optimal existence theorems for nonhomogeneous differential inclusions, *J. Funct. Anal.* **181** (2001), 447–475.
48. F. Murat, L'injection du cone positif de H^{-1} dans $W^{-1,q}$ est compacte pour tout $q < 2$, *J. Math. Pures Appl.* **60** (1981), 309–322.
49. J. Nash, C^1 isometric embeddings, *Ann. Math.* **60** (1954), 383–396.
50. V. Nesi and G.W. Milton, Polycrystalline configurations that maximize electrical resistivity, *J. Mech. Phys. Solids* **39** (1991), 525–542.
51. P. Pedregal, Laminates and microstructure, *Europ. J. Appl. Math.* **4** (1993), 121–149.
52. P. Pedregal, *Parametrized measures and variational principles*, Birkhäuser, 1997.
53. V. Scheffer, Regularity and irregularity of solutions to nonlinear second order elliptic systems of partial differential equations and inequalities, Dissertation, Princeton University, 1974.
54. D. Serre, Formes quadratiques et calcul des variations, *J. Math. Pures Appl.* **62** (1983), 177–196.
55. M. Šilhavý, Relaxation in a class of $SO(n)$-invariant energies related to nematic elastomers, Preprint.
56. D. Spring, *Convex integration theory*, Birkhäuser, 1998.
57. V. Šverák, Examples of rank-one convex functions, *Proc. Roy. Soc. Edinburgh* **A 114** (1990), 237–242.
58. V. Šverák, Rank-one convexity does not imply quasiconvexity, *Proc. Roy. Soc. Edinburgh* **120** (1992), 185–189.
59. V. Šverák, New examples of quasiconvex functions, *Arch. Rat. Mech. Anal.* **119** (1992), 293–300.
60. V. Šverák, On Tartar's conjecture, *Ann. Inst. H. Poincaré Anal. Non Linéaire* **10** (1993), 405–412.
61. V. Šverák, Lower semicontinuity of variational integrals and compensated compactness, in: Proc. ICM 1994 (S.D. Chatterji, ed.), vol. 2, Birkhäuser, 1995, 1153–1158.
62. V. Šverák and Xiadong Yan, A singular minimizer of a smooth strongly convex functional in three dimensions, *Calc. Var.* **10** (2000), 213–221.
63. M.A. Sychev, A new approach to Young measure theory, relaxation and convergence in energy, *Ann. Inst. H. Poincaré Anal. Non Linéaire* **16** (1999), 773–812.
64. M.A. Sychev, Comparing two methods of resolving homogeneous differential inclusions, *Calc. Var.* **13** (2001), 213–229.

65. L. Székelyhidi, PhD thesis, in preparation.
66. L. Tartar, Compensated compactness and partial differential equations, in: *Nonlinear Analysis and Mechanics: Heriot-Watt Symposium* Vol. IV, (R. Knops, ed.), Pitman, 1979, 136–212.
67. L. Tartar, The compensated compactness method applied to systems of conservations laws, in: *Systems of Nonlinear Partial Differential Equations*, (J.M. Ball, ed.), NATO ASI Series, Vol. C111, Reidel, 1983, 263–285.
68. L. Tartar, H-measures, a new approach for studying homogenisation, oscillations and concentration effects in partial differential equations, *Proc. Roy. Soc. Edinburgh Sect. A* **115** (1990), 193–230.
69. L. Tartar, Some remarks on separately convex functions, in: *Microstructure and phase transitions*, IMA Vol. Math. Appl. **54**, (D. Kinderlehrer, R.D. James, M. Luskin and J.L. Ericksen, eds.), Springer, 1993, 191–204.
70. Kewei Zhang, On the Dirichlet problem for a class of quasilinear elliptic systems of partial differential equations in divergence form, in: *Partial differential equations* (Tianjun, 1986), Lecture Notes in Mathematics **1306**, Springer, 1988.
71. Kewei Zhang, On the structure of quasiconvex hulls, *Ann. Inst. H. Poincaré Analyse Non Linéaire* **15** (1998), 663–686.

On the Korteweg - de Vries Equation and KAM Theory

Thomas Kappeler and Jürgen Pöschel

[1] Universität Zürich, Institut für Mathematik, Winterthurerstrasse 190, CH-8057 Zürich
tk@math.unizh.ch
[2] Mathematisches Institut A, Universität Stuttgart, Pfaffenwaldring 57, D-70569 Stuttgart
poschel@mathematik.uni-stuttgart.de

In this note we give an overview of results concerning the Korteweg-de Vries equation

$$u_t = -u_{xxx} + 6uu_x$$

and small perturbations of it. All the technical details will be contained in our forthcoming book [27].

The KdV equation is an evolution equation in one space dimension which is named after the two Dutch mathematicians Korteweg and de Vries [29], but was apparently derived even earlier by Boussinesq [10, 57]. It was proposed as a model equation for long surface waves of water in a narrow and shallow channel. Their aim was to obtain as solutions solitary waves of the type discovered in nature by Russell [58] in 1834. Later it became clear that this equation also models waves in other homogeneous, weakly nonlinear and weakly dispersive media. Since the mid-sixties the KdV equation received a lot of attention in the aftermath of the computational experiments of Kruskal and Zabusky [32], which lead to the discovery of the interaction properties of the solitary wave solutions and in turn to the understanding of KdV as an infinite dimensional integrable Hamiltonian system.

Our purpose here is to study small Hamiltonian perturbations of the KdV equation with periodic boundary conditions. In the unperturbed system *all* solutions are periodic, quasi-periodic, or almost periodic in time. The aim is to show that large families of periodic and quasi-periodic solutions persist under such perturbations. This is true not only for the KdV equation itself, but in principle for all equations in the KdV hierarchy. As an example, the second KdV equation will also be considered.

1 The KdV Equation

Let us recall those features of the KdV equation that are essential for our purposes. It was observed by Gardner [21] and Faddeev & Zakharov [18] that the KdV equation can be written in the Hamiltonian form

$$\frac{\partial u}{\partial t} = \frac{\mathrm{d}}{\mathrm{d}x}\frac{\partial H}{\partial u}$$

with the Hamiltonian

$$H(u) = \int_{S^1} \left(\tfrac{1}{2}u_x^2 + u^3\right) \mathrm{d}x,$$

where $\partial H/\partial u$ denotes the L^2-gradient of H. Since we are interested in *spatially periodic* solutions, we take as the underlying phase space the Sobolev space

$$\mathcal{H}^N = H^N(S^1;\mathbb{R}), \qquad S^1 = \mathbb{R}/\mathbb{Z},$$

of real valued functions with period 1, where $N \geq 1$ is an integer, and endow it with the Poisson bracket proposed by Gardner,

$$\{F,G\} = \int_{S^1} \frac{\partial F}{\partial u(x)} \frac{\mathrm{d}}{\mathrm{d}x} \frac{\partial G}{\partial u(x)} \mathrm{d}x.$$

Here, F and G are differentiable functions on $\mathcal{H}^N$ with L^2-gradients in $\mathcal{H}^1$. This makes $\mathcal{H}^N$ a Poisson manifold, on which the KdV equation may also be represented in the form $u_t = \{u,H\}$ familiar from classical mechanics.

The initial value problem for the KdV equation on the circle S^1 is well posed on every Sobolev space $\mathcal{H}^N$ with $N \geq 1$: for initial data $u^o \in \mathcal{H}^N$ it has been shown by Temam for $N = 1, 2$ [61] and by Saut & Temam for any real $N \geq 2$ [59] that there exists a unique solution evolving in $\mathcal{H}^N$ and defined globally in time. For further results on the initial value problem see for instance [60, 41, 44] as well as the more recent results [6, 7, 28].

The KdV equation admits infinitely many conserved quantities, or *integrals*, and there are many ways to construct such integrals [50, 51]. Lax [40] obtained a complete set of integrals in a particularly elegant way by considering the spectrum of an associated Schrödinger operator. For

$$u \in \mathcal{H}^0 = L^2 = L^2(S^1, \mathbb{R})$$

consider the differential operator

$$L = -\frac{\mathrm{d}^2}{\mathrm{d}x^2} + u$$

on the interval $[0, 2]$ of *twice* the length of the period of u with periodic boundary conditions. It is well known [42] that its spectrum, denoted $\mathrm{spec}(u)$, is pure point and consists of an unbounded sequence of *periodic eigenvalues*

$$\lambda_0(u) < \lambda_1(u) \leq \lambda_2(u) < \lambda_3(u) \leq \lambda_4(u) < \ldots \,.$$

Equality or inequality may occur in every place with a Ô≤Õ-sign, and one speaks of the *gaps* $(\lambda_{2n-1}(u), \lambda_{2n}(u))$ of the *potential* u and its *gap lengths*

$$\gamma_n(u) = \lambda_{2n}(u) - \lambda_{2n-1}(u), \qquad n \geq 1.$$

If some gap length is zero, then one speaks of a *collapsed gap*.

For $u = u(t,\cdot)$ depending also on t define the corresponding operator

$$L(t) = -\frac{\mathrm{d}^2}{\mathrm{d}x^2} + u(t,\cdot).$$

Lax observed that u is a solution of the KdV equation if and only if

$$\frac{\mathrm{d}}{\mathrm{d}t}L = [B,L],$$

where $[B,L] = BL - LB$ denotes the commutator of L with the *anti-symmetric* operator

$$B = -4\frac{\mathrm{d}^3}{\mathrm{d}x^3} + 3u\frac{\mathrm{d}}{\mathrm{d}x} + 3\frac{\mathrm{d}}{\mathrm{d}x}u.$$

It follows by an elementary calculation that the flow of

$$\frac{\mathrm{d}}{\mathrm{d}t}U = BU, \qquad U(0) = I,$$

defines a family of *unitary* operators $U(t)$ such that $U^*(t)L(t)U(t) = L(0)$. Consequently, the spectrum of $L(t)$ is independent of t, and so the periodic eigenvalues $\lambda_n = \lambda_n(u)$ are conserved quantities under the evolution of the KdV equation. In other words, the flow of the KdV equation defines an *isospectral deformation* on the space of all potentials in $\mathcal{H}^N$.

From an analytical point of view, however, the periodic eigenvalues are not satisfactory as integrals, as λ_n is not a smooth function of u whenever the corresponding gap collapses. But McKean & Trubowitz [45] showed that the *squared gap lengths*

$$\gamma_n^2(u), \qquad n \geq 1,$$

together with the average

$$[u] = \int_{S^1} u(x)\,\mathrm{d}x$$

form another set of integrals, which are *real analytic* on all of L^2 and Poisson commute with each other. Moreover, the squared gap lengths together with the average determine uniquely the periodic spectrum of a potential [22].

The space L^2 thus decomposes into the isospectral sets

$$\mathrm{Iso}(u) = \left\{v \in L^2 \colon \operatorname{spec}(v) = \operatorname{spec}(u)\right\},$$

which are invariant under the KdV flow and may also be characterized as

$$\mathrm{Iso}(u) = \left\{v \in L^2 \colon \text{gap lengths}(v) = \text{gap lengths}(u),\ [v] = [u]\right\}.$$

As shown by McKean & Trubowitz [45] these are compact connected *tori*, whose dimension equals the number of positive gap lengths and is infinite generically.

Moreover, as the asymptotic behavior of the gap lengths characterizes the regularity of a potential in exactly the same way as its Fourier coefficients do [43], we have

$$u \in \mathcal{H}^N \quad \Leftrightarrow \quad \operatorname{Iso}(u) \subset \mathcal{H}^N$$

for each $N \geq 1$. Hence also the phase space $\mathcal{H}^N$ decomposes into a collection of tori of varying dimension which are *invariant* under the KdV flow.

2 Action-Angle and Birkhoff Coordinates

In classical mechanics the existence of a foliation of the phase space into Lagrangian invariant tori is tantamount, at least locally, to the existence of action-angle coordinates. This is the content of the Liouville-Arnold-Jost theorem. In the infinite dimensional setting of the KdV equation, however, the existence of such coordinates is far less clear as the dimension of the foliation is *nowhere* locally constant. Invariant tori of infinite and finite dimension each form dense subsets of the foliation. Nevertheless, action-angle coordinates *can* be introduced *globally* as we describe now. They will form the basis of our study of perturbations of the KdV equation.

To formulate the statement we define the phase spaces more precisely. For any integer $N \geq 0$, let

$$\mathcal{H}^N = \left\{ u \in L^2(S^1, \mathbb{R}) \colon \; \|u\|_N < \infty \right\},$$

where

$$\|u\|_N^2 = |\hat{u}(0)|^2 + \sum_{k \in \mathbb{Z}} |k|^{2N} \, |\hat{u}(k)|^2$$

is defined in terms of the discrete Fourier transform $\hat{u}$ of u. The Poisson structure $\{\cdot,\cdot\}$ is degenerate on $\mathcal{H}^N$ and admits the average $[\cdot]$ as a Casimir functional. The leaves of the corresponding symplectic foliation are given by $[u] = \text{const}$. Instead of restricting the KdV Hamiltonian to each leaf, it is more convenient to fix one such leaf, namely

$$\mathcal{H}_0^N = \left\{ u \in \mathcal{H}^N \colon \; [u] = 0 \right\},$$

which is symplectomorphic to each other leaf by a simple translation, and consider the mean value as a parameter. On $\mathcal{H}_0^N$ the Poisson structure is nondegenerate and induces a symplectic structure. Writing $u = v + c$ with $[v] = 0$ and $c = [u]$, the Hamiltonian then takes the form

$$H(u) = H_c(v) + c^3$$

with

$$H_c(v) = \int_{S^1} \left(\tfrac{1}{2} v_x^2 + v^3 \right) \mathrm{d}x + 6c \int_{S^1} \tfrac{1}{2} v^2 \, \mathrm{d}x.$$

We consider H_c as a 1-parameter family of Hamiltonians on $\mathcal{H}_0^N$.

We remark that

$$H^0 = \frac{1}{2}\int_{S^1} v^2 \,\mathrm{d}x$$

corresponds to translation and is the zero-th Hamiltonian of the KdV hierarchy.

To describe the action-angle variables on $\mathcal{H}_0^N$ we introduce the model space

$$\mathscr{h}_r = \ell_r^2 \times \ell_r^2$$

with elements (x, y), where

$$\ell_r^2 = \left\{ x \in \ell^2(\mathbb{N}, \mathbb{R}) : \ \|x\|_r^2 = \sum_{n\geq 1} n^{2r} |x_n|^2 < \infty \right\}.$$

We endow $\mathscr{h}_r$ with the standard Poisson structure, for which $\{x_n, y_m\} = \delta_{nm}$, while all other brackets vanish.

The following theorem was first proven in [1] and [2]. A quite different approach for this result – and the one we expand on here – was first presented in [26].

Theorem 2.1 *There exists a diffeomorphism $\Psi \colon \mathscr{h}_{1/2} \to \mathcal{H}_0^0$ with the following properties.*

(i) *Ψ is one-to-one, onto, bi-analytic, and preserves the Poisson bracket.*
(ii) *For each $N \geq 0$, the restriction of Ψ to $\mathscr{h}_{N+1/2}$, denoted by the same symbol, is a map $\Psi \colon \mathscr{h}_{N+1/2} \to \mathcal{H}_0^N$, which is one-to-one, onto, and bi-analytic as well.*
(iii) *The coordinates (x, y) in $\mathscr{h}_{N+1/2}$ are global Birkhoff coordinates for KdV. That is, for any $c \in \mathbb{R}$ the transformed Hamiltonian $H_c \circ \Psi$ depends only on $x_n^2 + y_n^2$, $n \geq 1$, with (x, y) being canonical coordinates.*

In the coordinate system (x, y) the KdV Hamiltonian H_c is a real analytic function of the actions I alone, where $I = (I_n)_{n\geq 1}$ with

$$I_n = \frac{1}{2}(x_n^2 + y_n^2).$$

This allows us to define its frequencies ω_c in the usual fashion:

$$\omega_c = (\omega_{c,n})_{n\geq 1}, \qquad \omega_{c,n} = \frac{\partial H_c}{\partial I_n},$$

and to establish them as real analytic functions of I.

These results are not restricted to the KdV Hamiltonian. They simultaneously apply to *every* real analytic Hamiltonian in the Poisson algebra of all Hamiltonians which Poisson commute with all action variables $I_1, I_2, \ldots$. In particular, one obtains action-angle coordinates for every equation in the KdV hierarchy. As an example, we will also consider the second KdV Hamiltonian later.

The existence of action-angle coordinates makes it evident that every solution of the KdV equation is *almost periodic* in time. In the coordinates of the model space every solution is given by

$$I(t) = I^{\mathrm{o}}, \qquad \theta(t) = \theta^{\mathrm{o}} + \omega(I^{\mathrm{o}})t,$$

where $(\theta^{\mathrm{o}}, I^{\mathrm{o}})$ corresponds to the initial data u^{o}, and $\omega(I^{\mathrm{o}}) = \omega_c(I^{\mathrm{o}})$ are the frequencies associated with I^{o} defined above. Hence, in the model space every solution winds around some underlying invariant torus

$$T_I = \left\{(x,y) \colon x_n^2 + y_n^2 = 2I_n, \;\; n \geq 1\right\},$$

determined by the initial actions I^{o}. Putting aside for the moment questions of convergence, the solution in the space $\mathcal{H}_0^N$ is thus winding around the embedded torus $\Psi(T_I)$ and is of the form

$$\begin{aligned} u(t) &= \Xi(\theta^{\mathrm{o}} + \omega(I^{\mathrm{o}})t, I^{\mathrm{o}}) \\ &= \sum_{k \in \mathbb{Z}^\infty, |k| < \infty} \Xi_k(I^{\mathrm{o}}) e^{2\pi \mathrm{i} \langle k, \theta^{\mathrm{o}} \rangle} e^{2\pi \mathrm{i} \langle k, \omega(I^{\mathrm{o}}) \rangle t}. \end{aligned}$$

Here, $|k| = \sum_n |k_n|$, $\langle k, \theta \rangle = \sum_n k_n \theta_n$, and Ξ denotes the composition of the transformation from action-angle coordinates (θ, I) to Birkhoff coordinates (x, y) with Ψ, with each $\Xi_k(I^{\mathrm{o}})$ being an element of $\mathcal{H}_0^N$. Thus, every solution is almost-periodic in time.

We remark that the solution above can also be represented in terms of the Riemann theta function. The corresponding formula is due to Its & Matveev [13].

Among all almost-periodic solutions there is a dense subset of quasi-periodic solutions, which are characterized by a *finite* number of frequencies and correspond to finite gap potentials. To describe them more precisely, let $A \subset \mathbb{N}$ be a finite index set, and consider the set of *A-gap potentials*

$$\mathcal{G}_A = \left\{u \in \mathcal{H}_0^0 \colon \gamma_n(u) > 0 \Leftrightarrow n \in A\right\}.$$

That is, $u \in \mathcal{G}_A$ if and only if precisely the gaps $(\lambda_{2n-1}(u), \lambda_{2n}(u))$ with $n \in A$ are open. Clearly,

$$u \in \mathcal{G}_A \quad \Leftrightarrow \quad \mathrm{Iso}(u) \subset \mathcal{G}_A,$$

and all finite gap potentials are smooth, in fact real analytic, as almost all gap lengths are zero. As might be expected there is a close connection between the set $\mathcal{G}_A$ and the subspace

$$\hbar_A = \left\{(x,y) \in \hbar_0 \colon x_n^2 + y_n^2 > 0 \Leftrightarrow n \in A\right\}.$$

Addendum 1 (to Theorem 2.1). *The canonical transformation Ψ also has the following property.*

(iv) *For every finite index set $A \subset \mathbb{N}$, the restriction Ψ_A of Ψ to $\hbar_A$ is a map*

$$\Psi_A\colon\ \hbar_A \to \mathcal{G}_A,$$

which is one-to-one, onto, and bi-analytic.

It follows that $\mathcal{G}_A$ is a real analytic, invariant submanifold of $\mathcal{H}_0^0$, which is completely and analytically foliated into invariant tori $\mathrm{Iso}(u)$ of the same dimension $|A|$. The KdV flow consists of a quasi-periodic winding around each such torus which is characterized by $|A|$ frequencies ω_n, $n \in A$. The above representation reduces to a convergent, real analytic representation

$$u(t) = \sum_{k\in\mathbb{Z}^A} \Xi_k(I_A^{\circ})e^{2\pi\mathrm{i}\langle k,\theta_A^{\circ}\rangle}e^{2\pi\mathrm{i}\langle k,\omega_A(I_A^{\circ})\rangle t},$$

with $I_A = (I_n)_{n\in A}$, and similarly defined θ_A and ω_A.

3 Outline of the Proof of Theorem 2.1

The proof of the theorem splits into four parts. First we define actions I_n and angles θ_n for a potential q following a procedure well known for finite dimensional integrable systems. The formula for the actions I_n, due to Flaschka & McLaughlin [19], is given entirely in terms of the periodic spectrum of the potential. The angles θ_n, which linearize the KdV equation, were introduced even earlier by a number of authors, namely Dubrovin, Its, Krichever, Matveev, Novikov [12, 13, 15, 16, 25] (see also [14]), McKean & van Moerbeke [44], and McKean & Trubowitz [45, 46]. They are defined in terms of the Riemann surface $\Sigma(q)$ associated with the periodic spectrum of a potential. We show that each I_n is real analytic on L_0^2, while each θ_n, taken modulo 2π, is real analytic on the dense open domain $L_0^2 \setminus D_n$, where D_n denotes the subvariety of potentials with collapsed n-th gap.

Next, we define the cartesian coordinates x_n and y_n canonically associated to I_n and θ_n. Although defined originally only on $L_0^2 \setminus D_n$, we show that they extend real analytically to a complex neighbourhood W of L_0^2. Surely, the angle θ_n blows up when γ_n collapses, but this blow up is compensated by the rate at which I_n vanishes in the process. In particular, for real q the resulting limit will vanish.

Then we show that the thus defined map $\Omega\colon q \mapsto (x, y)$ is a diffeomorphism between L_0^2 and $\hbar_{1/2}$. The main problem here is to verify that $d_q\Omega$ is a linear isomorphism at *every* point q. This is done with the help of orthogonality relations among the coordinates, which are in fact their Poisson brackets. For the nonlinear Schrödinger equation the corresponding orthogonality relations have first been established by McKean & Vaninsky [47, 48]. It turned out that many of their ideas can also be used in the case of KdV.

We also verify that each Hamiltonian in the KdV hierarchy becomes a function of the actions alone, using their characterization in terms of the asymptotic expansion of the discriminant Δ as $\lambda \to -\infty$ and hence as spectral invariants.

Finally, we verify that Ω preserves the Poisson bracket. As it happens, it is more convenient to look at the associated symplectic structure. This way, we only need to establish the regularity of the gradient of θ_n at special points, not everywhere. Thus, we equivalently show that Ω is a symplectomorphism. This will complete the proof of the main results of Theorem 2.1.

4 Perturbations of the KdV Equation

Our aim is to investigate whether sufficiently small Hamiltonian perturbations of the KdV equation,

$$\frac{\partial u}{\partial t} = \frac{\mathrm{d}}{\mathrm{d}x}\left(\frac{\partial H_c}{\partial u} + \varepsilon\frac{\partial K}{\partial u}\right),$$

admit almost-periodic solutions as well, which wind around invariant tori in phase space.

In the classical setting of integrable Hamiltonian systems of finitely many degrees of freedom this question is answered by the theory of Kolmogorov, Arnold and Moser, known as *KAM theory*. It states that with respect to Lebesgue measure the majority of the invariant tori of a real analytic, nondegenerate integrable system persist under sufficiently small, real analytic Hamiltonian perturbations. They are only slightly deformed and still completely filled with quasi-periodic motions. The base of this partial foliation of the phase space into invariant tori, however, is no longer open, but has the structure of a *Cantor set*: it is a nowhere dense, closed set with no isolated points.

Till now, there is no general infinite dimensional KAM theory to establish the persistence of infinite dimensional tori with almost periodic solutions for Hamiltonian systems arising from partial differential equations, such as the KdV equation. There does exist a KAM theory to this effect, but it is restricted to systems, in which the coupling between the different modes of oscillations is of short range type. See [53] and the references therein. Such a theory does not apply here, and we can not make any statement about the persistence of almost-periodic solutions.

But as noted above, there are also families of finite dimensional tori on real analytic submanifolds, corresponding to finite gap solutions and filling the space densely. In the classical setting a KAM theorem about the persistence of such lower dimensional tori was first formulated by Melnikov [49], and proven later by Eliasson [17]. It was independently extended to the infinite dimensional setting of partial differential equations by Kuksin [33, 35, 37]. In the following we describe this kind of perturbation theory for finite gap solutions, following [52, 54].

Again, let $A \subset \mathbb{N}$ be a *finite* index, and let $\Gamma \subset \mathbb{R}_+^A$ be a compact set of positive Lebesgue measure. We then set

$$\mathcal{T}_\Gamma = \bigcup_{I\in\Gamma} \mathcal{T}_I \subset \mathcal{G}_A, \qquad \mathcal{T}_I = \Psi_A(T_I),$$

where

$$T_I = \{(x,y) \in \mathscr{h}_A : x_n^2 + y_n^2 = 2I_n,\ \ n \in A\} \cong \mathbb{T}^A \times \{I\},$$

with $\mathbb{T} = \mathbb{R}/2\pi\mathbb{Z}$ the circle of length 2π. Notice that $\mathcal{T}_\Gamma \subset \bigcap_{N\geq 0} \mathcal{H}_0^N$ in view of the Addendum to Theorem 2.1.

We show that under sufficiently small real analytic Hamiltonian perturbations of the KdV equation the majority of these tori persists together with their translational flows, the tori being only slightly deformed. The following theorem was first proven by S. Kuksin for the case $c = 0$.

Theorem 4.2 *Let $A \subset \mathbb{N}$ be a finite index set, $\Gamma \subset \mathbb{R}_+^A$ a compact subset of positive Lebesgue measure, and $N \geq 1$. Assume that the Hamiltonian K is real analytic in a complex neighbourhood U of $\mathcal{T}_\Gamma$ in $\mathcal{H}_{0,\mathbb{C}}^N$ and satisfies the regularity condition*

$$\frac{\partial K}{\partial u} \colon\ U \to \mathcal{H}_{0,\mathbb{C}}^N, \qquad \left\| \frac{\partial K}{\partial u} \right\|_{N;U}^{\mathrm{sup}} = \sup_{u\in U} \left\| \frac{\partial K}{\partial u} \right\|_N \leq 1.$$

Then, for any real c, there exists an $\varepsilon_0 > 0$ depending only on A, N, c and the size of U such that for $|\varepsilon| < \varepsilon_0$ the following holds. There exist

(i) *a nonempty Cantor set $\Gamma_\varepsilon \subset \Gamma$ with* $\operatorname{meas}(\Gamma - \Gamma_\varepsilon) \to 0$ *as* $\varepsilon \to 0$,
(ii) *a Lipschitz family of real analytic torus embeddings*

$$\Xi\colon\ \mathbb{T}^n \times \Gamma_\varepsilon \to U \cap \mathcal{H}_0^N,$$

(iii) *a Lipschitz map $\chi\colon \Gamma_\varepsilon \to \mathbb{R}^n$,*

such that for each $(\theta, I) \in \mathbb{T}^n \times \Gamma_\varepsilon$, the curve $u(t) = \Xi(\theta + \chi(I)t, I)$ is a quasi-periodic solution of

$$\frac{\partial u}{\partial t} = \frac{\mathrm{d}}{\mathrm{d}x}\left(\frac{\partial H_c}{\partial u} + \varepsilon \frac{\partial K}{\partial u}\right)$$

winding around the invariant torus $\Xi(\mathbb{T}^n \times \{I\})$. Moreover, each such torus is linearly stable.

Remark 4.3 Note that the L^2-gradient of a function on $\mathcal{H}_0^N$ has mean value zero by the definition of the gradient. On the other hand, the L^2-gradient of a function on the larger space $\mathcal{H}^N$ usually has man value different from zero, and the gradient of its restriction to $\mathcal{H}_0^N$ is the *projection* of the former onto $\mathcal{H}_0^N$:

$$\nabla \left(K|_{H_0^N}\right) = \operatorname{Proj}_{\mathcal{H}_0^N} \nabla K = \nabla K - [\nabla K],$$

with $\nabla = \partial/\partial u$. This, however, does not affect their Hamiltonian equations, since the derivative of the constant function $[\nabla K]$ vanishes. Therefore, we will not explicitly distinguish between these two gradients.

Remark 4.4 We already mentioned that $\mathcal{G}_A \subset \bigcap_{N\geq 0} \mathcal{H}_0^N$. Thus, the perturbed quasi-periodic solutions remain in $\mathcal{H}_0^N$ if the gradient of K is in $\mathcal{H}_0^N$.

Remark 4.5 The regularity assumption on K entails that K depends only on u, but not on its derivatives. So the perturbation effected by K is of *lower order* than the unperturbed KdV equation. In view of the derivation of the KdV equation as a model equation for surface waves of water in a certain regime expansion – see for example [63] – it would be interesting to obtain perturbation results which also include terms of higher order, at least in the region where the KdV approximation is valid. However, results of this type are still out of reach, if true at all.

Remark 4.6 We point out that the perturbing term $\varepsilon \partial K/\partial u$ need not be a differential operator. For example,

$$K(u) = \left(\int_{S^1} u^3 \, \mathrm{d}x \right)^2$$

has L^2-gradient

$$\frac{\partial K}{\partial u} = 6u^2 \int_{S^1} u^3 \, \mathrm{d}x,$$

to which the theorem applies as well.

Remark 4.7 The invariant embedded tori are linearly stable in the sense that the variational equations of motion along such a torus are reducible to constant coefficient form whose spectrum is located on the imaginary axis. Hence, all Lyapunov exponents of such a torus vanish.

Similar results hold for any equation in the KdV hierarchy. Consequently, for any given finite index set A, the manifold of A-gap potentials is foliated into the same family of invariant tori. The difference is only in the frequencies of the quasi-periodic motions on each of these tori. Therefore, similar results should also hold for the higher order KdV equations, once we can establish the corresponding nonresonance conditions.

As an example we consider the second KdV equation, which reads

$$\partial_t u = \partial_x^5 u - 10u\partial_x^3 u - 20\partial_x u \partial_x^2 u + 30u^2 \partial_x u.$$

Its Hamiltonian is

$$H^2(u) = \int_{S^1} \left(\tfrac{1}{2} u_{xx}^2 + 5uu_x^2 + \tfrac{5}{2} u^4 \right) \mathrm{d}x,$$

which is defined on $\mathcal{H}^2$. Again, with $u = v + c$, where $[v] = 0$, we get

$$H^2(u) = H_c^2(v) + \frac{5}{2} c^4$$

with

$$H_c^2(v) = H^2(v) + 10cH^1(v) + 30c^2H^0(v).$$

Here, $H^1(v) = \int_{S^1} \left(\frac{1}{2}v_x^2 + v^3\right) \mathrm{d}x$ is the familiar KdV Hamiltonian, and $H^0(v) = \frac{1}{2}\int_{S^1} v^2 \,\mathrm{d}x$ is the zero-th Hamiltonian of translation. We study this Hamiltonian on the space $\mathcal{H}_0^N$ with $N \geq 2$, considering c as a real parameter.

Theorem 4.8 *Let $A \subset \mathbb{N}$ be a finite index set, $\Gamma \subset \mathbb{R}_+^A$ a compact subset of positive Lebesgue measure, and $N \geq 3$. Assume that the Hamiltonian K is real analytic in a complex neighbourhood U of $\mathcal{T}_\Gamma$ in $\mathcal{H}_{0,\mathbb{C}}^N$ and satisfies the regularity condition*

$$\frac{\partial K}{\partial u}\colon\ U \to \mathcal{H}_{0,\mathbb{C}}^{N-2}, \qquad \left\| \frac{\partial K}{\partial u} \right\|_{N-2;U}^{\sup} \leq 1.$$

If $c \notin \mathcal{E}_A^2$, where the exceptional set $\mathcal{E}_A^2$ is an at most countable subset of the real line not containing 0 and with at most $|A|$ accumulation points, then the same conclusions as in Theorem 4.2 hold for the system with Hamiltonian $H_c^2 + \varepsilon K$.

Remark 4.9 The gradient $\partial K/\partial u$ is only required to be in $\mathcal{H}_0^{N-2}$. Still, the regularity assumption ensures that the perturbation is of *lower order* than the unperturbed equation.

We now give two simple examples of perturbations to which the preceding theorems apply. As a first example let

$$K(u) = \int_{S^1} F(x,u)\,\mathrm{d}x,$$

where F defines a real analytic map

$$\{\lambda \in \mathbb{R}\colon\ |\lambda| < R\} \to \mathcal{H}_0^N, \qquad \lambda \mapsto F(\,\cdot\,,\lambda),$$

for some $R > 0$ and $N \geq 1$. Then, with $f = \partial F/\partial \lambda$,

$$\frac{\partial K}{\partial u} = f(x,u) - [f(x,u)]$$

belongs to $\mathcal{H}_0^N$, and the perturbed KdV equation is

$$u_t = -u_{xxx} + 6uu_x + \varepsilon \frac{\mathrm{d}}{\mathrm{d}x} f(x,u).$$

Theorem 4.2 applies after fixing Γ and c for all sufficiently small ε. Of course, F may also depend on ε, if the dependence is, say, continuous.

We remark that perturbations of the KdV equation with K as above can be characterized equivalently as local perturbations given by $\frac{\mathrm{d}}{\mathrm{d}x} f(x,u(x))$, where f admits a power series expansion in the second argument,

$$f(x,\lambda) = \sum_{k\geq 0} f_k(x)\lambda^k,$$

convergent in $\mathcal{H}_0^N$. In this case, the perturbed equation is again a partial differential equation, and its Hamiltonian K is given by $K(u) = \int_{S^1} F(x,u)\,\mathrm{d}x$, where F is a primitive of f with respect to λ,

$$F(x,\lambda) = \sum_{k\geq 0} \frac{f_k(x)}{k+1}\lambda^{k+1}.$$

As a second example let

$$K(u) = \int_{S^1} F(x,u_x)\,\mathrm{d}x,$$

where F is as above with $N \geq 3$. More generally, F could also depend on u, but this adds nothing new. Then

$$\frac{\partial K}{\partial u} = -\frac{\mathrm{d}}{\mathrm{d}x} f(x,u_x)$$

belongs to $\mathcal{H}_0^{N-2}$, and the perturbed second KdV equation is

$$u_t = \cdots - \varepsilon \frac{\mathrm{d}^2}{\mathrm{d}x^2} f(x,u_x).$$

To this second example, Theorem 4.8 applies.

5 Outline of Proof of Theorems 4.2 and 4.8

A prerequisite for developing a perturbation theory of KAM type is the existence of coordinates with respect to which the variational equations along the unperturbed motions on the invariant tori reduce to constant coefficient form. Often, such coordinates are difficult to construct even locally. Here, they are provided *globally* by Theorem 2.1.

According to Theorem 2.1 the Hamiltonian of the KdV equation on the model space $\mathbf{h}_{3/2}$ is of the form

$$H_c = H_c(I_1, I_2, \ldots), \qquad I_n = \frac{1}{2}(x_n^2 + y_n^2).$$

The equations of motion are thus

$$\dot{x}_n = \omega_n(I) y_n, \qquad \dot{y}_n = -\omega_n(I) x_n,$$

with frequencies

$$\omega_n = \omega_{c,n} = \frac{\partial H_c}{\partial I_n}(I), \qquad I = (I_n)_{n\geq 1},$$

that are constant along each orbit. So each orbit is winding around some invariant torus $T_I = \{(x,y)\colon x_n^2 + y_n^2 = 2I_n,\ n \geq 1\}$, where the parameters $I = (I_n)_{n\geq 1}$ are the actions of its initial data.

We are interested in a perturbation theory for families of *finite*-dimensional tori T_I. So we fix an index set $A \subset \mathbb{N}$ of finite cardinality $|A|$, and consider tori with

$$I_n > 0 \quad \Leftrightarrow \quad n \in A.$$

The linearized equations of motion along any such torus have now constant coefficients and are determined by $|A|$ *internal frequencies* $\omega = (\omega_n)_{n\in A}$ and infinitely many *external frequencies* $\Omega = (\omega_n)_{n\notin A}$. Both depend on the $|A|$-dimensional parameter

$$\xi = (I_n^{\circ})_{n\in A},$$

since all other components of I vanish in this family.

The KAM theorem for such families of finite dimensional tori requires a number of assumptions, among which the most notorious and unpleasant ones are the so called *nondegeneracy* and *nonresonance conditions*. In this case, they essentially amount to the following. First, the map

$$\xi \mapsto \omega(\xi)$$

from the parameters to the internal frequencies has to be a local homeomorphism, which is Lipschitz in both directions. In the classical theory this is known as *Kolmogorov's condition*. Second, the zero set of any of the frequency combinations

$$\langle k, \omega(\xi)\rangle + \langle l, \Omega(\xi)\rangle$$

has to be a set of measure zero, for each $k \in \mathbb{Z}^A$ and $l \in \mathbb{Z}^{\mathbb{N}\setminus A}$ with $1 \leq |l| \leq 2$. This is sometimes called *Melnikov's condition*.

The verification of these conditions for the KdV Hamiltonian requires some knowledge of its frequencies. One way to obtain this knowledge is to use Riemann surface theory: Krichever proved that the frequency map $\xi \mapsto \omega(\xi)$ is a local diffeomorphism everywhere on the space of A-gap potentials, see [3, 31], and Bobenko & Kuksin showed that the second condition is satisfied in the case $c = 0$ using Schottky uniformization [4]. Here, however, we follow a different and more elementary route to verify these conditions by computing the first coefficients of the Birkhoff normal form of the KdV Hamiltonian, which we explain now.

In classical mechanics the Birkhoff normal form allows to view a Hamiltonian system near an elliptic equilibrium as a small perturbation of an integrable system. This tool is also applicable in an infinite dimensional setting as ours. Writing

$$u = \sum_{n \neq 0} \gamma_n q_n e^{2\pi i n x}$$

with weights $\gamma_n = \sqrt{2\pi |n|}$ and complex coefficients $q_{\pm n} = (x_n \mp i y_n)/\sqrt{2}$, the KdV Hamiltonian becomes

$$H_c = \sum_{n \geq 1} \lambda_n |q_n|^2 + \sum_{k+l+m=0} \gamma_k \gamma_l \gamma_m q_k q_l q_m$$

on $\hbar_{3/2}$ with

$$\lambda_n = (2\pi n)^3 + 6c \cdot 2\pi n.$$

Thus, at the origin we have an elliptic equilibrium with characteristic frequencies $\lambda_n, n \geq 1$.

To transform this Hamiltonian into its Birkhoff normal form up to order four two coordinate transformations are required: one to eliminate the cubic term, and one to normalize the resulting fourth order term. Both calculations are elementary. Expressed in real coordinates (x, y) the result is the following.

Theorem 5.10 *There exists a real analytic, symplectic coordinate transformation Φ in a neighbourhood of the origin in $\hbar_{3/2}$, which transforms the KdV Hamiltonian on $\hbar_{3/2}$ into*

$$H_c \circ \Phi = \frac{1}{2} \sum_{n \geq 1} \lambda_n (x_n^2 + y_n^2) - \frac{3}{4} \sum_{n \geq 1} (x_n^2 + y_n^2)^2 + \ldots,$$

where the dots stand for terms of higher order in x and y.

The important fact about the non-resonant Birkhoff normal form is that its coefficients are uniquely determined independently of the normalizing transformation, as long as it is of the form Òidentity + higher order termsÓ. For this reason, these coefficients are also called *Birkhoff invariants*. Comparing Theorem 5.10 with Theorem 2.1 and viewing Ψ as a *global* transformation into a *complete* Birkhoff normal form we thus conclude that the two resulting Hamiltonians on $\hbar_{3/2}$ must agree up to terms of order four. In other words, the local result provides us with the first terms of the Taylor series expansion of the globally integrable KdV Hamiltonian.

Corollary 5.11 *The canonical transformation Ψ of Theorem 2.1 transforms the KdV Hamiltonian into the Hamiltonian*

$$H_c(I) = \sum_{n \geq 1} \lambda_n I_n - 3 \sum_{n \geq 1} I_n^2 + \ldots,$$

where $I_n = \frac{1}{2}(x_n^2 + y_n^2)$, and the dots stand for higher order terms in (x, y). Consequently,

$$\omega_n(I) = \frac{\partial H_c}{\partial I_n}(I) = \lambda_n - 6 I_n + \ldots.$$

Here, λ_n and hence ω_n also depend on c.

By further computing some additional terms of order six in the expansion above, we gain sufficient control over the frequencies ω to verify all nondegeneracy and nonresonance conditions for any c.

Incidentally, the normal form of Theorem 5.10 already suffices to prove the persistence of quasi-periodic solutions of the KdV equation of sufficiently *small amplitude* under small Hamiltonian perturbations. In addition, if the perturbing term $\partial K/\partial u$ is of degree three or more in u, then no small parameter ε is needed to make the perturbing terms small, as it suffices to work in a sufficiently small neighbourhood of the equilibrium solution $u \equiv 0$. We will not expand on this point.

6 A Remark on the KAM Proof

Previous versions of the KAM theorem for partial differential equations such as [35, 54] were concerned with perturbations that were given by *bounded* nonlinear operators. This was sufficient to handle, among others, nonlinear Schrödinger and wave equations on a bounded interval, see for example [5, 39, 55]. This is *not* sufficient, however, to deal with perturbations of the KdV equation, as here the term

$$\frac{\mathrm{d}}{\mathrm{d}x}\frac{\partial K}{\partial u}$$

is an *unbounded* operator. This entails some subtle difficulties in the proof of the KAM theorem, as we outline now.

Write the perturbed Hamiltonian as

$$H = N + P,$$

where N denotes some integrable normal form and P a general perturbation. The KAM proof employs a rapidly converging iteration scheme of Newton type to handle small divisor problems, and involves an infinite sequence of coordinate transformations. At each step a transformation $\varPhi$ is constructed as the time-1-map $X_F^t|_{t=1}$ of a Hamiltonian vector field X_F that brings the perturbed Hamiltonian $H = N + P$ closer to some new normal form N_+. Its generating Hamiltonian F as well as the correction $\hat{N}$ to the given normal form N are a solution of the linearized equation

$$\{F,N\} + \hat{N} = R,$$

where R is some suitable truncation of the Taylor and Fourier expansion of P. Then $\varPhi$ takes the truncated Hamiltonian $H' = N + R$ into $H' \circ \varPhi = N_+ + R_+$, where $N_+ = N + \hat{N}$ is the new normal form and

$$R_+ = \int_0^1 \{(1-t)\hat{N} + tR,F\} \circ X_F^t \,\mathrm{d}t$$

the new error term arising from R. Accordingly, the full Hamiltonian $H = N + P$ is transformed into $H \circ \varPhi = N_+ + R_+ + (P - R) \circ \varPhi$.

What makes this scheme more complicated than previous ones is the fact that the vector field X_R generated by R represents an *unbounded* operator, whereas the vector field X_F generated by the solution F of the linearized equation has to represent a *bounded* operator to define a *bona fide* coordinate transformation. For most terms in F this presents no problem, because they are obtained from the corresponding terms in R by dividing with a *large divisor*. There is no such smoothing effect, however, for that part of R of the form

$$\frac{1}{2}\sum_{n\notin A} R_n(\theta;\xi)(x_n^2+y_n^2),$$

where $\theta = (\theta_n)_{n\in A}$ are the coordinates on the torus $\mathbb{T}^A$, and ξ the parameters mentioned above. We therefore include these terms in $\hat{N}$ and hence in the new normal form N_+. However, subsequently we have to deal with a generalized, *θ-dependent* normal form

$$N = \sum_{n\in A}\omega_n(\xi)I_n + \frac{1}{2}\sum_{n\notin A}\Omega_n(\theta;\xi)(x_n^2+y_n^2).$$

This, in turn, makes it difficult to obtain solutions of the linearized equation with useful estimates.

In [37] Kuksin obtained such estimates and thus rendered the iterative construction convergent. It requires a delicate discussion of a linear small divisor equation with *large, variable* coefficients.

7 Existing Literature

Theorem 2.1 was first given in [1] and [2]. A quite different approach to this result, and the one detailed here, was first presented in [26] and extended to the nonlinear Schrödinger equation in [23]. At the heart of the argument are orthogonality relations which first have been established in the case of the nonlinear Schrödinger equation by McKean & Vaninsky [47, 48].

A version of Theorem 4.2 in the case $c = 0$ is due to Kuksin [34, 37]. In the second paper, he proves a KAM theorem of the type discussed above which is needed to deal with perturbations given by unbounded operators, and combines it with earlier results [3, 4] concerning nonresonance properties of the KdV frequencies and the construction of local coordinates so that the linearized equations of motions along a given torus of finite gap potentials reduce to constant coefficients [34].

The proof of Theorem 4.2 presented in [27] is different from the approach in [34, 37]. Instead of the local coordinates constructed in [34] we use the global, real analytic action-angle coordinates given by Theorem 2.1 to obtain quasi-periodic solutions of arbitrary size for sufficiently small and sufficiently regular perturbations of the KdV equation. To verify the relevant nonresonance conditions we follow the line of arguments used in [39] where small quasi-periodic solutions for nonlinear

Schrödinger equations were obtained, and explicitly compute the Birkhoff normal form of the KdV Hamiltonian up to order 4 and a few terms of order 6.

We stress again that our results are concerned exclusively with the existence of *quasi-periodic* solutions. Nothing is known about the persistence of almost-periodic solutions. The KAM theory of [53] concerning such solutions is not applicable here, since the nonlinearities effect a strong, long range coupling among all "modes" in the KdV equation.

There are, however, existence results for simplified problems. Bourgain [8, 9] considered the Schrödinger equation

$$iu_t = u_{xx} - V(x)u - |u|^2 u$$

on $[0, \pi]$ with Dirichlet boundary conditions, depending on some analytic potential V. Given an almost-periodic solution of the linear equation with *very* rapidly decreasing amplitudes and nonresonant frequencies, he showed that the potential V may be modified so that this solution persists for the nonlinear equation. The potential serves as an infinite dimensional parameter, which has to be chosen properly for each initial choice of amplitudes. This result is obtained by iterating the Lyapunov-Schmidt reduction introduced by Craig & Wayne [11].

A similar result was obtained independently in [56] by iterating the KAM theorem about the existence of quasi-periodic solutions. As a result, one obtains for – in a suitable sense – almost all potentials V a set of almost periodic solutions, which – again in a suitable sense – has density one at the origin. See [56] for more details.

References

1. D. Bättig, A. M. Bloch, J.-C. Guillot & T. Kappeler, On the symplectic structure of the phase space for periodic KdV, Toda, and defocusing NLS. *Duke Math. J.* **79** (1995), 549–604.
2. D. Bättig, T. Kappeler & B. Mityagin, On the Korteweg-de Vries equation: Convergent Birkhoff normal form. *J. Funct. Anal.* **140** (1996), 335–358.
3. R. F. Bikbaev & S. B. Kuksin, On the parametrization of finite gap solutions by frequency vector and wave number vector and a theorem of I. Krichever. *Lett. Math. Phys.* **28** (1993), 115–122.
4. A. Bobenko & S. B. Kuksin, Finite-gap periodic solutions of the KdV equation are non-degenerate. *Physics Letters A* **161** (1991), 274–276.
5. A. I. Bobenko & S. B. Kuksin, The nonlinear Klein-Gordon equation on an interval as a perturbed sine-Gordon equation. *Comm. Math. Helv.* **70** (1995), 63–112.
6. J. Bourgain, Fourier transform restriction phenomena for certain lattice subsets and applications to nonlinear evolution equations. II. The KdV-equation. *Geom. Funct. Anal.* **3** (1993), 209–262.
7. J. Bourgain, On the Cauchy problem for periodic KdV-type equations. Proceedings of the Conference in Honor of Jean-Pierre Kahane (Orsay, 1993). *J. Fourier Anal. Appl.* (1995), Special Issue, 17–86.

8. J. Bourgain, Construction of approximative and almost periodic solutions of perturbed linear Schrödinger and wave equations. *Geom. Funct. Anal.* **6** (1996), 201–230.

9. J. Bourgain, *Global Solutions of Nonlinear Schrödinger Equations.* Colloquium Publications, American Mathematical Society, 1999.

10. J. Boussinesq, Théorie de l'intumescence liquid appelée onde solitaire ou de translation, se propageant dans un canal rectangulaire. *Comptes Rend. Acad. Sci. (Paris)* **72** (1871), 755–759.

11. W. Craig & C. E. Wayne, Newton's method and periodic solutions of nonlinear wave equations. *Comm. Pure Appl. Math.* **46** (1993), 1409–1498.

12. B. A. Dubrovin, Periodic problems for the Korteweg-de Vries equation in the class of finite band potentials. *Funct. Anal. Appl.* **9** (1975), 215–223.

13. B. A. Dubrovin, I. M. Krichever & S. P. Novikov, The Schrödinger equation in a periodic field and Riemann surfaces. *Sov. Math. Dokl.* **17** (1976), 947–951.

14. B. A. Dubrovin, I. M. Krichever & S. P. Novikov, Integrable Systems I. In: *Dynamical Systems IV,* Encyclopedia of Mathematical Sciences vol. 4, V. I. Arnold & S. P. Novikov (eds.). Springer, 1990, 173–280.

15. B. A. Dubrovin, V. B. Matveev & S. P. Novikov, Nonlinear equations of Korteweg-de Vries type, finite-zone linear operators and Abelian varieties. *Russ. Math. Surv.* **31** (1976), 59–146.

16. B. A. Dubrovin & S. P. Novikov, Periodic and conditionally periodic analogues of the many-soliton solutions of the Kortweg-de Vries equation. *Sov. Phys.-JETP* **40** (1974), 1058–1063.

17. L. H. Eliasson, Perturbations of stable invariant tori for Hamiltonian systems. *Ann. Sc. Norm. Sup. Pisa* **15** (1988), 115–147.

18. L. D. Faddeev & V. E. Zakharov, Kortweg-de Vries equation: a completely integrable Hamiltonian system. *Funct. Anal. Appl.* **5** (1971), 280–287.

19. H. Flaschka & D. McLaughlin, Canonically conjugate variables for the Korteweg-de Vries equation and Toda lattices with periodic boundary conditions. *Progress Theor. Phys.* **55** (1976), 438–456.

20. J. P. Françoise, The Arnol'd formula for algebraically completely integrable systems. *Bull. Amer. Math. Soc. (N.S.)* **17** (1987), 301–303.

21. C. S. Gardner, Korteweg-de Vries equation and generalizations. IV. The Korteweg-de Vries equation as a Hamiltonian system. *J. Math. Phys.* **12** (1971), 1548–1551.

22. J. Garnett & E. T. Trubowitz, Gaps and bands of one dimensional periodic Schrödinger operators. *Comment. Math. Helv.* **59** (1984), 258–312.

23. B. Grébert, T. Kappeler & J. Pöschel, Normal form theory for the NLS equation. *In preparation.*

24. Ph. Griffiths & J. Harris, *Principles of Algebraic Geometry.* John Wiley & Sons, New York, 1978.

25. A. R. Its & V. B. Matveev, A class of solutions of the Korteweg-de Vries equation. *Probl. Mat. Fiz. 8,* Leningrad State University, Leningrad, 1976, 70–92.

26. T. Kappeler & M. Makarov, On Birkhoff coordinates for KdV. *Ann. Henri Poincaré* **2** (2001), 807–856.

27. T. Kappeler & J. Pöschel, *KdV& KAM.* Springer, to appear.

28. C. KENIG, G. PONCE & L. VEGA, A bilinear estimate with applications to the KdV equation. *J. Amer. Math. Soc.* **9** (1996), 573–603.

29. D. J. KORTEWEG & G. DE. VRIES, On the change of form of long waves advancing in a rectangular canal, and on a new type of long stationary waves. *Phil. Mag. Ser. 5* **39** (1895), 422–443.

30. I. KRICHEVER, Integration of nonlinear equations by methods of algebraic geometry. *Funkt. Anal. Appl.* **11** (1977), 15–31.

31. I. KRICHEVER, "Hessians" of integrals of the Korteweg-de Vries equation and perturbations of finite-gap solutions. *Dokl. Akad. Nauk SSSR* **270** (1983), 1312–1317 [Russian]. English translation in *Sov. Math. Dokl.* **27** (1983), 757–761.

32. M. D. KRUSKAL & N. J. ZABUSKY, Interaction of "solitons" in a collisionless plasma and the recurrence of initial states. *Phys. Rev. Lett.* **15** (1965), 240–243.

33. S. B. KUKSIN, Hamiltonian perturbations of infinite-dimensional linear systems with an imaginary spectrum. *Funts. Anal. Prilozh.* **21** (1987), 22–37 [Russian]. English translation in *Funct. Anal. Appl.* **21** (1987), 192–205.

34. S. B. KUKSIN, Perturbation theory for quasiperiodic solutions of infinite-dimensional Hamiltonian systems, and its application to the Korteweg-de Vries equation. *Matem. Sbornik* **136** (1988) [Russian]. English translation in *Math. USSR Sbornik* **64** (1989), 397–413.

35. S. B. KUKSIN, *Nearly integrable infinite-dimensional Hamiltonian systems.* Lecture Notes in Mathematics 1556, Springer, 1993.

36. S. B. KUKSIN, On small-denominator equations with large variable coefficients. Preprint, 1995.

37. S. B. KUKSIN, A KAM-theorem for equations of the Korteweg-deVries type. *Rev. Math. Phys.* **10** (1998), 1–64.

38. S. B. KUKSIN, *Analysis of Hamiltonian PDEs.* Oxford University Press, Oxford, 2000.

39. S. B. KUKSIN & J. PÖSCHEL, Invariant Cantor manifolds of quasi-periodic oscillations for a nonlinear Schrödinger equation. *Ann. Math.* **143** (1996), 149–179.

40. P. LAX, Integrals of nonlinear equations of evolution and solitary waves. *Comm. Pure Appl. Math.* **21** (1968), 467–490.

41. P. LAX, Periodic solutions of the KdV equation. *Comm. Pure Appl Math.* **28** (1975), 141–188.

42. W. MAGNUS & S. WINKLER, *Hill's Equation.* Second edition, Dover, New York, 1979.

43. V. A. MARČHENKO, *Sturm-Liouville Operators and Applications.* Birkhäuser, Basel, 1986.

44. H. P. MCKEAN & P. VAN MOERBECKE, The spectrum of Hill's equation. *Invent. Math.* **30** (1975), 217–274.

45. H. P. MCKEAN & E. TRUBOWITZ, Hill's operator and hyperelliptic function theory in the presence of infinitely many branch points. *Comm. Pure Appl. Math.* **29** (1976), 143–226.

46. H. P. MCKEAN & E. TRUBOWITZ, Hill's surfaces and their theta functions. *Bull. Am. Math. Soc.* **84** (1978), 1042–1085.

47. H. P. MCKEAN & K. L. VANINSKY, Action-angle variables for the cubic Schroedinger equation. *Comm. Pure Appl. Math.* **50** (1997), 489–562.

48. H. P. McKean & K. L. Vaninsky, Cubic Schroedinger: The petit canonical ensemble in action-angle variables. *Comm. Pure Appl. Math.* **50** (1997), 594–622.

49. V. K. Melnikov, On some cases of conservation of conditionally periodic motions under a small change of the Hamilton function. *Soviet Math. Doklady* **6** (1965), 1592–1596.

50. R. M. Miura, Korteweg-de Vries equation and generalizations. I. A remarkable explicit nonlinear transformation. *J. Math. Phys.* **9** (1968), 1202–1204.

51. R. M. Miura, C. S. Gardner & M. D. Kruskal, Korteweg-de Vries equation and generalizations. II. Existence of conservation laws and constants of motion. *J. Math. Phys.* **9** (1968), 1204–1209.

52. J. Pöschel, On elliptic lower dimensional tori in Hamiltonian systems. *Math. Z.* **202** (1989), 559–608.

53. J. Pöschel, Small divisors with spatial structure in infinite dimensional Hamiltonian systems. *Comm. Math. Phys.* **127** (1990), 351–393.

54. J. Pöschel, A KAM-theorem for some nonlinear partial differential equations. *Ann. Sc. Norm. Sup. Pisa* **23** (1996), 119–148.

55. J. Pöschel, Quasi-periodic solutions for a nonlinear wave equation. *Comment. Math. Helv.* **71** (1996), 269–296.

56. J. Pöschel, On the construction of almost periodic solutions for a nonlinear Schrödinger equation. Preprint, 1996.

57. Lord Rayleigh, On waves. *Phil. Mag. Ser. 5* **1** (1876), 257-279.

58. J. S. Russell, Report on waves. In: *Report of the Fourteenth Meeting of the British Association for the Advancement of Sciences,* John Murray, London, 1844, 311–390.

59. J. Saut & R. Temam, Remarks on the Korteweg-de Vries equation. *Israel J. Math.* **24** (1976), 78–87.

60. A. Sjöberg, On the Korteweg-de Vries equation: existence and uniqueness. *J. Math. Anal. Appl.* **29** (1970), 569–579.

61. R. Temam, Sur un problème non linéaire. *J. Math. Pures Appl.* **48** (1969), 159–172.

62. P. van Moerbeke, The spectrum of Jacobi matrices. *Invent. Math.* **37** (1976), 45–81.

63. G. B. Witham, *Linear and Nonlinear Waves.* Wiley, New York, 1974.

Convergence of Approximate Solutions of Conservation Laws

Sebastian Noelle[1] and Michael Westdickenberg[2*]

[1] Institut für Geometrie und Praktische Mathematik, Rheinisch–Westfälische–Technische Hochschule Aachen, Templergraben 55, 52056 Aachen, Germany, `noelle@igpm.rwth-aachen.de`
[2] Institut für Angewandte Mathematik, Rheinische Friedrich-Wilhelms-Universität Bonn, Wegelerstrasse 10, 53115 Bonn, Germany
`mwest@iam.uni-bonn.de`

Summary. In this paper we consider convergence of approximate solutions of conservation laws. We start with an overview over the historical developments since the 1950s, and the analytical tools used in this context. Then we present some of our own results on the convergence of numerical approximations, discuss recent related work and open problems.

1 Exact and Approximate Solutions of Conservation Laws

Hyperbolic conservation laws are used widely to model the local conservation of physical quantities like density, momentum and energy. Applications include gasdynamics, ideal magnetohydrodynamics, shallow water and traffic flows. The equations take the form

$$\frac{\partial u}{\partial t} + \nabla \cdot f(u) = 0, \quad \text{for } (t,x) \in [0,T] \times \mathbb{R}^d, \tag{1}$$

where $u(t,x) := (u_1(t,x), \dots, u_m(t,x))$ is the m-vector of conservative variables with initial data $u(0,\cdot) =: u_0$, and $f(u) = (f_1(u), \dots, f_d(u))$ is the flux function. It is well known that for nonlinear fluxes, even for smooth initial data, the solution of (1) may cease to exist in the classical sense due to the formation of shocks in finite time (cf. Riemann 1859 [43]). Therefore it is necessary to consider weak solutions, i.e. functions u that are bounded and satisfy (1) in the distributional sense. Existence of weak solutions for general systems of conservation laws is largely unknown, and weak solutions are not unique. However, for scalar equations (with $m = 1$) the Cauchy problem is well understood.

As usual, a possible strategy to establish existence of weak solutions of the conservation law (1) is first to regularize the problem and obtain a sequence of approximate solutions $\{u^\epsilon\}_\epsilon$ for $\epsilon > 0$, then to show that a subsequence of u^ϵ converges to a limit function u which is a weak solution of (1). The regularization mostly used is the vanishing viscosity method, where u^ϵ solves the parabolic problem

* This work was supported by SFB 256 at Bonn University.

$\partial_t u^\epsilon + \nabla \cdot f(u^\epsilon) = \epsilon \Delta u^\epsilon$. Since in general exact solutions of conservation laws are not known, approximations through numerical computations are of utmost importance in the applications. As the grid size tends to zero, again a sequence of approximate solutions is created, and then a convergence analysis is important to make sure that the results are reliable. We say more on that issue in Section 3.

We will assume that the sequence $\{u^\epsilon\}_\epsilon$ of approximate solutions is uniformly bounded in $L^\infty([0,T] \times \mathbb{R}^d)$. For the vanishing viscosity method and for many numerical schemes this can be obtained from a (discrete) maximum principle or an invariant region argument. Then it is possible to extract a subsequence which converges weak* to some limit function $u \in L^\infty([0,T] \times \mathbb{R}^d)$. But in order to prove that u is a weak solution of (1), it is necessary to show that the weak limit of $f(u^\epsilon)$ coincides with $f(u)$. For a weakly converging sequence u^ϵ and a nonlinear flux f this is not true in general because oscillations may occur. A sufficient condition, however, would be strong L^1_{loc}-convergence. So the question is how this additional information can be obtained.

We already mentioned that for a general nonlinear hyperbolic conservation law there may exist many different weak solutions corresponding to the same initial data. In order to ensure uniqueness one imposes an additional condition, often called an entropy condition because of its analogy with the second law of thermodynamics, which selects the physically relevant weak solution out of all possible ones. For scalar conservation laws several related formulations are in use. We give the one that relies on convex entropies.

We call a pair of functions $\eta\colon \mathbb{R}^m \to \mathbb{R}$ and $q\colon \mathbb{R}^m \to \mathbb{R}^d$ an entropy-entropy flux pair if η is convex and if the compatibility relations

$$q_k'(u) = \eta'(u) f_k'(u) \quad \text{for all } u \in \mathbb{R}^m \text{ and } k = 1, \dots, d \tag{2}$$

hold. Then the entropy condition reads as follows: Out of all weak solutions of the conservation law (1) given above, we only select those satisfying

$$\frac{\partial \eta(u)}{\partial t} + \nabla \cdot q(u) \leq 0 \quad \text{in distributional sense} \tag{3}$$

for all entropy-entropy flux pairs (η, q). We also assume that the initial entropy is bounded: $\int_{\mathbb{R}^d} \eta(u_0)\, dx < \infty$. These solutions are called weak entropy solutions. For scalar conservation laws all convex functions η can serve as entropies, and it turns out that weak entropy solutions are unique. This is exceptional. For systems of conservation laws there often exists only a limited number or even only one single (the physical) entropy, because the corresponding compatibility relations are much more restrictive. The entropy inequalities provide crucial estimates in the convergence analysis for $\{u^\epsilon\}_\epsilon$, and the lack of enough entropy-entropy flux pairs is the main reason why the existence theory for systems of conservation laws is not as well developed as for the scalar case. For more on the entropy condition we refer to Lax [30] and Liu [34]. Further information on hyperbolic conservation laws can be found in Lax [29], Smoller [46] and Dafermos [12].

Our paper is organized as follows. In Section 2, we give an overview over the developments in the theory of conservation laws since the 1950s, and the analytical tools used to study the convergence of approximate solutions. These include classical compactness arguments, approximation theory, weak convergence methods (compensated compactness and measure-valued solutions), and the kinetic approach. In Section 3, we present some of our own work on the convergence of numerical approximations. Finally, in Section 4, we discuss recent related work and open problems.

2 Historical Remarks on Compactness Arguments

In this section we give an overview over some analytical tools used to study convergence of sequences of approximate solutions for conservation laws.

2.1 Regularity Estimates and Compactness

The classical approach to prove strong L^1_{loc}-convergence of a sequence $\{u^\epsilon\}_\epsilon$ of functions is to prove that the u^ϵ possess some positive regularity, i.e. the sequence is uniformly bounded in some appropriate function space. If the function space is BV, the space of functions with bounded variation, then strong compactness in L^1_{loc} follows from Helly's Theorem.

In 1957 Ladyzhenskaya [27] and Oleinik [39], both former students of Petrovskii in Moscow, independently published results on a scalar conservation law with strictly convex flux in one space dimension. The entropy condition they used to select the physical solution (today known as "condition E") can be interpreted as a one-sided Lipschitz condition: For any time $t > 0$, the solution $u(t, \cdot)$ is Lipschitz continuous where it is increasing (in x). There is no restriction where it is decreasing. In particular, decreasing jump discontinuities are allowed. From this a BV bound can be derived. We would like to remark that the one-sided Lipschitz continuity was already observed by E. Hopf [20] in his pioneering paper on the Burgers equation, but he did not use it systematically as selection criterion. While Ladyzhenskaya and Oleinik both studied the vanishing viscosity limit, Oleinik together with her student Vvedenskaya [54, 39] also proved convergence of the Lax-Friedrichs scheme [28]. In 1963, Oleinik's paper was translated into English, while Ladyzhenskaya and Vvedenskaya's work was not. Oleinik's contribution is also discussed in full detail in Smoller's famous book [46]. This may explain why condition E is nowadays usually attributed to Oleinik.

The multidimensional scalar problem was studied by Conway and Smoller [9] and Vol'pert [53] in the 1960s, and by Kruzkov [25] in 1970 who proved existence and uniqueness of weak entropy solutions for the scalar conservation law. Kruzkov's main observation was that the solution operator of the scalar conservation law is an L^1-contraction: He used the special convex entropies

$$\eta(u,k) := |u-k| \quad \text{and} \quad q(u,k) := \operatorname{sgn}(u-k)|f(u)-f(k)| \tag{4}$$

(nowadays known as Kruzkov entropies), where $k \in \mathbb{R}$ is a real parameter, and a "doubling of variables"-argument to conclude that

$$\int_{\mathbb{R}^d} |u(t,x) - v(t,x)|dx \le \int_{\mathbb{R}^d} |u_0(x) - v_0(x)|dx \tag{5}$$

for all times $t > 0$. Here u and v are weak entropy solutions of (1) corresponding to initial data u_0 and v_0, respectively. Uniqueness of weak entropy solutions of (1) is an immediate consequence of this estimate. And since the solution operator of (1) is invariant under translations in x, we can apply (5) also to $v(t,x) := u(t,x+h)$ for any $h \in \mathbb{R}^d$ and obtain from this a BV-bound $\|u(t,\cdot)\|_{BV(\mathbb{R}^d)} \le \|u_0\|_{BV(\mathbb{R}^d)}$. An analogous estimate for temporal regularity can then be derived using (1). So Helly's theorem can be applied.

Let us briefly mention that there is also a well-developed theory for systems of conservation laws in one space dimension relying on BV bounds. This theory started with the work of Glimm [17] in 1965 who introduced a numerical scheme that produces approximate solutions with bounded variation. Later on, Bressan and his coworkers extended Glimm's ideas considerably and set up a very powerful framework for studying existence and uniqueness of systems of conservation laws in 1-d. We refer to Bressan [3].

BV bounds also play a major role in the design of higher order accurate numerical algorithms. In 1983, Harten [19] introduced the class of TVD (total variation diminishing) schemes and proved their convergence in the linear case. Osher [40] introduced the class of "E-schemes", the most general class of schemes which satisfy a discrete entropy inequality, and proved convergence of second order accurate semidiscrete TVD schemes for nonlinear scalar conservation laws. Convergence of a second order accurate fully discrete scheme was proved by Osher and Tadmor in 1988 [41].

2.2 Approximation Theory

Approximation theory in the context of scalar conservation laws means to estimate the distance between an approximate solution (at some time $t > 0$, say) and the uniquely defined weak entropy solution which is already known to exist. This approach gives error estimates and convergence rates. Applied to numerical schemes it tells you how close an approximate solution is to the exact one.

The first who gave error estimates for approximate solutions of scalar conservation laws in several space dimensions was Kuznetsov [26], who studied the vanishing viscosity method and the Lax-Friedrichs scheme. His analysis relies on the L^1-contraction property of the solution operator of (1) we discussed in the previous section. In fact, it is possible to obtain an estimate like (5) even if v is not an exact, but only a (suitable) approximate solution. Then additional terms measuring the approximation error enter on the right hand side which can be controlled if the initial data is in BV. More precisely, one can estimate the L^1-distance between the weak entropy solution u and the approximate solution u^ϵ at a given time t in terms of the distance and BV-norm of the initial data, the residual etc. These estimates depend on ϵ (which could be the gridsize in numerical schemes, for example), and the convergence rate proved by Kuznetsov is $\epsilon^{1/2}$.

Kuznetsov's ideas were used later on to give convergence rate estimates for more sophisticated numerical schemes, as well. We refer to Sanders [45], Vila [52], Cockburn, Coquel and LeFloch [7], Noelle [37], and Cockburn, Gremaud and Yang [8]. We also refer to Bouchut and Perthame [1] who reformulated Kruzkov and Kuznetsov's approximation theory in a general, versatile form.

We also mention that recently, Tadmor [50] and coworkers developed a somewhat different approach to proving error estimates and convergence rates for scalar conservation laws in 1-d with strictly convex flux, the so-called Lip$'$-theory. The idea is to measure the distance between the exact and the approximate solution not in the L^1-norm, but in the much weaker topology of Lip$'(\mathbb{R})$. This is the topological dual of the space of Lipschitz continuous functions. The approach can also give pointwise error estimates: The theory shows that the convergence rate at some given point $x \in \mathbb{R}$ only depends on the regularity of the exact solution u in a small neighborhood around x.

2.3 Weak Convergence Methods

For modern numerical schemes using unstructured grids it may be too hard or even impossible to prove positive regularity in terms of boundedness of a sequence $\{u^\epsilon\}_\epsilon$ in appropriate function spaces. So the classical approach of proving compactness in L^1_{loc} by using Helly's theorem or Sobolev embeddings breaks down. There are other tools, which run under the name of Weak Convergence Methods, that can prove strong compactness without showing regularity first. These methods are specialized in the sense that they rely heavily on the structure of the problem at hand.

In 1954, Lax [28] proved compactness in L^1 of the solution operator of a one-dimensional, not necessarily strictly convex, scalar conservation law using a weak topology. More than twenty years later, Tartar [51] and Murat [35] introduced an even more powerful method, applicable also for systems of conservation laws, the compensated compactness theory. A typical result is the div-curl-Lemma: One assumes that two sequences of vector-valued functions $\{U^\epsilon\}_\epsilon$ and $\{V^\epsilon\}_\epsilon$ are bounded in $(L^2_{loc}(\mathbb{R}^n))^N$ and that the divergence of U^ϵ and the curl of V^ϵ are both precompact in $H^{-1}_{loc}(\mathbb{R}^n)$. Then there exist subsequences such that

$$U^\epsilon \cdot V^\epsilon \longrightarrow U \cdot V \quad \text{in distributional sense.} \tag{6}$$

Note that the assumptions on the differentiability of U^ϵ and V^ϵ are not sufficient to obtain strong compactness in $L^2_{loc}(\mathbb{R}^n)$ (from which (6) would follow trivially). But in combination they give enough information to have weak continuity of the scalar product.

The div-curl-Lemma has been used by Tartar to study the convergence of sequences of approximate solutions $\{u^\epsilon\}_\epsilon$ of the scalar conservation law (1) in one space dimension. Compensated Compactness arguments can also be used to study certain systems of conservation laws in 1-d. As an example, we refer to the work of DiPerna [13] and Chen [5] on isentropic Euler equations.

Another weak convergence method is DiPerna's theory of measure-valued solutions for multidimensional scalar conservation laws. As mentioned before, we assume that the sequence $\{u^\epsilon\}_\epsilon$ of approximate solutions is uniformly bounded in L^∞ and extract a subsequence converging weak* in L^∞. Then the sequence converges strongly in L^1_{loc} if and only if no oscillations occur, i.e. if there is convergence pointwise a.e. The concept of measure-valued solution was first introduced by Tartar [51] in 1975, who used Young measures to describe the oscillations occurring in $\{u^\epsilon\}_\epsilon$. A Young measure associated to $\{u^\epsilon\}_\epsilon$ is a weakly measurable mapping ν from $[0,T] \times \mathbb{R}^d$ into the space of probability measures $\mathrm{Prob}(\mathbb{R})$, such that for all continuous functions f we have $f(u^\epsilon) \to <\nu, f>$ in distributional sense, where the pairing $<\nu, f>$ is given by

$$<\nu, f> (t,x) := \int_{\mathbb{R}} f(\lambda)\, d\nu_{(t,x)}(\lambda).$$

So the distributional limit of any nonlinear function of u^ϵ can be written down using one single ν. A Young measure is called a measure-valued solution of the Cauchy problem (1) if the following identity holds

$$\frac{\partial}{\partial t} <\nu, \mathrm{id}> + \nabla \cdot <\nu, f> = 0 \quad \text{in distributional sense.} \tag{7}$$

Consistency with the entropy condition then means: For all convex η

$$\frac{\partial}{\partial t} <\nu, \eta> + \nabla \cdot <\nu, q> \leq 0 \quad \text{in distributional sense,} \tag{8}$$

where q is the corresponding entropy flux.

Note that a measure-valued solution is even weaker than a weak (distributional) solution. One can show that for any bounded sequence of approximate solutions $\{u^\epsilon\}_\epsilon$ there exists a Young measure ν and a subsequence converging to it in an appropriate sense. That ν is a measure-valued solution of (1) then follows from the consistency of the approximation.

One has $<\nu, \mathrm{id}> = u$, where u is the distributional limit of $\{u^\epsilon\}_\epsilon$. But in general $<\nu, f> \neq f(u)$. Equality holds for arbitrary nonlinear f if and only if the Young measure reduces to a Dirac measure, i.e. if ν is of the form $\nu_{(t,x)} = \delta_{u(t,x)}$ with δ_u the Dirac distribution on $\mathbb{R}$ centered at u. This is equivalent to saying that the sequence $\{u^\epsilon\}_\epsilon$ converges strongly in L^1_{loc}.

The following result is due to DiPerna [14] (see also Szepessy [47])

Theorem 2.1 *Assume that initial data $u_0 \in L^1 \cap L^\infty(\mathbb{R}^d)$ is given and that there is exists a Young measure ν satisfying the following properties*
(i) the function $(t,x) \mapsto <\nu_{(t,x)}, |\mathrm{id}|>$ is in $L^\infty([0,T], L^1(\mathbb{R}^d))$,
(ii) ν is a measure-valued solution of (1),
(iii) ν is consistent with the entropy condition,
(iv) ν assumes the initial data u_0 in the following sense

$$\lim_{t \to 0} \frac{1}{t} \int_0^t \int_{\mathbb{R}^d} <\nu_{(s,x)}, |\mathrm{id} - u_0(x)|> ds\, dx = 0. \tag{9}$$

Then the Young measure reduces to a Dirac measure, i.e. $\nu_{(t,x)} = \delta_{u(t,x)}$ for a.e. $(t,x) \in [0,T] \times \mathbb{R}^d$, where u is the unique entropy solution of (1).

Assumption (iv) says that the initial data must be attained in a stronger sense than just the sense of distributions. It excludes the occurrence of oscillations in the sequence of approximate initial data $\{u_0^\epsilon\}_\epsilon$. Then condition (iii) assures that no oscillations can develop at later times: the only oscillations that can exist in the sequence $\{u^\epsilon\}_\epsilon$ are those transported into the system from the initial data. If the initial data converge strongly, so does $\{u^\epsilon\}_\epsilon$ at any later time.

Theorem 2.1 was used to prove convergence of the streamline diffusion shock-capturing method by Szepessy [48], finite difference methods by Coquel and LeFloch [10], spectral viscosity approximations by Chen, Du and Tadmor [6] and finite volume schemes on unstructured polygonal grids by Kröner and Rokyta, [24], Kröner, Noelle and Rokyta [22], Noelle [36], and others. Rohde [44] extended the method to weakly coupled systems of conservation laws.

2.4 Kinetic Formulation and Velocity Averaging

The kinetic formulation was introduced by Lions, Perthame, and Tadmor [32]. They show that there is a one-to-one correspondence between weak entropy solutions of the scalar conservation law (1) and solutions of a linear transport equation with source term, for which a certain nonlinear constraint holds. More precisely, one considers a "density-like" function ρ depending on $(t,x) \in [0,t] \times \mathbb{R}^d$, and on an additional variable $v \in \mathbb{R}$, which is a solution of the transport equation

$$\frac{\partial \rho}{\partial t} + f'(v) \cdot \nabla \rho = \frac{\partial m}{\partial v} \quad \text{in distributional sense.} \tag{10}$$

Here m is a nonnegative bounded measure. Equation (10) is supplemented with an assumption on the structure of ρ. If $\chi\colon \mathbb{R} \to \mathbb{R}$ is defined by

$$\chi(v|\alpha) := \begin{cases} +1 & \text{if } 0 < v < \alpha \\ -1 & \text{if } \alpha < v < 0 \\ 0 & \text{otherwise} \end{cases} \tag{11}$$

for $\alpha \in \mathbb{R}$, then ρ should have the form

$$\rho(t,x,v) = \chi(v|u(t,x)) \tag{12}$$

for some scalar function u. If ρ is a solution of the kinetic equation (10) and satisfies the nonlinear constraint (12), then the function u is the unique weak entropy solution of (1). Vice versa, if u is the unique entropy solution of (1), then there exists a nonnegative bounded measure m such that the function ρ defined by (12) is a solution of the kinetic equation (10).

One big advantage of the kinetic formulation for scalar conservation laws is the possibility to apply velocity averaging lemmas to obtain regularity and compactness for (sequences of approximate) weak entropy solutions of (1). One can show that the macroscopic quantity u has more regularity than the ρ whose v-average it is, see Golse, Lions, Perthame and Sentis [18] and DiPerna, Lions and Meyer [15].

In Westdickenberg and Noelle [55] we used the kinetic formulation together with the velocity averaging technique to prove the convergence of a class of finite volume schemes for scalar conservation laws in several space dimensions. We will discuss this result in more detail in Section 3.3 below.

We remark that already in 1984, Brenier [2] constructed and analyzed an approximate evolution operator for (1) based on a kinetic decomposition of u. In 1998 Perthame [42] used the kinetic formulation to study again uniqueness of weak entropy solutions and error estimates in the sense of Kuznetsov's approximation theory. It is also possible to give a kinetic formulation for certain systems of conservation laws, e.g. the isentropic Euler equations in 1-d. In Lions, Perthame, Tadmor [33] and Lions, Perthame, Souganidis [31] this kinetic formulation was used together with compensated compactness to prove existence of weak entropy solutions, thereby extending DiPerna [13] and Chen's [5] results mentioned above.

3 Some Convergence Results for Finite Volume Schemes

In this section we give an overview over some of our own results on the convergence of finite volume schemes. These schemes define approximate solutions which are piecewise polynomial on a given polygonal grid of $\mathbb{R}^d$, and in general discontinuous at the cell interfaces. Integrating the conservation law (1) over a cell K with faces $e \subset \partial K$ from time t^n to time $t^{n+1} = t^n + \Delta t$ gives the following update formula for the cell average u_K^{n+1}:

$$u_K^{n+1} = u_K^n - \frac{\Delta t}{|K|} \sum_{e \subset \partial K} |e| \, g_{K,e}^n. \tag{13}$$

Here $g_{K,e}^n$ is a numerical flux which is consistent with $f(u) \cdot n_{K,e}$, where $n_{K,e}$ is the outward unit normal of face e. This flux has to be conservative, satisfy some upwinding property (e.g. an E-flux), and it may be higher order accurate in space and time, see e.g. [22, 36]. We consider explicit time discretizations, which means that the fluxes $g_{K,e}^n$ can be computed directly from the cell averages $u_{K'}^n$ in a neighborhood of K. Such discretizations are only stable for timesteps Δt which are so small that waves originating from face $e = K \cap K'$, where the approximation is discontinuous, do not cross the neighboring cells K and K' completely during one timestep. If one imposes this CFL-condition (named after Courant, Friedrichs and Lewy [11]), then it is possible to prove a discrete maximum principle.

3.1 Discrete Entropy Inequalities

The discrete maximum principle is enough to assure that a subsequence converges weak* to some limit function, but it does not guarantee that this limit is a weak solution of the conservation law. The crucial additional estimate needed to prove convergence is a discrete entropy inequality of the form

$$\eta(u_K^{n+1}) - \eta(u_K^n) + \frac{\Delta t}{|K|} \sum_{e \subset \partial K} |e| \, G_{K,e}^n \leq C h^{2\alpha}. \tag{14}$$

Here h is the maximal diameter of a cell of the computational grid, $\alpha \in (\frac{1}{2}, 1]$ is a constant depending on the triangulation and the polynomial reconstruction, the constant C depends on the entropy η and its derivatives, and $G_{K,e}^n$ is a numerical entropy flux which is consistent with the entropy flux $q \cdot n_{K,e}$ and is closely related to the numerical flux $g_{K,e}^n$.

In [22] we proved an entropy inequality of this type for a class of higher order schemes based on the Lax-Friedrichs and the Engquist-Osher schemes, and in [36] we could treat schemes which extend Godunov's exact Riemann solver to higher order accuracy (see [22, 36] for the definitions and the original references).

There are several points in the derivation of estimate (14) which are worth recalling. In [36] we first rewrite the update (13) as a convex combination of one-dimensional contributions. This is natural, since the numerical fluxes $g_{K,e}^n$ approximate one-dimensional fluxes $f \cdot n_{K,e}$ propagating in the normal direction to the cell-faces. Adapting the weights of the decomposition to the local wave speeds, our analysis admits larger time steps than previous work, even for the classical first order schemes in 1-d [49].

Next we use Tadmor's decomposition [49] of any E-flux as a convex combination of Lax-Friedrichs and Godunov's fluxes. If one uses a piecewise constant reconstruction, corresponding to a first order accurate discretization, then the discrete entropy inequality with $C = 0$ follows naturally from the upwinding properties and the CFL-condition. For higher order polynomial reconstructions, we have to generalize Harten's TVD conditions discussed in Section 2.1 above to multidimensional reconstructions. The estimate for the higher order Lax-Friedrichs numerical flux follows readily, but for the higher order Godunov's flux, which yields the best resolution of shock discontinuities, the structure of the Riemann solution has to be exploited in full detail, see [36]. Once the one-dimensional discrete entropy inequalities are obtained, convexity arguments yield the multidimensional inequality (14).

3.2 Convergence and Error Estimates via L^1-Contraction

Equipped with the discrete maximum principle and entropy inequality (14), we can now prove convergence and error estimates. In [22] we gave the first prove of convergence for higher order accurate finite volume schemes on unstructured grids for initial data which are merely in L^∞, using DiPerna's theory of measure-valued solutions. Subsequently, Kröner and collaborators designed practical algorithms along the lines of our convergence results.

In closely related work, Cockburn, Coquel and LeFloch [7] used Kuznetsov's approximation theory to prove convergence and error estimates for BV initial data in 1994. In [37], we could generalize these error estimates, and simplify their proof, using our discrete entropy inequalities.

Even though inequality (14) is the crucial estimate, careful work is required to finish the convergence proof. In this process, many papers (including [22]) assume that the triangulation is regular, meaning that the ratio of the outer diameter to the inner diameter of a cell is bounded above. In [36], we removed that restriction and treated grids where the cells may become flat in the limit as $h \to 0$. In [37], we quantified the influence of irregular grids on the convergence rate. An example in [36] shows that the rate with which the grid may degenerate is optimal.

3.3 Convergence via Kinetic Formulation

In Westdickenberg and Noelle [55] we gave a new convergence proof for sequences of approximate solutions of the scalar conservation law, produced by higher order finite volume schemes on unstructured grids in several space dimensions. We did not use the theory of measure-valued solutions, but relied on the kinetic formulation and velocity averaging lemmas instead. The key tools in our analysis were once more the discrete entropy condition (14) and the following compactness theorem.

Theorem 3.2 *Let $1 < p \leq 2$ and $0 < \gamma < 1$. Fix a compact set $K \subset \mathbb{R}$ and assume that there are sequences $\{\rho^\epsilon\}_\epsilon, \{m^\epsilon\}_\epsilon$, and $\{\pi^\epsilon\}_\epsilon$ uniformly bounded in $L^p(\mathbb{R}^{d+1} \times K), L^1(K, \mathcal{M}(\mathbb{R}^{d+1}))$, and $L^1(K, C^*_\gamma(\mathbb{R}^{d+1}))$, resp., such that*

$$\frac{\partial \rho^\epsilon}{\partial t} + f'(v) \cdot \nabla \rho^\epsilon = \frac{\partial m^\epsilon}{\partial v} + \pi^\epsilon \quad \text{in distributional sense} \tag{15}$$

$((t,x) \in \mathbb{R}^{d+1}, v \in K)$. If the following nondegeneracy condition holds

$$\sup_{(\tau,\xi)\in\mathbb{R}^{d+1}} \operatorname{meas}\{v \in \Lambda \colon \tau + f'(v) \cdot \xi = 0\} = 0, \tag{16}$$

then the sequence $\{z^\epsilon\}_\epsilon$ belongs to a compact subset of $L^1_{loc}(\mathbb{R}^{d+1})$, where

$$z^\epsilon := \int_K \rho^\epsilon(\cdot, v)\, dv. \tag{17}$$

Here, $\mathcal{M}(\mathbb{R}^{d+1})$ is the space of bounded Radon measures, and $C^*_\gamma(\mathbb{R}^{d+1})$ is the topological dual of the space of Hölder continuous functions. This theorem is a variant of the compactness theorem in [32]. Assumption (16) means that the flux function must be nonlinear enough. The regularizing effect stated in Theorem 3.2 does not exist for linearly degenerate problems (e.g. for advection equations). Note that the m^ϵ on the right hand side of (15) is a bounded measure, as we should expect from the kinetic formulation (10). The quantity π^ϵ is an error term (e.g. the numerical residual). It is measured in a function space with negative regularity, hence in a rather weak topology.

To prove Theorem 3.2 we decompose z^ϵ into two parts and show that one part can be made arbitrarily small in L^1_{loc} using the nondegeneracy of the flux (16), while the other part has some positive regularity in terms of Sobolev (in fact Besov) spaces and is therefore strongly compact in L^1_{loc}. Applying Theorem 3.2 to numerical approximations we obtain the strong compactness needed to pass to the limits in nonlinear quantities as explained in Section 1.

4 Related Work and Open Problems

In [23, 38] Kröner and Ohlberger derived a posteriori error estimates based on Kuznetsov's approximation theory and recent related work of Gallouet, Herbin, Chainais-Hillairet et al. [4]. Ohlberger [38] developed a fully adaptive, implicit finite volume scheme for scalar convection-reaction-diffusion equations based on these estimates.

In a recent preprint, Hwang and Tzavaras [21] used the kinetic formulation together with the velocity averaging technique to study the convergence of approximate solutions of scalar conservation laws that are produced by either a relaxation method or by a diffusion-dispersion approximation.

To conclude this note, we would like to sketch some important open problems concerning the convergence of numerical approximations. The convergence rate of $h^{1/2}$ proved by Kuznetsov for the Lax-Friedrichs scheme is only optimal for first order approximations of discontinuous solutions of linear advection equations. The rate obtained so far for unstructured grids is only $h^{1/4}$. A first attempt to overcome this barrier may be found in [8]. For strictly convex scalar conservation laws one expects a rate of $h \log h$, but this conjecture is so far only backed up by numerical experiments. For schemes which are formally higher order accurate one expects higher convergence rates away from discontinuities. For strictly convex scalar conservation laws a result in this direction is proven in [16]. It is also shown there that for systems of conservation laws, characteristics of one family crossing a numerical shock layer belonging to a different family may carry first order errors into the smooth postshock region, so the convergence rate deteriorates there even for higher order schemes. In ongoing work G. Kreiss et al. are beginning to design special numerical schemes which attempt to filter out this pollution effect.

References

1. F. Bouchut, B. Perthame: Kruzkov's estimates for scalar conservation laws revisited. Trans. AMS 350 (1998), pp. 2847–2870
2. Y. Brenier: Averaged multivalued solutions for scalar conservation laws. SIAM J. Numer. Anal. 21 (1984), pp. 1013–1037
3. A. Bressan: Hyperbolic systems of conservation laws. The one-dimensional Cauchy problem. Oxford Lecture Series in Mathematics and its Applications 20, Oxford University Press, Oxford, 2000
4. C. Chainais-Hillairet: Finite volume schemes for a nonlinear hyperbolic equation. Convergence towards the entropy solution and error estimate. M2AN Math. Model. Numer. Anal. 33 (1999), pp. 129–156
5. G.-Q. Chen: Convergence of the Lax-Friedrichs scheme for isentropic gas dynamics III. Acta Math. Sci. 6 (1986), pp. 75–120
6. G.-Q. Chen, Q. Du, E. Tadmor: Spectral viscosity approximations to multidimensional scalar conservation laws. Math. Comp. 61 (1993), pp. 629–643
7. B. Cockburn, F. Coquel and P. LeFloch: An error estimate for finite volume methods for multidimensional conservation laws. Math. Comp. 63 (1994), pp. 77–103

8. B. Cockburn, P.-A. Gremaud and J.X. Yang: A priori error estimates for numerical methods for scalar conservation laws. III. Multidimensional flux-splitting monotone schemes on non-Cartesian grids. SIAM J. Numer. Anal. 35 (1998), pp. 1775–1803
9. E. Conway and J. Smoller: Global solutions of the Cauchy problem for quasilinear first-order equations in several space variables. Comm. Pure Appl. Math. 19 (1966), pp. 95–105
10. F. Coquel and P. LeFloch: Convergence of finite difference schemes for conservation laws in several space dimensions: the corrected antidiffusive flux approach. Math. Comp. 57 (1991), pp. 169–210
11. R. Courant, K.O. Friedrichs and H. Lewy: Über die partiellen Differentialgleichungen der mathematischen Physik. Math. Annal. 100 (1928), pp. 32–74
12. C. Dafermos: Hyperbolic conservation laws in continuum physics. Grundlehren der Mathematischen Wissenschaften (Fundamental Principles of Mathematical Sciences) 325, Springer, Berlin, 2000
13. R.DiPerna: Convergence of the viscosity method for isentropic gas dynamics. Comm. Math. Phys. 91 (1983), pp. 1–30
14. R. DiPerna: Measure-valued solutions to conservation laws, Arch. Rat. Mech. Anal. 88 (1985), pp. 223-270
15. R.J. DiPerna, P.-L. Lions and Y. Meyer: L^p regularity of velocity averages. Ann. I.H.P. Analyse Nonlinéaire, 8 (1991), pp. 271–287
16. B. Engquist and B. Sjögreen: The convergence rate of finite difference schemes in the presence of shocks. SIAM J. Numer. Anal. 35 (1998), pp. 2464–2485
17. J. Glimm: Solutions in the large for nonlinear hyperbolic systems of equations. Comm. Pure Appl. Math. 18 (1965), pp. 697–715
18. F. Golse, P.-L. Lions, B. Perthame and R. Sentis: Regularity of the moments of the solution of a transport equation. J. Funct. Anal. 76 (1988), pp. 110–125
19. A. Harten: High resolution schemes for hyperbolic conservation laws, J. Comp. Phys. 49 (1983), pp. 357–393
20. E. Hopf: The partial differential equation $u_t + uu_x = \mu u_{xx}$. Comm. Pure Appl. Math. 3 (1950), pp. 201–230
21. S. Hwang and A.E. Tzavaras: Kinetic decomposition of approximate solutions to conservation laws: application to relaxation and diffusion-dispersion approximations. Preprint, University of Wisconsin-Madison, 2002
22. D. Kröner, S. Noelle and M. Rokyta: Convergence of higher order upwind finite volume schemes on unstructured grids for scalar conservation laws in several space dimensions. Numer. Math. 71 (1995), pp. 527–560
23. D. Kröner and M. Ohlberger: A posteriori error estimates for upwind finite volume schemes for nonlinear conservation laws in multidimensions. Math. Comp. 69 (2000), pp. 25–39
24. D. Kröner and M. Rokyta: Convergence of upwind finite volume schemes for scalar conservation laws in two dimensions. SIAM J. Numer. Anal. 31 (1994), pp. 324–343
25. S.N. Kruzkov: First order quasilinear equations in several independent variables. Math. Sb. 123 (1970), pp. 228–255; English transl. in Math. USSR Sbornik 10 (1970), pp. 217–243
26. N.N. Kuznetsov: Accuracy of some approximate methods for computing the weak solutions of a first-order quasi-linear equation. USSR Comp. Math. Math. Phys. 16/6 (1976), pp. 105–119
27. O.A. Ladyzhenskaya: On the construction of discontinuous solutions of quasi-linear hyperbolic equations as limits of solutions of the corresponding parabolic equations when

the "coefficient of viscosity" tends towards zero (in Russian). Trudy Moscov. Mat. Obsc. 6 (1957), pp. 465–480
28. P. Lax: Weak solutions of nonlinear hyperbolic equations and their numerical approximation. Comm. Pure Appl. Math. 7 (1954), pp. 159–193
29. P. Lax: Hyperbolic Systems of Conservation Laws and the Mathematical Theory of Shock Waves. SIAM, Philadelphia, 1973
30. P. Lax: Shock waves and entropy. In: E.A. Zarantonello (ed) Contributions to nonlinear functional analysis. Academic Press, New York, 1971
31. P.-L. Lions, B. Perthame, P.E. Souganidis: Existence and stability of entropy solutions for the hyperbolic systems of isentropic gas dynamics in Eulerian and Lagrangian coordinates. Comm. Pure Appl. Math. 49 (1996), pp. 599-638
32. P.-L. Lions, B. Perthame, E. Tadmor: A kinetic formulation of multidimensional scalar conservation laws and related equations. J. AMS, 7 (1994), pp. 169–191
33. P.-L. Lions, B. Perthame, E. Tadmor: Kinetic formulation of the isentropic gas dynamics and p-systems. Comm. Math. Phys. 163 (1994), pp. 415–431
34. T.-P. Liu: The entropy condition and the admissibility of shocks. J. Math. Anal. Appl. 53 (1976), pp. 78–88
35. F. Murat: Compacité par compensation. Ann. Sc. Norm. Sup. Pisa 5 (1978), pp. 489–509
36. S. Noelle: Convergence of higher order finite volume schemes on irregular grids. Adv. Comp. Math. 3 (1995), pp. 197–218
37. S. Noelle: A note on entropy inequalities and error estimates for higher order accurate finite volume schemes on irregular grids. Math. Comp. 65 (1996), pp. 1155–1163
38. M. Ohlberger: A posteriori error estimates for vertex centered finite volume approximations of convection-diffusion-reaction equations. M2AN Math. Model. Numer. Anal. 35 (2001), pp. 355–387
39. O.A. Oleinik: Discontinuous solutions of nonlinear differential equations. Usp. Mat. Nauk. 12 (1957), pp. 3–73; English transl. in AMS Transl. 26 (1963), pp. 95–172
40. S. Osher: Convergence of generalized MUSCL schemes. SIAM J. Numer. Anal. 22 (1985), pp. 947–961
41. S. Osher, E. Tadmor: On the convergence of difference approximations to scalar conservation laws. Math. Comp. 50 (1988), pp. 19–51
42. B. Perthame: Uniqueness and error estimates in first order quasilinear conservation laws via the kinetic entropy defect measure. J. Math. Pures Appl. 77 (1998), pp. 1055–1064
43. B. Riemann: Über die Fortpflanzung ebener Luftwellen von endlicher Schwingungsweite. Abh. d. Königl. Ges. d. Wiss. z. Göttingen, Bd. 8 (1858/59) (Math. Cl.), pp. 43–65
44. C. Rohde: Upwind finite volume schemes for weakly coupled hyperbolic systems of conservation laws in 2D. Numer. Math. 81 (1998), pp. 85–123
45. R. Sanders: On convergence of monotone finite difference schemes with variable spatial differencing. Math. Comp. 40 (1983), pp. 91–106
46. J. Smoller: Shock Waves and Reaction-Diffusion Equations. Springer, New York, 1983
47. A. Szepessy: An existence result for scalar conservation laws using measure valued solutions. Comm. Part. Diff. Equations 14 (1989), pp. 1329–1350
48. A. Szepessy: Convergence of a shock-capturing streamline diffusion finite element method for scalar conservation laws in two space dimensions. Math. Comp. 50 (1989), pp. 527–545
49. E. Tadmor: Numerical viscosity and the entropy condition for conservative difference schemes. Math. Comp. 43 (1984), pp. 369–381
50. E. Tadmor: Local error estimates for discontinuous solutions of nonlinear hyperbolic equations. SIAM J. Numer. Anal. 28 (1991), pp. 811–906

51. L. Tartar: Compensated compactness and applications to partial differential equations. In: R.J. Knopps (ed) Research Notes in Mathematics 39, Nonlinear Analysis and Mechanics, Herriot-Watt Symposium, Pittman Press, 1975
52. J.P. Vila: Convergence and error estimates in finite volume schemes for general multidimensional scalar conservation laws I. Explicit monotone schemes. RAIRO Anal. Numér. 28 (1994), pp. 267–295
53. A.I. Vol'pert: The space BV and quasilinear equations. Math. USSR Sbornik 2 (1967), pp. 225-267
54. N.D. Vvedenskaya: Solution of the Cauchy-problem for a nonlinear equation with discontinuous initial conditions by the method of finite differences. Dokl. Akad. Nauk SSSR 111 (1956), pp. 517-520. (in Russian)
55. M. Westdickenberg, S. Noelle: A new convergence proof for finite volume schemes using the kinetic formulation of conservation laws. SIAM J. Num. Anal. 37 (2000), pp. 742–757

Nonlinear Hyperbolic Systems of Generalized Navier-Stokes Type for Interactive Motion in Biology

Wolfgang Alt

Abteilung Theoretische Biologie, Botanisches Institut, Universität Bonn, D-53105 Germany

Summary. We review two modelling approaches to obtain genuinely nonlinear systems of one hyperbolic transport equation (for density) accompanied by parabolic or elliptic equations (for mean velocity and, eventually, pressure), namely generalized Navier-Stokes or (pseudo-stationary) Stokes equations. Background and applications are related to models of interactive biological motion, namely for contractile polymer networks in intra-cellular motility, for cell movement and tissue formation during wound healing as well as for cohorts of migrating birds.

One approach is to derive, after suitable scaling, a formal continuum limit of (stochastic) Hamiltonian equations for 'visco-elastic' multi-particle networks with specific interaction laws. The other consists in the derivation of (highly) viscous two-phase flow equations by minimization of a corresponding energy-loss functional. In both procedures there remain convergence or existence problems to be solved analytically. Some results and a few numerical simulations are shown, particularly for the 1-dimensional case. For further results, technical details and for comparison with other methods we give corresponding references.

1 'Visco-Elastic' Multi-particle Networks in Biological Systems

1.1 Model Setup

Let us describe the Hamiltonian dynamics of N 'particles' constituting a deformable network in $\mathfrak{R}^m$(meant are biochemical or biological entities as 'polymers', 'cells' or 'birds') by standard (stochastic) evolution equations for *positions* $X_i(t) \in \mathfrak{R}^m$ and *velocities* $V_i(t) \in \mathfrak{R}^m$, $i = 1, ..., N$:

$$dX_i = V_i \cdot dt \tag{1}$$

$$dV_i = (A_i + \sum_{j \neq i} A_{ij}) \cdot dt + \widetilde{\beta}_i \cdot d\zeta \tag{2}$$

where A_i resp. A_{ij} denote the deterministic acceleration vectors for the i-th 'particle', either induced by its own activity or by interaction with other 'particles', respectively, and where $\widetilde{\beta}_i$ measures the amplitude of small stochastic perturbations, e.g. due to 'thermodynamics' or individual arbitrariness, represented as mutually independent Brownian increments $d\zeta \in \mathfrak{R}^m$ during infinitesimally small time steps dt.

The first deterministic *acceleration* term in equation (2) might describe a self-induced adjustment to a given *preferred speed* S^*, common for all 'particles', say

$$A_i = \Phi(V_i) = \gamma(\| V_i \|) \cdot (S^* / \| V_i \| - 1) \cdot V_i(t). \tag{3}$$

with an *adjustment rate* $\gamma = \gamma(S_i)$ that could depend on the 'particle's actual speed $S_i = \| V_i \|$. In case of $S^* = 0$, the vector $A_i = -\gamma \cdot V_i$ could just describe physical friction force divided by mass. Analogously, the acceleration term A_{ij} could describe the physical interaction force, which the j-th 'particle' exerts onto the i-th one, divided by its body mass. However, for polymer networks, and even more for biological entities like cells or birds, which physiologically respond to interactions with neighbors and then adjust their moving velocity in an active manner, a simple linear superposition in $\sum_{j \neq i} A_{ij}$ is questionable. Therefore, later we will allow the term A_{ij} to depend not only on the variables of the $i \div j$ pair alone. Clearly, in more elaborate models, interaction could be indirectly transmitted by the dynamics of an 'external field' in-between the network particles (see e.g. [7]).

Here we would like to present a model class that considers **nearest neighbor interactions** only. To achieve this in a standard way, we use the well-known **Voronoi tesselation** and complementary **Delaunay triangulation**. For each network node $X_i(t)$ an open *Voronoi neighborhood* $\mathfrak{V}_i(t)$ is defined, the so-called *Voronoi polyeder*, whose boundary $\partial\mathfrak{V}_i(t) = \bigcup_{j \in N_i} \overline{\Delta_{ij}(t)}$ is the concatenation of open $(m-1)$-dimensional *contact surfaces* (or *Voronoi surfaces*)

$$\Delta_{ij}(t) := \{x \in \mathfrak{R}^m : |x - X_i(t)| = |x - X_j(t)| < |x - X_k(t)| \, \forall k \neq i,j\} \neq \emptyset$$

of all thereby defined *nearest neighbors of* i, namely $j \in N_i$. This contact surface Δ_{ij} orthogonally cuts the connecting vector $R_{ij} := X_j - X_i$ exactly at half the *neighbor distance* $r_{ij} := |R_{ij}|$, while the outer surface normal vector is $\nu_{ij} := R_{ij}/r_{ij}$ and the surface content is given by $|\Delta_{ij}| = \tau_{ij} \left(\frac{r_{ij}}{2}\right)^{m-1}$, where τ_{ij} is a measure for the *'space angle'*, under which the i-th 'particle' gets in contact with its j-neighbor. For example, in the 2-dimensional case this angle is $\psi_{ij} = \psi_{ij}^+ + \psi_{ij}^-$, looking to the left and to the right, with $\tau_{ij} = |\tan \psi_{ij}^+ + \tan \psi_{ij}^-|$, see Fig. 1. Obviously, $\bigcup_{i=1}^N \overline{\mathfrak{V}_i(t)}$ constitutes the Voronoi tesselation around the network 'particles'. Dual to this is the Delaunay triangulation of the network hull $\mathfrak{D}(t) = \bigcup_{k=1}^K \overline{\mathfrak{D}_k(t)}$, consisting of all non-empty m-dimensional *Delaunay polyeders* $\mathfrak{D}_k(t)$ generated by network nodes $X_i(t)$: If, for instance, $X_i(t) \in \overline{\mathfrak{D}_k(t)}$ is one of its corner points, then the edges emanating from this point are the adjacent connecting vectors R_{ij} to all those nearest neighbors j, whose Voronoi surfaces $\overline{\Delta_{ij}(t)}$ have a non-void mutual intersection; for this constellation we will write $i, j \in N_k^{\mathfrak{D}}$ and $k \in K_{ij}^{\mathfrak{D}}$. Thus, $|\mathfrak{D}_k(t)|$ measures part of the *'free space' volume* between these neighboring network 'particles', see Definition 1.1 below.

We introduce a proper scaling procedure, particularly suited for modelling **nearest neighbor interactions,** the only case that we are going to consider here. Let the **macroscopic length scale** $\mathfrak{L}$ be defined as a quantity in the order of the *network diameter*, and let us assume that it stays bounded, $\mathfrak{L} \sim 1$, on a non-dimensional scale. Furthermore, let δ be the **microscopic length scale** denoting the *mean neighbor distance* within the network. Then, for high *number of network particles*, $N \to \infty$, in a bounded network this mean free distance becomes infinitesimally small, $\delta \simeq \mathfrak{L}/N^{\frac{1}{m}} \to 0$, according to the assumption that in m spatial dimensions

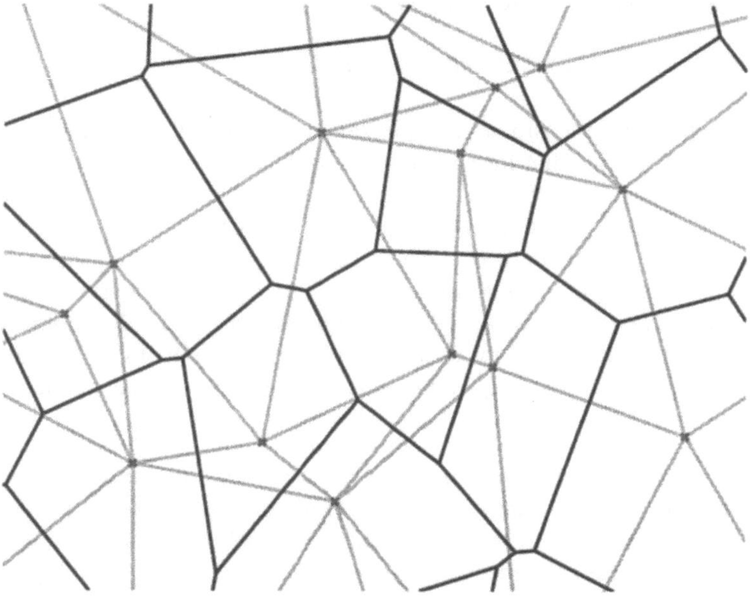

Fig. 1. Part of a 2-dimensional Voronoi-Delaunay triangulation induced by a set of points $X_i(t)$ representing the network of nearest neighbors in a simulated bird swarm; by courtesy from [31]. Obviously, there are regions with lower or higher 'network density ' where accordingly, at least on a local stochastic average, both the Delaunay triangles $\mathfrak{D}_k$ (with grey lines) and the Voronoi polygons $\mathfrak{V}_i$ (black) have larger or smaller size, respectively. Also, nearest neighbor distance r_{ij} and mean size of the two adjacent Delaunay triangles $\mathfrak{D}_k$ with $i, j \in N_k^{\mathfrak{D}}$ seem to be correlated. Furthermore notice, that the number of neighbors varies between 4 and 7; related statistics for larger swarms and longer runs can be found in [31].

the 'free space' volumes scale as $|\mathfrak{D}_k| \sim \delta^m$ on the average. Then we can introduce two measures of discrete, piecewise constant volume densities on $\mathfrak{D}(t)$, by the following definitions.

Definition 1.1 'Specific Delaunay volume' *function* $E_{\mathfrak{D}}(t,x) := \frac{|\mathfrak{D}_k(t)|}{\delta^m}$ *for* $x \in \mathfrak{D}_k(t)$, $k = 1, ..., K$,

and its inverse function, namely the

Definition 1.2 ***Delaunay density*** *function* $u_{\mathfrak{D}}(t,x) = \frac{1}{E_{\mathfrak{D}}(t,x)}$,

both being measures *'between' network particles*, and, dual to this,

Definition 1.3 'Specific Voronoi volume' *function* $E_{\mathfrak{V}}(t,x) := \frac{|\mathfrak{V}_i(t)|}{\delta^m}$ *for* $x \in \mathfrak{V}_i(t) \subset \mathfrak{D}(t)$, *i.e. only for interior network points* $i = 1, ..., N$, *whose Voronoi neighborhood lies in the hull* $\overline{\mathfrak{D}(t)}$,

and its inverse function, namely the

Definition 1.4 *Voronoi density function* $u_{\mathfrak{V}}(t,x) = \frac{1}{E_{\mathfrak{V}}(t,x)}$,

both being measures *'around' network particles.*

For a typical 2-dimensional example, taken as snapshot from a model simulation run, see Fig. 1. Notice that for a regular network, e.g. a hexagonal one, where each particle has 6 equally distant neighbors, e.g. chosen as an initial configuration for $t = 0$, both density functions would be constant and equal: $u_{\mathfrak{D}}(0,\cdot) = u_{\mathfrak{V}}(0,\cdot)$. Also, for anisotropically deformed but still quite regularly spaced networks, the two densities will have quite similar values, at least on a local stochastic average (as can be observed in Fig. 1). However, depending on the 'particle's dynamics, sometimes they could vary quite much, for instance, in degenerate situations described below; but still the concept of 'density' in the sense of measures can be maintained.

Remark 1.5 *As long as network 'particles' are disjoint, both densities $u_{\mathfrak{D}}$ and $u_{\mathfrak{V}}$ stay finite step functions, since the Delaunay polyeders do not degenerate (see e.g. [9]). The i-th 'particle' cannot cross the boundary of any other polyeder $\mathfrak{D}_k(t)$ except at one of its corners; then it collides with another 'particle' j. In case of such collisions there appear degenerate polyeders with vanishing volume $E_{\mathfrak{D}}$, but still the 'between particle' density $u_{\mathfrak{D}}$ is well-defined as n-dimensional Hausdorff measure with $(m-l) \leq n$ for l-fold collisions, such that the subsequent arguments and limit procedures could still be valid in suitably generalized function or measure spaces.*

Generically, i.e. in any cases of non-degenerate collisions, the Voronoi density $u_{\mathfrak{V}}$ stays bounded. Degeneration would occur, for example in spatial dimensions $m \geq 2$, when $l \geq 2$ 'particles' in a line simultaneously collide.

For the 1-dimensional case, with an ordered chain $X_1(t) < X_2(t) < \dots <$ $X_N(t)$, the 'free space' around the i-th interior 'particle' has size $|\mathfrak{V}_i| = \frac{1}{2}(R_{i(i+1)} + R_{(i-1)i})$, the mean value of both nearest neighbor distances, with $|\mathfrak{D}_i| = R_{i(i+1)}$, $i = 1, \dots, N-1$. Notice that therefore

$$u_{\mathfrak{V}}(t,x) = \frac{2\, u_{\mathfrak{D}}(t,x^+)\, u_{\mathfrak{D}}(t,x^-)}{u_{\mathfrak{D}}(t,x^+) + u_{\mathfrak{D}}(t,x^-)}$$

with $x^- = X_i(t)$ and $x^+ = X_{i+1}(t)$ for interval points satisfying $X_i(t) < x <$ $X_{i+1}(t)$. Then, if l 'particles' collide at an interior point $X_i(t)$, $u_{\mathfrak{D}}$ would there become a Dirac measure proportional to $(l-1)\delta_{X_i(t)}$, while $u_{\mathfrak{V}}$ would stay bounded for $l = 2$ and proportional to $(l-2)\delta_{X_i(t)}$ only for $l \geq 3$.

Finally, we construct the following linear interpolations on $\overline{\mathfrak{D}(t)}$

Definition 1.6 ***(Delaunay velocity)*** Piecewise linearly interpolation $\mathbf{v}(t,x)$ *with* $\mathbf{v}(t, X_i(t)) = V_i(t)$ *, for $i = 1, \dots, N$, i.e.* $\mathbf{v}$ *denotes the 1-st order spline interpolation of the velocity vector, by which the Delaunay 'triangulation' of the network is deformed with time,*

and

Definition 1.7 ***(Interpolated Delaunay characteristics)*** *For each point $y \in \overline{\mathfrak{D}(0)}$, the initial network hull, we define the* characteristics $x_y(t)$ *to satisfy*

$$\dot{x}_y(t) = \mathbf{v}\,(t, x_y(t))\ , \qquad x_y(0) = y\,. \tag{4}$$

Since $\mathbf{v}(t,\cdot)$ is piecewise linear, for any $y \in \mathfrak{D}_k(0)$ the characteristics $x_y(t)$ are corresponding linear interpolations of the 'particle's trajectories $X_i(t)$ with $i \in N_k^{\mathfrak{D}}$. Notice, that this definition even holds for stochastic swarm processes (i.e. equation (4) can be written as a stochastic differential equation and thereby establishing stochastic characteristics), provided that products of $V_i(t)$ with $X_i(t)$ are well-defined.

Let us, for any (sufficiently differentiable) function $h(t,\cdot)$ on $\overline{\mathfrak{D}(t)}$, denote the *derivative along characteristics* as

$$d_t h(t,x)\ :\ = \frac{d}{dt} h\,(t, x_y(t))\,, \text{ with } x = x_y(t) \tag{5}$$

$$= \partial_t h(t,x) + \mathbf{v}\,(t,x)\cdot \nabla_x h(t,x) \tag{6}$$

then even for the piecewise constant volume and density functions introduced above, the derivative (5) is well-defined and, by using (1) and Gauss' integration law, we can easily derive

Proposition 1.8 ***(Delaunay equations)*** *The 'discrete' volume and density functions 'between particles' defined in Definitions 1.1 and 1.2 satisfy (for any non-zero scaling parameter $\delta > 0$) the following standard constitutive 'volume' or 'mass balance' continuity equations on each $\mathfrak{D}_k(t)$, for $k = 1, ..., K$, thus in a generalized sense on the whole network hull $\overline{\mathfrak{D}(t)}$:*

$$d_t E_{\mathfrak{D}} = E_{\mathfrak{D}}\ \nabla_x \cdot \mathbf{v}\,, \tag{7}$$

$$d_t u_{\mathfrak{D}} = -u_{\mathfrak{D}}\ \nabla_x \cdot \mathbf{v}\,. \tag{8}$$

Furthermore, since for the edge vectors $R_{ij} = X_j - X_i$, $i, j \in N_k^{\mathfrak{D}}$, the ordinary differential equation holds

$$\frac{dR_{ij}}{dt} = V_j - V_i = |R_{ij}|\,\frac{V_j - V_i}{|X_j - X_i|} = R_{ij}\ \cdot \nabla_x \mathbf{v}\,, \tag{9}$$

then at any point $x \in \mathfrak{D}_k(t)$ every small 'tangential' vector $R_x = \widetilde{x} - x \in \mathfrak{R}^m$ can be written by a fixed linear combination of the edge vectors, and we derive for its evolutionary 'transport' along characteristics

$$d_t R_x = R_x \cdot \nabla_x \mathbf{v}\,. \tag{10}$$

Similarly, Gauss' integration law applied to the Voronoi neigborhoods $\mathfrak{V}_i(t)$ yields analogous continuity equations, though with a more refined interpolation of the velocities $V_i(t)$:

Proposition 1.9 ***(Voronoi equations)*** *The 'discrete' volume and density functions 'around particles' defined in Definitions 1.3 and 1.4 satisfy (for any non-zero scaling parameter $\delta > 0$) the following standard constitutive 'volume' or 'mass balance' continuity equations on each $\mathfrak{V}_i(t) \subset \mathfrak{D}(t)$, therefore in a generalized sense on the whole 'interior' network domain $\overline{\mathfrak{D}_{\mathfrak{V}}(t)} := \bigcup_{i \in \check{N}} \overline{\mathfrak{V}_i(t)}$*

$$d_t E_{\mathfrak{V}} = E_{\mathfrak{V}} \, \nabla_x \cdot \widetilde{\mathbf{v}} \,, \tag{11}$$

$$d_t u_{\mathfrak{V}} = -u_{\mathfrak{V}} \, \nabla_x \cdot \widetilde{\mathbf{v}} \,, \tag{12}$$

where we use the following

Definition 1.10 ***(Voronoi velocity)*** Piecewise linear 'refined' interpolation $\widetilde{\mathbf{v}}(t,x)$ *with $\widetilde{\mathbf{v}}(t, X_i(t)) = V_i(t) = \mathbf{v}(t, X_i(t))$ for $i = 1, ..., N$, as in Definition 1.6 but with the additional requirement $\widetilde{\mathbf{v}}(t, Z_l(t)) = \dot{Z}_l(t)$ for every corner point $Z_l(t)$ of any Voronoi polyeder $\partial\mathfrak{V}_i(t)$, i.e. $\widetilde{\mathbf{v}}$ denotes the velocity, by which the Voronoi tesselation of the network is deformed with time.*

By elementary calculations one can easily prove, at least in the 2-dimensional case, the

Proposition 1.11 ***(Relation between Delaunay and Voronoi velocities, explicitly given for*** $\mathrm{m} = 2$***)***

For any triple $i, j, k \in N_l^{\mathfrak{D}}$, whose positions define the corners of a Delaunay triangle $\mathfrak{D}_l(t)$, its geometric center $z = Z_l(t)$ with equal distance $|z - X_\iota(t)|$ for all $\iota \in N_l^{\mathfrak{D}}$, thus representing a potential boundary node $\{z\} = \Delta_{ij} \cap \Delta_{ik}$ of the Voronoi polyeder $\mathfrak{V}_i(t)$, moves according to the following differential equation, linear in the Delaunay velocities

$$\dot{Z}_l(t) = \widetilde{\mathfrak{P}}_{\mathfrak{D}_l}[V_i, V_j, V_k] = V_i + \frac{1}{2}\mathfrak{P}_{R_{ij}^\perp}(V_k - V_i) + \frac{1}{2}\mathfrak{P}_{R_{ik}^\perp}(V_j - V_i)$$

where $\mathfrak{P}_{R_{ij}^\perp}$ denotes the orthogonal projection onto the line $R_{ij}^\perp$ orthogonal to a boundary edge vector R_{ij}.

Notice that the center point $Z_l(t)$ can move outside the triangle $\mathfrak{D}_l(t)$, depending on how flat this is. Several such cases can be seen in Fig. 1, however, the center point always stays relatively near to the triangle's boundaries.

Thus generally, also for higher space dimensions, an estimation how far the Voronoi velocity $\widetilde{\mathbf{v}} = \dot{Z}_l(t)$ deviates from the Delaunay velocity $\mathbf{v}$ depends on the degree of degeneracy of the Delaunay polyeders in the network. Therefore, in the following we suppose a non-degeneracy condition which would guarantee that the Voronoi velocity $\widetilde{\mathbf{v}} = \widetilde{\mathfrak{P}}_{\mathfrak{D}_l}[\mathbf{v}]$ is *subordinate* to $\mathbf{v}$ in the sense of a uniform estimate $|\widetilde{\mathbf{v}}| = O(|\mathbf{v}|)$ i.e. boundedness of the linear operator $\widetilde{\mathfrak{P}}_{\mathfrak{D}_l}$ for each Delaunay polyeder $\mathfrak{D}_l$. Let us call this the **Delaunay uniformity criterium**.

Finally, in order to reformulate the dV_i model equations (2) in terms of the density and velocity interpolations defined above (Definitions 1.1–1.4) we assume a special rescaled **model for mutual acceleration**, namely

$$A_{ij} = \frac{|\Delta_{ij}|}{\delta^m\,|R_{ij}|} \cdot \left[\widetilde{\alpha}_{ij}\; R_{ij} + \widetilde{\mu}_{ij}\; \frac{dR_{ij}}{dt}\right] \text{, for } i,j = 1,...,N \tag{13}$$

reminding that $|\Delta_{ij}| = \tau_{ij}\,(|R_{ij}|\,/2)^{m-1}$. Thus, up to the first factor, the interaction between the i-th and the j-th 'particle' is supposed to depend linearly on the 'sensed' mutual distance R_{ij} and its temporal derivative $\frac{dR_{ij}}{dt}$, which is a reasonable, though the simplest assumption for biologically interactive entities. However, the coefficients $\widetilde{\alpha}_{ij}(t)$ and $\widetilde{\mu}_{ij}(t)$ will depend on other interactive conditions, in particular other nearby neighbors, see the discussion below. In any case, let us think of them as constant functions on any Voronoi contact surface $\Delta_{ij}(t)$. Moreover, notice that the factor in front of the brackets in (13) provides a weight term inversely proportional to neighbor distance $|R_{ij}|$ and proportional to $|\Delta_{ij}(t)|$, the latter measuring the degree of contact between the two neighbors. Both model assumptions are very reasonable, because thereby a newly appearing nearest neighbor j will first have a vanishingly small influence that will increase, the more it enters the 'interaction scope' of the i-th 'particle' being quantified by $\tau_{ij} = (2/\,|R_{ij}|)^{m-1}\,|\Delta_{ij}(t)|$. Thus, using (9) and observing that $R_{ij}\cdot\nabla_x\mathbf{v}$ is the same constant vector on each $\mathfrak{D}_k(t)$ with $k \in K_{ij}^{\mathfrak{D}}$, we can write the sum in the second term of equation (2) as a boundary integral over the Voronoi polyeders $\partial\mathfrak{V}_i = \bigcup_{j\in N_i}\overline{\Delta_{ij}}$ with outer normal vector $\nu = \nu_{ij} = R_{ij}/\,|R_{ij}|$ on each boundary surface Δ_{ij}, and then use Gauss' law to obtain a divergence term:

$$\sum_{j\neq i} A_{ij} = \frac{1}{\delta^m}\sum_{j\in N_i} |\Delta_{ij}|\,\nu_{ij}\cdot[\widetilde{\alpha}_{ij}\;\mathbf{I} + \widetilde{\mu}_{ij}\;\nabla_x\mathbf{v}] \tag{14}$$

$$= \frac{1}{\delta^m}\int\limits_{\partial\mathfrak{V}_i} \nu\cdot[\widetilde{\alpha}_{ij}\;\mathbf{I} + \widetilde{\mu}_{ij}\;\nabla_x\mathbf{v}] \tag{15}$$

$$= \frac{1}{\delta^m}\int\limits_{\mathfrak{V}_i} \nabla_x\cdot[\widetilde{\alpha}_{ij}\;\mathbf{I} + \widetilde{\mu}_{ij}\;\nabla_x\mathbf{v}] \tag{16}$$

$$= \frac{1}{u_{\mathfrak{V}}}\nabla_x\cdot[\widetilde{\alpha}_{ij}\;\mathbf{I} + \widetilde{\mu}_{ij}\;\nabla_x\mathbf{v}] + O(\delta) \tag{17}$$

where equality (16) with derivatives and integration only holds in a weak (distributional) sense and where the smallness of the error term in (17) would strongly depend on the kind of convergence in the continuum limit $\delta \to 0$, see part of the following discussion.

Modelling assumptions. Before we turn to this important point, let us discuss the modelling coefficients $\widetilde{\alpha}_{ij}$ and $\widetilde{\mu}_{ij}$ in formula (13) above. If the network 'particles' would just be physically connected by elastic springs with a common equilibrium elongation r_*, then we would have $\widetilde{\alpha}_{ij} \simeq \widehat{\alpha}(r_{ij})(r_{ij}-r_*)$ with some $\widehat{\alpha} > 0$, meaning repulsion for $r_{ij} < r_*$ and attraction for $r_{ij} > r_*$. However, biological entities, or even biopolymers, can develop much more complex interactions in a connected network. For instance, in space dimensions $m \geq 2$, repulsion or attraction could depend on the *'free interaction volume'* in the space around the connecting vector R_{ij}, e.g. expressed as a mean value of all adjacent Delaunay polyeder volumes $|\mathfrak{D}_k|$, namely

$\widetilde{\alpha}_{ij} = \widetilde{\alpha}_{ij}\left(|\mathfrak{D}_k|,\ k \in K_{ij}^{\mathfrak{D}}\right) = \alpha\left(\delta^m \left|K_{ij}^{\mathfrak{D}}\right| / \sum_{k \in K_{ij}^{\mathfrak{D}}} |\mathfrak{D}_k|\right)$, where $\alpha(u)$ would denote a monotone decreasing function of an averaged Delaunay density, $u = \overline{u_{\mathfrak{D}}}^{ij}$, positive (i.e. attracting) for low densities u, and negative (i.e. repulsing) for higher densities, with an equilibrium density u_*. For instance, in the 2-dimensional case

$$\overline{u_{\mathfrak{D}}}^{ij} = \frac{2\, {u_{\mathfrak{D}}}^{k+}\, {u_{\mathfrak{D}}}^{k-}}{{u_{\mathfrak{D}}}^{k+} + {u_{\mathfrak{D}}}^{k-}}$$

is a generalized mean between the Delaunay densities ${u_{\mathfrak{D}}}^k := u_{\mathfrak{D}}|_{\mathfrak{D}_k}$ regarding the two joining neigbors $k_+, k_- \in K_{ij}^{\mathfrak{D}}$. Similarly, we could assume that the 'viscosity' coefficients $\widetilde{\mu}_{ij}$ also depend on the surrounding 'free interaction volume', $\widetilde{\mu}_{ij} = \mu\left(\delta^m \left|K_{ij}^{\mathfrak{D}}\right| / \sum_{k \in K_{ij}^{\mathfrak{D}}} |\mathfrak{D}_k|\right)$, now with $\mu(u) > 0$ and eventually increasing for higher densities $u = \overline{u_{\mathfrak{D}}}^{ij}$.

1.2 Continuum Limit

Assuming that in the continuum limit, $\delta \to 0$, the above defined Delaunay and Voronoi density functions (or measures) $u_{\mathfrak{D}}$ and $u_{\mathfrak{V}}$ as well as the Delaunay and Voronoi velocities $\mathbf{v}$ and $\widetilde{\mathbf{v}}$ converge, in a suitably defined stochastic sense (e.g. in some configuration space [25]) and form a limit network domain $\mathfrak{D}(t) = \mathrm{supp}[\ u_{\mathfrak{D}}(t.\cdot)]$, then we could formulate the following

Conjecture 2. In a formal limit as $\delta \to 0$, the density measures and velocity functions defined above, as realizations of a suitably scaled and averaged stochastic limit process, fulfill the **stochastic Lagrange equations of Navier Stokes type** along stochastic characteristics as in Def. 6

$$d_t u_{\mathfrak{D}} = -u_{\mathfrak{D}}\ \nabla_x \cdot \mathbf{v}\,, \qquad (18)$$

$$d_t u_{\mathfrak{V}} = -u_{\mathfrak{V}}\ \nabla_x \cdot \widetilde{\mathbf{v}}\,, \qquad (19)$$

$$d_t \mathbf{v} = \Phi(\mathbf{v}) + \frac{1}{u_{\mathfrak{V}}} \nabla_x \cdot \left[\alpha(u_{\mathfrak{D}})\,\mathbf{I} + \mu(u_{\mathfrak{D}})\ \nabla_x \mathbf{v}\right] + \beta\,\Xi. \qquad (20)$$

Here the properties of the additive stochastic term Ξ (e.g. the standard $(m+1)$-dimensional white noise of the so-called Brownian sheet process) and expressions for its amplitude, β, would depend on corresponding stochastic convergence theorems that still have to be proven. In the simple 1-dimensional case first results in this direction have been obtained in [6]:

Remark 1.12 ***(Results for space dimension*** $m = 1$***)*** *For an ordered* 1*-dimensional chain (as in Remark 1.5) the model assumptions in Section 1.1 above would be consistent with nearest neighbor distance dependent coefficients* $\widetilde{\alpha}_{ij} = \alpha\left(\frac{\delta}{r_{ij}}\right)$ *and* $\widetilde{\mu}_{ij} = \mu\left(\frac{\delta}{r_{ij}}\right)$. *Under suitable conditions on the behavior of* $\Phi(\mathbf{v})$, $\alpha(u)$ *and* $\mu(u)$, *as well as for suitable scaling of the stochastic amplitude, namely* $\widetilde{\beta}_i \sim \sqrt{\delta}$ *in model* *equations (1–2), certain transformations and estimations can be performed yielding the continuum limit equations (18–20) above, with $u_{\mathfrak{V}} = u_{\mathfrak{D}} = u$, $\mathbf{v} = \widetilde{\mathbf{v}}$, the stochastic term $\Xi = \Xi_{t,x}$ representing independent infinitesimal increments of the standard temporal-spatial Brownian sheet and the stochastic amplitude scaling as $\beta \sim \sqrt{u}$, for details see [6].*

Convergence Assumption: Generally, let us assume that the initial network configuration, at $t = 0$, had been chosen so that in the limit $\delta \to 0$ the two densities coincide, $u_{\mathfrak{V}}(0,\dot{}) = u_{\mathfrak{D}}(0.\dot{})$. Moreover, let us suppose that the stochastic network fulfills a **uniformity criterium** (as defined in section 1.1) such that the more refined Voronoi velocity $\widetilde{\mathbf{v}} = \widetilde{\mathfrak{P}}_{\mathfrak{D}_l}[\mathbf{v}]$, by inducing relatively fast oscillations of the geometric centers $Z_l(t)$ around the Delaunay polyeders $\mathfrak{D}_l(t)$, after local averaging (on a relatively fast time scale) converges to the same limit as the Delaunay velocity $\mathbf{v}$; thus we would have $\widetilde{\mathbf{v}} = \mathbf{v}$ for $\delta \to 0$, as it holds for the 1-dimensional case. Clearly, then equations (18,19) would also imply the asymptotic equality $u_{\mathfrak{V}} = u_{\mathfrak{D}} =: u$.

Furthermore, let us suppose that, for a certain time interval, these convergences are valid in suitable function or measure spaces including the stochastics (e.g. again in some configuration space, cf. [25]). Then, by computing the derivative along characteristics (6), i.e. by performing the standard transformation from Lagrangian into Euler coordinates, we can formulate the

Conjecture 3. Under the *convergence assumptions* stated above, the stochastic differential equation system (18–20) would in the formal limit $\delta \to 0$ lead to the following **stochastic Navier-Stokes equations** for u and $\mathbf{v}$:

Hyperbolic continuity equation (mass balance)

$$\partial_t u + \nabla_x \cdot (u\mathbf{v}) = 0 \tag{21}$$

Stochastic parabolic velocity equation (impulse balance)

$$\partial_t \mathbf{v} + \mathbf{v} \cdot \partial_x \mathbf{v} = \Phi(\mathbf{v}) + \frac{1}{u} \nabla_x \cdot [\alpha(u)\, \mathbf{I} + \mu(u)\, \nabla_x \mathbf{v}] + \beta(u)\, \Xi. \tag{22}$$

If, in contrast to the modelling assumptions above, for dimensions $m \geq 2$ the 'elasticity' and 'viscosity' coefficients $\widetilde{\alpha}_{ij}$ and $\widetilde{\mu}_{ij}$ explicitely depend on nearest neighbor distance r_{ij} or, in general, on some other derived quantities of the network configuration, then there are two possibilities to proceed:

Remark 1.13 *(a) In case the stochastic perturbations are strong enough, so that the network configuration is isotropically mixed on macroscopic time scales, then similar ideas as in mean field approximations (still to be worked out in this situation) could be used to argue that the mean distance 'between particles' satisfies an asymptotic relation*

$$\frac{r}{\delta} = (u_{\mathfrak{D}})^{\frac{1}{m}} \textit{ for } \delta \to 0.$$

Then we would again obtain a but modified functional dependence $\alpha = \alpha(u)$ and $\mu = \mu(u)$ in the limit equation (22).

(b) For small stochastic perturbations and relatively mild elasticity and viscosity coefficients, one might be able to prove the uniformity criterium (defined in section 1.1), such that the linear deformation equation (10) along characteristics holds in the continuum limit and is solvable, at least for a certain time interval. Thus, starting with an initial network deformation tensor, $D_0(y)$, say isotropic, then for any $y \in \mathfrak{D}(t)$ there would exist a generalized semigroup $U_y(t) \in \mathfrak{R}^{m \times m}$, with $d_t U_y(t) = U_y(t) \cdot \nabla_x \mathbf{v}(t, X_y(t))$ which would yield for each point $x = X_y(t)$ at time t a deformation tensor $D(t, x)$. It describes the anisotropy in the continuum network so that in each direction $\theta \in \mathfrak{S}^{m-1}$ the 'rescaled particle distance' in this direction could be computed as $\widehat{r}_\theta(t, x) = \theta \cdot D(t, x) \cdot \theta^T$. Also, we would regain the Delaunay density $u_\mathfrak{D}$ from $\det D(t, x) = E_\mathfrak{D}(t, x) = \frac{\delta^m}{u_\mathfrak{D}(t,x)}$. Details of this procedure, which requires the derivation of constitutive continuum equations for the deformation tensor $D(t, x)$, still have to be elaborated.

1.3 Boundary Conditions

So far, our formal derivation of the differential equation system (21,22) is valid only in the interior of the limit continuum network domain $\mathfrak{D}(t)$. Thus, it can be directly extended to situations, where a 'locally uniform' network spans the whole space $\mathfrak{R}^m$. However, for domains with finite *moving boundary* $\partial\mathfrak{D}(t)$ one has to prove sufficient smoothness and derive certain induced 'natural' boundary conditions. For the 1-dimensional case of a *finite visco-elastic chain* this has been worked out in [6], where 'natural' Neumann boundary conditions for the velocity $\mathbf{v}$ appear, corresponding to a continuum model assumption of 'zero tension' at the two chain ends $x = X_\pm(t)$. Clearly, the second 'natural' boundary condition means, that the chain ends move with boundary velocity, $\dot{X}_\pm(t) = \mathbf{v}(t, X_\pm(t))$.

The standard generalization of this latter boundary condition for spatial dimensions $m > 1$, namely

$$d\Gamma/dt = \nu \cdot \mathbf{v}(t, \cdot)|_\Gamma \tag{23}$$

for the free boundary $\Gamma(t) := \partial\mathfrak{D}(t)$ with outer normal vector ν, would directly follow from the fact that the mass balance equation (21), for suitable definition of $u = 0|_{\mathfrak{R}^m \setminus \mathfrak{D}(t)}$, holds on the whole space $\mathfrak{R}^m$ in a distributional sense. However, the definition of the outer normal ν will strongly depend on a necessary, not at all straight-forward algorithm to define appropriate 'closures' of the discrete network domain $\mathfrak{D}(t)$, see [9, 31] for the 2-dimensional case.

1.4 Examples

Visco-elastic polymer networks. When considering the dynamics of, say 2-dimensional, polymer networks (e.g. thin films of polyacrylamid, used as substrates for experimental cell crawling on top of it), then after transforming the original model system to dimensionless coordinates one realizes that, for time and space scales of interest, all non-dimensional rate parameters in model equations (2,3 and

13) with $S^* = 0$, namely γ, $\widetilde{\alpha}$ and $\widetilde{\mu}$, are relatively large, meaning that friction, elasticity and viscosity forces all dominate inertial forces. Thus (2) might be replaced by an explicit pseudo-steady state equation for the i-th 'particle's velocity, $V_i = \frac{1}{\gamma}\left(\sum_{j\neq i} A_{ij} + F_i\right)$, thereby neglecting stochastic perturbations and assuming an additional outer forces F_i acting on each polymer. In the continuum limit (22) this corresponds to the pseudo-steady elliptic Stokes equation

$$\nabla_x \cdot [\alpha(u)\,\mathbf{I} + \mu(u)\,\nabla_x \mathbf{v}] + F = \gamma u \mathbf{v} \tag{24}$$

with eventual boundary conditions and a given outer force vector field F. Notice that this derived model equation for a highly *visco-elastic medium* consists in a simple force balance equation with viscosity and friction coefficients appearing as usual, but the elasticity being described by an isotropic tension term, for example the simplest linear case $\alpha(u) = \alpha_0\,(u_* - u)$. This clearly means that for densities u higher than the equilibrium density u_* the network elasticity (repulsion between polymers) induces a positive pressure, whereas a positive tension for lower densities, quite plausible phenomenological model assumptions. However, we should emphasize, that our derivation of (24) as a continuum limit of a discrete polymer network model is (formally) valid only with the special *modelling assumptions* proposed at the end of Section 1.1, meaning that elastic and viscous forces between connected polymers have a very special form in dependence of 'free space volume' and nearest neighbor distance, see equations (13). For more general modeling dependences probably one has to rely on additional anisotropic tensor equations as indicated in Remark 1.13(b).

For the linear elasticity parameter function (as above) and for constant viscosity parameter $\mu = \mu_0$, Till Bretschneider [11] has used these equations to simulate the deformations of visco-elastic polymer substrates under the influence of two or more migrating cells, which are distributed over a substratum, apply traction forces onto it and (by responding to the experienced substrate tension) move towards each other. This corresponds to experimental observations, which have been studied in detail [19],[28]. In these papers Micah Dembo even solves the inverse problem of computing the cell traction forces from observed displacements of fluorescent beads embedded into the substratum. For results of numerical simulations (as well as analytic investigations for $m = 1$) we refer to [11] and [7].

Bird swarms. For the 1-dimensional case the stochastic model equations (1–2) well describe the dynamics of ordered chains, which we find in migrating bird as cranes or wild geese. For reasonable choices of the distance dependent parameter functions, i.e. $\widetilde{\alpha} = \alpha(\delta/r)$ and $\widetilde{\mu} = \mu(\delta/r)$, and properly scaled stochastic noise, $\widetilde{\beta}_i \sim \sqrt{\delta}$, see Remark 1.12, numerical simulations reveal characteristic phenomena of so-called condensation waves travelling in both directions with a remarkably constant wave speed, see [6]. It turns out, that for a series of decreasing $\delta \simeq \mathfrak{L}/N \to 0$, say with particle number $N = 30, ..., 150$, for example, these condensation waves remain and carry about the same wave speed, see Fig. 2. Therefore the challenging question was, whether the limiting stochastic Navier-Stokes equations (21–22) with natural free boundary conditions would reproduce the same phenomenon of condensation

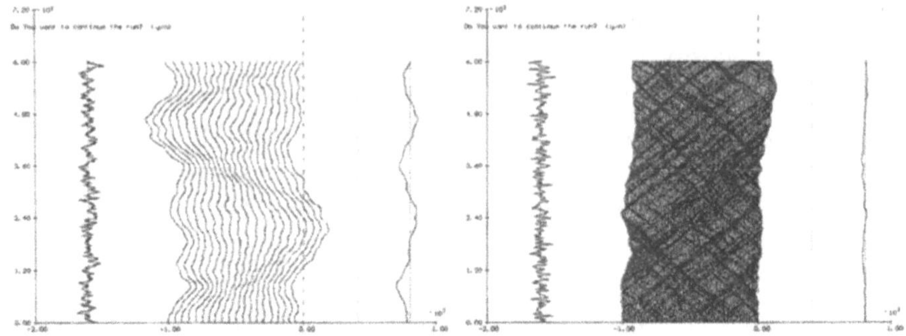

Fig. 2. For a 1-dimensional stochastic bird swarm model with correct scaling of the noise amplitude, $\widetilde{\beta}_i \sim \sqrt{\delta} \sim 1/\sqrt{N}$, the two pictures plot the positions of N birds ($N = 30$ in the left hand picture, $N = 150$ in the right hand one) over time (from 0 to 60 sec, in vertical direction). Here the uniform motion of the migrating birds with positive speed S^* has been substracted. Moreover, to the very left of each picture the fluctuating acceleration is plotted, which the leftmost bird experiences over time, whereas to the very right the fluctuations of the swarm diameter over time can be read off. Both simulations show characteristic condensation waves running through the swarm in both directions with a quite constant speed. From [6].

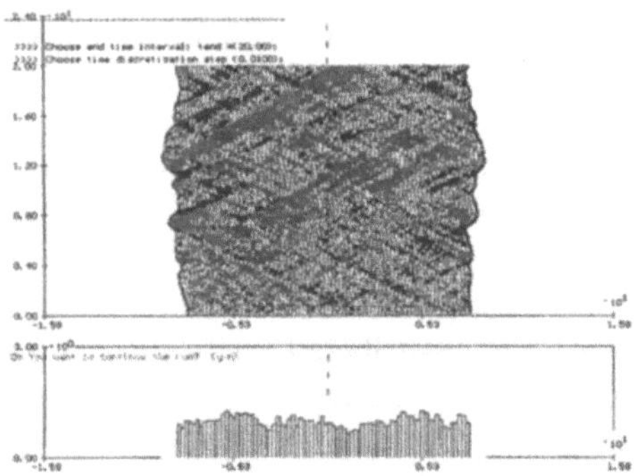

Fig. 3. For the 1-dimensional stochastic Navier-Stokes system arising as continuum limit equations for $\delta \to 0$ from the multi-particle model in Fig. 2, a suitable discretization of the temporally varying swarm interval (divided into $N = 70$ compartments) leads to a typical density distribution (at time $t = 20$), plotted in the lower panel, whereas the upper panel plots the whole density evolution with time (from 0 to 20 sec, in vertical direction). Comparison with the right hand picture in Fig. 2 reveals a striking similarity in the appearance of stochastic waves with constant speed as well as stochastic fluctuations of the two swarm boundaries in both the continuum model and the multi-particle model with $N = 150$ birds. For more details consult [6].

waves. Indeed, a suitable discretization procedure (with appropriate modifications of the Neumann boundary conditions mentioned in Section 1.3 above) produces numerical simulations with quite similar types of stochastically induced waves, see Fig. 3.

Also for higher dimensional cases the stochastic model equations (1–2) have been considered, but so far only with distance dependent 'elasticity' and 'viscosity', see [24] for $m = 3$, and [13, 22, 31] for $m = 2$. Nevertheless, the performed stochastic simulations of 2-dimensional swarms (Figs. 4 and 5) show characteristic behavioral properties resembling 'oscillating clouds' with quite smooth boundaries and quite persistent positional arrangement of 'birds', even in configurations with remarkable variations of swarm shape and internal density distributions. Therefore, it seems to be worth to check whether (with the type of used parameters) these discrete stochastic systems satisfy the *uniformity condition* which was assumed in Section 1.2 above, in order to derive a continuum limit of generalized Navier-Stokes type, although the purely distance dependence of the coefficients (without the 'interaction scope' factor in (13)) creates a difficulty to subsume the hitherto studied models in higher spatial dimensions under the presented model class.

Even independent of such multi-particle models and their analysis, also purely phenomenological continuum models for *cooperative motion* (as in bird swarms or in colonies of gliding bacteria) have been proposed and investigated, which – merely a surprise – again are nonlinear hyperbolic-parabolic or -elliptic systems of Navier-Stokes or Stokes type, respectively, with white noise stochastics, see [12],[13] and [35],[36]. Physical scaling arguments and numerical simulations show characteristic features of phase transitions from a trivial 'disordered' state to 'cooperative' states if, for example, the noise amplitude is successively increased.

However, so far there has been no trial to expand the analysis of such types of continuum models to biologically and mathematically interesting situations with *free boundaries* as in the 1-dimensional case (see Fig. 3).

1.5 Discussion

The here presented construction of 'densities' for dynamic multi-particle systems, the proposed scaling procedure and the subsequent (formal) derivation of a generalized Navier-Stokes system as continuum limit equations (furtheron called the *Voronoi-Delaunay limit*) are, to my knowledge, new and clearly contrast with the usually applied construction and scaling procedure of the so-called *hydrodynamical limit*, cf. [34],[32],[33].

1. **(Densities)** For any open test domain $\mathfrak{U}_\delta \subset \mathfrak{R}^m$ with $|\mathfrak{U}_\delta| \sim \delta^m$ the *hydrodynamical limit* procedure defines a discrete 'particle density' as the sum of Dirac measures $u|_{\mathfrak{U}_\delta} := \sum_{X_i \in \mathfrak{U}_\delta} \delta_{X_i}$ of all particles contained in $\mathfrak{U}_\delta$, thus, as an element in the space $\mathfrak{D}_0$ of distributions on $\mathfrak{R}^m$. Therefore its generic regularity is m degrees lower when compared to the *Voronoi and Delaunay densities* $u_{\mathfrak{V}}$ and $u_{\mathfrak{D}}$ which generically are bounded measurable step functions on $\mathfrak{R}^m$ and

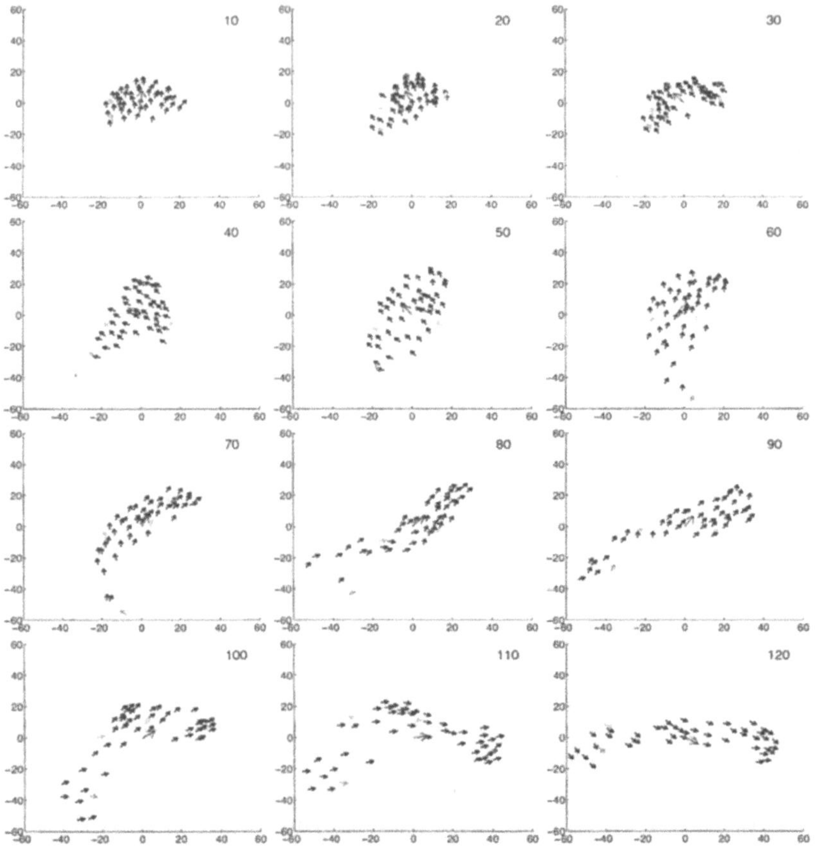

Fig. 4. Simulation of the 2-dimensional bird swarm model with center of mass held fixed and $N = 30$ birds. (12 frames plotted each 10 sec, in 4 lines starting from left). Observe the almost persistent connectivity of the deforming 'network' in spite of extensive elongation of its droplike shape. Also, during its deformation and shearing rotation (i.e. rotation of birds' mean orientation), the relative positions of single birds within the swarm and with respect to each other are, up to small stochastic fluctuations, merely changed (for example, follow the two grey arrows in the very left part of the swarm). From [31] by courtesy of Ralf Müller.

(because of the stochastics only transiently) might degenerate into successively lower dimensional Hausdorff measures, see Remark 1.5.

2. **(Scaling)** In the limit $\delta \to 0$ the *hydrodynamical limit* procedure requires that the test domains $\mathfrak{U}_\delta$, in which e.g. particle collisions are evaluated on the average, still contain (with $N \to \infty$) an infinitely increasing number of particles. Thus, the appropriate scaling has to be chosen as $\delta \simeq \mathfrak{L}/N^\kappa \succ \mathfrak{L}/N^{\frac{1}{m}}$ with an exponent $\kappa > \frac{1}{m}$, contrasting to the generic properties of the *Voronoi-Delaunay limit*, where $\delta \simeq \mathfrak{L}/N^{\frac{1}{m}}$ is chosen as the mean nearest neighbor distance, therefore

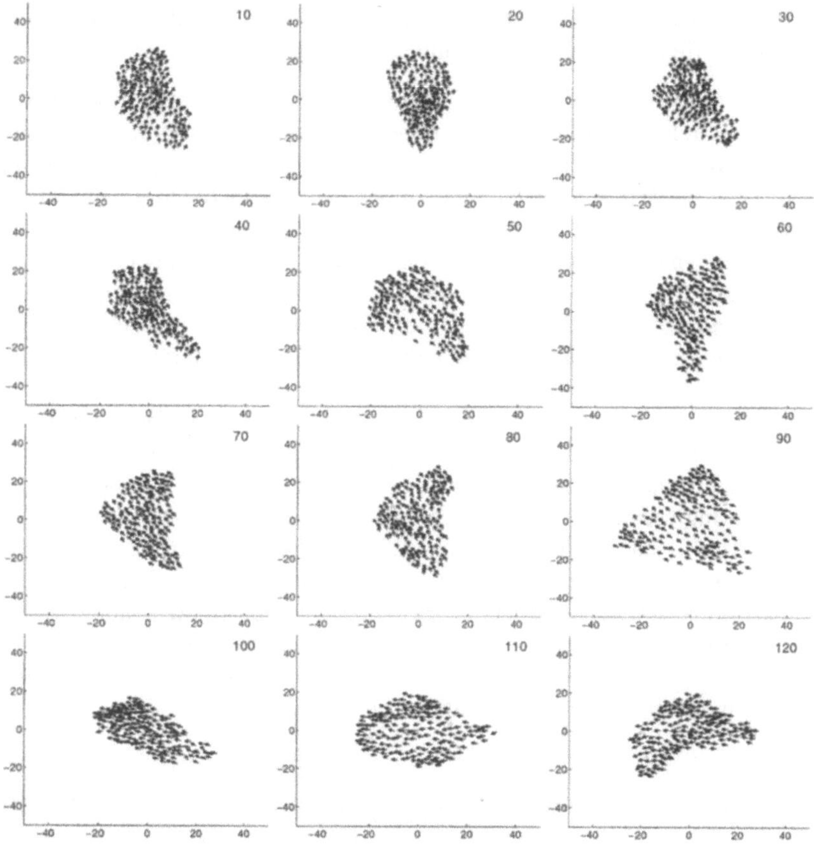

Fig. 5. Analogous plot of model simulations as in Fig. 4, but with $N = 200$ birds and appropriate rescaling of model parameters. Observe here, in addition, the remarkably smooth 'free boundary' of the swarm during extensive shape changes (droplike, triangular, oval), contraction or expansion, and clearly visible transient local accumulations or dilutions, i.e changes in local 'density' (both 'Voronoi' or 'Delaunay' depending on the observer's view, see Definitions 1.1–1.4 in Section 1.1). From [31].

any ball $\mathfrak{U}_\delta := \mathfrak{B}_{C\delta}$ with radius $C\delta$ would generically contain only a uniformly bounded number of 'particles'.

3. **(Stochastics)** As a typical 'mean field approximation' of a stochastic multi-particle process the *hydrodynamical limit* supposes that, at a time scale which is short relative to modelling time, positions of particles are well-mixed, so that averaging can be performed. In contrast, the *Voronoi-Delaunay limit* has to suppose some kind of 'uniformity criterium' (as assumed in Section 1.2); this allows such deformations of the multi-particle network that, at least for an 'interior particle', relative position and identity of its nearest neighbors may change, but the overall shape of its Voronoi neighborhood as well as the number of nearest neighbors (generically identical to the number of corner points of the

Voronoi polyeder) varies only within certain limits. For the 2-dimensional case compare Fig. 1 as well as the simulations in Figs. 4 and 5.

Thus, the required convergence estimates which still have to be performed in order to conclude a rigorous validity of the formally derived *Voronoi-Delaunay limit* equations, would rely on proving sufficient regularity even of higher derivatives of the velocity (e.g. the deformation tensor), similar to but probably more intricate as in the simpler 1-dimensional case. Such estimates could even be generalized to modelling situations, in which the limiting equations are not of simple Navier-Stokes type, see Remark 1.13. Moreover, let me emphasize that already in the simpler case of the *diffusion limit*, stochastic convergence of a discrete particle process to a Brownian sheet is quite easy to be proven in the 1-dimensional case, but more difficult in higher space dimension, see e.g. [25].

In spite of these remaining mathematical problems in analysis and stochastics, the presented technique of Delaunay triangulation and Voronoi tesselation for multi-particle systems has, in the mean time, been used by several authors for modelling and simulating cooperative motion of biopolymers in cells and of cells in tissues, see [9],[10],[30] and the review article [20]. On the other hand, simulations of discretized continuum models of Navier-Stokes types have been studied, e.g. in [12]. The ideas presented here, namely to formulate an appropriate framework for using the *Voronoi-Delaunay limit*, might stimulate further efforts to bring both modelling approaches into closer contact, with the aim to better understand the essential dynamics of interacting biological entities, see also [7].

2 Two-Phase Flow Models for Intracellular Polymer Dynamics

The following derivation is valid for general two-phase composite materials, where a *complex polymer phase*, e.g. in applications to biological cell motility the so-called '*cytoskeleton*' (more or less cross-linked filaments of actin polymers), is embedded within a more fluid material, subsequently called the *aqueous phase* (mainly the 'solvent' in a cell containing all kinds of monomers and polymer fragments). Both phases will be regarded as connected, cohesive and fluid-like. By the latter requirement we simply mean that the interconnections within a given phase are not permanent as in an elastic solid but rather are breaking and rearranging so that over some time scale elastic stresses are eventually relaxed.

For the essential continuum variables, scalar or vector functions of space $\mathbf{x}$ and time t, we use a simplifying notation:

- volume fraction of the *complex polymer phase*: $\theta_c = \theta$ with mean velocity $\mathbf{v}$
- volume fraction of the *aqueous phase*: $\theta_a = (1 - \theta_c)$ with mean velocity $\mathbf{w}$

2.1 Mass Balances

The first mass balance holds for the *volume fraction of the complex polymer phase*, θ_c and is expressed as a hyperbolic equation

$$\partial_t \theta_c + \nabla \cdot (\theta_c \mathbf{v}) = R_c(\theta_c, \theta_a) \ . \tag{25}$$

Here the source term R_c denotes the rate of chemical production of cytoskeletal (e.g. actin filaments) per unit volume. Obviously, there is a corresponding loss of mass from the solvent phase since small precursors of the filament system are regarded as being part of the aqueous phase (e.g. actin monomers). This yields a corresponding second mass balance equation for the *volume fraction θ_a of the aqueous phase*

$$\partial_t \theta_a + \nabla \cdot (\theta_a \mathbf{w}) = R_a(\theta_c, \theta_a) \ . \tag{26}$$

As long as the partial specific volume of various molecular components remain constant when they are added or substracted from cytoskeleton or solvent, the local net volume growth rate of the total cytoplasm must vanish;

$$R_c(\theta_c, \theta_a) + R_a(\theta_c, \theta_a) \ = 0.$$

Furthermore, since, $\theta_c + \theta_a = 1$ (by definition), we can surmise that $\partial_t(\theta_c + \theta_a) = 0$. Combining this with the two balance equations (25,26), we finally obtain the following mass balance laws for $\theta = \theta_c$ with $f(\theta) := R_c(\theta, 1 - \theta)$, namely a

– **hyperbolic transport equation**

$$\partial_t \theta + \nabla \cdot (\theta \ \mathbf{v}) = f(\theta) \ . \tag{27}$$

and the

free divergence for the total volume flux

$$\nabla \cdot (\theta \ \mathbf{v} + (\mathbf{1} - \theta) \ \mathbf{w}) = 0 \ . \tag{28}$$

At any time instant the two velocity fields, $\mathbf{v}$ and $\mathbf{w}$, must jointly satisfy the above divergence condition and also, as the most important constitutional relation for a 'creeping flow' of high viscosity (low Reynolds number), certain pseudo-stationary force balances within an, eventually time dependent, domain $\Omega(t)$ and on its boundary $\partial\Omega(t)$.

2.2 Power Functional and Force Balances for $\Omega(t) = R^m$

Previously we have stated and discussed a system of elliptic differential equations and suitable boundary conditions [14],[15],[17], which describe the pseudostationary **force balances of the two phases** described above. These equations were proposed to hold for creeping flow limit and for isotropic materials. We now show that, for given concentration $\theta = \theta(\mathbf{x}, t)$ at any fixed time t, these equations can also be derived (as Hamilton-Jacobi equations) from minimizing a global functional $J[\mathbf{v}, \mathbf{w}]$ that is a certain measure of the *total work performed per unit time ('power')*:

$$J[\mathbf{v},\mathbf{w}] := \frac{1}{2}\int_\Omega \nabla\mathbf{v}\,\mathbf{M}_c\,\nabla\mathbf{v} + \nabla\mathbf{w}\,\mathbf{M}_a\,\nabla\mathbf{w} + \Phi_c\theta|\mathbf{v}|^2 \tag{29}$$

$$+\frac{1}{2}\int_\Omega \varphi\theta(1-\theta)|\mathbf{v}-\mathbf{w}|^2 \tag{30}$$

$$-\int_\Omega \theta P_c\nabla\cdot\mathbf{v} + (1-\theta)P_a\nabla\cdot\mathbf{w} + P_{ac}(\mathbf{w}-\mathbf{v})\cdot\nabla\theta \tag{31}$$

$$-\int_\Omega \theta\mathbf{F}_c\cdot\mathbf{v} + (1-\theta)\mathbf{F}_a\cdot\mathbf{w} \tag{32}$$

This power functional ought to be minimized in a suitable Sobolev space $\mathfrak{V}$ of all velocity fields $(\mathbf{v},\mathbf{w})$ satisfying the divergence relation (28). The exact formulation is given below.

Before, let us explain the various terms of this functional. The principle is easy and general: Any linear *force balance* vector equation of the form $\Phi(\mathbf{v}) = \mathbf{F}$, where frictional forces equilibrate with inner potential forces or outer field forces, yields, after scalar multiplication by $\mathbf{v}$, a *power balance equation* $\mathbf{D}[\mathbf{v}] = \mathbf{F}\cdot\mathbf{v}$. Here, the *dissipative energy loss rate,* the symmetrical form $\mathbf{D}[\mathbf{v}] := \Phi(\mathbf{v})\cdot\mathbf{v}$, is in balance with the *energy gain rate* due to the effective flow in direction of the applied force field. However, minimizing the *power functional* $J[\mathbf{v}] := \frac{1}{2}\mathbf{D}[\mathbf{v}] - \mathbf{F}\cdot\mathbf{v}$, with the energy gain rate substracted from half the dissipation loss rate, gives back the full force balance vector equation $\Phi(\mathbf{v}_*) = \mathbf{F}$ for the unique minimizer $\mathbf{v}_*$. The minimal 'power' that is attained, $J[\mathbf{v}_*] := \frac{1}{2}\mathbf{D}[\mathbf{v}_*] - \mathbf{F}\cdot\mathbf{v}_* = -\frac{1}{2}\mathbf{D}[\mathbf{v}_*]$, always turns out to be negative.

Analogously, in the power functional above, the two Ω-integrals (29) and (30), which are positive definite, quantify the total rate of dissipation, i.e. conversion of kinematic energy into heat, whereas the following two Ω-integrals (31) and (32) measure the total energy gain rate (or loss rate, if negative) due to the action of internal pressures (P_c , P_a, and P_{ac}) or external forces ($\mathbf{F}_c$ and $\mathbf{F}_a$) on the various phases of the fluid, respectively. More detailed, we now comment the used parameters or parameter functions in the order of their appearance:

- $\mathbf{M}_c = \mathbf{M}_c(\theta)$ and $\mathbf{M}_a = \mathbf{M}_a(\theta)$ denote the viscosities of the complex polymer and aqueous phase, respectively. Both might depend on the volume fraction θ. In particular, we can suppose that $\mathbf{M}_c(\theta) \to 0$ for vanishing polymer concentration $\theta \to 0$. Moreover, notice the tensor notation which, e.g. for $m = 2$, reads $\nabla\mathbf{u}\,\mathbf{M}\,\nabla\mathbf{u} = \mu|\nabla\mathbf{u}|^2 + \lambda(\nabla\cdot\mathbf{u})^2 + \nu(\nabla\times\mathbf{u})^2$.
- The coefficient Φ_c describes a possibly present, distributed body friction (in the 3 dimensional case, for instance induced by a distributed small third phase) or a friction between cytoskeleton and adhesive substrata (important in 2- or 1-dimensional models), see also Section 1.4.1.
- The coefficient φ describes the interphase friction due to the drag between the two velocities with difference vector $(\mathbf{v}-\mathbf{w})$.

The subsequent integral term (31), with the minus sign in front, is the most important term as it describes the volumetric rate of energy supply within the domain Ω. When energy is converted from potential energy material interactions or from molecular motors or from any other processes so as to increase the organized kinematic energy, then these terms are positive. We claim that these simple terms are very general and can include all isotropic constitutive possibilities for supply of mechanical power merely by taking appropriate choices of the three scalar 'static pressures' P_c , P_a, and P_{ac} , as functions of the 'state' θ.

- The first of these (P_c) is a pressure associated purely with the complex polymer phase, e.g. negative if the effect of the microscopic distributed energy supply is to induce an attraction between complex particles (cytoskeletal filaments). If this results in a local contraction of the polymer phase, $\nabla \cdot \mathbf{v} < 0$, then the corresponding integrand is positive, meaning that locally kinematic energy is gained. The second (P_a) is an analogous term associated with the aqueous phase. For example, it is positive if aqueous particles tend to separate from each other (e.g. swelling of extracellular matrix). Usually this would induce a local expansion of the aqueous phase, $\nabla \cdot \mathbf{w} > 0$, so that the corresponding integrand is again positive, and kinematic is energy is gained locally (e.g. by the use of thermal energy). The third function ($P_{ac} = -P_{ca}$) is a pressure associated with the interface between the two phases. By sign convention the interphase pressure is positive when the two phases tend to avoid each other and negative when they tend to mix with each other. For example, the first case would again describe the process of swelling, in which water could be sucked into the complex polymer phase with a relative velocity 'up-gradient', i.e. $(\mathbf{w} - \mathbf{v}) \cdot \nabla\theta > 0$, leading to kinematic energy gain.

However, the fact that the model interpretations given for the last two pressure integral terms in (31) both rely on the same phenomenon 'swelling', might induce a feeling of redundancy. Indeed, it turns out that the three 'pressures' can be redefined by just adding a joint pressure function

$$P_r(\theta) = \tilde{P}_r(\theta) + \tilde{P}(\theta) \,, \quad r = c, a, ac \tag{33}$$

yielding the additional term $\int \tilde{P}(\theta)\nabla \cdot (\theta\mathbf{v} + (1-\theta)\mathbf{w}) = 0$ in (31). Therefore, because of (31) this redefinition would not change the value of the power functional J. Thus, the solution of the minimization problem does not depend on this redefinition, meaning that a *consistency condition* between the three pressure state functions must hold. For instance, by the particular choice

$$\tilde{P}(\theta) := -\tilde{P}_a(\theta) + \int_0^\theta \frac{\tilde{P}_a(\tilde{\theta}) + (\tilde{\theta})\tilde{P}_{ac}}{(1-\tilde{\theta})}\, d\tilde{\theta},$$

we could impose the following differential equation as one possible consistency condition

$$P_a'(\theta) = \frac{1}{1-\theta}(P_a(\theta) + P_{ac}(\theta))$$

implying

$$P_{ac}(\theta)\nabla\theta = \nabla((1-\theta)P_a(\theta)) \tag{34}$$

This equation describes a 'microscopic' force balance, where the aqueous pressure $P_a = P_a(\theta)$ appears to be induced only by an interphase pressure P_{ac} (e.g. hydrophilic swelling forces) such that no other 'intrinsic' pressures in the aqueous phase exist. Notice that (34) implies that the P_r-integrals in (31) after integration by parts (assuming that all terms at $|\mathbf{x}| \to \infty$ vanish) can be simply rewritten as

$$-\int_\Omega \bar{P}(\theta)\nabla \cdot \mathbf{v}$$

yiedling an energy gain/loss term, such as if the *mean intrinsic pressure of the two-phase fluid*

$$\bar{P}(\theta) := \theta P_c(\theta) + (1-\theta)P_a(\theta)\,, \tag{35}$$

would act only on the complex polymer phase flowing with mean velocity $\mathbf{v}$.

Remark 2.14 *A further crucial advantage of this choice of representation for the 'pressure triple' is that the integral on the left hand side is now generally defined for velocities, which are Sobolev functions only, even in occurring cases where the concentration function $\theta = \theta(x,t)$ carries jump discontinuities, so that $\nabla\theta$ would be a measure only, making a definition of the power functional J difficult. Thus, for J to be well defined also for such discontinuous concentrations θ, the condition (33) is required.*

Analogously, we get the following interpretation of the last term in the power functional:

- The integral term (32) describes the supply of mechanical power by body forces ($\mathbf{F}_c$ and $\mathbf{F}_a$) acting on the complex and aqueous phase (this may include gravitational fields, centrifugal fields etc).

From expressions (33) and the argumentation above the question remains, what is the total *pressure acting, for example, in the aqueous phase*? The answer is given by the classical *Lagrange multiplier* technique. Let us assume that $\mathfrak{V}$ is the space of all vector functions $(\mathbf{v}, \mathbf{w})$ satisfying (28) such that $\int_\Omega \theta|\nabla\mathbf{v}|^2 + \varepsilon|\nabla\mathbf{w}|^2 < \infty$ and $\int \theta|\mathbf{v}|^2 + \theta(1-\theta)|\mathbf{v}-\mathbf{w}|^2 < \infty$, for given concentration $\theta = \theta(x,t)$. This weighted Sobolev norm is the natural norm making J a coercive functional on $\mathfrak{V}$ (clearly in case of positive body friction Φ_c; otherwise imbedding theorems have to be used).

Then, under suitable conditions on the force fields F, one should be able to proof the following

Proposition 2.15 ***(Definition of pressure p as Lagrange multiplier)*** *For the uniquely determined minimization solution* $(\mathbf{v}_*, \mathbf{w}_*) \in \mathfrak{V}$ *there exists a* unique 'pressure' function p_* *satisfying* $\int \frac{1-\theta}{\theta}|p_*|^2 < \infty$*, such that*

$$DJ[\mathbf{v}_*, \mathbf{w}_*](\mathbf{v}, \mathbf{w}) = -\int_\Omega p_* \nabla \cdot (\theta \mathbf{v} + (1-\theta)\mathbf{w}).$$

We remark that this solution $(\mathbf{v}_*, \mathbf{w}_*, p_*)$ is the unique stationary point of the extended *saddle point type functional*

$$J[\mathbf{v}, \mathbf{w}, p] := \mathbf{J}[\mathbf{v}, \mathbf{w}] - \int_\Omega p \nabla \cdot (\theta \mathbf{v} + (1-\theta)\mathbf{w}).$$

Then, by denoting

$$p_r(x,t) = P_r(\theta(x,t)) + p(x,t)\ , \quad r = c, a, ac$$

the *total pressure* in each phase or at the interphase, respectively, and by writing down the corresponding Hamilton-Jacobi equations, the unique solution $(\mathbf{v}_*, \mathbf{w}_*, p_*)$ of the minimization problem satisfies the following system of elliptic differential equations (written for smooth solutions).

- **Linear elliptic system** for velocities $\mathbf{v}$ and $\mathbf{w}$

$$\nabla \cdot (\mathbf{M}_c \nabla \mathbf{v}) - \nabla(\theta p_c) - p_{ac} \nabla \theta = \varphi \theta (1-\theta)(\mathbf{v} - \mathbf{w}) + \Phi_c \theta \mathbf{v} - \theta \mathbf{F}_c \tag{36}$$

$$\nabla \cdot (\mathbf{M}_a \nabla \mathbf{w}) - \nabla((1-\theta) p_a) + p_{ac} \nabla \theta = -\varphi \theta (1-\theta)(\mathbf{v} - \mathbf{w}) - (1-\theta)\mathbf{F}_a \tag{37}$$

together with the

- **Divergence relation**

$$\nabla \cdot (\theta \mathbf{v} + (1-\theta)\mathbf{w}) = 0\,. \tag{38}$$

Again, here we have used, e.g. for $m = 2$, the following tensor convention $\mathbf{M}\ \nabla \mathbf{u} = \mu \nabla \mathbf{u} + \lambda(\nabla \cdot \mathbf{u})\mathbf{I} + \nu(\nabla \cdot \mathbf{u})\mathbf{I}^\perp$.

These equations together with the **hyperbolic equation** (27) for θ are exactly the classical two-phase flow continuum equations as they were originally derived from a more general modelling setup in [21] by [15], only that the latter authors used quite particular definitions and notations for the pressure terms.

2.3 Vanishing Solvent Viscocity

For further applications let us assume vanishing solvent viscosity $\mathbf{M}_a = 0$, which is reasonable in comparison to the much stronger polymer phase viscosity $\mathbf{M}_a$. Then, under the consistency condition (33) and equation (37) we can derive *Darcy's law for the 'effective pressure'* p, however, only at those points for which $\theta < 1$, that is where still aqueous phase is present:

$$\nabla p = \varphi\theta(\mathbf{v} - \mathbf{w}) - \mathbf{F}_a.$$

Thus the particular choice of the 'pressure triple' reveals the Lagrange multiplier p as a *hydrostatic pressure*, which is mainly related to the draft flow between the tho phases. Thus, we can explicitly express the *solvent velocity* $\mathbf{w}$ in terms of the other variables:

$$\mathbf{w} = \mathbf{v} - \frac{1}{\varphi\theta}\nabla p + \frac{1}{\varphi\theta}\mathbf{F}_a. \tag{39}$$

Finally, summation of eqns. (36) and (37) plus using the divergence relation (38) gives the following

- **Linear Stokes system** for velocity $\mathbf{v}$ and pressure p.

$$\nabla \cdot (\mathbf{M}_c \nabla \mathbf{v}) - \nabla(\bar{P}(\theta) + p) - \Phi\theta\mathbf{v} = \overline{\mathbf{F}} \tag{40}$$

$$\nabla \cdot (\frac{1-\theta}{\varphi\theta}\nabla p - \mathbf{v} + \frac{1-\theta}{\varphi\theta}\mathbf{F}_a) = 0 \tag{41}$$

where the *mean external force* is defined as

$$\overline{\mathbf{F}} := \theta\mathbf{F}_c + (1-\theta)\mathbf{F}_a$$

2.4 Power Functional and Boundary Conditions for Domains $\Omega(t)$ with Piecewise Smooth Boundary

In general, allowing for domains $\Omega(t)$ with boundaries, there will appear boundary integrals in the power functional. Here we restrict to the case that the boundary $\partial\Omega(t)$ can be divided into several parts, a fixed *impermeable boundary* Γ_C, a fixed *permeable boundary* Γ_B, and a *free impermeable boundary* Γ; the first boundary part is assumed to be inert, that is, to carry no extra forces or pressures. (For visualization and application to intracellular polymer dynamics see Fig. 4). Then the *extended power functional* $\widetilde{J}$ has volume integrals as J in (29–32) and, in addition, boundary integrals over Γ and Γ_B, but not over Γ_C:

$$\begin{aligned}\widetilde{J}[\mathbf{v},\mathbf{w}] :&= J[\mathbf{v},\mathbf{w}] \\ &+ \int_\Gamma \theta P_c^\Gamma(\mathbf{n}\cdot\mathbf{v}) + (1-\theta)P_a^\Gamma(\mathbf{n}\cdot\mathbf{w}) &(42)\\ &+\frac{1}{2}\int_{\Gamma_B}(1-\theta)\varphi_B(\mathbf{n}\cdot\mathbf{w})^2 + \int_{\Gamma_B}(1-\theta)P_B(\mathbf{n}\cdot\mathbf{w}). &(43)\end{aligned}$$

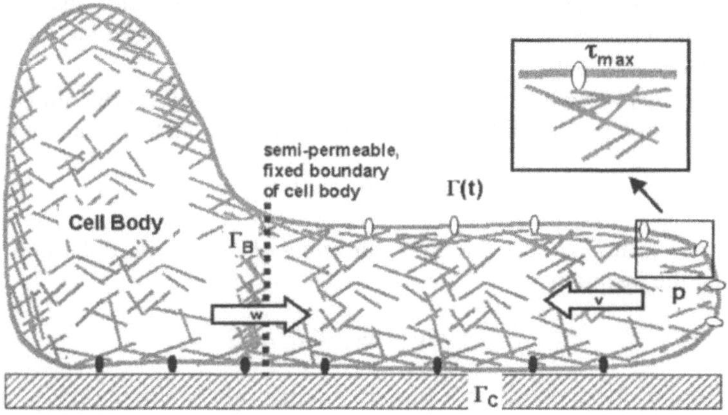

Fig. 6. Sketch of a 2-dimensional cross–sectional model for the two-phase flow dynamics of a cellular pseudopod. It is represented by a time variable domain $\Omega(t)$ with three boundaries: a fixed impermeable boundary Γ_C, where the pseudopod touches a substratum expressing adhesion proteins; another fixed but semi–permeable boundary is Γ_B, where the actin filaments are connected to the stable cytoskeletal cortex of the cell body, while the solvent can pour through with velocity $\mathbf{w}$; finally, a free boundary $\Gamma = \Gamma(t)$ constituting the cortical membrane–protein complex surrounding the pseudopod. The dynamics at the free boundary is determined by a force balance between hydrostatic pressure p and tension τ of the actin filament network that might be pulled inside with inward velocity $\mathbf{v}$, see eq. (56). From [5].

This functional is going to be minimized over all velocity fields $(\mathbf{v}, \mathbf{w}) \in\mathfrak{V}$ defined as in Section 2.2 but now, in addition, satisfying the following conditions at any boundary point of the respective type:

1. At the *fixed impermeable boundary* Γ_C, possibly representing the inert wall of a container in contact with the cell membrane, where the outer normal component of the aqueous flow has to vanish but that of the complex phase flow has to stay non-negative, meaning no outward flow of polymer phase allowed (however, newly assembled polymers near the boundary might flow inward with nonzero velocity as we suppose a 'non sticky' boundary)

$$\mathbf{n} \cdot \mathbf{w} = 0, \; \mathbf{n} \cdot \mathbf{v} \leq 0 \;\; on \; \Gamma_C \tag{44}$$

   with no additional integral term over this part of $\partial\Omega$ appearing in the functional; furthermore
2. At the *fixed permeable boundary* Γ_B, possibly representing the cortical border between a pseudopod and the cell body, where the polymers 'stick'

$$\mathbf{v} = 0 \;\; on \; \Gamma_B \tag{45}$$

   but the aqueous phase is allowed to permeate, thereby experiencing a drag resistance φ_B and an opposing *outer pressure* P_B expressed by the two integral terms over Γ_B in (43), the first quantifying energy dissipation analogous to the volume dissipation integrand (30), the second analogous to the volume force integrand (32). Finally,

3. at the *free impermeable boundary* $\Gamma = \Gamma(t)$, possibly representing the upper membrane of a pseudopod, which has to move with the total volume flux, i.e.

$$\dot{\Gamma} := \mathbf{n} \cdot (\theta \mathbf{v} + (1-\theta)\mathbf{w}) \tag{46}$$

but again, as on Γ_B, allowing for inward flow of the complex phase, $\mathbf{n} \cdot \mathbf{v} \leq \dot{\Gamma}$, yielding the inequality

$$\mathbf{n} \cdot (\mathbf{v} - \mathbf{w}) \leq 0 \ \ on\ \Gamma \tag{47}$$

Furthermore, there could be a flow resistance due to *outer surface pressures* effecting the complex polymer phase, P_c^{Γ}, or the aqueous phase, P_a^{Γ}, at the free boundary. The corresponding integral terms appear in the functional (42).

Now, when deriving the corresponding Hamilton-Jacobi equations, variations by vector functions $\tilde{\mathbf{v}}, \tilde{\mathbf{w}} \in \mathfrak{V}$, integration by parts and observation of the imposed boundary conditions (44,45,47) for $\mathbf{v}$ and $\mathbf{w}$, as well as for $\tilde{\mathbf{v}}$ and $\tilde{\mathbf{w}}$, yield the following boundary integrals to appear in the functional derivative

$$\int_{\Gamma_C} (\mathbf{n} \cdot \tilde{\mathbf{v}})(\mathbf{n} \cdot \mathbf{M}_c \nabla \mathbf{v} \cdot \mathbf{n} - \theta p_c) \tag{48}$$

$$\int_{\Gamma_B} (\mathbf{n} \cdot \tilde{\mathbf{w}})(\mathbf{n} \cdot \mathbf{M}_a \nabla \mathbf{w} \cdot \mathbf{n} - (1-\theta) p_a + (1-\theta)(\varphi_B(\mathbf{n} \cdot \mathbf{w}) + P_B)) \tag{49}$$

$$\begin{aligned} \int_{\Gamma} & (\mathbf{n} \cdot \tilde{\mathbf{v}})(\mathbf{n} \cdot \mathbf{M}_c \nabla \mathbf{v} \cdot \mathbf{n} - \theta p_a \theta P_a^{\Gamma}) \\ & + (\mathbf{n} \cdot \tilde{\mathbf{w}})(\mathbf{n} \cdot \mathbf{M}_a \nabla \mathbf{w} \cdot \mathbf{n} - (1-\theta) p_a + (1-\theta) P_a^{\Gamma}) \end{aligned} \tag{50}$$

in addition to terms containing the corresponding tangential components of $\mathbf{M}_c$ and $\mathbf{M}_a$, yielding *standard tangential Neumann boundary conditions* for $\mathbf{v}$ and $\mathbf{w}$.

1. Now, at such boundary points of Γ_C where $\mathbf{n} \cdot \mathbf{v} < 0$ holds, $\mathbf{n} \cdot \tilde{\mathbf{v}}$ can be varied arbitrarily with vanishing integral (48), whereas else the integral has only to be nonnegative for all $\mathbf{n} \cdot \tilde{\mathbf{v}} \leq 0$. Thus, we derive the following *Neumann conditions for* $\mathbf{v}$ *on* Γ_C

$$\mathbf{n} \cdot \mathbf{M}_c \nabla \mathbf{v} \cdot \mathbf{n} - \theta p_c \leq 0, \quad \text{and} \tag{51}$$

$$\mathbf{n} \cdot \mathbf{M}_c \nabla \mathbf{v} \cdot \mathbf{n} - \theta p_c \ = 0 \ \ \text{for points with } \mathbf{n} \cdot \mathbf{v} < 0 \tag{52}$$

Boundary points in the latter case we characterize by the property *'disrupture'* of the complex phase from the fixed boundary Γ_C, compared to the alternative case *'attachment'*. Notice that these boundary conditions include the effective pressure p because of $p_c = p + P_c(\theta)$.

From such *'disruptive' boundary points* with $\mathbf{n} \cdot \mathbf{v} < 0$, say $y \in \Gamma_C^d$, characteristics x_y for the hyperbolic transport equation (27), i.e. curves satisfying $\dot{x}_y(t) = \mathbf{v}(t, x_y(t))$ and $x_y(t_0) = y$, emanate into the domain $\Omega(t)$, so that we need an additional condition which (for applications) is chosen as a homogeneous

Dirichlet boundary condition for the *hyperbolic equation*:

$$\theta(t, y) = 0 \text{ for } y \in \Gamma_C^d(t) \, . \tag{53}$$

The expression $\tau_c := \mathbf{n} \cdot \mathbf{M}_c \nabla \mathbf{v} \cdot \mathbf{n} - \theta p_c$ appearing in (51,52), can be regarded as the *tension (or stress) within the complex polymer phase*. This polymer stress at the fixed boundary could become negative in the case of attachment (e.g. during expansive swelling) but never positive (i.e. contractile), because then disrupture of polymers would occur, whence the stress becomes zero. Indeed, this condition is consistent with (52) since in case of disrupture the polymer concentration θ at the boundary vanishes (53) and with it the viscosity tensor $M_c(\theta)$.

2. Since variations of $\mathbf{n} \cdot \tilde{\mathbf{w}}$ in (49) are free on the permeable part of the fixed boundary, we conclude the following *generalized Neumann conditions for* $\mathbf{w}$ *on* Γ_B

$$\mathbf{n} \cdot \mathbf{M}_a \nabla \mathbf{w} \cdot \mathbf{n}/(1-\theta) + \varphi_B(\mathbf{n} \cdot \mathbf{w}) = p + P_a(\theta) - P_B \tag{54}$$

or, in the limiting case of relatively small aqueous viscosity $\mathbf{M}_a$ and cortical boundary flow resistance φ_B, just the simple *Dirichlet condition for* p on Γ_B

$$p = P_B - P_a(\theta). \tag{55}$$

3. Finally, on the free boundary Γ we first perform variations in $\tilde{\mathbf{v}}$ and $\tilde{\mathbf{w}}$ satisfying the equality $\mathbf{n} \cdot (\tilde{\mathbf{v}} - \tilde{\mathbf{w}}) = 0$ and obtain the following *Neumann condition* for $\mathbf{v}$ and $\mathbf{w}$ on Γ

$$\mathbf{n} \cdot \mathbf{M}_c \nabla \mathbf{v} \cdot \mathbf{n} + \mathbf{n} \cdot \mathbf{M}_a \nabla \mathbf{w} \cdot \mathbf{n} = p + \theta(P_c(\theta) - P_c^\Gamma) + (1-\theta)(P_a(\theta) - P_a^\Gamma). \tag{56}$$

In addition, variations satisfying $\tilde{\mathbf{w}} = -\tilde{\mathbf{v}}$ and, in case of a boundary point with $\mathbf{n} \cdot (\mathbf{v} - \mathbf{w}) = 0$ also $n \cdot (\tilde{v} - \tilde{w}) \le 0$, induce the additional properties

$$\mathbf{n} \cdot \mathbf{M}_a(\nabla \mathbf{w}) \cdot \mathbf{n}/(1-\theta) \ge p + P_a(\theta) - P_a^\Gamma \text{ and} \tag{57}$$

$$\mathbf{n} \cdot \mathbf{M}_a(\nabla \mathbf{w}) \cdot \mathbf{n}/(1-\theta) = p + P_a(\theta) - P_a^\Gamma \text{ if } \mathbf{n} \cdot (\mathbf{v} - \mathbf{w}) < 0. \tag{58}$$

In the vanishing viscosity case $\mathbf{M}_a = 0$, assumed in the previous Section 2.3, these turn out to be *Dirichlet conditions for* p on Γ_B

$$p \le P_a^\Gamma - P_a(\theta) \quad \text{and} \tag{59}$$

$$p = P_a^\Gamma - P_a(\theta) \quad \text{for } \mathbf{n} \cdot (\mathbf{v} - \mathbf{w}) < 0. \tag{60}$$

Thus, in this case we obtain the following

Derived model system describing a *strongly viscous two-phase polymer flow with relatively low aqueous viscosity:* For each minimal solution of the power functional $\widetilde{J}[\mathrm{v},\mathrm{w}]$ defined at the beginning of this Section, in the limit $\mathrm{M}_a = 0$, we have derived, at least in a weak sense, the equations (27) and (40,41) as well as the derived boundary conditions above, that is a

– **hyperbolic-elliptic system of generalized Stokes type** on $\Omega(t)$

$$\partial_t\theta + \nabla\cdot(\theta\,\mathbf{v}) = f(\theta) \tag{61}$$

$$\nabla\cdot(\mathbf{M}_c(\theta)\nabla\mathbf{v}) - \nabla(\bar{P}(\theta)+p) - \Phi_c\theta\mathbf{v} = \overline{\mathbf{F}} \tag{62}$$

(63)

$$\nabla\cdot\left(\frac{1-\theta}{\varphi\theta}\nabla p - \mathbf{v} + \frac{1-\theta}{\varphi\theta}\mathbf{F}_a\right) = 0 \tag{64}$$

with **boundary conditions** for velocity $\mathbf{v}$ and pressure p, being inequalities or equalities on different parts of the boundary $\partial\Omega(t)$, namely in addition to 'natural' *tangential boundary conditions* mentioned above, the Neumann conditions (51,52) for $\mathbf{v}$ on the fixed impermeable boundary Γ_C with hyperbolic 'Dirichlet' condition (53) for θ, the Dirichlet condition (55) for p on the fixed impermeable boundary Γ_B as well as the Dirichlet conditions (59,60) for p and the Neumann condition (56) for $\mathbf{v}$ on the free impermeable boundary Γ. The normal speed $\dot{\Gamma}$ of the free boundary is explicitly given by equation (46) with $\mathbf{w}$ substituted as in (39).

2.5 Applications to Cytoplasm Flow and 1-Dimensional Examples

These derived two-phase flow equations of generalized Stokes type have been applied to several biological problems of cell motility, whereby particularly the two pressure terms $P_c(\theta)$ and $P_a(\theta)$ had to be chosen appropriately, for further discussions we refer to [17] and [5]. For example, the following results have been obtained: 3-dimensional rotationally symmetric model simulations for cell division, with detailed reproduction of experimental data for *Xenopus* eggs [23], 3-dimensional model for cytoplasm streaming in *Amoeba proteus* [18], 2-dimensional simulations of local cytoplasm protrusions [16], 2-dimensional movement of *Listeria* bacteria in the cytoplasm [26] and, as application in the simple 1-dimensional case, model simulations for lamellipodial protrusions and retractions [3],[5]. Here the domain $\Omega(t)$ is a finite interval with fixed (permeable) boundary point on one side (towards the cell body) and free (impermeable) boundary at the other side (the lamellipodial tip). In order to reproduce corresponding experimental observations from motility of human epidermal cells [2], the conditions for attachment and disrupture had to be slightly generalized by including stress dependent release of polymer binding to surface molecules; for more details and numerical simulations see [5].

However, already on fixed domains further simulations of the 1-dimensional hyperbolic-elliptic Stokes system reveal interesting dynamical properties of these

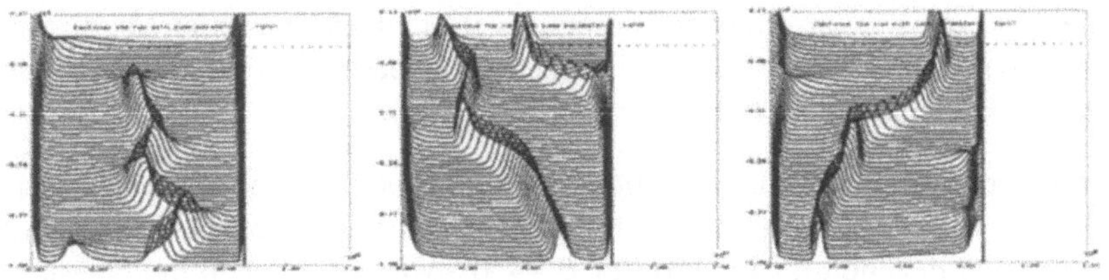

Fig. 7. 1-dimensional model simulation of a contractile, reactive and highly viscous two-phase fluid showing chaotic wave dynamics. Plot of actin polymer density $\theta(t,x)$ over a fixed spatial interval (horizontal) and over time in three successive panels (time running from 0 to 45 min, vertically from top to bottom in each panel)

viscous two-phase polymer flows. Indeed, for parameter choices such that the trivial constant state is dynamically unstable and dissipative friction is large enough, there appear solutions which are of chaotic character in time. For periodic boundary conditions in space (say a 1-dimensional polymer ring, e.g. actin filaments around a cell spread on a 2-dimensional substratum) they had been observed already by [14], where a standing wave showed (though very slight) irregular oscillations in amplitude and position. Moreover, simulations with fixed 'sticky' boundaries, i.e. zero flow conditions on both sides, have shown ongoing chaotic waves appearing in the interior, oscillating in position until finally they drift to one of the boundaries and melt there with persistent boundary layers, see Fig. 6 in [5] as well as Fig. 7.

2.6 Discussion

The appearance of such chaotic dynamics in the two-phase flow system strongly depends on the fact, that the constitutive law for the 'mean pressure' $\bar{P}(\theta) := \theta P_c(\theta) + (1-\theta)P_a(\theta)$ (35) in Stokes equation (40) for the velocity provides a **cubic non-linearity** which distinguishes regions of low polymer concentration θ with 'swelling' properties from regions of medium concentration θ having 'contractile' properties, i.e. where $\bar{P}'(\theta) < 0$. Thus, with such types of non-linearities, the polymer flow equations above model a **contractile and reactive, highly viscous two-phase fluid.**

Therefore it becomes clearly evident, why the chaotic wave phenomena in these 'deterministic fluids' have such different features when compared to the very constant speed waves appearing in the 'stochastic bird swarm fluids' of Chapter 1: There the derived continuum (Navier-Stokes) equations have a **monotone** 'attraction-repulsion function' as constitutive law for the 'elastic tension' $\alpha(u)$ in the infinitesimal network of (nearest) neighbors, see (22), and the globally stable constant equilibrium state has to be fed by continuing noise to produce such waves.

For the minimization problem presented in this Chapter 2, there remain several challenging difficulties to overcome in order to prove theorems on unique existence. One difficulty concerns the choice of suitably weighted Sobolev spaces $\mathfrak{V}$ in domains with boundaries, particularly in situations where the polymer concentration degenerates, $\theta \to 0$, at boundary points, as it occurs in cases of 'disrupture' (see Section 2.4). Even for proofs in the 1-dimensional case one probably needs uniform estimates of θ from below like $\theta(t,x) \succeq dist_{\partial\Omega}(x)$.

Furthermore, spatial discretizations of the elliptic Stokes system (say in week form) have to regard the fact that the leading (elliptic) coefficient in the velocity equation (63) degenerates to zero, for $\theta \to 0$, whereas that in the pressure equation (64) degenerates to infinity. Suitable discretization procedures at the boundaries, similar to those that have been used in 1-d [5], should be able to master this problem. However, with regard to the second equation it is essential for this class of solution procedures to guarantee a uniform bound $\theta \leq \theta_{\max} < 1$. Violation leads to a singular perturbed problem: For $\theta \to 1$ the hyperbolic two-phase flow system above degenerates into the incompressible stationary Stokes equations. Indeed, under the assumption $f(1) = 0$, i.e. if no polymer assembly or disassembly take place in the saturated domain $\Omega_1(t) := \{x \in \Omega(t) | \theta(t,x) = 1\}$, then there both equations, the hyperbolic (62) and the elliptic (64), change their type and merge into the simple incompressibility condition $\nabla \cdot \mathrm{v} = 0$. Recently, our Egyptian doctor student Mahmoud Mansour [29] could prove (by local singularity analysis and by using two different numerical simulation techniques) the existence of travelling front solutions $0 < \theta(x - ct) \leq 1$ for these 1-dimensional degenerate equations, with a uniquely determined minimal travelling speed c_* corresponding to the well-known minimal Fisher's diffusion speed; in fact, supposing a suitable behavior of the viscosity term $|\mathrm{M}_c(\theta)| \sim \theta^2$, for $\theta \to 0$ the hyperbolic two-phase flow system converges to a diffusion equation.

The last mentioned result is not only applicable to polymer dynamics, but also to the dynamics of interactively moving cells in the process of tissue formation (particularly for simulating wound healing). In contrast to the stochastic multi-particle approach pursued in Chapter 1, here – by assuming an already performed homogenization process – the more or less connected cells (eventually including extracellular matrix filaments connecting them) are modelled as the *complex phase*, whereas the surrounding *aqueous medium* is modelled as the other phase. Indeed, as a further generalization of the presented model class of two-phase flows, Sharon Lubkin [27] used a three-phase model to simulate the dynamical development of embryonal lung tissues, where two types of cells and the intercellular medium play their role. Moreover, I would like to emphasize and stimulate potential ideas to generalize the Voronoi-Delaunay concepts in order to derive such 'multi-phase' continuum equations for polymer networks and cell tissues directly from suitable interactive stochastic 'multi-particle' processes.

Thus, by taking a synopsis of Chapters 1 and 2 and realizing the various modelling derivations and simulation techniques that can lead to hyperbolic continuum equations of generalized Navier-Stokes type, we might express an increasing confidence towards a successful mathematical treatment of the rich spectrum of current and future complex problems offered by interdisciplinary research on interactive motion in biological systems.

Acknowledgements. The presented thoughts and formulations are the synergetic result of (since more than 15 years) ongoing scientific exchange between working groups in the Division of Theoretical Biology and the Institute of Applied Mathematics at the University of Bonn, mainly within the DFG special research program on 'Nonlinear Partial Differential Equations' (SFB 256). For Chapter 1 my particular thanks go to Martin Grothaus, Juri Kondratjev and Serge Albeverio for discussing the stochastic aspects of continuum limits, as well as to Hans-Wilhelm Alt for jointly looking at various analytic aspects, also for Chapter 2. The implementation of Voronoi-Delaunay methods into modelling of bird swarms and corresponding 2-dimensional simulations (including an extended statistical analysis) have been carefully and gratefully performed by Ralf Müller within his diploma theses. Chapter 2 essentially is a revision of a hitherto unpublished manuscript jointly written, over a longer time of fruitful collaboration and visit exchanges, with Micah Dembo in the Department of Biomedical Engineering at Boston University; special thanks to him for joint authorship with the bunches of good ideas and critical remarks.

Several authors of important work and publications related to both Chapters and partially mentioned in the list of references, who have been of grateful influence upon shaping my own modelling concepts, were active participants in a series of SFB-Workshops which I could organize together with colleagues and friends from the worldwide Mathematical Biology and Applied Mathematics community. There was the early Workshop in 1989 on 'Biological Motion' with Leah Edelstein-Keshet, Andreas Huth, Bob Tranquillo, beside others [1], a subsequent Workshop in 1995 on 'Cell and Tissue Motion' with Micah Dembo, Hans G. Othmer, Rich Dickinson [4] and a recent Workshop on 'Polymer and Cell Dynamics' again with Micah Dembo, then with Michael Griebel, Mark Chaplain, Dirk Drasdo, Eirikur Palsson and others [8]; to all of them my thankful acknowledgment as well as to Anke Thiedemann for being the effective and nice soul of the whole organization. Finally and most closely, I want to thank all the younger collaborators and students in our Theoretical Biology group but also my colleagues from mathematics and biosciences in Bonn, who helped to built and develop the lively mental and social nest that became the essential base for my own scientific work on modelling, simulation and analysis in biology; among these and in connection with the presented topic, I would particularly like to mention Jürgen Lenz, Beate Pfistner, Thomas Pohl, Guido Scholz, Volker Lendowski, Edith Geigant, Andreas Deutsch, Michael Wurzel, Ralf Müller, Arne Rüter, Dieter Felix, Boris Hinz, Till Bretschneider, Thorsten Libotte, Ulrike Schuldenzucker, Claudia Mulder and Mahmoud Mansour.

References

1. W. Alt, G. Hoffmann (eds.) (1990): Biological Motion. Lecture Notes in Biomath. vol. 89, Springer-Verlag, Berlin
2. W. Alt, O. Brosteanu, B. Hinz and H.W. Kaiser (1995): Patterns of spontaneous motility in videomicrographs of human epidermal keratinocytes (HEK). Biochemistry and Cell Biology, **73**, 441–459

3. W. Alt (1996): Biomechanics of actymyosin mediated motility of keratinocytes. Biophysics, **41**, 181–188
4. W. Alt, A. Deutsch, G. Dunn (eds.) (1997): Dynamics of Cell and Tissue Motion. Birkhäuser-Verlag, Basel, Boston, Berlin
5. W. Alt, M. Dembo (1999): Cytoplasm dynamics and cell motion: two-phase flow models. Math. Biosciences, **156**, 207–228
6. W. Alt (2002): One-dimensional dynamics of stochastic skeins: condensation waves and continuum limits. Submitted to J. Math. Biol.
7. W. Alt, T. Bretschneider, R. Müller (2002): Interactive movement, aggregation and swarm dynamics. In: W.Alt, M.Chaplain, M.Griebel, J.Lenz (eds.) Polymer and Cell Dynamics – Multiscale Modelling and Numerical Simulation. To appear in Birkhäuser-Verlag, Basel
8. W. Alt, M. Chaplain, M. Griebel, J. Lenz (eds.) (2002): Polymer and Cell Dynamics – Multiscale Modelling and Numerical Simulation. To appear in Birkhäuser-Verlag, Basel
9. D.C. Bottino (1998): Modeling viscoelastic networks and cell deformation in the context of the immersed boundary method. J. Comp. Phys., **147**, 86–113
10. D.C. Bottino (2001): Computer simulations of mechanochemical coupling in a deforming domain: Applications to cell motion. Preprint
11. T. Bretschneider (2002): Reinforcement of cytoskeleton-matrix bounds and tensiotaxis: a cell based model. In: A.Deutsch et al. (eds) Function and Regulation of Cellular Systems: Experiments and Models. To appear in Birkhäuser-Verlag, Basel
12. Z. Csahok, A. Czirok (1997): Hydrodynamics of bacterial motion. Physica A, **243**, 304–318
13. A. Czirok, T. Vicsek (2000): Collective behaviour of interacting self-propelled particles. J. Phys. A, **281**,17–29
14. M. Dembo, F.W. Harlow, W. Alt (1984): The biophysics of cell surface mobility. In: (A.D.Perelson, Ch.DelLisi, F.W.Wiegel eds.) *Cell Surface Dynamics, Concepts and Models*. Marcel Dekker, New York, Basel, 495–542
15. M. Dembo, F.W. Harlow (1986): Cell motion, contractile networks, and the physics of interpenetrating reactive flow. Biophys. J., **50**, 109–121
16. M. Dembo (1986): The mechanics of motility in dissociated cytoplasm. Biophys. J., **50**, 1165–1183
17. M. Dembo (1989): Field theories of the cytoplasm. Comments theor. Biol., **1**, 159
18. M. Dembo (1994): On free boundary problems and amoeboid motion. In: N. Akkas (ed) Biomechanics of Active Movement and Division of Cells. NATO ASI Ser. H84. Springer, Berlin, p. 231
19. M. Dembo, Y. Wang (1999): Stresses at the cell-to-substrate interface during locomotion of fibroblasts. Biophys. J., **76**, 2307–2316
20. D. Drasdo (2002): On selected individual-based approaches to the dynamics in multicellular systems. In: W.Alt, M.Chaplain, M.Griebel, J.Lenz (eds) Polymer and Cell Dynamics – Multiscale Modelling and Numerical Simulation. To appear in Birkhäuser-Verlag, Basel
21. D.A. Drew, L.A. Segel (1991): Averaged equations for two phase flow. Stud. Appl. Math., **50**, 205–231
22. W. Ebeling, F. Schweitzer (2001): Swarms of particle agents with harmonic interactions. Theory in Biosciences, **120**, 207–224
23. X. He, M. Dembo (1997): On the mechanics of the first cleavage division of the sea urchin egg. Exper. Cell Res., pp. 233–252

24. F.H. Heppner, U. Grenander (1990): A stochastic nonlinear model for coordinated bird flocks. In: S. Krasner (ed) The Ubiquity of Chaos. Washington, DC: American Association for the Advancement of Science.
25. H. Holden, B. Oksendal, J. Uboe, T. Zhang (1996): Stochastic Partial Differential Equations. A modeling white noise functional approach, Birkhäuser, Basel
26. V. Lendowski, A. Mogilner (1997): Origin of actin-induced locomotion of *Listeria*. In: W.Alt, A.Deutsch, G.Dunn (eds) Dynamics of Cell and Tissue Motion. Birkhäuser-Verlag, Basel, Boston, Berlin, 93–99
27. S. R. Lubkin, T. Jackson (2001): Multiphase Mechanics of Capsule Formation in Tumors. J. Biomech Eng, in press
28. T. Oliver, M. Dembo, K. Jacobson (1999): Seperation of propulsive and adhesive traction stresses in locomoting keratocytes. J. Cell Biology, **145**, 589–604
29. M. Mansour (2001): Singularity analysis of travelling wave solutions for degenerate diffusion and transport equations modelling cell motility. PhD thesis, Univ. Bonn
30. F.A. Meinicke, S.C. Potten, M. Loeffler (2001): Cell migration and organization in the intestinal crypt using a lattice-free model. Cell Prolif., **34**, 253–266
31. R. Müller (2002): Dynamik in stochastischen Vielteilchensystemen zur Modellierung von Vogelschwärmen. Diploma thesis Univ. Bonn
32. K. Oehlschläger (1989): On the derivation of reaction-diffusion equations as limit dynamics of systems of moderately interacting stochastic processes. Th. Rel. Fields, **82**, 565–589
33. K. Oehlschläger (1991) On the connection between Hamiltonian many-particle systems and the hydrodynamical equations. Arch. Rat. Mech. Anal., **115**, 297–310
34. S. Olla, S.R.S. Varadhan, H.T. Yau (1993): Hydrodynamical limit for a Hamiltonian system with weak noise. Comm. Math. Phys., **155**, 523–560
35. J. Toner, Y. Tu (1995): How birds fly together: Long range oder in a two-dimensional dynamical XY model. Phys. Rev. Lett. **75**, 4326–4329
36. J. Toner, Y. Tu (1998): Flocks, herds and schools: A quantitative theory of flocking. Phys. Rev. E, **58**, 4828–4858

On Peak and Periodic Solutions of an Integro-Differential Equation on S^1

Edith Geigant

Abteilung Theoretische Biologie, Universität Bonn, D-53105 Bonn, Germany
edith.geigant@uni-bonn.de

1 Orientational Models

For the dynamics of an elongated object its orientation is as important as its spatial position; think of a school of fish. Sometimes the orientational pattern alone attracts our attention. In our case it was the phenomenon of parallel bundling and orthogonal arrangement in networks of actin filaments. Actin is a filamentous (i.e. lengthy) protein which forms one of the main components of the cytoskeleton, giving the cell mechanical stability and the ability to crawl ([1]).

Orientational aggregation of filaments can be modeled by a stochastic jump process with interaction ([12]). Here we analyze the approximating mean field equation for orientational dynamics. Let $f(t,\theta)$ be the mean density distribution of filaments (e.g. actin) at time $t \geq 0$ over their orientation $\theta \in S^1 = \mathbb{R}/\mathbb{Z}$ (note our normalization $\mathrm{vol}S^1 = 1$). We consider the following integro-differential equation for the temporal evolution of f, using the notation $\dot{f}(\theta) = \partial_t f(.,\theta)$:

$$\dot{f}(\theta) = -\int_{S^1} W[f](\theta,\psi) f(\theta)\, d\psi + \int_{S^1} W[f](\psi,\theta) f(\psi)\, d\psi \tag{1}$$

where

$$W[f](\theta,\psi) = \int_{S^1} h(\phi-\theta)\, G_\sigma(\psi-\theta-v(\phi-\theta))\, f(\phi)\, d\phi\,.$$

For a filament with orientation θ the 'turning rate' $W[f](\theta,\psi)$ is the probability of jumping to location ψ, given the distribution f. So the first term in (1) (with a minus sign) accumulates all the filaments jumping away from orientation θ, and the second term (with a plus sign) accumulates all filaments jumping into orientation θ.

A filament at θ interacting with a filament at ϕ changes its orientation instantly to ψ, see Figure 1. Since the turning mechanism is invariant to rotations, ψ is chosen relative to θ and ϕ. The preferred value of ψ is the *optimal turning angle* $v(\phi-\theta)$. The variation around this angle is described by a periodic Gaussian $G_\sigma : S^1 \to \mathbb{R}_+$ with standard deviation $\sigma > 0$,

$$G_\sigma(\psi) = \frac{1}{\sqrt{2\pi}\sigma} \sum_{n\in\mathbb{Z}} e^{-\frac{1}{2}(\frac{\psi+n}{\sigma})^2},$$

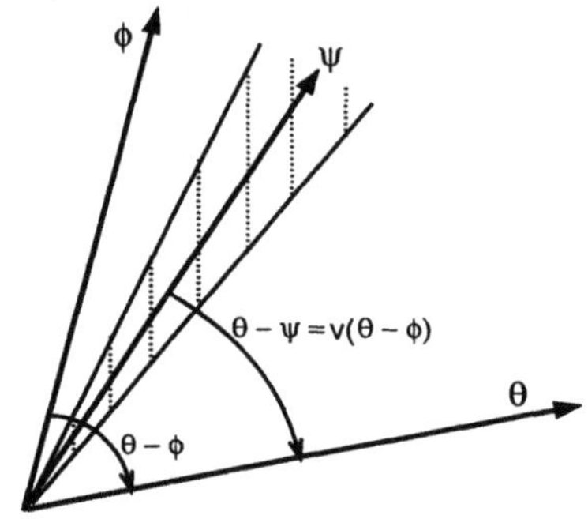

Fig. 1. Sketch of orienting filaments.

i.e. σ measures the *inaccuracy* of turning. The *interaction rate* $h : S^1 \to \mathbb{R}_+$ is an even and the optimal turning angle $v : S^1 \to S^1$ is an odd function of the interaction angle $\phi - \theta$, which follows from the assumption that the filaments cannot distinguish between left and right; they only can take into account angular differences. In mathematical terms: We have not only $SO(2)$ as symmetry group, i.e. solutions are invariant to rotations, but even $O(2)$, i.e. solutions may be also reflected.

The biological motivation is that orientational patterns of actin filaments are mediated by so-called actin binding proteins. Thus, h represents the binding rate of these proteins to actin, v describes the turning mechanism, e.g. favoring parallel alignment or orthogonal configuration, and G_σ models random deviations from the optimal resultant angle. The original non-periodic version of this equation was invented to model the distribution of dominance in insect colonies, [8]. It was applied to actin dynamics in [4]. Other models on actin dynamics use partial integro-differential equations, where the partial derivative is (orientational) diffusion and the integrals describe jumping and binding to filaments in favorable (relative) orientations, see [2] and [13] for modeling aspects, [9] and [10] for mathematical analysis (also in 3D), and [11] for a simplified spatio-angular model.

2 Bundles and Networks

The mass $\int f = \int_{S^1} f(\theta)\, d\theta$ is preserved. We rescale time and f to get $\int f = 1$ and $\int h = 1$. We assume that $h \in L^2(S^1)$ and v is measurable. Then there exist unique square-integrable solutions of (1), see [5]. Given a non-negative initial distribution, the solution is strictly positive for all times $t > 0$, because the deviation $G_\sigma > 0$ distributes filaments to any orientation (see [5] for a proof for continuous solutions).

Fourier transformation of equation (1) leads to an infinite system of ODEs (subscripts denote Fourier coefficients):

$$\dot{f}_l = \sum_{k\in\mathbb{Z}} w_{l,k}\, f_k f_{l-k} \tag{2}$$

with

$$w_{l,k} = -h_k + e^{-2\pi^2 l^2\sigma^2} \int_{S^1} h(\psi)\, \cos(2\pi(k\psi - lv(\psi)))\, d\psi.$$

The total mass is $f_0 = 1$. Because f is real-valued, we have $\bar{f}_l = f_{-l}$. Note that the inaccuracy σ occurs in the formula for $w_{l,k}$ only in the exponential factor in front of the integral.

Numerical algorithm. System (2) leads to a very fast numerical solution algorithm for the integro-differential equation (1). We cut the Fourier transformed system (2) at some fixed bound N for the Fourier indices:

$$\dot{f}_l = \sum_{\max(0,l)-N\le k\le N+\min(0,l)} w_{l,k}\, f_k f_{l-k} \quad \text{for} \quad -N \le l \le N,$$

and we increase N if for the higher Fourier coefficients f_l with $N < |l| \le 2N$ the growth $\dot{f}_l = \sum_{\max(0,l)-N\le k\le N+\min(0,l)} w_{l,k}\, f_k f_{l-k}$ is larger than some small error.

Eigenvalues of the linearization. From formula (2) we easily read off the eigenvalues of the linearization of (1) at the homogeneous solution $f = 1$:

$$\begin{aligned} c_l &= w_{l,0} + w_{0,l} \\ &= -1 - h_l + 2e^{-2\pi^2 l^2\sigma^2} \int h(\psi)\, \cos(\pi l\psi)\, \cos(2\pi l(v(\psi) - \tfrac{1}{2}\psi))\, d\psi, \end{aligned}$$

where we have used the trigonometric addition formula for the cosine. Since $c_l \to -1$ for $l \to \infty$, only finitely many eigenvalues can become unstable (i.e. positive). This holds true even for the limiting case $\sigma = 0$ if v is continuously differentiable and the points ψ such that $v'(\psi) = 0$ or $v'(\psi) = 1$ are isolated; because then the integral tends to 0 for $l \to \infty$.

Figures 2 shows stationary solutions of (1) where only first (c_1 and c_{-1}), second (c_2 and c_{-2}) and fourth eigenvalues (c_4 and c_{-4}) are unstable, respectively. Thus, Figure 2 shows (from left to right) unidirectional bundling (with $v(\psi) = \frac{1}{2}\psi, h = 1$), bidirectional bundling in the middle (with $v(\psi) = \frac{1}{2}\psi$ on $]-\frac{1}{4}, \frac{1}{4}[$, continued $\frac{1}{2}$-periodically, $h = 1$), and orthogonal arrangement (with $v(\psi) = -\frac{1}{8\pi}\sin(4\pi\psi)$ also $\frac{1}{2}$-periodic, but with $h(\psi) = \frac{1}{2}\left(G_\tau(\psi - \frac{1}{4}) + G_\tau(\psi + \frac{1}{4})\right)$, $\tau = 0.06$). The underlying optimal turning angle is sketched below each figure.

3 Hysteresis and Periodic Solutions

3.1 Codimension 1 Bifurcations

The examples above show that orientational aggregation is possible. However, if the inaccuracy is large, then inhomogeneities are smeared out by a diffusion-like

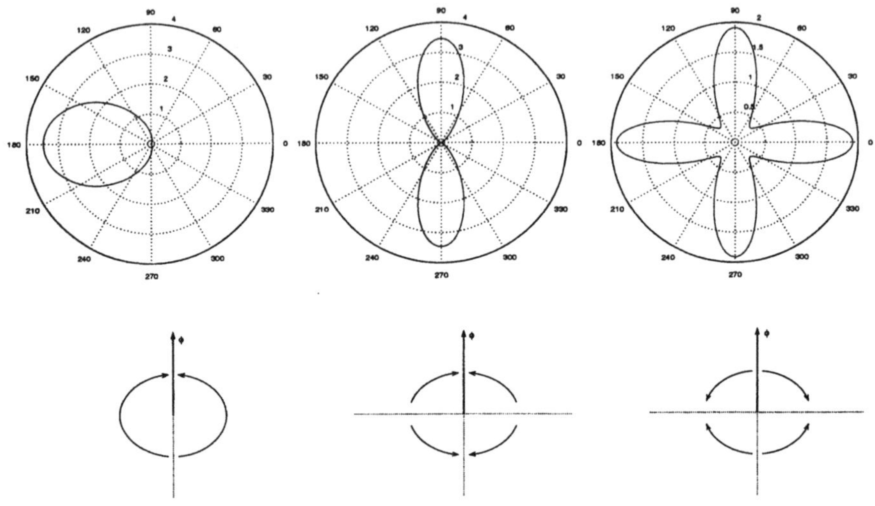

Fig. 2. Stationary solutions (upper row) and underlying optimal turning angle for equation (1). From left to right: a unidirectional bundle for $\sigma = 0.08$, a bidirectional bundle for $\sigma = 0.04$ and an orthogonal network for $\sigma = 0.02$ (parameter functions are given in the text). Depicted are polar plots in which the radius represents density $f(\theta)$.

mechanism, i.e. that for large times any solution converges to the constant solution $f = 1$ in L^2-norm. This is proved in [5] under some restricting assumptions on the interaction rate h; the statement holds e.g. for $h = 1$ or for periodic Gaussians.

It is now interesting to see how non-trivial solutions branch off the trivial solution, e.g. when the inaccuracy σ is decreased. If a single eigenvalue becomes unstable, then there is a pitchfork bifurcation which is either forward or backward. Examples for both types have been found, see [7] and Figure 3. Forward bifurcations are the rule, while backward bifurcations are rare. In the latter case a single (modulo rotation) unstable stationary solution branches off; in a secondary saddle-node-bifurcation a stable stationary solution appears, see Figure 3. The stable branch can be computed numerically (using the algorithm following equation (2)), the unstable branch in Figure 3 comes from a power series expansion based on a Lyapunov-Schmidt reduction, namely $f_l = \sum_{k\geq l} a_{l,k} f_1^k$ and $|f_1|^2 = \sum_{m\geq 1} b_m \lambda^m$, where $\lambda = \sigma - \sigma_{\text{crit}}$ is the bifurcation parameter. There is no obvious way to get it numerically — reversing time will not help, since there are infinitely many other directions in which this branch is stable.

In fact, the power series in the Lyapunov-Schmidt reduction converge, see [7]. One has to assume that the parameter functions h, v and σ depend analytically on a bifurcation parameter λ (e.g. $\sigma - \sigma_{\text{crit}}$), that $c_1(0) = 0$, $\frac{d}{d\lambda} c_1(0) = 0$, and that $|c_l(\lambda)| \geq \gamma > 0$ for $k \neq \pm 1$, λ small. Our proof that the series $f_l(f_1)$ $(l \geq 2)$ converge is based on the fact that the series coefficients can be calculated via an (explicitly given) recursion formula and can be bounded from above. This bound

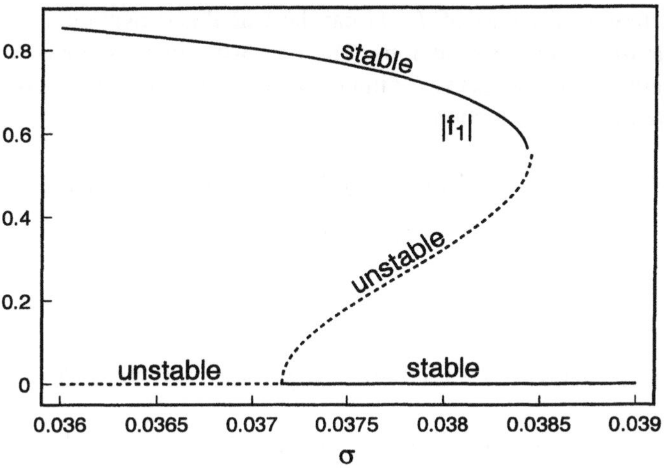

Fig. 3. Pitchfork bifurcation of an unstable branch from the constant solution and the secondary saddle-node-bifurcation. We use $v(\psi) = \frac{1}{2}\,\psi$ for $|\psi| < \frac{1}{6}$, $v(\psi) = 0$ otherwise, and $h(\psi) = 1$. For these values of σ higher eigenvalues than the first are unstable. Shown is the modulus of the first Fourier mode f_1.

yields a new system of fixed point equations. An Ansatz for solutions of the latter leads (unfortunately) not to solutions but to a second system of inequalities. Now, convergent solutions for these can be found and are upper bounds for the original series; the claim follows from this. The proof that the series $f_1(\lambda)$ converges works in a similar way but is even more technical.

3.2 Codimension 2 Bifurcations

The dynamics get more interesting when there are more eigenmodes that become unstable. If this happens at nearby points in parameter space, both bifurcations interact, and we get a great variety of different kinds of solution branches, see [3]. We assume that the first and second modes become unstable and perform a Lyapunov-Schmidt reduction. To do this we have to make some genericity assumptions similar to these for codimension one bifurcations (remember that c_l is the l-th eigenvalue of the linearization of (1) near the constant solution):

1.) h, v, σ depend analytically on two bifurcation parameters λ, μ.
2.) $c_1(0,0) = c_2(0,0) = 0$, and $\nabla_{\lambda\mu}\, c_1(0,0)$, $\nabla_{\lambda\mu}\, c_2(0,0)$ are independent.
3.) $|c_k(\lambda,\mu)| \geq \gamma > 0$ for $k \neq \pm 1, \pm 2$, (λ,μ) small.

We expand higher Fourier modes f_l (for $l \geq 3$) as power series in f_1, $\bar{f}_1$, f_2, $\bar{f}_2$ and the bifurcation parameters near the point where both eigenvalues vanish. Since solutions are invariant to turnings, the l-th series contains only terms of the form

$$f_1^{j_1}\,\bar{f}_1^{\,k_1}\,f_2^{j_2}\,\bar{f}_2^{\,k_2} \text{ where } j_1 - k_1 + 2j_2 - 2k_2 = l.$$

The first few coefficients of f_3, f_4 can be calculated explicitly when we plug the series into the stationary equations (2). Then, we substitute the power series back to get the reduced (complex) two-dimensional system of ODEs corresponding to the Lyapunov-Schmidt reduction:

$$\dot{f}_1 = c_1\, f_1 + \alpha_1\, \bar{f}_1 f_2 + \alpha_2\, f_1 |f_2|^2 + texthigherorderterms$$
$$\dot{f}_2 = c_2\, f_2 + \beta_1 f_1^2 + \beta_2\, |f_1|^2 f_2 + \beta_3\, f_2 |f_2|^2 + \text{higher order terms}$$

where

$$\alpha_1 = (w_{1,-1} + w_{1,2}), \quad \alpha_2 = -\frac{1}{c_3}\,(w_{1,-2} + w_{1,3})\,(w_{3,1} + w_{3,2}),$$
$$\beta_1 = w_{2,1}, \quad \beta_2 = -\frac{1}{c_3}\,(w_{2,-1} + w_{2,3})\,(w_{3,1} + w_{3,2}),$$
$$\beta_3 = -\frac{1}{c_4}\, w_{4,2}\,(w_{2,-2} + w_{2,4}).$$

From this reduced system, we can again get the bifurcating branches as (convergent) power series in the bifurcation parameters. In this instance, however, we have not really done the full power series expansions.

The dynamics near the codimension two bifurcation point is determined by the first few coefficients in the expansions of $\dot{f}_1$ and $\dot{f}_2$. The most important distinction is whether α_1 and β_1 have opposite signs or not, see [3]. An example for the latter case (same sign) is easy to find: one adds unipolar and bipolar bundling activity linearly. The results are biologically interesting: They show that transitions between unipolar and bipolar bundles are either via the homogeneous state (i.e. bundles dissolve and then aggregate anew) or via a mixed mode state which is neither unipolar nor bipolar (see Figure 4 for characteristics of a mixed mode solution although in a different model). Moreover, it becomes obvious that backward bifurcations (i.e. hysteresis) are a common phenomenon for non-constant interaction rate h (recall Figure 3 where $h = 1$ and the hysteresis effect is tiny).

If α_1, β_1 have opposite signs, there are much more different and interesting solutions. This is illustrated by the following example where $v(\psi) = \frac{1}{2}\psi$ for $\psi < \frac{1}{2}$ is attracting and $h(\psi) = \frac{1}{2}\,(G_\tau(\psi + 0.2) + G_\tau(\psi - 0.2))$ is maximal at interaction angles ± 0.2. At $\sigma \approx 0.103$ and $\tau \approx 0.033$ first and second eigenvalue are zero. Along with uni- and bipolar bundles and mixed mode solutions, we find traveling waves, modulated waves, oscillating bundles, and standing waves, see Figures 4 and 5. Because the equation is mirror symmetric, waves can move to the left (counter clockwise) or to the right (clockwise). Note that some of the patterns are clearly seen only for very large times (simulations have been started with some unstructured, randomly chosen distribution). In [7] a map of stable solutions near the codimension 2 bifurcation is given. It shows that oscillating bundles and traveling waves are the most common types of periodic solutions. One can also find a list of bifurcation types there.

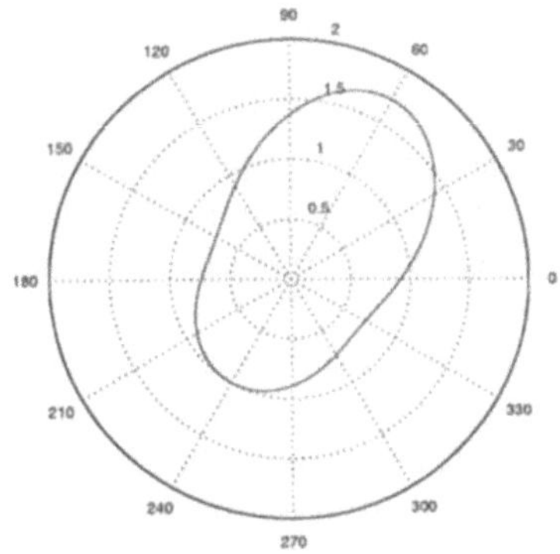

Fig. 4. A mixed mode solution for parameters $\sigma = 0.105$ and $\tau = 0.0315$.

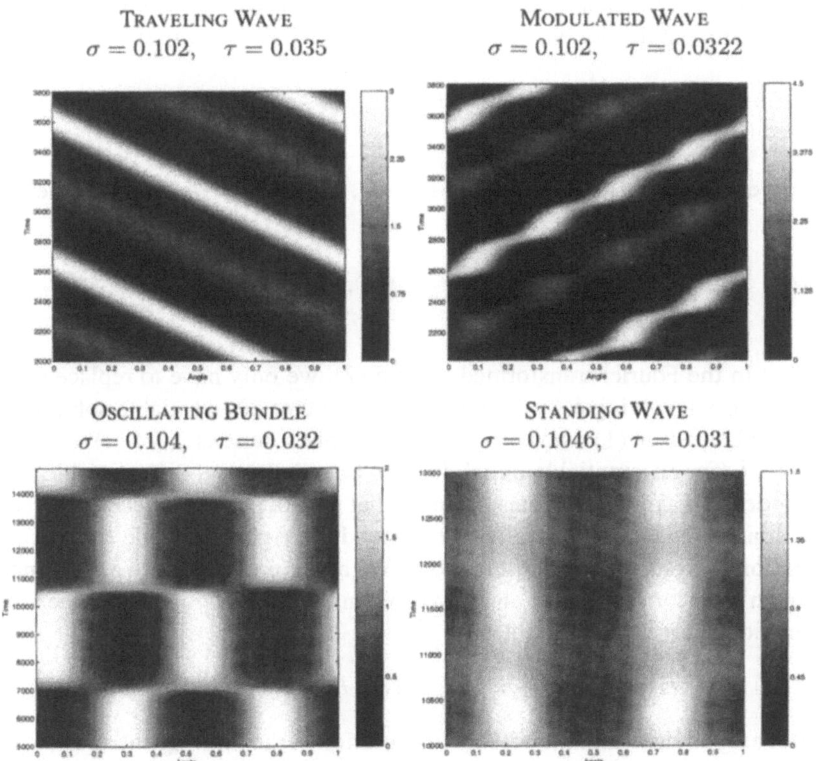

Fig. 5. Various kinds of periodic solutions near the codimension 2 bifurcation. Shown is angle θ versus time; the density is color coded, where the code is given in color bars on the right hand side. (cf. color plate 11, Page 671)

4 Peak Solutions

We call the optimal turning angle *attracting* if $0 < v(\psi) < \psi$ for $0 < \psi < \frac{1}{2}$, i.e. after turning the angle between two filaments is, on the average, narrower. If v is attracting, one can show that the first eigenvalue becomes positive for decreasing inaccuracy σ and one observes stable, unipolar stationary solutions (in some cases among other stationary solutions).

When σ tends to zero, these peaks become narrower. It is therefore natural to conjecture that in the limiting case $\sigma = 0$, there is a delta peak solution which is stable. For small σ, one can use an approximation by considering equation (1) on the real line instead of the circle (but still with functions f of unit mass). Then an exact Gaussian peak solution can be found that fits perfectly with the numerical data when σ is fairly small, see [4].

In order to approach the stability conjecture, we have to give sense to the limiting case. We write out the double integrals in (1) and perform some substitutions to get:

$$\dot{f}(\theta) = -(h * f)(\theta)\, f(\theta) + G_\sigma * \int_{S^1} h(\psi)\, f(\theta - v(\psi)) f(\theta + \psi - v(\psi))\, d\psi$$

Finally we replace G_σ by the delta distribution, i.e. we let $\sigma \to 0$, and get the limiting equation:

$$\dot{f}(\theta) = -(h * f)(\theta)\, f(\theta) + \int_{S^1} h(\psi)\, f(\theta - v(\psi)) f(\theta + \psi - v(\psi))\, d\psi. \qquad (3)$$

Note that in the Fourier transformed system (2), we only have to replace the exponential factors involving σ by 1. In [6] we prove that differentiable solutions f_σ of the original equation (1) converge to differentiable solutions f_0 of (3) uniformly on $[0, T] \times S^1$ (for differentiable initial density, bounded h and differentiable v).

In order to make sense of the statement that a delta peak is a steady state solution of the limiting equation (3), we have to extend the right hand side to measures (or distributions) $f \in \mathcal{D}'$ instead of functions. We apply the integral to a test function ϕ and obtain after some substitutions a two-fold integration of f against some modified test function (now in two variables):

$$\langle \dot{f}, \phi \rangle = -\langle f, (h * f)\, \varphi \rangle + \langle f(\theta), \langle f(\psi), h(\psi - \theta)\, \varphi(\theta + v(\psi - \theta)) \rangle_\psi \rangle_\theta.$$

Now it is easy to show that δ is a stationary solution of the limiting equation (3). Moreover, any delta peak $\delta_\theta := \delta(. - \theta)$ is stationary because equation (3) is equivariant under rotations (as in (1)).

Now we can look at the linearization of the limiting equation (3) at such a steady state peak solution, namely δ. This means that we perturb by small measures $\tilde{f}$, where 'small' means that we can safely ignore terms which are quadratic in $\tilde{f}$. There are more assumptions to make:

i) In order to stay in the space of measures of unit mass , $\tilde{f}$ must have mass zero ((3) is mass conserving as well as (1)).

ii) Since there is no result from interaction with filaments located exactly opposite of our peak ($v(\frac{1}{2}) = 0$ because of periodicity and oddness), we must only consider perturbations $\tilde{f}$ which have (compact) support contained in $]-\frac{1}{2}, \frac{1}{2}[$. In this case, we can use the open interval $]-\frac{1}{2}, \frac{1}{2}[$ instead of the circle as domain for our measures, and it makes sense to speak of the barycenter $\int_{-1/2}^{1/2} \tilde{f}(\theta)\, d\theta$ of a measure.
iii) It turns out that the linearized equation not only preserves mass but also the barycenter. This comes from the fact that interaction is symmetric — the probability of jumping from θ to ψ because of interaction with a filament at ϕ is the same as that of jumping from ϕ to $\phi - (\psi - \theta)$ because of interaction with a filament at θ. Hence our peak cannot be stable against perturbation with a measure of non-vanishing barycenter.

Since all shifted peaks must also be steady state solutions, the peak cannot be stable under all possible perturbations. Perturbation with a measure that has some non-zero barycenter will result in the peak being shifted to this barycenter. As it turns out, these three conditions are also sufficient for stability:

Theorem 4.1 *If the interaction is attracting, i.e. $0 < v(\psi) < \psi$ for $0 < \psi < \frac{1}{2}$, then the delta peak is a stationary solution of (3) and it is stable under perturbation by measures $\tilde{f}$ such that*

i) $\tilde{f}$ has compact support in $]-\frac{1}{2}, \frac{1}{2}[$,
ii) $\tilde{f}$ has zero mass, $\int_{-1/2}^{1/2} \tilde{f}(\psi)\, d\psi = 0$,
iii) $\tilde{f}$ has vanishing barycenter, $\int_{-1/2}^{1/2} \psi \tilde{f}(\psi)\, d\psi = 0$.

The Theorem is proved by writing down an explicit series development of the solution of the linearized equation. Some fairly involved manipulations then lead to the conclusion that the solution tends to zero, see [6]. Note that this stability result holds independently of in/stability of higher modes than the first.

5 Higher Nonlinearities

One might think that aggregation is strong if filaments turn to the orientation of their interaction partner, i.e. if $v(\psi) = \psi$. Actually, this is wrong: because both partners interact and turn, all solutions converge for large times to the homogeneous distribution (this is the same argument as on page 471; see also Figure 1). In fact, the integro-differential equation (1) simplifies to $\dot{f} = -f + G_\sigma * f$, which is a diffusion-like equation, and it is not difficult to prove convergence to the constant, e.g. by using Fourier transformation. Here and in the following, we assume that the interaction rate is homogeneous, $h = 1$.

Now we introduce non-symmetric interaction, namely

$$W[f](\theta, \psi) = \alpha \int_{S^1} G_\sigma(\psi - \phi)\, p(f(\phi))\, d\phi, \tag{4}$$

where $\alpha > 0$ is the rate constant and $p : \mathbb{R}_+ \to \mathbb{R}_+$, describes some non-proportional interaction of the filament at orientation θ with filaments at orientation ϕ. We use a typical sigmoid function

$$p(f) = f^2/(1 + \beta f^2),$$

At low filament density, a single filament that turns must interact with two filaments, because $p(f) \approx f^2$. At high density, differences in filament distribution at different orientations cannot be sensed by the filament, because $p(f) \approx 1/\beta$ is constant.

We rescale such that $\alpha = \beta = 1$. The eigenvalues of the linearization near the homogeneous distribution, $f = M > 0$, are

$$c_l = -p(M) + Mp'(M)\exp(-2\pi^2 l^2 \sigma^2) = \tfrac{M^2}{1+M^2}\left(-1 + \tfrac{2}{1+M^2} e^{-2\pi^2 l^2 \sigma^2}\right).$$

This shows that aggregation is now mass dependent. Only sufficiently small mass, namely $M < 1$ (for $M = 1$, $-p + Mp'$ vanishes), and decreasing inaccuracy σ lead to instability. In the following we are interested in cases with very small inaccuracy where the number of unstable eigenvalues becomes large; indeed, for $\sigma = 0$ and $M < 1$, all eigenvalues are positive ($c_l > 0$ for $|l| \neq 0$).

Theorem 5.2 *Let $\sigma = 0$, i.e. $G_\sigma = \delta$. The stationary solutions of (1) with (4),*

$$\dot{f} = -\int_{-1/2}^{1/2} p(f(\phi))\, d\phi\, f + M\, p(f),$$

are exactly the step functions

$$f = 0 \cdot \chi_0 + f_1 \cdot \chi_1 + f_2 \cdot \chi_2,$$

where χ_j are characteristic functions with volumes V_j, $V_0 + V_1 + V_2 = 1$,

$$\frac{p(f_1)}{f_1} = \frac{p(f_2)}{f_2} \quad \textit{and} \quad f_1 V_1 + f_2 V_2 = M.$$

Stationary solutions are unstable if $0 < f_1 < f_2$ or $f_1 = 0, f_2 = M/V_2 < 1$. Stationary solutions are stable (in the space of step functions) if $f_1 = 0$ and $f_2 = M/V_2 > 1$.

For a proof, we note that the function $P : 0 \leq f \mapsto p(f)/f$ is monotonically increasing for $f < 1$ and decreasing for $f > 1$. Any horizontal line, $y = p_0 \in]0, \max P[$, crosses the graph of P two times, namely in f_1 and f_2; moreover, the derivative $\dot{f}$ is negative below that line and positive above that line. Note that stability is a complicated concept because these stationary solutions depend only on the volumes V_j; the steps need not be connected. We use the fact that for any fixed partition the space of step functions is invariant. Then the integro-differential equation can be reduced to a finite system of ordinary differential equations. Stability holds for any such system with appropriate, i.e. fine enough, steps.

For small inaccuracy σ stationary solutions (in numerical simulations) consist of areas with equal density which are bounded by steep (continuous) descents to 0. To approximate these transitions layers we use a Kramers–Moyal–approximation of equation (1) with (4):

$$\dot{f} = -\int_{-1/2}^{1/2} p(f(\phi))\, d\phi\, f + M\, p(f) + \tfrac{M}{2}\, \sigma^2\, p(f)'',$$

which results from the fact that for small deviation the periodic Gaussian can be approximated on $]-1/2, 1/2[$ by $G_\sigma \approx \delta + \frac{\sigma^2}{2}\delta''$. The substitution $y = p(f)$ and rescaling with $x = \frac{\sigma^2}{2}\theta$ lead to a second order ordinary differential equation:

$$y'' = c\sqrt{\frac{y}{1-y}} - y \qquad \text{where} \qquad c = \frac{1}{M}\int_{-1/2}^{1/2} p(f(\phi))\, d\phi.$$

To find the heteroclinic orbit with $y'(0) = y'(\infty) = 0$ and $y(0) = 0$ we use the energy functional $E(y, y') = \frac{1}{2}(y^2 + y'^2) - c\int_0^y \sqrt{\eta/(1-\eta)}\, d\eta$. We get $f \approx 1.515$ for the density of aggregates and $V = M/f \approx M/1.515$ for the entire volume of the aggregate. These values and the (numerically calculated) transition layer coincide very well with the results of simulations.

Thus, for small inaccuracy the integro-differential equation is basically a reaction-diffusion-system in which the diffusion term comes from the deviation from the optimal orientation (v) and the reaction term is due to interaction and turning. The diffusion not only smoothes the boundaries of homogeneous aggregates but also determines their density. In the original equation (1) numerical simulations lead to several peak-like aggregation centers when turnings take a filament almost to the interaction partner (e.g. $v(\psi) = (1-\kappa)\psi$ with $0 < \kappa \ll 1$) and $\sigma \ll 1$, see [5]. Similarly, non-symmetric and non-saturating interaction, $p(f) = f^2$, leads to peak formation at small inaccuracy. With saturation, these peaks are flattened out to aggregation areas with homogeneous density in the interior. The basic reason is that the turning rate $W[f](\theta, .)$ is (more or less) constant toward all orientations where filament density is high enough.

6 Discussion

We have presented a fairly complete analysis of a jump process with a linear jump rate on the circle. Bifurcation analysis, based on the invariance to turning (i.e. $O(2)$-symmetry), is used to analyze the emergence of aggregates (i.e. non-homogeneous solutions), to find hysteresis (i.e. backward bifurcations), and to construct all kinds of time-periodic solutions. The existence of the latter was by no means obvious. Bifurcation analysis as well as linear stability analysis of the constant solution are based on spectral decomposition. Thus, they are restricted to systems with $O(2)$-symmetry. This is not the case for the linear stability analysis of a peak which was performed in Section 4. This method is more generally applicable because there are methods for solving linear equations, e.g. using semigroup theory. In the final section we have given an example of a jump process with a sigmoid jump rate. This leads to step function-like solutions whose stability can be clarified and which can be approximated by reducing the integro-differential equation to a partial differential equation. Thus it has become quite clear that these integro-differential equations share similarities with ordinary differential equations — many methods are applicable to both kinds of equations — as well as with partial differential equations.

References

1. Alberts, B., Bray, D., Lewis, J., Raff, M., Roberts, K., Watson, J.D. (1994): Molecular Biology of the Cell. Third Edition. Garland Publishing Inc., New York and London
2. Civelekoglu, G., Edelstein-Keshet, L. (1994): Modeling the dynamics of F-actin in the cell. Bull. Math. Biol., **56**, 587–616
3. Dangelmayr, G. (1986): Steady-state mode interaction in the presence of $O(2)$–symmetry. Dyn. Stab. Sys., **1**, 159–185
4. Geigant, E., Ladizhansky, K., Mogilner, A. (1999): An integro–differential model for orientational distributions of F-actin in cells. SIAM J. Appl. Math., **59**, 787–809
5. Geigant, E. (1999): Nichtlineare Integro–Differential–Gleichungen zur Modellierung interaktiver Musterbildungsprozesse auf S^1. PhD Thesis, Rheinische Friedrich-Wilhelms-Universität, Bonn
6. Geigant, E. (2000): Stability of a peak solution to an orientational aggregation model. In: Fiedler, B., Gröger, K., Sprekels, J.: Differential Equations. World Scientific, Singapore
7. Geigant, E., Stoll, M. (2002): Bifurcation analysis of an orientational aggregation model. Submitted to J. Math. Biol.
8. Jäger, E., Segel, L. (1992): On the distribution of dominance in populations of social organisms. SIAM J. Appl. Math., **29**, 33–58
9. Mogilner, A., Edelstein-Keshet, L. (1995): Selecting a common direction I: How orientational order can arise from simple contact responses between interacting cells. J. Math. Biol., **33**, 619–660
10. Mogilner, A., Edelstein-Keshet, L., Ermentrout, G.B. (1996): Selecting a common direction II: Peak-like solutions representing total alignment of cell clusters. J. Math. Biol., **34**, 811–842
11. Mogilner, A., Edelstein-Keshet, L. (1996): Spatio–angular order in populations of self-aligning objects: Formation of oriented patches. Physica D, **89**, 346–367
12. Othmer, H., Dunbar, S., Alt W. (1988): Models of dispersal in biological systems. J. Math. Biol., **26**, 263–298
13. Spiros, A., Edelstein-Keshet, L. (1998): Testing a model for the dynamics of actin structures with biological parameter values. Bull. Math. Biol., **60**, 275–305

Symmetrizing Measures for Infinite Dimensional Diffusions: An Analytic Approach

Sergio Albeverio[1], Yuri Kondratiev[2] and Michael Röckner[3]

[1] Institut für Angewandte Mathematik, Universität Bonn, 53115 Bonn, Germany; Research Center BiBoS, Universität Bielefefeld, 33615 Bielefeld, Germany
albeverio@uni-bonn.de

[2] Fakultät für Mathematik, Universität Bielefefeld, 33615 Bielefeld, Germany; Research Center BiBoS, Universität Bielefefeld, 33615 Bielefeld, Germany; Institute of Mathematics, Kiev, Ukraine
kondrat@mathematik.uni-bielefeld.de

[3] Fakultät für Mathematik, Universität Bielefefeld, 33615 Bielefeld, Germany; Research Center BiBoS, Universität Bielefefeld, 33615 Bielefeld, Germany
roeckner@mathematik.uni-bielefeld.de

Summary. The existence problem, characterizations and uniform a priori estimates for symmetrizing measures of lattice stochastic dynamics are discussed. A constructive criterion for the uniqueness of such measures is proven.

1 Introduction

The study of stochastic differential equations (SDE) for processes taking values in infinite dimensional spaces is not only an interesting generalization of the classical theory of SDE, but also has been strongly motivated by applications in Physics, Biology, Economics and Engineering. The theory of such stochastic processes (considered also as infinite dimensional diffusions) presents many new specific aspects as compared with the finite dimensional case. We refer the reader to the paper [5] for a review which contains more than 150 references to works discussing such stochastic processes. The connections between finite dimensional diffusions and certain aspects of the theory of partial differential equations (PDE) provide, in particular, very powerful analytic techniques in the study of the properties of these processes. However, most of the methods of the finite dimensional theory of PDE do not admit a direct extension to the infinite dimensional case. In this connection, it is very important to develop analytic techniques which are applicable also in the infinite dimensional situation. In this work we describe one of such methods. More precisely, we shall study the problem of existence and uniqueness of symmetrizing measures for infinite dimensional diffusions. In fact, our approach will be entirely analytic. The underlying SDE is only motivation and the reader is not expected to know anything about stochastic equations. The reason is that the measures in question are exactly the symmetrizing measures for the generator of the diffusion which is a second order partial differential operator in infinitely many variables.

The class of processes we consider corresponds to infinite systems of coupled SDE's. To explain the ideas, we deal with the simplest case of linear interactions between components in the drift term and constant diffusion coefficients. The gradient structure of the drift term is also essential for us. Then the symmetrizing measures we are looking for can be characterized by their Radon-Nikodym derivatives w.r.t. local shifts [2], but in applications an infinitesimal form of this characterization via an integration by parts formula is more useful. In Chapter 2 we discuss this characterization (see Proposition 2.3 below) which describes the symmetrizing measures as the solutions to a first order, but infinite system of PDE's. This gives rise to apply ideas from PDE's to construct these solutions via a priori estimates and tightness criteria, particularly suitable for measures. E.g., one can prove a priori estimates on the moments of symmetrizing measures which we discuss briefly in Section 2 (cf. Theorem 2.5). There exists an alternative constructive description of symmetrizing measures as Gibbs measures. This concept comes from statistical physics (see, e.g., [12]). Together with the mentioned a priori bounds it gives a possibility to show the existence of symmetrizing measures. In [4] this approach is realized in very general cases.

Another practical consequence of the Gibbs characterization is related to the uniqueness problem for symmetrizing measures. This problem is discussed in Section 3. We show there how the uniqueness conditions can be reduced to spectral gap estimates for Dirichlet operators associated with one-point conditional Gibbs measures. As a consequence we prove an easy to check criterion for the uniqueness of symmetrizing measures expressed in terms of parameters of the considered system.

We emphasize that, while Chapter 2 is is a review of our previous results on existence in [4] (formulated for pedagogical reasons in a concrete special case), the results about uniqueness of symmetrizing measures presented in Chapter 3 are new for the class of models considered in this paper.

Finally we would like to point out some important analytic consequences of our results for the study of partial differential operators in infinitely many variables. As mentioned above the generator L of a diffusion of the type studied in this paper is a second order elliptic partial differential operator which can be written as the sum of an infinite dimensional Laplacian and a vector field (called "drift") which is an infinite dimensional (generalized) gradient (cf. (9) below for the precise definition). Furthermore, L with the smooth cylinder functions as its domain is symmetric on $L^2(S', \mu)$, where S' is the infinite dimensional state space of the diffusion process and μ its symmetrizing measure (discussed above). In the analysis of L, μ now takes the role of the Lebesgue measure on $\mathbb{R}^d$ and one can consider corresponding Sobolev spaces $W^{r,p}(S', \mu)$. Though in contrast to the finite dimensional case functions belonging to all such Sobolev spaces are not smooth or not even continuous, in a number of cases regularity of solutions to the generalized heat equation corresponding to L or regularity properties of its resolvent can still be proved on the bases of suitable finite dimensional approximations and making full use of classical regularity theory on $\mathbb{R}^d$. We refer to [4] and for more recent work on this question to [9] and [14]. The latter treat the case where the above mentioned measure does not even have to be symmetrizing for L, but just has to be L-coharmonic, i.e. has to solve the equation $L^*\mu = 0$.

2 Symmetrizing Measures for Stochastic Gradient Systems

We shall consider anharmonic systems on the integer lattice $\mathbb{Z}^d$, $d \in \mathbb{N}$, with the Euclidean distance $|k-j|$, $k, j \in \mathbb{Z}^d \subset \mathbb{R}^d$. With any lattice point $k \in \mathbb{Z}^d$ there is associated a particle (oscillator) with the single spin space $\mathbb{R} \ni x_k$. In physical terminology the variable x_k describes the displacement of the particle from the equilibrium position at the point k and only for simplicity of notations it is assumed to be one-dimensional (which corresponds to having polarized oscillations). In fact, all our considerations below can be easily modified to the case of vector spins $x_k \in \mathbb{R}^l$, $l \leq d$. The configuration space of our system $\Omega = \mathbb{R}^{\mathbb{Z}^d}$ contains all real sequences $x = (x_k)_{k \in \mathbb{Z}^d}$ and is equipped with the product topology and with the corresponding Borel σ-algebra $\mathcal{B}(\Omega)$. The subset of finite sequences will be denoted by $\Omega_0 := \mathbb{R}_0^{\mathbb{Z}^d}$. In the following we will need also some topological subspaces of the configuration space Ω. Namely, for any $p \in \mathbb{Z}$ we define the Hilbert space

$$S_p := \left\{ x \in \Omega : \; |x|_p^2 = \sum_{k \in \mathbb{Z}^d} (1+|k|)^{2p} x_k^2 < \infty \right\}$$

and the mutually dual spaces

$$S := S(\mathbb{Z}^d) = \bigcap_{p \geq 1} S_p(\mathbb{Z}^d), \quad S' := S'(\mathbb{Z}^d) = \bigcup_{p \geq 1} S_{-p}(\mathbb{Z}^d)$$

of fastly decreasing resp. slowly increasing sequences over $\mathbb{Z}^d$. This gives a nuclear triplet

$$S(\mathbb{Z}^d) \subset l^2(\mathbb{Z}^d) \subset S'(\mathbb{Z}^d)$$

for the tangent Hilbert space $S_0 = l^2(\mathbb{Z}^d)$ with the scalar product $\langle x, x \rangle := (x, x)_0, x \in l^2(\mathbb{Z}^d)$. The duality between S and S' can be expressed in terms of the following scalar product:

$$\langle \varphi, x \rangle = \langle x, \varphi \rangle := \sum_{k \in \mathbb{Z}^d} \varphi_k x_k, \quad \varphi \in S, \, x \in S'. \tag{1}$$

We define $\mathcal{F}C_b^l\left(\mathbb{R}^{\mathbb{Z}^d}\right)$, $l \in \mathbb{N} \cup \{0, \infty\}$, as the set of all cylinder functions $f : \mathbb{R}^{\mathbb{Z}^d} \to \mathbb{R}$ which can be represented as

$$f(x) = f_N(x_{k_1}, \ldots, x_{k_N}), \quad x \in \mathbb{R}^{\mathbb{Z}^d}, \tag{2}$$

for some $N \in \mathbb{N}$, $\{k_1, ..., k_N\} \subset \mathbb{Z}^d$ and $f_N \in C_b^l(\mathbb{R}^N)$. Correspondingly, $\mathcal{F}C^l\left(\mathbb{R}^{\mathbb{Z}^d}\right)$ shall denote the space of cylinder functions with $f_N \in C^l(\mathbb{R}^N)$. For $f \in \mathcal{F}C_b^\infty\left(\mathbb{R}^{\mathbb{Z}^d}\right)$ we set $\nabla_k f := \frac{\partial f}{\partial x_k} \in \mathcal{F}C_b^\infty\left(\mathbb{R}^{\mathbb{Z}^d}\right)$. Then the gradient of f in a point $x \in \mathbb{R}^{\mathbb{Z}^d}$ is defined as the vector $\nabla f(x) := (\nabla_k f(x))_{k \in \mathbb{Z}^d} \in \mathbb{R}_0^{\mathbb{Z}^d}$.

For $p \in \mathbb{N}$, $C_b^1(S_{-p})$ will denote the set of all bounded Fréchet differentiable functions $f : S_{-p} \to \mathbb{R}$ with bounded continuous derivatives $f' : S_{-p} \to S_{+p}$ given by

$$\langle f'(x), h\rangle := \lim_{t\to 0} \frac{f(x+th) - f(x)}{t}, \qquad x, h \in S_{-p}.$$

As usual, $C_0^1(S_{-p}) \subset C_b^1(S_{-p})$ denotes the subset of functions with bounded support in S_{-p}.

Let us consider a system of anharmonic oscillators which can be described by means of the following heuristic potential energy functional

$$E(x) := \frac{J}{2} \sum_{\langle k,,j\rangle} (x_k - x_j)^2 + \sum_{k\in\mathbb{Z}^d} V(x_k). \tag{3}$$

Here $\sum_{\langle k,,j\rangle}$ is taken over all nearest-neighbor pairs $\langle k, j\rangle$ in $\mathbb{Z}^d$ and gives the potential energy of a quadratic interaction with the constant intensity $J > 0$. The self-interaction potential $V \in C^1(\mathbb{R})$ is assumed to satisfy the polynomial growth condition

$$|V(q)| \le C(1+|q|)^L, \quad |V'(q)| \le C(1+|q|)^{L-1}, \tag{4}$$

and the coercivity estimate

$$V'(q) \cdot q \ge A|q|^{2+\sigma} - B \tag{5}$$

(with some constants $A, B, C, \sigma > 0, L \ge 1$). *In all considerations below we assume that these conditions are satisfied without any additional references to them in the formulations of our statements.*

Note that the energy functional is well defined as a function on $\mathbb{R}_0^{\mathbb{Z}^d}$ but it has no meaning as a function on $S'(\mathbb{Z}^d)$. On the other hand, for any $k \in \mathbb{Z}^d$ we have heuristically

$$\nabla_k E(x) = J \sum_{j:|k-j|=1} (x_k - x_j) + V'(x_k), \nabla_k E \in \mathcal{F}C\left(\mathbb{R}^{\mathbb{Z}^d}\right) \tag{6}$$

with its right hand side being well-defined. Taking this as a definition of $\nabla_k E$, we conclude from (4) for the vector field $\nabla E := (\nabla_k E)_{k\in\mathbb{Z}^d}$ that

$$\nabla E \in C(S' \to S').$$

Let us consider the following infinite system of stochastic differential equations:

$$d\xi_k(t) = -\nabla_k E(\xi(t))\, dt + \sqrt{\frac{2}{\beta}} dw_k(t),$$
$$\xi_k(0) = x_k, \quad k \in \mathbb{Z}^d, t \ge 0. \tag{7}$$

Here $\xi(t) = (\xi_k(t))_{k\in\mathbb{Z}^d}$, $w(t) = (w_k(t))_{k\in\mathbb{Z}^d}$ is a family of independent standard one-dimensional Wiener processes, $\beta > 0$ denotes the inverse (absolute) temperature. This system of equations can be written in vector form as

$$d\xi(t) = -\nabla E(\xi(t))\,dt + \sqrt{\frac{2}{\beta}}dw(t), \xi(0) = x \in \Omega, \tag{8}$$

and describes an infinite dimensional dynamical system with gradient type vector field perturbed by Gaussian white noise with intensity depending on the temperature (so-called thermal noise).This system is called a stochastic gradient system, see e.g. [11], or a lattice Brownian motion with drift $b := -\nabla E$. The problem of existence and uniqueness of solutions for the system (7) has been intensively studied, for references and historical comments we refer the reader to [5]. In particular, for any $p \in \mathbb{N}$ and any given initial condition $x \in S_{-p}(\mathbb{Z}^d)$ there exists $p' \in \mathbb{N}$ s.t. (7) has a unique strong solution $\xi(t) = \xi^x(t) \in S_{-p'}(\mathbb{Z}^d), t \geq 0$, which is called the stochastic dynamics associated with our classical anharmonic lattice system.

A direct applications of the Ito's formula to the considered stochastic dynamics shows that the generator of the diffusion process $\xi(t)$ is the following infinite dimensional partial differential operator:

$$L_\beta f := \frac{1}{\beta}\sum_{k\in\mathbb{Z}^d} \Delta_k f + \sum_{k\in\mathbb{Z}^d} b_k \nabla_k f, \quad f \in \mathcal{F}C_b^\infty\left(\mathbb{R}^{\mathbb{Z}^d}\right), \tag{9}$$

where $\Delta_k f := \frac{\partial^2 f}{\partial x_k^2}$ and $b_k := -\nabla_k E$. We are interested in the study of symmetrizing measures μ for such stochastic dynamics, i.e., in measures for which the generator L_β become a symmetric operator in the corresponding $L^2(\mu)$-space. Considering such measures as initial distributions for the process we obtain time reversible infinite dimensional diffusions defined for all $t \in \mathbb{R}$. We first restrict the class of admissible measures.

Definition 2.1 *A probability measure μ on $(\Omega, \mathcal{B}(\Omega))$ is called tempered if*

$$\exists p = p(\mu) > d/2 : \qquad \mu\left(S_{-p}(\mathbb{Z}^d)\right) = 1.$$

We shall denote by $\mathcal{M}_t$ the set of all tempered probability measures on $(\Omega, \mathcal{B}(\Omega))$.

Definition 2.2 *Let $\mu \in \mathcal{M}_t$, $\mu\left(S_{-p}(\mathbb{Z}^d)\right) = 1$. The measure μ is called a symmetrizing measure for the generator L_β if*

$$\int_\Omega g \cdot L_\beta f\, d\mu = \int_\Omega f \cdot L_\beta g\, d\mu, \quad f, g \in C_0^2(S_{-p}). \tag{10}$$

Note that for any $f \in C_0^2(S_{-p})$ we have $L_\beta f \in C_0(S_{-p})$ (and, therefore, the integrals in (10) are well defined).

Now we can formulate a characterization result for symmetrizing measures. For the proof (even in the case where much more general interaction potentials in classical lattice systems are considered) we refer to [4].

Proposition 2.3 *Denote by $\mathcal{M}_t^b$ the set of all tempered probability measures on $(\Omega, \mathcal{B}(\Omega))$ such that the following integration by parts (IbP) formula*

$$\int_\Omega \partial_k f(x)\, d\mu(x) = -\beta \int_\Omega f(x) b_k(x)\, d\mu(x) \tag{11}$$

holds for all functions $f \in C_0^1(S_{-p})$ *and all* $k \in \mathbb{Z}^d$. *Then* $\mathcal{M}_t^b$ *coincides with the set of all symmetrizing measures for the generator* L_β.

Remark 2.4 *1. The (IbP)-formula (11) shows that the vector field* $\beta b(\cdot)$ *plays the role of the vector logarithmic derivative of the measure* μ.

2. The relations (11) can be considered as an infinite system of first order PDE for the unknown measures μ.

3. Different versions of Proposition 2.3 are known essentially from the beginning of the study of gradient dynamics in several particular cases. But in all of them, before the work [4], very restrictive a priori assumptions ensuring the integrability of the components b_k *of the logarithmic derivative were imposed on the solutions* μ, *see [4] for details and references.*

The (IbP)-characterization of symmetrizing measures gives a powerful tool for the analysis of such measures. Namely, one can show their existence by solving the first order system of PDE's given by (11). The first step is to prove a priori estimates, e.g. for the moments of the measure μ, as formulated in the following theorem proved in [4]. In fact, the proof of this theorem is based on a direct application of (11) to a proper class of functions $f \in C_0^1(S_{-p})$.

Theorem 2.5 (a priori estimates) *For any* $M \geq 1$ *there exists* $C_M > 0$ *s.t. for all* $k \in \mathbb{Z}^d$ *and all* $\mu \in \mathcal{M}_t^b$

$$\int |x_k|^M d\mu(x) \leq C_M. \tag{12}$$

As a consequence of the a priori estimates and (4) we have the following: every $\mu \in \mathcal{M}_t^b$ is supported by $\cap_{p>d/2} S_{-p}(\mathbb{Z}^d)$ and for any $M \geq 1$

$$\sup_{\mu \in \mathcal{M}_t^b} \sup_{k \in \mathbb{Z}^d} \int |b_k(x)|^M d\mu(x) < \infty. \tag{13}$$

The latter a priori integrability bound permits to extend the (IbP)-formula (11) to all $f \in \mathcal{F}C_b^\infty\left(\mathbb{R}^{\mathbb{Z}^d}\right)$ and to prove the symmetry of the generator L_β on the domain $\mathcal{F}C_b^\infty\left(\mathbb{R}^{\mathbb{Z}^d}\right)$ in $L^2(\mu)$ for any symmetrizing measure μ. Moreover, L_β is related with the classical Dirichlet form (see, e.g., [6]) corresponding to the measure μ from the (IbP)-formula it follows that

$$-(f, L_\beta g)_{L^2(\mu)} = \frac{1}{\beta} \int \langle \nabla f, \nabla g \rangle \, d\mu, \quad f, g \in \mathcal{F}C_b^\infty\left(\mathbb{R}^{\mathbb{Z}^d}\right). \tag{14}$$

As a consequence, we have a certain a priori information about symmetrizing measures for the stochastic dynamics we have considered, however, the question of the existence of such measures remains open. To solve this problem, we need a more constructive description of these measures.

For any finite subset $\Lambda \subset \mathbb{Z}^d$ (finite volume) we define a family of local specifications via stochastic kernels $\mu_\Lambda : \mathcal{B}(\Omega) \times \Omega \to [0,1]$ by the following formula, $\forall \Delta \in \mathcal{B}(\Omega) \quad \forall y \in \Omega$,

$$\mu_\Lambda(\Delta|y) := \frac{1}{Z_\Lambda(y)} \int\limits_{\mathbb{R}^\Lambda} \exp\{-\beta E_\Lambda(x_\Lambda \times y_{\Lambda^c})\} \mathbf{1}_\Delta(x_\Lambda \times y_{\Lambda^c}) \times_{k\in\Lambda} dx_k, \quad (15)$$

Here $x_\Lambda := (x_k)_{k\in\Lambda}$, $y_{\Lambda^c} := (y_j)_{j\in\Lambda^c}$, $\Lambda^c := \mathbb{Z}^d \setminus \Lambda$,

$$E_\Lambda(x_\Lambda \times y_{\Lambda^c}) := \frac{J}{2} \sum_{\{k,j\}\subset\Lambda} (x_k - x_j)^2 + \frac{J}{2} \sum_{k\in\Lambda, j\in\Lambda^c} (x_k - y_j)^2 + \sum_{k\in\Lambda} V_k(x_k) \quad (16)$$

is the energy in the volume Λ corresponding to the boundary condition y, and

$$Z_\Lambda(y) := \int\limits_{\mathbb{R}^\Lambda} \exp\{-\beta E_\Lambda(x_\Lambda \times y_{\Lambda^c})\} \times_{k\in\Lambda} dx_k \quad (17)$$

is the normalization factor (which is called partition function in the terminology of Statistical Physics). Because of our assumptions on the potentials, (17) is indeed well defined. Moreover, the family of stochastic kernels $\mu_\Lambda(\cdot|y)$ satisfies the consistency condition (cf. [12]):

$$\mu_{\Lambda'}\, \mu_\Lambda = \mu_{\Lambda'}, \quad \Lambda \subset \Lambda'. \quad (18)$$

Definition 2.6 *A probability measure μ on $(\Omega, \mathcal{B}(\Omega))$ is called a Gibbs measure for the local specification $\{\mu_\Lambda,\ \Lambda \subset \mathbb{Z}^d\}$ iff it satisfies the equation*

$$\forall \Lambda \subset \mathbb{Z}^d, \forall \Delta \in \mathcal{B}(\Omega) \quad \int\limits_\Omega \mu(dx)\mu_\Lambda(\Delta|x) = \mu(\Delta). \quad (19)$$

Equation (18) is called the Dobrushin-Lanford-Ruelle (DLR) equilibrium equation.

Let $\mathcal{G} = \mathcal{G}(J, V, \beta)$ denote the set of all Gibbs measures for our system. As before, we restrict our considerations to the set $\mathcal{G}_t = \mathcal{M}_t \cap \mathcal{G}$ of tempered Gibbs measures. The following proposition is a particular version of results from [1, 2], see also [4] for more general statements and historical comments.

Proposition 2.7 *We have*

$$\mathcal{M}_t^b = \mathcal{G}_t.$$

Proposition 2.7 asserts that the symmetrizing measures for the gradient stochastic dynamics can be described constructively as Gibbs measures. Combining this fact with the a priori bounds from Theorem 2.5 (which can be established in the same way for any measure $\mu_\Lambda(\cdot|x)$ of the specification) we can prove the following existence result, see [4].

Theorem 2.8 *There exists at least one symmetrizing measure, i.e.,*

$$\mathcal{M}_t^b = \mathcal{G}_t \neq \emptyset.$$

Previous proofs of existence of Gibbs measures for the case of unbounded spin systems have been based on a quite delicate analysis of the specifications for the given particular type of interactions, see, e.g. [7], [8]. The (IbP)-characterization leads to a (comparing with other methods) relatively easily proof for existence of Gibbs measures for a large class of interaction potentials, including the case of multiparticle interactions, see [4].

3 Uniqueness Problem

One of the specifically infinite dimensional effects in the considered stochastic dynamics is the possibility of having several symmetrizing measures. Due to Proposition 2.7 this phenomenon corresponds to what is called phase transition in Statistical Physics. In particular, if the lattice dimension $d \geq 3$ and the one-particle potential V has a symmetric two-wells shape, then for big enough β (low temperature regime) or J (strong coupling regime) our system has phase transitions, i.e., $|\mathcal{G}_t| > 1$. The latter follows by an application of the reflection positivity method and related infrared bounds to the periodic Gibbs state of the discussed model, see, e.g., [15] .

On the other hand, if the potential V is uniformly convex, then for all $J, \beta > 0$ we have $|\mathcal{G}_t| = 1$, see, e.g., [5]. Note that even for $V \equiv 0$ in the case where $d \geq 3$ one has $|\mathcal{G}_t| = \infty$ for all $J, \beta > 0$. Therefore, we should expect that the uniqueness conditions can be expressed in terms of the deviation of the one-particle potential from a convex one. To realize this observation, assume that the potential V has the form

$$V(q) = V_0(q) + W(q), \tag{20}$$

where V_0 is a uniformly convex function:

$$V_0 \in C^2(\mathbb{R}), \quad V''(q) \geq a^2 > 0, q \in \mathbb{R}, \tag{21}$$

which is together with its derivatives up to the second order polynomially bounded. The perturbation W is given by a bounded function

$$W \in C_b(\mathbb{R}). \tag{22}$$

We define the deviation $\delta(W)$ of the function W by

$$\delta(W) = \sup_{\mathbb{R}} W - \inf_{\mathbb{R}} W. \tag{23}$$

Theorem 3.9 *Suppose that the parameters of the considered system satisfy the inequality*

$$\frac{e^{\beta\delta(W)}}{2d + J^{-1}a^2} < \frac{1}{2d}. \tag{24}$$

Then there exists exactly one symmetrizing measure, i.e.,

$$|\mathcal{M}_t^b| = |\mathcal{G}_t| = 1.$$

Proof. For the proof of this fact, we shall use a modification of an approach developed in [3] for the derivation of a corresponding result for quantum lattice systems. The existence of a symmetrizing measure was already established in Section 1. To show uniqueness we will a well-known general uniqueness criterion by R.L. Dobrushin, see, e.g.,[10], [12] for a detailed discussion of this approach and related notions which we use below without giving additional references.

Let us consider the one-dimensional Gibbs distributions which correspond to the specification (15). Namely, for any $k \in \mathbb{Z}^d$ and any boundary condition $\xi \in \mathbb{R}^{\mathbb{Z}^d}$ we define a probability measure on $\mathcal{B}(\mathbb{R})$ as follows:

$$\nu_k\left(B|\xi\right) = \mu_{\{k\}}\left(\left\{x \in \mathbb{R}^{\mathbb{Z}^d} : x_k \in B\right\}|\xi\right), \quad B \in \mathcal{B}(\mathbb{R}).$$

Using (15) and (16) we have

$$\nu_k\left(dx_k|\xi\right) = \frac{1}{Z_k(\xi)} \exp\left\{-dJ\beta x_k^2 - \beta V\left(x_k\right) + J\beta x_k \sum_{|j-k|=1} \xi_j\right\} dx_k, \quad (25)$$

where as before $Z_k\left(\xi\right)$ is a normalizing factor. The Dobrushin approach is based on a comparison of these measures for different boundary conditions in the Kantorovich(–Ornstein–Rubinstein–Vasserstein) metric which is defined for two probability measures μ_1, μ_2 on $\mathcal{B}(\mathbb{R})$ by

$$R\left(\mu_1, \mu_2\right) := \sup_{f \in \mathrm{Lip}_1(\mathbb{R})} \left|\int f d\mu_1 - \int f d\mu_2\right|,$$

where

$$\mathrm{Lip}_1\left(\mathbb{R}\right) := \left\{f : \mathbb{R} \to \mathbb{R} \mid [f]_{\mathrm{Lip}} := \sup_{x \neq y} \frac{|f\left(x\right) - f\left(y\right)|}{|x-y|} \leq 1\right\}. \quad (26)$$

The Dobrushin matrix $C = \left(C_{kj}\right)_{k,j \in \mathbb{Z}^d}$ is defined by the relations

$$C_{kj} = \sup_{\substack{\xi,\eta \\ \forall i \neq j : \xi_i = \eta_i}} \left\{\frac{R\left(\nu_k\left(\cdot|\xi\right), \nu_k\left(\cdot|\eta\right)\right)}{|\xi_j - \eta_j|}\right\}. \quad (27)$$

In the considered case of translation invariant nearest-neighbors interactions on the lattice $\mathbb{Z}^d$, (25) implies $C_{kj} = 0$ if $|k-j| > 1$. Due to the Dobrushin criterion [12] the uniqueness of tempered Gibbs measures follows from the inequality

$$\sup_{k \in \mathbb{Z}^d} \sum_{j \in \mathbb{Z}^d, j \neq k} C_{kj} < 1 \quad (28)$$

and an obvious sufficient condition is then

$$C_{kj} < \frac{1}{2d}. \quad (29)$$

Note that due to (25) the measure $\nu_k(dx_k|\xi)$ depends on the boundary condition ξ only through the value $s_k = \sum_{|j-k|=1} \xi_j$.

Let us introduce a family of probability measures on $\mathcal{B}(\mathbb{R})$ parametrized by $s \in \mathbb{R}$ and given by

$$\nu^s(dx) := \frac{1}{Z_s} \exp\left\{-J\beta d x^2 - \beta V(x) + J\beta x s\right\} dx. \tag{30}$$

Then for all $k, j \in \mathbb{Z}^d$, $|k-j| = 1$ one has

$$C_{kj} = \sup_{\substack{s.t \\ s \neq t}} \left\{ \frac{R(\nu^s, \nu^t)}{|s-t|} \right\}. \tag{31}$$

Consider the mapping

$$\mathbb{R} \ni s \longmapsto \langle f \rangle_s := \int f(x)\, d\nu^s(x) \in \mathbb{R}$$

with fixed $f \in \mathrm{Lip}_1(\mathbb{R})$. Then

$$\frac{d}{ds} \langle f \rangle_s = J\beta \left(\langle f(\cdot) \cdot \rangle_s - \langle f(\cdot) \rangle_s \langle \cdot \rangle_s \right) = J\beta Cov_{\nu^s}(f(\cdot), \cdot), \tag{32}$$

and, therefore,

$$|\frac{d}{ds} \langle f \rangle_s| \leq J\beta \left(\mathrm{Var}_{\nu^s} f(\cdot)\right)^{\frac{1}{2}} \left(\mathrm{Var}_{\nu^s} \cdot\right)^{\frac{1}{2}}, \tag{33}$$

where

$$\mathrm{Var}_{\nu^s} f = \int \left(f - \langle f \rangle_s\right)^2 d\nu^s.$$

As will be shown separately below, this variance satisfies the following bound:

$$\mathrm{Var}_{\nu^s} f \leq C := \frac{e^{\beta\delta(W)}}{2J\beta d + a^2\beta}. \tag{34}$$

Hence

$$|\frac{d}{ds} \langle f \rangle_s| \leq J\beta C \tag{35}$$

and the mean-value theorem gives

$$\left| \int f d\nu^s - \int f d\nu^t \right| \leq J\beta C|s-t|, \quad f \in \mathrm{Lip}_1(\mathbb{R}), \tag{36}$$

or

$$R(\nu^s, \nu^t) \leq J\beta C|s-t|. \tag{37}$$

Using (30) and the assumption (24) of the theorem we obtain the estimate (28) and conclude then that $|\mathcal{G}_t| = 1$.

So, only the proof of the estimate (34) remains to be done. Let us consider the classical Dirichlet form corresponding to the probability measure ν^s:

$$\mathcal{E}_{\nu^s}(u,u) := \int |\nabla u|^2 d\nu^s, \quad u \in C_b^\infty(\mathbb{R}). \tag{38}$$

We rewrite the measure ν^s in the form

$$d\nu^s(x) = \frac{1}{N} e^{-\beta W(x)} d\nu_0^s(x), \tag{39}$$

where the probability measure ν_0^s is defined by formula (29) in which V is changed to V_0 :

$$\begin{aligned}\nu_0^s(dx) &:= \frac{1}{Z_s'} \exp\left\{-J\beta d x^2 - \beta V_0(x) + J\beta x s\right\} dx \\ &= \frac{1}{Z_s'} \exp\left\{-U_s(x)\right\} dx.\end{aligned} \tag{40}$$

Because $U_s''(x) \geq 2J\beta d + a^2\beta$ uniformly in x, $s \in \mathbb{R}$, we have the following spectral gap inequality, see, e.g., [13]:

$$\mathcal{E}_{\nu^s}(u,u) \geq \frac{2J\beta d + a^2\beta}{e^{\beta\delta(W)}} \|u\|_{L^2(\nu^s)}^2 \tag{41}$$

for all $u \in C_b^\infty(\mathbb{R})$ such that $\int u d\nu^s = 0$. After an obvious approximation the spectral gap inequality can be written as

$$\begin{aligned}\int \left(u - \langle u\rangle_{\nu^s}\right)^2 d\nu^s &= \frac{1}{2} \int\int \left(u(x) - u(y)\right)^2 d\nu^s(x)\, d\nu^s(y) \\ &\leq \frac{e^{\beta\delta(W)}}{2J\beta d + a^2\beta} \int |\nabla u|^2 d\nu^s\end{aligned} \tag{42}$$

for all $u \in C^1(\mathbb{R}) \cap \mathrm{Lip}(\mathbb{R})$. Thus

$$\mathrm{Var}_{\nu^s} u \leq \frac{e^{\beta\delta(W)}}{2J\beta d + a^2\beta} \sup |\nabla u|^2 = \frac{e^{\beta\delta(W)}}{2J\beta d + a^2\beta} [u]_{\mathrm{Lip}}^2. \tag{43}$$

Now let us take any $f \in \mathrm{Lip}(\mathbb{R})$. Then there exists a sequence $(u_n)_{n\geq 1} \subset C^1(\mathbb{R}) \cap \mathrm{Lip}(\mathbb{R})$ s.t.

$$[u_n]_{\mathrm{Lip}} \leq [f]_{\mathrm{Lip}}$$

and $u_n \to f, n \to \infty$, pointwise (to this end we can use a Friedrichs mollifier). Using such an approximation we obtain from (41) the desired inequality

$$\mathrm{Var}_{\nu^s} f = \int \left(f - \langle f\rangle_s\right)^2 d\nu^s \leq \frac{e^{\beta\delta(W)}}{2J\beta d + a^2\beta} [f]_{\mathrm{Lip}}^2. \tag{44}$$

The proof of the theorem is thus completed. □

We remark that the uniqueness criterion in Theorem 3.9 has a simple physical interpretation. Namely, we always can get to the "uniqueness regime" using high temperature or small coupling conditions.

References

1. Albeverio, S., Kondratiev, Yu.G., Röckner, M. (1997): Ergodicity of L^2-semigroups and extremality of Gibbs states. J. Funct. Anal., **144** (1997), 394-242
2. Albeverio, S., Kondratiev, Yu.G., Röckner, M. (1997): Ergodicity of the stochastic dynamics of quasi-invariant measures and applications to Gibbs states. J. Funct. Anal., **149**, 415–469
3. Albeverio, S., Kondratiev, Yu.G., Röckner, M., Tsikalenko, T.V. (1997): Uniqueness of Gibbs states for quantum lattice systems. Prob. Theory Related Fields, **108**, 193–218
4. Albeverio, S., Kondratiev, Yu.G., Röckner, M., Tsikalenko, T.V. (2000): A priori estimates on symmetrizing measures and their applications to Gibbs states. J. Funct. Anal., **171**, 366–400
5. Albeverio, S., Kondratiev, Yu.G., Röckner, M., Tsikalenko, T.V. (2001): Glauber dynamics for quantum lattice systems. Rev. Math. Phys., **13**, 51–124
6. Albeverio, S., Röckner, M. (1990): Dirichlet forms on topological vector spaces – Closability and a Cameron-Martin formula. J. Funct. Anal., **88**, 395–436
7. Bellissard, J., Høegh-Krohn, R. (1982): Compactness and the maximal Gibbs states for random Gibbs fields on a lattice. Comm. Math. Phys., **84**, 297–327
8. Cassandro, M., Olivieri, E., Pellegrinotti, A., Presutti, E. (1978): Existence and uniqueness of DLR measures for unbounded spin systems. Z. Wahrsch. verw. Gebiete, **41**, 313–334
9. Da Prato, G., Röckner, M. (2001): Singular dissipative stochastic equations in Hilbert space. BiBoS-Preprint 2001
10. Föllmer, H. (1982): A covariance estimate for Gibbs measures. J. Funct. Anal., **46**, 387–395
11. Fritz, J. (1982): Stationary measures of stochastic gradient systems, infinite lattice models. Z. Wahrsch. verw. Gebiete, **59**, 479–490
12. Georgii, H.-O. (1988): Gibbs Measures and Phase Transitions. Studies in Mathematics, 9, Walter de Gruyter, Berlin, New York
13. Gross, L. (1993): Logarithmic Sobolev inequalities and contractivity properties of semogroups. In Dirichlet Forms. Lect. Notes in Math., 1563, Springer, Berlin, New York
14. Röckner, M., Sobol, Z. (2002): Generalized stochastic Burgers equations: regularity theory for their Kolmogorov operators and applications. BiBoS-Preprint 2002
15. Sinai, Ya. (1982): Theory of Phase Trransitions. Rigorous Results. Pergamon Press, Oxford

Markov Semigroups and Harmonic Maps

Karl-Theodor Sturm

Institut für Angewandte Mathematik, Universität Bonn, D-53115 Bonn, Germany
sturm@uni-bonn.de

Summary. We present a semigroup approach to harmonic maps between metric spaces. Our basic assumption on the target space (N, d) is that it admits a "barycenter contraction", i.e. a contracting map which assigns to each probability measure q on N a point $b(q)$ in N. This includes all metric spaces with globally nonpositive curvature in the sense of Alexandrov as well as all Banach spaces.

The analytic input comes from the domain space (M, ρ) where we assume that we are given a Markov semigroup $(p_t)_{t>0}$. Typical examples come from elliptic or parabolic second order operators on $\mathbb{R}^n$, from Lévy type operators, from Laplacians on manifolds or on metric measure spaces, and from convolution operators on groups. In contrast to the work of [25], [26], [21], [22], [8], our semigroups are not required to be symmetric.

The linear semigroup acting e.g. on the space of bounded measurable functions $u : M \to \mathbb{R}$ gives rise to a nonlinear semigroup $(P_t^*)_t$ acting on certain classes of measurable maps $f : M \to N$. We will show that contraction and smoothing properties of the linear semigroup $(p_t)_t$ can be extended to the nonlinear semigroup $(P_t^*)_t$. Among others, we state existence, uniqueness and Lipschitz continuity of the solution to the Dirichlet problem for harmonic maps between metric spaces.

Introduction

A smooth map $f : M \to N$ between Riemannian manifolds is called harmonic iff its tension field $\tau(f) := \text{trace}\nabla(df)$ vanishes. Well known examples are harmonic functions ($N = \mathbb{R}$), geodesics ($M \subset \mathbb{R}$) and minimal surfaces. Harmonic maps play an important role in many areas of mathematics, see [5], [6] for a survey.

The first existence and regularity results for harmonic maps have been derived by [7], considering the parabolic equation $\frac{\partial}{\partial t} f(t, x) = \tau(f)(t, x)$ with given initial map $f(0, .)$ and then letting t go to ∞. An important assumption here is that the target N has nonpositive curvature. Otherwise, the solution may blow up, see e.g. [33], [34].

The elliptic approach, based on the fact that classical harmonic maps are critical values of the energy $E(f) := \frac{1}{2}\int_M \|df(x)\|^2 \text{vol}(dx)$, was initiated by [15], [16].

[17] characterized harmonic maps $f : M \to N$ by the fact that for each convex function φ, defined on some open $N_0 \subset N$, the function $\varphi \circ f$, defined on $f^{-1}(N_0) \subset M$, is subharmonic.

In the last decade, for several reasons it was found necessary also to study maps into more general target spaces, e.g. [14]. Ishihara's characterization indicates that any framework for such an extension will require an appropriate notion of subharmonic functions on the domain space and the notion of convex functions on the target space.

[25], [26] and [21], [22] independently began to develop a theory of harmonic maps into metric spaces of nonpositive curvature in the sense of Alexandrov (briefly: NPC spaces). These developments are based on the fact that a canonical extension of the energy functional can be defined for maps with values in NPC spaces. In the approach by KOREVAAR, SCHOEN, the domain space is still a Riemannian manifold. Generalizations to Lipschitz manifolds and Riemannian polyhedra are due to [12] and [8]. In JOST's approach, the domain space is a locally compact metric space with a Dirichlet form on it.

[22] succeeded to prove Hölder continuity of harmonic maps provided a scale invariant Poincaré inequality holds true on balls all the domain space. For the more specific case of Riemannian domain spaces, [25] could prove Lipschitz continuity of harmonic maps.

Our approach admits a more general class of target spaces than the class of NPC spaces. For instance, also Banach spaces and l_p-products of NPC spaces are included. We assume that the target space is a complete metric space (N, d) equipped with a map b which assigns to each probability measure p on N (with bounded support, say) a point $b(p) \in N$, called barycenter or center of mass. The intuitive meaning is that $b(p) = \int_N z\, p(dz)$. Indeed, for Banach spaces this may be used as a definition for b. For NPC spaces we may choose $b(p) := \text{argmin}_{y \in N} \int_N d^2(y, z)\, p(dz)$.

Instead of curvature conditions we require that

$$d(b(q_1), b(q_2)) \le \int_{N \times N} d(y_1, y_2)\, q(dy) \tag{1}$$

for each probability measure q on $N \times N$ with marginals q_1 and q_2. Our basic point of view is that curvature conditions (on domain as well as on target spaces) should be replaced by coupling properties.

Our domain space will be a metric space (M, ρ) with a semigroup of Markov kernels $p_t(x, dy)$ on it. In the classical case, it is given in terms of the heat kernel: $p_t(x, dy) = k_t(x, y)\text{vol}(dy)$. Other examples are derived from SDEs, from elliptic or subelliptic PDEs, from pseudodifferential operators as well as from operators on infinite dimensional spaces. Each Dirichlet form gives rise to such a Markov semigroup. However, we do not require that our semigroups are symmetric whereas in previous approaches symmetry is essential since everything is defined in terms of the energy.

As in the classical approach by [7], we first consider the solution to the parabolic problem. Given a map $f : M \to N$, we define its evolution after time t by

$$P_t^* f := \lim_{\delta_n \to 0} P_{\delta_n}^{\lfloor t/\delta_n \rfloor} f$$

(provided this limit exists for some sequence $(\delta_n)_n$) where

$$P_t f(x) := b\left(p_t(x, f^{-1}(.))\right)$$

denotes the barycenter of the push forward of the probability measure $p_t(x,.)$ under the map f. The intuitive meaning is that $P_t f(x) = \int_M f(y)\, p_t(x, dy)$.

Our main observation is that, under (1), regularizing properties of the linear semigroup $(p_t)_{t>0}$ carry over to the nonlinear semigroup $(P_t^*)_{t>0}$. For instance, if $\mathrm{dil} p_t u \le e^{-\kappa t}\mathrm{dil} u$ for all Lipschitz functions $u : M \to \mathbb{R}$ then $P_t^* f$ exists for all Lipschitz maps $f : M \to N$ and

$$\mathrm{dil} P_t^* f \le e^{-\kappa t}\mathrm{dil} f.$$

More involved assumptions on $(p_t)_{t>0}$ will imply that $P_t^* f$ exists for all bounded maps $f : M \to N$ and

$$\mathrm{dil} P_t^* f \le C_t \cdot \mathrm{osc} f.$$

We present many examples, including heat semigroups on manifolds and Alexandrov spaces, convolution semigroups on Lie groups, and Ornstein-Uhlenbeck semigroups on Wiener spaces.

Under weak assumptions, again on $(p_t)_{t>0}$, the maps $P_t^* f$ will converge as $t \to \infty$ to a map h with $P_t^* h = h$ ("invariance"), in particular, with

$$\lim_{t\to 0} \frac{1}{t} d(h, P_t^* h) = 0 \tag{2}$$

("harmonicity"). The solution to the Dirichlet problem on a set $D \subset M$ will be obtained in a similar manner, just replacing the original semigroup by the stopped semigroup $(p_{D,t})_{t>0}$ which preserves boundary data and, in the local case, leads to the same notion of harmonic maps.

In order to see the relation between our notion of harmonic maps and the classical one, let the target N be either a Riemannian manifold or a metric tree or a Banach space. Then (again under some minimal technical assumptions) a map $f : M \to N$ will be harmonic in the sense of (2) if and only if the function $\varphi \circ f$ is subharmonic (w.r.t. $(p_t)_t$) for each Lipschitz continuous convex function $\varphi : N \to \mathbb{R}$.

The discrete time version of our approach has been developed in [36, 37]. The relation to the energy approach in the symmetric case has been exploited in [35]. For more details and proofs of the results presented here, we refer to the forthcoming paper [38].

1 The Domain Space

Our domain space will be a measurable space $(M, \mathcal{M})$ with a given Markov semigroup $p = (p_t)_{t>0}$ on it.

That is, M is an arbitrary set, $\mathcal{M}$ is a σ-field on M and $p :]0, \infty[\times M \times \mathcal{M} \to [0, \infty]$ satisfies $\forall s, t > 0, \forall x \in M, \forall A \in \mathcal{M}$:

- $x \mapsto p_t(x, A)$ is a $\mathcal{M}$-measurable function on M;
- $A \mapsto p_t(x, A)$ is a probability measure on $(M, \mathcal{M})$;
- $p_{s+t}(x, A) = \int_M p_t(y, A) p_s(x, dy)$.

Occasionally, we require $(M, \mathcal{M})$ to be a *Radon measurable space*. All locally compact spaces with countable bases as well as all Polish spaces (= complete separable metric spaces) – equipped with their Borel σ-fields – are Radon measurable spaces.

Example 1.1 *Let M be a Riemannian manifold, $\mathcal{M}$ its Borel σ-field, m the Riemannian volume measure and $k :]0, \infty[\times M \times M \to [0, \infty]$ be the minimal heat kernel on M (= fundamental solution of $\frac{1}{2}\Delta - \frac{\partial}{\partial t}$). Then $p_t(x, dy) := k_t(x, y) m(dy)$ defines a Markov semigroup provided M is* stochastically complete, *i.e. provided $p_t(x, M) = 1$ for all $x \in M$ and some (hence all) $t > 0$.*
The latter is always satisfied if M is connected and complete and if the Ricci curvature of M is bounded from below or, more generally, if $Ric|_{B(r,x_0)} \geq C(r^2 + 1)$ or, even more generally, if $m(B_r(x_0)) \leq \exp[C(r^2 + 1)]$ (for some $x_0 \in M, C \in \mathbb{R}$ and all $r > 0$), cf. [13].

Example 1.2 *Let $M = \mathbb{R}^{d+1}$ equipped with its Borel σ-field $\mathcal{M}$ and let $k(s, x, t, dy)$ be the transition kernel for the parabolic partial differential equation*

$$\frac{\partial}{\partial s} u(x, s) = \sum_{i,j=1}^{d} a_{ij}(x, s) \frac{\partial^2}{\partial x_i \partial x_j} u(x, s) + \sum_{i=1}^{d} b_i(x, s) \frac{\partial}{\partial x_i} u(x, s)$$

on $\mathbb{R}^d$, where a_{ij} and b_i are bounded measurable functions on $\mathbb{R}^{d+1}$ and (a_{ij}) is locally uniformly elliptic, symmetric and continuous. Then $p_t((x, s), A) := \int 1_A((y, s+t)) k(s, x, s+t, dy)$ defines a Markov semigroup on $\mathbb{R}^{d+1}$. If the coefficients a_{ij} and b_i do not depend on time then $p_t(x, B) := k(0, x, t, B)$ defines a Markov semigroup on $\mathbb{R}^d$. See [32].

Similar results hold true for hypo- and subelliptic operators (cf. [11], [20]) as well as for certain pseudodifferential operators, for instance for $(-\Delta)^{\alpha/2}$ with $\alpha < 2$ which is covered by the next result.

Example 1.3 *Given a symmetric matrix $a \in \mathbb{R}^{d \times d}$, a vector $b \in \mathbb{R}^d$ and a measure μ on $\mathbb{R}^d$ satisfying $\int_{\mathbb{R}^d} \|y\|^2/(1+\|y\|^2)\mu(dy) < \infty$ there exists a unique convolution semigroup $(q_t)_{t>0}$ of probability measures on $\mathbb{R}^d$ such that $p_t(x, B) := q_t(B - x)$ defines the Markov semigroup for the Lévy operator*

$$\sum_{i,j=1}^{d} a_{ij} \frac{\partial^2}{\partial x_i \partial x_j} u(x) + \sum_{i=1}^{d} b_i \frac{\partial}{\partial x_i} u(x) + \int \left(u(x+y) - u(x) - \frac{y \cdot \nabla u(x)}{1 + \|y\|^2} \right) \mu(dy).$$

See e.g. [10], [18], [19], [39].

Lemma 1.4 *Each quasi-regular conservative Dirichlet form $(\mathcal{E}, \mathcal{D}(\mathcal{E}))$ on a σ-finite measure space $(M, \mathcal{M}, m)$ defines a Markov semigroup $(p_t)_{t>0}$ such that $\forall u \in L^2(M) \cap L^\infty(M)$ and for m-a.e. $x \in M$*

$$e^{ta}u(x) = \int_M u(y)p_t(x,dy). \tag{3}$$

Here a denotes the generator of $(\mathcal{E}, \mathcal{D}(\mathcal{E}))$. See [29].

Standard examples here are Dirichlet forms associated with elliptic differential operators in divergence form (with bounded measurable coefficients) on $\mathbb{R}^d$. Let us mention some non-classical examples of quasi-regular conservative Dirichlet forms:

- Dirichlet form on the Wiener space $C(\mathbb{R}_+, \mathbb{R}^n)$ and Ornstein-Uhlenbeck semigroup;
- Dirichlet forms on path or loop spaces $C(R_+, M)$ or $C(S^1, M)$, resp., over Riemannian manifolds.

The "quasi-regularity" of the Dirichlet form is not really essential here since in the sequel we only use p_t for $t \in \mathbb{T} := \{k \cdot 2^{-n} : k, n \in \mathbb{N}\}$ and for each conservative Dirichlet form on a Radon space $(M, \mathcal{M})$ there exists a Markov semigroup $(p_t)_{t\in\mathbb{T}}$ satisfying (3).

Lemma 1.5 *Each Markov process $(\Omega, \mathcal{A}, \mathbb{P}, X_t^x)$ with values in some measurable space $(M, \mathcal{M})$ defines a Markov semigroup on that space by*

$$p_t(x, A) := \mathbb{P}(X_t^x \in A). \tag{4}$$

If $(M, \mathcal{M})$ is a Radon measurable space then vice versa: each Markov semigroup on $(M, \mathcal{M})$ defines via (4) a Markov process (unique up to equivalence). See e.g. [2].

One of the main examples for such Markov processes are solutions of stochastic differential equations

$$dX_t^x = x + b(X_t^x)dt + \sigma(X_t^x)dW_t$$

on $\mathbb{R}^d$ where (W_t) denotes Brownian motion and $b : \mathbb{R}^d \to \mathbb{R}$ and $\sigma : \mathbb{R}^d \to \mathbb{R}^d$ are locally Lipschitz and bounded. Other remarkable examples are

- Super-Brownian motion on the space of measures on $\mathbb{R}^n$;
- Fleming-Viot processes on the space of probability measures on $\mathbb{R}^n$;
- Interacting particle systems as processes on the configuration space over $\mathbb{R}^n$.

In all the above mentioned examples, the Markov processes can be chosen to be *right Markov processes* which means that they have some additional minimal regularity properties.

Definition 1.6 *Let $(\Omega, \mathcal{A}, \mathbb{P}, X_t^x)$ be a right Markov process associated with a Markov semigroup $(p_t)_{t>0}$ on a Radon measurable space $(M, \mathcal{M})$. Then for each measurable subset $D \subset M$ the stopped semigroup $(p_{D,t})_{t>0}$ is the Markov semigroup on $(M, \mathcal{M})$ defined by*

$$p_{D,t}(x, A) := \mathbb{P}(X_{t\wedge\tau(D,x)}^x \in A)$$

where $\tau(D, x) := \inf\{t > 0 : X_t^x \notin D\}$ denotes the first exit time of D

Definition 1.7 *Given a Markov semigroup $(p_t)_{t>0}$ on a measurable space $(M, \mathcal{M})$ we define for each $t > 0$ the matching operator p_t^∇ acting on symmetric functions $\rho : M \times M \to \mathbb{R}_+$ by*

$$p_t^\nabla \rho(x_1, x_2) := \sup_u |p_t u(x_1) - p_t u(x_2)|$$

where the supremum is over all bounded measurable $u : M \to \mathbb{R}$ satisfying $|u(y_1) - u(y_2)| \le \rho(y_1, y_2)$ for all $y_1, y_2 \in M$. The coupling semigroup $(p_t^\diamond)_{t>0}$ acting on symmetric functions $\rho : M \times M \to \mathbb{R}_+$ is defined by

$$p_t^\diamond \rho(x_1, x_2) := \sup \left\{ p_{t_n}^\nabla \circ \ldots \circ p_{t_1}^\nabla \rho(x_1, x_2) : n \in \mathbb{N}, t_i > 0, \sum_{i=1}^n t_i = t \right\}.$$

Remark 1.8 *For each Markov semigroup $(p_t)_{t>0}$ on a measurable space $(M, \mathcal{M})$*

$$p_t^\diamond \rho(x_1, x_2) \le \inf \mathbb{E} \rho(X_t^{x_1}, X_t^{x_2})$$

where the infimum is over all Markov processes $(\Omega, \mathcal{A}, \mathbb{P}, X_t^x)$ associated with $(p_t)_{t>0}$. Cf. [30], [24].

2 The Target Space

Our target space will be a complete metric space (N, d) with a given barycenter contraction b on it.

We denote by $\mathcal{N}$ the Borel σ-field of N, and for each $\theta > 0$, by $\mathcal{P}^\theta(N)$ the set of all probability measures p on $(N, \mathcal{N})$ with separable support and with $\int_N d^\theta(z, x) p(dx) < \infty$ for some/all $z \in N$. Given two measures $p, q \in \mathcal{P}^1(N)$, a measure $\mu \in \mathcal{P}^1(N \times N)$ is called *coupling* of p and q iff

$$\mu(A \times N) = p(A), \ \mu(N \times A) = q(A). \qquad (\forall A \in \mathcal{N})$$

The L^1-*Wasserstein distance* or *Kantorovich-Rubinstein distance* of $p, q \in \mathcal{P}^1(N)$ is defined as

$$d^W(p, q) = \inf \left\{ \int_{N^2} d(x_1, x_2) \mu(dx) : \ \mu \in \mathcal{P}^1(N^2) \text{ is a coupling of } p \text{ and } q \right\}.$$

Definition 2.9 *A barycenter contraction is a map $b : \mathcal{P}^1(N) \to N$ such that*

- *$b(\delta_x) = x$ for all $x \in N$;*
- *$d(b(p), b(q)) \le d^W(p, q)$ for all $p, q \in \mathcal{P}^1(N)$.*

Remark 2.10 *If there exists a barycenter contraction on (N, d) then (N, d) is a geodesic space: For each pair of points $x_0, x_1 \in N$ we can define one geodesic $t \mapsto x_t$ connecting x_0 and x_1 by $x_t := b((1-t)\delta_{x_0} + t\delta_{x_1})$.*

Given any for points $x_0, x_1, y_0, y_1 \in N$, the function $t \mapsto d(x_t, y_t)$ is convex. In particular, the geodesic $t \mapsto x_t$ depends continuously on x_0 and x_1. However, it is not necessarily the only geodesic connecting x_0 and x_1.

If geodesics in N are unique then the existence of a barycenter contraction implies that $d : N \times N \to \mathbb{R}$ is convex. Thus N has (globally) "nonpositive curvature" in the sense of Busemann.

Example 2.11 *Let (N, d) be a complete metric space with (globally) "nonpositive curvature" in the sense of A.D. Alexandrov. Then for each $p \in \mathcal{P}^2(N)$ there exists a unique $b(p) \in N$ which minimizes the uniformly convex function*

$$z \mapsto \int_N d^2(z,x)p(dx)$$

on N. The map $b : \mathcal{P}^2(N) \to N$ extends to a barycenter contraction $\mathcal{P}^1(N) \to N$. See [35].
Equivalently, $b(p)$ can be defined via the law of large numbers as the unique accumulation point of the sequence

$$\frac{1}{n}\sum_{i=1}^{n} X_i(w)$$

for a.e. ω where $(X_i)_i$ is a sequence of independent random variables with distribution p. The point $\frac{1}{n}\sum_{i=1}^{n} X_i(w)$ is defined by induction on n as the point $\gamma_{1/n}$ on the geodesic from $\gamma_0 := \frac{1}{n-1}\sum_{i=1}^{n-1} X_i(w)$ to $\gamma_1 := X_n(w)$. See [36]. Examples of spaces with nonpositive curvature in the sense of A.D. Alexandrov are

- *complete, simply connected Riemannian manifolds with nonpositive sectional curvature;*
- *trees and, more generally, Euclidean Bruhat-Tits buildings;*
- *Hilbert spaces;*
- *L^2-spaces of maps into such spaces;*
- *Finite or infinite (weighted) products of such spaces;*
- *Gromov-Hausdorff limits of such spaces.*

See e.g. [1], [8], [21], [23], [25], [26].

Example 2.12 *Let N be a complete, simply connected Riemannian manifold and let d be a Riemannian distance.* Then (N, d) admits a barycenter contraction b if and only if N has nonpositive sectional curvature.
Indeed, if (N, d) admits a barycenter contraction then so does (N_0, d) for each closed convex $N_0 \subset N$. Hence, geodesics in N_0 are unique and thus $t \mapsto d(\gamma_t, \eta_t)$ is convex for any pair of geodesics γ and η in N_0. This implies that N has nonpositive curvature ([23]).
Conversely, if N has nonpositive curvature then it admits a barycenter contraction by the previous Example 2.11. It is worthwhile to mention that also the martingale mean defines a barycenter contraction (different from the previous one, cf. [40],

[41]). Here for each $p \in \mathcal{P}^1(N)$ the martingale mean $b(p)$ is defined to be the starting point Y_0 of a N-valued martingale $(Y_t)_{t\in[0,\infty]}$ whose terminal variable Y_∞ is distributed according to p.

Example 2.13 *Let (N, d) be a locally compact separable complete metric space with* negative curvature in the sense of Busemann. *Then [9] have constructed a barycenter contraction. For Riemannian manifolds, this is different from those in Examples 2.11 and 2.12, and also for trees, it is different from that in Example 2.11.*

Example 2.14 *Let $(N, \|.\|)$ be a (real or complex) Banach space and put $d(x, y) := \|x-y\|$. Then $\mathcal{P}^1(N)$ is the set of Radon measures p on N satisfying $\int_N \|x\|\, p(dx) < \infty$. For each $p \in \mathcal{P}^1(N)$, the identity $x \mapsto x$ on N is Bochner integrable and*

$$b(p) := \int_N x\, p(dx)$$

defines a barycenter contraction on (N, d). Cf. [28] for instance.

Lemma 2.15 *Let I be a countable set and for each $i \in I$, let (N_i, d_i) be a complete metric space with barycenter contraction b_i and "base" point $o_i \in N_i$. Given $\theta \in [1, \infty]$, define a complete metric space (N, d) with base point $o = (o_i)_{i\in I}$ by*

$$N := \left\{ x = (x_i)_{i\in I} \in \bigotimes_{i\in I} N_i : d(x, o) < \infty \right\}, \quad d(x, y) := \left[\sum_{i\in I} d_i^\theta(x_i, y_i) \right]^{\frac{1}{\theta}}$$

provided $\theta < \infty$ or by $d(x, y) = \sup_{i\in I} d_i(x_i, y_i)$ if $\theta = \infty$. One can define a barycenter contraction b on $\mathcal{P}^1(N)$ by

$$b(p) := (b_i(p_i))_{i\in I}$$

where $p_i \in \mathcal{P}^1(N_i)$ with $p_i : A \mapsto p(\{x = (x_j)_{j\in I} \in N : x_i \in A\})$ denotes the projection of $p \in \mathcal{P}^1(N)$ onto the i-th factor of N.

For instance, this applies to $N = \mathbb{R}^n, n \geq 2$ with the usual notion of barycenter but with "unusual" metric $d(x, y) = \sup\{|x_i - y_i| : i = 1, .., n\}$. In this case, geodesics are not unique, e.g. each curve $t \mapsto (t, \varphi_2(t), ..., \varphi_n(t))$ with $\varphi \in C^1(\mathbb{R}), \varphi_i(0) = \varphi_i(1) = 0$ and let $|\varphi_i'| \leq 1$ is a geodesic connecting $(0, 0, ..., 0)$ and $(1, 0, ..., 0)$.

3 The Nonlinear Heat Semigroup

Let $(M, \mathcal{M})$, $p = (p_t)_{t>0}$ and (N, d, b) be as in Chapters 1, 2 and let $\mathcal{L}(M, N, p)$ denote the set of all measurable maps $f : M \to N$ with separable ranges and with

$$\partial_t f(x) := \int_M d(f(x), f(y)) p_t(x, dy) < \infty$$

for all $t > 0$ and all $x \in M$. For each such f, t and x, the probability measure $p_t(x, f^{-1}(.))$ lies in $\mathcal{P}^1(N)$ and thus

$$P_t f(x) := b(p_t(x, f^{-1}(.)))$$

is well-defined.

Lemma 3.16 *For all $f, g \in \mathcal{L}(M, N, p)$, all $s, t > 0$, and all $x, y \in M$*

(i) $d(P_t f(x), P_t g(x)) \leq \int d(f(y), g(y)) p_t(x, dy)$;
(ii) $d(P_t f(x), f(x)) \leq \partial_t f(x)$;
(iii) $\partial_s(P_t f)(x) \leq \partial_{s+t} f(x) + \partial_t f(x)$;
(iv) $d(P_t f(x), P_t f(y)) \leq p_t^{\nabla} d_f(x, y)$ with p_t^{∇} from Definition 1.7 and d_f denoting the function $(x, y) \mapsto d(f(x), f(y))$ on $M \times M$;
(v) $P_t f \in \mathcal{L}(M, N, p)$.

The map $P_t : \mathcal{L}(M, N, p) \to \mathcal{L}(M, N, p)$ is called nonlinear Markov operator associated with the kernel p_t.

For the sequel, fix once for all a subsequence $(\delta_n)_{n \in \mathbb{N}}$ of $(2^{-n})_{n \in \mathbb{N}}$ and put $\mathbb{T} = \{k \cdot 2^{-n} : k, n \in \mathbb{N}\}$. Let $\mathcal{L}^*(M, N, p)$ denote the set of all $f \in \mathcal{L}(M, N, p)$ for which

$$P_t^* f(x) := \lim_{n \to \infty} (P_{\delta_n})^{t/\delta_n} f(x)$$

exists for all $t \in \mathbb{T}$ and all $x \in M$. Here $(P_\delta)^k$ denotes the k-th iteration of the nonlinear Markov operator P_δ.

Note that if $N = \mathbb{R}$ (equipped with the usual d and b) then $\mathcal{L}^*(M, \mathbb{R}, p) = \mathcal{L}(M, \mathbb{R}, p)$ is the set of all measurable $f : M \to \mathbb{R}$ with $\int |f(y)| p_t(x, dy) < \infty$ $(\forall t > 0, x \in M)$, and

$$P_t^* f(x) = P_t f(x) = \int_M f(y) p_t(x, dy).$$

Lemma 3.17 *For all $f, g \in \mathcal{L}^*(M, N, p)$, all $s, t \in \mathbb{T}$ and all $x, y \in M$*

(i) $d(P_t^* f(x), P_t^* g(x)) \leq \int d(f(y), g(y)) p_t(x, dy)$;
(ii) $d(P_t^* f(x), f(x) \leq \partial_t f(x))$ and $d(P_t^* f(x), P_{t+s}^* f(x)) \leq p_t(\partial_s f)(x)$;
(iii) $\partial_s(P_t^*)(x) \leq \partial_{s+t} f(x) + \partial_t f(x)$;
(iv) $d(P_t f(x), P_t f(y)) \leq p_t^{\diamond} d_f(x, y)$ with $p_t^{\diamond}$ from Definition 1.7 and d_f denoting the function $(x, y) \mapsto d(f(x), f(y))$ on $M \times M$;
(v) $P_t^* f \in \mathcal{L}^*(M, N, p)$ and $P_s^*(P_t^* f)(x) = P_{s+t}^* f(x)$.

The operator P_t^* on $\mathcal{L}^*(M, N, p)$ is called nonlinear heat operator associated with the "linear heat semigroup" $(p_t)_{t>0}$. The semigroup $(P_t^*)_{t \in \mathbb{T}}$ of operators on $\mathcal{L}^*(M, N, p)$ is called nonlinear heat semigroup.

Up to now, we have presented the pointwise approach to nonlinear Markov operators and nonlinear heat semigroups. In the rest of this Chapter, we will present the L_θ-approach. As before (N, d, b) will be a complete metric space with barycenter contraction and $(M, \mathcal{M})$ will be a measurable space with a Markov semigroup $(p_t)_{t>0}$

on it. In addition, we now fix a measure m on $(M, \mathcal{M})$ and a number $\theta \in [1, \infty]$ and we assume that $\exists C, \lambda \in \mathbb{R}$:

$$\|p_t u\|_\theta \leq C \cdot e^{-\lambda t} \cdot \|u\|_\theta \tag{5}$$

for all $u \in L_\theta(M)$ and $t > 0$. Here $L_\theta(M) = L_\theta(M, \mathcal{M}, m)$ denotes the Lebesgue space of (m-equivalence classes of) measurable functions $u : M \to \mathbb{R}$ with norm $\|u\|_\theta < \infty$.

For measurable $f, g : M \to N$ put $d_\theta(f, g) := \|d(f, g)\|_\theta$ where $d(f, g)$ denotes the function $x \mapsto d(f(x), g(x))$ on M. In particular, for $\theta < \infty$

$$d_\theta(f, g) = \left(\int_M d^\theta(f(x), g(x))\, m(dx) \right)^{1/\theta}.$$

Let $L_\theta(M, N, p)$ denote the set of (equivalence classes of) measurable $f : M \to N$ with separable ranges and with $\partial_t f \in L_\theta(M)$ for all $t > 0$. One easily verifies that $(L_\theta(M, N, p), d_\theta)$ is a complete metric space and each constant map lies in $\mathcal{L}_\theta(M, N, p)$. If $f \in \mathcal{L}_\theta(M, N)$ and $t > 0$ then $\partial_t f(x) < \infty$ for m-a.e. $x \in M$. Hence, $P_t f(x)$ is well-defined for m-a.e. $x \in M$ and $P_t f$ defines an m-equivalence class of measurable maps with separable ranges. Moreover,

$$f \in \mathcal{L}_\theta(M, N, p) \Longrightarrow P_t f \in \mathcal{L}_\theta(M, N, p) \quad \text{and} \quad d_\theta(f, P_t f) \leq \|\partial_t f\|_\theta.$$

Let $L^*_\theta(M, N, p)$ denote the set of $f \in L_\theta(M, N, p)$ for which the d_θ-limit

$$P^*_t f := \lim_{n \to \infty} (P_{\delta_n})^{t/\delta_n} f$$

exists for all $t \in \mathbb{T}$.

With exactly the same arguments as for Lemma 3.17 we deduce

Proposition 3.18 *For all $f, g \in \mathcal{L}_\theta{}^*(M, N, p)$ and all $s, t \in \mathbb{T}$:*

(i) $P^*_t f \in \mathcal{L}_\theta{}^*(M, N, p)$ and $P^*_s(P^*_t f) = P^*_{s+t} f$;
(ii) $d_\theta(P^*_t f, P^*_t g) \leq C \cdot e^{-\lambda t} \cdot d_\theta(f, g)$;
(iii) $d_\theta(P^*_t f, P^*_{t+s} f) \leq C \cdot e^{-\lambda t} \cdot \|\partial_s f\|_\theta$.

Remark 3.19 *Assume that $p_t(x, .) \ll m$ for all $x \in M$ and all $t \in \mathbb{T}$ ("absolute continuity of p_t"). Then $f(x) = g(x)$ for m-a.e. $x \in M$ implies $P^*_t f(x) = P^*_t g(x)$ for all $x \in M$ and $t \in \mathbb{T}$. In particular, for each $f \in L^*_\theta(M, N, p)$ and each $t \in \mathbb{T}$ the map $P^*_t f$ is pointwise well-defined on M.*

Theorem 3.20 *Assume that $\lambda > 0$ in (5). Then for each $f \in L^*_\theta(M, N, p)$ there exists a unique $h \in L^*_\theta(M, N, p)$ with $d_\theta(h, f) < \infty$ and*

$$P^*_t h = h \tag{6}$$

for all $t \in \mathbb{T}$. Indeed, $h = \lim_{\mathbb{T} \ni t \to \infty} P^*_t f$ and for $t \to \infty$

$$d_\theta(h, P^*_t f) \leq C \cdot e^{-\lambda t} \cdot d_\theta(h, f) \longrightarrow 0.$$

4 Lipschitz Continuity

Let $(M, \mathcal{M})$, $p = (p_t)_{t>0}$ and (N, d, b) be as before. In addition, throughout this Chapter we fix a nonnegative symmetric function ρ on $M \times M$. (Typically, ρ will be a metric on M. But $\mathcal{M}$ will not necessarily be the Borel σ-field of ρ.) For $f, g : M \to N$ put $d_\infty(f,g) := \sup_{x\in M} d(f(x), g(x))$,

$$\mathrm{dil}_\rho f := \sup_{x,y\in M} \frac{d(f(x), f(y))}{\rho(x,y)}$$

(with $\frac{0}{0} := 0$) and let $\mathcal{C}_\rho^{Lip}(M,N)$ denote the set of measurable $f : M \to N$ with separable range and $\mathrm{dil}_\rho f < \infty$. Moreover, let $\mathcal{L}_\infty(M,N)$ denote the set of bounded measurable $f : M \to N$ with separable range.

Theorem 4.21 *Assume that (N,d) has nonpositive and lower bounded curvature (in the sense of Alexandrov). Moreover, assume $\exists C, \beta > 0$ and $\forall t > 0 : \exists c_t$ such that $\sup_{s\le t} c_s < \infty$ and $\forall x, y \in M$:*

$$p_t^\circ \rho(x,y) \le c_t \cdot \rho(x,y) \tag{7}$$

$$\int \rho^4(x,z) p_t(x,dz) \le C \cdot t^{1+\beta} \tag{8}$$

Then $\mathcal{C}_\rho^{Lip}(M,N) \subset \mathcal{L}^*(M,N,p)$ and $(P_t^*)_{t\in\mathbb{R}_+}$ is a strongly continuous semigroup on $\mathcal{C}_\rho^{Lip}(M,N)$. More precisely, for all $x \in M, t \in \mathbb{R}_+$ and $f \in \mathcal{C}_\rho^{Lip}(M,N)$

$$P_t^* f(x) = \lim_{s\to 0} P_s^{\lfloor t/s \rfloor} f(x)$$

exists and the limit is continuous in each variable:

$$\mathrm{dil}_\rho P_t^* f \le c_t \cdot \mathrm{dil}_\rho f \tag{9}$$

$$d_\infty(P_t^* f, P_t^* g) \le d_\infty(f,g) \tag{10}$$

$$d_\infty(P_s^* f, P_t^* f) \le C^{1/4} \cdot \mathrm{dil}_\rho f \cdot |t-s|^{\frac{1+\beta}{4}}. \tag{11}$$

Remark 4.22 *(i) If $c_t = e^{-\kappa t}$ for some $\kappa \in \mathbb{R}$ and $N = \mathbb{R}$ then condition (7) is already necessary for (9). Indeed, for any Markov semigroup $(p_t)_t$ on a metric space (M,ρ) and any $\kappa \in \mathbb{R}$ the following are equivalent:*

- $\mathrm{dil}_\rho p_t u \le e^{-\kappa t} \cdot \mathrm{dil}_\rho u \qquad (\forall t, \forall u \in \mathcal{C}_\rho^{Lip}(M,\mathbb{R}))$
- $p_t^\circ \rho(x,y) \le e^{-\kappa t} \rho(x,y). \qquad (\forall t)$

(ii) Condition (8) excludes jump-diffusion processes. However, for diffusions it is a very weak assumption. E.g. for solutions of SDEs with bounded measurable coefficients on $\mathbb{R}^d$ or for the Markov semigroups from Example 1.2, $\int |x-y|^4 p_t(x,dy) \le C \cdot t^2$ for all $x, y \in \mathbb{R}^d$ and all t.

(iii) The assumption on the lower bounded curvature of (N,d) can be weakened in order to include also "spaces with a reverse variance inequality of some order > 2", e.g. the result of gluing together two copies of the set $\{z = (x,t) \in \mathbb{R}^k : t \le \psi(x)\}$ along their boundary $\{z = (x,t) \in \mathbb{R}^k : t = \psi(x)\}$ where $\psi : \mathbb{R}^{k-1} \to \mathbb{R}$ is any smooth convex function.

For typical examples satisfying (7), see Examples 4.25–4.29 below.

The main point in Theorem 4.21 is that it yields convergence independent of the choice of the sequence $(\delta_n)_n$. Our next result will give convergence for suitable choices of sequences $(\delta_n)_n$. Here the advantages will be:
- it does not require any kind of lower curvature bound for (N,d);
- it applies also to jump processes and to nonlocal equations;
- it also yields smoothing from $\mathcal{L}_\infty(M,N)$ to $\mathcal{C}_\rho^{Lip}(M,N)$.

In order to formulate the latter, let ρ_0 be another nonnegative symmetric function on $M \times M$ (besides ρ) and define dil_{ρ_0} and $\mathcal{C}_{\rho_0}^{Lip}(M,N)$ in an analogous way.

Theorem 4.23 *Assume that $(M,\mathcal{M})$ is a Radon measurable space, (M,ρ) is separable and (N,d) is locally compact. Moreover, assume that*
$\forall t \in \mathbb{T} : \exists C_t : \forall x,y \in M$:

$$p_t^\diamond \rho_0(x,y) \leq C_t \cdot \rho(x,y) \tag{12}$$

$$\int \rho_0(x,z)p_t(x,dz) < \infty. \tag{13}$$

Then for each sequence $(s_k)_k \subset (2^{-k})_k$ there exists a subsequence $(\delta_k)_k = (s_{n_k})_k$ such that $\mathcal{C}_{\rho_0}^{Lip}(M,N) \subset \mathcal{L}^*(M,N,p)$ and for each $t \in \mathbb{T}$

$$P_t^* : \mathcal{C}_{\rho_0}^{Lip}(M,N) \to \mathcal{C}_\rho^{Lip}(M,N).$$

More precisely, $\forall f \in \mathcal{C}_{\rho_0}^{Lip}(M,N), \forall x \in M, \forall t \in \mathbb{T}$ the limit

$$P_t^* f(x) = \lim_{k\to\infty} P_{\delta_k}^{t/\delta_k} f(x)$$

exists and

$$\mathrm{dil}_\rho P_t^* f \leq C_t \cdot \mathrm{dil}_{\rho_0} f. \tag{14}$$

Remark 4.24 *(i) The most important choices for ρ_0 are either $\rho_0 \equiv \rho$ or $\rho_0 \equiv 1$. In the latter case, $\mathrm{dil}_{\rho_0} f = \mathrm{osc} f := \sup_{x,y\in M} d(f(x),f(y))$ and thus $\mathcal{C}_{\rho_0}^{Lip}(M,N) = \mathcal{L}_\infty(M,N)$. Theorem 4.23 then proves existence of the limit $P_t^* f$ for all $f \in \mathcal{L}_\infty(M,N)$ and smoothing $P_t^* f : \mathcal{L}_\infty(M,N) \to \mathcal{C}_\rho^{Lip}(M,N)$.*

(ii) If in addition to the assumptions of the previous Theorem $\gamma_s(x) := \int \rho(x,z)p_s(x,dz) \to 0$ for $s \to 0$ then the limit

$$P_t^* f(x) = \lim_{k\to\infty} P_{\delta_k}^{\lfloor t/\delta_k \rfloor} f(x)$$

exists for all $t \in \mathbb{R}_+$ and for $s \to 0$

$$d(P_t^* f(x), P_{t+s}^* f(x)) \leq C_t \cdot \mathrm{dil}_{\rho_0} f \cdot \gamma_s(x) \to 0. \tag{15}$$

(iii) Assume that (12) only holds true for all x,y in a ρ-open set $M_1 \subset M$. Then the limit $P_t^* f(x) = \lim_{k\to\infty} P_{\delta_k}^{t/\delta_k} f(x)$ exists $\forall f \in \mathcal{C}_{\rho_0}^{Lip}(M,N), \forall x \in M_1, \forall t \in \mathbb{T}$ and is ρ-Lipschitz continuous on M_1.

Example 4.25 *Let $(q_t)_{t>0}$ be a convolution semigroup of probability measures on an Abelian group M and define a translation invariant Markov semigroup by $p_t(x,A) := q_t(x^{-1}A)$. Then for each symmetric $\rho : M \times M \to \mathbb{R}_+$*

$$p_t^\diamond \rho(x,y) \leq \int \rho(x+z, y+z)\, q_t(dz).$$

In particular, if ρ is translation invariant then

$$p_t^\diamond \rho(x,y) \leq \rho(x,y).$$

For instance, this applies to all Lévy semigroups on $\mathbb{R}^n$ as introduced in Example 1.3. For various other examples, see [3] and [4].

Example 4.26 *Let $(p_t)_{t>0}$ be a Markov semigroup on $M = \mathbb{R}^n$ such that*

$$p_t(x,B) = \int_{\mathbb{R}^n} 1_B(y) k_t(\|x-y\|) dy$$

for all $t > 0$ with some nonincreasing function $r \mapsto k_t(r)$ on $\mathbb{R}_+$ with

$$C_t := \int_{\mathbb{R}^{n-1}} k_t(\|z\|) dz < \infty.$$

Put $\rho(x,y) = \|x-y\|$. Then

$$p_t^\diamond 1(x,y) \leq C_t \cdot \rho(x,y).$$

For instance, if $(p_t)_{t>0}$ is the classical heat semigroup on $\mathbb{R}^n$, i.e. $k_t(r) = (2\pi t)^{-n/2} \exp(-r^2/(2t))$, this yields

$$C_t = \frac{1}{\sqrt{2\pi t}}.$$

More generally, for $\alpha < 2$ let $(p_t)_{t>0}$ be the symmetric α-stable semigroup *on $\mathbb{R}^n$, i.e. the Markov semigroup associated with the Lévy operator $(-\frac{1}{2}\Delta)^{\alpha/2}$. Then $\int_{\mathbb{R}^n} \exp(ix\xi) k_t(\|x\|) dx = \exp(-\|\xi\|^\alpha \cdot t/2)$ and thus*

$$C_t = \frac{1}{2\pi} \int_{\mathbb{R}} \exp(-s^\alpha \cdot t/2) ds = C'_\alpha \cdot t^{-1/\alpha}.$$

Example 4.27 *Let $(p_t)_{t>0}$ be the heat semigroup on a complete Riemannian manifold M of Ricci curvature $\geq \kappa$ and dimension n, and let ρ be the Riemannian distance ρ on M. Then $\exists C = C(\kappa, n) : \forall x, y, t :$*

- $p_t^\diamond \rho(x,y) \leq e^{-\kappa t} \cdot \rho(x,y)$
- $p_t^\diamond 1(x,y) \leq C \cdot t^{-1/2} \cdot e^{-\kappa t} \cdot \rho(x,y)$
- $\int \rho^4(x,y) p_t(x,dy) \leq C \cdot t^2.$

For examples of Alexandrov spaces M (= complete metric spaces of lower bounded "sectional" curvature) where the same estimates hold true, see [31].

Example 4.28 *Let M and $(p_t)_{t>0}$ as in the previous Example 4.27 and fix an open subset $D \subset M$. Let $(p_{D,t})_{t>0}$ be the stopped semigroup as introduced in Definition 1.6. Then for this semigroup, (12) with $\rho_0 = 1$ is satisfied on each relatively compact open subset $M_1 \subset D$.*

Example 4.29 *Let $(p_t)_{t>0}$ be a strongly continuous, symmetric semigroup on a σ-finite measure space $(M, \mathcal{M}, m)$ and assume a "curvature-dimension condition" in the sense of Bakry-Emery (see e.g. [27]) holds true with curvature bound κ and dimension bound n. Moreover, let ρ be a symmetric nonnegative function on $M \times M$ with the "Rademacher property"*

$$\mathrm{dil}_\rho u \leq 1 \quad \Longleftrightarrow \quad u \in \mathcal{D}(\Gamma), \Gamma(u) \leq 1 \; m-a.e.$$

where Γ denotes the square field operator associated with $(p_t)_{t>0}$. Then

$$p_t^\diamond \rho(x,y) \leq e^{-\kappa t} \cdot \rho(x,y).$$

For instance, this applies to the Ornstein-Uhlenbeck semigroup on the Wiener space $M = C(\mathbb{R}_+, \mathbb{R}^n)$. Here m=Wiener measure, ρ=Cameron-Martin distance, $\kappa = 1$ and $n = \infty$.

5 Invariant and Harmonic Maps

Definition 5.30 *A map $f : M \to N$ is called invariant iff $f \in \mathcal{L}^*(M, N, p)$ and for all $t \in \mathbb{T}$ and $x \in M$*

$$P_t^* f(x) = f(x).$$

It is called harmonic on a set $D \subset M$ iff for all $x \in D$

$$A^* f(x) := \limsup_{\mathbb{T} \ni t \to 0} \limsup_{n \to \infty} \frac{1}{t} d(f(x), P_{\delta_n}^{t/\delta_n} f(x)) = 0.$$

Given a set $D \subset M$ and a map $g : M \to N$, we say that f is a solution to the nonlinear Dirichlet problem iff f is harmonic on D and $f = g$ on $M \setminus D$.

Of course, each invariant map is harmonic on M (and thus on each set $D \subset M$).

Lemma 5.31 *Let $(p_t)_t$ be a right Markov semigroup and assume that for given $x \in M$ and $D \in \mathcal{M}$ the following locality condition is satisfied:*

$$\lim_{\mathbb{T} \ni t \to 0} \frac{1}{t} \mathbb{P}(\tau(D, x) < t) = 0.$$

Then a bounded map $f : M \to N$ is harmonic in x w.r.t. the semigroup $(p_t)_t$ if and only if it is harmonic in x w.r.t. the stopped semigroup $(p_{D,t})_t$ as introduced in Chapter 1. In particular, $P_{D,t}^* f(x) = f(x)$ implies $A^* f(x) = 0$.

This Lemma together with Theorem 3.20 may be used to deduce existence and uniqueness for the solution to the Dirichlet problem. For sake of simplicity, we restrict ourselves to the pointwise version (with uniform convergence). Similar results hold true in $L_\theta(M, N)$.

Corollary 5.32 *Fix $D \subset M$ and assume that there exist numbers $C, \lambda_D > 0$such that for all $t > 0$*

$$\sup_{x \in D} p_{D,t}(x, D) \leq C \cdot e^{-\lambda_D \cdot t}. \tag{16}$$

Then for each bounded $f \in \mathcal{L}^*(M, N, p)$ there exists a unique $h \in \mathcal{L}^*(M, N, p)$ with $d_\infty(h, f) < \infty$ and $h = f$ on $M \setminus D$ and $P^*_{D,t}h = h$ for all $t \in \mathbb{T}$.

In all typical examples, M is a metric space and $\lambda_D > 0$ for each bounded open subset $D \subset M$.

A general uniqueness result for the Dirichlet problem may be deduced from the following Lemma which in turn follows easily from Lemma 3.17.

Lemma 5.33 *If $f, g : M \to N$ are invariant then $u : M \to \mathbb{R}, x \mapsto d(f(x), g(x))$ is subinvariant in the following sense: $p_t u(x) \geq u(x)$ for all $t \in \mathbb{T}$ and $x \in M$.*

If f, g are harmonic on $D \subset M$ then $u := d(f, g)$ is subharmonic on D in the following sense:

$$a^* u(x) := \limsup_{\mathbb{T} \ni t \to 0} \frac{1}{t} (p_t u(x) - u(x)) \geq 0.$$

Corollary 5.34 *Assume that a Maximum Principle holds true for subharmonic functions on D in the following sense: if $u : M \to \mathbb{R}_+$ is subharmonic on D and vanishes on $M \setminus D$ then $u = 0$. Then solutions to the nonlinear Dirichlet problem for harmonic maps on D are unique.*

A consequence of Jensen's inequality is

Lemma 5.35 *Let (N, d) be either a complete metric space of (globally) nonpositive curvature or a Banach space and let $\varphi : N \to \mathbb{R}$ be convex and Lipschitz continuous. If $f \in \mathcal{L}^*(M, N, p)$ is invariant then $\varphi \circ f : M \to \mathbb{R}$ is subinvariant. If f is harmonic on some set D then $\varphi \circ f : M \to \mathbb{R}$ is subharmonic on this set D.*

A complete characterization of harmonic maps in terms of convex and subharmonic functions is possible for some of the most important target spaces.

Theorem 5.36 *Let (N, d) be either a smooth, simply connected, complete Riemannian manifold of nonpositive curvature or a metric tree with discrete set of vertices or a Banach space. Let $f : M \to N$ be measurable with separable range and fix $x \in M$ with*

$$\limsup_{t \to 0} \frac{1}{t} \int d^2(f(x), f(y))\, p_t(x, dy) < \infty \tag{17}$$

$$\lim_{t \to 0} \frac{1}{t} \int [d(f(x), f(y)) - r]_+ \, p_t(x, dy) = 0 \tag{18}$$

($\forall r > 0$). Then the following are equivalent

(i) $A^* f(x) = 0$;
(ii) $a^*(\varphi \circ f)(x) \geq 0$ *for all Lipschitz continuous, convex* $\varphi : N \to \mathbb{R}$.

Remark 5.37 *Let ρ be a metric on M. Then conditions (17) and (18) are satisfied for all $f \in C_\rho^{Lip}(M, N)$ provided*

$$\limsup_{t \to 0} \frac{1}{t} \int \rho^2(x, y)\, p_t(x, dy) < \infty \tag{19}$$

$$\lim_{t \to 0} \frac{1}{t} \int [\rho(x, y) - r]_+\, p_t(x, dy) = 0 \tag{20}$$

($\forall r > 0$). The first one, is a linear bound for the quadratic variation, the second one a continuity (or locality) condition. Under (19) it is equivalent to

$$\lim_{t \to 0} \frac{1}{t} p_t(x, M \setminus B_r(x)) = 0$$

($\forall r > 0$) which is a well known sufficient condition for continuity of paths of the associated Markov process.

An even more complete picture is obtained for invariant maps in an L_θ-context.

Theorem 5.38 *Let (N, d) be as in the previous Theorem, let $(M, \mathcal{M}, m)$ be a measure space and let $(p_t)_t$ be a Markov semigroup on $(M, \mathcal{M})$ such that m is excessive w.r.t. $(p_t)_t$, i.e. $\int_M p_t(x, A) m(dx) \leq m(A)$ for all $A \in \mathcal{M}$. Finally, let $f : M \to N$ be a measurable map with separable range and with*

$$\limsup_{t \to 0} \frac{1}{t} \int\int d^2(f(x), f(y)) p_t(x, dy) m(dx) < \infty \tag{21}$$

$$\lim_{t \to 0} \frac{1}{t} \int\int [d(f(x), f(y)) - r]_+\, p_t(x, dy) m(dx) = 0 \tag{22}$$

($\forall r > 0$). Then the following assertions are equivalent

(i) For each $t > 0$

$$P_t^* f := \lim_{s \to 0} P_s^{\lfloor t/s \rfloor} f$$

exists in $L^1(M, N)$ and $P_t^* f = f$.

(ii) There exist subsequences $(t_n)_n$ and $(s_k)_k$ of $(2^{-k})_k$ such that for m-a.e. $x \in M$

$$P_{t_n}^* := \lim_{k \to \infty} P_{s_k}^{\lfloor t_n / s_k \rfloor} f(x)$$

exists in N and $P_{t_n}^* f(x) = f(x)$.

(iii) For each $t > 0$, for m-a.e. $x \in M$ and for each convex, Lipschitz continuous $\varphi : N \to \mathbb{R}$

$$(\varphi \circ f)(x) \leq p_t(\varphi \circ f)(x).$$

(iv) $\lim_{t\to 0} \frac{1}{t} d_1(f, P_t f) = 0.$

(v) There exist subsequences $(t_n)_n$ *and* $(s_k)_k$ *of* $(2^{-k})_k$ *such that*

$$P^*_{t_n} f := \lim_{k\to\infty} P_{s_k}^{\lfloor t_n/s_k \rfloor} f$$

exists in $L^1(M, N)$ *and* $\liminf_{n\to\infty} \frac{1}{t_n} d_1(f, P^*_{t_n} f) = 0.$

References

1. W. Ballmann (1995): Lectures on spaces of nonpositive curvature. DMV Seminar Band **25**, Birkhäuser
2. H. Bauer (1996): Probability theory. de Gruyter Studies in Mathematics, **23**. Walter de Gruyter & Co., Berlin
3. A. Bendikov (1995): Potential theory on infinite-dimensional Abelian groups. de Gruyter Studies in Mathematics, **21.** Walter de Gruyter, Berlin
4. A. Bendikov, L. Saloff-Coste (2001): On the absolute continuity of Gaussian measures on locally compact groups. J. Theor. Probab. **14**, No. 3, 887–898
5. J. Eells, L. Lemaire (1978): A report on harmonic maps. Bull. London Math. Soc. **10**, 1–68
6. J. Eells, L. Lemaire (1988): Another report on harmonic maps. Bull. London Math. Soc.**20**, 385–524
7. J. Eells, J.H. Sampson (1964): Harmonic mappings of Riemannian manifolds. Amer. J. Math. **86**, 109–160
8. J. Eells, B. Fuglede (2001): Harmonic maps between Riemannian polyhedra. Cambridge Tracts in Mathematics, **142.** Cambridge University Press, Cambridge
9. A. Es-Sahib, H. Heinich (1999): Barycentre canonique pour un espace métrique à courbure négative. Séminaire de Probabilités, XXXIII, 355–370, Lecture Notes in Math., **1709**, Springer, Berlin
10. S. N. Ethier, T. G. Kurtz (1986): Markov processes. Characterization and convergence. Wiley Series in Probability and Mathematical Statistics: Probability and Mathematical Statistics. John Wiley, New York
11. C. Fefferman, D. H. Phong (1983): Subelliptic eigenvalue problems. Conference on harmonic analysis, 590–606, Chicago, Wadsworth Math. Ser.
12. G. Gregori (1998): Sobolev spaces and harmonic maps between singular spaces. Calc. Var. Partial Differ. Equ. **7**, 1–18
13. A. Grigoryan (1999): Estimates of heat kernels on Riemannian manifolds. Spectral theory and geometry, 140–225, London Math. Soc. Lecture Note Ser., **273**, Cambridge Univ. Press
14. M. Gromov, R. Schoen (1992): Harmonic maps into singular spaces and p-adic superrigidity for lattices in groups of rank one. Publ. Math., Inst. Hautes Étud. Sci., 76, 165-246
15. S. Hildebrandt, H. Kaul, K.-O. Widman (1975): Harmonic mappings into Riemannian manifolds with non-positive sectional curvature. Math. Scandinav. **37**, 257–263
16. S. Hildebrandt, H. Kaul, K.-O. Widman (1977): An existence theorem for harmonic mappings of Riemannian manifolds. Acta Math. **138**, 1–16
17. T. Ishihara (1979): A mapping of Riemannian manifolds which preserves harmonic functions. J. Math. Kyoto Univ. **19**, 215–229

18. N. Jacob (1996): Pseudo-differential operators and Markov processes. Mathematical Research, **94**. Akademie Verlag, Berlin
19. N. Jacob (2001): Pseudo differential operators and Markov processes. Vol. I. Fourier analysis and semigroups. Imperial College Press, London
20. D. S. Jerison, A. Sanchez-Calle (1986): Estimates for the heat kernel for a sum of squares of vector fields. Indiana Univ. Math. J. **35**, no. 4, 835–854
21. J. Jost (1994): Equilibrium maps between metric spaces. Calc. Var. Partial Differential Equations 2, **2**, 173–204
22. J. Jost (1997): Generalized Dirichlet forms and harmonic maps. Calc. Var. Partial Differential Equations 5, no. 1, 1–19
23. J. Jost (1997a): Nonpositive curvature: geometric and analytic aspects. Lectures in Mathematics, ETH Zürich, Birkhäuser, Basel
24. W. S. Kendall (1990): Probability, convexity, and harmonic maps with small images. I. Uniqueness and fine existence. Proc. London Math. Soc. **61**, 371–406
25. N. Korevaar, R. Schoen (1994): Sobolev spaces and harmonic maps for metric space targets. Comm. Anal. Geom. **1**, 561–569
26. N. Korevaar, R. Schoen (1997): Global existence theorems for harmonic maps to non-locally compact spaces. Comm. Anal. Geom. **5**, 333–387
27. M. Ledoux (2000): The geometry of Markov diffusion generators. Probability theory. Ann. Fac. Sci. Toulouse Math. **9**, 305–366
28. M. Ledoux, Talagrand (1991): Probability in Banach spaces. Isoperimetry and processes. Ergebnisse der Mathematik und ihrer Grenzgebiete **23**. Springer, Berlin
29. Z. M. Ma, M. Röckner (1992): Introduction to the theory of (nonsymmetric) Dirichlet forms. Universitext. Springer, Berlin
30. S. T. Rachev, L. Rüschendorf (1998): Mass transportation problems. Vol. I. Theory. Probability and its Applications. Springer, New York
31. M.-K. von Renesse (2002): Comparison properties of diffusion semigroups on spaces with lower curvature bounds. PhD Thesis, Bonn
32. D. W. Stroock, S. R. S. Varadhan (1979): Multidimensional diffusion processes. Grundlehren der Mathematischen Wissenschaften **233**. Springer-Verlag, Berlin-New York
33. M. Struwe (1985): On the evolution of harmonic mappings of Riemannian surfaces. Comment. Math. Helv. **60**, 558–581
34. M. Struwe (1988): On the evolution of harmonic maps in higher dimensions. J. Differential Geom. **28**, 485–502
35. K.T. Sturm (2001): Nonlinear Markov operators associated with symmetric Markov kernels and energy minimizing maps between singular spaces. Calc. Var. **12**, 317-357
36. K.T. Sturm (2002): Nonlinear martingale theory for processes with values in metric spaces of nonpositive curvature. To appear in Ann. Prob.
37. K.T. Sturm (2002a): Nonlinear Markov operators, discrete heat flow, and harmonic maps between singular spaces. Potential Analysis **16**, 305–340
38. K.T. Sturm (2002b): A semigroup approach to harmonic maps. Preprint, Bonn
39. K. Taira (1991): Boundary value problems and Markov processes. Lecture Notes in Mathematics, 1499. Springer-Verlag, Berlin
40. A. Thalmaier (1996): Martingales on Riemannian manifolds and the nonlinear heat equation. Stochastic analysis and applications, 429–440, World Sci. Publishing, River Edge, NJ
41. A. Thalmaier (1996a): Brownian motion and the formation of singularities in the heat flow for harmonic maps. Probab. Theory Related Fields 105, no. 3, 335–367

Boundary Regularity for Nonlinear Elliptic Systems: Applications to the Transmission Problem

Carsten Ebmeyer[1], Jens Frehse[2], and Moritz Kassmann[2]

[1] Mathematisches Seminar, Universität Bonn, Nussallee 15, D-53115 Bonn, Germany
[2] Institut für Angewandte Mathematik, Universität Bonn, Beringstr. 6, D-53115 Bonn, Germany

Summary. The Dirichlet problem for nonlinear elliptic systems with piecewise continuous coefficients is studied. The domain can be a n-dimensional polyhedron, possibly non-convex and exhibiting slits. Regularity in Nikolskii spaces up to the transmission surface and the boundary is proved.

1 Introduction

We are concerned with the nonlinear elliptic system

$$-\sum_{i=1}^{n} \partial_i F_i(x, \nabla u) = f(x) \qquad \text{in } \Omega \tag{1}$$

on a bounded domain $\Omega \subset \mathbb{R}^n$ $(n \geq 2)$, where $u : \Omega \to \mathbb{R}^N$ $(N \geq 1)$ and $F_i : \Omega \times \mathbb{R}^{nN} \to \mathbb{R}^N$. A transmission surface is given by a hyperplane across which the functions $F_i(\cdot, p)$ have discontinuities. The aim of this work is to investigate the regularity of weak solutions u in the joint neighborhood of the transmission surface and the boundary of Ω. Full details are given only for the simplest situations.

It is well-known that a weak solution u of (1) may have singularities if the data are non-smooth, such as piecewise continuous coefficients $F(\cdot, p)$ (transmission problem), changing boundary value conditions, or corner points of $\partial\Omega$. Then the weak solution u can only be expected to belong to some fractional order space, for instance, a Sobolev space $W^{s,p}(\Omega; \mathbb{R}^N)$ or a Nikolskii space $\mathcal{N}^{s,p}(\Omega; \mathbb{R}^N)$, where $1 < s < 2$.

In order to prove regularity in a Nikolskii space, say $u \in \mathcal{N}^{1+s,2}(\Omega; \mathbb{R}^N)$ and $0 < s < 1$, we have to estimate the difference quotients of ∇u of order s. More precisely, we have to show that

$$\sup_{1 \leq i \leq n} \int_{\Omega_\delta} \left| \frac{\nabla u(x + \delta e_i) - \nabla u(x)}{|\delta|^s} \right|^2 dx \leq c \qquad \text{for all } \delta > 0, \tag{2}$$

where e_i is an unit vector of $\mathbb{R}^n$. Typically, there are certain directions e_i for which estimate (2) can easily be proved. For instance, let us consider the transmission problem. If the coefficients have a jump along the hyperplane $\{x_1 = 0\}$, estimate

(2) can be shown for the $x_2-, \ldots, x_n-$ directions using some difference quotient method. Or, in the case of a mixed boundary value problem, we obtain (2) for all directions e_i parallel to $\partial\Omega$, but not across $\partial\Omega$. Thus, a typical situation is that estimate (2) is known for $n-1$ directions, say, the $x_2-, \ldots, x_n-$ directions. In this paper we are concerned with this situation. Our aim is to show how to obtain then regularity in the x_1- direction. Therefore, we use Fourier analysis. In a similar manner, in [2, 3] such a technique has been utilized, where mixed boundary value problems are treated. Here, we have improved this method. In particular, we do not consider the equation in the reflected domains. A further idea of our proof is to use *first* order difference quotients of solutions as test functions, such as in [2, 3].

We confine our short discussion about known results to transmission problems characterized by piecewise smooth coefficients and a single weak formulation. Especially, we do not discuss cases where different growth conditions are considered in different regions. An overview of regularity results for linear systems is given together with an exhaustive list of references in [5]. As explained in [5] one obtains in the linear case fine regularity results like asymptotic expansions with singular functions. In the case of nonlinear elliptic systems the accessible results are much weaker since even smoothness properties in interior regions need not hold. We obtain here the almost optimal order of differentiation near the set where the transmission surface hits the boundary. We consider our approach as a convenient method to prove $\mathcal{N}^{\frac{3}{2}-\varepsilon,2}$-regularity also for nonlinear scalar equations or even linear equations in multidimensional domains. Classical results concerning the regularity in the interior of the domain but up to the transmission surface can be found in [6] and [7], see also the references therein.

This paper is organized as follows. In section 2 we estimate difference quotients of periodic functions using Fourier series. In section 3 we apply these results to the transmission problem and prove

$$u \in \mathcal{N}^{\frac{3}{2}-\varepsilon,2} \qquad (\varepsilon > 0)$$

up to the boundary of the region, where $F(\cdot, p)$ is smooth, – in particular up to $\partial\Omega$.

2 Difference Quotients of Periodic Functions

The aim of this section is to prove two results for difference quotients of periodic functions; see Lemma 1 and Lemma 2 below.

Let $Q := \{x \in \mathbb{R}^n; -\frac{l}{2} < x_1, \ldots, x_n < \frac{l}{2}\}$ be a cube and $w \in L^2(\mathbb{R}^n/(l\mathbb{Z})^n)$ a periodic function. That is, $w(x+le_j) = w(x)$ for all $x \in \mathbb{R}^n$ and $1 \le j \le n$, where e_j is the j-th unit vector in $\mathbb{R}^n$. In order to prove the main results of this section we consider the Fourier series of w,

$$w(x) = \sum_m c_m e^{imx}.$$

Here, $m = (m_1, m_2, \ldots, m_n)$ is a multi-index of length n, $mx = \sum_{k=1}^n m_k x_k$ for $x \in \mathbb{R}^n$,

$$c_m = l^{-n} \int_Q w(x)\, e^{imx}\, dx$$

are the Fourier coefficients of w, and

$$\sum_m = \sum_{m_1,\dots,m_n} = \sum_{\pm m_1 \in \mathbb{N}_0} \cdots \sum_{\pm m_n \in \mathbb{N}_0} .$$

Let $h > 0$ and $j \in \{1, \dots, n\}$. We set

$$(D_j^{\pm h} w)(x) = h^{-1}(w(x \pm h e_j) - w(x)) ,$$

$$(M_j^h w)(x) = h^{-1} \int_0^h w(x + t e_j)\, dt .$$

We will utilize the following well-known statements [2]:

(A) Assume there is a constant C such that for given $1 \le j \le n$ and $\mu > 0$ the following estimate holds:

$$\sup_{0<h<\mu} \int_Q |h^{\frac{1}{2}} D_j^h w(x)|^2\, dx \le C .$$

Then there exists a $\delta_0 \in (0,1)$ such that, for any $\delta \in (0, \delta_0)$ there exists a constant C_1 with

$$\sum_m (c_m)^2 |m_j|^{1-\delta} \le C_1 .$$

(B) Assume there is a constant C such that for given $\delta_0 \in (0,1), \delta \in (0, \delta_0)$ $1 \le j \le n$ and $\mu > 0$ the following estimate holds:

$$\sum_m (c_m)^2 |m_j|^{1-\delta} \le C .$$

Then there exists a constant C_2 such that

$$\sup_{0<h<\mu} \int_Q |h^{\frac{1}{2}+\frac{\delta}{2}} D_j^h w(x)|^2\, dx \le C_2 .$$

Now we prove the following two lemmas.

Lemma 2.1 *Let $l > 0$ and $w \in L^2(\mathbb{R}^n/(l\mathbb{Z})^n)$. Assume that the following two assertions hold:*

$$\sup_{h>0} h^{\frac{1}{2}} \|D_j^h w\|_{L^2(Q)} \le C \quad \forall j \in \{2, \dots, n\} , \tag{3}$$

$$\sup_{h>0} h^{\frac{1}{2}} \|D_1^h (M_2^h \circ \dots \circ M_n^h) w\|_{L^2(Q)} \le C . \tag{4}$$

Then, for any $\delta > 0$ there exists a constant C such that

$$\sup_{h>0} h^{\frac{1}{2}+\delta} \|D_1^h w\|_{L^2(Q)} \le C . \tag{5}$$

Proof. A direct consequence of (3) and (A) is the fact that, for any $\delta \in (0,1)$ there exists a constant C such that

$$\sum_m |m_j|^{1-\delta}|c_m|^2 \le C \quad \forall j \in \{2,\dots,n\}\,.$$

Define $\mathcal{I} := \{m \in (\pm\mathbb{N}_0)^n : |m_1| < \max_{j=1,.,n} |m_j|\}$ and $\mathcal{I}^c := (\pm\mathbb{N}_0)^n \backslash \mathcal{I}$. In view of (B) it is sufficient to prove

$$\sum_{m\in\mathcal{I}^c} |m_1|^{1-\delta}|c_m|^2 \le C \quad \forall j \in \{2,\dots,n\}\,.$$

Let us deduce properties of w from (4) via the representation of w by its Fourier series. Define $\mathcal{I}_0 : \{m \in \mathcal{I} : (\prod_{j=2}^n m_j) = 0\}$ and $\mathcal{I}_+ := \mathcal{I}\backslash\mathcal{I}_0 = \{m \in \mathcal{I} : m_j \neq 0\ \forall j = 2,\dots,n\}$. We have

$$\begin{aligned}
M_j^h w &= \sum_{m,m_j\neq 0} c_m \frac{1}{ihlm_j}(e^{ilm_jh}-1)e^{ilmx} + \sum_{m,m_j=0} c_m e^{ilmx},\\
(M_2^h \circ\dots\circ M_n^h)w &= \sum_{\mathcal{I}_+} c_m \Big(\prod_{j=2}^n \frac{1}{ihlm_j}(e^{ihlm_j}-1)\Big)e^{ilmx} + \sum_{\mathcal{I}_0} c_m e^{ilmx},\\
D_1^h(M_2^h \circ\dots\circ M_n^h)w &= \frac{1}{h}\sum_{\mathcal{I}_+} c_m \Big(\prod_{j=2}^n \frac{1}{ihlm_j}(e^{ihlm_j}-1)\Big)e^{ilmx}(e^{ilm_1h}-1)\\
&\quad + \frac{1}{h}\sum_{\mathcal{I}_0} c_m e^{ilmx}(e^{ilm_1h}-1)\,.
\end{aligned}$$

Taking into account $|e^{ix}-1|^2 = (2\sin(\frac{x}{2}))^2$ we can conclude from (4)

$$\sup_{h>0}\sum_{\mathcal{I}_+} |c_m|^2 \frac{4\sin^2(\frac{lm_1h}{2})}{h}\Big(\prod_{j=2}^n \frac{4\sin^2(\frac{lm_jh}{2})}{|hlm_j|^2}\Big) + \sum_{\mathcal{I}_0} \frac{4\sin^2(\frac{lm_1h}{2})}{h}|c_m|^2 \le C\,.$$

Restricting h to be of the form $h = 2^{-s}$, $s \in \mathbb{N}_0$, we can estimate for given small $\delta > 0$:

$$\sum_{s\in\mathbb{N}_0}\Big(\frac{1}{2}\Big)^{\delta s}\sum_{\mathcal{I}_+}|c_m|^2\frac{4\sin^2(lm_12^{-s-1})}{2^{-s}}\Big(\prod_{j=2}^n \frac{1}{|2^{-s}lm_j|^2}4\sin^2(lm_j2^{-s-1})\Big) \le C\,,$$

$$\sum_{s\in\mathbb{N}_0}\Big(\frac{1}{2}\Big)^{\delta s}\sum_{\mathcal{I}_0}|c_m|^2\frac{4\sin^2(lm_12^{-s-1})}{2^{-s}} \le C\,.$$

Consider the following subset $\mathcal{I}_s^c$ of $\mathcal{I}^c$:

$$\mathcal{I}_s^c := \{m \in \mathcal{I}^c; |m_1| \in [\frac{2^{s-1}}{l}, \frac{2^s}{l})\}\,.$$

For $m \in \mathcal{I}_s^c$ we have :

$$\frac{4\sin^2(lm_1 2^{-s-1})}{2^{-s}} \geq Cl^2|m_1|^2 2^{-s} \geq C|m_1| \,,$$

$$\frac{4\sin^2(lm_j 2^{-s-1})}{|2^{-s}lm_j|^2} \geq \frac{\sin^2(lm_j 2^{-s-1})}{|lm_j 2^{-s-1}|^2} \geq C \quad \text{because of } lm_j \leq 2^{-1} \,.$$

Therefore, we deduce from (6):

$$\sum_{m\in\mathcal{I}^c} |m_1|^{1-\delta}|c_m|^2 \leq C \sum_{s\in\mathbb{N}_0} (\frac{1}{2})^{\delta s} (\sum_{m\in\mathcal{I}_s^c} (c_m)^2|m_1|) \leq C \,.$$

The lemma has been proved completely. □

Lemma 2.2 *Consider the upper halfspace* $\mathbb{R}^n_+ := \{x \in \mathbb{R}^n : x_n > 0\}$. *Let* $f \in L^2(\mathbb{R}^n_+)$. *Consider functions* $W_k \in L^2(\mathbb{R}^n_+)$, $k = 1, \ldots, n$, *satisfying*

$$\sum_{j=1}^{n} \int_{\mathbb{R}^n_+} W_j(x)\, \partial_j\varphi(x)\, dx = \int_{\mathbb{R}^n_+} f(x)\, \varphi(x)\, dx \quad \forall \varphi \in W_0^{1,p}(\mathbb{R}^n_+) \,. \tag{6}$$

Assume that there exists $R > 0$ *such that*

$$W_k(x) = 0 \quad \forall x : |x| > R \quad \forall k \in \{1, \ldots, n\} \,.$$

Assume further that the following two estimates hold:

$$\sup_{h>0} h^{\frac{1}{2}} \|D_j^h W_k\|_{L^2(\mathbb{R}^n_+)} \leq C \quad \forall j \in \{2, \ldots, n\} \quad \forall k \in \{1, \ldots, n\} \,. \tag{7}$$

$$\sup_{h>0} h^{-1} \int_{-\infty}^{\infty} \cdots \int_{-\infty}^{\infty} \int_0^h |W_k(x)|^2 \, dx_n \, dx_{n-1} \ldots dx_1 \leq C \quad \forall k \in \{1, \ldots, n\} \,. \tag{8}$$

Then, for any $\delta > 0$ *there exists a constant* C *such that*

$$\sup_{0<h<R} h^{\frac{1}{2}+\delta} \|D_1^h W_1\|_{L^2(\mathbb{R}^n_+)} \leq C \tag{9}$$

Corollary 2.3 *The lemma holds, as well, if* $\mathbb{R}^n_+$ *is replaced by* $\mathbb{R}^n_+ \cap \{x_1 > 0\}$ *everywhere.*

Proof (of Lemma 2.2). In order to apply Lemma 2.1 to the function W_1 one has to extend W_1 to a periodic function in $\mathbb{R}^n$. First, we define $\widehat{W_1} : B_R(0) \to \mathbb{R}$ in the following way:

$$\widehat{W_1}(x) := \begin{cases} W_1(x_1, \ldots, x_{n-1}, x_n) & ; \; x \in \{x_n > 0\} \\ W_1(x_1, \ldots, x_{n-1}, -x_n) & ; \; x \in \{x_n \leq 0\} \end{cases}$$

Considering the cube $K := [-2R, 2R]^n$ it is clear that $\widehat{W_1}$ can be extended to a periodic function $\widetilde{W}_1 \in L^2(\mathbb{R}^n/((4R)\mathbb{Z})^n)$. Set $w(x) := \widetilde{W}_1(x)$. We first have to prove that assumption (7) extends to (3), i.e.

$$\sup_{h>0} h^{\frac{1}{2}} \|D_j^h w\|_{L^2(K)} \leq C \quad \forall j \in \{2, \dots, n\}. \tag{10}$$

Due to (7), estimate (10) is trivial in the case $j \neq n$. The case $j = n$ is treated as follows. One has to find a constant C, independent of h, such that

$$h^{-1} \int_K |\widetilde{W}_1(x_1, \dots, x_{n-1}, x_n + h) - \widetilde{W}_1(x)|^2 \, dx \leq C \, .$$

Consider the decomposition

$$\begin{aligned} K &= (K \cap \{x_n > 0\}) \cup (K \cap \{x_n \leq -h\}) \cup (K \cap \{-h < x_n \leq 0\}) \\ &=: M_1 \cup M_2 \cup M_3 \, . \end{aligned}$$

The only set of interest in view of (7) is M_3. Introducing the following settings

$$\tilde{h} := -2|x_n| + h, \quad \tilde{z}_n := |x_n|, \quad \hat{h} := 2|x_n| - h, \quad \hat{z}_n := -|x_n| + h,$$

we can derive

$$\begin{aligned} h^{-1} &\int_{M_3} |W_1(x_1, \dots, x_{n-1}, x_n + h) - W_1(x_1, \dots, x_{n-1}, -x_n)|^2 dx \\ &= h^{-1} \int_{K \cap \{0 \leq |x_n| < \frac{h}{2}\}} |W_1(x_1, \dots, x_{n-1}, \tilde{z}_n + \tilde{h}) - W_1(x_1, \dots, x_{n-1}, \tilde{z}_n)|^2 dx \\ &\quad + h^{-1} \int_{K \cap \{\frac{h}{2} \leq |x_n| < h\}} |W_1(x_1, \dots, x_{n-1}, \hat{z}_n) - W_1(x_1, \dots, x_{n-1}, \hat{z}_n + \hat{h})|^2 dx \\ &\leq h^{-1} \int_{K \cap \{z_n > 0\}} |w(z + \tilde{h} e_n) - w(z)|^2 dz \\ &\quad + h^{-1} \int_{K \cap \{z_n > 0\}} |w(z + \hat{h} e_n) - w(z)|^2 dz \leq C \, , \end{aligned}$$

where we applied assumption (7) and standard arguments. Thus, we have shown estimate (10). Next, we have to prove that (4) holds. Equation (6) can be written as the following equality in the sense of distributions, in $\mathbb{R}^n_+$:

$$-\partial_1 W_1 = \sum_{j=2}^{n} \partial_j W_j + f \, .$$

Applying the operator $(M_1^h \circ \dots \circ M_n^h)$ to both sides we obtain

$$\begin{aligned} &D_1^h (M_2^h \circ \dots \circ M_n^h) W_1 \\ &= \sum_{j=2}^{n} D_j^h (M_1^h \circ \dots \circ M_{j-1}^h \circ M_{j+1}^h \circ \dots \circ M_n^h) W_j + (M_1^h \circ \dots \circ M_n^h) f \, . \end{aligned}$$

Multiplying by $h^{\frac{1}{2}}$ and utilizing $f \in L^2(\mathbb{R}^n_+)$ and assumption (7) we deduce

$$\sup_{h>0} h^{\frac{1}{2}} \|D_1^h(M_2^h \circ \ldots \circ M_n^h) W_1\|_{L^2(\mathbb{R}^n_+)} \leq C\,.$$

Let us try to prove (4) by reflecting W_1 at the hypersurface $\{x_n = 0\}$. Assumption (8) will become important now. Defining w by $w(x) := \widetilde{W}_1(x)$ as above we have to prove

$$\sup_{h>0} h^{-1} \|M_n^h(M_2^h \circ \ldots \circ M_{n-1}^h)(D_1^h w)\|^2_{L^2(K)} \leq C\,.$$

Considering the same decomposition of the cube K as above, namely $K = M_1 \cup M_2 \cup M_3$ it is clear that the only nontrivial task is to estimate the L^2-norm on M_3. We deduce

$$\begin{aligned}
&h^{-1} \int_{M_3} \left| \frac{1}{h} \int_0^h (M_2^h \circ \ldots \circ M_{n-1}^h)(D_1^h w)(x + te_n)\, dt \right|^2 dx \\
&\leq C(M_2^h \circ \ldots \circ M_{n-1}^h) \left(\frac{1}{h^2} \int_0^h \int_{M_3} |w(x + te_n + he_1) - w(x + te_n)|^2 \right) \\
&\leq C(M_2^h \circ \ldots \circ M_{n-1}^h) \left(\frac{1}{h^2} \int_0^h \int_{M_3} (|w(x + te_n + he_1)|^2 + |w(x + te_n)|^2) \right) \\
&\leq C(M_2^h \circ \ldots \circ M_{n-1}^h)(1) \leq C\,,
\end{aligned}$$

where me made use of assumption (8). Altogether,we have verified assumptions (3) and (4) of Lemma 2.1 for the function $w(x) = \widetilde{W}_1(x)$. Therefore, we conclude (5). Notice that $W_1(x) = w(x)\ \forall x \in \mathbb{R}^n_+ \cap K$. Thus, (5) implies the assertion (9). □

3 The Transmission Problem

Let us consider the nonlinear elliptic Dirichlet problem

$$\begin{aligned}
-\textstyle\sum_{i=1}^n \partial_i F_i(x, \nabla u) &= f(x) \quad \text{in } \Omega, \\
u(x) &= 0 \quad\quad \text{on } \partial\Omega,
\end{aligned} \tag{11}$$

where $\Omega \subset \mathbb{R}^n$ is a bounded domain, and the functions $F_i(x,p)$ $(x \in \mathbb{R}^n, p \in \mathbb{R}^{nN})$ are piecewise continuous with respect to x. More precisely, let

$$\Omega^1 = \{x \in \Omega : |x_1| \geq 0\} \quad \text{and} \quad \Omega^2 = \{x \in \Omega : |x_1| < 0\}.$$

Thus, $\Omega = \Omega^1 \cup \Omega^2$. We assume there are functions $G_i(p)$ and $H_i(p)$ such that

$$F_i(x,p) = \begin{cases} G_i(p) \text{ if } x \in \Omega^1, \\ H_i(p) \text{ if } x \in \Omega^2, \end{cases}$$

for all $1 \le i \le n$. Just for simplicity, the functions G_i and H_i do not depend on x. A smooth dependence on x would not affect our results. Further, we suppose there is a function $F(x,p) \in \mathbb{R}$, that is a C^2–function with respect to p, and fulfills $\frac{\partial}{\partial p_i^\nu} F(x,p) = F_i^\nu(x,p)$ for $1 \le i \le n$ and $1 \le \mu \le N$ (here, F_i^ν denotes the ν-th component of F_i). Further, let $F_{i,j}^{\nu\mu}(x,p) = \frac{\partial}{\partial p_j^\mu} F_i^\nu(x,p)$. We assume there are constants $c_1, \dots, c_6$ such that for all $x \in \mathbb{R}^n$, $p \in \mathbb{R}^{nN}$, $1 \le i,j \le n$, and $1 \le \nu, \mu \le N$, the following conditions are satisfied:

(H1) $c_1 + c_2|p|^2 \le F(x,p) \le c_3(1+|p|^2)$ and $c_2 > 0$.
(H2) $|F_i^\nu(x,p)| \le c_4(1+|p|)$.
(H3) $|F_{i,j}^{\nu\mu}(x,p)| \le c_5$.
(H4) $c_6|\xi|^2 \le \sum_{\nu,\mu=1}^N \sum_{i,j=1}^n F_{i,j}^{\nu\mu}(x,p)\xi_i^\nu \xi_k^\mu$ for all $\xi \in \mathbb{R}^{nN}$ and $c_6 > 0$.
(H5) $f(x) \in L^2(\Omega; \mathbb{R}^N)$.

Note, that the pointwise coercivity condition (H4) can be replaced by an integral one in order to treat systems where only the Legendre-Hadamard condition holds.

Furthermore, $\Omega \in \mathbb{R}^n$, $n \ge 2$, is a polyhedral domain which may be non-convex and may have slits. We only assume that Ω^1 and Ω^2 are convex in a neighborhood of $\partial\Omega^1 \cap \partial\Omega^2$. That is, there is a constant $\gamma > 0$ such that the two sets $\Omega^k \cap \{x \in \mathbb{R}^n : |x_1| < \gamma\}$, $k = 1,2$, are convex. (Clearly, we can also consider situations which arise after a $W^{2,\infty}$–mapping.)

We call $u \in W_0^{1,2}(\Omega; \mathbb{R}^N)$ a weak solution of (11), if

$$\sum_{i=1}^n \int_\Omega F_i(x, \nabla u) \cdot \partial_i \varphi = \int_\Omega f \cdot \varphi \qquad \text{for all } \varphi \in W_0^{1,2}(\Omega; \mathbb{R}^N). \tag{12}$$

It is well-known that there exists a unique weak solution. We now state our main result.

Theorem 3.4 *Let u be the weak solution of (11). Then there holds*

$$u \in \mathcal{N}^{\frac{3}{2}-\varepsilon,2}(\Omega^k; \mathbb{R}^N) \qquad \textit{for all } \varepsilon > 0 \textit{ and } k = 1,2. \tag{13}$$

Here, $\mathcal{N}^{\frac{3}{2}-\varepsilon,2}(\Omega^k)$ is a Nikolskii space. It consists of all $L^2(\Omega^k)$-functions w for which the norm

$$\|w\|^2_{\mathcal{N}^{\frac{3}{2}-\varepsilon,2}(\Omega^k)} = \|w\|^2_{L^2(\Omega^k)} + \sup_{\substack{\eta>0 \\ 0<|z|<\eta}} \int_{\Omega^k_\eta} \left| \frac{\nabla w(x+z) - \nabla w(x)}{|z|^{\frac{1}{2}-\varepsilon}} \right|^2 dx$$

is finite, where $\Omega^k_\eta = \{x \in \Omega^k : \text{dist}(x, \partial\Omega^k) \ge \eta\}$.

Remark 3.5 *In the case where $u \in C^{\alpha}(\bar{\Omega})$ as in the scalar case the results can be improved considerably. Namely, one obtains that u belongs to a Nikolskii-Morrey space with the same degree of differentiation and additional Morrey exponent $n - 3 + 2\alpha$. This implies $\nabla u \in L^{3+\delta}$ for small $\delta > 0$ in any dimension $n \geq 3$.*

Remark 3.6 *(i) Our proof yields that the norms $\|u\|_{\mathcal{N}^{\frac{3}{2}-\varepsilon,2}(\Omega^k;\mathbb{R}^N)}$ are uniformly bounded by the data.*
(ii) The embedding of Nikolskii spaces into Sobolev spaces (cf. [4]) implies that $u \in W^{s,2}(\Omega^k;\mathbb{R}^N)$ for all $s < \frac{3}{2}$ and $k = 1, 2$.
(iii) Let us emphasize that we obtain regularity of u up to the boundary.
(iv) It is well-known that $u \in \mathcal{N}^{\frac{3}{2}-\varepsilon,2}(\Omega;\mathbb{R}^N)$ if the coefficients $F(x,p)$ are smooth with respect to x; see [1, 2].

Now we give the proof of Theorem 3.4. As a starting point we assume that $\partial\Omega$ is plane around the set where the transmission surface $\{x_1 = 0\}$ hits $\partial\Omega$. More precisely, we assume that

$$\Omega \cap B_1 = \mathbb{R}^n_+ \cap B_1,$$

where $B_r = \{x \in \mathbb{R}^n : |x| < r\}$. At the end of the proof we shortly mention how to treat various geometrical situations. Let us define $B(r) = B_r \cap \Omega$, $R = \frac{1}{8}$, and $\Omega_h = \{x \in B(R) : 0 < x_n < h\}$. Our aim is to show $u \in \mathcal{N}^{\frac{3}{2}-\varepsilon,2}(B(R) \cap \Omega^k;\mathbb{R}^N)$. First, we estimate difference quotients of ∇u with respect to the $x_2-, \ldots, x_n-$ directions. These directions are parallel to the face $\partial\Omega^1 \cap \partial\Omega^2$.

Lemma 3.7 *Assume u is a weak solution of (11). Then there exist constants C_1 and C_2 such that*

$$\sup_{0<h<R} \int_{B(R)} h|D_j^h \nabla u|^2 dx \leq C_1 \qquad \mathit{for} 2 \leq j \leq n \tag{14}$$

and

$$\sup_{0<h<R} h^{-1} \int_{\Omega_h} |\nabla u|^2 dx \leq C_2. \tag{15}$$

Proof. First, we show estimate (14) for $j = n$. Let $\mathbb{R}^n_- = \{x \in \mathbb{R}^n : x_n < 0\}$. We extend the functions $F(\cdot, p)$ and $f(\cdot)$ onto $B_1 \cap \mathbb{R}^n_-$ by reflection at the hyperplane $\{x_n = 0\}$. Moreover, we set $u(x) = 0$ for all $x \in B_1 \cap \mathbb{R}^n_-$. Let $\tau \in W^{1,\infty}(\mathbb{R}^n)$ be a positive cut-off function with supp $\tau = B_{\frac{1}{2}}$ and $\tau \equiv 1$ in $B_{\frac{1}{4}}$. Further, let $0 < h < R$. It is easy to see that $\tau D_n^{-h} u$ is an admissible test function in equation (12). We find

$$\begin{aligned} J_1 &:= -\sum_{i=1}^{n} \int_{B(\frac{1}{2})} F_i(x, \nabla u) \cdot \tau \partial_i D_n^{-h} u \\ &= \sum_{i=1}^{n} \int_{B(\frac{1}{2})} F_i(x, \nabla u) \cdot \partial_i \tau D_n^{-h} u - \int_{B(\frac{1}{2})} f \cdot \tau D_n^{-h} u \,. \end{aligned} \tag{16}$$

Let us estimate the left-hand side from below. The Taylor expansion of $F(x,\cdot)$ and the ellipticity condition (H4) entail

$$\begin{aligned} F(x,p') - F(x,p) &= \sum_{\nu=1}^{N}\sum_{i=1}^{n}(p'-p)_i^\nu F_i^\nu(x,p) \\ &\quad + \sum_{\nu,\mu=1}^{N}\sum_{i,k=1}^{n}(p'-p)_i^\nu (p'-p)_k^\mu \, F_{ik}^{\nu\mu} \\ &\geq \sum_{i=1}^{n}(p'-p)_i \cdot F_i(x,p) + \frac{c_6}{2}|p'-p|^2, \end{aligned}$$

where $F_{ik}^{\nu\mu} = \int_0^1 (1-t)F_{i,k}^{\nu\mu}(x, tp' + (1-t)p)\,dt$. We put $p = \nabla u(x)$ and $p' = \nabla u(x - he_n)$. This yields $h^{-1}(p'-p)_i = D_n^{-h}\partial_i u$. Multiplying the inequality by τh^{-1} and integrating over $B(\frac{1}{2})$ we get

$$J_1 \geq \frac{c_6}{2}\int_{B(\frac{1}{2})} \tau h\left|D_n^{-h}\nabla u\right|^2 - \int_{B(\frac{1}{2})} \frac{\tau}{h}\left[F(x,\nabla u(x-he_n)) - F(x,\nabla u(x))\right].$$

Using this estimate and (16) we may conclude that

$$\int_{B(\frac{1}{2})} \tau h\left|D_n^{-h}\nabla u\right|^2 \leq c. \tag{17}$$

Therefore, the integral $\int_{B(\frac{1}{2})} \tau h^{-1}\left[F(x,\nabla u(x-he_n)) - F(x,\nabla u(x))\right]$ has to be estimated. Due to the fact that $\tau = 0$ in $B(1)\setminus B(\frac{1}{2})$ we may integrate over $B(1)$ instead $B(\frac{1}{2})$. Then we use the estimate

$$\int_{B(1)} D_n^{-h}(\tau F(x,\nabla u)) = h^{-1}\int_{\Omega_{-h}} \tau F(x,\nabla u) \leq c_3 h^{-1}|\Omega_{-h}| \leq c,$$

where $\Omega_{-h} = \{x \in B(1) : -h < x_n < 0\}$. Here, we have utilized (H1) and the fact that $\nabla u = 0$ in $B_1 \cap \mathbb{R}^n_-$. Now, it is not hard to obtain estimate (17). Notice that $\tau = 1$ in $B(\frac{1}{4})$ and $R = \frac{1}{8} > h$. Thus, the assertion (14) follows for $j = n$. Further, the proof of (14) for $2 \leq j \leq n-1$ is similar. (More details about this difference quotient method can be found in [1, 2].)

Finally, recalling again that $\nabla u = 0$ in $B_1 \cap \mathbb{R}^n_-$ we obtain from (17)

$$c \geq \int_{\Omega_h} h\left|D_n^{-h}\nabla u\right|^2 = \frac{1}{h}\int_{\Omega_h} |\nabla u(x-he_n) - \nabla u(x)|^2 = \frac{1}{h}\int_{\Omega_h} |\nabla u|^2.$$

Hence, we have shown the assertion (15). Thus, the lemma has been proved completely. □

Next, we prove regularity in the x_1–direction.

Lemma 3.8 *Assume u is a weak solution of (11). Further, let $K := [-R_0, R_0]^n$ be a cube such that $K \subset B_{\frac{R}{2}}$. For each $\delta > 0$ there is a constant C such that*

$$\sup_{0<h<R} \int_{K\cap\Omega^1} h^{1+\delta} |D_1^{-h} \nabla u|^2 dx + \sup_{0<h<R} \int_{K\cap\Omega^2} h^{1+\delta} |D_1^{h} \nabla u|^2 dx \le C\,. \quad (18)$$

Proof. We show the assertion for Ω^k, say $k = 2$. Recall that $F_1(x, \nabla u) = H_1(\nabla u)$ in Ω^2. The Taylor expansion yields

$$\sum_{\nu=1}^{N} H_1^\nu(p') - H_1^\nu(p) = \sum_{\nu,\mu=1}^{N} H_{11}^{\nu\mu}(p'-p)_1^\mu + \sum_{j=2}^{n} \sum_{\nu,\mu=1}^{N} H_{1j}^{\nu\mu}(p'-p)_j^\mu,$$

where $H_{1j}^{\nu\mu} = \int_0^1 H_{1,j}^{\nu\mu}(tp' + (1-t)p)\, dt$. Let us note that (H4) entails

$$\left| \sum_{\nu,\mu=1}^{N} H_{11}^{\nu\mu}(p'-p)_1^\mu \right|^2 \ge c\,|p_1' - p_1|^2.$$

We put $p = \nabla u(x)$ and $p' = \nabla u(x + he_1)$. Integrating over $K_0 := K \cap \Omega^2$ and utilizing (H3) we find

$$\int_{K_0} h^{1+\delta} \left| D_1^h \partial_1 u \right|^2 \le c \left(\int_{K_0} h^{1+\delta} \left| D_1^h H_1(\nabla u) \right|^2 + \sum_{j=2}^{n} \int_{K_0} h^{1+\delta} \left| D_1^h \partial_j u \right|^2 \right)$$
$$:= J_2 + J_3.$$

We set $W_k(x) := \tau H_k(\nabla u)$ for $1 \le k \le n$, where τ is a smooth cut-off function with supp $\tau = B_R$ and $\tau \equiv 1$ in $B_{\frac{R}{2}}$. Due to (14) and (15) the assumptions (7) and (8) of Lemma 2.2 are satisfied. Hence, the estimate (9) provides a constant c_δ such that

$$J_2 \le c_\delta.$$

Moreover, let us extend the functions $\partial_j u$, $1 \le j \le n$, by reflecting Ω^2 at the hyperplanes $\{x_1 = 0\}$ and $\{x_n = 0\}$. In view of Lemma 3.7 the assumptions (3) and (4) of Lemma 2.1 are satisfied. Thus, (5) yields

$$J_3 \le c_\delta.$$

Collecting results the assertion follows. □

Proof (of Theorem 3.4). If $\Omega \cap B_1 = \mathbb{R}_+^n \cap B_1$, the estimates (14) and (18) yield

$$\sup_{0<h<R} \int_{K\cap\Omega^1} h^{1+\delta} |D_j^{-h} \nabla u|^2 dx + \sup_{0<h<R} \int_{K\cap\Omega^2} h^{1+\delta} |D_j^{h} \nabla u|^2 dx \le C$$

for $1 \leq j \leq n$, where K is the cube defined in Lemma 3.8. Thus, covering Ω by a finite set of appropriate cubes we may conclude that

$$u \in \mathcal{N}^{\frac{3}{2}-\varepsilon,2}(\Omega^1; \mathbb{R}^N) \quad \text{and} \quad u \in \mathcal{N}^{\frac{3}{2}-\varepsilon,2}(\Omega^2; \mathbb{R}^N) \qquad \forall \varepsilon > 0.$$

If $\Omega \cap B_1 \neq \mathbb{R}^n_+ \cap B_1$, we consider suitable parallelograms instead of cubes. Clearly, using a coordinate transformation we can map a parallelogram onto a cube. Then we can apply the technique based on Fourier analysis as before. This yields the assertion. The accurate study of these general situations would be too technical to be included here. The pictures below shall illustrate the geometrical range of application.

Example 3.9 *Let us consider some examples of admissible two-dimensional domains. In* **Fig. 1** *a domain with two slits is given. Notice that Ω^1 is not convex. However, $\Omega^1 \cap \{x \in \mathbb{R}^n : |x_1| < \gamma\}$ is convex if $\gamma > 0$ is small. Next,* **Fig. 2-4** *show three domains. Possible choices of cubes and parallelograms to apply the Fourier analysis are given.*

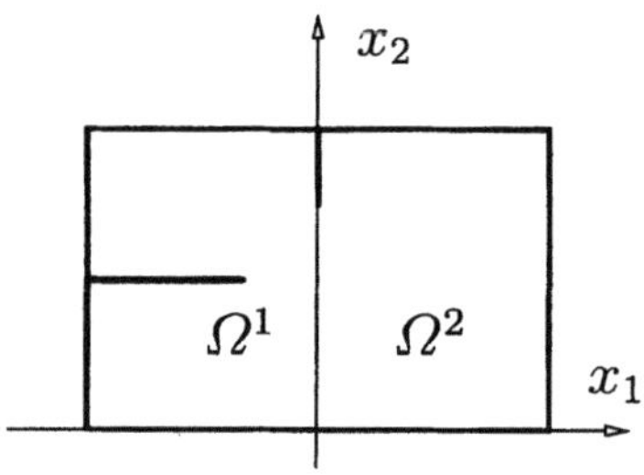

Fig. 1

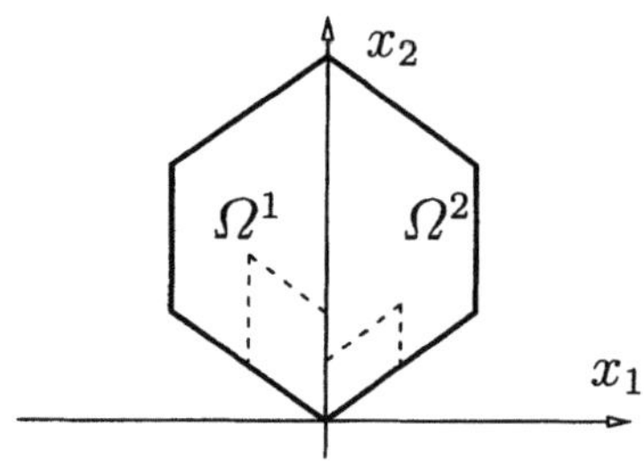

Fig. 2

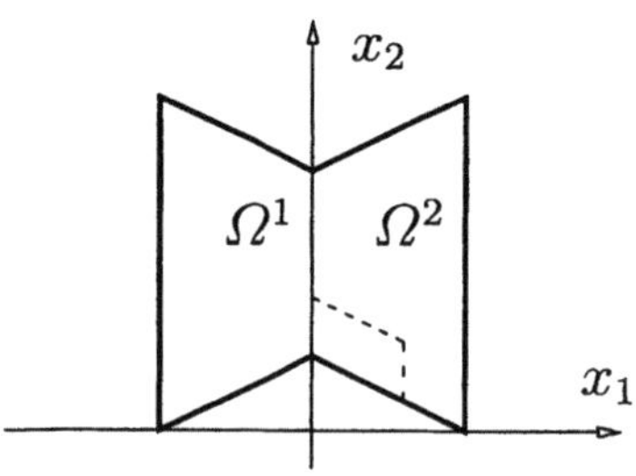

Fig. 3a

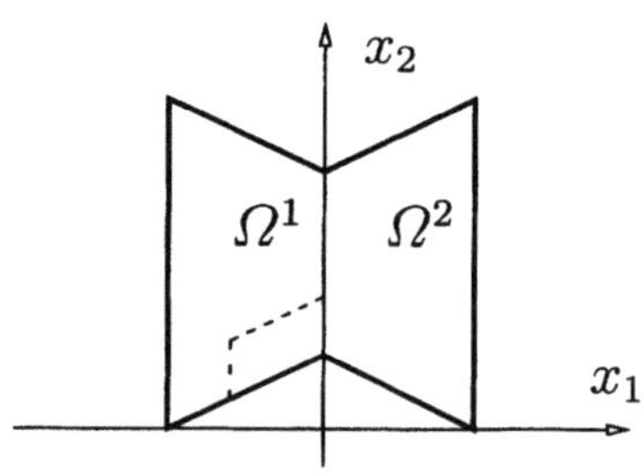

Fig. 3b

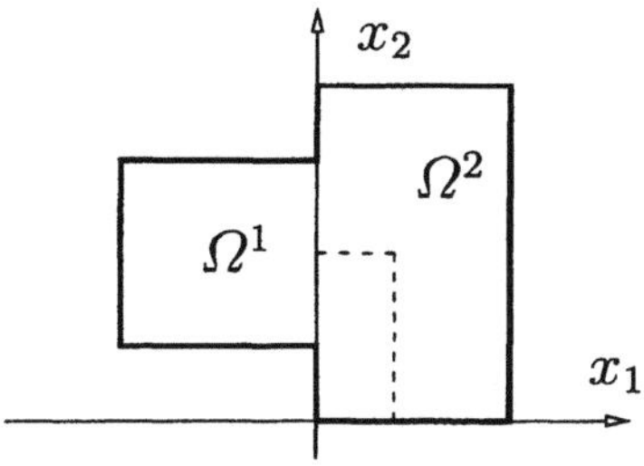

Fig. 4a

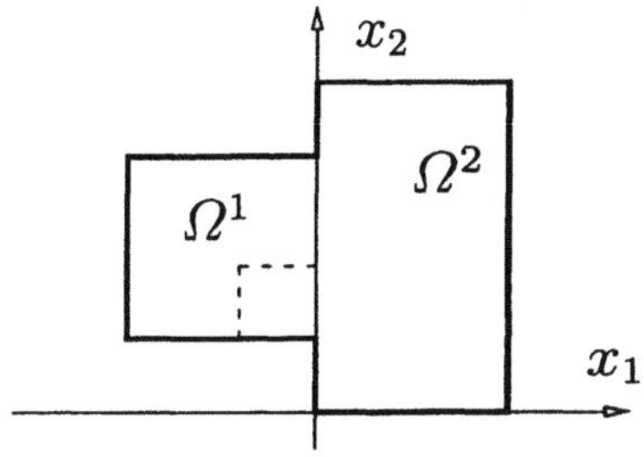

Fig. 4b

References

1. Ebmeyer, C. (1999): Mixed boundary value problems for nonlinear elliptic systems in n-dimensional Lipschitzian domains. Zeit. Anal. Anwend. **18**, 539–555
2. Ebmeyer, C., Frehse, J. (1999): Mixed boundary value problems for nonlinear elliptic equations in multidimensional non-smooth domains. Math. Nachr. **203**, pp. 47–74
3. Ebmeyer, C., Frehse, J. (2001): Steady Navier-Stokes equations with mixed boundary value conditions in three-dimensional Lipschitzian domains. Math. Ann. **319**, pp. 349–381
4. Kufner, A., John, O., Fučik, S. (1977): Function Spaces. Academia, Prague
5. Nicaise, S., Sändig, A.-M., (1999): Transmission problems for the Laplace and elasticity operators: regularity and boundary integral formulation. Math. Models Methods Appl. Sci. **9**, pp. 855-898
6. Oleinik, O.A. (1961): Boundary-value problems for linear equations of elliptic parabolic type with discontinuous coefficients.Izv. Akad. Nauk SSSR Ser. Mat. **25**, 3–20
7. Rivkind, V. Ja., Ural'ceva, N. N. (1972): Classical solvability and linear schemes for the approximate solution of diffraction problems for quasilinear equations of parabolic and of elliptic type. In: Problems of mathematical analysis, No. 3: Integral and differential equations (Russian). Izdat. Leningrad. Univ., Leningrad

A Particle–Partition of Unity Method Part V: Boundary Conditions

M. Griebel and M.A. Schweitzer

Institut für Angewandte Mathematik, Universität Bonn, Wegelerstr. 6,
D-53115 Bonn, Germany

Summary. In this sequel to [12, 13, 14, 15] we focus on the implementation of Dirichlet boundary conditions in our partition of unity method. The treatment of essential boundary conditions with meshfree Galerkin methods is not an easy task due to the non-interpolatory character of the shape functions. Here, the use of an almost forgotten method due to Nitsche from the 1970's allows us to overcome these problems at virtually no extra computational costs. The method is applicable to general point distributions and leads to positive definite linear systems. The results of our numerical experiments, where we consider discretizations with several million degrees of freedom in two and three dimensions, clearly show that we achieve the optimal convergence rates for regular and singular solutions with the (adaptive) h-version and (augmented) p-version.

1 Introduction

Meshfree methods for the numerical treatment of partial differential equations gained much attention in recent years. These methods allow for the construction of shape functions without the need of a mesh or grid so that they are promising approaches to overcome the problem of mesh generation.

Since meshfree methods are independent of a mesh, their shape functions are in general more complex than finite element shape functions. In a meshfree method the shape functions are usually piecewise rational functions, whereas in a finite element method (FEM) they are piecewise polynomials. This makes the meshfree Galerkin discretization of a partial differential equation (PDE) more challenging than its discretization with a FEM. See [12, 13, 27, 28] for details on the discretization process with the partition of unity method (PUM). Furthermore, the shape functions are in general non-interpolatory which makes the design of optimal linear solvers [14] as well as the implementation of essential boundary conditions not an easy task.

Many different approaches toward the treatment of Dirichlet problems have been proposed over the years [4, 10, 12, 13, 16, 17, 19, 22, 27]. Most of them, however, suffer from one or several drawbacks (restrictions on the distribution of the discretization points, need for an additional boundary function space, saddle-point structure, non-optimal convergence rates, etc.). Yet, the problem was in fact already solved in the early 1970's when Nitsche [23] developed a general approach for the treatment of essential boundary conditions where the trial and test functions do not have to fulfill the boundary conditions. Nitsche's method, however, seems to be quite unknown. Recently, Stenberg [29] revived the interest in this non-standard method by showing its connection to some stabilization techniques for the Lagrange multiplier method.

In this paper we present the implementation of essential boundary conditions by Nitsche's method in the context of our partition of unity method for the numerical solution of elliptic PDE. The remainder of the paper is organized as follows. We begin with a short review of the construction of PUM spaces, the Galerkin discretization of an elliptic PDE and the multilevel solution of the arising linear block-system in §2. The treatment of essential boundary conditions is presented in §3. Here, we discuss the various approaches and their difficulties and give a detailed introduction to Nitsche's method. Then, we present the results of our numerical experiments where we apply our PUM Dirichlet problems with regular and singular solutions in two and three dimensions in §4. Finally, we conclude with some remarks in §5.

2 Partition of Unity Method

In the following, we shortly review the construction partition of unity spaces and the meshfree Galerkin discretization of an elliptic PDE, see [12, 13] for details. Furthermore, we give a summary of the efficient multilevel solution of the arising linear block-system, see [14] for details.

2.1 Construction of Partition of Unity Method

In a partition of unity method, we define a global approximation u^{PU} simply as a weighted sum of local approximations u_i,

$$u^{\mathrm{PU}}(x) := \sum_{i=1}^{N} \varphi_i(x) u_i(x). \tag{1}$$

These local approximations u_i are completely independent of each other, i.e., the local supports $\omega_i := \mathrm{supp}(u_i)$, the local basis $\{\psi_i^n\}$ and the order of approximation p_i for every single $u_i := \sum u_i^n \psi_i^n \in V_i^{p_i}$ can be chosen independently of all other u_j. Here, the functions φ_i form a partition of unity (PU). They are used to splice the local approximations u_i together in such a way that the global approximation u^{PU} benefits from the local approximation orders p_i yet it still fulfills global regularity conditions, see [12]. Hence, the global approximation space on Ω is defined as

$$V^{\mathrm{PU}} := \sum_i \varphi_i V_i^{p_i} = \sum_i \varphi_i \,\mathrm{span}\langle\{\psi_i^n\}\rangle = \mathrm{span}\langle\{\varphi_i \psi_i^n\}\rangle. \tag{2}$$

The starting point for any meshfree method is a collection of N independent points $P := \{x_i \in \mathbb{R}^d \,|\, x_i \in \overline{\Omega}, i = 1, \ldots, N\}$. In the PU approach we need to construct a partition of unity $\{\varphi_i\}$ on the domain of interest Ω to define an approximate solution (1) where the union of the supports $\mathrm{supp}(\varphi_i) = \overline{\omega_i}$ covers the domain $\overline{\Omega} \subset \bigcup_{i=1}^N \omega_i$

and $u_i \in V_i^{p_i}(\omega_i)$ is some locally defined approximation of order p_i to u on ω_i. Thus, the first (and most crucial) step in a PUM is the efficient construction of an appropriate cover C_Ω. This is not and easy task [13, 27]. Throughout this paper we use a tree-based construction algorithm for d-rectangular covers presented in [13]. Here, the cover patches ω_i are products of intervals $(x_i^l - h_i^l, x_i^l + h_i^l)$. With the help of weight functions W_k defined on these cover patches ω_i we can easily generate a partition of unity by *Shepard's method*, i.e., we define

$$\varphi_i(x) = \frac{W_i(x)}{\sum_{\omega_k \in C_\Omega^i} W_k(x)}, \tag{3}$$

where $C_i := \{\omega_j \in C_\Omega \,|\, \omega_i \cap \omega_j \neq \emptyset\}$ is the set of all geometric neighbors of a cover patch ω_i. Due to the use of d-rectangular patches ω_i, the most natural choice for a weight function W_i is a product of one-dimensional functions, i.e., $W_i(x) = \prod_{l=1}^d W_i^l(x^l) = \prod_{l=1}^d \mathcal{W}(\frac{x - x_i^l + h_i^l}{2h_i^l})$ with $\operatorname{supp}(\mathcal{W}) = [0,1]$ such that $\operatorname{supp}(W_i) = \overline{\omega_i}$. It is sufficient for this construction to choose a one-dimensional weight function $\mathcal{W}$ with the desired regularity which is non-negative. The partition of unity functions φ_i inherit the regularity of the generating weight function $\mathcal{W}$. We always use a normed B-spline [27] as the generating weight function $\mathcal{W}$.

In general, a partition of unity $\{\varphi_i\}$ can of course only recover the constant function on the domain Ω. Hence, we need to improve the approximation quality to use the method for the discretization of a PDE. To this end, we multiply the partition of unity functions φ_i locally with polynomials ψ_i^n. Since we use d-rectangular patches ω_i only, a local tensor product space is the most natural choice. Throughout this paper, we use products of univariate Legendre polynomials as local approximation spaces $V_i^{p_i}$, i.e., we choose

$$V_i^{p_i} = \operatorname{span}\langle\{\psi_i^n \,|\, \psi_i^n = \prod_{l=1}^d \mathcal{L}_i^{\hat{n}_l}, \|\hat{n}\|_1 = \sum_{l=1}^d \hat{n}_l \leq p_i\}\rangle,$$

where $\hat{n}$ is the multi-index of the polynomial degrees $\hat{n}_l$ of the univariate Legendre polynomials $\mathcal{L}_i^{\hat{n}_l} : [x_i^l - h_i^l, x_i^l + h_i^l] \to \mathbb{R}$, and n is the index associated with the product function $\psi_i^n = \prod_{l=1}^d \mathcal{L}_i^{\hat{n}_l}$.

In summary, we can view the construction given above as follows

$$\begin{pmatrix} \{x_i\} \\ \mathcal{W} \\ \{p_i\} \end{pmatrix} \to \begin{pmatrix} \{\omega_i\} \\ \{W_i\} \\ \{V_i^{p_i} = \operatorname{span}\langle\psi_i^n\rangle\} \end{pmatrix} \to \begin{pmatrix} \{\varphi_i\} \\ \{V_i^{p_i}\} \end{pmatrix} \to V^{\mathrm{PU}} = \sum \varphi_i V_i^{p_i},$$

where the set of points $P = \{x_i\}$, the generating weight function $\mathcal{W}$ and the local approximation orders p_i are assumed to be given.

2.2 Basic Convergence Theory

In this section we state the notation and basic theory of the partition of unity method as developed by Babuška and Melenk in [5, 6]. Crucial to the theory of the partition of unity method is the notion of a (M, C_∞, C_∇) partition of unity. The conditions formulated in the following definitions allow to show the basic approximation properties as given in Theorem 2.3 of a partition of unity space V^{PU} (2).

Definition 2.1 (Partition of Unity) *Let $\Omega \subset \mathbb{R}^d$ be an open set, let $C_\Omega = \{\omega_i\}$ be an open cover of Ω satisfying a point-wise overlap condition*

$$\exists M \in \mathbb{N} \quad \forall x \in \Omega \quad \mathrm{card}(\{i \,|\, x \in \omega_i\}) \le M\,.$$

Let $\{\varphi_i\}$ be a Lipschitz partition of unity subordinate to the cover C_Ω satisfying

$$\begin{aligned} &\mathrm{supp}(\varphi_i) \subset \overline{\omega_i} \quad \textit{for all } i, && \textstyle\sum_i \varphi_i \equiv 1 \textit{ on } \Omega, \\ &\|\varphi_i\|_{L^\infty(\mathbb{R}^d)} \le C_\infty, && \|\nabla\varphi_i\|_{L^\infty(\mathbb{R}^d)} \le \tfrac{C_\nabla}{\mathrm{diam}(\omega_i)}, \end{aligned}$$

where C_∞ and C_∇ are two positive constants. Then $\{\varphi_i\}$ is called a (M, C_∞, C_∇) partition of unity subordinate to the cover C_Ω. The partition of unity is said to be of degree $k \in \mathbb{N}_0$ if $\varphi_i \in C^k(\mathbb{R}^d)$ for all i. The covering sets ω_i are called patches.

Definition 2.2 (Partition of Unity Space) *Let $C_\Omega = \{\omega_i\}$ be an open cover of $\Omega \subset \mathbb{R}^d$ and let $\{\varphi_i\}$ be a (M, C_∞, C_∇) partition of unity subordinate to C_Ω. Let $V_i^{p_i} = \mathrm{span}\langle\{\psi_i^n\}\rangle \subset H^1(\Omega \cap \omega_i)$ be given. Then the space*

$$V^{\mathrm{PU}} := \sum_i \varphi_i V_i^{p_i} = \sum_i \varphi_i \,\mathrm{span}\langle\{\psi_i^n\}\rangle = \mathrm{span}\langle\{\varphi_i \psi_i^n\}\rangle$$

is called a partition of unity space. The space V^{PU} is said to be of degree $k \in \mathbb{N}$ if $V^{\mathrm{PU}} \subset C^k(\Omega)$. The spaces $V_i^{p_i}$ are referred to as the local approximation spaces.

Theorem 2.3 (Approximation Property) *Let $\Omega \subset \mathbb{R}^d$ be given. Let $C_\Omega = \{\omega_i\}$, $\{\varphi_i\}$ and $V_i^{p_i}$ be as in Definitions 2.1 and 2.2. Let $u \in H^1(\Omega)$ be the function to be approximated. Assume that the local approximation spaces $V_i^{p_i}$ have the following approximation properties: On each patch $\Omega \cap \omega_i$, the function u can be approximated by a function $v_i \in V_i^{p_i}$ such that $\|u - v_i\|_{L^2(\Omega\cap\omega_i)} \le \hat{\epsilon}_i$, and $\|\nabla(u - v_i)\|_{L^2(\Omega\cap\omega_i)} \le \tilde{\epsilon}_i$ hold. Then the function*

$$u^{\mathrm{PU}} := \sum_i \varphi_i v_i \in V \subset H^1(\Omega)$$

satisfies

$$\begin{aligned} \|u - u^{\mathrm{PU}}\|_{L^2(\Omega)} &\le \sqrt{M} C_\infty \Big(\sum_{i=1}^N \hat{\epsilon}_i^2\Big)^{\frac{1}{2}}, \\ \|\nabla(u - u^{\mathrm{PU}})\|_{L^2(\Omega)} &\le \sqrt{2M} \Big(\sum_{i=1}^N \Big(\frac{C_\nabla}{\mathrm{diam}(\omega_i)}\Big)^2 \hat{\epsilon}_i^2 + C_\infty^2 \tilde{\epsilon}_i^2\Big)^{\frac{1}{2}}. \end{aligned}$$

Proof. See [5, 6].

Due to this theorem we can use the partition of unity method as an h-version, a p-version or even an hp-version. From the above estimate we can see that the approximation property of the approximation space V^{PU} may be improved via the reduction of the diameters of the cover patches ω_i (the insertion of points x_i) or via the enhancement of the approximation qualities of the local spaces $V_i^{p_i}$ (the increment of the local approximation order p_i). These refinement strategies are independent of each other, thus we may use both at the same time. Since the PUM only employs a scattered data set, an adaptive h-refinement (by the insertion of new points x_i) does not have to cope with problems caused by the grid-structure like hanging nodes, etc. Furthermore, there are no compatibility restrictions on the local spaces $V_i^{p_i}$. Hence, we may also employ an adaptive p-refinement on the local spaces $V_i^{p_i}$, i.e., we may use a different order of approximation p_i and even different basis functions ψ_i^n in different parts of the domain.

2.3 Galerkin Discretization of Elliptic Equations

We want to solve elliptic boundary value problems of the type

$$\begin{aligned} Lu &= f \text{ in } \Omega \subset \mathbb{R}^d\,, \\ Bu &= g \text{ on } \partial\Omega\,, \end{aligned} \tag{4}$$

where L is a symmetric partial differential operator of second order and B expresses suitable boundary conditions.

In the following let $a\,(\cdot,\cdot)$ be the continuous and elliptic bilinear form induced by L on $V := H^1(\Omega)$. We discretize the partial differential equation using Galerkin's method. Then, we have to compute the stiffness matrix

$$A = \left(A_{(i,n),(j,m)}\right), \text{ with } A_{(i,n),(j,m)} = a\,(\varphi_j\psi_j^m, \varphi_i\psi_i^n) \in \mathbb{R}\,,$$

and the right hand side vector

$$\hat{f} = \left(f_{(i,n)}\right), \text{ with } f_{(i,n)} = \langle f, \varphi_i\psi_i^n\rangle_{L^2} = \int_\Omega f\varphi_i\psi_i^n \in \mathbb{R}\,.$$

If we restrict ourselves for reasons of simplicity to the case $L = -\Delta$ we have to compute the integrals $\int_\Omega \varphi_i\psi_i^n f$ for the right hand side and the integrals $\int_\Omega \nabla(\varphi_i\psi_i^n)\nabla(\varphi_j\psi_j^m)$ for the stiffness matrix. Recall that φ_i is defined by (3), i.e.

$$\varphi_i(x) = \frac{W_i(x)}{\sum W_k(x)}\,.$$

Now we carry out the differentiation in $\int_\Omega \nabla(\varphi_i\psi_i^n)\nabla(\varphi_j\psi_j^m)$. With the notation $\mathcal{S} := \sum W_k$, $\mathcal{T} := \sum \nabla W_k$ and $\mathcal{G}_i := \nabla W_i \mathcal{S} - W_i \mathcal{T}$ we end up with the integrals

$$\begin{aligned} a(\varphi_j\psi_j^m, \varphi_i\psi_i^n) = \int_\Omega \mathcal{S}^{-4}\mathcal{G}_i\psi_i^n\mathcal{G}_j\psi_j^m \quad &+ \textstyle\int_\Omega \mathcal{S}^{-2}W_i\nabla\psi_i^n W_j\nabla\psi_j^m\, + \\ \int_\Omega \mathcal{S}^{-3}\mathcal{G}_i\psi_i^n W_j\nabla\psi_j^m &+ \textstyle\int_\Omega \mathcal{S}^{-3}W_i\nabla\psi_i^n\mathcal{G}_j\psi_j^m \end{aligned} \tag{5}$$

for the stiffness matrix and the integrals

$$\langle f, \varphi_i \psi_i^n \rangle_{L^2} = \int_\Omega \mathcal{S}^{-1} W_i \psi_i^n f \tag{6}$$

for the right hand side. These integrands may have quite a number of jumps of significant size since we use piecewise polynomial weights W_i whose supports ω_i overlap in the Shepard construction (3). Therefore, the integrals of the weak form have to be computed with certain care using an appropriate numerical quadrature scheme, see [12, 13].

2.4 Multilevel Solution of Resulting Linear System

The product structure of the shape functions $\varphi_i \psi_i^n$ implies two natural block-partitions of the resulting linear system $A\tilde{u} = \hat{f}$, where $\tilde{u}$ denotes a coefficient vector and $\hat{f}$ denotes a moment vector.

1. The stiffness matrix A can be arranged in *spatial blocks*. A spatial block A_{nm} corresponds to a discretization of the PDE on the complete domain Ω using the trial functions $\varphi_j \psi_j^m$ and the test function $\varphi_i \psi_i^n$ with fixed n and m. Here, all blocks A_{nm} are sparse matrices and have the same row and column dimensions which corresponds to the number of partition of unity functions φ_i.
2. The stiffness matrix A may also be arranged in *polynomial blocks*. Here, a single block A_{ij} corresponds to a local discretization of the PDE on the domain $\omega_i \cap \omega_j \cap \Omega$. The polynomial blocks A_{ij} are dense matrices and may have different dimensions corresponding to the dimensions of the local approximation spaces $V_j^{p_j}$ and $V_i^{p_i}$.

This separation of the degrees of freedom into local approximation functions ψ_i^n and partition of unity functions φ_i can be used to define two different multilevel concepts [14]. Throughout this paper we assume that the stiffness matrix is given in polynomial block-form and we use the corresponding spatial multilevel solver developed in [14] for the fast and efficient solution of the resulting large sparse linear block-system $A\tilde{u} = \hat{f}$, where $\tilde{u}$ denotes a coefficient block-vector and $\hat{f}$ a moment block-vector.

In a multilevel method we need a sequence of discretization spaces V_k with $k = 0, \dots, J$ where J denotes the finest level. To this end we construct a sequence of PUM spaces V_k^{PU} as follows. We use a tree-based algorithm developed in [13, 14] to generate a sequence of point sets P_k and covers C_Ω^k from a given initial point set $\tilde{P}$. Following the construction given in §2.1 we can then define an associated sequence of PUM spaces V_k^{PU}. Note that these spaces are nonnested, i.e., $V_{k-1}^{\mathrm{PU}} \not\subset V_k^{\mathrm{PU}}$, and that the shape functions $\varphi_{i,k}\psi_{i,k}^n$ are non-interpolatory. Thus, we need to construct appropriate transfer operators $I_{k-1}^k : V_{k-1}^{\mathrm{PU}} \to V_k^{\mathrm{PU}}$ and $I_k^{k-1} : V_k^{\mathrm{PU}} \to V_{k-1}^{\mathrm{PU}}$. With such transfer operators I_{k-1}^k, I_k^{k-1} and the stiffness matrices A_k coming from the Galerkin discretization on each level k we can then set up a standard multiplicative multilevel iteration to solve the linear system $A_J \tilde{u}_J = \hat{f}_J$.

Our multilevel solver utilizes special localized L^2-projections for the interlevel transfers and a block-smoother to treat all local degrees of freedom ψ_i^n within a patch ω_i simultaneously. The convergence rates of this multilevel iteration are independent of the number of cover patches $\mathrm{card}(C_\Omega^J)$ and only slightly dependent on the polynomial degrees. The typical convergence rate of a $V(1,1)$-cycle is $\rho \leq 0.25$ for a PUM space with linear local approximation spaces. For further details see [14].

3 Boundary Conditions

Our PUM shape functions $\varphi_i \psi_i^n$ are non-interpolatory since the partition of unity functions φ_i are (in general) non-interpolatory, i.e., $\varphi_i(x_j) \neq \delta_{ij}$. Furthermore the usage of local approximation spaces $V_i^{p_i}$ with $\dim(V_i^{p_i}) > 1$ generates an approximation space $V^{\mathrm{PU}} = \sum_i \varphi_i V_i^{p_i}$ with more degrees of freedom than interpolation nodes x_i. Thus, we have to cope with the problem: How to fulfill boundary conditions?

First consider (4) with Neumann boundary conditions $Bu = u_n := \partial u / \partial n := \nabla u \cdot n = g$ on $\partial\Omega$, where n denotes the outer normal. Here, we learn from the variational formulation

$$F(v) := \frac{1}{2} a(v,v) - \langle f, v \rangle_{L^2} - \int_{\partial\Omega} gv \rightarrow \min\{v \in H^1(\Omega)\}, \tag{7}$$

that the trial functions v have to fulfill no additional constraint besides being from the definition space $H^1(\Omega)$ of the differential operator L in its weak form. The boundary conditions are not imposed explicitly on the function space. Thus, the basis of a finite-dimensional subspace $V \subset H^1(\Omega)$ used to approximate the solution of (7) may be compiled of arbitrary functions $v \in H^1(\Omega)$. The basis functions do not need to be interpolatory. Hence, we may use our functions $\varphi_i \psi_i^k$ as trial and test functions in a Galerkin procedure without any modification.

However, Dirichlet boundary conditions $Bu = u = g$ on $\partial\Omega$ explicitly impose the values of the solution u on the boundary $\partial\Omega$. Thus, the trial space of the usual weak formulation

$$\text{Find } u \in H_g^1(\Omega): \quad a(u,v) = \langle f, v \rangle_{L^2} \text{ for all } v \in H_0^1(\Omega)$$

is not the complete space $H^1(\Omega)$ but $H_g^1(\Omega) := \{v \in H^1(\Omega) \,|\, u = g \text{ on } \partial\Omega\}$, whereas the test space is $H_0^1(\Omega)$. Note that we can enforce vanishing Dirichlet boundary conditions within the PUM by the selection of appropriate local approximation spaces $V_i^{p_i}$, i.e., we need $\psi_i^k|_{\partial\Omega} \equiv 0$ for all local basis functions ψ_i^k [4, 10]. But the implementation of a trial space $V \subset H_g^1(\Omega)$ with $g \not\equiv 0$ by the selection of appropriate local approximation spaces is not feasible.

There are many different approaches toward the treatment of Dirichlet boundary conditions with meshfree methods [4, 10, 16, 27], e.g.

1. coupling to mesh-based methods close to the boundary [19],
2. penalty or perturbation methods [17, 22],
3. the Lagrange multiplier method [12, 27],
4. and Nitsche's method [4, 28].

Obviously the coupling of a meshfree method to a mesh-based method destroys the meshfree character of the original method just for the sake of the implementation of Dirichlet boundary conditions. This loss of generality and freedom can hardly be justified. Furthermore, the coupling itself must be consistent with the local approximation orders p_i of the local spaces or we also experience an adverse effect on the approximation quality of the overall method only due to the poor implementation of Dirichlet boundary conditions.

The penalty or perturbation approaches are very general concepts for the implementation of constraints in a variational problem. In our setting we would introduce an additional surface term in the variational formulation to enforce the boundary conditions. This penalty term may change the properties of the functional and we need to be concerned with the issues of existence and uniqueness of a solution. Furthermore, we usually do not achieve the maximal rate of convergence [2, 4, 23], i.e., again we would experience a reduction in the approximation quality of the overall method just because of the inappropriate treatment of boundary conditions.

The Lagrange multiplier method is a general approach toward the solution of constrained minimization problems which is also used in the finite element [3, 8] and wavelet [20] context to implement essential boundary conditions. It is well-known that the method converges with the optimal rates if the function spaces involved, in our setting the interior PUM approximation space and the multiplier space on the boundary, fulfill a (discrete) Ladyzhenskaya–Babuška–Brezzi (or inf-sup) condition. Here, the main problem is the design of an appropriate multiplier space on the boundary. Within the finite element context Pitkäranta [24, 25, 26] showed that there is not much freedom in the design of the multiplier space if the optimal convergence of the method is desired. Furthermore, the use of Lagrange multipliers (in general) leads to a saddle-point problem and the arising linear system is indefinite and the design of an optimal solver is not an easy task.

A different variational approach to Dirichlet problems due to Nitsche [23], however, allows for the use of subspaces $V_N \subset H^1(\Omega)$ which do not have to fulfill the boundary conditions explicitly, yet it gives the optimal rate of convergence. Let us consider the Poisson problem

$$\begin{aligned} -\Delta u &= f \text{ in } \Omega \subset \mathbb{R}^d, \\ u &= g \text{ on } \partial\Omega, \end{aligned} \tag{8}$$

for reasons of simplicity. We are interested in finding an approximate solution $u_N \in V_N \subset H^1(\Omega)$ to (8) — within optimal error bounds. In an early yet not widely known paper [23] Nitsche proposed to minimize the functional $J_N(v-u)$ among all $v \in V_N$ where u is the solution of (8),

$$J_N(w) := \int_\Omega |\nabla w|^2 - 2\int_{\partial\Omega} w w_n + \beta_N \int_{\partial\Omega} w^2,$$

and $\beta_N > 0$ depends only on the subspace V_N, i.e., the approximation u_N is given by $J(u_N - u) := \inf_{v \in V_N} J_N(v - u)$. Note that the subscript n denotes the normal derivative, i.e., $w_n = \nabla w \cdot n$, whereas the subscript N indicates a dependence on the discretization space $V_N \subset H^1(\Omega)$. The minimizer $u_N \in V_N$ can be computed from the input data f and g of (8) since

$$\begin{aligned} J_N(v-u) &= J_N(v) + J_N(u) - 2\left(\int_\Omega \nabla v \nabla u + \int_{\partial\Omega} u(\beta_N v - v_n) - v u_n\right), \\ &= J_N(v) + J_N(u) - 2\left(\int_\Omega f v + \int_{\partial\Omega} g(\beta_N v - v_n)\right). \end{aligned}$$

The corresponding weak formulation is given by $a_N(u_N, v) = \langle l_N, v\rangle$ for all $v \in V_N$ where

$$\begin{aligned} a_N(w,v) &:= \int_\Omega \nabla v \nabla w - \int_{\partial\Omega} v w_n - \int_{\partial\Omega} w v_n + \beta_N \int_{\partial\Omega} v w, \\ \langle l_N, v\rangle &:= \int_\Omega f v - \int_{\partial\Omega} g v_n + \beta_N \int_{\partial\Omega} g v. \end{aligned}$$

Although the bilinearform $a_N(\cdot,\cdot)$ is indefinite on the space $H^1(\Omega)$ it is symmetric positive definite on the subspace V_N under the assumptions that

$$\|v_n\|_{L^2(\partial\Omega)} \le C_N \|\nabla v\|_{L^2(\Omega)} \tag{9}$$

holds for all $v \in V_N$ with $C_N > 0$ and that $\beta_N > 2C_N^2$ since

$$\begin{aligned} a_N(v,v) &= \|\nabla v\|^2_{L^2(\Omega)} - 2\int_{\partial\Omega} v v_n + \beta_N \|v\|^2_{L^2(\partial\Omega)} \\ &\ge \|\nabla v\|^2_{L^2(\Omega)} - 2C_N \|v\|_{L^2(\partial\Omega)} \|\nabla v\|_{L^2(\Omega)} + \beta_N \|v\|^2_{L^2(\partial\Omega)} \\ &\ge \frac{1}{2}\|\nabla v\|^2_{L^2(\Omega)} + (\beta_N - 2C_N^2)\|v\|^2_{L^2(\partial\Omega)}. \end{aligned}$$

Nitsche furthermore proved optimal error estimates if the relation

$$C_N^2 \simeq \operatorname{diam}(\operatorname{supp}(\phi))^{-1} \tag{10}$$

for C_N in (9) holds for all basis functions $\phi \in V_N$ and if the respective approximation property is given in V_N. The proportionality (10) is valid e.g. if we have estimates of the form

$$\int_{\partial\Omega} |\phi_n|^2 \le C_{N,1} (\operatorname{diam}(\operatorname{supp}(\phi)))^{d-1} \tag{11}$$

and

$$\int_\Omega |\nabla\phi|^2 \ge C_{N,2} (\operatorname{diam}(\operatorname{supp}(\phi)))^{d} \tag{12}$$

for all basis functions $\phi \in V_N$ with $\text{supp}(\phi) \cap \partial\Omega \neq \emptyset$. Note that (11) and (12) essentially introduce some geometric constraints on the intersections $\text{supp}(\phi) \cap \Omega$ and $\text{supp}(\phi) \cap \partial\Omega$, i.e., in our meshfree context on the cover C_Ω or in the finite element context on the regularity of the mesh.

The proof of (10), i.e., of (11) and (12), for linear finite elements on a regular mesh is trivial since the first order derivatives $\partial_i \phi$ of a shape function ϕ are constant on an element. In general a proof of (10) is simplified when we only need to consider a regular reference configuration, i.e., where the map to the reference configuration is affine. Here, we find

$$\begin{aligned}\frac{\int_{\partial\,\text{supp}(\phi)} |\phi_n|^2}{\int_{\text{supp}(\phi)} |\nabla\phi|^2} &= \frac{\det(J_{\partial T}) \int_{\partial\omega_{\text{ref}}} |\phi_n \circ \partial T|^2}{\det(J_T) \int_{\omega_{\text{ref}}} |\nabla\phi \circ T|^2} \\ &= \frac{\det(J_{\partial T})}{\det(J_T)} C_{N,\text{ref}} \\ &\simeq (\text{diam}(\text{supp}(\phi)))^{-1} C_{N,\text{ref}}\end{aligned}$$

where $C_{N,\text{ref}}$ is only dependent on the polynomial degree of the shape function ϕ. Thus, if we limit ourselves to the use of uniform covers and a fixed local approximation space we only need to consider very few reference cases (depending on the number of edges of $\text{supp}(\varphi_i \psi_i^n) \cap \partial\Omega$ and the polynomial degree p_i of $V_i^{p_i}$). However, for the general situation where we have an irregular point distribution and locally varying approximation spaces this approach cannot be pursued. Furthermore, from a computational point of view we must be interested in the *value* of C_N in (9), not only the type of the proportionality (10). Only a large enough value C_N will lead to a definite problem formulation. Yet, a wrong choice of the regularization parameter may have an impact on the condition number of the linear system or other adverse effects on the applicability of certain linear solvers. The parameter C_N is not only dependent on the support sizes but also on the selected local basis functions ψ_i^k for the space $V_i^{p_i}$ and the local approximation orders p_i. Hence, we need to be concerned with the automatic computation of a reliable estimate of C_N. Here, we decided to approach the estimate (9) as a generalized eigenvalue problem

$$\mathcal{A}x = \lambda \mathcal{B}x \tag{13}$$

where

$$\mathcal{A}_{(i,k),(j,m)} := \int_{\partial\Omega} (\varphi_j \psi_j^m)_n (\varphi_i \psi_i^k)_n$$

and

$$\mathcal{B}_{(i,k),(j,m)} := \int_{\Omega} \nabla(\varphi_j \psi_j^m) \nabla(\varphi_i \psi_i^k)$$

for all index pairs $(i,k),(j,m)$ which correspond to a shape function which overlaps the boundary $\partial\Omega$, i.e., $\omega_i \cap \partial\Omega \neq \emptyset$ and $\omega_j \cap \partial\Omega \neq \emptyset$. Solving (13) for the maximal eigenvalue $\lambda_{\max}$ we get a good estimate for C_N^2.

Note that the assembly of the matrices $\mathcal{A}$ and $\mathcal{B}$ does not introduce a significant amount of additional computational costs. The entries of $\mathcal{B}$ are needed for the stiffness matrix anyway and can be re-used. The remaining additional costs associated with the computation of the regularization parameter β_N come from the solution of the eigenvalue problem (13). The eigenvalue $\lambda_{\max}$ can be computed very efficiently by a simultaneous Rayleigh-quotient minimization method [7, 11, 21] due to the similar structure of the matrices $\mathcal{A}$ and $\mathcal{B}$. Here, the minimization of $\frac{x^T \mathcal{B} x}{x^T \mathcal{A} x}$ only involves matrix-vector-products. We do not need to solve a linear system. Furthermore, the eigenvalue problem (13) involves only boundary degrees of freedom and is therefore of smaller dimension. Hence, the computational costs associated with the assembly of a Dirichlet problem are comparable to the costs associated with the respective Neumann problem.

Nitsche's method is closely related to some stabilized Lagrange multiplier techniques [29] which are used in Discontinuous Galerkin methods [1], Domain decomposition methods [9] and Mortar finite element methods [18, 30]. From our point of view, this approach provides the most natural implementation of Dirichlet boundary conditions for meshfree methods. The main advantages of Nitsche's method are:

1. It does *not* introduce constraints on the distribution of the points $x_i \in P$.
2. The problem formulation involves only the a single function space. There is no need of an additional appropriate function space on the boundary.
3. The method leads to symmetric definite linear systems. We do not need to be concerned with linear solvers for saddle-point problems.

Furthermore, we can automatically compute the regularization parameter β_N without significant extra computational costs. This is especially important for highly irregular point sets P, varying local basis functions ψ_i^k or varying local approximation orders p_i. Hence, we use Nitsche's method for the implementation of Dirichlet boundary conditions throughout this paper. In summary, the implementation of Neumann or Dirichlet boundary conditions involve only the evaluation of boundary integrals of the PUM shape functions $\varphi_i \psi_i^k$ and their gradients. We get the same rates of convergence for both types of boundary conditions and the arising linear systems are positive definite independent of the boundary conditions. The computational costs associated with the assembly of the stiffness matrix A and right-hand side vector $\hat{f}$ are comparable for both types of boundary conditions. Finally, we give the weak formulation of a Poisson problem

$$\begin{aligned} -\Delta u &= f \quad \text{in } \Omega \subset \mathbb{R}^d, \\ u &= g_D \text{ on } \Gamma_D \subset \partial\Omega, \\ u_n &= g_N \text{ on } \Gamma_N = \partial\Omega \setminus \Gamma_D, \end{aligned}$$

with mixed boundary conditions where the Dirichlet boundary conditions are realized with Nitsche's method and the Neumann boundary conditions are implemented in the standard fashion as an additional surface term on the right-hand side. This weak formulation

$$\int_{\Omega} \nabla u \nabla v + \int_{\Gamma_D} u(\beta v - v_n) - u_n v = \int_{\Omega} f v + \int_{\Gamma_D} g_D(\beta v - v_n) + \int_{\Gamma_N} g_N v,$$

where β now denotes the respective regularization parameter, is the problem formulation which we have implemented. Here, the volume and surface integrals are computed with the help of a sparse grid quadrature scheme presented in [13].

Note that the introduction of a space-dependent (i.e. level-dependent) regularization parameter may effect the quality of a multilevel solver for the resulting linear system. The regularization parameter grows with the number of levels due to the relation $\beta_l \sim h^{-1} = 2^l$ so that one approach could be the use of β_l from the finest level l also on the coarser levels $k < l$, i.e. $\beta_k = \beta := \beta_l$ for all $k = 0, \dots, l$. With this choice for β we have the same variational form on all levels k. This approach, however, leads to an imbalance between the approximation of the boundary conditions and the approximation of the PDE in the domain on the different levels which may adversely effect the overall convergence rate of the multilevel solver. Furthermore, if we use the regularization from the finest level on all levels we need to fix the finest discretization level in advance. We cannot re-use the discretized systems from previous computations or levels if we need to refine the underlying cover (uniformly or in an adaptive way). These issues make this approach less useful.

Thus, we use a different approach. On each level $k = 0, \dots, l$ we use the minimal regularization parameter β_k so that we have a different variational form on each level, i.e., now we are faced with a sequence of nonnested discretization spaces with non-interpolatory basis functions and a sequence of non-inherited variational forms. The results of our numerical experiments with these two multilevel approaches showed that this second approach with a level-dependent regularization parameter β_k leads to iterations with smaller convergence rates ρ than the first approach. In fact the convergence rates of a multilevel $V(1,1)$-cycle with a different variational form on each level are essentially the same as we have for a respective Neumann problem ($\rho \leq 0.25$), see [28] for details.

4 Numerical Experiments

We apply our PUM to the model problem

$$\begin{aligned} -\Delta u &= f \text{ in } \Omega, \\ u &= g \text{ on } \partial\Omega, \end{aligned} \tag{14}$$

in two and three dimensions. In all our experiments we use a linear spline as the generation weight function $\mathcal{W}$ for the partition of unity (3) and we use uniform, quasi Monte Carlo and graded point sets P for our cover construction algorithm [13]. Here, we use $\alpha = 1.3$ so that we have an overlap of 30% of two cover patches (in the uniform case). As local approximation spaces we use Legendre polynomials.

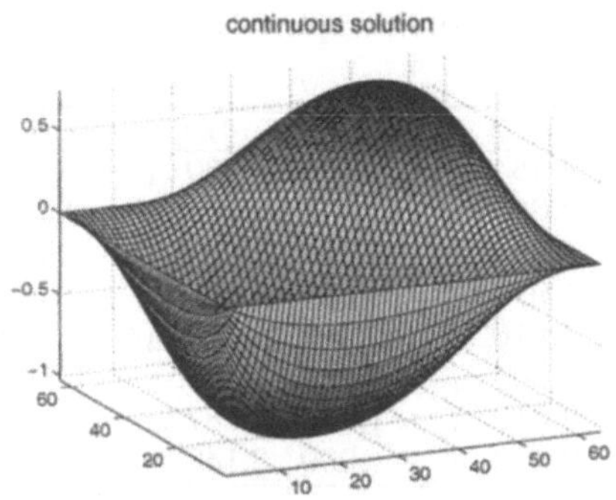

Fig. 1. Surface plot of the solution (15).

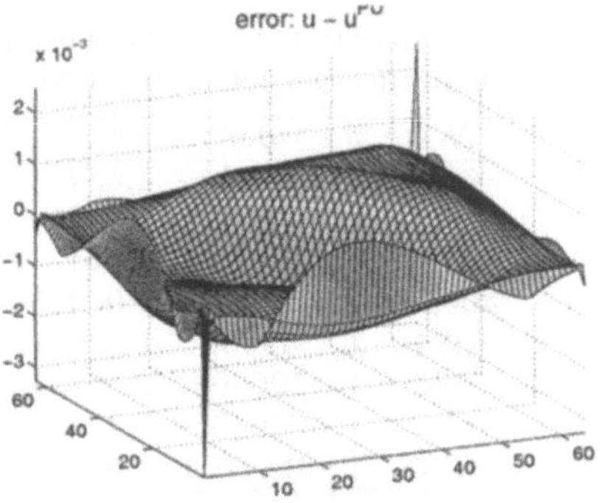

Fig. 2. Surface plot of the error $u - u_l^{\mathrm{PU}}$ with $l = 6$ for the h-version PUM.

Example 4.4 In our first example we consider the model problem (14) on the unit square $\Omega = (0,1)^2$ and choose f and g such that the solution to (14) is given by

$$u(x,y) = \arctan\left(100\left(\frac{x+y}{\sqrt{2}} - 0.8\right)\left(x - x^2\right)\left(y - y^2\right)\right), \tag{15}$$

see Figure 1.

We use the uniform h-version of our PUM for the approximation of (14), i.e., we use the cell centers of a uniform grid for the cover construction. Here, we increase the number of points $N = \mathrm{card}(P) = 2^{ld}$ for our PUM construction by increasing the refinement level l of the underlying uniform grid. Hence, we expect an algebraic error estimate of the form

$$\|u - u_l^{\mathrm{PU}}\| = O(\mathsf{dof}_l^{\rho})$$

and compute the respective convergence rates

$$\rho := \frac{\log\left(\frac{\|u - u_l^{\mathrm{PU}}\|}{\|u - u_{l-1}^{\mathrm{PU}}\|}\right)}{\log\left(\frac{\mathsf{dof}_l}{\mathsf{dof}_{l-1}}\right)} \tag{16}$$

from the errors $u - u_l^{\mathrm{PU}}$ measured in various norms. These error norms $\|u - u_l^{\mathrm{PU}}\|$ are computed with the help of the numerical quadrature scheme presented in [13], i.e., the L^∞-norm is approximated in the quadrature points only.

Note that due to the use of such regular point sets P we have $N^{-1/d} = h$ and we can compute the standard $O(h^\gamma)$ convergence rates γ for an h-version PUM. We have $\mathsf{dof} \simeq Np^d = h^{-d}p^d$ and thus we find

$$\gamma = -\rho d.$$

Due to our choice of parameters we expect a quadratic h-convergence in the L^2-norm and a linear h-convergence in the H^1-norm, i.e., we expect to measure $\rho_{L^2} = -1$

Table 1. Relative errors e (17) and convergence rates ρ (16) for Example 4.4 and solution (15) with the uniform h-version of the PUM.

l	N	p	dof	e_{L^∞}	ρ_{L^∞}	e_{L^2}	ρ_{L^2}	e_{H^1}	ρ_{H^1}
3	64	1	192	1.554_{-1}		9.484_{-2}		3.880_{-1}	
4	256	1	768	5.059_{-2}	-0.810	2.806_{-2}	-0.879	1.997_{-1}	-0.479
5	1024	1	3072	1.444_{-2}	-0.904	7.503_{-3}	-0.951	1.003_{-1}	-0.497
6	4096	1	12288	3.969_{-3}	-0.932	1.913_{-3}	-0.986	5.028_{-2}	-0.498
7	16384	1	49152	1.020_{-3}	-0.980	4.816_{-4}	-0.995	2.515_{-2}	-0.500
8	65536	1	196608	2.567_{-4}	-0.995	1.205_{-4}	-0.999	1.258_{-2}	-0.500
9	262144	1	786432	6.423_{-5}	-0.999	3.012_{-5}	-1.000	6.287_{-3}	-0.500
10	1048576	1	3145728	1.606_{-5}	-1.000	7.529_{-6}	-1.000	3.143_{-3}	-0.500

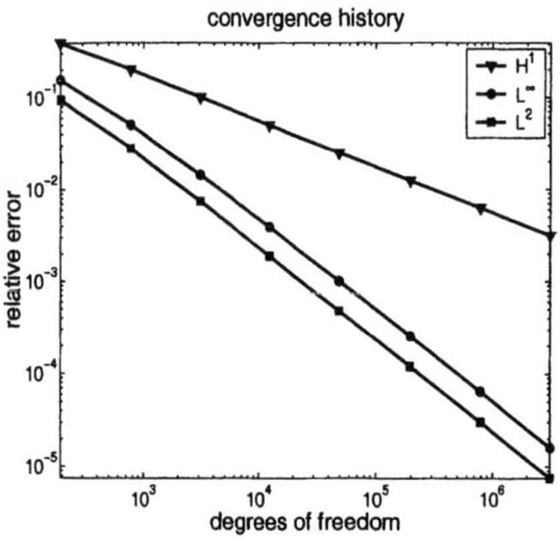

Fig. 3. Convergence history of the h-version of the PUM for Example 4.4 and solution (15).

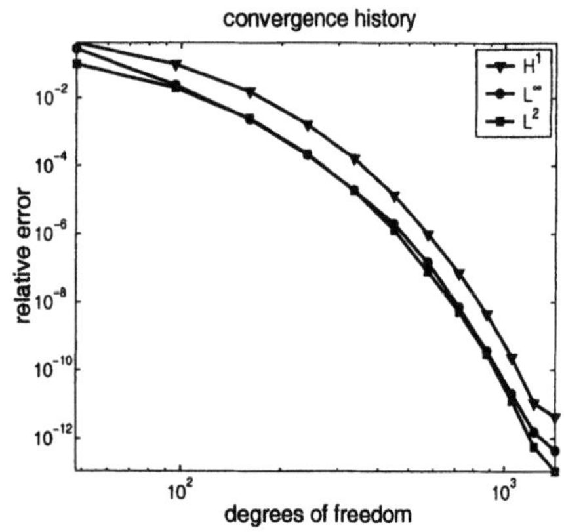

Fig. 4. Convergence history of the p-version of the PUM for Example 4.4 and solution (18).

and $\rho_{H^1} = -0.5$ in this two dimensional example. This anticipated convergence behavior can be observed from the relative errors

$$e_{L^\infty} := \frac{\|u - u_l^{\mathrm{PU}}\|_{L^\infty}}{\|u\|_{L^\infty}}, \quad e_{L^2} := \frac{\|u - u_l^{\mathrm{PU}}\|_{L^2}}{\|u\|_{L^2}}, \quad \text{and} \quad e_{H^1} := \frac{\|\nabla(u - u_l^{\mathrm{PU}})\|_{L^2}}{\|\nabla u\|_{L^2}} \tag{17}$$

displayed in Table 1 and the respective convergence rates ρ_{L^∞}, ρ_{L^2} and ρ_{H^1}. Also the graphs depicted in Figure 3 clearly show the expected convergence behavior of the h-version of our PUM with linear polynomials.

Let us now consider higher order discretizations, i.e., we now use Legendre polynomials of degree $p = 1, \dots, 12$ for the approximation of (14). Here, we choose f and g such that the solution is given by the analytic function

$$u(x, y) = \exp(4(x + y)). \tag{18}$$

Table 2. Relative errors e (17) and convergence rates ρ (16) for Example 4.4 and solution (18) with the p-version of the PUM.

l	N	p	dof	e_{L^∞}	ρ_{L^∞}	e_{L^2}	ρ_{L^2}	e_{H^1}	ρ_{H^1}
1	4	2	24	1.069_{-1}	-0.808	1.287_{-1}	-0.974	3.045_{-1}	-0.536
2	16	2	96	2.373_{-2}	-1.086	1.979_{-2}	-1.350	9.773_{-2}	-0.820
3	64	2	384	4.324_{-3}	-1.228	2.196_{-3}	-1.586	2.530_{-2}	-0.975
4	256	2	1536	6.411_{-4}	-1.377	2.398_{-4}	-1.597	6.258_{-3}	-1.008
1	4	3	40	1.991_{-2}	-1.233	3.235_{-2}	-1.856	9.496_{-2}	-1.097
2	16	3	160	2.348_{-3}	-1.542	2.474_{-3}	-1.855	1.522_{-2}	-1.321
3	64	3	640	2.257_{-4}	-1.689	1.532_{-4}	-2.007	2.000_{-3}	-1.464
4	256	3	2560	1.824_{-5}	-1.815	8.843_{-6}	-2.057	2.455_{-4}	-1.513
1	4	4	60	3.584_{-3}	-1.675	5.667_{-3}	-2.255	2.096_{-2}	-1.586
2	16	4	240	2.095_{-4}	-2.048	2.169_{-4}	-2.354	1.656_{-3}	-1.831
3	64	4	960	9.712_{-6}	-2.216	6.193_{-6}	-2.565	1.004_{-4}	-2.022
4	256	4	3840	3.881_{-7}	-2.323	1.661_{-7}	-2.610	5.781_{-6}	-2.059
1	4	5	84	8.455_{-4}	-1.884	7.812_{-4}	-2.691	3.706_{-3}	-2.042
2	16	5	336	1.878_{-5}	-2.746	1.848_{-5}	-2.701	1.658_{-4}	-2.241
3	64	5	1344	4.617_{-7}	-2.673	3.087_{-7}	-2.952	5.507_{-6}	-2.456
4	256	5	5376	9.590_{-9}	-2.795	4.724_{-9}	-3.015	1.693_{-7}	-2.512
1	4	6	112	1.158_{-4}	-2.525	1.151_{-4}	-2.960	6.213_{-4}	-2.400
2	16	6	448	1.962_{-6}	-2.942	1.282_{-6}	-3.244	1.368_{-5}	-2.753
3	64	6	1792	2.278_{-8}	-3.214	1.124_{-8}	-3.417	2.385_{-7}	-2.921
4	256	6	7168	2.117_{-10}	-3.375	9.043_{-11}	-3.479	3.851_{-9}	-2.976
1	4	7	144	1.739_{-5}	-2.745	1.397_{-5}	-3.452	8.879_{-5}	-2.858
2	16	7	576	1.483_{-7}	-3.436	7.875_{-8}	-3.736	1.000_{-6}	-3.236
3	64	7	2304	8.300_{-10}	-3.741	3.393_{-10}	-3.929	8.690_{-9}	-3.423
4	256	7	9216	4.167_{-12}	-3.819	1.303_{-12}	-4.012	6.865_{-11}	-3.492
1	4	8	180	1.594_{-6}	-3.385	1.731_{-6}	-3.790	1.206_{-5}	-3.234
2	16	8	720	7.233_{-9}	-3.892	5.162_{-9}	-4.195	7.233_{-8}	-3.691
3	64	8	2880	2.378_{-11}	-4.124	1.215_{-11}	-4.365	3.372_{-10}	-3.872
4	256	8	11520	1.155_{-13}	-3.843	4.033_{-14}	-4.117	2.044_{-12}	-3.683
1	4	9	220	1.558_{-7}	-3.848	1.886_{-7}	-4.180	1.439_{-6}	-3.638
2	16	9	880	3.698_{-10}	-4.359	2.963_{-10}	-4.657	4.577_{-9}	-4.148
3	64	9	3520	1.052_{-12}	-4.229	8.197_{-13}	-4.249	2.898_{-11}	-3.652
1	4	10	264	1.747_{-8}	-4.498	1.705_{-8}	-4.516	1.482_{-7}	-4.070
2	16	10	1056	2.048_{-11}	-4.868	1.263_{-11}	-5.199	2.315_{-10}	-4.661
3	64	10	4224	9.127_{-14}	-3.905	2.960_{-14}	-4.369	1.197_{-12}	-3.798
1	4	11	312	1.963_{-9}	-4.788	1.479_{-9}	-4.991	1.408_{-8}	-4.541
2	16	11	1248	1.562_{-12}	-5.148	5.720_{-13}	-5.668	1.107_{-11}	-5.156
1	4	12	364	1.793_{-10}	-5.003	1.226_{-10}	-5.420	1.273_{-9}	-4.908
2	16	12	1456	4.401_{-13}	-4.335	1.107_{-13}	-5.057	4.271_{-12}	-4.110

Due to the unlimited smoothness of (18) we expect that the algebraic convergence rates (16) for the L^2-norm is given by $\rho_{L^2} = -\frac{p+1}{d}$ and by $\rho_{H^1} = -\frac{p}{d}$ for the energy norm if we use an h-refinement. This anticipated convergence behavior can be observed from the numbers displayed in Table 2. We see some fluctuations in the convergence rates for higher polynomial degrees ($p > 7$) where we do not fully achieve the anticipated convergence rates. Here, the errors are already so small

Table 3. Relative errors e (17) and convergence rates ρ (16) for Example 4.5 with the uniform h-version of the PUM.

l	N	p	dof	e_{L^∞}	ρ_{L^∞}	e_{L^2}	ρ_{L^2}	e_{H^1}	ρ_{H^1}
1	8	1	32	2.054_{-1}	-0.450	1.349_{-1}	-0.617	4.967_{-1}	-0.189
2	64	1	256	8.970_{-2}	-0.398	3.239_{-2}	-0.686	2.537_{-1}	-0.323
3	512	1	2048	2.448_{-2}	-0.625	9.225_{-3}	-0.604	1.303_{-1}	-0.321
4	4096	1	16384	5.974_{-3}	-0.678	2.640_{-3}	-0.602	6.565_{-2}	-0.330
5	32768	1	131072	1.573_{-3}	-0.642	7.183_{-4}	-0.626	3.286_{-2}	-0.333
6	262144	1	1048576	3.872_{-4}	-0.674	1.886_{-4}	-0.643	1.643_{-2}	-0.333
7	2097152	1	8388608	9.616_{-5}	-0.670	4.836_{-5}	-0.654	8.213_{-3}	-0.333

($\|e\|_{L^\infty} \leq 10^{-12}$) that we experience some effects from the limited accuracy of floating-point arithmetic. If we use a p-refinement where we increase the polynomial degree successively but keep the number of points fixed we anticipate an exponential convergence behavior. From the plots of the relative errors against the number of degrees of freedom given in Figure 4, we can observed the exponential convergence of the p-version of our PUM for smooth functions as expected. Again, we see a drop-off in the convergence rates for very small errors ($\|e\|_{L^\infty} \leq 10^{-12}$).

In summary, the results of this example clearly show that the implementation of Dirichlet boundary conditions by Nitsche's method preserves the approximation properties of the h-version as well as the p-version of our PUM.

Example 4.5 In our second example we consider the model problem (14) on the unit cube $\Omega = (0,1)^3$ where we choose f and g such that the solution to (14) is given by

$$u(x) = \|x\|_2^{\frac{5}{2}}. \tag{19}$$

Again, we begin with a uniformly refined sequence of point sets before we discuss the results of the h-version of our PUM based on randomly chosen points. We expect to measure convergence rates $\rho_{L^2} = -\frac{2}{3}$ and $\rho_{H^1} = -\frac{1}{3}$ which correspond to the usual $O(h^2)$ and $O(h)$ behavior in three dimensions. The measured errors and convergence rates for the uniform h-version of our PUM are given in Table 3. From these we can clearly observe the anticipated convergence behavior.

Up to now we have considered uniform point sets only to avoid any fluctuations in the measured convergence rates due to an irregular point distribution. Now that we have seen that our PUM does indeed converge with the expected properties we turn to the use of general point sets. To this end, we employ a Halton$(2,3,5)$ sequence[1] to generate an initial point set for our tree-based cover construction algorithm [13]. Here, we use the same number of points as we had for the uniform point set (see N in

[1] Halton sequences are quasi Monte Carlo sequences, which are used in sampling and numerical integration. Consider $n \in \mathbb{N}_0$ given as $\sum_j n_j p^j = n$ for some prime p. We can define the transformation H_p from $\mathbb{N}_0$ to $[0,1]$ with $n \mapsto H_p(n) = \sum_j n_j p^{-j-1}$. Then, the Halton$(p,q)$ sequence with N points in two dimensions is given by $\{(H_p(n), H_q(n)) \mid n = 0, \ldots, N-1\}$.

Table 4. Relative errors e (17) and convergence rates ρ (16) for Example 4.5 with the h-version of the PUM using Halton$(2,3,5)$ point sets.

l	N	p	dof	e_{L^∞}	ρ_{L^∞}	e_{L^2}	ρ_{L^2}	e_{H^1}	ρ_{H^1}
2	64	1	256	2.676_{-2}	-0.653	1.034_{-2}	-0.824	2.000_{-1}	-0.290
4	176	1	704	1.999_{-2}	-0.288	9.083_{-3}	-0.128	1.832_{-1}	-0.087
6	1373	1	5492	5.142_{-3}	-0.661	2.724_{-3}	-0.586	9.289_{-2}	-0.331
6	10606	1	42424	1.621_{-3}	-0.565	7.277_{-4}	-0.646	4.686_{-2}	-0.335
7	91820	1	367280	3.763_{-4}	-0.676	1.847_{-4}	-0.635	2.322_{-2}	-0.325

Table 3) but since our cover construction automatically introduces additional points to insure the shape regularity of the cover patches ω_i we find a larger number of points N than before. Hence, the number points cannot be prescribed exactly so that N does not grow by a constant factor of 2^d from row to row in Table 4. The values of l now correspond to the maximal refinement level of the cover tree, i.e., the minimal support size is given by $\mathrm{diam}(\omega_i) \simeq \alpha 2^{-l}$. This fluctuation in the number of degrees of freedom, however, should not effect the measured convergence rates substantially since the Halton point sets are uniformly distributed. The results presented in Table 4 clearly support this assertion. Again, we find $\rho_{L^2} = -\frac{2}{3}$ and $\rho_{H^1} = -\frac{1}{3}$.

Example 4.6 So far all considered problems were H^2-regular and hence the solutions of these problems could be approximated with the usual optimal convergence rates ρ. Let us now turn to the treatment of problems with singular solutions. To this end, we consider our model problem (14) on an L-shaped domain $\Omega = (-1,1)^2 \setminus [0,1)^2$ where we choose f and g such that the solution (see Figure 5) is given by

$$u(r,\theta) = r^{\frac{2}{3}} \sin\left(\frac{2\theta - \pi}{3}\right) \tag{20}$$

which is in H^s with $s < 1 + \frac{2}{3}$ only. Hence, we cannot expect to measure the same convergence rates as we did in the previous examples. From finite element theory

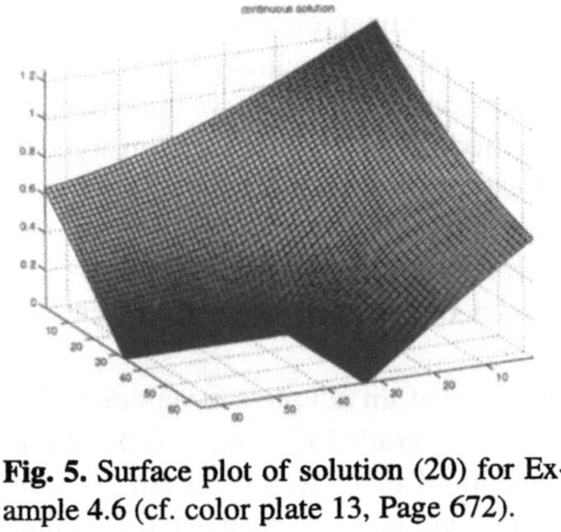

Fig. 5. Surface plot of solution (20) for Example 4.6 (cf. color plate 13, Page 672).

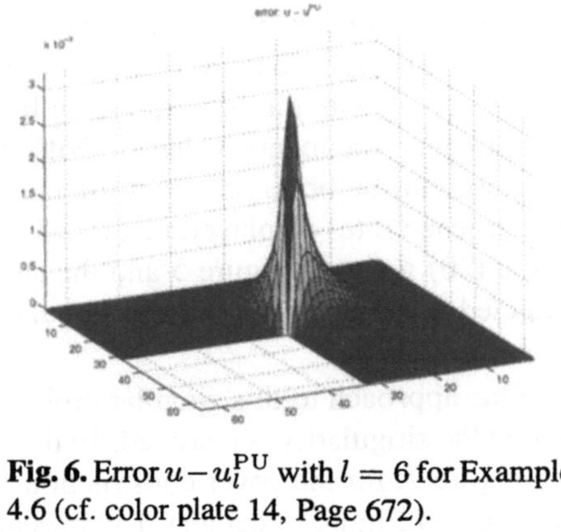

Fig. 6. Error $u - u_l^{\mathrm{PU}}$ with $l = 6$ for Example 4.6 (cf. color plate 14, Page 672).

Table 5. Relative errors e (17) and convergence rates ρ (16) for Example 4.6 with the uniform h-version of the PUM.

l	N	p	dof	e_{L^∞}	ρ_{L^∞}	e_{L^2}	ρ_{L^2}	e_{H^1}	ρ_{H^1}
3	48	1	144	5.792_{-2}		6.487_{-3}		1.112_{-1}	
4	192	1	576	3.122_{-2}	-0.446	3.195_{-3}	-0.511	7.296_{-2}	-0.304
5	768	1	2304	1.816_{-2}	-0.391	1.434_{-3}	-0.578	4.723_{-2}	-0.314
6	3072	1	9216	1.160_{-2}	-0.324	5.659_{-4}	-0.671	3.003_{-2}	-0.327
7	12288	1	36864	7.275_{-3}	-0.336	2.272_{-4}	-0.658	1.906_{-2}	-0.328
8	49152	1	147456	4.507_{-3}	-0.345	9.244_{-5}	-0.649	1.208_{-2}	-0.329
9	196608	1	589824	2.869_{-3}	-0.326	3.629_{-5}	-0.674	7.622_{-3}	-0.332
10	786432	1	2359296	1.826_{-3}	-0.326	1.424_{-5}	-0.675	4.804_{-3}	-0.333

Table 6. Relative errors e (17) and convergence rates ρ (16) for Example 4.6 with the h-version of the PUM using graded Halton(2, 3) point sets.

l	N	p	dof	e_{L^∞}	ρ_{L^∞}	e_{L^2}	ρ_{L^2}	e_{H^1}	ρ_{H^1}
5	186	1	558	8.075_{-2}	-0.398	2.896_{-2}	-0.560	1.877_{-1}	-0.265
7	300	1	900	2.727_{-2}	-2.271	1.059_{-2}	-2.104	1.261_{-1}	-0.833
10	1293	1	3879	7.265_{-3}	-0.905	3.173_{-3}	-0.825	6.569_{-2}	-0.446
11	4968	1	14904	1.945_{-3}	-0.979	7.927_{-4}	-1.030	3.268_{-2}	-0.519
14	20142	1	60426	5.029_{-4}	-0.966	2.048_{-4}	-0.967	1.677_{-2}	-0.476
17	80574	1	241722	1.394_{-4}	-0.925	5.030_{-5}	-1.013	8.239_{-3}	-0.513
17	317637	1	952911	3.820_{-5}	-0.944	1.272_{-5}	-1.002	4.158_{-3}	-0.499
20	1310847	1	3932541	9.223_{-6}	-1.003	3.224_{-6}	-0.969	2.092_{-3}	-0.485

[31, Chapter 8] we know that a solution u to a homogeneous Dirichlet problem on an L-shaped domain can be split into a regular part $u_R \in H^2$ and a singular part u_S where this singular part is given by the our solution (20). Hence, this example represents the hard part of the general situation. If we can approximate solution (20) with a particular convergence rate then we can approximate any solution to a homogeneous Dirichlet problem on an L-shaped domain using the same PUM space with the same convergence rate.

It is well-known that a uniform h-version with linear elements will give an $O(h^{2/3})$ convergence in the energy norm only. Thus, we may achieve a convergence of $\rho_{H^1} < -\frac{1}{3}$ with the corresponding linear h-version PUM where we approximate the (singular) boundary value by Nitsche's method. This anticipated convergence behavior can be observed from the measured errors (see also Figure 9 (left)) and convergence rates displayed in Table 5. From the surface plot of the continuous solution (20) given in Figure 5 and the surface plot of the error $u - u_l^{\mathrm{PU}}$ with $l = 6$ depicted in Figure 6 we see the singular character of the solution at the re-entrant corner.

One approach to this singular problem is the use of an adaptive h-version PUM so that the singularity is resolved. To this end we use a graded Halton(2, 3) point set, see Figure 7, for the cover construction. The results of this experiment are given in Table 6 and Figure 9 (center). The measured convergence rates clearly indicate that

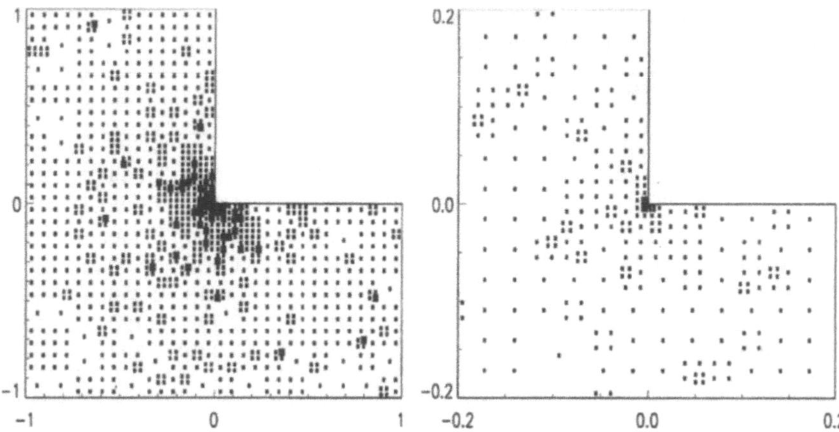

Fig. 7. Fine level point set P_{10} generated from a graded Halton$(2,3)$ point set with 1024 points (left), zoom-in (right).

we can resolve the singularity at the re-entrant corner by local (h-type) refinement of the cover patches. Again, we measure the optimal rates $\rho_{L^2} = -1$ and $\rho_{H^1} = -\frac{1}{2}$.

The PUM approach, however, also allows for a different p-type adaptation process. One benefit of the PUM approach is the independence of the local approximation spaces, i.e., we may include singular functions in a local approximation space without the need to pay any attention to neighboring (overlapping) local spaces. Hence, the introduction of a singular function into the local approximation spaces $V_i^{p_i}$ in the vicinity of the re-entrant corner can be realized very easily within the PUM. From finite element theory we know that this may improve the convergence from $O(h^{2/3})$ in the energy norm to the usual $O(h)$ behavior.

To this end, consider the sub-domains $\Omega_p := \Omega \setminus [-a,a]^2 \subset \Omega$ and $\Omega_a := \Omega \cap [-a,a]^2 \subset \Omega$. On the cover patches ω_i whose associated center point x_i is sufficiently far away from the singularity at the origin, i.e. $x_i \in \Omega_p$, we use the linear Legendre polynomials as before to approximate the (smooth part of the) solution. But close to the singularity, $x_i \in \Omega_a$, we use augmented local approximation spaces $V_i^{p_i^a} := \operatorname{span}\langle V_i^{p_i}, \Phi\rangle$ where Φ is the (global) singular function (20). Similar constructions have also been used in the GFEM [32] and in the FEM [31] context.

The numerical results of this experiment with two different values for a, namely $a = 0.5$ and $a = 0.25$, are presented in Table 7 and Figure 9 (right). From the numbers given there we can clearly observe that the measured convergence rates are now $\rho_{L^2} = -1$ and $\rho_{H^1} = -\frac{1}{2}$ which correspond to the usual $O(h^2)$ and $O(h)$ convergence behavior respectively. This improvement in the error evolution can also be observed from a comparison of the plots depicted in Figure 9. The size of the sub-domain Ω_a where we use augmented local spaces $V_i^{p_i^a}$ essentially determines the location and the absolute value of the maximal error which can be observed from the surface plots given in Figure 8. But it has no effect on the (asymptotic) convergence rates as long as it is fixed on all levels.

Table 7. Relative errors e (17) and convergence rates ρ (16) for Example 4.6 with the uniform h-version of the PUM using augmented local spaces.

l	N	p	dof	e_{L^∞}	ρ_{L^∞}	e_{L^2}	ρ_{L^2}	e_{H^1}	ρ_{H^1}
				$a = 0.5$					
2	12	1	39	2.186_{-2}		8.939_{-3}		8.802_{-2}	
3	48	1	156	5.897_{-3}	-0.945	2.141_{-3}	-1.031	4.044_{-2}	-0.561
4	192	1	624	1.727_{-3}	-0.886	5.182_{-4}	-1.023	1.925_{-2}	-0.535
5	768	1	2496	4.896_{-4}	-0.909	1.284_{-4}	-1.006	9.330_{-3}	-0.523
6	3072	1	9984	1.278_{-4}	-0.969	3.204_{-5}	-1.002	4.559_{-3}	-0.517
7	12288	1	39936	3.310_{-5}	-0.974	8.035_{-6}	-0.998	2.256_{-3}	-0.508
8	49152	1	159744	8.483_{-6}	-0.982	2.014_{-6}	-0.998	1.122_{-3}	-0.504
9	196608	1	638976	2.122_{-6}	-1.000	5.043_{-7}	-0.999	5.594_{-4}	-0.502
				$a = 0.25$					
2	12	1	39	2.186_{-2}		8.939_{-3}		8.802_{-2}	
3	48	1	147	1.054_{-2}	-0.550	4.121_{-3}	-0.584	6.207_{-2}	-0.263
4	192	1	588	3.635_{-3}	-0.768	1.043_{-3}	-0.991	3.066_{-2}	-0.509
5	768	1	2352	1.117_{-3}	-0.851	2.511_{-4}	-1.027	1.497_{-2}	-0.517
6	3072	1	9408	3.053_{-4}	-0.936	6.051_{-5}	-1.027	7.335_{-3}	-0.515
7	12288	1	37632	8.095_{-5}	-0.957	1.486_{-5}	-1.013	3.630_{-3}	-0.507
8	49152	1	150528	2.102_{-5}	-0.973	3.683_{-6}	-1.006	1.805_{-3}	-0.504
9	196608	1	602112	5.305_{-6}	-0.993	9.169_{-7}	-1.003	8.999_{-4}	-0.502

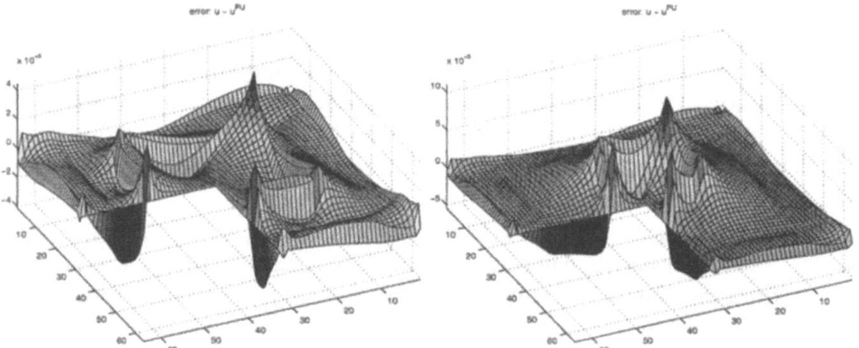

Fig. 8. Error $u - u_l^{\mathrm{PU}}$ with $l = 6$ for the h-version PUM with augmented local spaces (left: $a = 0.5$, right: $a = 0.25$) (cf. color plate 15, Page 673).

Finally, we use the augmented p-version of our PUM to approximate (14) on the L-shaped domain where we choose f and g such that the solution is given by

$$u(r,\theta) = (1 - (r\cos(\theta))^2)(1 - (r\sin(\theta))^2)r^{\frac{2}{3}} \sin\left(\frac{2\theta - \pi}{3}\right), \quad (21)$$

see Figure 10. Again, we use augmented spaces in Ω_a with $a = 0.25$ but we now use a p-version refinement. Note that we use the singular function (20) as before for the augmentation of the local spaces, we do not use the solution (21). Since the introduction of the singularity into the PUM space V^{PU} should resolve the singularity

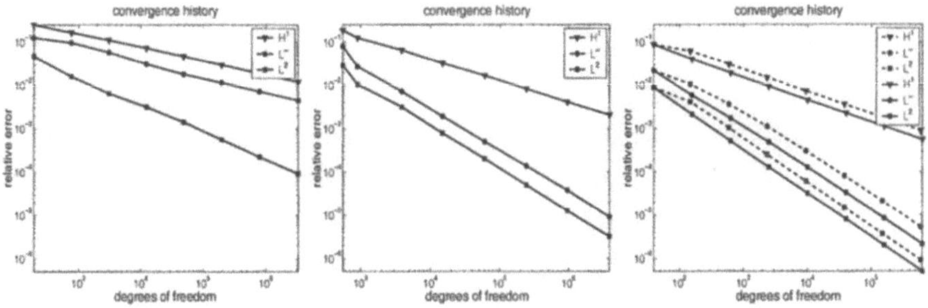

Fig. 9. Convergence history of the h-version of the PUM with linear local spaces (left: uniform point set, center: graded Halton$(2,3)$ point set) and augmented local spaces (right: solid lines $a = 0.5$, dashed lines $a = 0.25$) for Example 4.6.

at the re-entrant corner, we expect an exponential convergence of the p-version for the singular solution (21). This anticipated convergence behavior can be observed from the plots given in Figure 14 for larger errors ($\|e\|_{L^\infty} \geq 10^{-4}$). Then, we experience some pollution effect within Ω_a which comes from the couplings between overlapping polynomial spaces in Ω_p and augmented spaces in Ω_a, see Figures 11, 12, and 13. Here, a smaller overlap between augmented and polynomial spaces may overcome this pollution effect and could extend the exponential convergence behavior even further. Nonetheless, we are able to approximate the singular solution (21) with a relative accuracy of five digits with about 1000 degrees of freedom only.

In summary, the results of our numerical experiments clearly show the expected convergence properties of our PUM independent of the boundary conditions. The implementation of Dirichlet boundary conditions by Nitsche's method has no adverse effect on the convergence rates for regular as well as singular solutions. We achieve the optimal convergence rates for regular as well as irregular point sets, polynomial basis functions and singular basis functions.

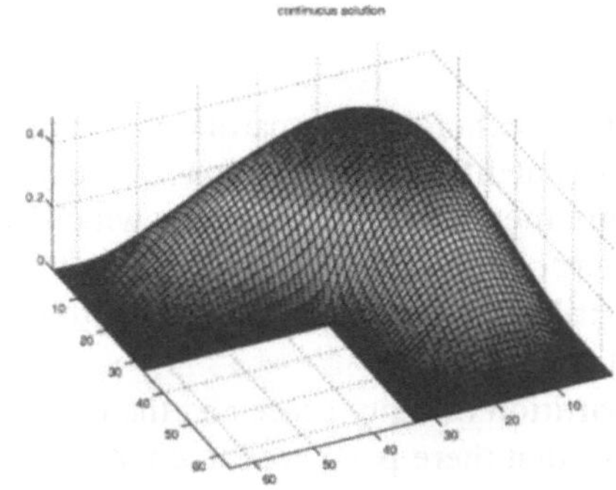

Fig. 10. Surface plot of solution (21) for Example 4.6 (cf. color plate 16, Page 673).

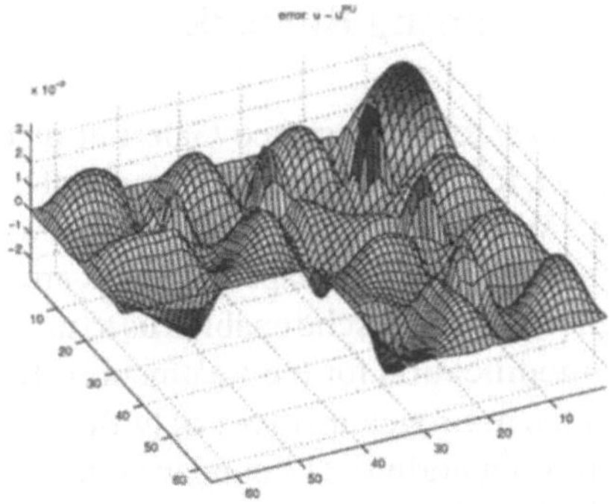

Fig. 11. Error $u - u_l^{\mathrm{PU}}$ with $l = 2, p = 3$ for Example 4.6 (cf. color plate 17, Page 673).

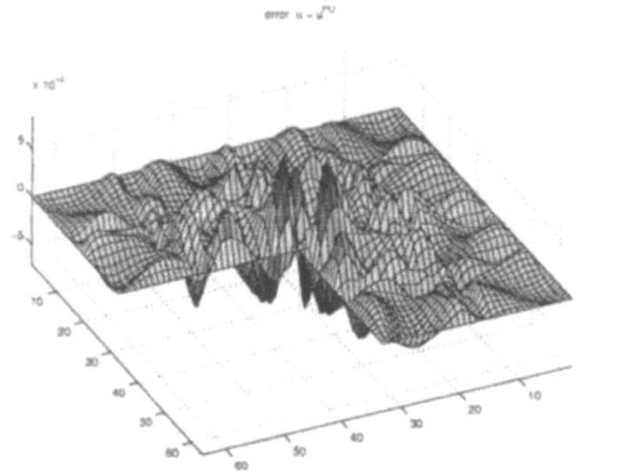

Fig. 12. Error $u - u_l^{\mathrm{PU}}$ with $l = 2, p = 5$ for Example 4.6 (cf. color plate 18, Page 673).

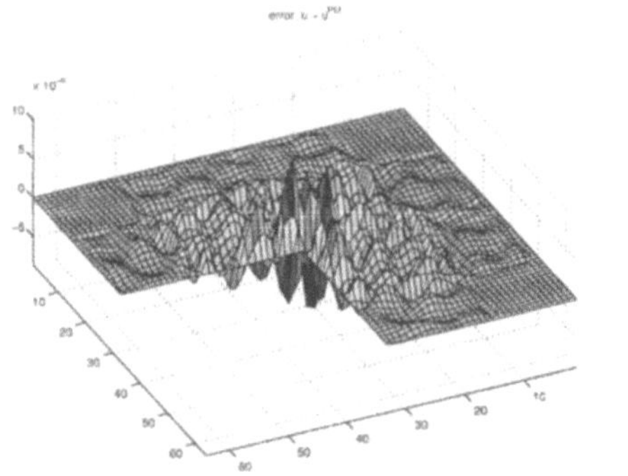

Fig. 13. Error $u - u_l^{\mathrm{PU}}$ with $l = 2, p = 8$ for Example 4.6 (cf. color plate 19, Page 673).

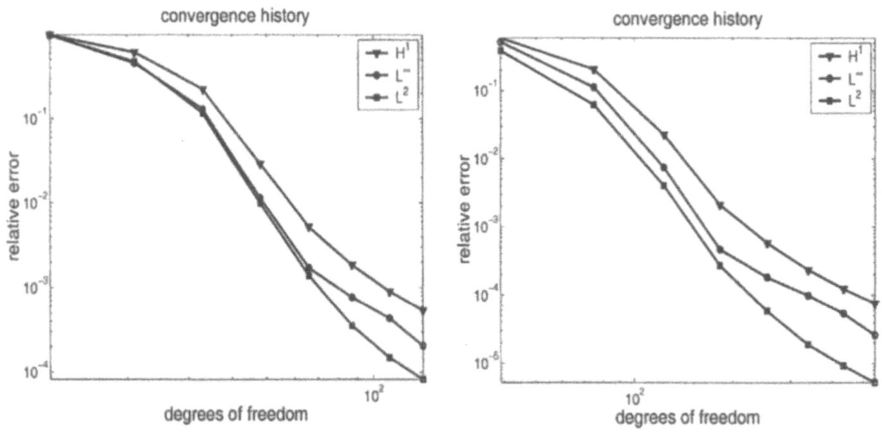

Fig. 14. Convergence history of the augmented p-version of the PUM with linear local spaces $l = 1$ (left) and $l = 2$ (right) for Example 4.6 with solution (21).

5 Concluding Remarks

We presented a meshfree Galerkin discretization technique, the partition of unity method, for the numerical treatment of elliptic partial differential equations of second order. We focused on the implementation of Dirichlet boundary conditions which is not trivial since the shape functions are non-interpolatory. Here, an almost forgotten method due to Nitsche enabled us to use our partition of unity shape functions without any modification for the treatment of Dirichlet problems.

This approach involves only the original partition of unity space and the evaluation of an additional boundary integral term so that there is no need to change the method or the implementation in any way. We can use our general cover construction algorithm [13], our sparse grid integration scheme [13] as well as our multilevel solver [14] without any modification for the treatment of Dirichlet problems with our partition of unity method. Hence, also the parallelization and parallel efficiency of the method [15] is not effected by the boundary conditions. Furthermore, Nitsche's method is also applicable to general domains, it was in fact developed for general domains.

The results of our numerical experiments where we considered two- and three-dimensional problems with several million degrees of freedom clearly showed that we achieve optimal convergence rates of the h-version and p-version of our PUM for Dirichlet problems. This observation holds not only for regular point sets and regular solutions but also for highly adaptive discretizations and singular solutions.

References

1. D. N. Arnold, F. Brezzi, B. Cockburn, D. Marini: Discontinuous Galerkin methods for elliptic problems. In: B. Cockburn, G. Karniadakis, and C.-W. Shu (eds) Discontinuous Galerkin Methods: Theory, Computation and Applications. Springer, 89–101, 1999
2. I. Babuška: Numerical Solution of Boundary Value Problems by the Perturbated Variational Principle. Tech. Note BN-624, University of Maryland, 1969
3. I. Babuška: The finite element method with Lagrangian multipliers, Numer. Math., 20, 179–192, 1973
4. I. Babuška, U. Banerjee, J. E. Osborn: Meshless and generalized finite element methods: A Survey of some major results. In: M. Griebel, M. A. Schweitzer (eds) Meshfree Methods for Partial Differential Equations, Lecture Notes in Computational Science and Engineering, Vol. 26, Springer, 1–20, 2002
5. I. Babuška, J. M. Melenk: The partition of unity finite element method: basic theory and applications. Comput. Meth. Appl. Mech. Engrg., 139, 289–314, 1996. Special Issue on Meshless Methods
6. I. Babuška, J. M. Melenk: The partition of unity method. Int. J. Numer. Meth. Engrg., 40, 727–758, 1997
7. W. W. Bradburry, R. Fletcher: New iterative methods for the solution of the Eigenproblem. Numer. Math., 9, 259–267, 1966
8. J. H. Bramble: The Lagrange multiplier method for Dirichlet's problem. Math. Comp., 37, 1-11, 1981
9. F. Brezzi, L. P. Franca, D. Marini, A. Russo: Stabilization techniques for domain decomposition methods with non-matching grids. In: P. E. Bjørstad, M. S. Espedal, D. E. Keyes (eds) Domain Decomposition Methods in Science and Engineering. Domain Decomposition Press, 1–11, 1998
10. C. A. M. Duarte, J. T. Oden: hp clouds – a meshless method to solve boundary value problems. Numer. Meth. for PDE, 12, 673–705, 1996
11. J. Garcke: Berechnung von Eigenwerten der stationären Schrödingergleichung mit der Kombinationstechnik. Diplomarbeit, Institut für Angewandte Mathematik, Universität Bonn, 1998
12. M. Griebel, M. A. Schweitzer: A particle-partition of unity method for the solution of elliptic, parabolic and hyperbolic PDE. SIAM J. Sci. Comput., 22, 853–890, 2000
13. M. Griebel, M. A. Schweitzer: A particle-partition of unity method—Part II: Efficient cover construction and reliable integration. SIAM J. Sci. Comput., 23, 1655–1682, 2002
14. M. Griebel, M. A. Schweitzer: A particle-partition of unity method—Part III: A multilevel solver. SIAM J. Sci. Comput., 24, 377–409, 2002

15. M. Griebel, M. A. Schweitzer: A particle-partition of unity method—Part IV: Parallelization. In: M. Griebel, M. A. Schweitzer (eds) Meshfree Methods for Partial Differential Equations. Lecture Notes in Computational Science and Engineering, Vol. 26, Springer, 161–192, 2002
16. F. C. Günther, W. K. Liu: Implementation of boundary conditions for meshless methods. Comput. Meth. Appl. Mech. Engrg., 163, 205–230, 1998
17. W. Han, X. Meng: Some studies of the reproducing kernel particle method. In: M. Griebel, M. A. Schweitzer (eds) Meshfree Methods for Partial Differential Equations. Lecture Notes in Computational Science and Engineering, Vol. 26, Springer, 191–208, 2002
18. B. Heinrich, K. Pietsch: Nitsche Type Mortaring for some Elliptic Problems with Corner Singularities. Tech. Rep. SFB393/00-34, Sonderforschungsbereich 393, 2000
19. Y. Krongauz, T. Belytschko: Enforcement of essential boundary conditions in meshless approximations using finite elements. Comput. Meth. Appl. Mech. Engrg., 131,133–145, 1996
20. A. Kunoth: Multilevel preconditioning – appending boundary conditions by Lagrange multipliers. Adv. Comput. Math., 4, 145–170, 1995. Special Issue on Multiscale Methods
21. D. E. Longsine, S. F. McCormick: Simultaneous Rayleigh-quotient minimization methods for $Ax = \lambda Bx$. Lin. Alg. Appl., 34, 195–234, 1980
22. Y. Y. Lu, T. Belytschko, L. Gu: A new implementation of the element free Galerkin method. Comput. Math. Appl. Mech. Engrg., 113, 397–414, 1994
23. J. Nitsche: Über ein Variationsprinzip zur Lösung von Dirichlet-Problemen bei Verwendung von Teilräumen, die keinen Randbedingungen unterworfen sind. Abh. Math. Sem. Univ. Hamburg, 36, 9–15, 1970–1971
24. J. Pitkäranta: Boundary subspaces for the finite element method with Lagrange multipliers. Numer. Math., 33, 273–289, 1979
25. J. Pitkäranta: Local stability conditions for the Babuška method of Lagrange multipliers. Math. Comp., 35, 1113–1129, 1980
26. J. Pitkäranta: The finite element method with Lagrange multipliers for domains with corners. Math. Comp., 37, 13–30, 1981
27. M. A. Schweitzer: Ein Partikel–Galerkin–Verfahren mit Ansatzfunktionen der Partition of Unity Method. Diplomarbeit, Institut für Angewandte Mathematik, Universität Bonn, 1997
28. M. A. Schweitzer: A Parallel Multilevel Partition of Unity Method for Elliptic Partial Differential Equations. Dissertation, Institut für Angewandte Mathematik, Universität Bonn, 2002
29. R. Stenberg: On some techniques for approximating boundary conditions in the finite element method. J. Comput. Appl. Math., 63, 139–148, 1995
30. R. Stenberg: Mortaring by a method of J. A. Nitsche. In: S. Idelsohn, E. Onate, and E. Dvorkin (eds) Computational Mechanics, New Trends and Applications. 1–6, 1998
31. G. Strang, G. J. Fix: An Analysis of the Finite Element Method. Prentice–Hall, 1973
32. T. Strouboulis, I. Babuška, K. Copps: The design and analysis of the generalized finite element method. Comput. Meth. Appl. Mech. Engrg., 181, 43–69, 1998

On Uniqueness- and Regularity Criteria for the Navier-Stokes Equations

Mark Steinhauer

Mathematisches Seminar, Universität Bonn, Nussallee 15, D-53115 Bonn
mark@ms1.uni-bonn.de

Summary. We survey and improve some results concerning uniqueness and regularity of solutions to the instationary Navier-Stokes equations in three (and higher) dimensions. In particular we show that the class of weak solutions which additionally belong to the space $L^2(0,T;BMO)$ guarantees uniqueness as well as regularity. The method of proof which we present is elementary and depends deeply on the "div-curl" structure of the nonlinear convective term $u\cdot\nabla u$ of the Navier-Stokes equations together with $\operatorname{div} u = 0$ and according to Coifman, Lions, Meyer & Semmes it belongs to the Hardy space $\mathcal{H}^1$. This also shows that it is applicable to other equations in hydrodynamics as for example the Boussinesq equations, the equations of Magneto-Hydrodynamics and the equations of higher grade type fluids.

1 Introduction and Problem Formulation

Let us consider the initial value problem for the Navier-Stokes equations in $(0,T)\times\mathbb{R}^n$ with $0<T<+\infty$ and $n\geq 3$:

$$\begin{aligned} &\partial_t u^i - \Delta u^i + u^j D_j u^i + D_i\pi = f^i \quad \text{in } (0,T)\times\mathbb{R}^n,\\ &\operatorname{div} u = 0 \quad \text{in } (0,T)\times\mathbb{R}^n, \qquad u^i(0,x) = a^i(x), \end{aligned} \tag{1}$$

where $u = u(t,x) = (u^1(t,x),\ldots,u^n(t,x))$ and $\pi = \pi(t,x)$ denote the unknown velocity vector and pressure of the fluid at the point $(t,x)\in(0,T)\times\mathbb{R}^n$, while $a = a(x) = (a^1(x),\ldots,a^n(x))$ is the given initial velocity vector and $f = f(t,x) = (f^1(t,x),\ldots,f^n(t,x))$ is a given external force.

We are interested in the classical problem of finding sufficient conditions for weak solutions of (1) such that they become unique and/or regular.

In the following we use some notations for function spaces which will be explained in section 2. We know that for every $a\in L^2_\sigma$ and every $f\in L^1(0,T;L^2)$, there exists at least one weak solution u of (1) satisfying the energy inequality :

$$||u(t)||^2_{0,2} + 2\int_0^t ||\nabla u(s)||^2_{0,2}\,ds \leq ||a||^2_{0,2} + 2\int_0^t (f(s),u(s))ds \quad . \tag{2}$$

By a weak solution we mean a function u in $L^\infty(0,T;L^2_\sigma)\cap L^2(0,T;H^{1,2}_\sigma)$ which satisfies (1) in the sense of distributions (for a more precise definition see section 2). This existence result was long ago proven by Leray in 1934 [18] (in the case n =

3) and Hopf in 1951 [13] (general case). Unfortunately we do not know up to now whether such weak solutions are unique and/or regular or not in the case $n \geq 3$. This is in contrast to the two dimensional case, in which we know that weak solutions are unique and regular (see for instance: Constantin, Foias [3], Ladyzhenskaya [17] or J. Lions [19] and references cited therein). Therefore the question arises: under which additional assumption(s) on the weak solution are we able to deduce uniqueness and regularity?

Introducing the class $L^\alpha(0,T;L^\beta)$, Prodi [22] showed 1959 that if u is a weak solution of (1) belonging to $L^\alpha(0,T;L^\beta)$ with α, β satisfying

$$PS(\alpha,\beta) \equiv \frac{2}{\alpha} + \frac{n}{\beta} = 1\,,\ \beta > n \tag{3}$$

and $n = 3$, then u is unique. Two years later Foias [10] proved the same result for general n but under the slightly more restrictive assumption $PS(\alpha,\beta) < 1$. For $n \leq 4$ Serrin [24] extended Prodi's and Foias' result to $\frac{2}{\alpha} + \frac{n}{\beta} = 1\,,\ \beta > n$ for bounded domains Ω in $\mathbb{R}^n$. More precisely, Serrin proved that if u is a weak solution belonging to $L^\alpha(0,T;L^\beta)$ for α and β as above and if v is another weak solution with the same data satisfying the energy inequality, then u and v coincide: $u \equiv v$. This criterion of Serrin has the advantage that it guarantees uniqueness of weak solutions where only one solution – say u – is required to be of class $L^\alpha(0,T;L^\beta)$ and the other one belongs to the larger class satisfying the usual energy inequality.

It is worth to remark that every weak solution of (1) in the class $L^4(0,T;L^4)$ fulfills the energy inequality, which turns in this case even into an identity. Therefore we can rephrase Serrin's result as follows: $L^\alpha(0,T;L^\beta)$ solutions are unique in the larger class of $L^4(0,T;L^4)$-solutions. One simply has to observe that functions which belong to $L^\infty(0,T;L^2) \cap L^2(0,T;H^{1,2}) \cap L^\alpha(0,T;L^\beta)$ are automatically also elements of $L^4(0,T;L^4)$, which follows from some interpolation inequalities (see also section 2).

Further important contributions to this question were made by Masuda [21], Sohr-von Wahl [30] and finally by Kozono-Sohr [15].

Let us finish our discussion of the uniqueness problem with the hint that one can find further details in Kozono-Sohr [15].

Concerning the regularity problem the starting paper is Serrin [23]. There it is shown that a weak solution u is of class $C^\infty((0,T) \times \mathbb{R}^n)$ if it belongs to the space $L^\alpha(0,T;L^\beta)$ with exponents α, β satisfying $\frac{2}{\alpha} + \frac{n}{\beta} < 1$. In proving this Serrin considers the vorticity equation

$$\partial_t \omega^i - \Delta\omega^i + u^j\, D_j\omega^i - \omega^j\, D_j u^i = (\operatorname{curl} f)^i$$

where ω is the vorticity, i.e. $\omega \equiv \operatorname{curl} u$. Reading this equation as a heat equation and using the assumption on u he improved step by step the regularity of ω and using the fact that $\operatorname{div} u = 0$ also that of u.

Ten years later Fabes-Jones-Rivière [8] treated the case of data in L^p and proved regularity and uniqueness under the same condition as Serrin supposed.

Sohr [25] succeeded in proving that the class $L^\alpha(0,T;L^\beta)$ with $\frac{2}{\alpha}+\frac{n}{\beta}=1$, $\beta>n$ is also a regularity class extending Serrin's result to the "limit case" in which the critical quantity $\frac{2}{\alpha}+\frac{n}{\beta}$ can be equal to 1 but (still) β must be strictly larger than $n=$ dimension. Essential tools in Sohr's approach are the use of the Yosida-approximation and the potential theoretic estimates of Solonnikov [26], [27]. He also investigated the limit case $L^\infty(0,T;L^n)$ and formulated a nice criterion for this case.

Independently of each other Giga [12], Struwe [29] and Takahashi [31] proved also the latter result with quite different methods. In fact Giga applies "abstract semigroup-theory" (i.e. results like that of Fujita-Kato [9]), Struwe goes back to Serrin's idea of using the vorticity equation, but then be proceeds differently (employing test function techniques in stead of integral representations). Takahashi finally refines results of Ladyženskaya-Solonnikov-Ural'ceva [20] on parabolic systems via a cut-off technique.

Another regularity criterion was treated by Beirão Da Veiga [1]. He extended Serrin's regularity criterion to gradients showing that if

$$\nabla u \in L^\alpha(0,T;L^\beta) \quad \text{with} \quad \frac{2}{\alpha}+\frac{n}{\beta}=2\,,$$

then

$$\nabla u \in C^0([0,T);L^{\frac{\alpha}{\alpha-1}}) \cap L^{\frac{\alpha}{\alpha-1}}(0,T;L^{\frac{2\beta}{n-2}})\,,$$

in particular u is a (strong and) regular solution.

The purpose of this note is to study the following four limit cases of the above mentioned regularity classes:

$$\begin{aligned} &1)\quad u\in L^2(0,T;L^\infty)\,,\ 2)\quad \nabla u\in L^2(0,T;L^n)\,,\\ &3)\ \ \nabla u\in L^1(0,T;L^\infty)\,,\ 4)\ \ \nabla^2 u\in L^1(0,T;L^n)\,, \end{aligned}$$

and to weaken/sharpen the assumption concerning the regularity with respect to the space variables. Instead of 1) – 4) we will only assume

$$\begin{aligned} &1')\quad u\in L^2(0,T;BMO)\,,\ 2')\quad \nabla u\in L^2(0,T;L^{2,n-2})\,,\\ &3')\ \ \nabla u\in L^1(0,T;BMO)\,,\ 4')\ \ \nabla^2 u\in L^1(0,T;L^{2,n-2})\,, \end{aligned}$$

where BMO denotes the John-Nirenberg space of functions with bounded mean oscillation and $L^{2,n-2}$ denotes a certain Morrey space related to BMO via the Poincaré inequality (see section 2 for details). Our aim will be that under these weaker assumptions still the classical assertions are true, i.e. all four classes are regularity classes and the first two classes are also uniqueness classes.

2 Statements of the Results

Before stating our results, we introduce some function spaces, recall some of their properties and then we give the definition of a weak solution.

Let $C^\infty_{0,\sigma}$ denote the set of all C^∞ vector functions $\phi = (\phi^1, \dots, \phi^n)$ with compact support such that $\operatorname{div} \phi = 0$.
L^γ_σ is the closure of $C^\infty_{0,\sigma}$ with respect to the L^γ-norm $\|\cdot\|_{0,\gamma}$; $(\cdot,\cdot)$ denotes the duality pairing between L^γ and $L^{\gamma'}$, where $\frac{1}{\gamma} + \frac{1}{\gamma'} = 1$. $\mathring{W}^{1,\gamma}_\sigma$ denotes the closure of $C^\infty_{0,\sigma}$ with respect to the norm

$$\|\phi\|_{1,\gamma} \equiv \|\phi\|_{0,\gamma} + \|\nabla\phi\|_{0,\gamma}\,,$$

where $\nabla\phi = \left(\frac{\partial\phi^i}{\partial x_j}\right)$; $i, j = 1, \dots, n$.

For an interval I in $\mathbb{R}$ and a Banach space X, $L^p(I;X)$ and $C^m(I;X)$ denote the usual Banach spaces, where $1 \le p \le +\infty$, $m = 0, 1, 2, \dots$.

BMO denotes the John-Nirenberg space of functions with bounded mean oscillation i.e.

$$BMO \equiv \left\{ f \in L^1_{\text{loc}}(\mathbb{R}^n) : \sup_B \frac{1}{|B|} \int_B |f(y) - f_B|\, dy < +\infty \right\},$$

where $f_B \equiv \fint_B f(y)\, dy = \frac{1}{|B|}\int_B f(y)\, dy$ is the mean value of f over B and B denotes an arbitrary ball in $\mathbb{R}^n$. As a norm on BMO we take

$$\|f\|_{BMO} = \|f\|_* \equiv \sup_B \frac{1}{|B|} \int_B |f(y) - f_B|\, dy\,.$$

With this norm BMO becomes a Banach space, if we identify functions which differ by a constant:

$$\|f\|_{BMO} = 0 \Leftrightarrow f \equiv \text{const}\,.$$

So in fact BMO is a quotient space. To imagine how "large" the space BMO is we recall from the literature the following facts (see for example [28]):

i) $L^\infty \subset BMO \subset L^p$ for any $p < +\infty$;
ii) $\log|x| \in BMO$, but $\log|x| \notin L^\infty$, so $L^\infty \subsetneqq BMO$;
iii) $\log|P(x)| \in BMO$, P some homogeneous polymonial.

$L^{2,n-2}$ denotes the Morrey space

$$L^{2,n-2} \equiv \left\{ f \in L^2(\mathbb{R}^n) : \|f \mid L^{2,n-2}\|^2 := \sup_{z,r} \frac{1}{r^{n-2}} \int_{B_r(z)} |f(y)|^2\, dy < +\infty \right\},$$

where the supremum is taken over all $z \in \mathbb{R}^n$, $r > 0$. $L^{2,n-2}$ is a Banach space with respect to the norm $\|f|L^{2,n-2}\|$ and it is related to the space BMO via Poincaré's inequality:

$$u \in W^{1,2}\,,\ \nabla u \in L^{2,n-2} \Rightarrow u \in BMO$$

because of

$$\int\limits_{B_r(z)} |u(y) - u_{B_r(z)}|^2\,dy \le Cr^n \|\nabla u | L^{2,n-2}\|^2\,.$$

Our definition of weak solutions of (1) is as follows:

Definition 2.1 *Let $a \in L^2_\sigma$ and $f \in L^1(0,T;L^2)$. A measurable function u is called a weak solution of (1) on $(0,T)$ if*

i) $u \in L^\infty(0,T;L^2_\sigma) \cap L^2(0,T;\mathring{W}^{1,2}_\sigma)$ *and*
ii)

$$\int\limits_0^T \{-(u,\partial_t\Phi) + (\nabla u,\nabla\Phi) + (u\cdot\nabla u,\Phi)\}\,dt = (a,\Phi(0)) + \int\limits_0^T (f,\Phi)\,dt \quad (4)$$

for all $\Phi \in C^1([0,T);\Upsilon)$ with $\Phi(T) = 0$, where $\Upsilon \equiv \mathring{W}^{1,2}_\sigma \cap L^n$.

Remark 2.2 *1. For u and Φ as above, the integral $\int\limits_0^T (u\cdot\nabla u,\Phi)\,dt$ is well defined since we have by the Sobolev inequality $\|u\|_{0,\frac{2n}{n-2}} \le C\|\nabla u\|_{0,2}$ (for $n \ge 3$) that*

$$\int\limits_0^T |(u\cdot\nabla u,\Phi)|\,dt \le C\big(\sup_{0<t<T}\|\Phi(t)\|_{0,n}\big)\int\limits_0^T \|\nabla u\|^2_{0,2}\,dt\,.$$

2. The space Υ is a Banach space with respect to the norm $\|\Phi\|_\Upsilon \equiv \|\Phi\|_{1,2} + \|\Phi\|_{0,n}$. It is known that $C^\infty_{0,\sigma}$ is dense in Υ, if the underlying domain Ω is either the whole space, or a half space, or a bounded domain with $C^{2,\mu}$ boundary, or a domain having a compact complement with $C^{2,\mu}$ boundary (see Masuda [21], Proposition 1, Giga [12], Appendix and Kozono-Sohr [15], Theorem 2).
Hence we may take Φ as the test function of the above definition having the form $\Phi(t,x) = h(t)\phi(x)$, where $\phi \in C^\infty_{0,\sigma}$ and $h \in C^1([0,T))$ with $h(T) = 0$. See also Masuda [21], Lemma 2.2.
3. After redefinition of $u(t)$ on a set of measure zero on $(0,T)$, we may assume that $u(t)$ is weakly continuous in L^2_σ (see Prodi [22]).

Our theorem concerning uniqueness now reads:

Theorem 2.3 *There is at most one solution of (1) in the sense of Definition 2.1 such that*

i) $u \in L^\infty(0,T;L^2_\sigma) \cap L^2(0,T;\mathring{W}^{1,2}_\sigma) \cap L^2(0,T;BMO)$ *or*
ii) $u \in L^\infty(0,T;L^2_\sigma) \cap L^2(0,T;\mathring{W}^{1,2}_\sigma)$ *and* $\nabla u \in L^2(0,T;L^{2,n-2})$.

Such a solution would be continuous from $[0,T]$ into L^∞_σ.

Remark 2.4 *The above uniqueness-classes weaken/sharpen the assumption concerning the regularity with respect to the space variables of the following "known" uniqueness-classes:*

$$i)\ u \in L^2(0,T;L^\infty)\ , \qquad ii)\ \nabla u \in L^2(0,T;L^n)\ .$$

This will be clear from the proof of our theorem, but is mostly known to experts in this field of PDE-theory. We refer to von Wahl [32], Beirao da Veiga [1] for corresponding results in the literature.

Concerning regularity we prove in the sequel the following a priori estimates:

Theorem 2.5 *Assume that u is a regular solution of (1) in some interval $[0,T]$ (with $f \equiv 0$ for simplicity). Then, if*

$$i)\ \nabla u \in L^2(0,T;L^{2,n-2}) \quad or \quad ii)\ u \in L^2(0,T;BMO) \quad or$$
$$iii)\ \nabla^2 u \in L^1(0,T;L^{2,n-2}) \quad or \quad iv)\ \nabla u \in L^1(0,T;BMO)\ ,$$

one has

$$u \in C(0,T;\mathring{W}^{1,2}_\sigma) \cap L^2(0,T;\mathring{W}^{1,2}_\sigma \cap W^{2,2})\,. \tag{5}$$

Moreover,

$$\operatorname*{ess\,sup}_{0\le t<T} \|\nabla u(t)\|_{0,2}^2 + \int_0^T \|\nabla^2 u(\tau)\|_{0,2}^2\,d\tau \tag{6}$$

$$\le C\,\|\nabla u(0)\|_{0,2}^2 \left\{ 1 + \exp\left(C \int_0^T \|\nabla^{\alpha_i} u(\tau)|X_i\|^{\beta_i}\,d\tau \right) \right\}\ ,$$

where $i = 1,2,3,4$ and

$$\alpha_1 = 1;\quad X_1 = L^{2,n-2};\quad \beta_1 = 2; \tag{7}$$
$$\alpha_2 = 0;\quad X_2 = BMO;\quad \beta_2 = 2; \tag{8}$$
$$\alpha_3 = 2;\quad X_3 = L^{2,n-2};\quad \beta_3 = 1; \tag{9}$$
$$\alpha_4 = 1;\quad X_4 = BMO;\quad \beta_4 = 1; \tag{10}$$

according to the cases (i)–(iv) above.

Here and in the sequel we denote by C (or $C_0, C_1, \dots$) positive constants that depend, at most, on the dimension n and other absolute constants like the constant in the Sobolev inequality. The symbol C may be used, even in the same equation, to denote distinct constants.

In order to show how the above a priori estimates apply in the framework of classical (weak) Leray-Hopf solutions (see [18] and [13]), we state the following Theorem which is an easy adaption to our setting of a corresponding Theorem of Beirão da Veiga [1].

Theorem 2.6 *Suppose $a \in W^{1,2}_\sigma$ and u is a Leray-Hopf solution of (1) in the sense of Definition 2.1 in $[0,T)$. If u belongs to one of the four classes specified in Theorem 2.5 (i)–(iv), then*

$$u \in C(0,T;W^{1,2}_\sigma) \cap L^2(0,T;W^{2,2}_\sigma) \tag{11}$$

and

$$\operatorname*{ess\,sup}_{0\le t<T} \|\nabla u(t)\|^2_{0,2} + \int_0^T \|\nabla^2 u(t)\|^2_{0,2}\,dt \tag{12}$$

$$\le C\,\|\nabla a\|^2_{0,2}\left\{1+\exp\left(c\int_0^T \|\nabla^{\alpha_i}u(\tau)|X_i\|^{\beta_i}\,d\tau\right)\right\}, \tag{13}$$

where α_i, X_i and β_i have the same meaning as in Theorem 2.5. In particular u is regular and unique solution in $[0,T]$.

3 Proofs of the Theorems

We start by preparing the proof of Theorem 2.3. If u and v are weak solutions of (1) with the same exterior force f and the same initial velocity a, which additionally belong to one of the classes specified there, we get for the difference $w := u - v$ the system of equations

$$\begin{aligned} \partial_t w^i - \Delta w^i + D_i\pi_w &= -w^j D_j v^i - u^j D_j w^i\,, \\ \operatorname{div} w = 0\,, &\quad w(0,x) = 0\,. \end{aligned} \tag{14}$$

By testing with w^i we obtain formally the identity

$$\frac{1}{2}\frac{d}{dt}\|w(t)\|^2_{0,2} + \|\nabla w(t)\|^2_{0,2} = \int_{\mathbb{R}^n} w^j D_j w^i v^i\,dx\,, \tag{15}$$

because the pressure cancels due to $\operatorname{div} w = 0$ and the term

$$-\int_{\mathbb{R}^n} u^j D_j w^i w^i\,dx = -\int_{\mathbb{R}^n} u^j D_j \frac{|w|^2}{2}\,dx = 0 \tag{16}$$

vanishes also due to $\operatorname{div} u = 0$. We also used integration by parts together with $\operatorname{div} w = 0$ to transform the right hand side in (15). From (15) we would like to estimate the convective term $(w\cdot\nabla w, v) = \int_{\mathbb{R}^n} w^j D_j w^i v^i\,dx$ in such a way that on applying Gronwall's inequality we could finish the proof because of $w(0,x) = 0$ respective $\|w(0)\|_{0,2} = 0$. For this purpose we need several auxiliary results:

Proposition 3.7 *(S. Chanillo 1991) If $f \in W^{1,2}(\mathbb{R}^n)$, $g \in L^2_\sigma(\mathbb{R}^n)$ and $h \in W^{1,2}(\mathbb{R}^n)$ such that $\nabla h \in L^{2,n-2}(\mathbb{R}^n)$, then there exists a constant $C = C(n) > 0$ independent of f, g and h such that*

$$\Big|\int_{\mathbb{R}^n} f g^j D_j h\, dx\Big| = \Big|\int_{\mathbb{R}^n} g^j D_j f h\, dx\Big| \le C\, \|\nabla h|L^{2,n-2}\|\, \|\nabla f\|_{0,2}\, \|g\|_{0,2}\,. \quad (17)$$

We refer to [4] for a proof of this proposition, but we shall make some comments to the proposition and its proof after completing the proof of Theorem 2.3 in case of $\nabla v \in L^2(0,T; L^{2,n-2})$ (case (ii)). This will be an immediate consequence of Proposition (3.7). Applying the proposition with $g^j = w^j$, $D_j f = D_j w^i$ and $h = v^i$ yields directly

$$\begin{aligned}\Big|\int_{\mathbb{R}^n} w^j D_j w^i v^i\, dx\Big| &\le C\|\nabla v|L^{2,n-2}\|\, \|\nabla w\|_{0,2}\, \|w\|_{0,2} \\ &\le \frac{1}{2}\, \|\nabla w\|^2_{0,2} + \frac{C}{2}\, \|\nabla v|L^{2,n-2}\|^2\, \|w\|^2_{0,2} \end{aligned} \quad (18)$$

with little help of Young's inequality. Therefore (15) shows that

$$\frac{d}{dt}\|w(t)\|^2_{0,2} + \|\nabla w(t)\|^2_{0,2} \le C\, \|\nabla v(t)|L^{2,n-2}\|^2\, \|w(t)\|^2_{0,2} \quad (19)$$

and according to Gronwall's inequality in conjunction with $\|w(0)\|_{0,2} = 0$ we conclude by the assumption $\nabla v \in L^2(0,T; L^{2,n-2})$:

$$\|w(t)\|^2_{0,2} = 0 \qquad \text{for } 0 \le t < T\,, \quad (20)$$

from which the desired uniqueness follows. Therefore Theorem 2.3 (ii) is proven.

Remark 3.8 *1. Chanillo's inequality (17) was used by L. C. Evans to establish partial regularity of weakly harmonic stationary maps which are valued in spheres $S^{m-1} \subset \mathbb{R}^m$ (see [6] for details). The underlying system of partial differential equations is a diagonal elliptic system of critical growth:*

$$-\Delta u^i = u^i|\nabla u|^2\,, \quad |u|^2 = 1\,.$$

2. It is tempting to prove Chanillo's inequality by employing the following scheme:

$$\Big|\int_{\mathbb{R}^n} fg\cdot\nabla h\, dx\Big| \le \Big(\int_{\mathbb{R}^n} |f|^2|\nabla h|^2\, dx\Big)^{1/2} \|g\|_{0,2}$$

using the Cauchy-Schwarz inequality. Now try to show

$$\Big(\int_{\mathbb{R}^n} |f|^2|\nabla h|^2\, dx\Big)^{1/2} \le C\Big(\int_{\mathbb{R}^n} |\nabla f|^2\, dx\Big)^{1/2},$$

where the constant C depends only on "natural" quantities and on $\|\nabla h|L^{2,n-2}\| < +\infty$. This just about fails for general $h \in W^{1,2}$, $\nabla h \in L^{2,n-2}$, but it is in fact true for $h(x) = \log|x|$ because of $|\nabla h|^2 \le \frac{1}{|x|^2}$ and the well known inequality

$$\int_{\mathbb{R}^n} \frac{|f|^2}{|x|^2}\,dx \le K \int_{\mathbb{R}^n} |\nabla f|^2\,dx \qquad \forall f \in W^{1,2}$$

(see for example [17]). Remember: $h(x) = \log|x| \in BMO$! This strongly suggests that Chanillo's inequality will also be true for $h \in W^{1,2} \cap BMO$.

The preceding remarks suggest that Chanillo's inequality will also be true by replacing the $L^{2,n-2}$-norm of ∇h with the BMO-norm of h. In fact we have

Proposition 3.9 *If $f \in W^{1,2}(\mathbb{R}^n)$, $g \in L^2(\mathbb{R}^n, \mathbb{R}^n)$, $\operatorname{div} g = 0$ and $h \in W^{1,2} \cap BMO(\mathbb{R}^n)$, then there exists a positive constant $C = C(n)$ independent of f, g and h such that*

$$\Big|\int_{\mathbb{R}^n} g^j D_j f h\,dx\Big| \le C\,\|g\|_{0,2}\,\|\nabla f\|_{0,2}\,\|h|BMO\|\,. \tag{21}$$

Before proving Proposition 3.9 – the proof requires some preparation – let us quickly finish the proof of Theorem 2.3 case (i): Setting $g^j = w^j$, $D_j f = D_j w^i$ and $h = v^i$ the proposition shows

$$\begin{aligned}\Big|\int_{\mathbb{R}^n} w^j D_j w^i v^i\,dx\Big| &\le C\,\|w\|_{0,2}\,\|\nabla w\|_{0,2}\,\|v|BMO\| \\ &\le \frac{1}{2}\,\|\nabla w\|_{0,2}^2 + \frac{C}{2}\,\|v|BMO\|^2\,\|w\|_{0,2}^2 \end{aligned}\tag{22}$$

and we can conclude in the same way as in case (ii) that $\|w(t)\|_{0,2}^2 = 0$ for $0 \le t < T$ (see (19), (20)).

Next we review the definition and some properties of the Hardy space $\mathcal{H}^1 = \mathcal{H}^1(\mathbb{R}^n)$. Assume that $g \in L^1(\mathbb{R}^n)$ and let Φ be any smooth function with support in the unit ball:

$$\Phi \in C_0^\infty(B_1(0))\,, \quad \int_{\mathbb{R}^n} \Phi(x)\,dx = 1\,. \tag{23}$$

We set

$$g^*(x) := \sup_{r>0}\Big|\frac{1}{r^n}\int g(y)\Phi\Big(\frac{x-y}{r}\Big)\,dy\Big| \tag{24}$$

and say that g belongs to the Hardy space $\mathcal{H}^1$ if $g^* \in L^1(\mathbb{R}^n)$. We write:

$$\|g|\mathcal{H}^1(\mathbb{R}^n)\| = \|g^*\|_{0,1}\,. \tag{25}$$

Observe that $g \in \mathcal{H}^1$ implies $\int_{\mathbb{R}^n} g(x)\,dx = 0$. An easy consequence of the definition is the fact that $\mathcal{H}^1(\mathbb{R}^n)$ is a separable Banach space. A fundamental theorem of C. Fefferman [7], [11] asserts that

$$(\mathcal{H}^1)^* = BMO \tag{26}$$

and in particular provides the duality inequality

$$\left| \int_{\mathbb{R}^n} f(x)g(x)\,dx \right| \le C \|f|BMO\|\ \|g|\mathcal{H}^1(\mathbb{R}^n)\| \tag{27}$$

for example for $f \in L^\infty(\mathbb{R}^n)$, $g \in \mathcal{H}^1(\mathbb{R}^n)$ or $f \in BMO$, $g \in \mathcal{H}^1_a$, where $\mathcal{H}^1_a$ is the dense subspace of $\mathcal{H}^1$ consisting of finite linear combinations of $\mathcal{H}^1$ atoms (see Stein [28], chapt. IV). The constant C in (27) depends only on the dimension n.

Finally we recall for the reader's convenience a result of Coifman, Lions, Meyer and Semmes (see [5], [6] for a nice proof).

Proposition 3.10 *(i) Assume* $E \in L^p(\mathbb{R}^n, \mathbb{R}^n)$, $B \in L^{p'}(\mathbb{R}^n, \mathbb{R}^n)$ *with* $1 < p < +\infty$ *and* $\frac{1}{p} + \frac{1}{p'} = 1$ *and*

$$\operatorname{div} B = 0\,, \qquad \operatorname{curl} E = W(E) = \nabla E - (\nabla E)^T = 0 \tag{28}$$

in the distribution sense. Then $B \cdot E = B_j E_j \in \mathcal{H}^1(\mathbb{R}^n)$, *with the bound*

$$\|B \cdot E\|_{\mathcal{H}^1(\mathbb{R}^n)} \le C \|B\|_{0,p'} \|E\|_{0,p}\,. \tag{29}$$

(ii) If $v \in L^{p'}_\sigma(\mathbb{R}^n)$, $u \in W^{1,p}(\mathbb{R}^n)$ *with* $1 < p < +\infty$, $\frac{1}{p'} + \frac{1}{p} = 1$, *then* $v \cdot \nabla u = v^j D_j u \in \mathcal{H}^1(\mathbb{R}^n)$ *with the bound*

$$\|v \cdot \nabla u\|_{\mathcal{H}^1(\mathbb{R}^n)} \le C \|v\|_{0,p'} \|\nabla u\|_{0,p'}\,. \tag{30}$$

Now the proof of Proposition 3.9 is an immediate consequence of the last proposition and the duality inequality (27):

Proof.

$$\left| \int_{\mathbb{R}^n} g^j D_j f h\,dx \right| \le C \|g \cdot \nabla f|\mathcal{H}^1\|\ \ \|h|BMO\| \le C \|g\|_{0,2}\ \|\nabla f\|_{0,2}\ \ \|h|BMO\|\,. \tag{31}$$

□

Let us interrupt our discussion for a short moment for inserting a comment on the duality inequality (27):

For general $f \in BMO$ and $g \in \mathcal{H}^1$, the integral $\int_{\mathbb{R}^n} fg\,dx$ does not converge absolutely, which means that fg in general is **not** in $L^1(\mathbb{R}^n)$. There is an "explicit" counterexample in Stein's book [28, chapt. IV 6.2, p. 178]. Thus one needs to define this "integral" initially by taking g to be in an appropriate dense linear subspace of $\mathcal{H}^1$ or by taking f to be bounded, then the integrand belongs to L^1, the corresponding inequality (27) is true and one can extend this functional by continuity (see [28, chapt. IV, p. 142 ff.]). So we "conclude" that (27) is also satisfied even with fg in L^1 for $f \in L^1 \cap BMO$ and $g \in \mathcal{H}^1 \cap L^p$ for some $p > 1$. We will show now that in our applications this extra assumption is always satisfied:

Lemma 3.11 *(i) If $u \in L^\infty(0,T;L^2_\sigma) \cap L^2(0,T;\mathring{H}^1_\sigma \cap BMO)$, then $u \in L^4(0,T;L^4)$ and*

$$\|u|L^4(0,T;L^4)\| \leq C\|u|L^\infty(0,T;L^2_\sigma)\|^{1/2} \ \|u|L^2(0,T;BMO)\|^{1/2}. \qquad (32)$$

(ii) If $u \in L^\infty(0,T;L^2) \cap L^2(0,T;W^{1,2})$ and $\nabla u \in L^1(0,T;BMO)$, then $\nabla u \in L^{4/3}(0,T;L^4)$ and

$$\|\nabla u|L^{4/3}(0,T;L^4)\| \leq C\|\nabla u|L^2(0,T;L^2)\|^{1/2} \ \|\nabla u|L^1(0,T;BMO)\|^{1/2}. \qquad (33)$$

Proof. The proof consists of one line:

$$\|f\|_{0,4} \leq A_4\|f^\#\|_{0,4} \leq A_4\|f^\#\|_{0,\infty}^{1/2}\|f^\#\|_{0,2}^{1/2} \leq C\|f|BMO\|^{1/2}\|f\|_{0,2}^{1/2} \qquad (34)$$

for all $f \in L^2 \cap BMO$, by using first that $f \in BMO \Leftrightarrow f^\# \in L^\infty$, where $f^\#(x) := \sup_{x\in B}\frac{1}{|B|}\int_B |f(y) - f_B|\,dy$ denotes the Fefferman-Stein sharp function, second $f \in L^{p_0}$ for some $1 < p_0 < +\infty$ and $f^\# \in L^p$ for some $p > p_0$ imply $f \in L^p$ and $\|f\|_{0,p} \leq A_p\|f^\#\|_{0,p}$, and third Hölder's inequality (see Stein [28, p. 146 f.]). □

An immediate consequence of the lemma is the fact that for weak solutions u of (1) in the class $u \in L^\infty(0,T;L^2_\sigma) \cap L^2(0,T;W^{1,2} \cap BMO)$ the energy inequality turns into an identity and $u \in C^0(0,T;L^2_\sigma)$. This observation leads us to the following

Remark 3.12 *The uniqueness criterion of Theorem 2.3 can also be formulated in the following way: If u and v are two weak solutions of (1) on $(0,T)$ with the same initial velocity $a \in L^2_\sigma$ such that u belongs additionally to $L^2(0,T;BMO)$ and v satisfies the energy inequality*

$$\|v(t)\|_{0,2}^2 + 2\int_0^t \|\nabla v(\tau)\|_{0,2}^2\,d\tau \leq \|a\|_{0,2}^2$$

for $0 < t \leq T$, then $u(t) = v(t)$ for all $t \in [0,T]$. For simplicity we took here $f \equiv 0$, but this is no serious restriction.

For two reasons we do not give a proof of this version of the uniqueness criterion here: Firstly, the above stated version coincides with Theorem 2(1) of a recent paper by Kozono-Taniuchi [16] and the reader can find there a detailed proof. Secondly, we shall not apply this version of the uniqueness criterion and we feel that it is not necessary to do this because the (formal) idea of the proof is clear.[1]

Next we pass over to the proof of Theorem 2.5, which is now easy:

Proof. Since we assume that u is a regular solution (for instance a strong one) on $[0,T)$ we are allowed to test by $-\Delta u^i$ providing

$$\frac{1}{2}\frac{d}{dt}\|\nabla u(t)\|_{0,2}^2 + \|\nabla^2 u(t)\|_{0,2}^2 = \int_{\mathbb{R}^n} D_k u^j D_j D_k u^i u^i \, dx - \int_{\mathbb{R}^n} f^i \Delta u^i \, dx \,, \tag{35}$$

by using integration by parts

$$\int_{\mathbb{R}^n} u^j D_j u^i \Delta u^i \, dx = \int_{\mathbb{R}^n} D_k u^j D_j D_k u^i u^i \, dx + \int_{\mathbb{R}^n} \Delta u^j D_j u^i u^i \, dx \,, \tag{36}$$

observing that the pressure cancels due to $0 = \operatorname{div} u = \operatorname{div} \Delta u$ and that for the same reason

$$\begin{aligned} \int_{\mathbb{R}^n} u^j D_j D_k u^i D_k u^i \, dx &= \int_{\mathbb{R}^n} u^j D_j \frac{|\nabla u|^2}{2} \, dx = 0 \\ \int_{\mathbb{R}^n} \Delta u^j D_j u^i u^i \, dx &= \int_{\mathbb{R}^n} \Delta u^j D_j \frac{|u|^2}{2} \, dx = 0 \,. \end{aligned} \tag{37}$$

In case (i) we use Chanillo's inequality (17) of Proposition 3.7 to estimate:

$$\left| \int_{\mathbb{R}^n} D_k u^j D_j D_k u^i u^i \, dx \right| \leq C \, \|\nabla u | L^{2,n-2}\| \, \|\nabla^2 u\|_{0,2} \, \|\nabla u\|_{0,2} \,. \tag{38}$$

Moving the L^2-norm of the second derivatives with Young's inequality to the left we obtain

$$\frac{d}{dt}\|\nabla u(t)\|_{0,2}^2 + \|\nabla^2 u(t)\|_{0,2}^2 \leq C \, \|\nabla u(t) | L^{2,n-2}\|^2 \, \|\nabla u(t)\|_{0,2}^2 + C \, \|f(t)\|_{0,2}^2 \tag{39}$$

and Gronwall's inequality delivers

[1] The results/remarks which are described in this chapter were obtained by the author in 1996/97 and he gave a talk on this topic in the spring 1997 on an international conference in honour of Prof. Dr. J. Nečas, which took place in Lisbon. So they are obtained independently from the paper of Kozono-Taniuchi.

$$\operatorname*{ess\,sup}_{t\in(0,T)}\|\nabla u(t)\|_{0,2}^2 \le \exp\left(C\int_0^T \|\nabla u(s)|L^{2,n-2}\|^2\,ds\right)\times$$
$$\left\{\|\nabla a\|_{0,2}^2 + C\int_0^T \|f(s)\|_{0,2}^2\,ds\right\}, \tag{40}$$

which by integrating (39) with respect to t finally leads to

$$\operatorname*{ess\,sup}_{t\in(0,T)}\|\nabla u(t)\|_{0,2}^2 + \int_0^T \|\nabla^2 u(t)\|_{0,2}^2\,dt \le \left(\|\nabla a\|_{0,2}^2 + C\int_0^T \|f(s)\|_{0,2}^2\,ds\right)\times$$
$$\left\{1+\exp\left(C\int_0^T \|\nabla u(s)|L^{2,n-2}\|^2\,ds\right)\right\}. \tag{41}$$

Case (i) is proven. To prove case (ii) we use instead of (17) (Proposition 3.7) inequality (21) (Proposition 3.9) and proceed in completely the same way as before, i. e. replacing in the formula above $\|\nabla u(t)|L^{2,n-2}\|$ by $\|u(t)|BMO\|$. Case (iii) and (iv) are proven in the same manner, but by starting to estimate the right hand side from the "stage" before:

$$\left|\int_{\mathbb{R}^n} D_k u^j D_j u^i D_k u^i\,dx\right| \le C\,\|\nabla^2 u|L^{2,n-2}\|\,\|\nabla u\|_{0,2}^2 \tag{42}$$

by Chanillo's inequality (17) ($\nabla f \sim D_j u^i, g \sim D_k u^j, h \sim D_k u^i$) or by (21)

$$\left|\int_{\mathbb{R}^n} D_k u^j D_j u^i D_k u^i\,dx\right| \le C\,\|\nabla u|BMO\|\,\|\nabla u\|_{0,2}^2\,, \tag{43}$$

where we used the fact that $D_k u^j D_j u^i$ belongs to $\mathcal{H}^1$ and fulfills

$$\|D_k u^j D_j u^i|\mathcal{H}^1\| \le C\,\|\nabla u\|_{0,2}^2$$

for all $1 \le i, k \le n$ as a consequence of Proposition 3.10, so Theorem 2.5 is proven. □

Let us finish this section with some remarks:

Remark 3.13 *1) The proof of Theorem 2.5 in cases (iii) and (iv) shows that the same assertion holds for solutions of the Euler equations. This comes from the fact that in these cases we did not use the "dissipation-term" $-\Delta u$ during the proof. So we have proven also a regularity class criterion for Euler's equations which replaces the class $L^1(0,T;W^{1,\infty})$, which is known to be a regularity class for the Euler equations. In* *fact it is the sole class among the classes "*$\nabla u \in L^{\alpha}(0,T;L^{\beta})$ *with* $\frac{2}{\alpha}+\frac{n}{\beta}=2$*" to be a regularity class for these equations (at least according to what is known at present).*

2) In all statements of Theorem 2.5 one can replace ∇u *by* $W(u)=\nabla u-(\nabla u)^T$ *or* $A(u)=\nabla u+(\nabla u)^T$*. This observation goes back to Kato-Ponce [14] and relies on the fact that the gradient (=Jacobian matrix here) can be recovered from the Spin tensor* $W(u)$ *or the deformation tensor* $A(u)$ *via some Riesz transforms provided* $\operatorname{div} u = 0$*:*

$$D_j u^k = R_j R_i W_{ik}(u) = R_j R_i (D_k u^i - D_i u^k)\,. \tag{44}$$

Similarly we have

$$D_j u^k = -R_j R_i A_{ik}(u) = -R_j R_i (D_k u^i + D_j u^k)\,. \tag{45}$$

Since all Riesz transforms are bounded operators on BMO*, we see by (45), (44) that*

$$\nabla u \in BMO \Leftrightarrow A(u) \in BMO \Leftrightarrow W(u) \in BMO \tag{46}$$

for solenoidal vector fields u*.*

3) The results of Theorem 2.5/2.6 can be reformulated in the spirit of the paper "Remarks on the break down of smooth solutions for the 3-D Euler equations" of Beale, Kato and Majda from 1984 (see [2], [14] and [16]).

References

1. Beirão da Veiga, H. (1995): A new regularity class for the Navier-Stokes equations in $\mathbb{R}^n$. Chinese Ann. Math. Ser. B, **16(4)**, 407–412
2. Beale, J.T., Kato, T., Majda, A. (1984): Remarks on the breakdown of smooth solutions for the 3-D Euler equations. Comm. Math. Phys., **94(1)**, 61–66
3. Constantin, P., Foias, C. (1988): Navier-Stokes equations. University of Chicago Press, Chicago, IL
4. Chanillo, S. (1991): Sobolev inequalities involving divergence free maps. Comm. Partial Differential Equations, **16(12)**, 1969–1994
5. Coifman, R., Lions, P.L., Meyer, Y., Semmes, S. (1993): Compensated compactness and Hardy spaces. J. Math. Pures Appl. (9), **72(3)**, 247–286
6. Evans, L.C. (1991): Partial regularity for stationary harmonic maps into spheres. Arch. Rational Mech. Anal., **116(2)**, 101–113
7. Fefferman, C. (1971): Characterizations of bounded mean oscillation. Bull. Amer. Math. Soc., **77**, 587–588
8. Fabes, E.B., Jones, B.F., Rivière, N.M. (1972): The initial value problem for the Navier-Stokes equations with data in L^p. Arch. Rational Mech. Anal., **45**, 222–240
9. Fujita, H., Kato, T. (1964): On the Navier-Stokes initial value problem. I. Arch. Rational Mech. Anal., **16**, 269–315
10. Foias, C. (1961): Une remarque sur l'unicité des solutions des équations de Navier-Stokes en dimension n. Bull. Soc. Math. France, **89**, 1–8

11. Fefferman, C., Stein, E.M. (1972): H^p spaces of several variables. Acta Math., **129(3-4)**, 137–193
12. Giga, Y. (1986): Solutions for semilinear parabolic equations in L^p and regularity of weak solutions of the Navier-Stokes system. J. Differential Equations, **62(2)**, 186–212
13. Hopf, E. (1951): Über die Anfangswertaufgabe für die hydrodynamischen Grundgleichungen. Math. Nachr., **4**, 213–231
14. Kato, T., Ponce, G. (1988): Commutator estimates and the Euler and Navier-Stokes equations. Comm. Pure Appl. Math., **41(7)**, 891–907
15. Kozono, H., Sohr, H. (1996): Remark on uniqueness of weak solutions to the Navier-Stokes equations. Analysis, **16(3)**, 255–271
16. Kozono, H., Taniuchi, Y. (2000): Bilinear estimates in BMO and the Navier-Stokes equations. Math. Z., **235(1)**, 173–194
17. Ladyzhenskaya, O.A. (1963): The mathematical theory of viscous incompressible flow. Gordon and Breach Science Publishers, New York
18. Leray, J. (1934): Sur le mouvement d'un liquide visquex emplissant l'espace. Acta Mathematica, **63**, 93–248
19. Lions, J.L. (1969): Quelques méthodes de résolution des problèmes aux limites non linéaires. Dunod, Paris
20. Ladyženskaja, O.A., Solonnikov, V. A., Ural′ceva, N.N. (1967): Linear and quasilinear equations of parabolic type. American Mathematical Society, Providence, R.I.
21. Masuda, K. (1984): Weak solutions of Navier-Stokes equations. Tohoku Math. J. (2), **36(4)**, 623–646
22. Prodi; G. (1959): Un teorema di unicità per le equazioni di Navier-Stokes. Ann. Mat. Pura Appl. (4), **48**, 173–182
23. Serrin, J. (1962): On the interior regularity of weak solutions of the Navier-Stokes equations. Arch. Rational Mech. Anal., **9**, 187–195
24. Serrin, J. (1963): The initial value problem for the Navier-Stokes equations. In: Nonlinear Problems (Proc. Sympos., Madison, Wis.), Univ. of Wisconsin Press, Madison, Wis.
25. Sohr, H. (1983): Zur Regularitätstheorie der instationären Gleichungen von Navier-Stokes. Math. Z., **184(3)**, 359–375
26. Solonnikov, V.A. (1968): Estimates of the solutions of a nonstationary linearized system of Navier-Stokes equations. Am. Math. Soc., Transl., II. Ser. 75, **14**, 1–116
27. Solonnikov, V.A. (1977): Estimates for solutions of nonstationary Navier-Stokes equations. J. Sov. Math., **8**, 467–529
28. Stein, E.M. (1993): Harmonic analysis: real-variable methods, orthogonality, and oscillatory integrals. Princeton University Press, Princeton, NJ
29. Struwe, M. (1988): On partial regularity results for the Navier-Stokes equations. Comm. Pure Appl. Math., **41(4)**, 437–458
30. Sohr, H., von Wahl, W. (1984): On the singular set and the uniqueness of weak solutions of the Navier-Stokes equations. Manuscripta Math., **49(1)**, 27–59
31. Takahashi, S. (1990): On interior regularity criteria for weak solutions of the Navier-Stokes equations. Manuscripta Math., **69(3)**, 237–254
32. von Wahl, W. (1985): The equations of Navier-Stokes and abstract parabolic equations. Friedr. Vieweg & Sohn, Braunschweig

Problems Due to the No-Slip Boundary in Incompressible Fluid Dynamics*

Jens Frehse[1] and Josef Málek[2]

[1] Institut für Angewandte Mathematik, Universität Bonn, Beringstr. 6, D-53115 Bonn, Germany
[2] Mathematical Institute of Charles University, Sokolovská 83, 186 75 Prague 8, Czech Republic

Summary. Dealing with the analysis of the systems of PDE's describing unsteady flows of incompressible fluids, we focus on open problems that occur if the equations are supplemented by the homogeneous Dirichlet (no-slip) boundary conditions, but that are successfully solvable in the spatially periodic setting, for example. Within this framework we also present four different approaches to obtain compactness of weakly converging quantities and discuss how these tools can be used in the global existence theory for the power-law fluids and their various generalizations.

1 Introduction

There is a saying *"The boundary is the invention of the devil"* attributed to Werner Karl Heisenberg in connection with his research on turbulence modelling. We paraphrase Heisenberg's words and formulate the question: *"Is the no-slip boundary in incompressible fluid dynamics an invention of the devil?"* In this paper, we justify this question by presenting several problems concerning the analysis of evolutionary PDE's stemming from fluid mechanics that are supplemented by the no-slip boundary conditions. To avoid any confusion, we stress at the beginning that the analysis of turbulence models is not the subject of this paper at all. We wish to remark however that it is commonly accepted that turbulent flows are generated by the (no-slip) boundary.

The purpose of this article is threefold.

First, focusing on the evolutionary Navier-Stokes equations in three spatial dimensions (3d NSEs for short) we discuss three different problems that are still open for the no-slip boundary, but which have been successfully solved for the Cauchy or spatially periodic problem, or even for other type of the boundary conditions. More precisely, we discuss the following questions related to weak solutions of 3d NSEs:

- Regularity of the pressure and the time derivative of the velocity.
- The existence of Leray's self-similar solutions.
- Shift of the 3d NSEs from supercritical problems to critical ones.

These issues are addressed in Section 2.

* J. Málek thanks the Grant Agency of the Czech republic (GACR 201/00/0768) and the Czech Ministry of Education (MSM 113200007) for their support.

Then, in Section 3, we will be mainly concerned with the existence theory "in large" for the power-law fluids. We present four different approaches that provide the compactness of the velocity gradient (for different ranges of the power-law exponent) which is the key step in passage to the limit from suitable chosen approximations to the original problem. We also show briefly how these methods are incorporated in analyzing the power-law fluids pointing out the difficulties/differences that are due to no-slip boundary conditions.

Finally, the third aim is to expose a part of the results obtained in the last decade at the Division of Applied Analysis in the Institute for Applied Mathematics of the University of Bonn, supported in part by the Sonderforschungsbereich 256.

In the rest of this introductory section we describe the models and the boundary conditions we deal with. Throughout the whole paper we assume that the fluid occupies an open, three-dimensional set Ω with a smooth boundary $\partial\Omega$; $T \in (0,\infty)$ denotes the length of a considered time interval, and $Q_T = (0,T) \times \Omega$.

Restricting to isothermal flows, incompressible fluid dynamics is captured by the following system of equations in Q_T

$$\operatorname{div} \boldsymbol{v} = 0\,, \qquad \boldsymbol{v}_t + \operatorname{div}(\boldsymbol{v} \otimes \boldsymbol{v}) - \operatorname{div} \mathbf{S} = -\nabla p + \boldsymbol{f}\,. \tag{1}$$

Here, we look for a couple $(\boldsymbol{v}, p)$ consisting of the velocity $\boldsymbol{v} = (v_1, v_2, v_3)$ and the pressure p, while the density of the external body forces $\boldsymbol{f} = (f_1, f_2, f_3)$ and the constitutive equation for the extra stress $\mathbf{S}$ have to be given. Assuming that $\mathbf{S}$ is a tensorial function of the velocity gradient, the most general model *supported* by experimental observations (see [20], [15] and [7] for more details and further references) is given by

$$\mathbf{S} = \nu(p, |\mathbf{D}(\boldsymbol{v})|^2)\mathbf{D}(\boldsymbol{v})\,. \tag{2}$$

Here, $\mathbf{D}(\boldsymbol{v})$ denotes the symmetric part of the velocity gradient with the components $D_{ij}(\boldsymbol{v}) = \frac{1}{2}(\partial_{x_j} v_i + \partial_{x_i} v_j)$.

We will consider two particular cases of (2), namely,

– the Navier-Stokes fluid with

$$\mathbf{S} = 2\nu_0 \mathbf{D}(\boldsymbol{v})\,. \qquad (\nu_0 > 0) \tag{3}$$

Then the governing equations (1) reduce to the Navier-Stokes equations

$$\operatorname{div} \boldsymbol{v} = 0\,, \qquad \boldsymbol{v}_t + \operatorname{div}(\boldsymbol{v} \otimes \boldsymbol{v}) - \nu_0 \Delta \boldsymbol{v} = -\nabla p + \boldsymbol{f}\,. \tag{4}$$

– the power-law fluids with

$$\mathbf{S} = 2\nu_0 |\mathbf{D}(\boldsymbol{v})|^{r-2} \mathbf{D}(\boldsymbol{v})\,, \qquad (r \in (1,\infty)) \tag{5}$$

where r is the power-law exponent. From a physical point of view, it is of interest to deal with both $r > 2$ and $r \in \langle 1, 2)$. The former captures the fluids that have the ability to shear thicken, i.e. ν is increasing as $|\mathbf{D}(\boldsymbol{v})|$ increases. The latter captures shear thinning fluids where the generalized viscosity ν is a decreasing function of $|\mathbf{D}(\boldsymbol{v})|$. Note that if $r = 2$ in (5) we obtain the Navier-Stokes fluid. The governing equations for the model given by (5) are

$$\operatorname{div} \boldsymbol{v} = 0\,, \qquad \boldsymbol{v}_t + \operatorname{div}(\boldsymbol{v} \otimes \boldsymbol{v}) - 2\nu_0 \operatorname{div}(|\mathbf{D}(\boldsymbol{v})|^{r-2}\mathbf{D}(\boldsymbol{v})) = -\nabla p + \boldsymbol{f}\,. \tag{6}$$

These equations are completed by convenient initial and boundary conditions for the velocity. Regarding the latter we distinguish three cases:

- *No-slip boundary.* It has been experimentally observed that a viscous fluid ($\mathbf{S} \neq \mathbf{0}$) can adhere to a given boundary. If $\partial\Omega$ is fixed, it means that the velocity vanishes at the boundary and the fluid exhibits no-slip there. Thus,

$$\boldsymbol{v} = \mathbf{0} \quad \text{at } [0,T] \times \partial\Omega\,. \tag{7}$$

- *No boundary.* It happens if the fluid occupies the whole three-dimensional space and the velocity is vanishing at the infinity ($\Omega = \mathbb{R}^3$); or all considered fields are spatially periodic with period L_i at each variable x_i (then $\Omega = (0, L_1) \times (0, L_2) \times (0, L_3)$).
- *Navier's boundary.* This is described by the conditions

$$\boldsymbol{v} \cdot \boldsymbol{n} = 0 \quad \text{and} \quad \mathbf{S}\boldsymbol{n} \cdot \boldsymbol{t}^s + \alpha\, \boldsymbol{v} \cdot \boldsymbol{t}^s = 0 \text{ at } [0,T] \times \partial\Omega \ \text{ for } s = 1, 2\,, \tag{8}$$

where $\boldsymbol{n}$ is the outward normal vector and $\boldsymbol{t}^1$, $\boldsymbol{t}^2$ are the corresponding tangent vectors. The first condition in (8) expresses the impermeability of the boundary, while the other conditions relates the tangent component of the velocity at the boundary with forces exerted by the fluid onto the boundary. The coefficient α is supposed to be non-negative and it may depend on $|\boldsymbol{v}|$. If $\alpha = 0$ one talks about the no-stick or slippery boundary.

2 No-Slip Boundary and Navier-Stokes Equations

2.1 Regularity of the Pressure and the Time Derivative for Weak Solutions

Let us consider a globally existing weak solution to 3d NSEs fulfilling

$$\boldsymbol{v} \in L^\infty(0,T; L^2(\Omega)) \cap L^2(0,T; W^{1,2}_{div}(\Omega)) \cap L^{10/3}(Q_T)\,, \tag{9}$$

and discuss the question to which spaces $\boldsymbol{v}_t$ and p belong. Clearly, from (4) one concludes (at least formally) that

$$\boldsymbol{v}_t + \nabla p \in (L^{5/2}(0,T; W^{1,5/2}(\Omega))^*\,, \tag{10}$$

assuming that $\boldsymbol{f}$ also belongs to this space. To separate information for p from $\boldsymbol{v}_t$ or vice versa one can choose one of the following ways:

1. To apply div to (4) and obtain the equation for the pressure eliminating the time derivative and the Laplace operator thanks to $\operatorname{div} \boldsymbol{v} = 0$. This yields

$$-\Delta p = \operatorname{div} \operatorname{div}(\boldsymbol{v} \otimes \boldsymbol{v}) - \operatorname{div} \boldsymbol{f} = \operatorname{div}(v_k \partial_{x_k} \boldsymbol{v}) - \operatorname{div} \boldsymbol{f}\,, \tag{11}$$

   which is very useful in no boundary case, while the other boundary conditions bring a nonhomogeneous Neumann condition for p, difficult to deal with.
2. To observe directly from (4) that for any convenient $s > 1$

$$\boldsymbol{v}_t \in (L^s(0,T; W^{1,s}(\Omega)))^*\,. \tag{12}$$

   Note that it is easy to observe that $\boldsymbol{v}_t$ belongs to a dual of the space of divergenceless functions, which is however of no use in identifying the pressure. (For example, it is well known that $\boldsymbol{v}_t \in L^{4/3}(0,T; (W^{1,2}_{div}(\Omega))^*)$.)
   To achieve (12), we apply the Helmholtz decomposition to any function $\boldsymbol{\varphi} \in L^s(0,T; W^{1,s}(\Omega))$ and obtain

$$\boldsymbol{\varphi} = \boldsymbol{\varphi}_0 + \nabla z \quad \text{with} \quad \operatorname{div} \boldsymbol{\varphi}_0 = 0 \text{ in } \Omega \quad \text{and} \quad \boldsymbol{\varphi}_0 \cdot \boldsymbol{n} = 0 \text{ at } \partial\Omega\,. \tag{13}$$

   For all considered types of the boundaries, $\boldsymbol{v}_t$ is orthogonal to ∇z. Hence we have

$$\langle \boldsymbol{v}_t, \boldsymbol{\varphi} \rangle = \langle \boldsymbol{v}_t, \boldsymbol{\varphi}_0 \rangle$$

   which can be used to derive (12) provided $\boldsymbol{\varphi}_0$ is a suitable test function. Thanks to (13) it happens not only for no boundary case, but also for Navier's boundary. It is however the incorrect test function for the no-slip boundary.
3. To apply the results known for the evolutionary Stokes system. One can identify the Navier-Stokes equations with evolutionary Stokes system

$$\boldsymbol{v}_t - \nu_0 \Delta \boldsymbol{v} - \nabla p = \boldsymbol{f} - v_k \partial_{x_k} \boldsymbol{v} \tag{14}$$

   supplemented by the corresponding boundary conditions. Due to uniqueness of the evolutionary Stokes system, one incorporates the L^q-theory to (14) (see [27]) noticing that (9) implies $v_k \partial_{x_k} \boldsymbol{v} \in L^{5/4}(Q_T)$. Therefore, if $\boldsymbol{f}$ is regular one concludes that

$$\begin{aligned} &\|\boldsymbol{v}_t\|_{L^{5/4}(Q_T)} + \|\nabla^2 \boldsymbol{v}\|_{L^{5/4}(Q_T)} + \|\nabla p\|_{L^{5/4}(Q_T)} \\ &\qquad \le C(\|v_k \partial_{x_k} \boldsymbol{v}\|_{L^{5/4}(Q_T)} + \|\boldsymbol{f}\|_{L^{5/4}(Q_T)}) \le K, \end{aligned} \tag{15}$$

   where K depends on the data. This is true for all types boundaries we deal with. A remarkable difference between the results for no-slip and no boundary case occur if the right-hand side $\boldsymbol{f}$ is of the form

$$\boldsymbol{f} = \operatorname{div} \mathsf{G} \quad \text{with } \mathsf{G} \in L^q(0,T; L^q(\Omega))\,.$$

   In case of no boundary or Navier's boundary, the above presented approaches imply that

$$\|\nabla \boldsymbol{v}\|_{L^s(Q_T)} + \|p\|_{L^s(Q_T)} \leq C\|\mathbf{G}\|_{L^s(Q_T)} \quad \text{with } s = \min\{2, q\}\,, \tag{16}$$

whereas the recent studies of H. Koch and V.A. Solonnikov [8], [9] show that in case of no-slip boundary the estimate for the pressure is not any more true. In fact, one has, see [8], that $p = p^1 + \partial_t p^2$, where p^1 fulfils the estimate (16), but p^2 being a harmonic function is such that in general $\partial_t p^2$ does not belong to $L^s(Q_T)$, see [9].

One can ask why we mentioned the first two approaches to detect the information on p and $\boldsymbol{v}_t$ if the third approach seems to be optimal. The reason comes from studying more general models where the Laplace operator is replaced by a nonlinear operator, compare (4) with (6). Then the results for the evolutionary Stokes system (14) based on the representation formula for linear operators are of a little use.

2.2 Leray's Selfsimilar Weak Solutions

It is well known that if $(\boldsymbol{v}, p)$ solves NSEs (4) then $(\boldsymbol{v}^\lambda, p^\lambda)$ defined through

$$\boldsymbol{v}^\lambda(t,x) = \lambda \boldsymbol{v}(\lambda^2 t, \lambda x)\,, \qquad p^\lambda = \lambda^2 p(\lambda^2 t, \lambda x) \tag{17}$$

solve NSEs too.

In 1934, Jean Leray [14] proposed to construct a weak solution exhibiting the singularity at prescribed time T^* in the form

$$\boldsymbol{v}(t,x) = \frac{1}{\sqrt{T^* - t}} \boldsymbol{U}\Big(\frac{x}{\sqrt{T^* - t}}\Big)\,, \quad p(t,x) = \frac{1}{T^* - t} P\Big(\frac{x}{\sqrt{T^* - t}}\Big)\,, \tag{18}$$

provided that there is a nontrivial solution $(\boldsymbol{U}, P)$ of the system

$$\operatorname{div} \boldsymbol{U} = 0\,, \qquad \frac{1}{2}\boldsymbol{U} + \frac{1}{2} y_k \partial_{y_k} \boldsymbol{U} + U_k \partial_{y_k} \boldsymbol{U} - \nu_0 \Delta \boldsymbol{U} = -\nabla P\,.$$

Note that (18) fits into the selfsimilar transformation (17). Observe also that the selfsimilar transformation (17) can be considered in any conical domain having the vertex at the origin.

In the no boundary case (i.e. $\Omega = \mathbb{R}^3$), the existence of $(\boldsymbol{U}, P)$ has been excluded in the series of papers: see [22] (assuming $\boldsymbol{U} \in L^3$), [16] (if $\boldsymbol{U} \in W^{1,2}$) and [28] (provided that $\boldsymbol{v}$ satisfies energy estimates locally). Remark that a generalization of (18) to so-called pseudo-selfsimilar solutions proposed in [16] was finally also excluded in [21].

In case of any conical domain with the no-slip boundary, particularly in the half space, the possibility of constructing the singular (weak) solution in the form (18) is completely open.

2.3 Supercriticality and Criticality of the NSEs

In this subsection we repeat the arguments presented by V. Šverák at the 7th School on Mathematical Theory in Fluid Mechanics, held at Paseky (Czech Republic) in June 2001.

The total amount of energy dissipated out of the system driven by (4) is given by

$$d(\boldsymbol{v}) = \int_{Q_T} \mathbf{S} \cdot \mathbf{D}(\boldsymbol{v})\, dx\, dt = 2\nu_0 \int_{Q_T} |\mathbf{D}(\boldsymbol{v})|^2\, dx\, dt\,.$$

Studying the behavior of $d(\boldsymbol{v}^\lambda)$ as $\lambda \to \infty$ for $\boldsymbol{v}^\lambda$ defined in (17), one can classify the problems as follows:

$$\text{If} \left\{ \begin{array}{l} d(\boldsymbol{v}^\lambda) \to 0 \\ d(\boldsymbol{v}^\lambda) = d(\boldsymbol{v}^1) \\ d(\boldsymbol{v}^\lambda) \to \infty \end{array} \right. \text{ then the problem is } \left\{ \begin{array}{l} \text{supercritical.} \\ \text{critical.} \\ \text{subcritical.} \end{array} \right. \tag{19}$$

Thus, 2d NSEs represent a critical problem, while 3d NSEs constitute a supercritical one. Recall that this type of classification can be assigned to various problems admitting selfsimilar scaling.

This classification helps to evaluate the difficulty of a given problem in dealing with regularity questions. Roughly speaking, regularity issues for the subcritical problems can be solved by standard methods, the critical problems requires to use subtle techniques and the supercritical problems can be handled only if an additional information is discovered, shifting the problem from supercritical to (sub)-critical. Axially symmetric flows (with zero angular component) represent the exemplary paradigm for (supercritical) 3d NSEs: the additional information comes from the vorticity which in the whole space satisfies the maximum principle, giving then the full regularity of the solution, see [10] or [13]. This approach does not work if Ω is the half space with no-slip boundary. Similarly, for the stationary NSEs in five dimensions that enjoys the same scaling of $d(\boldsymbol{v})$ as 3d evolutionary NSEs, the key role in establishing the full regularity (globally for no boundary problems and locally for the no-slip boundary) plays the quantity $(|\boldsymbol{v}|^2+2p)^+$ satisfying also the maximum principle, see [5], [6] for details.

In both examples, NSEs without the boundary are shifted from a supercritical problem to critical one, whereas it seems that the no-slip boundary keeps these problems supercritical.

3 No-Slip Boundary and Power-Law Fluids

Our main task sounds as follows: for arbitrary (reasonable) data $\Omega, T \in (0,\infty)$ and $\boldsymbol{v}_0$ satisfying

$$\operatorname{div} \boldsymbol{v}_0 = 0 \quad \text{in } \Omega \quad \text{and } \boldsymbol{v}_0 = \mathbf{0} \quad \text{at } \partial\Omega\,, \tag{20}$$

we find a couple $(\boldsymbol{v}, p) : Q_T \to \mathbb{R}^3 \times \mathbb{R}$ solving weakly (6), and satisfying the initial condition $\boldsymbol{v}(0,\cdot) = \boldsymbol{v}_0$ in Ω and the no-slip boundary conditions (7).

We first summarize the basic energy estimates and their consequences.

The balance of mechanical energy implies the following estimates for suitable approximations $(\boldsymbol{v}^\varepsilon, p^\varepsilon)$ of (6)

$$\|\boldsymbol{v}^\varepsilon(t)\|_2^2 + \int_0^T \|\nabla \boldsymbol{v}^\varepsilon\|_r^r \, ds + \int_0^T \|\boldsymbol{v}^\varepsilon\|_{\frac{5r}{3}}^{\frac{5r}{3}} \, ds \le K \,. \tag{21}$$

From here we have the following bound on $\partial_t \boldsymbol{v}^\varepsilon$ (see (5.2.25) in [15], p.207)

$$\|\partial_t \boldsymbol{v}^\varepsilon\|_{L^{r'}(0,T;(W^{s,2}_{div})^*)} \le K \qquad s > \frac{5}{2}\,,\; r' = \frac{r}{r-1}\,. \tag{22}$$

This immediately implies (at least for a subsequence) that

$$\begin{aligned} \nabla \boldsymbol{v}^\varepsilon &\to \nabla \boldsymbol{v} \quad \text{weakly in } L^r(0,t;L^r(\Omega))\,, \\ \boldsymbol{v}^\varepsilon_t &\to \boldsymbol{v}_t \quad \text{weakly in } L^{r'}(0,T;(W^{s,2}_{div})^*)\,, \end{aligned} \tag{23}$$

and if $r > \frac{6}{5}$ the Aubin-Lions compactness lemma gives

$$\boldsymbol{v}^\varepsilon \to \boldsymbol{v} \quad \text{strongly in } L^s(0,t;L^2(\Omega)) \qquad (\forall s \in \langle 1, \infty))\,. \tag{24}$$

It then easily follows from (23) and (24) that if $r > \frac{6}{5}$ one can pass to the limit both in the convective term and in the time derivative term.

The key point is the passage to the limit in nonlinear elliptic term involving $\mathbf{S}(\mathbf{D}(\boldsymbol{v})) = 2\nu_0 |\mathbf{D}(\boldsymbol{v})|^{r-2}\mathbf{D}(\boldsymbol{v})$. Since $|\mathbf{S}(\mathbf{D})| \le C\,|\mathbf{D}|^{r-1}$, this will follow from

$$\text{almost everywhere convergence of } \nabla \boldsymbol{v}^k \text{ to } \nabla \boldsymbol{v} \text{ in } Q_T\,. \; (\{\boldsymbol{v}^k\} \subset \{\boldsymbol{v}^\varepsilon\}) \tag{25}$$

We present four different ways how to obtain (25).

1. Higher regularity method. It is based on standard difference quotient techniques, which lead in no boundary case (see [15] and [23]) to the inequality

$$\begin{aligned} \frac{d}{dt}\|\nabla \boldsymbol{v}^\varepsilon\|_2^2 + J(\boldsymbol{v}^\varepsilon) &\le C\|\nabla \boldsymbol{v}\|_2^{2\lambda}\|\nabla \boldsymbol{v}\|_r^r \qquad \text{with } \lambda = \frac{2(3-r)}{3r-5} \\ \text{and } J(\boldsymbol{v}^\varepsilon) &= \int_\Omega |\mathbf{D}(\boldsymbol{v}^\varepsilon)|^{r-2}\,|\mathbf{D}(\nabla \boldsymbol{v}^\varepsilon)|^2 \, dx\, dt\,. \end{aligned} \tag{26}$$

If $\lambda \le 1$, which happens if $r \ge \frac{11}{5}$, then we obtain higher regularity from which (25) follows using the compact imbedding. See [15] for more details; note however that $J(\boldsymbol{v}^\varepsilon)$ is nondegenerate therein.

If $\lambda > 1$ (it means that $r < \frac{11}{5}$), the inequality (26) does not, in general, improve the smoothness of the solution. However, it follows from (26) and (21) that

$$\int_0^T \frac{J(\boldsymbol{v}^\varepsilon)}{(1 + \|\nabla \boldsymbol{v}\|_2^2)^\lambda} \, ds \le K\,. \tag{27}$$

The further arguments differ in dependence whether $r > 2$ or $r < 2$. For simplicity, let us consider the latter. Then (27) together with (21) implies that if $(2 \ge) r > \frac{9}{5}$ we have

$$\int_0^T \|\nabla^{(2)} \boldsymbol{v}^{\varepsilon_n}\|_r^{2\beta}\, dt \le K \qquad \text{with } \beta \in (0, \tfrac{1}{3}]\,. \tag{28}$$

All these estimates together with an interpolation technique for certain $\sigma > 0$ and $r_0 \in (1, r)$ lead then to the following

$$\left.\begin{array}{l} \int_0^T \|\boldsymbol{v}^{\varepsilon_n}\|_{1,r}^{r}\, dt \le K \\ \int_0^T \|\boldsymbol{v}^{\varepsilon_n}\|_{2,r}^{2\beta}\, dt \le K \end{array}\right. \Longrightarrow \int_0^T \|\boldsymbol{v}^{\varepsilon_n}\|_{1+\sigma,r}^{r_0}\, dt \le K\,. \tag{29}$$

Having (29) and (22) at hand, we apply the Aubin-Lions compactness lemma again and conclude the strong convergence of $\nabla \boldsymbol{v}^{\varepsilon_n}$ to $\nabla \boldsymbol{v}$ in $L^{r_0}(0, T; W^{1,r})$, which implies (25).

Although this method has been extended to the no-slip boundary condition (see [18] and [19] for details) the obtained results are worse than in the spatially-periodic case. On the other hand, it follows from [18] that there is a weak solution to 3d NSEs with the no-slip boundary satisfying

$$\int_0^T \|\nabla^2 \boldsymbol{v}\|_2^{2/3}\, dt \le C\,,$$

the result that has been known before only for the spatially periodic case, see [2].

The following three methods have a common goal: to show that for some $\theta \in (0, 1)$, say $\theta = \frac{1}{2}$, and for an arbitrarily chosen $\eta > 0$ there is $\{\boldsymbol{v}^n\} \subset \{\boldsymbol{v}^\varepsilon\}$ such that

$$\limsup_{n\to\infty} \int_0^T \int_\Omega \Big[\{\mathbf{S}(\mathbf{D}(\boldsymbol{v}^n)) - \mathbf{S}(\mathbf{D}(\boldsymbol{v}))\} : \mathbf{D}(\boldsymbol{v}^n - \boldsymbol{v})\Big]^\theta\, dx\, dt \le \eta\,. \tag{30}$$

Since the corresponding nonlinear operator is strictly monotone, once we obtain (30), the almost everywhere convergence (25) follows.

2. Standard monotone operator method. To achieve (30) it seems natural to multiply the approximate problem by $\boldsymbol{v}^n - \boldsymbol{v}$. This is possible for all type of boundaries, as shown by O.A. Ladyzhenskaya for the first time (see [11] for example), but it restricts ourselves to the range of r,

$$r \ge \frac{11}{5}\,, \tag{31}$$

due to required finiteness of the integral

$$\int_{Q_T} v_k^n \partial_{x_k} \boldsymbol{v}^n \cdot (\boldsymbol{v}^n - \boldsymbol{v})\, dx\, dt\,. \tag{32}$$

Once the integrability is assured, the term in (32) vanishes as $n \to \infty$ thanks to (24).

3. Method of truncated Sobolev functions. In order to obtain (30) for $r < \frac{11}{5}$, one can attempt to replace the multiplier $\boldsymbol{v}^n - \boldsymbol{v}$ by its bounded truncation

$$\boldsymbol{\psi}^n := (\boldsymbol{v}^n - \boldsymbol{v})(1 - \min\{1, \delta^{-1}|\boldsymbol{v}^n - \boldsymbol{v}|\}) . \qquad (\delta > 0) \tag{33}$$

That it is in principle possible the reader can verify in [3], where the existence of weak solution to no boundary and Navier's boundary is proved if

$$r > \frac{8}{5} . \tag{34}$$

Note however that δ has to be chosen in a suitable way so that the difference between the integral in (30) and the term

$$\int_0^T \int_\Omega \{\mathbf{S}(\mathbf{D}(\boldsymbol{v}^n)) - \mathbf{S}(\mathbf{D}(\boldsymbol{v}))\} : \mathbf{D}(\boldsymbol{\psi}^n)\, dx\, dt$$

can be made as small as needed (at least for a subsequence). This requirement leads to show that for all $\eta > 0$ there is $\{\boldsymbol{v}^k\} \subset \{\boldsymbol{v}^n\}$ and δ independent of k such that

$$\int_{Q_\delta^k} \{\mathbf{S}(\mathbf{D}(\boldsymbol{v}^k)) - \mathbf{S}(\mathbf{D}(\boldsymbol{v}))\} : \mathbf{D}((\boldsymbol{v}^k - \boldsymbol{v}) \min\{1, \delta^{-1}|\boldsymbol{v}^k - \boldsymbol{v}|\})\, dx\, dt \le \eta ,$$
$$\text{where } Q_\delta^k := \{(t,x) \in Q_T; |\boldsymbol{v}^k - \boldsymbol{v}| < \delta\} . \tag{35}$$

Note that functions behind the integral sign (restricted to Q_δ^k) are uniformly bounded in $L^1(Q_T)$. A proof of (35) is based on the following assertion:
Let $\{g^n\}$ be such that $\|g^n\|_{L^1(Q_T)} \le M$ for all n. Then for all $\eta > 0$ there are $\{\boldsymbol{v}^k\} \subset \{\boldsymbol{v}^n\}$, $\{g^k\} \subset \{g^n\}$ and $\delta < \frac{1}{M}$ independent of k such that

$$\int_{A_\delta^k} g^k\, dx\, dt \le \eta , \qquad \textit{where } A_\delta^k := \{(t,x) \in Q_T; \delta^2 \le |\boldsymbol{v}^k - \boldsymbol{v}| < \delta\} .$$

The bound (34) is due to convective term: it follows from (21) that $v_k^n \partial_{x_k} \boldsymbol{v}^n \in L^1(Q_T)$ uniformly if (34) holds; consequently for $r > \frac{8}{5}$ we have

$$\int_{Q_T} \left| v_k^n \partial_{x_k} \boldsymbol{v}^n \cdot \boldsymbol{\psi}^n \right| dx\, dt \le K\, \delta ,$$

which can be made arbitrarily small by our choice of δ.

The reason why the no-slip boundary is not covered in [3] has the origin in $\operatorname{div} \boldsymbol{\psi}^n \neq 0$. As indicated in Sect. 2.1, it is difficult to introduce the pressure to (6). Thus we can modify (33) by subtracting a suitable vector field so that the result is divergenceless. Subtracting $\boldsymbol{h}^n$ satisfying $\operatorname{div} \boldsymbol{h}^n = \operatorname{div} \boldsymbol{\psi}^n$ with no-slip boundary conditions we obtain test functions with correct boundary conditions but not enough regularity, while subtracting ∇z^n, where $-\Delta z^n = \operatorname{div} \boldsymbol{\psi}^n$ with homogeneous Neumann boundary conditions, we obtain suitable test functions only for problems with Navier's boundary or without boundary.

Despite these difficulties, we conjecture that even in the case of no boundary the existence of weak solution for $r > \frac{8}{5}$ can be proved via this method.

4. Method of Lipschitz approximations of Sobolev functions. This method has been applied to steady problem (i.e. $\boldsymbol{v}_t = \boldsymbol{0}$ in (6)) so far, and gives the existence of weak solution for (see [4])

$$r > \frac{6}{5}. \tag{36}$$

An extension of this approach to the evolutionary model is the subject of current investigation.

In stationary problems, this method stems from the following assertion (see [1] and [12] for example):
There exists a $C > 0$ such that for all $u^n \to 0$ weakly in $W_0^{1,r}(\Omega)$ and all $\lambda > 0$ there are $u_\lambda^n \in W^{1,\infty}(\Omega)$ such that

$$\|\nabla u_\lambda^n\|_{L^\infty(\Omega)} \le \lambda\,, \tag{37}$$

$$u_\lambda^n \to 0 \text{ strongly in } L^\infty(\Omega)\,, \tag{38}$$

$$u_\lambda^n \to 0 \text{ weakly in } W_0^{1,s}(\Omega)\,, \qquad (\forall s \in [1,\infty)) \tag{39}$$

$$|\{x \in \Omega;\ u^n(x) \ne u_\lambda^n(x)\}| \le C\lambda^{-r}\|u^n\|_{W^{1,r}(\Omega)}^r\,, \tag{40}$$

$$|\{x \in \Omega;\ M(|\nabla u^n(x)|) + M(u^n(x)) > \mu\}| \le C\mu^{-r}\|u^n\|_{W^{1,r}(\Omega)}^r\,, \tag{41}$$

where $M(f)$ denotes the Hardy-Littlewood maximal function to f. Note also that $|\{x \in \Omega;\ u^n(x) \ne u_\lambda^n(x)\}| = |\{x \in \Omega;\ M(|\nabla u^n(x)|) + M(u^n(x)) > \lambda\}|$.

Using this observation, we can replace the test function $\boldsymbol{v}^n - \boldsymbol{v}$ by its Lipschitz approximations $(\boldsymbol{v}^n - \boldsymbol{v})_\lambda$. The bound (36) is due to the convective term, surprisingly. We have

$$\int_\Omega \boldsymbol{v}^n \otimes \boldsymbol{v}^n : \nabla(\boldsymbol{v}^n - \boldsymbol{v})_\lambda\, dx = \int_\Omega \boldsymbol{v} \otimes \boldsymbol{v} : \nabla(\boldsymbol{v}^n - \boldsymbol{v})_\lambda\, dx$$
$$+ \int_\Omega [(\boldsymbol{v}^n - \boldsymbol{v}) \otimes \boldsymbol{v}^n + \boldsymbol{v} \otimes (\boldsymbol{v}^n - \boldsymbol{v})] : \nabla(\boldsymbol{v}^n - \boldsymbol{v})_\lambda\, dx\,.$$

Taking $\lambda > 0$ and $r > \frac{6}{5}$ fixed and letting $n \to \infty$ we observe that the first term at the right hand vanishes thanks to (39) applied to $u^n = (\boldsymbol{v}^n - \boldsymbol{v})$, while the second term tends to 0 thanks to (37) and the strong convergence of $\boldsymbol{v}^n \to \boldsymbol{v}$ in $L^2(\Omega)$.

Note again that λ has to be chosen in an appropriate way so that the integral

$$\int_{\Omega_\lambda^k} \{\mathbf{S}(\mathbf{D}(\boldsymbol{v}^k)) - \mathbf{S}(\mathbf{D}(\boldsymbol{v}))\} : \mathbf{D}((\boldsymbol{v}^k - \boldsymbol{v})_\lambda)\, dx$$
$$\text{with } \Omega_\lambda^k := \{x \in \Omega; (\boldsymbol{v}^k - \boldsymbol{v})_\lambda(x) \ne (\boldsymbol{v}^k - \boldsymbol{v})(x)\} \tag{42}$$

is small. Here, the following assertion is used (see [4]):

Let $\{g^n\}$ be such that $\|g^n\|_{L^1(\Omega)} \leq M$ for all n. Then for all $\eta > 0$ there are $\{g^k\} \subset \{g^n\}$, $\{v^k\} \subset \{v^n\}$ and $\lambda < \frac{1}{M}$ independent of k such that $\int_{A^k_\lambda} g^k\, dx \leq \eta$, where $A^k_\lambda := \{x \in \Omega; \lambda < M(|\nabla(v^k - v)(x)|) + M(v^k(x) - v(x)) \leq \lambda^2\}$.

Based on these results, we conjecture the global-in-time existence of weak solutions to (6) via the Lipschitz approximation method for the same bound as in the stationary case, i.e. $r > \frac{6}{5}$.

4 Conclusion

We have intended to show that the no-slip boundary even for the classical three-dimensional Navier-Stokes equations leave several open questions, which are however solvable for the spatially periodic boundary conditions, the Cauchy problem, or Navier's or no-stick boundary conditions. In fact, this peculiarity of the no-slip boundary occurs even in two-dimensional setting studying the question of convergence of the solutions of 2d NSEs to the solution of the 2d Euler equations as $\nu_0 \to 0+$. While the affirmative answer holds for problems without boundary, it is open even for the two-dimensional half-space with the no-slip boundary.

We have also presented four different approaches how to obtain almost everywhere convergence for weakly convergent gradients, and we sketched how these tools can be used in the existence theory for the power-law fluids. We remark that the power-law model has been considered here just for simplicity. The results mentioned in Sect. 3 however hold for any viscosity $\nu = \nu(|\mathbf{D}|^2)$ that behaves as $(\varepsilon + |\mathbf{D}|^2)^{\frac{r-2}{2}}$, with $\varepsilon \geq 0$. We have also tried to emphasize that these approaches encounter difficulties when dealing with the no-slip boundary.

Being supported by the results in [24] and [17] for example, we believe that the presented tools can help in analysis of more complex incompressible models with the nonconstant viscosity of the type $\nu(Q, |\mathbf{D}(v)|^2)$, where the quantity $Q : \Omega \to \mathbb{R}$ or $\mathbb{R}^3$ can have a different meaning. For example, Q can be the identity map ($Q(x) = x$) to include the inhomogenity of the viscosity, or Q can be the pressure p, or another quantity as the density (if the fluid is inhomogeneous), the temperature (if the process is not isothermal), the electric field (when dealing with electro-rheological fluids) and the magnetic field.

The no-slip boundary for the Navier-Stokes equations at small ν_0 initiates the presence of thin boundary layer in which the turbulence eddies originate. But, the boundary layers occur even for slow flows for power-law models if $r \to 1+$, see [25]. One may thus speculate that the difficulties in analysis of no-slip boundary problems are not only because of the lack of correct mathematical arguments, but are of a physical nature.

A question of different, but related character is to ask whether there are boundary conditions that may smooth out the solution. For a viscoelastic model such a feature has been observed in [26].

Acknowledgement. This text is certainly influenced by valuable and fruitful discussions with many our colleagues. In particular, we thank J. Nečas, K.R. Rajagopal, M. Růžička, and M. Steinhauer.

References

1. Acerbi, E. and Fusco, N. (1988): An approximation lemma for $W^{1,p}$-functions. In: Material instabilities and continuum mechanics, 1–5. Oxford University Press
2. Foias, C., Guillope, C., Temam, R. (1981): New a priori estimates for Navier-Stokes equations in dimension 3. Comm. Partial Differential equations, **6**, 329–359
3. Frehse, J., Málek, J., Steinhauer, M. (1999): On existence results for fluids with shear dependent viscosity – unsteady flows. In: Partial Differential Equations, Theory and Numerical Solution, Vol. 406, CRC Research Notes in Mathematics series, Boca Raton, CRC Press UK, 121–129
4. Frehse, J., Málek, J., Steinhauer, M. (2002): On analysis of steady flows of fluids with shear-dependent viscosity based on the Lipschitz truncation method. Submitted. Preprint SFB611, University of Bonn
5. Frehse, J., Růžička, M. (1994): On the regularity of the stationary Navier-Stokes equations. Ann. Scuola Norm. Sup. Pisa Cl. Sci., **21**, 63–95
6. Frehse, J., Růžička, M. (1994): Regularity for the stationary Navier-Stokes equations in bounded domains. Arch. Rational Mech. Anal., **128**, 361–381
7. Hron, J., Málek, J., Rajagopal, K. R. (2001): Simple flows of fluids with pressure dependent viscosities. Proceedings of The Royal Society London A: Mathematical, Physical and Engineering Sciences, **457**, 1603–1622.
8. Koch, H., Solonnikov, V. A. (2001): L^p-estimates for a solutions to the nonstationary Stokes equation. Journal of Mathematical Sciences, **106**, 3042–3072
9. Koch, H., Solonnikov, V. A. (2002): L^q-estimates of first-order derivatives of solutions to the nonstationary stokes problem. In: Nonlinear problems in mathematical physics and related topics I, International Mathematical series, Rozhkovskaya and Kluwer/Plenum publishers, 163–177
10. Ladyzhenskaya, O.A. (1968): On the unique solvability in large of a three-dimensional Cauchy problem for the Navier-Stokes equations in the presence of axial symmetry. Zapiski Naukchnych Sem. LOMI, **7**, 155–177
11. Ladyzhenskaya, O.A. (1970): The mathematical theory of viscous incompressible flow. Gordon and Breach, New York
12. Landes, R. (1996): Quasimonotone versus pseudomonotone. Proceedings of the Royal Society of Edinburgh, **126A**, 705–717
13. Leonardi, S., Málek, J., Nečas, J., Pokorný, M. (1999): On axially symmetric flows in $\mathbb{R}^3$. Z. Anal. Anwendungen, **31**, 52–61
14. Leray, J. (1934): Sur le mouvement d'un liquide visqueux emplissant l'espace. Acta Math., **63**, 193–248
15. Málek, J., Nečas, J., Rokyta, M., Růžička, M. (1996): Weak and measure-valued solutions to evolutionary PDEs. Chapman & Hall, London
16. Málek, J., Nečas, J., Pokorný, M., Schonbek, M. E. (1999): On possible singular solutions to the Navier-Stokes equations. Math. Nachr., **319**, 97–114
17. Málek, J., Nečas, J., Rajagopal, K. R. (2001): Global analysis of the flows of fluids with pressure dependent viscosities. Accepted to Arch. Rat. Mech. Anal. Preprint SFB256 No.737, University of Bonn

18. Málek, J., Nečas, J., Růžička, M. (2001): On weak solutions to a class of non-Newtonian incompressible fluids in bounded three-dimensional domains: the case $p \geq 2$. Advances in Differential Equations, **6(3)**, 257–302
19. Málek, J., Nečas, J., Růžička, M. (2002): On weak solutions to a class of non-Newtonian incompressible fluids in bounded three-dimensional domains: the case $p < 2$. In preparation
20. Málek, J., Rajagopal, K.R., Růžička, M. (1995): Existence and regularity of solutions and stability of the rest state for fluids with shear dependent viscosity. Math. Models Methods in Appl. Sci., **6**, 789–812
21. Miller, J., O'Leary, M., Schonbek, M. E. (2001): Nonexistence of singular pseudo-self-similar solutions of the Navier-Stokes system. Math. Ann., **319**, 809–815
22. Nečas, J., Růžička, M., Šverák, V. (1996): On Leray's self-similar solutions of the Navier-Stokes equations. Acta Math., **176**, 283–294
23. Pokorný, M. (1996): Cauchy problem for the non-Newtonian viscous incompressible fluid. Appl. Math., **41**, 169–201
24. Růžička, M. (2000): Electrorheological fluids: modeling and mathematical theory. Springer.
25. Rajagopal, K. R. (1995): Boundary layers in non-linear fluids. In: Trends in applications of mathematics to mechanics, vol. 77 of Pitman Monogr. Surveys Pure Appl. Math., Longman, Harlow, 209–218
26. Rao, I. J., Rajagopal, K. R. (1999): Some simple flows of a Johnson-Segelman fluid. Acta Mechanica, **132**, 209–219
27. Solonnikov, V. A. (1964): Estimates for solutions of a non-stationary linearized system of Navier-Stokes equations. Trudy Mat. Inst. Steklov. **70** 213–317, 1964. English translation:Am. Math. Soc. Transl. Ser. II, **75**, (1968), 1–116
28. Tsai, T.-P. (1998): On Leray's self-similar solutions of the Navier-Stokes equations satisfying local energy estimates. Arch. Rational Mech. Anal., **143**, 29–51

Comparison of Finite Volume and Discontinuous Galerkin Methods of Higher Order for Systems of Conservation Laws in Multiple Space Dimensions

A. Dedner, D. Kröner, C. Rohde, T. Schnitzer, and M. Wesenberg

Institut für Angewandte Mathematik, Universität Freiburg, Hermann-Herder-Str.10, D-79104 Freiburg, Germany
{dedner, dietmar, chris, schnitz, wesenber}@mathematik.uni-freiburg.de

1 Introduction

The methods most frequently used in computational fluid mechanics for solving the compressible Navier-Stokes or compressible Euler equations are finite volume schemes on structured or on unstructured grids. First order as well as higher order space discretizations of MUSCL type, including flux limiters and higher order Runge-Kutta methods for the time discretization, guarantee robust and accurate schemes. But there is an important difficulty. If one increases the order, the stencil for the space discretization increases too, and the scheme becomes very expensive. Therefore schemes with more compact stencils are necessary. Discontinuous Galerkin schemes in the sense of [3] are of this type. They are identical to finite volume schemes in the case of formal first order, and for higher order they use nonconformal ansatz functions whose restrictions to single cells are polynomials of higher order. Therefore they seem to be more efficient and it is of highest interest to compare finite volume and discontinuous Galerkin methods for real applications with respect to their efficiency. Experiences [1] with the Euler equations of gas dynamics indicate that the discontinuous Galerkin methods have some advantages. Since there are no systematic studies available in the literature, we will present in this paper some numerical experiments for hyperbolic conservation laws in multiple space dimensions to compare their efficiency for different situations. As important instances of hyperbolic conservation laws we consider the Euler equations of gas dynamics and Lundquist's equations of ideal magneto-hydrodynamics (MHD). Furthermore we have found a new limiter which improves the results from [14]. Similar studies have been done in [4].

The paper is structured as follows. First we will give some motivation and the definition of the discontinuous Galerkin method. Then we recall the definition of the finite volume schemes. Finally we will describe different testproblems and will compare the numerical results.

2 Motivation and Definition of the Discontinuous Galerkin Scheme

In this section we will derive the discontinuous Galerkin method [2], [3] for the initial value problem

$$\begin{aligned} \partial_t u + \operatorname{div} f(u) &= 0 \quad \text{in} \quad \mathbb{R}^2 \times (0,T),\ T>0\,, \\ u(x,0) &= u_0(x) \quad \text{in} \quad \mathbb{R}^2\,. \end{aligned} \tag{1}$$

Here we assume $f \in C^1(\mathbb{R}^2)$ and $u_0 \in L^\infty(\mathbb{R}^2)$. We restrict ourselves to the spatially two-dimensional case for the sake of notational simplicity. In order to define a numerical approximation we assume that we have a triangulation

$$\mathcal{T} = \{T_j \ :\ j \in \mathcal{I},\ T_j \text{ triangle}\}\,, \quad \mathcal{I} \text{ index set}\,.$$

For the triangulation $\mathcal{T}$ we assume all the conditions which are mentioned in [6]. Let $P^k(T_j)$ denote the space of polynomials of degree $\le k$ on the cell T_j. Then we look for a numerical approximation u_h of the exact solution u of (1) in the space

$$V_h = \{v_h \in L^\infty(\mathbb{R}^2) : v_h|_{T_j} \in P^k(T_j) \quad \text{for all} \quad j \in \mathcal{I}\}\,. \tag{2}$$

Let $(v_{j,i})_{i=1}^n$ be a basis of the space $P^k(T_j)$ and

$$u_h(x,t) = \sum_{i=1}^n u_{ji}(t) v_{j,i}(x) \quad x \in T_j, \quad 0 \le t \le T\,. \tag{3}$$

Now for the motivation of the scheme we multiply the partial differential equation in (1) by $v_h \in V_h$, integrate over the triangle T_j, and replace the exact solution u of (1) by the numerical approximation u_h

$$\begin{aligned} \frac{d}{dt}\int_{T_j} u_h(x,t) v_h(x)\,dx = &- \sum_{S_{jl}\in\partial T_j} \int_{S_{jl}} f(u_h(x,t))\cdot n_{jl} v_h(x)\,dx \\ &+ \int_{T_j} f(u_h(x,t))\cdot \nabla v_h(x)\,dx\,. \end{aligned} \tag{4}$$

Here S_{jl} is the joint edge with the neighbouring triangle T_{jl} for $1 \le l \le 3$ (see Fig. 1). The integrals in (4) will be replaced by the quadrature rules

$$\int_{S_{jl}} f(u_h(x,t))\cdot n_{jl} v_h(x)\,dx \approx \sum_{q=1}^L \omega_q f(u_h(x_q^{jl},t))\cdot n_{jl} v_h(x_q^{jl})|S_{jl}| \tag{5}$$

and

$$\int_{T_j} f(u_h(x,t))\cdot \nabla v_h(x)\,dx \approx \sum_{q=1}^M \alpha_q f(u_h(y_q^j,t))\cdot \nabla v_h(y_q^j)|T_j|\,. \tag{6}$$

Here $\omega_q, \alpha_q \in \mathbb{R}$ are suitable quadrature weights, x_q^{jl} and y_q^j are the integration points on S_{jl} and T_j, respectively, and n_{jl} denotes the outer unit normal on S_{jl} pointing from T_j to T_l.

In general $u_h(x_q^{jl}, t)$ is not defined for arbitrary functions $u_h \in V_h$, since u_h can be discontinuous across S_{jl}. Therefore the term $\sum_{q=1}^{L} \omega_q f(u_h(x_q^{jl}, t)) \cdot n_{jl} v_h(x_q^{jl}) |S_{jl}|$ has to be replaced by some well-defined expression. Instead of $\sum_{q=1}^{L} \omega_q f(u_h(x_q^{jl}, t)) \cdot n_{jl} v_h(x_q^{jl}) |S_{jl}|$ we will use

$$\sum_{q=1}^{L} \omega_q g_{jl}(u_h^j(x_q^{jl}, t), u_h^l(x_q^{jl}, t)) v_h(x_q^{jl}) . \tag{7}$$

Here the numerical flux g_{jl} has to satisfy the conditions in [7], page 159 and $u_h^j(x_q^{jl}, t)$ in g_{jl} is the numerical solution u_h on triangle T_j, evaluated in x_q^{jl} on the joint edge S_{jl}.

Summarizing the last steps of the motivation we are led to:

Definition 2.1 *(Discontinuous Galerkin method) On the triangle T_j, $j \in \mathcal{I}$ we define*

$$u_h(\cdot, 0) := P(u_0) , \tag{8}$$

where P is the L^2-projection on the space V_h. Then, for $t > 0$, find $u_h \in V_h$ such that

$$\frac{d}{dt} \int_{T_j} u_h(x,t) v_h(x)\, dx = Q_j(u_h, v_h) \tag{9}$$

for all $v_h \in P^k(T_j)$ and for all $j \in \mathcal{I}$, where

$$\begin{aligned} Q_j(u_h, v_h) = &- \sum_{l=1}^{3} \sum_{q=1}^{L} \omega_q g_{jl}(u_h^j(x_q^{jl}, t), u_h^l(x_q^{jl}, t)) v_h(x_q^{jl}) \\ &+ \sum_{q=1}^{M} \alpha_q f(u_h(y_q^j, t)) \cdot \nabla v_h(y_q^j) |T_j| \end{aligned} \tag{10}$$

and the notation is chosen according to Fig. 1.

If we replace u_h in (9) by its representation (3) we obtain on T_j

$$M_j \cdot \begin{pmatrix} \partial_t u_{j1} \\ \partial_t u_{j2} \\ \vdots \\ \partial_t u_{jn} \end{pmatrix} = \begin{pmatrix} Q_j(u_h, v_{j,1}) \\ Q_j(u_h, v_{j,2}) \\ \vdots \\ Q_j(u_h, v_{j,n}) \end{pmatrix} , \tag{11}$$

where M_j is given by

$$M_j = \begin{pmatrix} \int_{T_j} v_{j,1}^2 & \int_{T_j} v_{j,2}v_{j,1} & \cdots & \int_{T_j} v_{j,n}v_{j,1} \\ \int_{T_j} v_{j,1}v_{j,2} & \int_{T_j} v_{j,2}^2 & \cdots & \int_{T_j} v_{j,n}v_{j,2} \\ \vdots & \vdots & \ddots & \vdots \\ \int_{T_j} v_{j,1}v_{j,n} & \int_{T_j} v_{j,2}v_{j,n} & \cdots & \int_{T_j} v_{j,n}^2 \end{pmatrix} . \tag{12}$$

Up to now we have not defined the basis functions $(v_{j,i})_{i=1}^n$ on the triangle T_j. If we choose orthogonal basis functions, the matrix M_j will be a diagonal matrix. For $k = 1$ we choose on each triangle T_j the basis functions $v_{j,1}, v_{j,2}, v_{j,3}$, such that $v_{j,i}$ is equal to one in the midpoint of the edge S_{ji} and equal to zero in the midpoints of the other edges of T_j, that is, $v_{j,i}(z_{jk}) = \delta_{ik},\ i, k = 1, 2, 3$ (see Fig. 1). If we then choose a quadrature rule which is exact for polynomials in $P^2(T_j)$ in (6) (which happens to use the values of u_h in the midpoints of the edges z_{jk}, $k = 1, 2, 3$), we can easily see that the chosen basis $v_{j,1}$, $v_{j,2}$, $v_{j,3}$ is orthogonal. Therefore (9) reduces for these basis functions on the triangle T_j to

$$m_{ji}\partial_t u_{ji} = Q_j(u_h, v_{j,i}) \tag{13}$$

for $i = 1, 2, 3$ where

$$m_{ji} := \int_{T_j} v_{j,i}^2 .$$

Similar as for all MUSCL type schemes, in particular for finite volume schemes, one has to use limiters, which we will define in Definition 2.2 (see [14]).

Up to now the best results concerning the experimental order of convergence for higher order finite volume schemes, even for systems, have been obtained with the

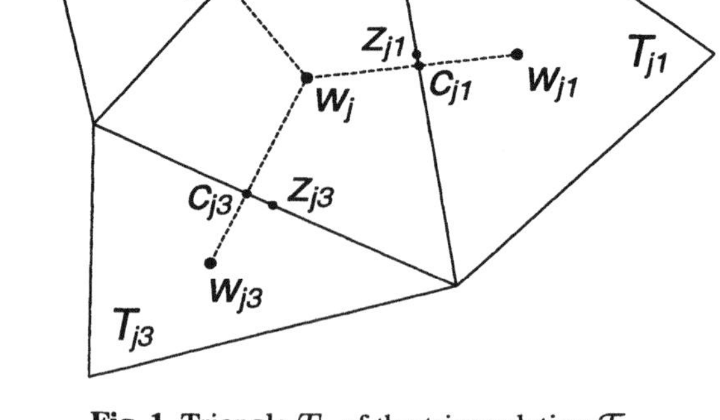

Fig. 1. Triangle T_j of the triangulation $\mathcal{T}$

Wierse limiter (Definition 2.2), defined in [14]. This limiter satisfies the conditions in [6] for the convergence proof. An important modification, which still improves the results, will be given in Definition 2.3.

Definition 2.2 *(Wierse-limiter for $P^k(T_j) = P^1(T_j)$) For any triangle T_j we have an affine function, which can be written as*

$$F_j^n(x) = u_j^n + g_j^n \cdot (x - w_j) \tag{14}$$

with $g_j^n \in \mathbb{R}^2$. Let $u_j^{\max}$ and $u_j^{\min}$ be the maximal and minimal values, respectively, of u_h on the triangle T_j and its direct neighbours (see Fig. 1). For the neighbouring triangle T_{jl} of T_j define

$$t(l) := F_j^n(z_{jl}) - u_j^n \quad \text{if} \quad (F_j^n(z_{jl}) - u_j^n)(u_j^n - u_l^n) \leq 0 \ , \tag{15}$$

$$t(l) := 0 \quad \text{otherwise} \ . \tag{16}$$

Now we distinguish two cases:

First case: *If $t(l) = 0$ for two neighbours T_{jl} of T_j then let $\beta := 0$.*

Second case: *Otherwise:*

For $l, m, n \in \{1, 2, 3\}$, if $t(l) = 0$, $t(m) \neq 0$, $t(n) \neq 0$ and $t(m)t(n) < 0$, then we change the values $t(m)$ and $t(n)$:

$$minvalue := \min\{|t(m)|, |t(n)|\} \ , \tag{17}$$

$$t(m) := \operatorname{sign} t(m)\, minvalue \ , \tag{18}$$

$$t(n) := \operatorname{sign} t(n)\, minvalue \ . \tag{19}$$

Let $\alpha \in]\frac{1}{2}, 1]$ be fixed. Now in this case determine the largest $\beta \in [0, 1]$ such that for all neighbouring triangles T_{jl} we have

$$\beta\, |t(l)| \leq C_1 h^\alpha \ , \tag{20}$$

$$u_j^{\min} \leq u_j^n + \beta\, t(l) \leq u_j^{\max} \ . \tag{21}$$

Then we define

$$L_j^n(z_{jl}) = u_j^n + \beta \cdot t(l) \tag{22}$$

and the limited reconstruction Λu_h is defined on T_j as the linear interpolation of $L_j^n(z_{j1})$, $L_j^n(z_{j2})$, $L_j^n(z_{j3})$.

Definition 2.3 *(modified Wierse-limiter for $P^k(T_j) = P^1(T_j)$) If we replace in (15) the term $F_j^n(z_{jl}) - u_j^n$ by $F_j^n(c_{jl}) - u_j^n$ where c_{jl} is the intersection between the line through w_j and w_{jl} with the edge S_{jl} (see Fig. 1) we will get the modified Wierse limiter.*

The discretization, mentioned in (13) together with the limiter guarantees a stable second order space discretization. In order to get also second order in time we will use a Runge-Kutta method (see [14], [3]).

Definition 2.4 *(Runge-Kutta time discretization of order k) Define* $u_h^0 := \Lambda u_h(.,0)$, *where* Λ *denotes the limiter from Definition 2.3. Assume that* u_h^n *is known already. Define* $u^0 := u_h^n$. *Choose the time step* Δt *such that the CFL condition (see [14]) is satisfied. Then, for* $s = 1, \dots, k+1$ *compute for* $i = 1,2,3$

$$r_{ji}^s := \sum_{l=0}^{s-1} \left(\alpha_{sl} u_{ji}^l + \beta_{sl} \Delta t \, Q_j(u^l, v_{j,i}) \right) , \tag{23}$$

where α_{sl}, β_{sl} *are the well-known Runge-Kutta coefficients [2], let*

$$\tilde{u}^s(x, t^n) := \sum_{i=1}^{3} r_{ji}^s v_{j,i}(x) , \quad x \in T_j \tag{24}$$

and

$$u^s := \Lambda \tilde{u}^s . \tag{25}$$

Finally define

$$u_h^{n+1} := u^{k+1} . \tag{26}$$

In [2] it is shown that this scheme is formally uniformly high-order accurate.

3 Definition of Finite Volume Schemes of Higher Order

In this section we will define finite volume schemes of higher order on unstructured grids (see [7]). On the same grid as in the preceding section we define

$$u_j^0 := \frac{1}{|T_j|} \int_{T_j} u_0 \tag{27}$$

for all $j \in \mathcal{I}$. Let us assume that u_j^n for all n and j are given already. Now we have to define a linear reconstruction on T_j.

Definition 3.5 *(Reconstruction and DEOmod limiter) For any* T_j *we have three neighbouring triangles* T_{j1}, T_{j2}, T_{j3}. *For any* T_j *we define three piecewise linear functions* K_{j1}, K_{j2}, K_{j3} *which interpolate the values* u_j^n, u_{j2}^n, u_{j3}^n *and* u_j^n, u_{j1}^n, u_{j3}^n *and* u_j^n, u_{j1}^n, u_{j2}^n, *respectively. Let*

$$\begin{aligned} a_{jl} &:= \nabla K_{jl} \cdot (w_{jl} - w_j) , \\ d_{jl} &:= u_{jl}^n - u_j^n , \\ m_{jl} &:= \begin{cases} 0 & \textit{if } \ d_{jl} a_{jl} \le 0 \\ d_{jl}/a_{jl} & \textit{if } \ d_{jl} a_{jl} > 0, \ |a_{jl}| > |d_{jl}| \\ 1 & \textit{otherwise} . \end{cases} \end{aligned}$$

Then for $l = 1, 2, 3$ we obtain linear functions

$$D_{jl}(x) := u_j^n + m_{jl}\nabla K_{jl} \cdot (x - w_j) \ . \qquad (28)$$

For the limited reconstruction L_j^n on T_j we choose the linear function D_{jr}, $r \in \{1, 2, 3\}$, such that

$$|\nabla D_{jr}| \geq |\nabla D_{js}| \quad \textit{for} \quad s \in \{1, 2, 3\} \ , \qquad (29)$$

i.e., $L_j^n := D_{jr}$.

Now the finite volume scheme of higher (second) order can be defined as follows.

Definition 3.6 *(Finite volume scheme of higher order) Let $g_{jl}(u, v)$ be a conservative numerical flux, let z_{jl} denote the midpoint of the edge between the triangle T_j and T_{jl} (see Fig. 1) and let L_j, L_{jl} be the reconstructed limited linear functions on the triangle T_j and T_{jl} as defined in Definition 3.5. Then the new value u_j^{n+1} is defined by*

$$u_j^{n+1} := u_j^n - \frac{\Delta t}{|T_j|} \sum_{l=1}^{3} g_{jl}(L_j(z_{jl}), L_{jl}(z_{jl})) \ . \qquad (30)$$

Higher order discretizations with respect to time are obtained by a Runge-Kutta method as in Definition 2.4.

For schemes using limited reconstructions we can prove the following convergence results and error estimates [6], [9].

Theorem 3.7 *(Convergence of finite volume schemes and error estimates) Let (1) be a scalar equation, g_{jl} be a conservative, Lipschitz continuous, consistent, and monotone numerical flux (see [6]), assume that the linear reconstruction is defined as in [14] and the limited reconstruction L_j is given as in Definition 2.2. Let*

$$u_h(x, t) = u_j^n$$

if $x \in T_j$ and $t \in [t^n, t^{n+1}[$ where u_j^n is defined in (27), (30) and assume that an CFL condition is satisfied as in [9]. Then u_h converges to the uniquely defined entropy solution of (1). The same result holds if the time discretization is replaced by a second order Runge-Kutta method (see [9]). Furthermore in this case we have the following a priori error estimate (see [9]).

$$||u - u_h||_{L^1} \leq ch^{\frac{1}{4}} + ||u_0 - u_h(0)||_{L^1} \ . \qquad (31)$$

Remark 3.8 *(A posteriori error estimate) Under the assumption of Theorem 3.7 an a posteriori error estimate has been proven for the first order version of the finite volume scheme in [8]. This can be used for the efficient control of local grid refinement.*

Remark 3.9 *(Generalization to systems) Although there are no corresponding theoretical results for systems in multiple space dimensions, there are accepted, efficient, and reliable schemes based on the finite volume (see [7]) and discontinuous Galerkin methods (see [2]) as described above.*

4 Comparison, Numerical Testproblems, and Results

We present a comparison of the efficiency of the finite volume and discontinuous Galerkin schemes applied to Lundquist's equatins of ideal MHD. They are a mathematical model for electrically conducting fluids, based on the conservation of mass, momentum, and energy and the Maxwell equations. The MHD equations can be written in the form

$$\begin{aligned}
\partial_t \rho + \nabla \cdot (\rho \mathbf{u}) &= 0 \quad \textit{(conservation of mass)},\\
\partial_t (\rho \mathbf{u}) + \nabla \cdot (\rho \mathbf{u}\mathbf{u}^t + \mathcal{P}) &= 0 \quad \textit{(conservation of momentum)},\\
\partial_t \mathbf{B} + \nabla \cdot (\mathbf{u}\mathbf{B}^t - \mathbf{B}\mathbf{u}^t) &= 0 \quad \textit{(induction equation)},\\
\partial_t (\rho e) + \nabla \cdot (\rho e \mathbf{u} + \mathcal{P}\mathbf{u}) &= 0 \quad \textit{(conservation of energy)},\\
\nabla \cdot \mathbf{B} &= 0 \quad \textit{(divergence constraint)}.
\end{aligned} \tag{32}$$

Here, ρ denotes the density, $\mathbf{u} = (u_x, u_y, u_z)$ the velocity, $\mathbf{B} = (B_x, B_y, B_z)$ the magnetic field and e the energy; the pressure tensor $\mathcal{P}$ is defined as

$$\mathcal{P} := \left(p + \frac{1}{8\pi}|\mathbf{B}|^2\right)\mathcal{I} - \frac{1}{4\pi}\mathbf{B}\mathbf{B}^t ,$$

where p denotes the gas pressure and $\mathcal{I}$ the unit tensor. All quantities depend on the space variables x, y, z and the time t. Since we assume the fluid to be an ideal polytropic gas with adiabatic exponent $\gamma > 1$, the system is closed by the equation of state

$$p = (\gamma - 1)\left(\rho e - \frac{1}{2}\rho|\mathbf{u}|^2 - \frac{1}{8\pi}|\mathbf{B}|^2\right) , \tag{33}$$

and suitable initial and boundary conditions.

Remark 4.10 *For (32) local existence results have been shown in [5]. The proof of global existence is still open. For* $\mathbf{B} = 0$ *this system reduces to the Euler equations of gas dynamics.*

Before we compare the efficiency of the discontinuous Galerkin and the finite volume schemes, let us consider the computational effort. If N denotes the number of cells in 1D, the computational costs for one time step are shown in Table 1. Since FV1 and DG1 are exactly the same schemes, the costs are the same. FV2 and DG2 are both formally of second order, but the effort for computing the fluxes is two times higher for DG2 than for FV2, and for DG3 more than four times higher than for FV2. The corresponding results in 2D are shown in Table 2. Here E denotes the number of edges and N the number of triangles. In this case DG2 requires six times as many flux calculations as FV2. However, in 1D and in 2D the linear reconstructions have to be computed only for FV2.

These considerations indicate that it is not obvious that the discontinuous Galerkin schemes are more efficient than the finite volume schemes. Therefore we will study their efficiency in the following examples.

Table 1. Comparison of the computational effort between the different schemes in 1D. N = number of cells.

	FV1	DG1	FV2	DG2	DG3
Flux	$N+1$	$N+1$	$2N+2$	$4N+4$	$9N+9$
Limiter	0	0	$2N$	$2N$	$3N$
Reconstruction	0	0	$2N$	0	0

Table 2. Comparison of the computational effort between the different schemes in 2D. E = number of edges, N = number of triangles.

	FV1	DG1	FV2	DG2
Flux	$1E$	$1E$	$2E$	$12E$
Limiter	0	0	$2N$	$2N$
Reconstruction	0	0	$2N$	0

For the numerical approximation in the following test cases we have used the discontinuous Galerkin (9) and the finite volume method (30). On the basis of a comprehensive study of several Riemann solvers in [12] we decided to choose the numerical flux of Dai-Woodward, which is one of the most efficient numerical fluxes for this class of problems. For the examples in 1D the limitation is based on the min-mod limiter [7]. In 2D we apply the modified Wierse-limiter (Definition 2.3) for the discontinuous Galerkin method and the DEOmod limiter (Definition 3.5) for the finite volume schemes. The numerical results presented in the following can be found in [10]. The finite volume algorithm is the same as used for the results in [12].

Example 4.11 *(MHD, advection for smooth solutions in 1D) We consider in this example the MHD system (32) in 1D. The smooth initial data are given by*

Initial data ($\gamma = 2.0, B_x = 0.0$)						
ρ	u_x	u_y	u_z	B_y	B_z	e
1.0	*1.0*	2.0	*3.0*	$r(x)$	$1 - r(x)$	*10.0*

where

$$r(x) = \begin{cases} \exp\left(\frac{1}{(x+1/2)^2 - 1/4} + 4\right) & -1 < x < 0 \\ 0 & \textit{otherwise} \end{cases} .$$

The initial B_y-component of the magnetic field is plotted in Fig. 2 (dashed line). The smooth exact solution consists of the transported initial data with velocity 1. On the interval $]-1, 1[$ we have used grids with 50 *up to* 3200 *cells. ($h = 0.04 \cdot 2^{-p}$, $p = 0, \dots, 6$). The B_y-component of the exact solution at time $t = 1$ is shown in Fig. 2.*

In Fig. 3, left, the experimental order of convergence (EOC) for the finite volume method of first order (FV1), of second order (FV2), of the discontinuous Galerkin method of first order (DG1), of second order (DG2) and of third order (DG3) are shown, respectively. Since FV1 and DG1 are the same schemes there is no special

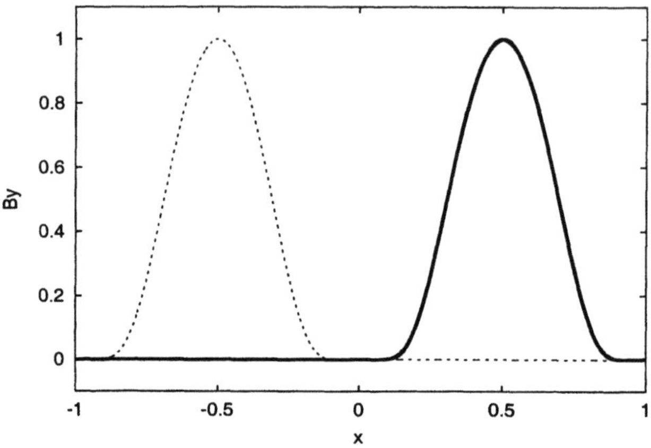

Fig. 2. Initial data ($T = 0$) and exact solution ($T = 1$) for the MHD advection

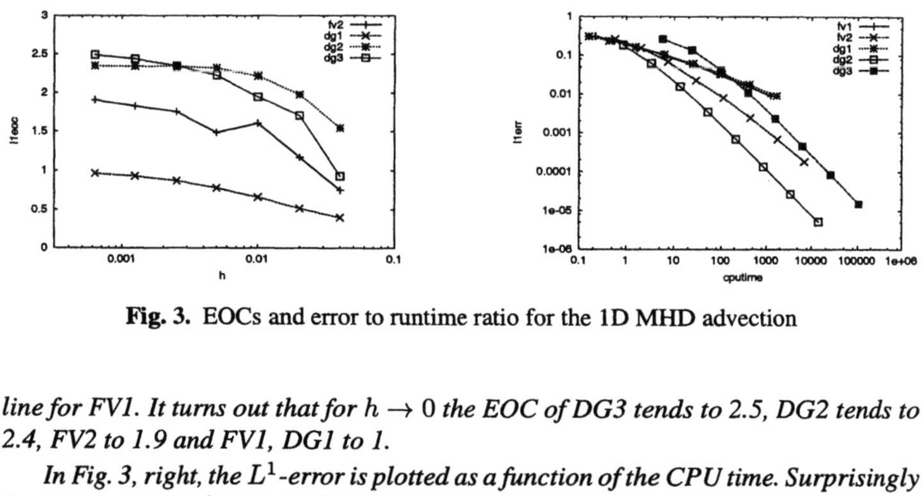

Fig. 3. EOCs and error to runtime ratio for the 1D MHD advection

line for FV1. It turns out that for $h \to 0$ the EOC of DG3 tends to 2.5, DG2 tends to 2.4, FV2 to 1.9 and FV1, DG1 to 1.

In Fig. 3, right, the L^1-error is plotted as a function of the CPU time. Surprisingly for this regime of h we see that the best results are obtained by the discontinuous Galerkin method of second order and not by the third order. Even the second order finite volume schemes are more efficient than the discontinuous Galerkin method of third order.

Example 4.12 *(MHD shock tube problem, discontinuous solutions in 1D) Now we consider a Riemann problem for the MHD system (32) in 1D. The initial data are given by*

Initial data ($\gamma = \frac{5}{3}, B_x = 4.0$)							
	ρ	u_x	u_y	u_z	B_y	B_z	p
$x < 0$	*0.8129*	*1.801*	*0.3672*	*0.1836*	*1.7856*	*0.8928*	*0.4809*
$x > 0$	*1.053402*	*-1.732090*	*-0.062036*	*-0.031019*	*2.043758*	*1.021879*	*0.157034*

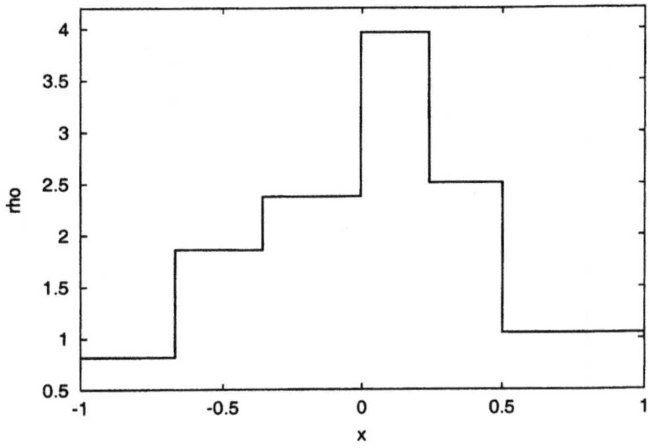

Fig. 4. Exact solution for the 1D MHD shocktube

The discontinuous exact solution is known (see [11] and Fig. 4). On the interval $]-1,1[$ *we have used grids with* 50 *up to* 6400 *cells (*$h = 0.04 \cdot 2^{-p},\ p = 0, \ldots, 7$*).*

As before, in Fig. 5, left, the experimental order of convergence (EOC) for the different finite volume and discontinuous Galerkin methods is shown. On the finest scale the best results have been obtained by the three discontinuous Galerkin methods.

In Fig. 5, right, the L^1*-errors for the finite volume and the discontinuous Galerkin methods are shown. While in the smooth case we have obtained the best results for the discontinuous Galerkin method of second order, now in the nonsmooth case the best results are obtained by the finite volume methods of second order. For smaller* h *the higher order methods should be better but for this moderate regime of* h *the higher numerical costs for the discontinuous Galerkin methods are dominating.*

The following three examples refer to testproblems in 2D. As a result of a study for different limiters we decided finally to choose the modified Wierse limiter (see Definition 2.3 and [14]) for which we have got the best results. In order to see the differences we discretized the smooth initial data from Example 4.13 according to

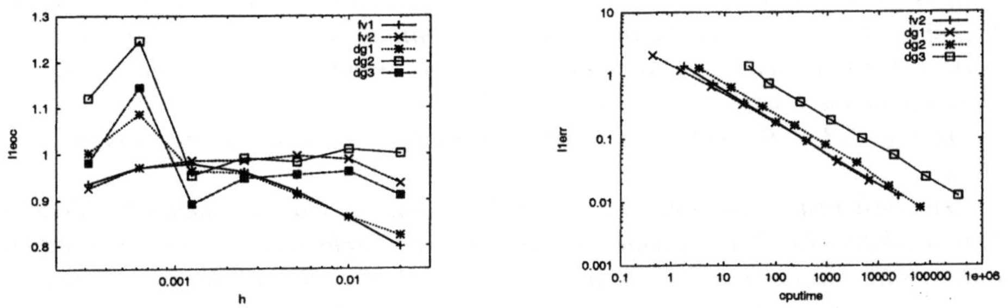

Fig. 5. EOCs and error to runtime ratio for the 1D MHD shocktube

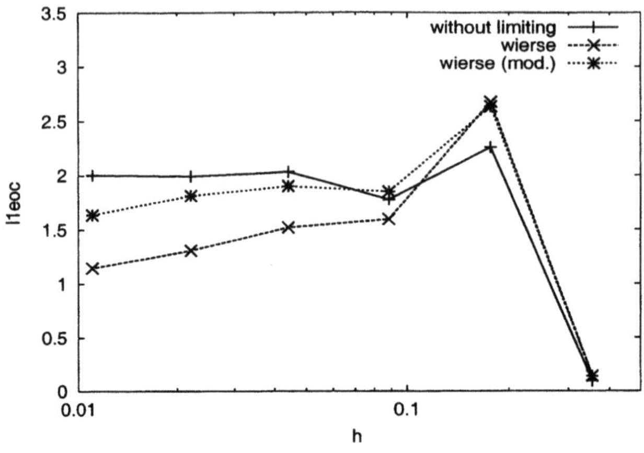

Fig. 6. The original vs. the modified Wierse limiter

(8) and applied to this discrete data the original Wierse limiter (Definition 2.2) and the modified Wierse limiter (Definition 2.3). The EOCs for the error between these results and the exact initial data and the error between the L^2-Projection and the exact initial data are plotted in Fig. 6. It is shown that the modified limiter is much better than the original one.

Example 4.13 *(MHD, advection for smooth solutions in 2D) The initial data are*

Initial data ($\gamma = 2.0$)							
ρ	u_x	u_y	u_z	B_x	B_y	B_z	e
1.0	*1.0*	*1.0*	*0.0*	*0.0*	*0.0*	*r(x)*	*10.0*

where

$$r(x) = \begin{cases} 4096x^4 - 128x^2 + 1 & |x| < 0.125 \\ 0 & \textit{otherwise} . \end{cases}$$

The domain $\Omega := [-0.5, 1.5] \times [-0.5, 1.5]$ *has been discretized by a grid with* 128 *to* 32768 *triangles (*$h = 0.4 \cdot 2^{-p}$, $p = 0, \ldots, 4$*). Since we have seen for the 1D Examples 4.11, 4.12 that the third order discontinuous Galerkin method does not perform better than the second order finite volume or the second order discontinuous Galerkin method, we have not used it in 2D.*

In Fig. 7, left, the EOCs against h, the maximal diameter of the triangles, are shown.

The best results are obtained by the discontinuous Galerkin method without a limiter (rkdg-w/o). This indicates that the limiter switches down the scheme to first order nearly everywhere. On the other hand the EOC of the scheme without limiter converges to the order two. Concerning the EOCs the finite volume schemes of first and second order are worse. The result for the efficiency is different (see Fig. 7,

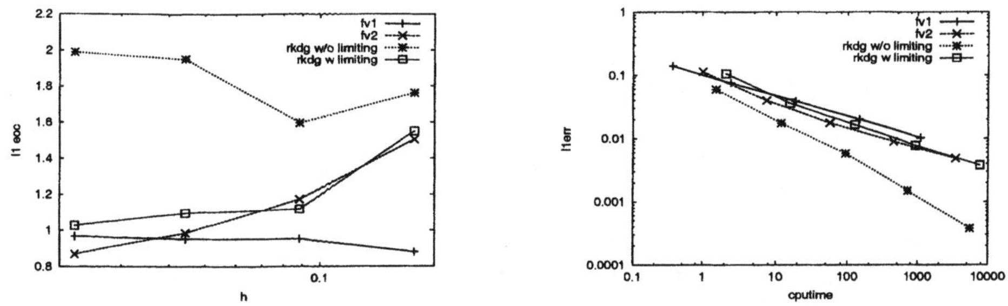

Fig. 7. EOCs and runtime to error ratio for the 2D MHD advection

right). The discontinuous Galerkin method without limiter (rkdg-w/o) is the best one but the finite volume is slightly better than the discontinuous Galerkin scheme with limiter. This is important for the approximation of discontinuous solutions, which we will consider in the next example. The CPU time which is necessary for obtaining the corresponding L^1*−error and the corresponding EOC for finite volume of first and second order and the discontinuous Galerkin schemes with and without limiter are also shown in Table 3. It turns out that in the group of schemes with limiters the finite volume methods are the most efficient schemes.*

Remark 4.14 *The results in Table 3 and Fig. 7 are based on triangular grids obtained by bisection. In [13] it is shown that the EOC for FV2 on grids obtained by quartering are much better.*

Table 3. 2D MHD advection: L^1-errors and eoc for the different schemes

Scheme: fv						
	Order: 1			**Order:** 2		
h	*time*	$\|\|u-u_h\|\|_{L^1}$	*EOC*	*time*	$\|\|u-u_h\|\|_{L^1}$	*EOC*
0.354	0.36	0.137620		1.01	0.112432	
0.177	2.43	0.074810	0.879398	7.50	0.039638	1.504088
0.088	18.91	0.038660	0.952386	**56.86**	**0.017581**	1.172894
0.044	**148.33**	**0.019978**	0.952445	443.17	0.008875	0.986110
0.022	1100.92	0.010215	0.967659	3437.55	0.004858	0.869425

Scheme: rkdg, **Order:** 2						
	Limiter: none			**Limiter:** wierse (mod.)		
h	*time*	$\|\|u-u_h\|\|_{L^1}$	*EOC*	*time*	$\|\|u-u_h\|\|_{L^1}$	*EOC*
0.354	1.54	0.059057		2.09	0.104836	
0.177	**11.92**	**0.017434**	1.760244	15.79	0.035794	1.550343
0.088	93.38	0.005765	1.596356	**126.25**	**0.016496**	1.117615
0.044	719.01	0.001491	1.950956	935.76	0.007713	1.096714
0.022	5516.49	0.000376	1.989243	7649.47	0.003784	1.027323

Table 4. Initial data for the shock tube problem in 2D

Initial data ($\gamma = 1.4$)								
	ρ	u_x	u_y	u_z	B_x	B_y	B_z	p
$x < 0$	1.0	0.0	0.0	0.0	0.0	0.0	0.0	1.0
$x > 0$	0.125	0.0	0.0	0.0	0.0	0.0	0.0	0.1

Example 4.15 *(Euler equations, shock tube problem in 2D) We set* $\mathbf{B} = 0$ *in (32) and obtain the Euler equations of gas dynamics. We use the 1D shock tube data (see Table 4) as initial data for the 2D calculation.*

As before the domain $\Omega :=]-1, 1[^2$ *is discretized by an grid with 32768 triangles. The resolution of the density is shown in Fig. 8. A zoom into the contact discontinuity and the shock indicates that the discontinuous Galerkin method is slightly more accurate than the finite volume scheme. Within the shock they are comparable. But for this shock tube problem in 2D the CPU-time for the discontinuous Galerkin method was much longer than for the finite volume result in order to get the same* L^1*-error.*

We should mention that in this example the MHD-fluxes have been used for solving the Euler equations. If we use simple Euler-fluxes, the results could be (slightly) different.

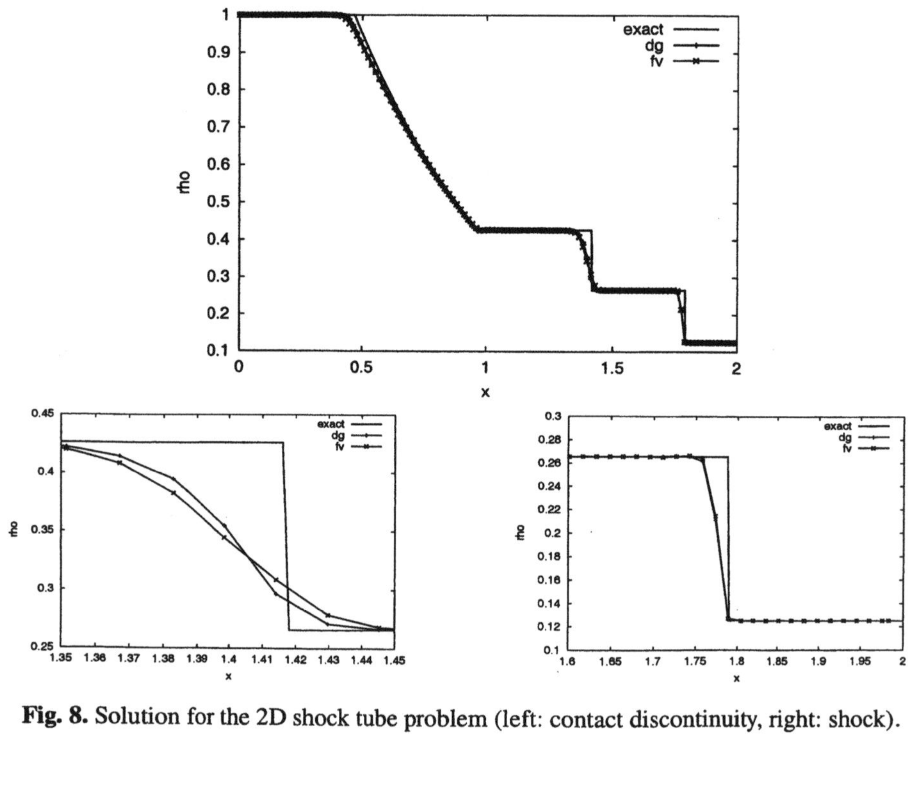

Fig. 8. Solution for the 2D shock tube problem (left: contact discontinuity, right: shock).

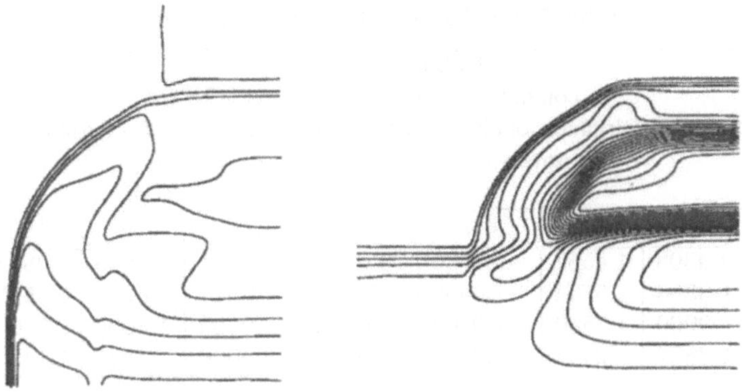

(a) Finite volume method of first order

(b) Finite volume method of second order

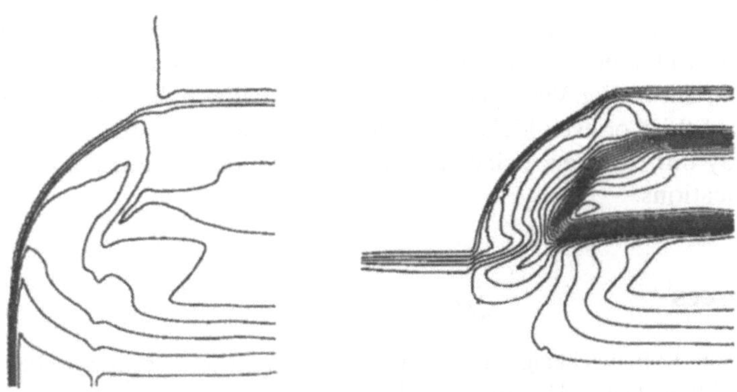

(c) Discontinuous Galerkin method of second order

Fig. 9. 2D Riemann problem (components shown: ρ, B_x)

Table 5. Initial data for the 2D Riemann problem

quadr	ρ	ρu_x	ρu_y	ρu_z	B_x	B_y	B_z	ρe
Initial data ($\gamma = 5/3$) computational domain: $[-1,1] \times [-1,1]$ boundaries: Dirichlet condition (numerical solution to 1d Riemann problems) quadrants : $1 : x > 0,\ y > 0$ / $2 : x < 0,\ y > 0$ / $3 : x < 0,\ y < 0$ / $4 : x > 0,\ y < 0$								
1	0.9308	1.4557	−0.4633	0.0575	1.2411	3.4845	1.0812	5.0838
2	1.0304	1.5774	−1.0455	−0.1016	1.2411	1.8000	0.5585	5.7813
3	1.0000	1.7500	−1.0000	0.0000	2.0000	1.8000	0.9000	6.0000
4	1.8887	0.2334	−1.7422	0.0733	2.0000	3.4845	1.7423	12.999

Example 4.16 *(MHD, 2D Riemann problem) Initial data in this case are given by the values in the four quadrants (see Table 5). The domain is discretized by 32768 triangles. In Fig. 9 level lines for finite volume of first and second order and for the discontinuous Galerkin method, respectively, are shown at time* $T = 0.25$. *On the left-hand side we see the density, on the right-hand side the* B_x*-component of the magnetic field. The finite volume method of second order yields the best resolution of the discontinuities. Because we don't know the exact solution in this case, we have no information about the efficiency.*

5 Conclusions

For the approximation of smooth solutions, if a limiter is not necessary, the discontinuous Galerkin method performs much better than the finite volume schemes. In the case of discontinuous solutions, when limiters are definitely necessary, the situation is much different. For a moderate regime of h, the finite volume schemes of second order seem to be the best choice, compared to discontinuous Galerkin schemes of second and third order. We expect that for much smaller h the discontinuous Galerkin methods of third or even higher order will be most successful. But this case is not covered by our experience since the case of moderate regimes of h is relevant for real applications.

References

1. Becker, J.: Entwicklung eines effizienten Verfahrens zur Lösung hyperbolischer Differentialgleichungen. Universität Freiburg, Dissertation (1999), http://www.freidok.uni-freiburg.de/volltexte/123/
2. Cockburn, B., Hou, S., Shu, C.-W.: TVB Runge-Kutta local projection discontinuous Galerkin finite element method for conservation laws IV: The multidimensional case. Math. Comp. 54 (1990), no. 190, 545–581
3. Cockburn, B., Karniadakis, E., Shu, C.-W.: The development of discontinuous Galerkin methods. Lecture Notes in Computational Science and Engineering 11 (2000), 3–52

4. Dolejsi, V., Feistauer, M., Schwab, C.: On some aspects of the discontinuous Galerkin finite element method for conservation laws. To appear in: Mathematics and Computers in Simulation
5. Friedrichs, K.O.: On the laws of relativistic electro-magneto-fluid dynamics. Comm. pure appl. Math. 27 (1974), 749–808
6. Kröner, D., Noelle, S., Rokyta, M.: Convergence of higher order upwind finite volume schemes on unstructured grids for scalar conservation laws in several space dimensions. Numer. Math., 71 (1995), no. 4, 527–560
7. Kröner, D.: Numerical schemes for conservation laws. Wiley-Teubner series advances in numerical mathematics. B. G. Teubner Verlagsgesellschaft mbH, Stuttgart, first edition (1997)
8. Kröner, D., Ohlberger, M.: A posteriori error estimates for upwind finite volume schemes for nonlinear conservation laws in multi dimensions. Math. Comput. 69 (2000), no. 229, 25–39
9. Küther, M.: Error estimates for second order finite volume schemes using a TVD-Runge-Kutta time discretization for a nonlinear scalar hyperbolic conservation law. East-West J. Numer. Math. 8 (2000), no. 4, 299–322
10. Schnitzer, T.: Discontinuous Galerkin Verfahren angewandt auf die MHD-Gleichungen, Diplomarbeit (2002)
11. Wesenberg, M.: Finite-Volumen-Verfahren für die Gleichungen der Magnetohydrodynamik in ein und zwei Raumdimensionen, Diplomarbeit (1998)
12. Wesenberg, M.: Efficient MHD Riemann solvers for simulations on unstructured triangular grids. To appear in: J. of Numer. Math.
13. Wesenberg, M.: Efficient higher–order finite volume schemes for (real gas) magnetohydrodynamics. PhD thesis (2002)
14. Wierse, M.: A new theoretically motivated higher order upwind scheme on unstructured grids of simplices. Adv. Comput. Math. 7 (1997), no. 3, 303–335

Existence of Strong Solutions for Electrorheological Fluids in Two Dimensions: Steady Dirichlet Problem

Frank Ettwein and Michael Růžička

Institute of Applied Mathematics, Albert-Ludwigs-University Freiburg, Eckerstr. 1, D-79104 Freiburg, Germany, {ettwein, rose}@mathematik.uni-freiburg.de

Summary. We prove the existence of local strong solutions to a system of nonlinear partial differential equations describing steady planar motions of electrorheological fluids with Dirichlet boundary conditions for $p_\infty > \frac{6}{5}$.

1 Introduction

Electrorheological fluids (ERF) are special viscous liquids that change their viscosity rapidly when an electric field is applied (cf. color plate 5, Page 668). This behaviour is explored and partly already used in engineering applications and in medical technology, for example in shock absorbers, clutches, force feedback devices, earthquake-proof buildings and artificial muscles.

In [8] a model was developed which is based on a non-dimensionalisation and a subsequent approximation taking into account the special properties of ERFs. Starting points were general conservation laws of electrodynamics and thermodynamics. The resulting model is a system of partial differential equations which reads[1]

$$\begin{aligned} \operatorname{div}(\mathbf{E}+\mathbf{P}) &= 0, \\ \operatorname{curl}\mathbf{E} &= 0, \end{aligned} \tag{1a}$$

$$\begin{aligned} \rho_0\frac{\partial \mathbf{v}}{\partial t} - \operatorname{div}\mathbf{S} + \rho_0\,[\nabla\mathbf{v}]\,\mathbf{v} + \nabla\phi &= \rho_0\mathbf{f} + [\nabla\mathbf{E}]\,\mathbf{P}, \\ \operatorname{div}\mathbf{v} &= 0, \end{aligned} \tag{1b}$$

where $\mathbf{E}$ is the electric field, $\mathbf{P}$ the polarisation, ρ_0 the density, $\mathbf{v}$ the velocity, $\mathbf{S}$ the extra stress tensor, ϕ the pressure and $\mathbf{f}$ the mechanical force.

The above system has to be supplemented with constitutive relations for $\mathbf{P}$ and $\mathbf{S}$ and with boundary and initial conditions. If we choose a linear relation for $\mathbf{P}$, i.e. $\mathbf{P} = \chi^E\mathbf{E}$, where χ^E is the dielectric susceptibility, the system decouples in the

[1] Here and in the following we use the notation $[\nabla\mathbf{u}]\,\mathbf{w} = \left(\frac{\partial u_i}{\partial x_j}w_j\right)_{i=1,2}$, where the summation convention over repeated indices is used and little Latin indices are either 1 or 2.

mechanical and electric variables. Since Maxwell's equations are well understood (cf. [10], [9]) we will only treat the system (1b). Moreover, we restrict ourselves to the steady planar case and Dirichlet boundary conditions. Since we assume constant density we divide through ρ_0 and obtain a new $\mathbf{S}$ and ϕ. Moreover, $\mathbf{E}$ can be made as regular as necessary and thus we introduce for brevity the notation $\mathbf{F} = \mathrm{f} + \chi^E\,[\nabla\mathbf{E}]\,\mathbf{E}$.

Hence we study the existence of strong solutions to the system

$$\begin{aligned} -\operatorname{div}\mathbf{S}(\mathbf{D}(\mathbf{v}),\mathbf{E}) + [\nabla\mathbf{v}]\,\mathbf{v} + \nabla\phi = \mathbf{F} &\quad \text{in } \Omega, \\ \operatorname{div}\mathbf{v} = 0 &\quad \text{in } \Omega, \\ \mathbf{v} = \mathbf{0} &\quad \text{on } \partial\Omega, \end{aligned} \tag{2}$$

where $\mathbf{S} : \mathbb{R}^{2\times 2}_{sym} \times \mathbb{R}^2 \to \mathbb{R}^{2\times 2}_{sym}$ is given by

$$\begin{aligned} \mathbf{S}(\mathbf{D},\mathbf{E}) :=&\alpha_{21}\Big(\big(1+|\mathbf{D}|^2\big)^{\frac{p(|\mathbf{E}|^2)-1}{2}} - 1\Big)\mathbf{E}\otimes\mathbf{E} \\ &+ \big(\alpha_{31} + \alpha_{33}|\mathbf{E}|^2\big)\big(1+|\mathbf{D}|^2\big)^{\frac{p(|\mathbf{E}|^2)-2}{2}}\mathbf{D} \\ &+ \alpha_{51}\big(1+|\mathbf{D}|^2\big)^{\frac{p(|\mathbf{E}|^2)-2}{2}}\big(\mathbf{D}\mathbf{E}\otimes\mathbf{E} + \mathbf{E}\otimes\mathbf{D}\mathbf{E}\big). \end{aligned} \tag{3}$$

Here p is a given material function with $1 < p_\infty \le p(|\mathbf{E}|^2) \le p_0$ and $\mathbf{D}$ means the symmetric velocity gradient, i.e. $\mathbf{D}(\mathbf{v}) = \frac{1}{2}\left(\nabla\mathbf{v} + (\nabla\mathbf{v})^t\right)$. We require the α_{ij} should be chosen such that[2]

$$\begin{aligned} \partial_{kl}S_{ij}(\mathbf{D},\mathbf{E})B_{ij}B_{kl} &\ge c\big(1+|\mathbf{D}|^2\big)^{\frac{p(|\mathbf{E}|^2)-2}{2}}|\mathbf{B}|^2, \\ \mathbf{S}(\mathbf{D},\mathbf{E}) : \mathbf{D} &\ge c\big(1+|\mathbf{D}|^2\big)^{\frac{p(|\mathbf{E}|^2)-2}{2}}|\mathbf{D}|^2, \\ |\partial_{kl}\mathbf{S}(\mathbf{D},\mathbf{E})| &\le C\big(1+|\mathbf{D}|^2\big)^{\frac{p(|\mathbf{E}|^2)-2}{2}}, \quad k,l = 1,2, \\ |\partial_n\mathbf{S}(\mathbf{D},\mathbf{E})| &\le C\big(1+|\mathbf{D}|^2\big)^{\frac{p(|\mathbf{E}|^2)-1}{2}}\left(1+\ln(1+|\mathbf{D}|^2)\right), \quad n = 1,2 \end{aligned} \tag{4}$$

for all $\mathbf{B},\mathbf{D} \in X := \{\mathbf{D} \in \mathbb{R}^{2\times 2}_{sym}, \operatorname{tr}\mathbf{D} = 0\}$ and $\mathbf{E} \in \mathbb{R}^2$. We call the first property of $\mathbf{S}$ (uniform) monotonicity and the second coercivity. The constants c, C may depend on $\mathbf{E}$. In [1], and [9], one can find conditions on α_{ij} in the two-dimensional, and in the three-dimensional case respectively, that ensure the validity of (4).

Let us denote $\mathcal{V}$ the space of divergence-free smooth functions with compact support, V_q the closure of $\mathcal{V}$ in the semi-norm of $W^{1,q}$ and

$$E_{p(\cdot)} := \overline{\mathcal{V}}^{\|\mathbf{D}(\cdot)\|_{p(\cdot)}}.$$

Here

[2] We use the following notations: $\mathbf{A} : \mathbf{B} = A_{ij}B_{ij}$ means the scalar product of matrices and we abbreviate $\partial_{kl}\mathbf{S} = \frac{\partial\mathbf{S}}{\partial D_{kl}}$ and $\partial_n\mathbf{S} = \frac{\partial\mathbf{S}}{\partial E_n}$.

$$\|f\|_{p(\cdot)} := \inf\{\lambda > 0 : \left|\frac{f}{\lambda}\right|_{p(\cdot)} \leq 1\}$$

is the norm of the generalised Lebesgue space

$$L^{p(\cdot)}(\Omega) := \{f \in L^1(\Omega) : \|f\|_{p(\cdot)} < \infty\}$$

with $|f|_{p(\cdot)} := \int_\Omega |f(x)|^{p(x)}\, dx$. Generalised Sobolev spaces are defined analogously. We write $p(x)$ for $p(|\mathbf{E}(x)|^2)$. For details of the generalised Lebesgue and Sobolev spaces see [5] or [9]. If p is constant the norm will coincide with the norm of the usual Lebesgue space L^p.

Let $\mathbf{F} \in (W_0^{1,p_\infty}(\Omega))'$. We say $\mathbf{v} \in E_{p(x)}$ is a *weak solution* to (2) if and only if

$$\int_\Omega \mathbf{S}(\mathbf{D}(\mathbf{v}), \mathbf{E}) : \mathbf{D}(\varphi)\, dx + \int_\Omega [\nabla \mathbf{v}]\, \mathbf{v} \cdot \varphi\, dx = \langle \mathbf{F}, \varphi \rangle_{1,p_\infty} \tag{5}$$

for all smooth divergence-free φ with compact support. Note that the second integral, called the convective term, will only be defined for $p_\infty \geq \frac{3d}{d+2}$ (here $d = 2 \Rightarrow p_\infty \geq \frac{3}{2}$) if one uses test functions $\varphi \in E_{p(x)}$ in equation (5). In the case of strict inequality the existence of weak solutions can be shown with the theory of pseudomonotone operators, see [9] for details.

In [9] the full system (1a) and (1b) is investigated in three dimensions, however the results/methods can be also applied to the planar case. It is shown that strong solutions (strong means that second derivatives are locally integrable) exists for $\frac{3}{2} < p_\infty \leq p_0 < p_\infty^*$ (in dimensions higher than 2 there are some extra conditions on p_0 additional to the requirement $p_\infty > \frac{3d}{d+2}$), where $p^* = \frac{dp}{d-p}$ for $p < d$ so that $W^{1,p} \hookrightarrow L^{p^*}$. Our goal is to enhance this result in two dimensions. In [4] (Dirichlet boundary conditions) or [3] (periodic case) the authors study a similar system with the extra stress tensor $\mathbf{S}$ given by a potential and with standard growth. They pursue further to obtain $C^{1,\alpha}$-solutions with Meyers' L^p-estimate for the gradient what will be a future task for us. This article rests upon a master thesis [1] where one can find the missing details of the computations.

Our scheme is the following: Consider $p \equiv 2$. Show existence of weak solutions with monotonicity methods and than show regularity with the help of the difference quotient technique. Then approximate the general case by the case $p \equiv 2$ through substituting $\mathbf{S}$ by $\mathbf{S}^\lambda$ and get uniform bounds for the approximate solutions $\mathbf{v}^\lambda$ and therefore a candidate $\mathbf{v}$. At last we will prove that the limit $\mathbf{v}$ is a solution to (5).

Thus we will prove:

Theorem 1.1 *Let* $\Omega \subset \mathbb{R}^2$ *be a bounded domain and* $\mathbf{E} \in C^1(\overline{\Omega})$ *and* $\mathbf{f} \in L^{p'_\infty}(\Omega)$ *(therewith* $\mathbf{F} \in L^{p'_\infty}(\Omega)$ *holds) be given. Further assume* $6/5 < p_\infty \leq p(|\mathbf{E}(x)|^2) \leq p_0 < \infty$ *and* $p \in C^1(\mathbb{R})$. *Then there exists a strong solution* $\mathbf{v}$ *to* (2) *with* $\mathbf{S}$ *as in* (3) *satisfying* (4). *This solution satisfies* (5) *and*

$$\mathbf{v} \in \bigcap_{0<\epsilon\leq 1} W^{2,2-\epsilon}_{loc}(\Omega) \cap E_{p(\cdot)} \cap V_{p_\infty}. \tag{6}$$

2 Proof of Theorem 1.1

Before we give the proof let us define our approximation and state its properties. For $\mathbf{S}$ given by (3) and satisfying (4) we define for $\lambda \in [0,1]$ a λ-*approximation* by

$$\begin{aligned}\mathbf{S}^\lambda(\mathbf{D},\mathbf{E}) :=& \alpha_{21}\Big((1+\lambda|\mathbf{D}|^2)^{\frac{1-p(|\mathbf{E}|^2)}{2}}(1+|\mathbf{D}|^2)^{\frac{p(|\mathbf{E}|^2)-1}{2}} - 1\Big)\mathbf{E}\otimes\mathbf{E} \\ &+ (1+|\mathbf{D}|^2)^{\frac{p(|\mathbf{E}|^2)-2}{2}}(1+\lambda|\mathbf{D}|^2)^{\frac{2-p(|\mathbf{E}|^2)}{2}} \times \\ &\times \Big((\alpha_{31}+\alpha_{33}|\mathbf{E}|^2)\mathbf{D} + \alpha_{51}(\mathbf{D}\mathbf{E}\otimes\mathbf{E}+\mathbf{E}\otimes\mathbf{D}\mathbf{E})\Big).\end{aligned} \tag{7}$$

According to the definition $\mathbf{S}^0 = \mathbf{S}$ holds.

One can show with some calculations that $\mathbf{S}^\lambda$ has the following properties which allow us to use, for fixed $\lambda > 0$, the theory with linear growth (cf. Appendix A).

Lemma 2.2 *Let* $\mathbf{S}^\lambda$ *be given by* (7). *Then we have for* $0 < \lambda \leq 1$

$$\partial_{kl} S^\lambda_{ij}(\mathbf{D},\mathbf{E}) B_{ij} B_{kl} \geq c\left(\frac{1+|\mathbf{D}|^2}{1+\lambda|\mathbf{D}|^2}\right)^{\frac{p-2}{2}} |\mathbf{B}|^2, \tag{8}$$

$$|\mathbf{S}^\lambda(\mathbf{D},\mathbf{E})| \leq C\left(\frac{1+|\mathbf{D}|^2}{1+\lambda|\mathbf{D}|^2}\right)^{\frac{p-2}{2}} |\mathbf{D}|, \tag{9}$$

$$\mathbf{S}^\lambda(\mathbf{D},\mathbf{E}) : \mathbf{D} \geq c\left(\frac{1+|\mathbf{D}|^2}{1+\lambda|\mathbf{D}|^2}\right)^{\frac{p-2}{2}} |\mathbf{D}|^2, \tag{10}$$

$$|\partial_{kl}\mathbf{S}^\lambda(\mathbf{D},\mathbf{E})| \leq C\left(\frac{1+|\mathbf{D}|^2}{1+\lambda|\mathbf{D}|^2}\right)^{\frac{p-2}{2}}, \tag{11}$$

$$|\partial_n\mathbf{S}^\lambda(\mathbf{D},\mathbf{E})| \leq C\left(\frac{1+|\mathbf{D}|^2}{1+\lambda|\mathbf{D}|^2}\right)^{\frac{p-2}{2}} \ln\left(\frac{1+|\mathbf{D}|^2}{1+\lambda|\mathbf{D}|^2}\right)(1+|\mathbf{D}|^2) \tag{12}$$

for all $\mathbf{B},\mathbf{D} \in X$. *The constants* c, C *are independent of* λ.

Now we can proof our theorem.

Proof (of the Theorem 1.1). Without loss of generality let $p_\infty < 2$, otherwise see [9]. For each fixed λ with $0 < \lambda \le 1$ the assumptions of Proposition A.2 are fulfilled since $1 \le \frac{1+|\mathbf{D}|^2}{1+\lambda|\mathbf{D}|^2} \le \frac{1}{\lambda}$. Thus we get a solution $\mathbf{v}^\lambda \in W^{2,2}_{loc}(\Omega) \cap V_2$.

(i) Energy Estimates:
We use $\mathbf{v}^\lambda$ as a test function in equation (5) and obtain

$$\int_\Omega \mathbf{S}^\lambda(\mathbf{D}(\mathbf{v}^\lambda),\mathbf{E}) : \mathbf{D}(\mathbf{v}^\lambda)\,dx + \int_\Omega [\nabla \mathbf{v}^\lambda]\,\mathbf{v}^\lambda \cdot \mathbf{v}^\lambda\,dx = \int_\Omega \mathbf{F}\cdot\mathbf{v}^\lambda\,dx. \quad (13)$$

The second integral vanishes because $\mathbf{v}^\lambda$ is divergence-free. If we use the coercivity (10) and Young's inequality with ϵ, we will conclude that

$$\begin{aligned} c\int_\Omega \left(\frac{1+|\mathbf{D}(\mathbf{v}^\lambda)|^2}{1+\lambda|\mathbf{D}(\mathbf{v}^\lambda)|^2}\right)^{\frac{p(|\mathbf{E}|^2)-2}{2}} & |\mathbf{D}(\mathbf{v}^\lambda)|^2\,dx \\ \le \int_\Omega \mathbf{S}^\lambda(\mathbf{D}(\mathbf{v}^\lambda),\mathbf{E}) &: \mathbf{D}(\mathbf{v}^\lambda)\,dx = \int_\Omega \mathbf{F}\cdot\mathbf{v}^\lambda\,dx \\ \le C_\epsilon \|\mathbf{F}\|_{p'_\infty}^{p'_\infty} &+ \epsilon\|\mathbf{v}^\lambda\|_{p_\infty}^{p_\infty}. \end{aligned}$$

Using $(1+x)^{p-2}x^2 \ge x^p - C_p$ we obtain

$$\begin{aligned} \int_\Omega \left(\frac{1+|\mathbf{D}(\mathbf{v}^\lambda)|^2}{1+\lambda|\mathbf{D}(\mathbf{v}^\lambda)|^2}\right)^{\frac{p(|\mathbf{E}|^2)-2}{2}} & |\mathbf{D}(\mathbf{v}^\lambda)|^2\,dx \\ & \ge \|\mathbf{D}(\mathbf{v}^\lambda)\|_{p_\infty}^{p_\infty} - C, \end{aligned}$$

where C depends on the measure of the domain Ω.
Combining the last two estimates we reach our first estimate employing also Korn's inequality (see for instance [7])

$$\|\mathbf{v}^\lambda\|_{W^{1,p_\infty}}^{p_\infty} \le C\left(\|\mathbf{F}\|_{p'_\infty}^{p'_\infty} + 1\right), \quad (14)$$

respectively

$$\int_\Omega \left(\frac{1+|\mathbf{D}(\mathbf{v}^\lambda)|^2}{1+\lambda|\mathbf{D}(\mathbf{v}^\lambda)|^2}\right)^{\frac{p(|\mathbf{E}|^2)-2}{2}} |\mathbf{D}(\mathbf{v}^\lambda)|^2\,dx \le C(\mathbf{F}). \quad (15)$$

(ii) Uniform local estimates in W^{2,p_∞}:
We would like to test with $\operatorname{rot}(\xi^{2\alpha}\operatorname{curl}\mathbf{v}^\lambda) = -\xi^{2\alpha}\Delta\mathbf{v}^\lambda + \operatorname{curl}\mathbf{v}^\lambda \operatorname{rot}(\xi^{2\alpha})$, where $\operatorname{rot}\psi = \left(\frac{\partial\psi}{\partial x_2}, -\frac{\partial\psi}{\partial x_1}\right)^t$ for scalar functions ψ and $\operatorname{curl}\mathbf{u} = \frac{\partial u_2}{\partial x_1} - \frac{\partial u_1}{\partial x_2}$ for vector-valued functions $\mathbf{u}$. ξ is an usual smooth cut-off function with compact support, $0 \le \xi \le 1$ and such that there exists a set $U \ne \emptyset$ with $\xi = 1$ in $U \subset\subset \Omega$. For all vector-valued functions $\operatorname{div}\operatorname{rot}\mathbf{u} = 0$ holds and further for divergence-free functions $\operatorname{rot}\operatorname{curl}\mathbf{u} = -\Delta\mathbf{u}$ holds.

Starting from the identity (5) we get the following equation by density arguments:

$$\begin{aligned} 0 = &- \int_\Omega \operatorname{div} \mathbf{S}^\lambda(\mathbf{D}(\mathbf{v}), \mathbf{E}) \cdot \operatorname{rot}(\xi^{2\alpha} \operatorname{curl} \mathbf{v}^\lambda)\, dx \\ &+ \int_\Omega [\nabla \mathbf{v}]\, \mathbf{v}^\lambda \cdot \operatorname{rot}(\xi^{2\alpha} \operatorname{curl} \mathbf{v}^\lambda)\, dx \\ &- \int_\Omega \mathbf{F} \cdot \operatorname{rot}(\xi^{2\alpha} \operatorname{curl} \mathbf{v}^\lambda)\, dx =: I_1 + I_2 + I_3\,. \end{aligned} \tag{16}$$

Let us examine the three integrals separately.

$$\begin{aligned} I_1 = &\int_\Omega \frac{\partial}{\partial x_k} S_{ij}^\lambda(\mathbf{D}(\mathbf{v}^\lambda), \mathbf{E}) D_{ij}\left(\tfrac{\partial \mathbf{v}^\lambda}{\partial x_k}\right) \xi^{2\alpha}\, dx \\ &+ 2\alpha \int_\Omega S_{ij}^\lambda(\mathbf{D}(\mathbf{v}^\lambda), \mathbf{E}) D_{ij}\left(\tfrac{\partial \mathbf{v}^\lambda}{\partial x_k}\right) \xi^{2\alpha-1} \frac{\partial \xi}{\partial x_k}\, dx \\ &- \alpha \int_\Omega S_{ij}^\lambda(\mathbf{D}(\mathbf{v}^\lambda), \mathbf{E}) (\Delta v_i^\lambda \xi^{2\alpha-1} \frac{\partial \xi}{\partial x_j} + \Delta v_j^\lambda \xi^{2\alpha-1} \frac{\partial \xi}{\partial x_i})\, dx \\ &+ \int_\Omega S_{ij}^\lambda(\mathbf{D}(\mathbf{v}^\lambda), \mathbf{E}) D_{ij}(\operatorname{rot}(\xi^{2\alpha}) \operatorname{curl} \mathbf{v}^\lambda)\, dx \\ =:&\ J_1 + J_2 + J_3 + J_4. \end{aligned}$$

We apply the chain rule and obtain for the first term

$$\begin{aligned} J_1 = &\int_\Omega \partial_{lm} S_{ij}^\lambda(\mathbf{D}(\mathbf{v}^\lambda), \mathbf{E}) D_{lm}\left(\tfrac{\partial \mathbf{v}^\lambda}{\partial x_k}\right) D_{ij}\left(\tfrac{\partial \mathbf{v}^\lambda}{\partial x_k}\right) \xi^{2\alpha}\, dx \\ &+ \int_\Omega \partial_n S_{ij}^\lambda(\mathbf{D}(\mathbf{v}^\lambda), \mathbf{E}) \frac{\partial E_n}{\partial x_k} D_{ij}\left(\tfrac{\partial \mathbf{v}^\lambda}{\partial x_k}\right) \xi^{2\alpha}\, dx \\ =:&\ J_{1,1} + J_{1,2}. \end{aligned}$$

The monotonicity (8) helps us to obtain

$$\begin{aligned} J_{1,1} \geq &\ c \int_\Omega \left(\frac{1 + |\mathbf{D}(\mathbf{v}^\lambda)|^2}{1 + \lambda |\mathbf{D}(\mathbf{v}^\lambda)|^2} \right)^{\frac{p(|\mathbf{E}|^2)-2}{2}} |\mathbf{D}(\nabla \mathbf{v}^\lambda)|^2 \xi^{2\alpha}\, dx \\ =:&\ c I_p^\lambda(\mathbf{D}(\nabla \mathbf{v}^\lambda), \xi^\alpha). \end{aligned}$$

With the lower bound for I_p^λ (cf. Lemma B.4), Hölder's inequality (cf. Lemma B.3) and the energy estimates we obtain bounds uniform in λ, namely

$$\|\nabla^2 \mathbf{v}^\lambda \xi^\alpha\|_{p_\infty}^2 + \|\nabla \mathbf{v} \xi^{\frac{2\alpha}{p_\infty}}\|_r^{p_\infty} \leq C(\mathbf{F}) \left(1 + I_p^\lambda(\mathbf{D}(\nabla \mathbf{v}^\lambda), \xi^\alpha)\right), \tag{17}$$

where $1 \leq r < \infty$ is arbitrary.
With the estimate (12) for $\partial_n \mathbf{S}^\lambda$ and the bound $\ln(1 + x) \leq C_\epsilon (1 + x)^\epsilon$, with arbitrary $\epsilon > 0$, we achieve after few calculations with a small δ

$$
\begin{aligned}
|J_{1,2}| \le c(\mathbf{E}) \int_\Omega \left(\frac{1+|\mathbf{D}(\mathbf{v}^\lambda)|^2}{1+\lambda|\mathbf{D}(\mathbf{v}^\lambda)|^2}\right)^{\frac{p-2}{2}} (1+|\mathbf{D}(\mathbf{v}^\lambda)|) \times \\
\times \ln\left(\frac{1+|\mathbf{D}(\mathbf{v}^\lambda)|^2}{1+\lambda|\mathbf{D}(\mathbf{v}^\lambda)|^2}\right) |\mathbf{D}(\nabla\mathbf{v}^\lambda)|\xi^{2\alpha}\,dx \\
\le c \int_\Omega \left(\frac{1+|\mathbf{D}(\mathbf{v}^\lambda)|^2}{1+\lambda|\mathbf{D}(\mathbf{v}^\lambda)|^2}\right)^{\frac{p-2}{2}+\epsilon} |\mathbf{D}(\mathbf{v}^\lambda)||\mathbf{D}(\nabla\mathbf{v}^\lambda)|\xi^{2\alpha}\,dx \\
+ \delta I_p^\lambda(\mathbf{D}(\nabla\mathbf{v}^\lambda), \xi^\alpha) + C_\delta.
\end{aligned}
$$

This term looks like the term we obtained from $J_{1,1}$, but it is a little bit harder to handle for $p_0 > 2$. However, after Hölder's $(2, a, \frac{2a}{a-2})$ and Young's $(2, b, \frac{2b}{b-2})$ inequalities we get for $a, b > 2$ and $A \in (0,1)$

$$
\begin{aligned}
\int_\Omega \left(\frac{1+|\mathbf{D}(\mathbf{v}^\lambda)|^2}{1+\lambda|\mathbf{D}(\mathbf{v}^\lambda)|^2}\right)^{\frac{p-2}{2}+\epsilon} |\mathbf{D}(\mathbf{v}^\lambda)||\mathbf{D}(\nabla\mathbf{v}^\lambda)|\xi^{2\alpha}\,dx \\
\le \delta \int_\Omega \left(\frac{(1+|\mathbf{D}(\mathbf{v}^\lambda)|^2)}{1+\lambda|\mathbf{D}(\mathbf{v}^\lambda)|^2}\right)^{\frac{p-2}{2}} |\mathbf{D}(\nabla\mathbf{v}^\lambda)|^2\xi^{2\alpha} \\
+ C(\delta)\left(\int_\Omega \left(\frac{(1+|\mathbf{D}(\mathbf{v}^\lambda)|^2)}{1+\lambda|\mathbf{D}(\mathbf{v}^\lambda)|^2}\right)^{\frac{p-2}{4}aA} |\mathbf{D}(\mathbf{v}^\lambda)|^{aA}\right)^{\frac{2b}{a(b-2)}} \\
+ \delta\left(\int_\Omega \left(C(p_0) + |\mathbf{D}(\mathbf{v}^\lambda)|^{(2\epsilon+\frac{p_0}{2}(1-A))\frac{2a}{a-2}}\right) \xi^{\alpha\frac{2a}{a-2}}\right)^{\frac{a-2}{2a}b}.
\end{aligned}
$$

We require that $aA = 2$ and $\frac{a-2}{2a}b = \frac{p_\infty}{r_1}$ with $r_1 = (2\epsilon + \frac{p_0}{2}(1-A))\frac{2a}{a-2} = p_0 + \epsilon\frac{4a}{a-2}$. One possible solution is $\epsilon = \frac{1}{8}, b = 4, a = 4\frac{p_0}{2p_0+1-p_\infty}$ and $A = \frac{2p_0+1-p_\infty}{2p_0}$. $0 < A < 1$ and $a > 2$ are true since $1 < p_\infty \le p_0$. We conclude with the energy estimate (15) and with another δ

$$
|J_{1,2}| \le +\delta I_p^\lambda(\mathbf{D}(\nabla\mathbf{v}^\lambda), \xi^\alpha) + \delta\left(\int_\Omega |\mathbf{D}(\mathbf{v}^\lambda)|^{r_1}\xi^{\frac{\alpha 4 r_1}{p_\infty}}\right)^{\frac{p_\infty}{r_1}} + C(\mathbf{F}), \quad (18)
$$

where r_1 is defined above. Note that the second integral can be absorbed by the second term of the left-hand side in (17).

We can bound J_2, J_3 and J_4 with the estimate (15) of step (i) and Hölder's inequality as follows:

$$
\begin{aligned}
|J_2+J_3+J_4| &\le C(\nabla\xi)\int_\Omega |\mathbf{S}^\lambda(\mathbf{D}(\mathbf{v}^\lambda),\mathbf{E})|\,|\mathbf{D}(\nabla\mathbf{v}^\lambda)|\xi^{2\alpha-1}\,dx\\
&\le C\int_\Omega\left(\frac{1+|\mathbf{D}(\mathbf{v}^\lambda)|^2}{1+\lambda|\mathbf{D}(\mathbf{v}^\lambda)|^2}\right)^{\frac{p-2}{2}}|\mathbf{D}(\mathbf{v})|\,|\mathbf{D}(\nabla\mathbf{v}^\lambda)|\xi^{2\alpha-1}\,dx\\
&\le \delta\int_\Omega\left(\frac{1+|\mathbf{D}(\mathbf{v}^\lambda)|^2}{1+\lambda|\mathbf{D}(\mathbf{v}^\lambda)|^2}\right)^{\frac{p-2}{2}}|\mathbf{D}(\nabla\mathbf{v}^\lambda)|^2\xi^{\alpha}\,dx \qquad (19)\\
&\quad + C_\delta\int_\Omega\left(\frac{1+|\mathbf{D}(\mathbf{v}^\lambda)|^2}{1+\lambda|\mathbf{D}(\mathbf{v}^\lambda)|^2}\right)^{\frac{p-2}{2}}|\mathbf{D}(\mathbf{v}^\lambda)|^2\xi^{2(\alpha-1)}\,dx\\
&\le \delta I_p^\lambda(\mathbf{D}(\nabla\mathbf{v}^\lambda),\xi^\alpha)+C(\mathbf{F}).
\end{aligned}
$$

Now we treat the convective term which is responsible for the lower bound of p_∞:

$$
\begin{aligned}
|I_2| &= \left|\int_\Omega[\nabla\mathbf{v}^\lambda]\mathbf{v}^\lambda\cdot(-\xi^{2\alpha}\Delta\mathbf{v}^\lambda+\operatorname{curl}\mathbf{v}^\lambda\operatorname{rot}(\xi^{2\alpha}))\,dx\right|\\
&\le \left|-\int_\Omega[\nabla\mathbf{v}^\lambda]\mathbf{v}^\lambda\cdot\Delta\mathbf{v}^\lambda\xi^{2\alpha}\,dx\right|+c(\nabla\xi)\int|\nabla\mathbf{v}^\lambda|^2|\mathbf{v}^\lambda|\xi^{2\alpha-1}\,dx.
\end{aligned}
$$

In two dimensions it is specific that the term with second derivatives can be avoided. We transform the first term by partial integration:

$$
\begin{aligned}
-\int_\Omega\nabla\mathbf{v}^\lambda\mathbf{v}^\lambda\cdot\Delta\mathbf{v}^\lambda\xi^{2\alpha}\,dx &= -\int_\Omega v_j^\lambda\frac{\partial v_i^\lambda}{\partial x_j}\frac{\partial^2 v_i^\lambda}{\partial x_k^2}\xi^{2\alpha}\,dx\\
=\int_\Omega\frac{\partial v_j^\lambda}{\partial x_k}\frac{\partial v_i^\lambda}{\partial x_j}\frac{\partial v_i^\lambda}{\partial x_k}\xi^{2\alpha}\,dx &+ \int_\Omega v_j^\lambda\frac{\partial^2 v_i^\lambda}{\partial x_k\partial x_j}\frac{\partial v_i^\lambda}{\partial x_k}\xi^{2\alpha}\,dx\\
+\int_\Omega v_j^\lambda\frac{\partial v_i^\lambda}{\partial x_j}\frac{\partial v_i^\lambda}{\partial x_k}2\alpha\xi^{2\alpha-1}\frac{\partial\xi}{\partial x_k}\,dx.&
\end{aligned}
$$

The first integral vanishes because of $A_{ij}A_{jk}A_{ik}=0$ for $\mathbf{A}\in\mathbb{R}^{2,2}$ with $\operatorname{tr}\mathbf{A}=0$. By partial integration we see that

$$
\int_\Omega v_j^\lambda\frac{\partial^2 v_i^\lambda}{\partial x_k\partial x_j}\frac{\partial v_i^\lambda}{\partial x_k}\xi^{2\alpha}\,dx=-\int_\Omega\mathbf{v}^\lambda\cdot\nabla\xi|\nabla\mathbf{v}^\lambda|^2\alpha\xi^{2\alpha-1}\,dx.
$$

So the convective term can be estimated by

$$
|I_2|\le c(\nabla\xi)\int|\nabla\mathbf{v}^\lambda|^2|\mathbf{v}^\lambda|\xi^{2\alpha-1}\,dx. \qquad (20)
$$

For the third term of (16) we use Hölder's inequality and the energy estimates to get

$$
\begin{aligned}
|I_3| &\le \delta\|\nabla^2\mathbf{v}^\lambda\xi^\alpha\|_{p_\infty}^2+\|\nabla\mathbf{v}^\lambda\|_{p_\infty}^2+C\|\mathbf{F}\|_{p'_\infty}^2\\
&\le \delta\|\nabla^2\mathbf{v}^\lambda\xi^\alpha\|_{p_\infty}^2+C(\mathbf{F}).
\end{aligned} \qquad (21)
$$

Summarizing (17)–(21), we obtain for arbitrary $r \geq 1$

$$\|\nabla^2 \mathbf{v}^\lambda \xi^\alpha\|_{p_\infty}^2 + \|\nabla \mathbf{v}^\lambda \xi^{\frac{2}{p_\infty}\alpha}\|_r^{p_\infty} \leq C\Big(1 + \int_\Omega |\nabla \mathbf{v}^\lambda|^2 |\mathbf{v}^\lambda| \xi^{2\alpha-1}\, dx\Big), \quad (22)$$

where the constant C depends on $\mathbf{F}, \mathbf{E}, \nabla\xi, p_0$. It remains only one integral to consider:

$$\begin{aligned} \int_\Omega |\nabla \mathbf{v}^\lambda|^2 |\mathbf{v}^\lambda| \xi^{2\alpha-1}\, dx &= \int_\Omega |\nabla \mathbf{v}^\lambda|^{2A} \xi^{2\alpha-1} |\nabla \mathbf{v}^\lambda|^{2(1-A)} |\mathbf{v}^\lambda|\, dx \qquad (23) \\ \leq \Big(\int_\Omega |\nabla \mathbf{v}^\lambda|^{2A\beta_1} \xi^{(2\alpha-1)\beta_1}\, dx\Big)^{\frac{1}{\beta_1}} &\Big(\int_\Omega |\nabla \mathbf{v}^\lambda|^{2(1-A)\beta_2}\, dx\Big)^{\frac{1}{\beta_2}} \times \\ \times \Big(\int_\Omega |\mathbf{v}^\lambda|^{\frac{2p_\infty}{2-p_\infty}}\, dx\Big)^{\frac{2-p_\infty}{2p_\infty}}&. \end{aligned}$$

The following equations for $\beta_1, \beta_2 > 1$ and $A \in (0,1)$ should be fulfilled:

$$\frac{1}{\beta_1} + \frac{1}{\beta_2} + \frac{2-p_\infty}{2p_\infty} = 1,$$
$$2A\beta_1 = \frac{2p_\infty}{2-p_\infty} \quad \text{and} \quad 2(1-A)\beta_2 = p_\infty.$$

For $p_\infty > \frac{6}{5}$ they are solvable and the solutions are

$$A = \frac{3(2-p_\infty)}{2p_\infty}, \quad \beta_1 = \frac{2}{3}\frac{p_\infty^2}{(2-p_\infty)^2} \quad \text{and} \quad \beta_2 = \frac{p_\infty^2}{5p_\infty - 6}.$$

The last two integrals in (23) are bounded by a constant due to the estimates in step (i). For α big enough we would like to absorb the first integral in the first term on the left-hand side of (22) using Young's inequality and Sobolev's embedding theorem and then we would obtain

$$\|\nabla^2 \mathbf{v}^\lambda \xi^\alpha\|_{p_\infty}^2 + \|\nabla \mathbf{v}^\lambda \xi^{\frac{2}{p_\infty}\alpha}\|_r^{p_\infty} \leq C.$$

This will be possible if $\frac{1}{\beta} < 2\frac{1}{p_\infty^*}$ holds (in two dimensions $p^* = \frac{2-p}{2p}$) which is valid for $p_\infty > \frac{6}{5}$. Thus we obtain for all $r \geq 1$

$$\|\nabla^2 \mathbf{v}^\lambda \xi\|_{p_\infty} \leq C(\mathbf{F}, \mathbf{E}, \nabla\xi, p_0), \qquad (24)$$
$$\|\nabla \mathbf{v}^\lambda \xi\|_r^{p_\infty} \leq C(\mathbf{F}, \mathbf{E}, \nabla\xi, p_0), \qquad (25)$$
$$\int_\Omega \Big(\frac{1 + |\mathbf{D}(\mathbf{v}^\lambda)|^2}{1 + \lambda|\mathbf{D}(\mathbf{v}^\lambda)|^2}\Big)^{\frac{p-2}{2}} |\mathbf{D}(\nabla \mathbf{v}^\lambda)|^2 \xi^2\, dx \leq C(\mathbf{F}, \mathbf{E}, \nabla\xi, p_0), \qquad (26)$$

where ξ is a cut-off function and the constants are independent of λ.

(iii) Uniform local estimates in $W^{2,2-\epsilon}$:
We write $2-\epsilon = 2s^{-1}$ with $s = s(\epsilon) > 1$. With the estimates (25) and (26) we get

$$\begin{aligned}
&\int_\Omega |\nabla^2 \mathbf{v}^\lambda|^{2-\epsilon} \xi \, dx \le C \int_\Omega |\mathbf{D}(\nabla \mathbf{v}^\lambda)|^{2s^{-1}} \xi \, dx \\
&\le C\Big(\int_\Omega \Big(\frac{1+|\mathbf{D}(\mathbf{v}^\lambda)|^2}{1+\lambda |\mathbf{D}(\mathbf{v}^\lambda)|^2} \Big)^{\frac{p_\infty - 2}{2}} |\mathbf{D}(\nabla \mathbf{v}^\lambda)|^2 \xi \, dx \Big)^{\frac{1}{s}} \times \\
&\qquad \times \Big(\int_\Omega (1+|\mathbf{D}(\mathbf{v}^\lambda)|^2)^{\frac{2-p_\infty}{2} \frac{1}{s-1}} \, dx \Big)^{\frac{s-1}{s}} \\
&\le C.
\end{aligned}$$

Thus

$$\begin{aligned}
&\mathbf{v}^\lambda \rightharpoonup \mathbf{v} \text{ weak in } W^{2,2-\epsilon}(\tilde{\Omega}) \cap W^{1,p_\infty}(\Omega), \\
&\mathbf{v}^\lambda \to \mathbf{v} \text{ strong in } \bigcap_{r \ge 1} W^{1,r}(\tilde{\Omega}) \cap L^{p_\infty^*}(\Omega)
\end{aligned} \tag{27}$$

for $\tilde{\Omega} \subset\subset \Omega$, $r \ge 1$ and $0 < \epsilon \le 1$ arbitrary.
With Vitali's convergence theorem one can show that $\mathbf{v} \in E_p$ and $\mathbf{v}$ is a solution of (5).

We want to remark that the existence of the pressure ϕ can be deduced with De Rham's theorem (see for instance [2]). Further one can show the uniqueness of the solution for small data $\mathbf{f}, \mathbf{E}$ with the same arguments as in [9] and with estimates similar to the estimates used here.

A The Case $p \equiv 2$

Here we consider $p \equiv 2$ as in the case of the Navier-Stokes equations.

Proposition A.1 *Assume* $\mathbf{S}^\lambda \in C^1(\mathbb{R}^{2\times 2}_{sym} \times \mathbb{R}^2; \mathbb{R}^{2\times 2}_{sym})$ *satisfies*

$$\begin{aligned}
\partial_{kl} S^\lambda_{ij}(\mathbf{D}, \mathbf{E}) B_{ij} B_{kl} &\ge c_\lambda |\mathbf{B}|^2, \\
\mathbf{S}^\lambda(\mathbf{D}, \mathbf{E}) : \mathbf{D} &\ge c_\lambda |\mathbf{D}|^2, \\
|\partial_n \mathbf{S}^\lambda(\mathbf{D}, \mathbf{E})| &\le C_\lambda (1+|\mathbf{D}|), \\
|\partial_{kl} \mathbf{S}^\lambda(\mathbf{D}, \mathbf{E})| &\le C_\lambda
\end{aligned}$$

for all $\mathbf{B}, \mathbf{D} \in X$ *and* $\mathbf{E} \in \mathbb{R}^2$.
Then a weak solution $\mathbf{v}^\lambda \in V_2$ *of* (5) *with* $\mathbf{S}$ *replaced by* $\mathbf{S}^\lambda$ *exists for* $\mathbf{F} \in W^{-1,2}(\Omega)$. *This solution* $\mathbf{v}^\lambda$ *satisfies the following estimate*

$$\|\mathbf{v}\|_{W^{1,2}(\Omega)} \le C \|\mathbf{F}\|_{W^{-1,2}(\Omega)}. \tag{28}$$

Proof. This proposition is a special case of what is proved in [9] chapter three. On the other hand one can prove it directly, namely with the help of Brézis' theory of pseudomonotone operators (compare [6] or [11] and [12]) we get $\mathbf{v}^\lambda \in V_2$. Notice that the perturbation of a monotone operator with a compact operator is pseudomonotone.

Proposition A.2 *Let* $\mathbf{F} \in L^2(\Omega)$ *and* $\mathbf{E} \in C^1(\overline{\Omega})$. *Then* $\mathbf{v}^\lambda \in W^{2,2}_{loc}(\Omega) \cap V_2$.

Proof. This is a standard difference quotient technique. It is done more generally in [9].

B Auxiliary Results

The next statement is true in every dimension.

Lemma B.3 *Let* $u \in W^{2,2}(\Omega, \mathbb{R}^2)$ *and* $q < 2$. *Then the following estimate holds:*

$$\|\nabla^2 \mathbf{u}\xi^\alpha\|_q^2 \le \|(1+|\mathbf{D}(\mathbf{u})|^2)^{\frac{1}{2}}\|_q^{2-q} \int_\Omega (1+|\mathbf{D}(\mathbf{u})|^2)^{\frac{q-2}{2}} |\nabla^2 \mathbf{u}|^2 \xi^{2\alpha}\, dx.$$

Proof. This follows by Hölder's inequality.

Next we will use that we are in the planar case.

Lemma B.4 *Let* $\mathbf{u} \in W^{2,2}(\Omega, \mathbb{R}^2)$, *then for arbitrary* $r \ge 1$, $\alpha > 0$ *and* $q \ge 0$ *holds*

$$\|\nabla \mathbf{u}\xi^{\alpha\frac{2}{q}}\|_r^{\frac{q}{2}} \le C\left(1 + \int_\Omega \left(\frac{1+|\mathbf{D}(\mathbf{u})|^2}{1+\lambda|\mathbf{D}(\mathbf{u})|^2}\right)^{\frac{q-2}{2}} |\mathbf{D}(\nabla \mathbf{u})|^2 \xi^{2\alpha}\, dx + \|\nabla \mathbf{u}\|_q^{\frac{q}{2}}\right).$$

Proof. The case $q \ge 2$ is trivial. Assume therefore $q < 2$. Then the proof is based on the facts that

$$\frac{\partial}{\partial x_i}(1+|\mathbf{D}(\mathbf{u})|^2)^{\frac{q}{4}} = \frac{q}{2}(1+|\mathbf{D}(\mathbf{u})|^2)^{\frac{q-4}{4}} \mathbf{D}(\frac{\partial \mathbf{u}}{\partial x_i})$$

and that with Sobolev's embedding theorem holds

$$\|(1+|\mathbf{D}(\mathbf{u})|^2)^{\frac{q}{4}}\xi^\alpha\|_r \le C\|\nabla\left((1+|\mathbf{D}(\mathbf{u})|^2)^{\frac{q}{4}}\xi^\alpha\right)\|_2.$$

Lemma B.5 *Let* $\mathbf{u}$ *be twice (weakly) differentiable, then* $|\mathbf{D}(\nabla \mathbf{u})|$ *and* $|\nabla^2 \mathbf{u}|$ *are equivalent, i.e. there exit constants* c, C *so that* $c|\mathbf{D}(\nabla \mathbf{u}(x))| \le |\nabla^2 \mathbf{u}(x)| \le C|\mathbf{D}(\nabla \mathbf{u}(x))|$ *for a.e.* x.

Proof. The statement holds because of the identity

$$\frac{\partial^2 u_i}{\partial x_k \partial x_l} = D_{ik}(\frac{\partial \mathbf{u}}{\partial x_l}) + D_{il}(\frac{\partial \mathbf{u}}{\partial x_k}) - D_{kl}(\frac{\partial \mathbf{u}}{\partial x_i}).$$

References

1. Frank Ettwein: Elektrorheologische Flüssigkeiten: Existenz starker Lösungen in zweidimensionalen Gebieten. Diplomarbeit, Universität Freiburg, 2002. In preparation.
2. Giovanni P. Galdi: An Introduction to the Mathematical Theory of the Navier-Stokes Equations: Linearised Steady Problems. Volume 1. Springer, Berlin, 1998
3. Petr Kaplický, Josef Málek, Jana Stará: Full regularity of weak solutions to a class of nonlinear fluids in two dimensions - stationary, periodic problem. Comment. Math. Univ. Carolinae, 38(4), 681–695, 1997
4. Petr Kaplický, Josef Málek, Jana Stará: $C^{1,\alpha}$- solutions to a class of nonlinear fluids in two dimensions- stationary Dirichlet problem. Zapiski Nauchnyh Seminarov POMI, 259, 122–144, 1999
5. Ondrej Kováčik and Jiří Rákosník: On spaces $L^{p(x)}$ and $W^{k,p(x)}$. Czechoslovak Math. J., 41(116)(4), 592–618, 1991
6. Jacques L. Lions: Quelques méthodes de résolution des problémes aux limites non linéaires. Springer, Paris, 1969
7. Josef Málek, Jindřich Nečas, Mirko Rokyta, Michael Růžička: Weak and measure-valued solutions to evolutionary PDEs. Chapman & Hall, London, 1996
8. K.R. Rajagopal and Michael Růžička: Mathematical modeling of electrorheological materials. Continuum Mechanics and Thermodynamics, 13, 59–78, 2001
9. Michael Růžička: Electrorheological fluids: modeling and mathematical theory. Vvolume 1748 of Lecture Notes in Mathematics, Springer-Verlag, Berlin, 2000
10. Günter Schwarz: Hodge Decomposition – A Method for Solving Boundary Value Problems. Volume 1607 of Lecture Notes in Mathematics. Springer, Berlin,Heidelberg, 1995
11. Eberhard Zeidler: Nonlinear Functional Analysis and its Applications. Volume 2B: Nonlinear Monotone Operators. Springer, New York, 1984
12. Eberhard Zeidler: Nonlinear Functional Analysis and its Applications. Volume 4: Applications to Mathematical Physics. Springer, New York, 1988

Spinodal Decomposition in the Presence of Elastic Interactions

Harald Garcke[1], Stanislaus Maier-Paape[2], and Ulrich Weikard[3]

[1] Naturwissenschaftliche Fakultät I - Mathematik, Universität Regensburg, D-93040 Regensburg, Germany, harald.garcke@mathematik.uni-regensburg.de
[2] Institut für Mathematik, RWTH Aachen, D-52062 Aachen, Germany maier@instmath.RWTH-Aachen.DE
[3] Institut für Mathematik, Universität Duisburg, Lotharstraße 63, D-47048 Duisburg, Germany and Institut für Angewandte Mathematik, Universität Bonn, Wegelerstraße 6, 53115 Bonn, Germany, wkd@iam.uni-bonn.de

Summary. Spinodal decomposition, i.e., the separation of a homogeneous mixture into different phases, can be modeled by the Cahn-Hilliard equation - a fourth order semilinear parabolic equation. If elastic stresses due to a lattice misfit become important, the Cahn-Hilliard equation has to be coupled to an elasticity system to take this into account.

It is the goal of this paper to understand how elastic effects influence the formation of patterns during spinodal decomposition and to analyze what kind of morphologies one has to expect. It is shown that with a probability close to one, the dynamics of randomly chosen initial data in the neighborhood of a uniform mixture will be dominated by an invariant manifold which is tangential to the most unstable eigenfunctions of the linearized operator. For example in the case of cubic anisotropy it is shown that the most unstable eigenfunctions reflect the cubic anisotropy and the anisotropy will influence the dynamics quite drastically.

1 Introduction

In this paper we consider spinodal decomposition in binary alloys in the case where elastic effects become important. It is well known that complex patterns may form in the early stages of spinodal decomposition. We are interested to understand the effect elastic interactions may have on the formation of patterns.

The typical scenario of spinodal decomposition is as follows. At temperatures above a certain critical temperature a uniform mixture of the two components in the alloy is stable. After a rapid quenching this state can become unstable and regions where one or the other of the two components dominate occur. It is the goal of this paper to understand how these regions form and to analyze what kind of morphologies one has to expect.

Let c_1, c_2 denote the concentrations of the two alloy components. Since $c_1 + c_2 = 1$ the variable $c := c_1 - c_2$ completely determines the concentrations. Deformations of the reference configuration are described by the displacement field u, i.e., a material point x in the reference configuration will be found at the point $x + u(t, x)$ at time t.

Since displacement gradients in phase separating systems are small, the theory we consider will be based on the linearized strain tensor

$$\mathcal{E}(u) = \frac{1}{2}(\nabla u + (\nabla u)^T)$$

where ∇ is the gradient with respect to space and $(\nabla u)^T$ is the transpose of ∇u.

In this paper we consider a generalization of the Cahn-Hilliard model taking elastic effects into account. The model is due to Larché and Cahn [13] and Onuki [18] using ideas introduced by Eshelby [3] and Khachaturyan [12]. The theory is based on a free energy of the form

$$E(c,u) = \int_{\Omega} \left[\frac{\varepsilon^2}{2}|\nabla c|^2 + \psi(c) + W(c, \mathcal{E}(u))\right] dx$$

where $\Omega \subset \mathbb{R}^d$ is the domain under consideration, $c : \Omega \to \mathbb{R}$ is the concentration difference and $u : \Omega \to \mathbb{R}^d$ is the displacement field. The function $\psi : \mathbb{R} \to \mathbb{R}$ is the free energy density and is assumed here to be a non-convex function of c, e.g. a double well potential of the form $\psi(c) = (c^2 - 1)^2$. Without the term W this type of free energy goes back to van der Waals [21] and was introduced in the theory of spinodal decomposition by Cahn and Hilliard [2]. The third term is the elastic energy density $W : \mathbb{R} \times \mathbb{R}^{d\times d} \to \mathbb{R}$ which is chosen to be

$$W(c,\mathcal{E}) = \frac{1}{2}(\mathcal{E} - \tilde{\mathcal{E}}(c)) : C[\mathcal{E} - \tilde{\mathcal{E}}(c)]$$

(see Eshelby [3], Khachaturyan [12], Larché and Cahn [13], Fratzl, Penrose and Lebowitz [9]). Here, $\tilde{\mathcal{E}}(c) \in \mathbb{R}^{d\times d}$ is the stress free strain at concentration c, C is a fourth-rank elasticity tensor and the :–product of two matrices $A = (A_{ij})_{i,j=1,\dots,d}$ and $B = (B_{ij})_{i,j=1,\dots,d}$ is given by

$$A : B = \sum_{i,j=1}^{d} A_{ij}B_{ij} \quad \text{and} \quad |A|^2 = A : A\,.$$

Our standing assumptions are:

Assumption 1.1 *The stress free strains depend linearly on the concentration (Vegard's law), i.e.,*

$$\tilde{\mathcal{E}}(c) = c \cdot \mathcal{E}^*$$

with a fixed symmetric matrix

$$\mathcal{E}^* \in \mathbb{R}^{d\times d}.$$

The domain Ω is a assumed to have rectangular shape, i.e.,

$$\Omega = [0,\ell_1] \times \cdots \times [0,\ell_d]$$

with $\ell_1, \dots, \ell_d > 0$. The elasticity tensor

$$C = (C_{ijmn})_{i,j,m,n=1,\dots,d}$$

is assumed to be positive definite and to fulfill the symmetry conditions of linear elasticity, i.e.,

$$\begin{aligned} C_{ijmn} &= C_{ijnm} = C_{jimn}, \\ C_{ijmn} &= C_{mnij} \end{aligned} \tag{1}$$

and there exists some $d_0 > 0$, such that

$$\mathcal{E} : C[\mathcal{E}] \geq d_0 |\mathcal{E}|^2 \tag{2}$$

for all symmetric $\mathcal{E} \in \mathbb{R}^{d\times d}$.

We note that the symmetry condition (1) gives for all matrices A and B

$$C[A] : B = A : C[B].$$

Let us give two typical examples for the elasticity tensor C. In the **isotropic case** (cf. [11]) we have

$$C_{ijmn} = \lambda\delta_{ij}\delta_{mn} + \mu(\delta_{im}\delta_{jn} + \delta_{in}\delta_{jm}) \tag{3}$$

where λ, μ are the Lamé constants and δ_{ij} is the Kronecker symbol. This means

$$\begin{aligned} C[\mathcal{E}] = \left(\sum_{m,n=1}^{d} C_{ijmn}\, \mathcal{E}_{mn} \right)_{i,j=1,\dots,d} &= \lambda(tr\mathcal{E}) \cdot Id + \mu(\mathcal{E} + \mathcal{E}^T) \\ &= \lambda(tr\mathcal{E}) \cdot Id + 2\mu\mathcal{E} \end{aligned}$$

for all symmetric $\mathcal{E} \in \mathbb{R}^{d\times d}$. For many systems it is more realistic to assume **cubic symmetry** (cf. [9], p. 168). In this case there are three degrees of freedom for the elasticity tensor. One usually introduces the notation

$$\begin{aligned} C_{iiii} &:= C_{11}\ i = 1, \dots, d, \\ C_{iijj} &:= C_{12}\ i \neq j, \\ C_{ijij} &:= C_{44}\ i \neq j\ , \end{aligned}$$

where C_{11}, C_{12}, and C_{44} are given constants. All other entries of C are then either given due to the cubic symmetry, i.e., either there is a symmetry with respect to the coordinate axis or they are set to be zero. Formally this means

$$Q\, C[\mathcal{E}]\, Q^T = C[Q\mathcal{E}Q^T]$$

for all orthogonal $Q \in \mathbb{R}^{d\times d}$ with $\det Q = 1$ which let the d–dimensional cube $[-1, 1]^d$ invariant. The elasticity tensor can than be written as

$$\begin{aligned} C_{ijmn} = &(C_{11} - C_{12} - 2C_{44})\delta_{ij}\delta_{jm}\delta_{mn} \\ &+ C_{12}\delta_{ij}\delta_{mn} \\ &+ C_{44}(\delta_{im}\delta_{jn} + \delta_{in}\delta_{jm}). \end{aligned} \tag{4}$$

The differential equation for c is given by (see [5, 6, 7, 9])

$$\partial_t c = \Delta w,$$

where w is the chemical potential difference defined as

$$w = \frac{\delta E}{\delta c} = -\varepsilon^2 \Delta c + \psi_{,c}(c) + W_{,c}(c, \mathcal{E}(u))$$

and $W_{,c}(c, \mathcal{E}(u)) = -\mathcal{E}^* : C[\mathcal{E}(u) - c\mathcal{E}^*]$.

The system is then completed by the assumption of quasi-static equilibrium for the mechanical part. This is justified since mechanical equilibrium is attained on a much faster time scale than mass diffusion takes place. Therefore, we obtain

$$0 = \frac{\delta E}{\delta u} = -\nabla \cdot W_{,\mathcal{E}}(c, \mathcal{E}(u)) = -\nabla \cdot S,$$

where

$$S = W_{,\mathcal{E}}(c, \mathcal{E}(u)) = C[\mathcal{E}(u) - c\mathcal{E}^*]$$

is the stress tensor and $\nabla\cdot$ is the divergence operator acting on rows. Altogether we obtain a system of a scalar and a vector-valued equation

$$\begin{aligned} \partial_t c &= \Delta(-\varepsilon^2 \Delta c + \psi_{,c}(c) - \mathcal{E}^* : S), \\ 0 &= \nabla \cdot S = \nabla \cdot C[\mathcal{E}(u) - c\mathcal{E}^*] \,. \end{aligned} \tag{5}$$

For definiteness, we assume periodic boundary conditions for c and u. The above equations imply mass conservation for c, i.e.

$$\int_\Omega c(t,x)\, dx =: c_m = \text{const.} \tag{6}$$

Therefore the new variable $v := c - c_m$ satisfies $\int_\Omega v dx = 0$. We set

$$f(c) := -\psi_{,c}(c)$$

and, after replacing v by c again, we arrive at

$$\partial_t c = (-\Delta)(\varepsilon^2 \Delta c + f(c_m + c) + \mathcal{E}^* : S), \tag{7}$$

$$0 = \nabla \cdot C[\mathcal{E}(u) - c\mathcal{E}^*] \,. \tag{8}$$

The elasticity equation (8) is linear in u and c. In Section 2 it is shown that for all previously described elasticity tensors C equation (8) can be solved using Fourier transformation (see also [12, 17]). We obtain some $u = u(c)$ for any given c with $\int_\Omega c\, dx = 0$.

Secondly, in Section 2 we compute the term $\mathcal{E}^* : S = \mathcal{E}^* : C[\mathcal{E}(u) - c\mathcal{E}^*]$ with $u = u(c)$ which enters the equation (7) for c. This will give an operator $\mathcal{L}$

$$\mathcal{L}: X \to X,\ c \mapsto \mathcal{E}^* : S, \text{ where } X := \left\{c \in L^2(\Omega) : \int_\Omega c dx = 0\right\}, \tag{9}$$

which is linear in c. With the help of $\mathcal{L}$ we can rewrite the equation for c as follows

$$\partial_t c = (-\Delta)(\varepsilon^2 \Delta c + f(c_m + c) + \mathcal{L}(c)) . \tag{10}$$

The understand the behavior of (10) for $c \approx 0$ we linearize at $c = 0$ to obtain

$$\partial_t c = (-\Delta)\left(\varepsilon^2 \Delta c + f'(c_m)\, c + \mathcal{L}(c)\right) . \tag{11}$$

The eigenfunctions of the right hand side in (11) are

$$\varphi_\kappa(x) = e^{\mathbf{i}\kappa \cdot x} ,$$

where $\mathbf{i} = \sqrt{-1}$ and

$$\begin{aligned} \kappa &= (\kappa_1, ..., \kappa_d) = 2\pi(\tfrac{\nu_1}{\ell_1}, \ldots, \tfrac{\nu_d}{\ell_d}), \\ \nu &= (\nu_1, \ldots, \nu_d) \in \mathbb{Z}^d . \end{aligned} \tag{12}$$

The associated eigenvalues are

$$\lambda_{\kappa,\varepsilon} = |\kappa|^2\Big(- \varepsilon^2 |\kappa|^2 + f'(c_m) + L(\kappa)\Big) , \tag{13}$$

where we use that in our applications the linear operator $\mathcal{L}$ can be written as

$$\mathcal{L}(\varphi_\kappa) = L(\kappa) \cdot \varphi_\kappa \tag{14}$$

with some 0–homogeneous function $L : \mathbb{R}^d \backslash \{0\} \to \mathbb{R}$. We set $L_{\max}$ and $L_{\min}$ to be the maximal and minimal value of L.

Now the equilibrium $c = 0$ of (10) is unstable, if the maximal eigenvalue in (13) is positive. Therefore we assume:

Assumption 1.2 *Let the mean value c_m of c (cf. (6)) and the maximum of L from (14) satisfy*

$$f'(c_m) + L_{\max} > 0 .$$

It is easy to see that then the eigenvalues are bounded by

$$\lambda_\varepsilon^{\max} := \frac{(f'(c_m) + L_{\max})^2}{4\varepsilon^2} .$$

In this situation, we can make an adaption of the theory of Maier–Paape and Wanner on spinodal decomposition for the Cahn–Hilliard equation (cf. [14] and [15]). We will outline this in Section 3. In conclusion, we find that the behavior of (10) near $c = 0$ is dominated by a finite dimensional strongly unstable subspace

$$\mathcal{Y}_\varepsilon^+ := \text{span}\,\{\varphi_\kappa : \lambda_{\kappa,\varepsilon} > \gamma_0 \cdot \lambda_\varepsilon^{\max}\}$$

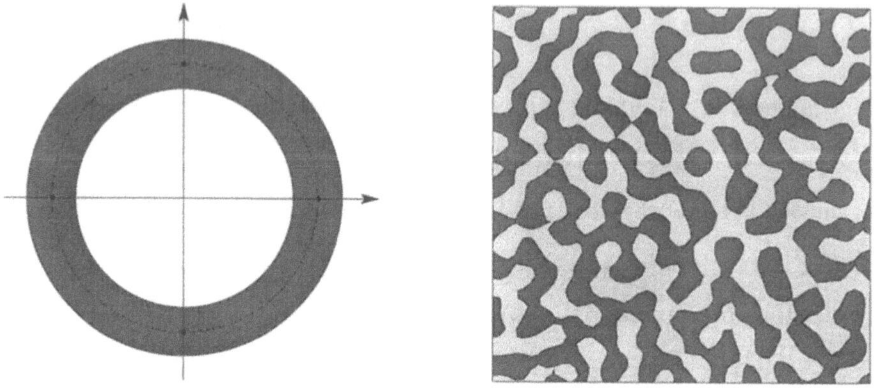

Fig. 1. Cahn–Hilliard situation

for some $\gamma_0 < 1$ close to one. It turns out that real parts and imaginary parts of this subspace are just a subset of the dominating subspace occurring for the Cahn–Hilliard equation. Therefore we inherit the small order $O(\varepsilon)$ wavelength estimate for the elements in $\mathcal{Y}_\varepsilon^+$ (cf. [14], Section 4). Figure 1 is a sketch of the situation for the Cahn–Hilliard equation ($\mathcal{L} = 0$) which we give for reference. To the left we have the Fourier vectors κ that are excited the most and to the right we have the nodal domains of a typical element of $\mathcal{Y}_\varepsilon^+$ for that case.
The next two situations show the dominant Fourier vectors and typical pattern for the case of cubic symmetry (4), $\mathcal{E}^* = q \cdot Id$ and Ω = square. We first have negative anisotropy, i.e. $\Delta C = C_{11} - C_{12} - 2C_{44} < 0$, and then in Figure 3 the situation for positive anisotropy, i.e. $\Delta C > 0$.
The last picture, Figure 4, shows the situation for isotropic symmetry (3) and $\mathcal{E}^*$ being not a multiple of the identity matrix (non-dilatational misfit).

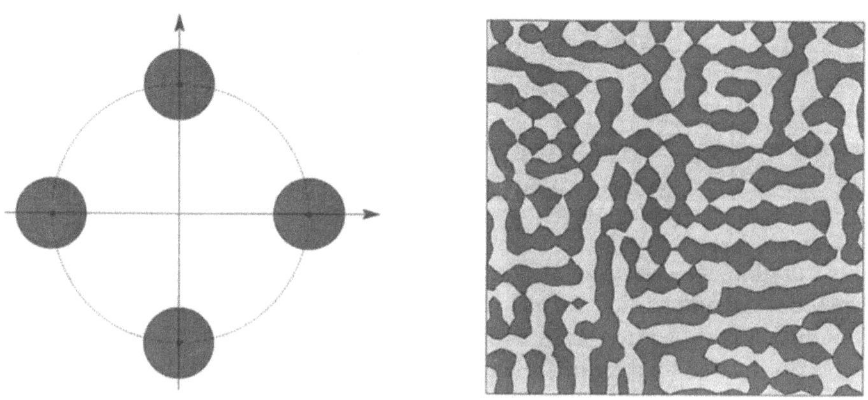

Fig. 2. Cubic symmetry with negative anisotropy

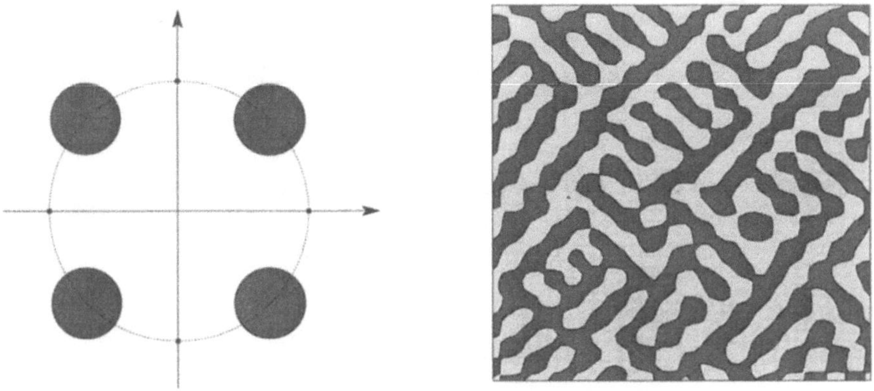

Fig. 3. Cubic symmetry with positive anisotropy

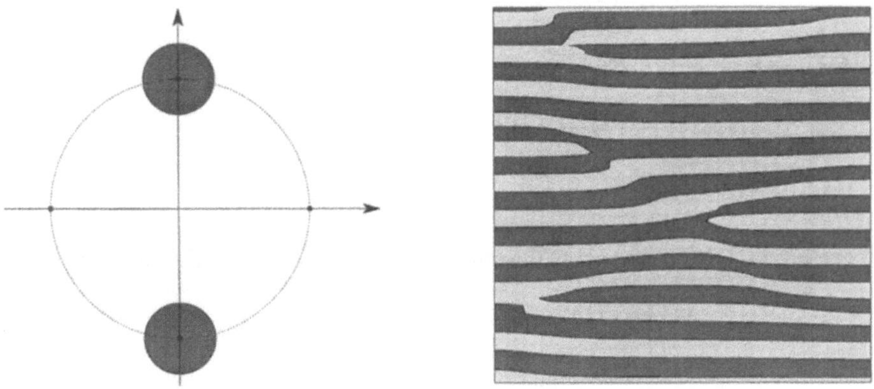

Fig. 4. Isotropic symmetry with anisotropic eigenstrains

In each of the Figures 2, 3 and 4 we see that certain directions of the nodal domains are selected.

Finally, in Section 4 we present results from numerical simulations of the elastically modified Cahn-Hilliard system which are in agreement with the theoretical predictions of Sections 2 and 3.

2 Solving the Elasticity System

We start this section by computing $\mathcal{L}(c) = \mathcal{E}^* : S$, where $S = C[\mathcal{E}(u) - c\mathcal{E}^*]$ and $u = u(c)$ is the unique solution of $\nabla \cdot S = 0$ for given c with $\int_\Omega c dx = 0$. Let us compute $\mathcal{L}$ for the Fourier mode

$$\varphi_\kappa(x) = e^{\mathrm{i}\kappa \cdot x}\,, \tag{15}$$

where $\mathbf{i} = \sqrt{-1}$ and

$$\kappa = (\kappa_1, ..., \kappa_d) = 2\pi(\tfrac{\nu_1}{\ell_1}, \dots, \tfrac{\nu_d}{\ell_d}),$$
$$\nu = (\nu_1, \dots, \nu_d) \in \mathbb{Z}^d \ .$$

For the solution u of (8) with $c = \varphi_\kappa$ as in (15) we make the ansatz

$$u(x) = (u_1(x), ..., u_d(x)) = \tilde{u} e^{\mathbf{i}\kappa \cdot x} \tag{16}$$

with $\tilde{u} \in \mathbb{R}^d$. We obtain

$$\partial_m u_n(x) = \tilde{u}_n \mathbf{i} \kappa_m e^{\mathbf{i}\kappa \cdot x}$$

and

$$\mathcal{E}(u) = \frac{1}{2}\mathbf{i}(\kappa \otimes \tilde{u} + \tilde{u} \otimes \kappa) e^{\mathbf{i}\kappa \cdot x}, \tag{17}$$

where $a \otimes b := ab^T \in \mathbb{R}^{d \times d}$ for all $a, b \in \mathbb{R}^d$. We calculate,

$$\begin{aligned} \nabla \cdot C[\varphi_\kappa \mathcal{E}^*] &= \nabla \cdot C[\mathcal{E}^* e^{\mathbf{i}\kappa \cdot x}] \\ &= (C[\mathcal{E}^*]) \mathbf{i} \kappa e^{\mathbf{i}\kappa \cdot x} \\ &= S^* \mathbf{i} \kappa e^{\mathbf{i}\kappa \cdot x} \in \mathbb{R}^d \ , \end{aligned}$$

where $S^* := C[\mathcal{E}^*]$, i.e., S^* is the stress induced by $\mathcal{E}^*$. In addition we compute

$$\begin{aligned} \nabla \cdot C[\mathcal{E}(u)] &= \nabla \cdot (C[\tfrac{1}{2}\mathbf{i}(\kappa \otimes \tilde{u} + \tilde{u} \otimes \kappa)] e^{\mathbf{i}\kappa \cdot x}) \\ &= -\tfrac{1}{2} e^{\mathbf{i}\kappa \cdot x} C[\kappa \otimes \tilde{u} + \tilde{u} \otimes \kappa] \kappa \ . \end{aligned}$$

The symmetries of the elasticity tensor $C_{ijmn} = C_{ijnm}$ (cf. (1)) can be used to obtain

$$C[\kappa \otimes \tilde{u}] = C[\tilde{u} \otimes \kappa]$$

and therefore

$$\nabla \cdot C[\mathcal{E}(u)] = -e^{\mathbf{i}\kappa \cdot x} C[\kappa \otimes \tilde{u}] \kappa \ .$$

Introducing the matrix

$$Z^{-1}(\kappa) = \left(\sum_{j,m=1}^{d} C_{ijmn} \kappa_j \kappa_m \right)_{i,n=1,\dots,d} , \tag{18}$$

we obtain

$$\nabla \cdot C[\mathcal{E}(u)] = -e^{\mathbf{i}\kappa \cdot x} Z^{-1}(\kappa) \tilde{u} \ .$$

The above used matrix $Z^{-1}(\kappa)$ is in fact the inverse of some matrix $Z(\kappa) \in \mathbb{R}^{d \times d}$. This follows from the next lemma.

Lemma 2.3 *The matrix $Z^{-1}(\kappa)$ is strictly positive definite for all $\kappa \in \mathbb{R}^d \setminus \{0\}$ and therefore invertible.*

Proof. Let $\zeta \in \mathbb{R}^d$. Then we obtain using the symmetry properties of C (1) and the fact that C is positive definite (2)

$$\begin{aligned}\zeta^T Z^{-1}(\kappa)\zeta &= \sum_{i,n=1}^{d} \zeta_i \sum_{j,m=1}^{d} C_{ijmn}\kappa_j\kappa_m\zeta_n \\ &= \sum_{i,j,m,n} C_{ijmn}\zeta_i\kappa_j\zeta_n\kappa_m \\ &= \sum_{i,j,m,n} C_{ijnm}\zeta_i\kappa_j\zeta_n\kappa_m \\ &\geq d_0|\zeta \otimes \kappa|^2 = d_0|\zeta|^2|\kappa|^2 .\end{aligned}$$

The above inequality shows that $Z^{-1}(\kappa)$ is positive definite whenever $\kappa \neq 0$. □

To obtain u we need to solve

$$\nabla \cdot C[\mathcal{E}(u)] = \nabla \cdot C[\varphi_\kappa \mathcal{E}^*] .$$

This is equivalent to

$$-Z^{-1}(\kappa)\tilde{u} = \mathbf{i}S^*\kappa .$$

Using that $Z^{-1}(\kappa)$ is invertible, we get

$$\tilde{u} = -\mathbf{i}Z(\kappa)S^*\kappa .$$

Hence from (16)

$$u(x) = -(\mathbf{i}Z(\kappa)S^*\kappa)e^{\mathbf{i}\kappa\cdot x} = -\mathbf{i}Z(\kappa)S^*\kappa\varphi_\kappa(x) . \tag{19}$$

Lemma 2.4 *For $\kappa \in \mathbb{R}^d \setminus \{0\}$ we obtain*

$$\mathcal{L}(\varphi_\kappa) = \mathcal{E}^* : (C[Z(\kappa)S^*\kappa\kappa^T] - S^*)\varphi_\kappa .$$

Proof. Using the formulas for u and $\tilde{u}$ above and (17) we compute

$$\begin{aligned}\mathcal{L}(\varphi_\kappa) = \mathcal{E}^* : S &= \mathcal{E}^* : C[\mathcal{E}(\tilde{u}e^{\mathbf{i}\kappa\cdot x}) - \varphi_\kappa\mathcal{E}^*] \\ &= \mathcal{E}^* : C[\tilde{u} \otimes \kappa\mathbf{i}e^{\mathbf{i}\kappa\cdot x} - \mathcal{E}^* e^{\mathbf{i}\kappa\cdot x}] \\ &= \mathcal{E}^* : C[Z(\kappa)S^*\kappa\kappa^T - \mathcal{E}^*]e^{\mathbf{i}\kappa\cdot x} \\ &= \mathcal{E}^* : (C[Z(\kappa)S^*\kappa\kappa^T] - S^*)e^{\mathbf{i}\kappa\cdot x}.\end{aligned}$$

This shows the lemma. □

Remark 2.5 *The operator $\mathcal{L}$ can be interpreted as a pseudo-differential operator of order* 0*. This follows from the fact that*

$$\mathcal{L}(\varphi_\kappa) = L(\kappa)\varphi_\kappa \tag{20}$$

with a function L. The function L is 0-homogeneous because $Z(\kappa)$ is (-2)-homogeneous which follows from (18) and Cramer's rule. With u taken from (19) and after a partial integration we obtain using (8)

$$0 \geq -\int_\Omega W(\varphi_\kappa, \mathcal{E}(u))\, dx = \frac{1}{2}\int_\Omega \varphi_\kappa \mathcal{E}^* : C[\mathcal{E}(u) - \varphi_\kappa \mathcal{E}^*] = \frac{1}{2} L(\kappa) \int_\Omega \varphi_\kappa^2 \,.$$

We conclude that in the elastically modified Cahn-Hilliard equation (10) the $\mathcal{L}$-term has a stabilizing effect.

2.1 The Isotropic Case

In this subsection we assume that

$$C[\mathcal{E}] = \lambda(tr\mathcal{E})Id + \mu(\mathcal{E} + \mathcal{E}^T)$$

with given λ, $\mu \in \mathbb{R}$ and for some $q \in \mathbb{R}$

$$\mathcal{E}^* = qId.$$

Lemma 2.6 *Under the above assumption on C and $\mathcal{E}^*$ we obtain*

$$\mathcal{L} = q^2(d\lambda + 2\mu)\frac{2\mu(1-d)}{2\mu + \lambda} \cdot Id.$$

Proof. Let $\kappa \in \mathbb{R}^d \setminus \{0\}$ be given. First we compute

$$\begin{aligned} Z^{-1}(\kappa) &= \lambda \left(\sum_{j,m=1}^{d} \delta_{ij}\delta_{mn}\kappa_j\kappa_m \right)_{i,n=1,\dots,d} \\ &\quad + \mu \left(\sum_{j,m=1}^{d} (\delta_{im}\delta_{jn} + \delta_{in}\delta_{jm})\kappa_j\kappa_m \right)_{i,n=1,\dots,d} \\ &= \lambda\kappa\kappa^T + \mu(\kappa\kappa^T + |\kappa|^2 Id) \\ &= (\lambda + \mu)\kappa\kappa^T + \mu|\kappa|^2 Id \,. \end{aligned}$$

We can easily invert $Z^{-1}(\kappa)$ to obtain $Z(\kappa)$ defined as follows

$$Z(\kappa)\kappa = \frac{1}{(2\mu + \lambda)} \frac{1}{|\kappa|^2}\kappa$$

and

$$Z(\kappa)e = \frac{1}{\mu|\kappa|^2} e \quad \text{for all } e \text{ with} \quad \kappa^T \cdot e = 0 \,.$$

Furthermore

$$S^* = C[\mathcal{E}^*] = qCId = q(d\lambda Id + 2\mu Id) = q(d\lambda + 2\mu)Id$$

and

$$Z(\kappa)S^*\kappa\kappa^T = \frac{q(d\lambda + 2\mu)}{(2\mu + \lambda)} \frac{1}{|\kappa|^2} \kappa\kappa^T .$$

To compute

$$\begin{aligned} \mathcal{E}^* : (C[Z(\kappa)S^*\kappa\kappa^T] - S^*) &= q\, Id : (C[Z(\kappa)S^*\kappa\kappa^T] - S^*) \\ &= q \cdot tr(C[Z(\kappa)S^*\kappa\kappa^T] - S^*) , \end{aligned}$$

we calculate

$$\begin{aligned} trC[\kappa\kappa^T] &= \sum_i \sum_{m,n} (\lambda\delta_{ii}\delta_{mn} + \mu(\delta_{im}\delta_{in} + \delta_{in}\delta_{im}))\kappa_m \cdot \kappa_n \\ &= (d\lambda + 2\mu)|\kappa|^2. \end{aligned}$$

This gives

$$\begin{aligned} \mathcal{E}^* : (C[Z(\kappa)S^*\kappa\kappa^T] - S^*) &= q \cdot tr(C[Z(\kappa)S^*\kappa\kappa^T] - S^*) \\ &= q^2(d\lambda + 2\mu)\left(\tfrac{d\lambda+2\mu}{(2\mu+\lambda)} - d\right) \\ &= q^2(d\lambda + 2\mu)\tfrac{2\mu(1-d)}{2\mu+\lambda} , \end{aligned}$$

which implies

$$\mathcal{L}(\varphi_\kappa) = q^2(d\lambda + 2\mu)\, \frac{2\mu(1-d)}{2\mu + \lambda}\varphi_\kappa$$

for all $\kappa \neq 0$. Therefore the claim is proved. □

Remark 2.7 *We have shown that in the isotropic case, the operator $\mathcal{L}$ is a multiple of the identity. C as in (3) is positive definite if and only if*

$$\mu > 0 \quad \text{and} \quad 2\mu + d\lambda > 0.$$

Then in particular $2\mu + \lambda > 0$. This implies that $\mathcal{L}$ is a negative multiple of the identity for $q \neq 0$ and $d > 1$. For the evolution equation we observe again that the operator $\mathcal{L}$ has a stabilizing effect when compared to the case without elasticity.

2.2 A Cubic Elasticity Tensor

In this subsection we consider the case $\mathcal{E}^* = q \cdot Id$ and we assume cubic symmetry for the elasticity tensor C (cf. (4)), i.e.,

$$\begin{aligned} C_{ijmn} = &(C_{11} - C_{12} - 2C_{44})\delta_{ij}\delta_{jm}\delta_{mn} \\ &+C_{12}\delta_{ij}\delta_{mn} \\ &+C_{44}(\delta_{im}\delta_{jn} + \delta_{in}\delta_{jm}). \end{aligned}$$

We remark that C is positive definite if and only if $C_{44} > 0$ and

$$\begin{aligned} &C_{11} > (d-1)|C_{12}| \,, &&\text{in case } C_{12} < 0 \\ &C_{11} > C_{12} \,, &&\text{in case } C_{12} \geq 0 \,. \end{aligned}$$

The following lemma is easy to verify.

Lemma 2.8 *The matrix* $Z^{-1}(\kappa)$ *has the entries*

$$\begin{aligned} (Z^{-1}(\kappa))_{ii} &= (C_{11} - C_{44})\kappa_i^2 + C_{44}|\kappa|^2, \quad & i = 1, \dots, d\,, \\ (Z^{-1}(\kappa))_{in} &= (C_{12} + C_{44})\kappa_i \kappa_n & i \neq n\,. \end{aligned}$$

Lemma 2.9 *With cubic symmetry for* C *and* $\mathcal{E}^* = q \cdot Id$ *we obtain*

$$\mathcal{L}(\varphi_\kappa) = q^2(C_{11} + (d-1)C_{12}) \left[(C_{11} + (d-1)C_{12}) \left(\sum_{m,p=1}^{d} Z_{mp}(\kappa)\kappa_p\kappa_m \right) - d \right] \varphi_\kappa\,.$$

Proof. According to Lemma 2.4 we have to compute

$$\mathcal{L}(\varphi_\kappa) = q \cdot tr(C[Z(\kappa)S^*\kappa\kappa^T] - S^*)\varphi_\kappa\,.$$

Under the assumptions on C and $\mathcal{E}^*$ we calculate

$$\begin{aligned} (S^*)_{ij} &= (C[\mathcal{E}^*])_{ij} = q(C[Id])_{ij} \\ &= q \left(\sum_{m=1}^{d} C_{ijmm} \right)_{ij} \\ &= q((C_{11} - C_{12} - 2C_{44} + dC_{12} + 2C_{44})\delta_{ij})_{ij}. \end{aligned}$$

This implies

$$S^* = q(C_{11} + (d-1)C_{12})Id$$

and

$$trS^* = q\, d(C_{11} + (d-1)C_{12}).$$

Furthermore, we have

$$\begin{aligned} & tr(C[Z(\kappa)S^*\kappa\kappa^T]) = \\ &= q(C_{11} + (d-1)C_{12}) tr(C[Z(\kappa)\kappa\kappa^T]) \\ &= q(C_{11} + (d-1)C_{12}) \sum_{i,m,n} C_{iimn} \sum_p Z_{mp}(\kappa)\kappa_p\kappa_n \\ &= q(C_{11} + (d-1)C_{12}) \left(\sum_{i,m,p} C_{iimm} Z_{mp}(\kappa)\kappa_p\kappa_m \right) \\ &= q(C_{11} + (d-1)C_{12}) \left(\sum_i (C_{11} - C_{12}) \sum_p Z_{ip}(\kappa)\kappa_p\kappa_i + C_{12} d \sum_{m,p} Z_{mp}(\kappa)\kappa_p\kappa_m \right) \\ &= q(C_{11} + (d-1)C_{12})^2 \sum_{m,p} Z_{mp}(\kappa)\kappa_p\kappa_m. \end{aligned}$$

□

For simplicity we only give the formulas for $\mathcal{L}$ in the physically interesting cases $d = 2, 3$.

Lemma 2.10 *(d=2) In two space dimensions we obtain*

$$\mathcal{L}(\varphi_\kappa) = -\frac{q^2 C_{44}(C_{11}^2 - C_{12}^2)|\kappa|^4}{C_{44}C_{11}|\kappa|^4 + (C_{11}+C_{12})(C_{11}-C_{12}-2C_{44})\kappa_1^2\kappa_2^2}\varphi_\kappa .$$

Proof. For $d = 2$ we obtain the inverse of $Z^{-1}(\kappa)$ as

$$Z(\kappa) = \frac{1}{\det Z^{-1}(\kappa)} \begin{pmatrix} |\kappa|^2 C_{44} + (C_{11} - C_{44})\kappa_2^2 & -(C_{12}+C_{44})\kappa_1\kappa_2 \\ -(C_{12}+C_{44})\kappa_1\kappa_2 & |\kappa|^2 C_{44} + (C_{11}-C_{44})\kappa_1^2 \end{pmatrix}$$

with

$$\det Z^{-1}(\kappa) = C_{11}C_{44}|\kappa|^4 + [(C_{11}-C_{44})^2 - (C_{12}+C_{44})^2]\kappa_1^2\kappa_2^2 .$$

Hence,

$$\sum_{m,p=1}^{2} Z_{mp}(\kappa)\kappa_p\kappa_m = \frac{|\kappa|^4 C_{44} + 2(C_{11}-C_{12}-2C_{44})\kappa_1^2\kappa_2^2}{\det Z^{-1}(\kappa)} .$$

Altogether, we obtain

$$\mathcal{L}(\varphi_\kappa) = q^2(C_{11}+C_{12}) \left[\frac{(C_{11}+C_{12}) \cdot [|\kappa|^4 C_{44} + 2(C_{11}-C_{12}-2C_{44})\kappa_1^2\kappa_2^2]}{|\kappa|^4 C_{44}C_{11} + (C_{11}+C_{12})(C_{11}-C_{12}-2C_{44})\kappa_1^2\kappa_2^2} - 2 \right] \varphi_\kappa$$

$$= -\frac{q^2 C_{44}(C_{11}^2 - C_{12}^2)|\kappa|^4}{C_{44}C_{11}|\kappa|^4 + (C_{11}+C_{12})(C_{11}-C_{12}-2C_{44})\kappa_1^2\kappa_2^2}\varphi_\kappa .$$

□

Remark 2.11 *In the situation of Lemma 2.10 the 0–homogeneous function L is given by*

$$L(\kappa) = -\frac{q^2 C_{44}(C_{11}^2 - C_{12}^2)|\kappa|^4}{C_{44}C_{11}|\kappa|^4 + (C_{11}+C_{12})(C_{11}-C_{12}-2C_{44})\kappa_1^2\kappa_2^2} .$$

Let us compute now for which directions $\kappa \in \mathbb{R}^d \setminus \{0\}$ $L(\kappa)$ becomes maximal. First of all we note that the fact that C is positive definite implies

$$C_{11} > |C_{12}| \quad \textit{and} \quad C_{44} > 0 .$$

Hence we obtain:

i) In the case of positive anisotropy, i.e., $\Delta C := C_{11} - C_{12} - 2C_{44} > 0$:

$$L(\kappa) \quad \textit{is maximal if} \quad \kappa_1^2 = \kappa_2^2 .$$

ii) In the case of negative anisotropy, i.e., $\Delta C < 0$:

$$L(\kappa) \quad \textit{is maximal if either } \kappa_1 \textit{ or } \kappa_2 \textit{ is equal to zero.}$$

For the evolution problem this will imply that elastic interactions will amplify wave numbers κ lying in segments around the coordinate axes stronger than other segments in the case that the anisotropy is negative. For positive anisotropy instead the diagonal directions will be stronger amplified.
In the case of three space dimensions we obtain:

Lemma 2.12 *For cubic elasticity and $\mathcal{E}^* = qId$ we obtain in three space dimensions for all $\varphi_\kappa(x) = e^{i\kappa\cdot x}$ with $\kappa \in \mathbb{R}^3 \setminus \{0\}$*

$$\mathcal{L}(\varphi_\kappa) = q^2(C_{11}+2C_{12})\Big[\tfrac{(C_{11}+2C_{12})}{det\, Z^{-1}(\kappa)}\Big(C_{44}^2|\kappa|^6 + 2C_{44}(C_{11}-C_{12}-2C_{44})|\kappa|^2$$
$$(\kappa_1^2\kappa_2^2+\kappa_2^2\kappa_3^2+\kappa_3^2\kappa_1^2) + 3(C_{11}-C_{12}-2C_{44})^2\kappa_1^2\kappa_2^2\kappa_3^2\Big) - 3\Big]\varphi_\kappa$$

where

$$det\, Z^{-1}(\kappa) = C_{44}^2C_{11}|\kappa|^6 + C_{44}(C_{11}-C_{12}-2C_{44})(C_{12}+C_{11})|\kappa|^2$$
$$(\kappa_1^2\kappa_2^2+\kappa_2^2\kappa_3^2+\kappa_3^2\kappa_1^2) + (C_{11}-C_{12}-2C_{44})^2$$
$$(2C_{12}+C_{44}+C_{11})\kappa_1^2\kappa_2^2\kappa_3^2.$$

Proof. A straightforward computation using Lemma 2.8 shows

$$Z_{ii}(\kappa) = \tfrac{1}{\det Z^{-1}(\kappa)}\,(C_{44}^2|\kappa|^4 + (C_{11}-C_{44})C_{44}|\kappa|^2(\kappa_j^2+\kappa_\ell^2)$$
$$+(C_{11}-C_{12}-2C_{44})(C_{12}+C_{11})\kappa_j^2\kappa_\ell^2)$$

for $i = 1,2,3$ with j and ℓ such that $j \neq i$, $j \neq \ell$, $\ell \neq i$ and

$$Z_{ij}(\kappa) = \frac{-1}{\det Z^{-1}(\kappa)}(C_{12}+C_{44})\kappa_i\kappa_j(C_{44}|\kappa|^2 + (C_{11}-C_{12}-2C_{44})\kappa_\ell^2)$$

for $i,j = 1,2,3$, $i \neq j$ and ℓ such that $\ell \neq j$ and $\ell \neq i$. We also have

$$\det Z^{-1}(\kappa) = C_{44}^2C_{11}|\kappa|^6 + C_{44}(C_{11}-C_{12}-2C_{44})(C_{12}+C_{11})|\kappa|^2$$
$$(\kappa_1^2\kappa_2^2+\kappa_2^2\kappa_3^2+\kappa_3^2\kappa_1^2)+$$
$$+(C_{11}-C_{12}-2C_{44})^2(2C_{12}+C_{44}+C_{11})\kappa_1^2\kappa_2^2\kappa_3^2\,.$$

Having computed $Z(\kappa)$ we obtain

$$\sum_{m,p} Z_{mp}(\kappa)\kappa_p\kappa_m = \tfrac{1}{\det Z^{-1}(\kappa)}\Big[C_{44}^2|\kappa|^6$$
$$+2|\kappa|^2C_{44}(C_{11}-C_{44})(\kappa_1^2\kappa_2^2+\kappa_2^2\kappa_3^2+\kappa_3^2\kappa_1^2)$$
$$+3(C_{11}-C_{12}-2C_{44})(C_{12}+C_{11})\kappa_1^2\kappa_2^2\kappa_3^2$$
$$-2|\kappa|^2C_{44}(C_{12}+C_{44})(\kappa_1^2\kappa_2^2+\kappa_2^2\kappa_3^2+\kappa_3^2\kappa_1^2)$$
$$-6(C_{12}+C_{44})(C_{11}-C_{12}-2C_{44})\kappa_1^2\kappa_2^2\kappa_3^2\Big]$$
$$= \tfrac{1}{\det Z^{-1}(\kappa)}\Big[C_{44}^2|\kappa|^6 + 2C_{44}(C_{11}-C_{12}-2C_{44})|\kappa|^2(\kappa_1^2\kappa_2^2+\kappa_2^2\kappa_3^2+\kappa_3^2\kappa_1^2)$$
$$+3(C_{11}-C_{12}-2C_{44})^2\kappa_1^2\kappa_2^2\kappa_3^2\Big]$$

Now Lemma 2.9 gives the assertion. □

We now discuss the result of Lemma 2.12. As in (20) we can write $\mathcal{L}$ with the help of a 0–homogeneous function L. We want to determine directions for which L becomes maximal.

We obtain the following result.

Lemma 2.13 *In the situation of Lemma 2.12 with $\mathcal{L}(\varphi_\kappa) = L(\kappa) \cdot \varphi_\kappa$ we find for L: from zero we have:*

Case A**:** *Positive anisotropy, i.e., $\Delta C > 0$. Here L is maximal in directions given by the eight points*

$$\Big(\pm\frac{1}{\sqrt{3}}, \pm\frac{1}{\sqrt{3}}, \pm\frac{1}{\sqrt{3}}\Big). \tag{21}$$

Case B**:** *Negative anisotropy, i.e., $\Delta C < 0$. Here L is maximal in directions given by the six points*

$$(\pm 1, 0, 0)\,, \;(0, \pm 1, 0), \quad \textit{and} \quad (0, 0, \pm 1)\,. \tag{22}$$

Proof. In both cases we start with the identity

$$\begin{aligned} &\tfrac{C_{11}}{q^2(C_{11}+2C_{12})^2} L(\kappa) + 2\tfrac{C_{11}-C_{12}}{C_{11}+2C_{12}} = \tfrac{\Delta C}{\det Z^{-1}(\kappa)}\Big[C_{44}\,(C_{11} - C_{12})\; |\kappa|^2 \\ &\textstyle\sum_{i,j,i<j} \kappa_i^2\kappa_j^2 + \Delta C(2C_{11} - 2C_{12} - C_{44})\kappa_1^2\kappa_2^2\kappa_3^2\Big]\,. \end{aligned} \tag{23}$$

For negative anisotropy we use this to find

$$\begin{aligned} &\tfrac{C_{11}}{q^2(C_{11}+2C_{12})^2} L(\kappa) + 2\tfrac{C_{11}-C_{12}}{C_{11}+2C_{12}} = \tfrac{\Delta C}{\det Z^{-1}(\kappa)}\Big[C_{44}\,(C_{11} - C_{12}) \textstyle\sum_{i,j,\,i\neq j} \kappa_i^4\kappa_j^2 + \\ &2\left((C_{11} - C_{12} - C_{44})^2 + C_{44}\,(C_{11} - C_{12})\right)\kappa_1^2\kappa_2^2\kappa_3^2\Big] \end{aligned}$$

and conclude that L is maximal in the directions corresponding to the coordinate axes. For κ on the coordinate axes the right hand side of the above identity is zero whereas for all other directions the right hand side is strictly negative. Here we use the fact that C is positive definite which yields $C_{44} > 0$, $C_{11} - C_{12} > 0$, $C_{11} > 0$, $C_{11} + 2C_{12} > 0$ and $det Z^{-1}(\kappa) > 0$.

To discuss the case of positive anisotropy, i.e., $\Delta C > 0$, we consider

$$M(\kappa) = \frac{\alpha}{\Delta C}\left(\frac{C_{11}}{q^2(C_{11} + 2C_{12})^2} L(\kappa) + 2\frac{C_{11} - C_{12}}{C_{11} + 2C_{12}}\right) - \beta\,,$$

where

$$\begin{aligned} \beta &= 9C_{44}(C_{11} - C_{12}) + \gamma_1, \\ \alpha &= 27C_{44}^2 C_{11} + 9\gamma_2 + \gamma_3, \\ \gamma_1 &= (\Delta C)\,(2C_{11} - 2C_{12} - C_{44})\,, \\ \gamma_2 &= (\Delta C)C_{44}(C_{11} + C_{12}), \\ \gamma_3 &= (\Delta C)^2\,(2C_{12} + C_{44} + C_{11}) \end{aligned}$$

which are all positive constants. Using (23) we get

$$M(\kappa) = \frac{1}{\det Z^{-1}(\kappa)} \Big[-(9a+b)|\kappa|^6 + (27a-c)|\kappa|^2 \sum_{i,j,i<j} \kappa_i^2 \kappa_j^2 + (27b+9c)\kappa_1^2\kappa_2^2\kappa_3^2 \Big] , \tag{24}$$

with

$$\begin{aligned} a &= C_{44}(C_{11}-C_{12})C_{44}^2 C_{11}, \\ b &= \gamma_1 C_{44}^2 C_{11}, \\ c &= \gamma_1\gamma_2 - \gamma_3\, C_{44}(C_{11}-C_{12}) = (\Delta C)^3 C_{44} C_{11} . \end{aligned}$$

Again the verification of the above identities needs some patience but is straightforward. We remark that $a, b, c > 0$.

The above relation between L and M and the following statement proves the lemma.

Claim: $M(\kappa) \le 0$ and $M(\kappa) = 0$ if and only if $\kappa_1^2 = \kappa_2^2 = \kappa_3^2$.

We have:

$$\begin{aligned} &\left(\det Z^{-1}(\kappa)\right) M(\kappa) \\ &\quad = 9a \left(3|\kappa|^2 \textstyle\sum_{i,j,i<j} \kappa_i^2\kappa_j^2 - |\kappa|^6\right) \\ &\quad + b\left(27\kappa_1^2\kappa_2^2\kappa_3^2 - |\kappa|^6\right) \\ &\quad + c\left(9\kappa_1^2\kappa_2^2\kappa_3^2 - |\kappa|^2 \textstyle\sum_{i,j,i<j} \kappa_i^2\kappa_j^2\right) . \end{aligned}$$

To show the claim we are going to show that the term on the right hand side in the above identity is not positive for κ with $|\kappa| = 1$. For the first two terms it is easy to verify that the maximum on S^2 is attained if and only if $\kappa_1^2 = \kappa_2^2 = \kappa_3^2$ and in this case the terms are zero. The last term is a bit more difficult to handle. W.l.o.g. we assume $\kappa_1^2 \le \kappa_2^2 \le \kappa_3^2$. Then we obtain

$$\begin{aligned} &9\kappa_1^2\kappa_2^2\kappa_3^2 - |\kappa|^2 \textstyle\sum_{i,j,i<j} \kappa_i^2\kappa_j^2 \\ &\quad = \kappa_1^2\kappa_2^2(2\kappa_3^2 - \kappa_1^2 - \kappa_2^2) + \kappa_2^2\kappa_3^2(2\kappa_1^2 - \kappa_2^2 - \kappa_3^2) + \kappa_1^2\kappa_3^2(2\kappa_2^2 - \kappa_1^2 - \kappa_3^2) \\ &\quad \le \kappa_1^2\kappa_3^2(2\kappa_3^2 - \kappa_1^2 - \kappa_2^2) + \kappa_1^2\kappa_3^2(2\kappa_1^2 - \kappa_2^2 - \kappa_3^2) + \kappa_1^2\kappa_3^2(2\kappa_2^2 - \kappa_1^2 - \kappa_3^2) \\ &\quad = 0 \end{aligned}$$

and this proves the lemma. □

2.3 Anisotropic Eigenstrains

In this section we consider the case that the eigenstrain $\mathcal{E}^*$ is not a multiple of the identity matrix. We restrict ourselves to isotropic elasticity in two dimensions and eigenstrains of the form

$$\mathcal{E}^* = \begin{pmatrix} a & 0 \\ 0 & b \end{pmatrix}, \quad a, b \in \mathbb{R} \quad \text{with } a \cdot b > 0 .$$

Lemma 2.14 *For an isotropic elasticity tensor C and $\mathcal{E}^*$ as above we obtain for all $\kappa \in \mathbb{R}^2 \setminus \{0\}$*

$$\mathcal{L}(\varphi_\kappa) = L(\kappa)\varphi_\kappa$$

with

$$\begin{aligned} L(\kappa) = {} & \tfrac{1}{\mu(\lambda+2\mu)|\kappa|^4}\Big[\mu e^2\kappa_1^4 + \mu f^2\kappa_2^4 \\ & + \left[(\lambda+\mu)(e-f)^2 + \mu(e^2+f^2)\right]\kappa_1^2\kappa_2^2\Big] \\ & - \Big[\lambda(a+b)^2 + 2\mu(a^2+b^2)\Big] \end{aligned}$$

where $e = \lambda(a+b) + 2\mu a$ *and* $f = \lambda(a+b) + 2\mu b$.

Proof. For all $\kappa \neq 0$ we have to compute

$$L(\kappa) = \mathcal{E}^* : (C[Z(\kappa)S^*\kappa\kappa^T] - S^*).$$

Using the symmetry of C and the definition $S^* = C[\mathcal{E}^*]$ we obtain

$$L(\kappa) = S^* : (Z(\kappa)S^*\kappa\kappa^T - \mathcal{E}^*).$$

Furthermore, we have

$$\begin{aligned} S^* = C[\mathcal{E}^*] &= \lambda tr\mathcal{E}^* Id + 2\mu\mathcal{E}^* \\ &= \lambda(a+b)Id + 2\mu\begin{pmatrix} a & 0 \\ 0 & b \end{pmatrix} \\ &= \begin{pmatrix} \lambda(a+b)+2\mu a & 0 \\ 0 & \lambda(a+b)+2\mu b \end{pmatrix}. \end{aligned}$$

We have (cf. the proof of Lemma 2.10 with $C_{12} = \lambda$, $C_{44} = \mu$ and $C_{11} = \lambda + 2\mu$)

$$Z(\kappa) = \frac{1}{\det Z^{-1}(\kappa)}\begin{pmatrix} \mu|\kappa|^2 + (\lambda+\mu)\kappa_2^2 & -(\lambda+\mu)\kappa_1\kappa_2 \\ -(\lambda+\mu)\kappa_1\kappa_2 & \mu|\kappa|^2 + (\lambda+\mu)\kappa_1^2 \end{pmatrix}$$

with

$$\det Z^{-1}(\kappa) = \mu(\lambda+2\mu)|\kappa|^4.$$

Therefore,

$$S^*Z(\kappa)S^* =$$
$$\begin{pmatrix} (\lambda(a+b)+2\mu a)^2 Z_{11}(\kappa) & (\lambda(a+b)+2\mu a)(\lambda(a+b)+2\mu b)Z_{21}(\kappa) \\ (\lambda(a+b)+2\mu a)(\lambda(a+b)+2\mu b)Z_{12}(\kappa) & (\lambda(a+b)+2\mu b)^2 Z_{22}(\kappa) \end{pmatrix}.$$

Altogether we obtain

$$\begin{aligned}
&tr(S^*Z(\kappa)S^*\kappa\kappa^T)\\
&= \tfrac{1}{\mu(\lambda+2\mu)|\kappa|^4}\Big[(\lambda(a+b)+2\mu a)^2(|\kappa|^2\mu+(\lambda+\mu)\kappa_2^2)\kappa_1^2\\
&\quad -2(\lambda(a+b)+2\mu a)(\lambda(a+b)+2\mu b)(\lambda+\mu)\kappa_1^2\kappa_2^2\\
&\quad +(\lambda(a+b)+2\mu b)^2(|\kappa|^2\mu+(\lambda+\mu)\kappa_1^2)\kappa_2^2\Big]\\
&= \tfrac{1}{\mu(\lambda+2\mu)|\kappa|^4}\Big[\mu(\lambda(a+b)+2\mu a)^2\kappa_1^4+\mu(\lambda(a+b)+2\mu b)^2\kappa_2^4\\
&\quad +\big[(\lambda+2\mu)(\lambda(a+b)+2\mu a)^2-2(\lambda+\mu)(\lambda(a+b)+2\mu a)\\
&\quad (\lambda(a+b)+2\mu b)+(\lambda(a+b)+2\mu b)^2(\lambda+2\mu)\big]\,\kappa_1^2\kappa_2^2\Big]\\
&= \tfrac{1}{\mu(\lambda+2\mu)|\kappa|^4}\Big[\mu(\lambda(a+b)+2\mu a)^2\kappa_1^4+\mu(\lambda(a+b)+2\mu b)^2\kappa_2^4\\
&\quad +\big[(\lambda+\mu)4\mu^2(a-b)^2+\mu((\lambda(a+b)+2\mu a)^2+(\lambda(a+b)+2\mu b)^2)\big]\,\kappa_1^2\kappa_2^2\Big]\,.
\end{aligned}$$

Also we compute

$$\begin{aligned}
S^* : \mathcal{E}^* &= (\lambda(a+b)+2\mu a)a+(\lambda(a+b)+2\mu b)b\\
&= \lambda(a+b)^2+2\mu(a^2+b^2)\,.
\end{aligned}$$

This proves the claim. □

Remark 2.15 *We want to determine the maximum of L on S^1. Setting $z=\kappa_1^2$ (i.e. $1-z=\kappa_2^2$) an analysis of the function*

$$g(z)=\mu e^2z^2+\mu f^2(1-z)^2+\big[(\lambda+\mu)(e-f)^2+\mu(e^2+f^2)\big]\,z(1-z)$$

shows that g can attain its maximum only for $z=0,1$. This means that L attains its maximum on the coordinate axes. Here one obtains that the maximum is attained on the x_1-axis if and only if $a^2>b^2$ and on the x_2-axis if and only if $b^2>a^2$.

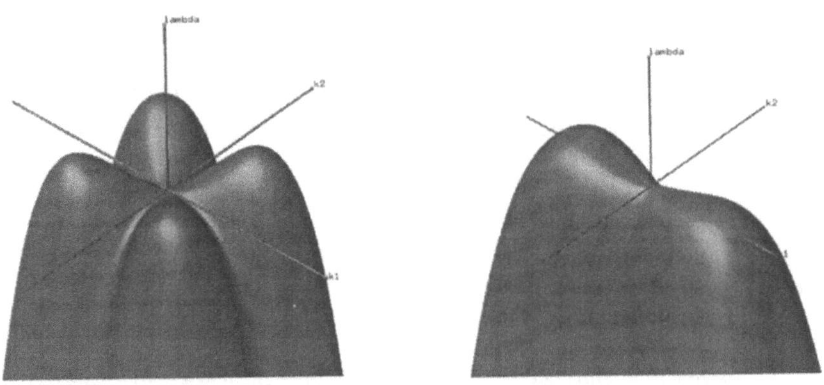

Fig. 5. $\lambda_{\kappa,\epsilon}$ in the two dimensional case as a function of κ_1 and κ_2; on the left: with positive anisotropy of the elasticity tensor, on the right: with anisotropic eigenstrains.

The wave numbers κ for which L attains its maximum are amplified most strongly by elastic interactions. In this case only wave numbers lying on one particular coordinate axis are most strongly amplified.

Remark 2.16 *In Figure 5 we show the graph of $\lambda_{\kappa,\varepsilon}$ in the two dimensional case as a function of κ_1 and κ_2 given by equation (13). On the left hand we consider the case of an elasticity tensor with positive anisotropy (cf. Lemma 2.10) whereas on the right the elasticity tensor is isotropic but the eigenstrain is not (cf. Lemma 2.14).*

3 Spinodal Decomposition and Elastic Interactions

In this section we will apply the abstract results of Section 2 of [15] to the Cahn-Hilliard equation with elasticity. This application is very much along the lines of the application of this theory given in Section 3 of [15] and Section 2 of [16], where the binary and multicomponent Cahn–Hilliard model were discussed. We consider the equation 10, i.e.,

$$\begin{aligned} \partial_t c &= (-\Delta)(\varepsilon^2 \Delta c + f(c_m + c) + \mathcal{L}(c)) \text{ in } \Omega\,, \\ &c \text{ is periodic}, \\ &\int_\Omega c dx = 0\,, \end{aligned} \tag{25}$$

where $\varepsilon > 0$ is a small parameter. We specify our assumptions as follows.

(A1) On top of our standing Assumption 1.1 we let Ω be of rectangular shape in $\mathbb{R}^d$, where $d \in \{1,2,3\}$.
(A2) Let $f : \mathbb{R} \to \mathbb{R}$ be a smooth function.

In this situation we defined the linear operator $\mathcal{L} : X \to X$ in Section 1 (cf. (9)), which, as we saw in Section 2, always has a representation on the eigenfunctions $\varphi_\kappa = e^{i\kappa x}$ as $\mathcal{L}(\varphi_\kappa) = L(\kappa)\, \varphi_\kappa$, with some 0–homogeneous function $L : \mathbb{R}^d \setminus \{0\} \to [L_{\min}, L_{\max}] \subset \mathbb{R}$ (cf. (14)).

For later reference let us also introduce the following notation.

Definition 3.17 *We denote all κ related to eigenvalues of $-\Delta$ by*

$$\Lambda := \{\kappa \in \mathbb{R}^d \setminus \{0\} : \kappa \textit{ is of the form (12)}\}$$

and $P\Lambda = \{\theta \in S^{d-1} : \kappa/|\kappa| = \theta \textit{ for some } \kappa \in \Lambda\}$. Furthermore, for given $\theta \in P\Lambda$ let

$$\Lambda(\theta) = \{\kappa \in \Lambda : \kappa/|\kappa| = \theta\} \subset \mathbb{R}^d \setminus \{0\}$$

be all κ corresponding to eigenvalues of $-\Delta$ lying on a half ray. We set for $N \in \mathbb{N}$

$$P\Lambda_N := \left\{\theta \in P\Lambda : \exists\, \kappa = 2\pi \left(\frac{\nu_1}{\ell_1}, \ldots, \frac{\nu_d}{\ell_d}\right) \in \Lambda(\theta)\,,\, \nu \in \mathbb{Z}^d \setminus \{0\}\,,\, |\nu| < N\right\}.$$

Our last assumption is:

(A3) (a) Let c_m and L satisfy Assumption 1.2 to make the equilibrium c_m unstable.

(b) Furthermore, we assume the existence of some $N_0 \in \mathbb{N}$ such that the maximum of L, $L_{\max}$, is assumed for some $\theta \in P\Lambda_{N_0}$. Any such $\theta \in P\Lambda$ satisfying $L(\theta) = L_{\max}$ is called $\theta_{\max}$ and $\theta_{\max}$ lies always in $P\Lambda_{N_0}$.

The part (b) of Assumption (A3) is not really necessary, but it simplifies the proofs. It is satisfied in all our applications, but in general not. To rewrite (25) we set

$$\tilde{f}(c) := f(c_m + c) - f'(c_m)c - f(c_m) , \tag{26}$$

so that $\tilde{f}(0) = \tilde{f}'(0) = 0$. Furthermore, let

$$\begin{aligned} A_\varepsilon c &:= (-\Delta)(\varepsilon^2 \Delta c + f'(c_m)c) \,, \; B_\varepsilon c := A_\varepsilon c + (-\Delta)\, \mathcal{L}(c) \,, \\ \text{and} \qquad F(c) &:= (-\Delta)\tilde{f}(c) \,. \end{aligned} \tag{27}$$

Then formally the first equation in (25) is of the form

$$\partial_t c = B_\varepsilon c + F(c) \,. \tag{28}$$

The evolution equation (28) is of the same form as the one used in the binary Cahn–Hilliard case considered in [15]. Thus, we can use the abstract theory developed there to prove the dominance properties mentioned in the introduction. For this we only have to verify hypotheses (H1) through (H3) in [15]. This will be done in the following subsections. Basically, we have to verify the following three claims.

(H1) The operator $-B_\varepsilon$ is sectorial in the Hilbert space X.

(H2) There exists a decomposition $X = X^{--} \oplus X^{-} \oplus X^{+} \oplus X^{++}$ into pairwise orthogonal subspaces, such that all subspaces are finite-dimensional except X^{--}, and such that the linear semigroup corresponding to $\partial_t c = B_\varepsilon c$ satisfies several dichotomy estimates, see Lemma 3.24(b) below.

(H3) The nonlinear mapping $F : X^\alpha \to X$ is C^1 with $F(0) = 0$ and $DF(0) = 0$. Furthermore, it satisfies a global Lipschitz condition with constant L_F, i.e., for all $c, \tilde{c} \in X^\alpha$ we have

$$\|F(c) - F(\tilde{c})\|_X \le L_F \|c - \tilde{c}\|_{X^\alpha} \,.$$

Here X^α denotes the fractional power space corresponding to B_ε and $\alpha \in (0, 1)$.

Hypothesis (H3) was already developed in [15], Section 3.3, because the nonlinearity in (25) is exactly the same as for the Cahn–Hilliard equation. In the course of verifying the hypotheses (H1) and (H2), we also calculate several constants introduced in [15], which in turn furnish an upper bound on the Lipschitz constant L_F. (Notice that since the nonlinearity in our example (see (27)) does not satisfy a global Lipschitz condition, we have to employ a standard cut-off technique.) This will eventually determine the size of the neighborhood on which our results are valid.

3.1 Spectral Properties

In the following lemma we collect several properties of the linear operator B_ε which will be needed later. These results are the obvious generalization of Lemma 3.1 in [15], where the operator A_ε with Neumann type boundary conditions was discussed. We will again use the eigenfunctions $\varphi_\kappa = e^{i\kappa x}$ of $-\Delta$ and their eigenvalues $|\kappa|^2$.

Lemma 3.18 *Assume that (A1), (A2), and (A3) are satisfied, and let X be defined as in (9). Define the operator $B_\varepsilon : X \to X$ by*

$$B_\varepsilon c = (-\Delta)\left(\varepsilon^2 \Delta c + f'(c_m)c + \mathcal{L}(c)\right) ,$$

with domain

$$D(B_\varepsilon) = \left\{c \in X \cap H^4(\Omega) : \ c \ \textit{is periodic}\right\} .$$

Then the following assertions hold.

(a) The spectrum of the operator $-\Delta : X \to X$ with domain

$$D(-\Delta) = \{c \in X \cap H^2(\Omega) : \ c \ \textit{is periodic}\}$$

consists of all $|\kappa|^2$, where $\kappa \in \mathbb{R}^d \setminus \{0\}$ satisfies (12). The corresponding normalized eigenfunctions $\tilde{\varphi}_\kappa = \varphi_\kappa / \|\varphi_\kappa\|$ form a complete $L^2(\Omega)$-orthonormal set in X. Furthermore, if $N_d(\mu)$ denotes the number of eigenvalues less than $\mu \in \mathbb{R}$ (counting multiplicities), then we have that

$$N_d(\mu) \sim \mu^{d/2} \ \textit{as} \ \mu \to \infty , \tag{29}$$

where the proportionality constant depends only on Ω.

(b) The operator $-B_\varepsilon$ defined above is selfadjoint and sectorial. The spectrum of B_ε consists of real eigenvalues $\lambda_{\kappa,\varepsilon}$ given in (13) with corresponding eigenfunctions φ_κ. Moreover, the largest eigenvalue is for small $\varepsilon > 0$ of the order

$$\lambda_\varepsilon^{\max} := \frac{\left(f'(c_m) + L_{\max}\right)^2}{4\varepsilon^2} , \quad \textit{and bounded by} \ \ \lambda_\varepsilon^{\max} . \tag{30}$$

Due to the above lemma, B_ε generates an analytic semigroup $S_\varepsilon(t)$ on X. Furthermore, for every B_ε the fractional power space $X^{1/2,\varepsilon} \subset X$ is defined; cf. the discussion in Subsection 2.1 of [15] following (H1). Although formally this fractional power space depends on ε, it will turn out in the next lemma that *algebraically*, i.e., as a vector space, we have $X^{1/2,\varepsilon} = H^2_{av}(\Omega)$, where

$$H^2_{av}(\Omega) := \left\{ c \in H^2(\Omega) : \int_\Omega c\,dx = 0 \quad \text{and} \quad c \ \text{is periodic} \right\} .$$

Only the *topological* structure given by the norm $\|\cdot\|_{1/2,\varepsilon}$ on $X^{1/2,\varepsilon}$ will depend on the parameter ε. Fortunately, the norms $\|\cdot\|_{1/2,\varepsilon}$ will turn out to be equivalent to the standard $H^2(\Omega)$-norm on $H^2_{av}(\Omega)$. More precisely, we have the following result.

Lemma 3.19 *Assume that (A1) and (A3) are satisfied, and let* $||\cdot||_{1/2,\varepsilon}$ *denote the norm on* $X^{1/2,\varepsilon}$ *defined by*

$$||c||_{1/2,\varepsilon} := \left|\left|(-B_\varepsilon + b_\varepsilon I)^{1/2} c\right|\right| \quad \textit{for all} \quad c \in X^{1/2,\varepsilon} ,$$

where $b_\varepsilon = \left(f'(c_m) + L_{\max}\right)^2 / \varepsilon^2$ *and* $||\cdot|| = ||\cdot||_{L^2(\Omega)}$. *Moreover, let* $||\cdot||_{H^2(\Omega)}$ *denote the standard* $H^2(\Omega)$*-norm, and define another norm* $||\cdot||_*$ *on* $H^2(\Omega)$ *by*

$$||c||_* := \sqrt{||c||^2 + ||\Delta c||^2} \quad \textit{for all} \quad c \in H^2(\Omega) .$$

For $\varepsilon^2 > 0$ *we assume* $\varepsilon^2 \leq f'(c_m) + L_{\max}$ *and, in case* $f'(c_m) + L_{\min} < 0$, *we assume additionally* $\varepsilon^2 < |f'(c_m) + L_{\min}|$. *Then we have*

$$X^{1/2,\varepsilon} = H^2_{av}(\Omega) ,$$

and the norm $||\cdot||_{1/2,\varepsilon}$ *is equivalent to both* $||\cdot||_{H^2(\Omega)}$ *and* $||\cdot||_*$. *More precisely, there exists a (*ε*-independent) constant* C *depending only on the domain* Ω *such that for all* $c \in H^2_{av}(\Omega)$ *the estimates*

$$\frac{\varepsilon}{\sqrt{2}} \cdot ||c||_* \leq ||c||_{1/2,\varepsilon} \leq \frac{\hat{C}}{\varepsilon} \cdot ||c||_* \tag{31}$$

and

$$\frac{\varepsilon}{C \cdot \sqrt{2}} \cdot ||c||_{H^2(\Omega)} \leq ||c||_{1/2,\varepsilon} \leq \frac{\hat{C} \cdot C}{\varepsilon} \cdot ||c||_{H^2(\Omega)}$$

hold. Here we set

$$\hat{C} := \begin{cases} f'(c_m) + L_{\max} , & \textit{in case} \;\; f'(c_m) + L_{\min} \geq 0 \\ \sqrt{\left(f'(c_m) + L_{\max}\right)^2 + \frac{1}{2}\left(f'(c_m) + L_{\min}\right)^2} , & \textit{otherwise} . \end{cases}$$

Proof. As already pointed out in Lemma 3.2 in [15] we have that the norm $||\cdot||_*$ is an equivalent norm on $H^2(\Omega)$, i.e., there is a constant C such that for all $c \in H^2(\Omega)$ the estimate

$$\frac{1}{C} \cdot ||c||_{H^2(\Omega)} \leq ||c||_* \leq C \cdot ||c||_{H^2(\Omega)} \tag{32}$$

holds, cf. Temam [20, p. 150] and Fife, Kielhöfer, Maier-Paape, Wanner [4, Lemma 6.1] for the necessary elliptic regularity results.

Therefore we only have to verify (31). According to Amann [1, Theorem 4.6.7] the fractional power space $X^{1/2,\varepsilon}$ consists of all functions $c \in X$ whose $L^2(\Omega)$-Fourier coefficients $\xi_\kappa = (c, \tilde{\varphi}_\kappa)$, $\kappa \in \Lambda$, satisfy

$$\sum_{\kappa \in \Lambda} (b_\varepsilon - \lambda_{\kappa,\varepsilon}) \xi_\kappa^2 ||\tilde{\varphi}_\kappa||^2 < \infty .$$

In that case, the norm $||c||_{1/2,\varepsilon}$ is given by

$$||c||_{1/2,\varepsilon} = \left(\sum_{\kappa \in \Lambda} (b_\varepsilon - \lambda_{\kappa,\varepsilon}) \xi_\kappa^2 ||\tilde{\varphi}_\kappa||^2 \right)^{1/2} .$$

On the other hand a function $c \in X$ is in $H^2_{av}(\Omega)$ if and only if

$$||c||_*^2 = \sum_{\kappa \in \Lambda} \left(1 + |\kappa|^4\right) \xi_\kappa^2 ||\tilde{\varphi}_\kappa||^2 < \infty . \tag{33}$$

These characterizations eventually furnish the stated results. Using the representation (13) of $\lambda_{\kappa,\varepsilon}$ and $L(\kappa) \in [L_{\min}, L_{\max}]$ we obtain

$$|\kappa|^2 \left(-\varepsilon^2 |\kappa|^2 + f'(c_m) + L_{\min}\right) \leq \lambda_{\kappa,\varepsilon} \leq |\kappa|^2 \left(-\varepsilon^2 |\kappa|^2 + f'(c_m) + L_{\max}\right) .$$

An easy calculation gives for all $s \geq 0$ and $0 < \varepsilon^2 < f'(c_m) + L_{\max}$

$$\frac{\varepsilon^2}{2} \left(1 + s^2\right) \leq b_\varepsilon + \varepsilon^2 s^2 - \left(f'(c_m) + L_{\max}\right) s$$

yielding the left hand side of (31). For the right hand side of (31) we observe for $s \geq 0$ and ε as assumed

$$b_\varepsilon + \varepsilon^2 s^2 - \left(f'(c_m) + L_{\min}\right) s \leq \frac{\hat{C}^2}{\varepsilon^2} \left(1 + s^2\right) .$$

These estimates immediately imply the assertions of the lemma. □

Since according to the above lemma the fractional power space $X^{1/2,\varepsilon}$ algebraically does not depend on ε, we will omit the superscript ε in the following and simply write $X^{1/2} = H^2_{av}(\Omega)$. Also, Lemma 3.19 shows that we may choose either the standard $H^2(\Omega)$-norm or the norm $|| \cdot ||_*$ on the fractional power space $X^{1/2}$, since both of them are equivalent to the graph norm $|| \cdot ||_{1/2,\varepsilon}$ of $(-B_\varepsilon + b_\varepsilon I)^{1/2}$. It will turn out to be very convenient to choose the norm $|| \cdot ||_*$ as a (non-standard) norm on $X^{1/2}$. Thus, from now on we identify the space $(X^{1/2}, || \cdot ||_{1/2,\varepsilon})$ with the space $(H^2_{av}(\Omega), || \cdot ||_*)$.

3.2 Spectral Gaps and Exponential Dichotomy Estimates

The following lemma proves the existence of suitable spectral gaps in the spectrum of B_ε. They will be used to define the decomposition of X mentioned in (H2), and therefore eventually furnish the dichotomy estimates, see Lemma 3.24(b). The size of these gaps turns out to be crucial, because it provides a restriction on the possible size of the global Lipschitz constant of the nonlinearity in (H3).

Lemma 3.20 *Assume that (A1) and (A3) are satisfied, so that in particular we have $d \in \{1,2,3\}$. Furthermore, fix two constants $v_* < v^* < 1$ and let $\lambda_\varepsilon^{\max}$ be defined as in (30).*

Then there exist constants $\varepsilon_0, \mu_0 > 0$ depending only on v_*, v^*, N_0, Ω, and $f'(c_m) + L_{\max}$ such that for arbitrary $0 < \varepsilon \leq \varepsilon_0$ the following holds. The linear operator B_ε has eigenvalues $\lambda_*(\varepsilon)$ and $\lambda^*(\varepsilon)$ satisfying both

$$v_* \cdot \lambda_\varepsilon^{\max} \leq \lambda_*(\varepsilon) < \lambda^*(\varepsilon) \leq v^* \cdot \lambda_\varepsilon^{\max}$$

and

$$\lambda^*(\varepsilon) - \lambda_*(\varepsilon) \geq \mu_0 \cdot \varepsilon^{d-2} .$$

Moreover, the whole interval $(\lambda_*(\varepsilon), \lambda^*(\varepsilon))$ is part of the resolvent set of B_ε. In fact, both the interval $(v_* \cdot \lambda_\varepsilon^{\max}, \lambda_*(\varepsilon))$ and $(\lambda^*(\varepsilon), v^* \cdot \lambda_\varepsilon^{\max})$ contains eigenvalues of the form $\lambda_{\kappa,\varepsilon}$ if $\theta := \frac{\kappa}{|\kappa|} \in PA_{N_0}$ and if

$$\lambda_{\varepsilon,\theta}^{\max} := \max\{s(f'(c_m) + L(\theta) - \varepsilon^2 s) \; : \; s \geq 0\} \geq v^* \cdot \lambda_\varepsilon^{\max} . \tag{34}$$

In particular, in the above intervals are always eigenvalues $\lambda_{\kappa,\varepsilon}$ with $\kappa \in \Lambda(\theta_{\max})$.

Proof. Fix two constants $v_* < v_{**} < v^{**} < v^*$. Due to Lemma 3.18, especially (29), we can choose $\varepsilon_0 > 0$ small enough such that for all $0 < \varepsilon \leq \varepsilon_0$ the following two assertions hold.

1. Both in the interval $[v^{**}, v^*] \cdot \lambda_\varepsilon^{\max}$ and in $[v_*, v_{**}] \cdot \lambda_\varepsilon^{\max}$ there is at least one eigenvalue of B_ε. Let $\lambda^{**}(\varepsilon)$ denote the smallest eigenvalue in the first interval, and $\lambda_{**}(\varepsilon)$ the largest one in the second interval.
2. The number of eigenvalues of B_ε in the interval $(v_{**}, v^{**}) \cdot \lambda_\varepsilon^{\max}$ is bounded above by $C \cdot \varepsilon^{-d} - 1$, where C depends only on v_*, v^*, Ω, and $f'(c_m) + L_{\max}$.

Let $\mu_0 := (f'(c_m) + L_{\max})^2 (v^{**} - v_{**})/(4C)$ and assume that any two consecutive eigenvalues of the operator B_ε in the interval $[\lambda_{**}(\varepsilon), \lambda^{**}(\varepsilon)]$ are strictly less than $\mu_0 \cdot \varepsilon^{d-2}$ apart. Then we get

$$\lambda^{**}(\varepsilon) - \lambda_{**}(\varepsilon) < \mu_0 \cdot \varepsilon^{d-2} \cdot C \cdot \varepsilon^{-d} = \frac{(f'(c_m) + L_{\max})^2 (v^{**} - v_{**})}{4\varepsilon^2} .$$

This however contradicts the fact that

$$\lambda^{**}(\varepsilon) - \lambda_{**}(\varepsilon) \geq (v^{**} - v_{**}) \cdot \lambda_\varepsilon^{\max} = \frac{(f'(c_m) + L_{\max})^2 (v^{**} - v_{**})}{4\varepsilon^2} .$$

The remaining is easy and therefore left to the reader. □

Loosely speaking, the above lemma states that it is possible to find gaps in the spectrum of B_ε whose size can be controlled as $\varepsilon \to 0$. We will use this fact to obtain the spectral gaps needed for assumption (H2). To this end, choose constants

$$\underline{v}^{--} < \overline{v}^{--} \ll 0 \ll \underline{v}^{-} < \overline{v}^{-} < \underline{v}^{+} < \overline{v}^{+} < 1\,, \tag{35}$$

where typically the differences $\overline{v}^{--} - \underline{v}^{--}$, $\overline{v}^{-} - \underline{v}^{-}$, and $\overline{v}^{+} - \underline{v}^{+}$ will be small. Using these constants the following results are an immediate consequence of Lemma 3.20. They give the appropriate choices of the spectral gaps for the operators B_ε needed for our application.

Corollary 3.21 *Assume that all the assumptions of Lemma 3.20 are satisfied. Then with the constants from (35) there exist intervals*

$$\begin{aligned} \hat{J}_\varepsilon^{--} &:= [\hat{a}_\varepsilon^{--}, b_\varepsilon^{--}] \subset [\underline{v}^{--}, \overline{v}^{--}] \cdot \lambda_\varepsilon^{\max}\,, \\ J_\varepsilon^{-} &:= [a_\varepsilon^{-}, b_\varepsilon^{-}] \subset [\underline{v}^{-}, \overline{v}^{-}] \cdot \lambda_\varepsilon^{\max}\,, \\ J_\varepsilon^{+} &:= [a_\varepsilon^{+}, b_\varepsilon^{+}] \subset [\underline{v}^{+}, \overline{v}^{+}] \cdot \lambda_\varepsilon^{\max} \end{aligned}$$

such that for sufficiently small $\varepsilon > 0$ the following holds.

(a) Each of the intervals $\hat{J}_\varepsilon^{--}$, J_ε^{-}, J_ε^{+} is contained in the resolvent set of B_ε.
(b) If we define $a_\varepsilon^{--} := (\hat{a}_\varepsilon^{--} + b_\varepsilon^{--})/2$ and let

$$J_\varepsilon^{--} := [a_\varepsilon^{--}, b_\varepsilon^{--}] \subset \hat{J}_\varepsilon^{--} \subset [\underline{v}^{--}, \overline{v}^{--}] \cdot \lambda_\varepsilon^{\max}\,,$$

then there is an ε-independent constant $\mu_1 > 0$ such that the length of each of the intervals J_ε^{--}, J_ε^{-}, and J_ε^{+} is at least $\mu_1 \cdot \varepsilon^{d-2}$. The constant μ_1 depends only on Ω, $f'(c_m) + L_{\max}$, and the constants in (35).
(c) The interval $[\underline{v}^{--} \cdot \lambda_\varepsilon^{\max}, a_\varepsilon^{--})$ is not contained in the resolvent set of B_ε. In fact this interval contains at least one eigenvalue $\lambda_{\kappa,\varepsilon}$ of B_ε with $\kappa \in \Lambda(\theta_{\max})$.

Using this result, we can now define the subspace decomposition of X needed for applying the results of [15].

Definition 3.22 *Using the constants introduced in Corollary 3.21, define the intervals $I_\varepsilon^{--} := (-\infty, a_\varepsilon^{--})$, $I_\varepsilon^{-} := (b_\varepsilon^{--}, a_\varepsilon^{-})$, $I_\varepsilon^{+} := (b_\varepsilon^{-}, a_\varepsilon^{+})$, and $I_\varepsilon^{++} := (b_\varepsilon^{+}, \lambda_\varepsilon^{\max}]$. Furthermore, let X_ε^{--}, X_ε^{-}, X_ε^{+}, and X_ε^{++} denote the sum of all eigenspaces of the operator B_ε corresponding to eigenvalues $\lambda_{\kappa,\varepsilon}$ in I_ε^{--}, I_ε^{-}, I_ε^{+}, and I_ε^{++}, respectively.*

For further reference we define

Definition 3.23 *For given $\theta \in P\Lambda$, $\varepsilon > 0$, we denote by $\lambda_{i,\theta,\varepsilon}$, $i \in \mathbb{N}$ the ordered eigenvalues $\lambda_{\kappa,\varepsilon}$ of B_ε with κ lying in $\Lambda(\theta)$. Hence we have $\lambda_{1,\theta,\varepsilon} \geq \lambda_{2,\theta,\varepsilon} \geq \ldots \to -\infty$. The corresponding normalized eigenfunctions $\psi_{i,\theta,\varepsilon}, i \in \mathbb{N}$ of B_ε are obtained form the eigenfunctions φ_κ through this ordering procedure in the obvious way. Furthermore, denote by $\tilde{\kappa}_{i,\theta,\varepsilon} \in \Lambda(\theta), i \in \mathbb{N}$ the related wave vector κ of $\psi_{i,\theta,\varepsilon}$. Note that $|\tilde{\kappa}_{i,\theta,\varepsilon}|$ is monotonically increasing for increasing i in the region $|\tilde{\kappa}_{i,\theta,\varepsilon}|^2 > (f'(c_m) + L(\theta))/\varepsilon^2$.*

The restrictions of B_ε or of the corresponding (linear) analytic semigroup $S_\varepsilon(t)$ to each of the subspaces defined above will be denoted by the appropriate superscript. With these definitions we can proceed to verifying the two hypotheses (H1) and (H2) for the linearization. Again we use the non-standard norm $||\cdot||_*$ (which is equivalent to the norm $||\cdot||_{1/2,\varepsilon}$) on $X^{1/2} = H^2_{av}(\Omega)$.

Lemma 3.24 *Assume that (A1) and (A3) hold. Let* $-B_\varepsilon : X \to X$ *denote the self-adjoint and sectorial operator defined in Lemma 3.18, let* $S_\varepsilon(t) : X \to X$, $t \geq 0$, *denote the corresponding analytic semigroup, and let* $X^{1/2} = H^2_{av}(\Omega)$ *denote the fractional power space of Subsection 3.1 with norm* $\|\cdot\|_*$. *Furthermore, consider the constants and intervals introduced in Corollary 3.21 and Definition 3.22. Then the following assertions hold for arbitrary* $0 < \varepsilon \leq \varepsilon_0$, *where* ε_0 *depends only on the domain* Ω, N_0, *the constant* $f'(c_m) + L_{\max}$, *and the constants in (35).*

(a) The subspaces X_ε^-, X_ε^+, *and* X_ε^{++} *are finite-dimensional subspaces of* $X^{1/2}$, *and their dimensions are proportional to* ε^{-d}, *where d denotes the dimension of the domain* Ω. *Furthermore, the spaces* X_ε^{--}, X_ε^-, X_ε^+, *and* X_ε^{++} *are orthogonal with respect to the* $L^2(\Omega)$*-scalar product* $(\cdot,\cdot)$, *and their restrictions to* $X^{1/2}$ *are orthogonal with respect to* $(\cdot,\cdot)_*$.

(b) There exists a constant $M_\varepsilon^{--} > 0$ *such that for every* $c^{++} \in X_\varepsilon^{++}$, $c^+ \in X_\varepsilon^+$, $c^- \in X_\varepsilon^-$, $c_*^{--} \in X_\varepsilon^{--} \cap X^{1/2}$, *and* $c^{--} \in X_\varepsilon^{--}$ *the estimates*

$$\begin{aligned}
\|S_\varepsilon^{++}(t)c^{++}\|_* &\leq e^{b_\varepsilon^+ t} \cdot \|c^{++}\|_* && \text{for } t \leq 0\,,\\
\|S_\varepsilon^{+}(t)c^{+}\|_* &\leq e^{a_\varepsilon^+ t} \cdot \|c^{+}\|_* && \text{for } t \geq 0\,,\\
\|S_\varepsilon^{+}(t)c^{+}\|_* &\leq e^{b_\varepsilon^- t} \cdot \|c^{+}\|_* && \text{for } t \leq 0\,,\\
\|S_\varepsilon^{-}(t)c^{-}\|_* &\leq e^{a_\varepsilon^- t} \cdot \|c^{-}\|_* && \text{for } t \geq 0\,,\\
\|S_\varepsilon^{-}(t)c^{-}\|_* &\leq e^{b_\varepsilon^{--} t} \cdot \|c^{-}\|_* && \text{for } t \leq 0\,,\\
\|S_\varepsilon^{--}(t)c_*^{--}\|_* &\leq e^{a_\varepsilon^{--} t} \cdot \|c_*^{--}\|_* && \text{for } t \geq 0\,,\\
\|S_\varepsilon^{--}(t)c^{--}\|_* &\leq M_\varepsilon^{--} \cdot t^{-1/2} \cdot e^{a_\varepsilon^{--} t} \cdot \|c^{--}\| && \text{for } t > 0
\end{aligned}$$

hold, and

$$M_\varepsilon^{--} \leq C \cdot \varepsilon^{-(1+d/2)} \quad \text{as} \quad \varepsilon \to 0\,,$$

where $C > 0$ *depends only on* Ω, $f'(c_m) + L_{\max}$, *and the constants in (35). Note that due to the finite dimensions of the subspaces* X_ε^-, X_ε^+, *and* X_ε^{++} *the semigroups* $S_\varepsilon^-(t)$, $S_\varepsilon^+(t)$, *and* $S_\varepsilon^{++}(t)$ *can be extended to groups.*

(c) There exists a constant $M_{1/2,\varepsilon} \geq 1$ *such that for all* $c \in X_\varepsilon^- \oplus X_\varepsilon^+ \oplus X_\varepsilon^{++}$ *we have*

$$\frac{1}{M_{1/2,\varepsilon}} \cdot \|c\| \leq \|c\| \leq \|c\|_* \leq M_{1/2,\varepsilon} \cdot \|c\|\,,$$

and

$$M_{1/2,\varepsilon} \cdot \varepsilon^2 \to C \quad \text{as} \quad \varepsilon \to 0\,,$$

where $C > 0$ *depends only on* $f'(c_m) + L_{\max}$, *and the constants in (35).*

Proof. The assertions of *(a)* follow easily from Lemma 3.18, (29), Corollary 3.21, and Definition 3.22.
As for the proof of *(b)*, let $c \in X$ be arbitrary, let $\psi_{i,\theta,\varepsilon}$ denote the eigenfunctions of B_ε according to Definition 3.23, and let $c = \sum_{\theta \in P\Lambda} \sum_{i=1}^\infty \xi_{i,\theta,\varepsilon} \psi_{i,\theta,\varepsilon}$ denote the

Fourier series representation of c in X, i.e., let $\xi_{i,\theta,\varepsilon} := (c, \psi_{i,\theta,\varepsilon})$, where $(\cdot,\cdot)$ denotes the standard $L^2(\Omega)$-scalar product. Then we have an explicit spectral representation of the semigroup $S_\varepsilon(t)$ given by

$$S_\varepsilon(t)c = \sum_{\theta\in P\Lambda} \sum_{i=1}^{\infty} e^{\lambda_{i,\theta,\varepsilon}\cdot t} \cdot \xi_{i,\theta,\varepsilon} \cdot \psi_{i,\theta,\varepsilon} \qquad \text{for} \qquad t > 0\,,$$

and if $c \in X^{1/2}$, then (33) furnishes

$$\|c\|_*^2 = \sum_{\theta\in P\Lambda} \sum_{i=1}^{\infty} (1 + |\tilde{\kappa}_{i,\theta,\varepsilon}|^4) \cdot \xi_{i,\theta,\varepsilon}^2 \cdot \|\psi_{i,\theta,\varepsilon}\|^2 < \infty\,.$$

These two identities already imply the first six inequalities in part *(b)*. For example, let $\lambda_{1,\theta,\varepsilon} \geq \ldots \geq \lambda_{n_0(\theta),\theta,\varepsilon}$, for $\theta \in P\Lambda$, denote all eigenvalues of B_ε in the interval I_ε^{++}, where $n_0(\theta) = 0$ is possible. Then an element $c^{++} \in X_\varepsilon^{++}$ has the Fourier series representation $c^{++} = \sum_{\theta\in P\Lambda} \sum_{i=1}^{n_0(\theta)} \xi_{i,\theta,\varepsilon}\psi_{i,\theta,\varepsilon}$, and for every $t \leq 0$ we obtain

$$\begin{aligned}\|S_\varepsilon^{++}(t)c^{++}\|_*^2 &= \sum_{\theta\in P\Lambda} \sum_{i=1}^{n_0(\theta)} (1 + |\tilde{\kappa}_{i,\theta,\varepsilon}|^4) \cdot e^{2\lambda_{i,\theta,\varepsilon}\cdot t} \cdot \xi_{i,\theta,\varepsilon}^2 \cdot \|\psi_{i,\theta,\varepsilon}\|^2 \\ &\leq \sum_{\theta\in P\Lambda} e^{2\lambda_{n_0(\theta),\theta,\varepsilon}\cdot t} \cdot \sum_{i=1}^{n_0(\theta)} (1 + |\tilde{\kappa}_{i,\theta,\varepsilon}|^4) \cdot \xi_{i,\theta,\varepsilon}^2 \cdot \|\psi_{i,\theta,\varepsilon}\|^2 \\ &\leq e^{2b_\varepsilon^+\cdot t} \cdot \|c^{++}\|_*^2\,,\end{aligned}$$

since $\lambda_{n_0(\theta),\theta,\varepsilon} \in I_\varepsilon^{++} = (b_\varepsilon^+, \lambda_\varepsilon^{\max}]$ according to Definition 3.22. The remaining five of the first six inequalities follow analogously.

In order to prove the seventh inequality, let $c^{--} \in X_\varepsilon^{--}$ be arbitrary. If $n_1(\theta) \geq 1$ is chosen in such a way that $\lambda_{i,\theta,\varepsilon}$ for $i \geq n_1(\theta)$ and $\theta \in P\Lambda$ denote all the eigenvalues of B_ε which are contained in I_ε^{--}, then c^{--} has the Fourier series representation $c^{--} = \sum_{\theta\in P\Lambda} \sum_{i=n_1(\theta)}^{\infty} \xi_{i,\theta,\varepsilon}\psi_{i,\theta,\varepsilon}$ in X, and for arbitrary $t > 0$ we actually have $S_\varepsilon^{--}(t)c^{--} \in X^{1/2}$. Due to the choice of the interval I_ε^{--} one further obtains $\lambda_{i,\theta,\varepsilon} \leq \lambda_{n_1(\theta),\theta,\varepsilon} < \hat{a}_\varepsilon^{--} < 0$ for all $i \geq n_1(\theta)$, and therefore $|\tilde{\kappa}_{i,\theta,\varepsilon}|^2 > \beta_\theta/\varepsilon^2$ for all $i \geq n_1(\theta)$, where we set $\beta_\theta = f'(c_m) + L(\theta)$.

Thus,

$$\|S_\varepsilon^{--}(t)c^{--}\|_*^2 = \sum_{\theta\in P\Lambda} \sum_{i=n_1(\theta)}^{\infty} (1 + |\tilde{\kappa}_{i,\theta,\varepsilon}|^4) \cdot e^{2\lambda_{i,\theta,\varepsilon}\cdot t} \cdot \xi_{i,\theta,\varepsilon}^2 \cdot \|\psi_{i,\theta,\varepsilon}\|^2\,.$$

Now it is easy to verify that for all $t > 0$, $i \geq n_1(\theta)$, and $\lambda > \lambda_{n_1(\theta),\theta,\varepsilon}$ we have

$$\begin{aligned}(1 + |\tilde{\kappa}_{i,\theta,\varepsilon}|^4) \cdot e^{2\lambda_{i,\theta,\varepsilon}\cdot t} &\leq \frac{1 + |\tilde{\kappa}_{i,\theta,\varepsilon}|^4}{2e(\lambda - \lambda_{i,\theta,\varepsilon})} \cdot t^{-1} \cdot e^{2\lambda\cdot t} \\ &= \frac{1 + |\tilde{\kappa}_{i,\theta,\varepsilon}|^4}{2e\left(\lambda - \beta_\theta|\tilde{\kappa}_{i,\theta,\varepsilon}|^2 + \varepsilon^2|\tilde{\kappa}_{i,\theta,\varepsilon}|^4\right)} \cdot t^{-1} \cdot e^{2\lambda\cdot t}\,.\end{aligned}$$

To continue, define the function $h(s) = (1+s^2)\cdot(\lambda - \beta s + \varepsilon^2 s^2)^{-1}$, where $\lambda = a_\varepsilon^{--} < 0$, $s \in Z := \{s > 0 \ : \ \lambda - \beta s + \varepsilon^2 s^2 > 0\}$, and $\beta = \beta_\theta$ for an arbitrary $\theta \in P\Lambda$. Assume first that θ is such that $\beta_\theta \geq 0$. Then h is monotonically decreasing on Z, and this implies

$$C_{i,\theta,\varepsilon}^2 := \frac{1+|\tilde{\kappa}_{i,\theta,\varepsilon}|^4}{2e(\lambda - \beta_\theta|\tilde{\kappa}_{i,\theta,\varepsilon}|^2 + \varepsilon^2|\tilde{\kappa}_{i,\theta,\varepsilon}|^4)} \leq \frac{1+|\tilde{\kappa}_{n_1(\theta),\theta,\varepsilon}|^4}{2e(\lambda - \lambda_{n_1(\theta),\theta,\varepsilon})}$$

$$\leq \frac{1+(s_{\theta,\varepsilon}^*)^2}{2e\left(a_\varepsilon^{--} - \hat{a}_\varepsilon^{--}\right)} \leq \frac{1+|\tilde{\kappa}_{n_1(\theta_{\max}),\theta_{\max},\varepsilon}|^4}{2e(a_\varepsilon^{--} - \hat{a}_\varepsilon^{--})} =: \left(\widetilde{M}_\varepsilon^{--}\right)^2,$$

where we set $s = s_{\theta,\varepsilon}^* > 0$ the unique positive solution of $-\varepsilon^2 s^2 + \beta_\theta\, s = \hat{a}_\varepsilon^{--}$. If $\beta_\theta < 0$, then h switches exactly once from decreasing to increasing in Z, and this yields

$$C_{i,\theta,\varepsilon} \leq \max\left\{\widetilde{M}_\varepsilon^{--}, \varepsilon^{-1}\right\} =: M_\varepsilon^{--}.$$

This proves the seventh estimate in (b). Finally, using a calculation which was already employed in the proof of Lemma 3.6(b) in [15] and which uses the estimate $\lambda_{n_1(\theta_{\max}),\theta_{\max},\varepsilon} \geq \underline{v}^{--}\cdot\lambda_\varepsilon^{\max} = \underline{v}^{--}\cdot(\beta_{\theta_{\max}})^2/(4\varepsilon^2)$ from Corollary 3.21(c), we obtain

$$M_\varepsilon^{--} \leq \varepsilon^{-(1+d/2)}\cdot\max\left\{\varepsilon^{d/2},\ \sqrt{\frac{\varepsilon^4 + (\beta_{\theta_{\max}})^2\cdot(1+\sqrt{1-\underline{v}^{--}})^2/4}{\mu_1}}\right\},$$

where $\mu_1 > 0$ is defined in Corollary 3.21(b). This finally proves the asymptotic behavior of the constant M_ε^{--} for $\varepsilon \to 0$.

The proof of part *(c)* follows the lines of the proof of Lemma 3.6(c) in [15] and is therefore omitted. □

According to the above lemma the linear part of the Cahn-Hilliard equation with elasticity (25) satisfies (with respect to the non-standard norm $\|\cdot\|_*$) both (H1) and (H2), as well as (7) from Section 2 in [15]. Moreover, the asymptotic behavior for $\varepsilon \to 0$ of certain spectral gaps in the spectrum of B_ε, and of the constants M_ε^{--} and $M_{1/2,\varepsilon}$ have been obtained. We close this subsection with the following remark.

Remark 3.25 With the results of Corollary 3.21 and Lemma 3.24 we can deduce the asymptotic behavior of certain constants introduced in [15, Section 2] for $\varepsilon \to 0$. Although the specific values of these constants are different if compared to the application in [15, Section 3], their dependence on ε is not. Hence, we obtain exactly the same asymptotics, i.e., $C_\varepsilon^{--} \geq C\cdot\varepsilon^d$ and $C_\varepsilon^{+} \geq C\cdot\varepsilon^{d-2}$ for $\varepsilon \to 0$, see Remark 3.7 in [15].

Even though we did not formally introduce these constants, we want to point out that their asymptotic behavior is used to prove that the abstract theory of Section 2 in [15] can be applied to B_ε and a nonlinear function F, whose Lipschitz constant satisfies $0 \leq L_F \leq C\cdot\varepsilon^d$. For more details see [15, Remark 2.11]. All of the constants C above depend only on Ω, $f'(c_m) + L_{\max}$, and the constants in (35).

3.3 Properties of the Nonlinearity

Hypothesis (H3) is valid for some function $\hat{F} : H^2_{av}(\Omega) \to L^2(\Omega)$ which coincides with F from (27) on a certain neighborhood of the origin. In order to obtain a global Lipschitz constant $L_{\hat{F}}$ of the order ε^d (as required by Remark 3.25), the size of this neighborhood has to be proportional to ε^d with respect to the $H^2(\Omega)$-norm. This can be proved by applying the results from [15, Section 3.3] to $\tilde{f}$ and F. This immediately furnishes the following result.

Corollary 3.26 *The nonlinear operator F defined in (27) satisfies (H3) with a Lipschitz constant L_F of the order ε^d on an $H^2(\Omega)$-neighborhood of* 0 *with size proportional to ε^d.*

3.4 Spinodal Decomposition

In the previous subsections we established all properties of (25) which are necessary to apply the abstract results of [15, Section 2] to the Cahn–Hilliard equation with elasticity — and this can be done exactly as in Subsection 3.4 of [15]. Moreover, since the asymptotic behavior of the involved constants remains basically unchanged, we obtain exactly the same result, of course after adopting the new notation (c_m instead of μ, B_ε instead of A_ε and so on). Therefore, we refrain from presenting our main theorem again in as detailed a form as in the binary case, and state only an intuitive abbreviated version.

Suppose that three constants $0 < r \ll \rho \ll R$ are given. We consider initial conditions from the ball

$$B_r(c_m) = \{v \in c_m + X_\varepsilon^- \oplus X_\varepsilon^+ \oplus X_\varepsilon^{++} : \|v - c_m\|_* < r\} \subset c_m + H^2_{av}(\Omega)$$

and their evolution under the dynamics of (25). Let M_r denote the set of all initial conditions $v \in B_r(c_m)$ whose corresponding solution of (25) either remains in the larger ball $B_R(c_m)$ for all time, or has distance greater than ρ from $X_\varepsilon^+ \oplus X_\varepsilon^{++}$ upon exiting $B_R(c_m)$. See also Figure 6 in [15]. In other words, the initial conditions in M_r cannot be considered as being dominated by the strongly unstable subspace $\mathcal{Y}_\varepsilon^+ := X_\varepsilon^+ \oplus X_\varepsilon^{++}$.

Our main theorem states that the volume (which is the canonical Lebesgue volume of the finite-dimensional space $X_\varepsilon^- \oplus X_\varepsilon^+ \oplus X_\varepsilon^{++}$) of these "bad"´initial conditions compared to the volume of all initial conditions in $B_r(c_m)$ is arbitrarily small, provided the constants $0 < r \ll \rho \ll R$ are chosen proportional to ε^d as $\varepsilon \to 0$.

Theorem 3.27. *We consider solutions of the Cahn–Hilliard equation with elasticity (25) and assume that hypotheses (A1), (A2), and (A3) are satisfied. Then there exists a positive constant ε_0 which depends only on Ω, N_0, $f'(c_m) + L_{\max}$,*

$|f'(c_m) + L_{\min}|$, and the constants in (35), such that for arbitrary $0 < \varepsilon \le \varepsilon_0$ the following holds.

For every $0 < p \ll 1$ there exist constants $0 < r \ll \rho \ll R$ which depend only on $f'(c_m) + L_{\max}$, Ω, and the constants in (35) (r depends additionally on p) and which are all proportional to ε^d as $\varepsilon \to 0$, such that

$$\frac{\text{vol}(M_r)}{\text{vol}(B_r(c_m))} \leq p\,. \tag{36}$$

Proof. One only has to apply the abstract theory of Section 2 in [15]. Hypotheses (H1) through (H3) have been established in Subsections 3.1 through 3.3, and the necessary constants have been calculated, furnishing the ball with size proportional to ε^d on which the result is valid. □

Remark 3.28 This theorem guarantees that the initial conditions near c_m are dominated by the subspace $\mathcal{Y}_\varepsilon^+$. This is why we call $\mathcal{Y}_\varepsilon^+$ the dominating subspace. Since the nonlinearity F is exactly the same as in [15] we expect that also the results of Sander and Wanner [19] on second phase spinodal decomposition hold true for the Cahn-Hilliard equation with elasticity.

4 Numerical Simulations

Finally, we would like to show that the patterns predicted by our analysis in Sections 2 and 3 are in fact observed in typical solutions having initial data close to an unstable homogeneous state c_m, i.e., $\psi_{,cc}(c_m) - L_{\max} = -(f'(c_m) + L_{\max}) < 0$. We made a series of numerical simulations based on a finite element method developed in Garcke, Rumpf, Weikard [10] and in this section we will present typical patterns seen in the numerical experiments. We point out that in [10] optimal error estimates are established, i.e. the method used is very reliable.

Before we present the numerical results, let us remark that the patterns shown are generic in the sense that the initial data have to be degenerate in a certain sense in order not to lead to similar patterns. In all simulations we were choosing $c_m = 0$ and were taking a random perturbation around c_m as initial data. All calculations have been performed on the unit square with $\varepsilon = 10^{-3}$, $\psi(c) = \frac{1}{4}(c^2 - 0.16)^2$ and the solutions are shown at time $t = 0.001$.

At first we consider the case without elasticity. The left hand side of Figure 6 shows the modulus of the Fourier coefficients of a solution to the Cahn-Hilliard equation during spinodal decomposition after a fixed time. On the right the sign of the concentration difference c (black denoting positive, white denoting negative values) is shown. The Fourier coefficients were calculated using the FFTW-software package (see http://www.fftw.org for details). Figure 6 corresponds to Figure 1 in that neither in the Fourier coefficients nor in the concentration picture there are any distinguished directions reflecting the isotropy of the Cahn-Hilliard model without elasticity. Essentially the same picture occurs in the case of isotropic elasticity.

In the Figures 7 and 8 we show numerical results with an anisotropic elasticity tensor. Parameters were $C_{11} = 10$, $C_{12} = 1$, $C_{44} = 1$ and $q = \frac{\sqrt{2}}{10}$ (this implies

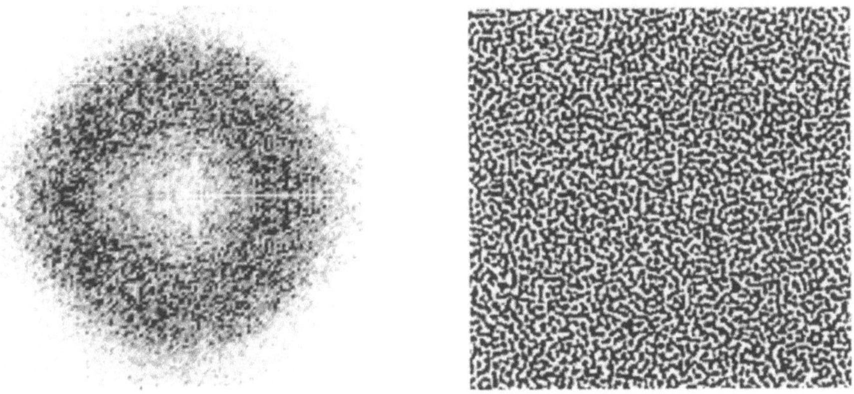

Fig. 6. The case without elasticity; modulus of the Fourier coefficients (left) and sign of the concentration difference c (right)

$L_{max} \approx -0.068$, $L_{min} = -0.2$ and $f'(c_m) = 0.16$) for the case with positive anisotropy and $C_{11} = 2$, $C_{12} = 1$, $C_{44} = 100$ and $q = 0.2$ (i.e., $L_{max} = -0.06$, $L_{min} \approx -0.236$ and $f'(c_m) = 0.16$) for the case with negative anisotropy. In both cases the anisotropy is clearly visible in the Fourier coefficients as well as in the concentration.

Finally we consider the case with isotropic elasticity tensor but anisotropic eigenstrains (see Figure 9). The parameter were $\lambda = 1$, $\mu = \frac{1}{2}$ and $a = 1$, $b = 0.1$ (i.e., $L_{max} = -0.015$, $L_{min} = -1.5$ and $f'(c_m) = 0.16$).

Acknowledgment. This work has been started while SMP was visiting the University of Bonn as a guest of the Sonderforschungsbereich 256 "Nichtlineare partielle Differentialgleichungen" (SFB256). HG and UW have been members of the SFB256

Fig. 7. Negative anisotropy of the elasticity tensor; modulus of the Fourier coefficient (left) and sign of the concentration difference c (right)

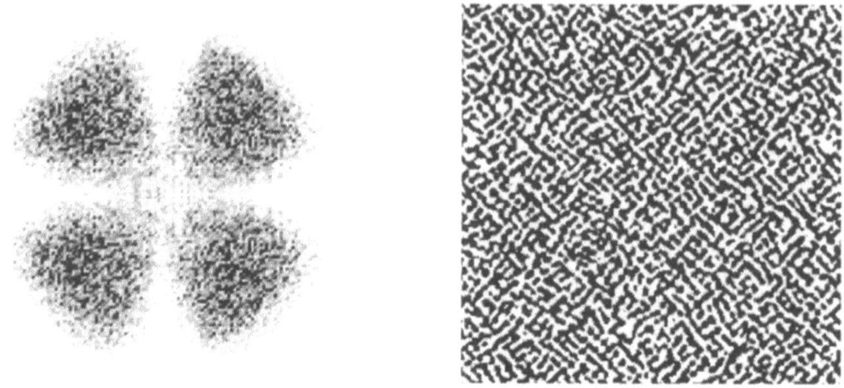

Fig. 8. Positive anisotropy of the elasticity tensor; modulus of the Fourier coefficient (left) and sign of the concentration difference c (right)

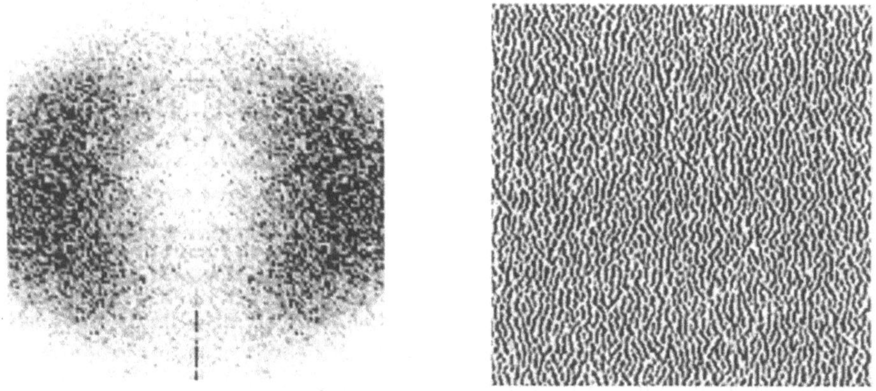

Fig. 9. Anisotropic eigenstrain and isotropic elasticity tensor; modulus of the Fourier coefficient (left) and sign of the concentration difference c (right)

for many years and all three authors gratefully acknowledge the support of the Sonderforschungsbereich.

References

1. H. Amann: Linear and Quasilinear Parabolic Problems. Volume I: Abstract Linear Theory. Birkhäuser, Basel – Boston – Berlin, 1995
2. J. W. Cahn, J. E. Hilliard: Free energy of a nonuniform system. I. Interfacial free energy. Journal of Chemical Physics, Vol. 28, 258–267, 1958
3. J.D. Eshelby: The continuum theory of lattice defects. Solid State Physics, 3, 79–144, 1956

4. P.C. Fife, H. Kielhöfer, S. Maier-Paape, T. Wanner: Perturbation of doubly periodic solution branches with applications to the Cahn-Hilliard equation. Physica D, 100(3-4), pp. 257–278, 1997
5. P. Fratzl, O. Penrose, J. L. Lebowitz: Modelling of phase separation in alloys with coherent elastic misfit. J. Stat. Physics, 95 5/6,1429–1503, 1999
6. H. Garcke: On mathematical models for phase separation in elastically stressed solids. Habilitation thesis, Bonn, 2000
7. H. Garcke: On Cahn–Hilliard systems with elasticity. Submitted
8. H. Garcke: The Γ–limit of the elastically modified Ginzburg–Landau energy. Manuscript
9. H. Garcke: On a Cahn-Hilliard model for phase separation with elastic misfit. Manuscript
10. H. Garcke, M. Rumpf, U. Weikard: The Cahn-Hilliard equation with elasticity: Finite element approximation and qualitative studies. Interfaces and Free Boundaries, 3, 101–118, 2001
11. M.E. Gurtin: An introduction to Continuum Mechanics. Academic Press, 1981
12. A. G. Khachaturyan: Theory of Structural Transformations in Solids. Wiley, New York, 1983
13. F. C. Larché, J. W. Cahn: The effect of self–stress on diffusion in solids. Acta Metall., 30, 1835–1845, 1982
14. S. Maier–Paape, T. Wanner: Spinodal decomposition for the Cahn-Hilliard equation in higher dimensions. Part I: Probability and wavelength estimate. Comm. Math. Phys., 195, 435–464, 1998
15. S. Maier–Paape, T. Wanner: Spinodal decomposition for the Cahn-Hilliard equation in higher dimensions. Nonlinear dynamics. Arch. Rat. Mech. Anal., 151, 187–219, 2000
16. S. Maier-Paape, B. Stoth, T. Wanner: Spinodal decomposition for multicomponent Cahn–Hilliard systems. J. Stat. Phys., Vol. 98, Nos. 3/4, 871–890, 2000
17. T. Mura: Micromechanics of Defects in Solids. Nijhoff, the Hague, 1982
18. A. Onuki: Ginzburg–Landau approach to elastic effects in the phase separation of solids. J. Phys. Soc. Jpn., 58, 3065–3068, 1989
19. E. Sander, T. Wanner: Unexpectedly linear behavior for the Cahn–Hilliard equation. SIAM J. Appl. Math., 60(6), 2182–2202, 2000
20. R. Temam: Infinite-Dimensional Dynamical Systems in Mechanics and Physics. Springer-Verlag, New York–Berlin–Heidelberg, 1988
21. J. D. van der Waals: The thermodynamic theory of capillarity under the hypothesis of a continuous variation of density (in Dutch). Verhaendel. Kronik. Akad. Weten. Amsterdam, Vol. 1, (1893). Engl. translation by J.S. Rowlinson, J. Stat. Phys., 20, 197–244, 1979

Waiting Time Phenomena for Degenerate Parabolic Equations – A Unifying Approach

Roberta Dal Passo[1], Lorenzo Giacomelli[2], and Günther Grün[3]

[1] Dipartimento di Matematica, Università di Roma "Tor Vergata", 00133 Roma, Italy
[2] Dipartimento Me.Mo.Mat., Università di Roma "La Sapienza", 00161 Roma, Italy
[3] Institut für Angewandte Mathematik, Universität Bonn, 53115 Bonn, Germany

1 Introduction

The aim of this contribution is to present a new approach to establish the occurrence of waiting time phenomena for solutions to degenerate parabolic equations. Originally developed by the authors in [15] for the thin film equation

$$u_t + \operatorname{div}(u^n \nabla \Delta u) = 0, \tag{1}$$

it may in fact be used for a broad class of degenerate parabolic equations having divergence structure. The model problems we have in mind are – besides the aforementioned thin film-type equations – doubly nonlinear second order equations

$$u_t - \operatorname{div}(u^{m-1}|\nabla u|^{p-2}\nabla u) = 0 \tag{2}$$

of porous media type as well as higher order doubly nonlinear equations with variational structure like the equation

$$(|u|^{q-2}u)_t + (-1)^k \sum_{|\lambda|=k} D^\lambda(|D^k u|^{p-2} D^\lambda u) = 0. \tag{3}$$

In fact, our method only requires integral inequalities of the kind

$$\begin{aligned}\int_{\mathbf{R}^N} \zeta^{kp}|w(t,x)|^q dx + C^{-1}\int_0^t \int_{\mathbf{R}^N} \zeta^{kp}|D^k w|^p dxdt \\ \leq \int_{\mathbf{R}^N} \zeta^{kp}|w_0|^q dx + C\int_0^t \int_{\{\zeta>0\}} |w|^p dxdt\end{aligned} \tag{4}$$

to be satisfied for some

$$k \in \mathbf{N}, \quad 0 < q < p, \quad p \geq 1. \tag{5}$$

In the applications to the aforementioned model problems, w is a suitable strictly monotone function of u such that $w = 0$ at $u = 0$. In this case, we write for simplicity $w(t,x)$ and $w_0(x)$ to denote $w(u(t,x))$ and $w(u_0(x))$, respectively. Whenever such inequalities hold for a suitable class of test functions (cf. (10) below), we can identify a general criterion on the growth of initial data which guarantees the occurrence of a waiting time phenomenon; that is, the existence of a positive time during which the support of w (locally in space) does not expand:

Definition 1.1 *A function* $w(t,x) : [0,\infty) \times \mathbf{R}^N \to \mathbf{R}$ *has positive waiting time at* $x_0 \in \partial\mathrm{supp}(w_0)$ *if there exists* $T^\star > 0$ *such that*

$$x_0 \in \overline{\mathbf{R}^N \setminus \mathrm{supp}(w(t))} \qquad \text{for a.e. } t \in [0, T^\star).$$

Note that this is a notion of "forward" waiting time: It does not exclude the possibility that the support of w retracts away from x_0.

The occurrence of a waiting time phenomenon for solutions u of degenerate parabolic equations can be associated with a growth condition on the initial datum u_0 in a neighbourhood of a Lipschitz–regular boundary point x_0. In the sequel, we will call $\gamma > 0$ the *critical exponent* if it is the infimum over all positive numbers σ with the property that a local growth condition

$$|u_0(x)| \le C|x - x_0|^\sigma \quad \forall\, |x - x_0| < \delta \tag{6}$$

for some $C > 0$, $\delta > 0$ entails $T^\star$ to be positive.

For non-negative solutions of the porous media equation ($m > 1$, $p = 2$), the occurrence of waiting time phenomena was first proven by Knerr [23] in one space dimension via comparison arguments. A necessary and sufficient condition in arbitrary space dimensions was later obtained by Alikakos [1]: $T^\star > 0$ if and only if

$$\limsup_{r\to 0} r^{-\frac{2}{m-1}} \fint_{B(x_0,r)} u_0(x)\,dx < \infty \tag{7}$$

(the reader should be aware of a repeated misprint concerning the exponent in the paper [1]). When translated into a "power–law–type" condition, (7) reads as follows:

$$\begin{aligned} &T^\star > 0 \text{ if } u_0(x) \le C|x - x_0|^{\frac{2}{m-1}} \\ &T^\star = 0 \text{ if } u_0(x) \ge C|x - x_0|^\gamma, \quad \gamma < \frac{2}{m-1}. \end{aligned} \tag{8}$$

Both the sufficient and the necessary part of Alikakos' proof ultimately rely upon the comparison principle, the former one via an estimate from below for Δu^m (cf. [8],[6]), the latter one via the Aronson–Caffarelli inequality [7]. This excludes the applicability of his technique to higher order equations. In [14], Chipot and Sideris obtained an upper bound for the waiting time $T^\star$ by energy methods, but the argument does not seem to carry over to operators in non-variational structure as (1).

A different energy method, first introduced by Antontsev [3], was used by Antontsev, Diaz and Shmarev in a series of papers ([4], [5] and the references therein) in the case of doubly nonlinear second order equations, considering also source and absorption terms. Their basic idea was to estimate the function on nested balls using an interpolation inequality for the trace [17]. They prove that a solution which is initially zero on a ball, remains zero on that ball provided the initial datum is sufficiently flat around it. Therefore, their results are not entirely local and — as a consequence — they do not yield the critical exponent $\gamma = \frac{2}{m-1}$ for $p = 2$ in higher space dimensions.

A third approach, introduced by Shishkov [25], was used by Shishkov and Shchelkov [26] to treat the case of higher order doubly nonlinear equations in variational structure, allowing for absorption terms, too. It is based on a–priori estimates which remind of Saint Venant's principle in elasticity. Due to the fact that estimates are worked out on nested half–spaces, in higher space dimensions their result applies only to convexity points of $\partial \text{supp}(u_0)$, and it does not yield the critical exponent (in the sense of (6)) for the porous medium equation, neither.

The new approach, which we present here, applies also to equations which, as (1) for $n \neq 1$, do not have variational structure, and is capable to capture the critical exponent for the porous medium equation in arbitrary space dimensions, even for a non-convex initial support: in fact, we only need an "exterior cone condition" to be satisfied locally in space.

The first main tool, with respect both to optimality and to non-convexity, is an appropriate choice of the class of test functions in (4), and in particular of their support. To introduce such a class, let e_N be the canonical basis vector parallel to the x_N-axis, and consider for given $\theta > 0$ and $r \geq 0$ functions

$$\tilde{\zeta}_{r,\theta}(x) := [r - x_N]_+ \left[(\tan\theta)^2 - \frac{|\overline{x}_N|^2}{(r - x_N)^2}\right]_+, \qquad \overline{x}_N = (x_1, \ldots, x_{N-1}).$$

These functions are supported on a half–cone in $\mathbf{R}^{N-1}$ with vertex in re_N, axis parallel to e_N (pointing downwards) and opening angle θ. We shall denote that half-cone by

$$\mathcal{C}_{r,\theta} := \text{supp}(\tilde{\zeta}_{r,\theta}). \tag{9}$$

From a technical point of view, it will be essential that the conical shape of supports permits the use of the homogeneous version of Gagliardo–Nirenberg's inequality (cf. Theorem 2.4 on page 641) on those sets. To localize the test–functions, we introduce a cut–off

$$\varphi(s) = \begin{cases} \left[1 - (|s| - 3)^2\right]_+ & |s| \geq 3 \\ 1 & |s| < 3, \end{cases}$$

and we consider for $r_0 > 0$ functions

$$\zeta_{r,\theta,r_0}(x) := \varphi\left(\tfrac{|x|}{r_0}\right) \tilde{\zeta}_{r,\theta}(x).$$

Taking also translations and rotations into account, the class of test functions to be referred to in the sequel can be defined as follows.

$$\mathcal{F} = \bigcup_{x_0 \in \mathbf{R}^N} \mathcal{F}_{x_0}, \tag{10}$$

where

$$\mathcal{F}_{x_0} = \{\zeta_{r,\theta,r_0}(x_0 + Ax);\ r > 0,\ \theta \in (0, \tfrac{\pi}{2}),\ r_0 > 0,\ A \in SO(N)\}.$$

In addition to (4), we will assume that w has finite speed of propagation in the sense that

$$\begin{aligned} &B(x_0, r_0) \cap \mathrm{supp}(w_0) = \emptyset \implies \\ &\quad B(x_0, r_0 - r(t)) \cap \mathrm{supp}(w(t)) = \emptyset \quad \forall\, t:\ r(t) < r_0/2 \end{aligned} \tag{11}$$

with r being a non-negative increasing function which is independent of x_0, r_0 and which satisfies $r(0) = 0$. We are now ready to state the main result.

Theorem 1.2 *Assume that a function $w \in L^p((0,T); L^p_{loc}(\mathbf{R}^N))$ satisfies (11) and (4) for all $\zeta \in \mathcal{F}$ and for some k, p and q such that (5) holds. Then w has positive waiting time at x_0 provided*

– $\mathrm{supp}(w_0)$ *has the "exterior cone property" locally at x_0, i.e.*

$$\mathrm{supp}(w_0) \cap \mathrm{supp}(\zeta) = \emptyset \quad \text{for some} \quad \zeta \in \mathcal{F}_{x_0}; \tag{12}$$

– w_0 *satisfies the following growth condition:*

$$\limsup_{r \to 0} r^{-\frac{kp}{p-q}q} \fint_{B(x_0,r)} |w_0(x)|^q \, dx < \infty. \tag{13}$$

When translated into a "power law–type" condition, (13) reads as follows:

$$T^\star > 0 \quad \text{if} \quad w_0(x) \le C|x - x_0|^{\frac{kp}{p-q}}. \tag{14}$$

For the porous medium equation, inequality (4) is satisfied by $w = u^{\frac{m+1}{2}}$ letting $p = 2$, $q = \frac{4}{m+1}$, and condition (14) becomes

$$T^\star > 0 \quad \text{if} \quad u_0(x) \le C|x - x_0|^{\frac{2}{m-1}},$$

which coincides with (8). The fact that we are able to recover the critical exponent for the porous medium equation is partly due to the choice of test functions — we estimate over a family of nested cones with same axis and same opening angle, thus capturing the local behaviour of the initial datum. Nevertheless, there is a second main tool, namely the following extension of the well–known Stampacchia Lemma:

Lemma 1.3 (An Extension of Stampacchia's Lemma) *Assume that a given non-negative, nonincreasing function $G : (0, \rho_0) \to \mathbf{R}$ satisfies:*

$$G(\xi) \le \frac{c_0}{(\xi - \eta)^\alpha} \left(G(\eta) + (\rho_0 - \eta)^\sigma\right)^\beta \tag{15}$$

for $0 \le \eta < \xi \le \rho_0$ and positive numbers $c_0, \alpha, \beta, \sigma$ such that

$$\beta > 1 \qquad \text{and} \qquad \sigma \geq \frac{\alpha}{\beta - 1}.$$

Assume further that

$$\rho_0^\alpha \geq 2^{\frac{\alpha\beta}{\beta-1}}(1 + 2^{\frac{\alpha}{\beta-1}-\sigma})^\beta \cdot c_0 \cdot (G(0) + \rho_0^\sigma)^{\beta-1}.$$

Then

$$G(\rho_0) = 0.$$

The proof can be found in [15]. We remark that if the term $(\rho_0 - \eta)^\sigma$ did not occur in (15), the result would be identical to Stampacchia's Lemma [27, Lemma 4.1 (i)].

The paper is organised as follows: In §2 we will prove Theorem 1.2, and in §3 we will give applications of this result to solutions of (2), (1) and (3).

2 Proof of Theorem 1.2

In the course of the proof we will make use of Gagliardo–Nirenberg's inequality [18], [24]. In the formulation we provide below, some of the summability powers are allowed to be less than one (see [16] for a proof), while homogeneity follows immediately from a rescaling argument.

Theorem 2.4 (Gagliardo-Nirenberg) *Let* $0 < q < p$, $1 \leq r \leq \infty$, $k \in \mathbf{N}$, $m > 0$. *Let* $\Omega \subset \mathbf{R}^N$ *be diffeomorphic to a cone. Suppose* u *belongs to* $L^q(\Omega)$ *and its derivatives of order* m *belong to* $L^r(\Omega)$. *Then the following inequalities hold (with the constant* C_1 *depending only on* Ω, k, q, r*):*

$$\|u\|_p \leq C_1 \|D^k u\|_r^\Theta \cdot \|u\|_q^{1-\Theta}$$

where

$$\tfrac{1}{p} = \Theta\left(\tfrac{1}{r} - \tfrac{k}{N}\right) + (1-\Theta)\tfrac{1}{q}$$

for all Θ *in the interval* $[0, 1)$.

Proof of Theorem 1.2 We assume, up to a translation and rotation, that $x_0 = 0$ and that the axis of the exterior cone is parallel to e_N, pointing downwards. Therefore

$$\operatorname{supp}(\zeta_{0,2\theta,2r_0}) \cap \operatorname{supp}(w_0) = \emptyset$$

for some $\theta > 0$ and $r_0 > 0$. It follows from the finite speed of propagation property (11) that there exists $T_0 > 0$ such that

$$\operatorname{supp}(\zeta_{-r_0,2\theta,r_0}) \cap \operatorname{supp}(w(t)) = \emptyset \qquad \forall t < T_0. \tag{16}$$

We choose the functions

$$\zeta_r(x) := \zeta_{r,\theta,r_0}(x)$$

as test functions in (4). It follows from (16) and simple geometric arguments that there exists $\epsilon > 0$ such that

$$\operatorname{supp}(\zeta_r) \cap \operatorname{supp}(w(t)) \subseteq \operatorname{supp}(\zeta_{r_0}) \setminus \operatorname{supp}(\zeta_{-r_0,2\theta,r_0}) \subset B(0, 3r_0 - \epsilon) \quad (17)$$

for all $r \in [0, r_0]$. Therefore, denoting by $\mathcal{C}(r) := \mathcal{C}_{r,\theta}$ the half–cone with vertex in re_N, axis parallel to e_N and opening angle θ (see equation (9)), we have in particular

$$\int_{\mathcal{C}(r)\cap B(0,3r_0)} \zeta_r^{kp} |w(t,x)|^q dx + C^{-1} \int_0^t \int_{\mathcal{C}(r)\cap B(0,3r_0)} \zeta_r^{kp} |D^k w|^p dx dt \quad (18)$$
$$\leq \int_{\mathcal{C}(r)\cap B(0,3r_0)} \zeta_r^{kp} |w_0|^q dx + C \int_0^t \int_{\mathcal{C}(r)\cap B(0,3r_0)} |w|^p dx dt.$$

Since by definition

$$\zeta_r(x) = \tilde{\zeta}_r(x) := [r - x_N]_+ \left[(\tan\theta)^2 - \frac{|\overline{x}_N|^2}{(r - x_N)^2} \right]_+ \quad \text{on } B(0, 3r_0),$$

we may substitute ζ_r by $\tilde{\zeta}_r$ in (18). In addition, property (17) allows to extend w by zero on $\mathcal{C}(r_0)$; we denote this extension again by w. Taking the supremum on the left-hand side of (18), we conclude that

$$\sup_{\tau\in(0,t)} \int_{\mathcal{C}(r)} \tilde{\zeta}_r^{kp} |w(\tau,x)|^q dx + C^{-1} \int_0^t \int_{\mathcal{C}(r)} \tilde{\zeta}_r^{kp} |D^k w|^p dx dt \quad (19)$$
$$\leq \int_{\mathcal{C}(r)} \tilde{\zeta}_r^{kp} |w_0|^q dx + C \int_0^t \int_{\mathcal{C}(r)} |w|^p dx dt.$$

It is not difficult to check that, for all $\rho \in [0, r)$,

$$\tilde{\zeta}_r(x) \geq (\tan\theta)^2 (r - \rho) \qquad \forall x \in \mathcal{C}(\rho).$$

We use this relation to minorize the left-hand side of (19), and we obtain

$$\sup_{\tau\in(0,t)} \int_{\mathcal{C}(\rho)} |w(\tau,x)|^q dx + C^{-1} \int_0^t \int_{\mathcal{C}(\rho)} |D^k w|^p dx dt \quad (20)$$
$$\leq \frac{1}{(r-\rho)^{kp}} \left(\int_{\mathcal{C}(r)} \tilde{\zeta}_r^{kp} |w_0|^q dx + C \int_0^t \int_{\mathcal{C}(r)} |w|^p dx dt \right)$$

for all $\rho \in (0, r)$. By virtue of Gagliardo–Nirenberg's inequality, we estimate the last term on the right–hand side of (20) with r replaced by ρ:

$$\|w\|_{L^p(\mathcal{C}(\rho))} \leq C \left\|D^k w\right\|^{\Theta}_{L^p(\mathcal{C}(\rho))} \|w\|^{1-\Theta}_{L^q(\mathcal{C}(\rho))},$$

where

$$\Theta = \tfrac{N(p-q)}{kpq+N(p-q)}.$$

Taking the pth-power and integrating in time, we obtain

$$\int_0^t \int_{\mathcal{C}(\rho)} |w|^p dx dt \leq C \int_0^t \left(\int_{\mathcal{C}(\rho)} |D^k w|^p dx \right)^{\Theta} \left(\int_{\mathcal{C}(\rho)} |w|^q dx \right)^{\frac{p(1-\Theta)}{q}} dt$$

$$\leq C t^{1-\Theta} \left(\sup_{\tau \in (0,t)} \int_{\mathcal{C}(\rho)} |w(\tau,x)|^q dx \right)^{\frac{p(1-\Theta)}{q}} \left(\int_0^t \int_{\mathcal{C}(\rho)} |D^k w|^p dx dt \right)^{\Theta}.$$

Combining this inequality with (20) yields

$$\int_0^t \int_{\mathcal{C}(\rho)} |w|^p \leq C t^{1-\Theta} \left(\frac{1}{(r-\rho)^{kp}} \left(\int_{\mathcal{C}(r)} \tilde{\zeta}_r^{kp} |w_0|^q + \int_0^t \int_{\mathcal{C}(r)} |w|^p \right) \right)^{1+\delta} \tag{21}$$

where

$$\delta = \frac{(1-\Theta)(p-q)}{q} > 0.$$

Observing that (by straightforward geometric considerations)

$$\mathcal{C}(r) \cap \operatorname{supp}(w_0) \subseteq B(0,r)$$

and using the growth condition on w_0, we find $C > 0$ and $\rho_0 \in (0, r_0]$ such that

$$\int_{\mathcal{C}(r)} \tilde{\zeta}_r^{kp} |w_0|^q dx \leq C r^{kp} \int_{B(0,r)} |w_0|^q dx \leq C r^{\sigma} \quad \forall\, r \in [0, \rho_0],$$

where

$$\sigma = kp + N + \frac{kpq}{p-q}.$$

Therefore

$$\int_0^t \int_{\mathcal{C}(\rho)} |w|^p dx dt \leq C t^{1-\Theta} \left(\frac{1}{(r-\rho)^{kp}} \left(C r^{\sigma} + \int_0^t \int_{\mathcal{C}(r)} |w|^p dx dt \right) \right)^{1+\delta}. \tag{22}$$

Let us introduce for $0 \leq \xi \leq \rho_0$ the non-negative decreasing function

$$G(\xi) = \int_0^t \int_{\mathcal{C}(\rho_0 - \xi)} |w|^p dx dt;$$

specifying $\xi = \rho_0 - \rho$ and $\eta = \rho_0 - r$, we may rewrite (22) as follows:

$$G(\xi) \leq \frac{C t^{1-\Theta}}{(\xi - \eta)^{kp(1+\delta)}} \left(G(\eta) + (\rho_0 - \eta)^{\sigma} \right)^{1+\delta}.$$

This inequality has exactly the same structure as that one in our iteration lemma 1.3. In order to apply it, we still have to convince ourselves that

$$\sigma \geq \frac{kp(1+\delta)}{\delta}$$

and that

$$\rho_0^\alpha \geq Ct^{1-\Theta}\left(G(0)+\rho_0^\sigma\right)^\delta .$$

The latter condition is satisfied for $t = T^\star$ sufficiently small, while the former one is actually an equality; indeed

$$\begin{aligned}\frac{kp(1+\delta)}{\delta} &= kp + \frac{kpq}{(1-\Theta)(p-q)} = kp + \frac{kpq + N(p-q)}{(p-q)}\\ &= kp + N + \frac{kpq}{p-q} = \sigma.\end{aligned}$$

Hence

$$\int_0^{T^\star}\int_{C(0)} |w|^p dxdt = 0.$$

This implies in particular that $x_0 = 0$ belongs to the closure of $\mathbf{R}^N \setminus \operatorname{supp}(w(t))$ for almost every $t \in (0, T^\star)$, and the proof is complete.

3 Applications

3.1 Second Order Doubly Nonlinear Equations

The equation

$$u_t - \operatorname{div}(u^{m-1}|\nabla u|^{p-2}\nabla u) = 0$$

is well known to model the flow of a non-Newtonian polytropic gas of density u through a homogeneous porous medium. The nonlinearities depend on the rheological properties of the gas: in particular, $m > 1$ for a polytropic gas and $p \neq 2$ for a non-Newtonian gas. The following statement summarizes results in [22] and [17]:

Proposition 3.5 *For any $p > 1$, $m \geq 1$, $m + p > 3$, and any non-negative $u_0 \in L^\infty(\mathbf{R}^N)$ with compact support, there exists a Hölder continuous solution of (2) (in the sense of distributions) with initial datum u_0; moreover u has finite speed of propagation in the sense of (11).*

Multiplying the equation by $u^\alpha\zeta^p$, $\alpha > 0$, integrating by parts and using Hölder's inequality, it is not difficult to show that the following estimate holds for all $\zeta \in W^{1,\infty}(\mathbf{R}^N)$, in particular for any $\zeta \in \mathcal{F}$:

$$\begin{aligned}&\tfrac{1}{\alpha+1}\int_{\mathbf{R}^N} \zeta^p u^{\alpha+1}(t,x)dx + C^{-1}\int_0^t \zeta^p |\nabla u^{\frac{\alpha+m+p-2}{p}}|^p dxdt\\ &\leq \tfrac{1}{\alpha+1}\int_{\mathbf{R}^N}\zeta^p u_0^{\alpha+1}dx + C\int_0^t\int_{\operatorname{supp}(\zeta)} u^{\alpha+m+p-2}|\nabla\zeta|^p dxdt.\end{aligned}$$

Therefore we can apply Theorem 1.2 with

$$w = u^{\frac{\alpha+m+p-2}{p}}, \quad q = \frac{p(\alpha+1)}{\alpha+m+p-2}$$

to obtain

Corollary 3.6 *Let u be a solution of (2) in the sense of Proposition 3.5. If* $\text{supp}(u_0)$ *has the exterior cone property (12) at a boundary point x_0 and if*

$$\limsup_{r\to 0} r^{-\frac{p}{m+p-3}(\alpha+1)} \int_{B(x_0,r)} u_0(x)^{\alpha+1}\, dx < \infty$$

for a positive number α, then u has positive waiting time at x_0.

Our conditions for the occurrence of a waiting time phenomenon are, in higher space dimensions, weaker than those formulated in [5] and in [26]; in particular, they yield the critical exponent $\gamma = \frac{p}{m+p-3}$ in the sense of (6):

$$T^\star > 0 \quad \text{for all } u_0 \quad \text{s.t.} \quad u_0(x) \le C|x-x_0|^{\frac{p}{m+p-3}} \quad \forall\, |x-x_0| < \delta$$

for some $C > 0$, $\delta > 0$.

3.2 The Thin Film Equation

The height of a thin film of viscous liquid, which spreads on a horizontal surface under the influence of surface tension, can be modeled by the equation

$$u_t + \text{div}(u^n \nabla \Delta u) = 0. \tag{23}$$

This equation is one of the few examples of fourth order parabolic equations which allow for globally non-negative solutions. Another peculiarity is the property that both qualitative behaviour of solutions and analytic tools to be applied are sensitive to the diffusion growth exponent $n > 0$. For $n \ge 3$, not only formal arguments suggest the solution's support to be constant in time. For $0 < n < 2$, *strong solutions* – characterized by a zero contact angle at the free boundary – exist that satisfy the integral estimate

$$\frac{1}{\alpha(\alpha+1)} \int_{\mathbf{R}^N} \zeta^4 u(t,x)^{\alpha+1} dx + C^{-1} \int_0^t \int_{\mathbf{R}^N} \zeta^4 |D^2 u^{\frac{\alpha+n+1}{2}}|^2 dx dt$$
$$\le \frac{1}{\alpha(\alpha+1)} \int_{\mathbf{R}^N} \zeta^4 u_0^{\alpha+1} dx + C \int_0^t \int_{\{\zeta>0\}} u^{\alpha+n+1} dx dt.$$

Note that the parameter α has to satisfy the condition

$$\alpha \in (\max(0, \frac{1}{2} - n), 2 - n). \tag{24}$$

By virtue of Theorem 1.2, choosing

$$w = u^{\frac{\alpha+n+1}{2}}, \quad p = 2, \quad q = \frac{2(\alpha+1)}{\alpha+n+1},$$

we immediately infer the following

Corollary 3.7 *Let* $0 < n < 2$, $N < 4$, *suppose* $u_0 \in H^1(\mathbf{R}^N; \mathbf{R}_0^+)$ *to be compactly supported and assume* u *to be a strong solution as constructed in [9] and in [13], respectively. Assume moreover that* $\text{supp}(u_0)$ *has the exterior cone property (12) at a boundary point* x_0. *If*

$$\limsup_{r\to 0} r^{-\frac{4}{n}(\alpha+1)} \fint_{B(x_0,r)} u_0^{\alpha+1}\,dx < \infty \tag{25}$$

for a number α *as in (24), then* u *exhibits a waiting time phenomenon at* x_0.

Note that (25) is satisfied if $|u_0(x)| \le C|x - x_0|^{4/n}$ with a positive constant C for values of x near x_0.

Surprisingly, for $n \in [2, 3)$ the structure of the operator changes once more. Due to the restriction (24), an estimate of type (4) declines to hold in that parameter regime. Hence, the argumentation has to be based on appropriate modifications of the only remaining integral estimate known so far which is the energy estimate

$$\frac{1}{2}\int_{\mathbf{R}^N} |\nabla u(t,x)|^2 dx + \int_0^t \int_{\mathbf{R}^N} u^n |\nabla \Delta u|^2 dx dt \le \frac{1}{2}\int_{\mathbf{R}^N} |\nabla u_0|^2 dx. \tag{26}$$

Combining the iteration method formulated in the present paper with appropriate versions of Hardy's inequality and with interpolation inequalities of the type

$$\int_\Omega |\nabla u^{\frac{n+2}{6}}|^6 dx + \int_\Omega |\nabla \Delta u^{\frac{n+2}{2}}|^2 dx \le C(n,N) \int_\Omega u^n |\nabla \Delta u|^2 dx$$

(cf. [12] for $N = 1$, [21, 20] for $N > 1$), the following result may be established. For the one-dimensional case, we refer to [15], the multi-dimensional case is treated in [19].

Theorem 3.8 *Let* $2 \le n < 3$, $N < 4$ *and assume* u *to be a strong solution of equation (23) with compactly supported initial data* $u_0 \in H^1(\mathbf{R}^N; \mathbf{R}_0^+)$ *as it has been constructed in [21]. Suppose that* $\text{supp}(u_0) \cap \mathcal{C} = \emptyset$, *where* $\mathcal{C}$ *is an exterior cone centered at a boundary point* x_0, *and assume that*

$$\limsup_{r\to 0} r^{-2(\frac{4}{n}-1)} \fint_{B(x_0,r)} |\nabla u_0(x)|^2 dx < \infty. \tag{27}$$

Then, u *exhibits a waiting time phenomenon locally at* x_0.

Note that condition (27) is stronger than condition (25) in the sense that it invokes a condition on the derivative of initial data which reads in a power-law-formulation as $|\nabla u_0(x)| \le C|x - x_0|^{\frac{4}{n}-1}$. It remains an open problem whether (27) can be weakened and substituted by (25) also for $n \in [2, 3)$.

3.3 Higher Order Doubly Nonlinear Equations with Variational structure

The following statement summarizes results of Bernis [10], [11]:

Proposition 3.9 *For any $1 < q < p$ and any $u_0 \in L^q(\mathbf{R}^N)$ with compact support, there exists a solution (in the sense of distributions)*

$$u \in C([0,T]; L^q(\mathbf{R}^N)) \cap L^p(0,T; W_c^{k,p}(\mathbf{R}^N))$$

of equation (3) with initial datum u_0.

Arguing as in [2] (cf. in particular formula (4.5)), it is possible to show that for $\zeta \in \mathcal{F}$

$$\int_{\mathbf{R}^N} \zeta^{kp}|u(t,x)|^q dx + C^{-1} \int_0^t \int_{\mathbf{R}^N} \zeta^{kp}|D^k u|^p dxdt \leq \int_{\mathbf{R}^N} \zeta^{kp}|u_0|^q dx + C \int_0^t \int_{\{\zeta>0\}} |u|^p dxdt,$$

which coincides with (4). Though finite speed of propagation has been considered by many authors (see for example [11, 26, 2]), to the best of our knowledge a local statement of the form (11) is not available for $k > 1$. Assuming that (11) holds, we can apply Theorem 1.2: A waiting time phenomenon occurs at x_0 provided the exterior cone property is satisfied and

$$\limsup_{r\to 0} r^{-\frac{kp}{p-q}q} \fint_{B(x_0,r)} |u_0(x)|^q \, dx < \infty.$$

which yields $\gamma = \frac{kp}{p-q}$ as critical exponent. For $k = 1, p = 2, q = \frac{m+1}{m}$ and $u \mapsto u^m$, we recover the critical exponent $\gamma = \frac{2}{m-1}$ for the porous medium equation.

References

1. Alikakos, N.D.: On the pointwise behaviour of solutions of the porous medium equation as t approaches zero or infinity. Nonlinear Anal., **9**, 1095–1113, 1985
2. Andreucci, D. Tedeev, A. F.: Finite speed of propagation for the thin-film equation and other higher-order parabolic equations with general nonlinearity. Interfaces Free Bound. 3, no. 3, 233–264, 2001
3. Antontsev, S.N.: On the localization of solutions of nonlinear degenerate elliptic and parabolic equations. Soviet Math. Dokl., **24**, 420–424, 1981
4. Antontsev, S.N., Diaz, J.I.: On space or time localization of solutions of nonlinear elliptic and parabolic equations via energy methods. In: Recent advances in nonlinear elliptic and parabolic problems (Nancy, 1988). Pitman Res. Notes Math. Ser., 208, 3–14, Longman Sci. Tech., Harlow, 1989
5. Antontsev, S.N., Diaz, J.I., Shmarev, S.I.: The support shrinking in solutions of parabolic equations with non-homogeneous absorption terms. In: Elliptic and parabolic problems (Pont-à-Mousson, 1994), Pitman Res. Notes Math. Ser., 325, 24–39, Longman Sci. Tech., Harlow, 1995
6. Aronson, D.G.; Bénilan, P.: Régularité des solutions de l'équation des milieux poreux dans $\mathbf{R}^N$. C. R. Acad. Sci. Paris, **288**, 103–105, 1979
7. Aronson, D.G., Caffarelli, L.A.: The initial trace of a solution of the porous medium equation. Trans. Amer. Math. Soc., **280**, 351–366, 1983

8. Bénilan, P., Crandall, G., Pierre, M.: Solutions of the porous medium equation in $\mathbf{R}^N$ under optimal conditions on initial values. Indiana Univ. Math. J., **33**, 51–87, 1984
9. Beretta, E., Bertsch, M., Dal Passo, R.: Nonnegative solutions of a fourth order nonlinear degenerate parabolic equation. Arch. Ration. Mech. Anal., **129**, 175–200, 1995
10. Bernis, F.: Existence results for "doubly" nonlinear higher order parabolic equations on unbounded domains. Math. Ann., **279**, 373–394, 1988
11. Bernis, F.: Qualitative properties for some nonlinear higher order degenerate parabolic equations. Houston J. Math., **14**, 319–352, 1988
12. Bernis, F.: Integral inequalities with applications to nonlinear degenerate parabolic equations. In: Angell, T.S., Cook, L.P., Kleinman, R.E., Olmstead, W.E. (eds) Nonlinear Boundary Value Problems. SIAM, Philadelphia, 1996
13. Bertsch, M., Dal Passo, R., Garcke, H., Grün, G.: The thin viscous flow equation in higher space dimensions. Adv. Differential Equations, **3**, 417–440, 1998
14. Chipot, M., Sideris, T.: An upper bound for the waiting time for nonlinear degenerate parabolic equations. Trans. Amer. Mat. Soc., **288**, 423–427, 1985
15. Dal Passo, R., Giacomelli, L., Grün, G.: A waiting time phenomenon for thin film equations. Ann. Scuola Norm. Sup. Pisa, **XXX**, 437–463, 2001
16. Dal Passo, R., Giacomelli, L., Shishkov, A.: The thin film equation with nonlinear diffusion. Comm. PDE, **26**, 1509–1557, 2001
17. Diaz, J.I., Veron, L.: Local vanishing properties of solutions of elliptic and parabolic quasilinear equations. Trans. Amer. Math. Soc., **290**, 787–814, 1985
18. Gagliardo, E.: Ulteriori properità di alcune classi di funzioni in piú variabili. Ricerche di Mat. Napoli, 24–51, 1959
19. Grün, G.: Droplet spreading under weak slippage: the waiting time phenomenon. Submitted for publication
20. Grün, G.: On Bernis' interpolation inequalities in multiple space dimensions. Z. Anal. Anwendungen, **20**, 987–998, 2001
21. Grün, G.: On free boundary problems arising in thin film flow. Habilitation thesis, University of Bonn (2001), accepted
22. Ivanov, A.V.: Existence and uniqueness of a regular solution of the Cauchy-Dirichlet problem for doubly nonlinear parabolic equations. Z. Anal. Anwendungen, **14**, 751–777, 1995
23. Knerr, B.F.: The porous medium equation in one dimension. Trans. Am. Math. Soc., **234**, 381–415, 1977
24. Nirenberg, L.: On elliptic partial differential equations. Ann. Scuola Norm. Sup. Pisa, **13**, 115–162, 1959
25. Shishkov, A.E.: On estimates of the speed of propagation of perturbations in quasilinear divergent degenerate parabolic equations of higher order. Ukranian Math. J., **44**, 1335–1340, 1992
26. Shishkov, A.E., Shchelkov, A.G.: Dynamics of the support of energy solutions of mixed problems for quasi–linear parabolic equations of arbitrary order. Izv. Math., **62**, 601–626, 1998
27. Stampacchia, G.: Équations elliptiques du second ordre à coefficients discontinus. Les presses de l'Université de Montréal, Montréal, 1966

The Mathematics of Ostwald Ripening

Barbara Niethammer

Institut für Angewandte Mathematik, Wegelerstr. 6, D-53115 Bonn, Germany
barbara@iam.uni-bonn.de

Summary. We review recent progress in the analysis of mean–field models for Ostwald ripening with particular emphasis on their regime of validity and their large–time dynamics.

1 Introduction

Ostwald ripening, a fundamental process in the aging of materials, occurs as the last stage of many first–order phase transformations. Consider for example a mixture consisting of two components A and B, which is quenched below a critical temperature into an unstable region. There the free energy favors two phases, each with a characteristic value of the relative concentration of the components. If the volume fraction of one component, say A, is sufficiently small, many small particles of the phase with higher equilibrium concentration of A atoms form, immersed in the second phase, also called background phase or matrix. Now, the concentration field has approximately reached equilibrium in each phase, but the interface between particles and matrix carries surface energy. In the last stage of the phase–transition the system evolves to reduce this surface energy by diffusion of A atoms from small to large particles. As a consequence large particles grow, while smaller ones shrink and eventually vanish, such that the number of particles decreases, whereas the typical length scales such as particle size and interparticle distance increase. This form of competitive growth is known as Ostwald ripening.

Due to the paradigmatic nature of Ostwald ripening, which can be observed in virtually all two–phase systems, there is considerable interest in a theory to predict the statistics of this process. In their pioneering works, Lifshitz and Slyozov [16] and Wagner [40] presented and analyzed a model which allows to make predictions on a system with a large number of particles. Their theory is nowadays known as the classical LSW–theory for Ostwald ripening and since then an enormous and still growing body of work appeared in the physics and metallurgical literature.

More recently also the attention of mathematicians was drawn to the subject; it is the purpose of this article to review some of the mathematical contributions. We will concentrate mainly on two issues:

- Rigorous derivation of reduced models from free boundary problems and the validity of the classical LSW–model
- Analysis of the large–time dynamics of the LSW–model

We do not aim to give an overview of the work in the applied literature. Instead, we refer to the excellent review articles [37, 38] and to the references in the articles we mention below.

2 Derivation of Mean–Field Models

2.1 The Mullins–Sekerka Model

The last stage of the phase transformation is well described by Stefan–type free boundary problems coupled with the Gibbs–Thomson law on the interface which accounts for surface tension. We will for simplicity concentrate on the simplest variant, which in the context of binary alloys is also known as the Mullins–Sekerka free boundary problem and can be derived as the sharp interface limit of the Cahn–Hilliard model (see e.g. [30]). Let the whole system cover a domain $\Omega \subset I\!R^d$, $d = 2, 3$, while the two phases are represented by a region $G \subset \Omega$ and its complement respectively. The Mullins–Sekerka model is based on the assumption that the diffusion field, which is given by the negative gradient $-\nabla u$ of the chemical potential u, is in quasi–stationary equilibrium given the phase distribution, i.e.

$$\Delta u = 0 \quad \text{in the bulk } \Omega \backslash \partial G, \tag{1}$$

$$u = \kappa \quad \text{at the interface } \partial G, \tag{2}$$

where κ denotes the mean curvature of the interface. Equation (2) is just the Gibbs–Thomson law for local equilibrium at the interface. The kinematic Stefan–condition prescribes the normal velocity of the interface via

$$V = [\nabla u \cdot \mathbf{n}] \quad \text{at the interface } \partial G. \tag{3}$$

Here $\mathbf{n}$ denotes the normal to the interface, $[\nabla u \cdot \mathbf{n}]$ the jump of the normal component of the gradient $\nabla u \cdot \mathbf{n}$ across the interface and V the normal velocity of the interface. Under Neumann– or periodic boundary condition, this interfacial motion reduces the area of ∂G and preserves the volume of G. It is well known that the evolution can also be interpreted as the gradient flow of the surface energy with respect to the H^{-1}–norm in the bulk.

2.2 Small Volume Fraction

In the following we consider the case when one phase consists of $N \gg 1$ particles and we are interested in their average dynamics. Reduced models can usually be derived if there is a clear separation of length scales and we consider in the following the case that the volume fraction ε of one phase is small, in other words,

$$\text{system size} \gg \text{typical distance} \gg \text{typical radius}. \tag{4}$$

One observes in experiments in this regime that particles quickly become radially symmetric and do essentially not drift in space. Also, in the Mullins–Sekerka model drifting of particles and deviations from sphericity are higher order effects compared with the variation of the radii, as has been argued by several authors (cf. eg. [17, 36]). More recently, this has also been formulated in a mathematically rigorous manner in [1, 2]. There, a configuration of a finite number of almost spherical particles is considered, such that the ratio between size of particles and distance as well as the deviation from sphericity compared to size is small. It is shown for the full Mullins–Sekerka evolution that the particles continue to be close to spherical for all times, which allows to associate to each particle a radius R_i and center X_i. In addition it is derived that to leading order the evolution of R_i satisfies the equation of the classical LSW–theory (see (10), (11) below), whereas the centers move much slower than the radii.

This analysis suggests to replace problem (1)-(3) by:

$$\Delta u = 0 \text{ in } \Omega \setminus \bigcup_i B_i\,(t)\,, \tag{5}$$

$$u = \frac{1}{R_i} \text{ in } \partial B_i\,(t)\,, \tag{6}$$

$$\dot{R}_i = \frac{1}{|\partial B_i\,(t)|}\int_{\partial B_i(t)} \frac{\partial u}{\partial n}, \tag{7}$$

where R_i is the radius of the ball $B_i\,(t)$. The balls have centers $\{X_i\}$ which are given initially and remain fixed in this model for $t > 0$.

It is easily derived that this evolution keeps the main features of the Mullins–Sekerka problem. The volume fraction of particles is preserved, i.e.

$$\sum_i R_i^3(t) = \sum_i R_i^3(0) \qquad \text{for all } t > 0 \tag{8}$$

and the surface energy is decreasing, i.e.

$$2\pi \sum_i R_i^2(t) + \int_0^t \int_\Omega |\nabla u|^2 \, dx\, ds = 2\pi \sum_i R_i^2(0). \tag{9}$$

2.3 The Classical LSW–Theory

In the case of small volume fraction and $d = 3$, Lifshitz, Slyozov and Wagner [16, 40] formally identified the leading order term for the evolution of $\{R_i\}_i$. Their main assumption is that due to small volume fraction the chemical potential is approximately constant in space at distances from the particles which are much larger than the typical sizes, i.e. $u(t,x) \approx \overline{u} = \overline{u}(t)$ for $\min_i |x - X_i| \gg \langle R \rangle$. Then the growth rate of a particle is determined by the growth rate of an isolated particle which evolves according to Mullins–Sekerka where u satisfies the boundary condition $u(t,\infty) = \overline{u}(t)$. The solution can be easily computed explicitly and one obtains from (7)

$$\dot{R}_i = \frac{1}{R_i^2}(R_i\,\overline{u}(t) - 1), \qquad \text{as long as } R_i > 0. \tag{10}$$

The so–called mean–field $\overline{u}(t)$ is determined by the conservation of total particle volume (8), which together with (10) gives

$$\overline{u}(t) = \frac{\sum_{i:R_i>0} 1}{\sum_i R_i(t)}, \tag{11}$$

i.e. $\overline{u}$ equals the mean radius in the systems. If one introduces the particle radii distribution $f(t,R)$ via

$$\int_{R_+}^{R_-} f(t,R)\,dR = \text{number of particles with radius in } (R_+, R_-),$$

the system of ordinary differential equations (10) translates into the kinetic equation

$$\partial_t f + \partial_R\Big(\frac{1}{R^2}\,(R\overline{u} - 1)\,f\Big) = 0, \tag{12}$$

and (11) can be written as

$$\overline{u} = \frac{\int f\,dR}{\int R\,f\,dR}. \tag{13}$$

Thus, LSW reduce the Mullins–Sekerka evolution to an evolution for the distribution ν of the particle radii — a tremendous reduction.

The non–local transport equation (12) has the scale invariance $R \sim t^{1/3}$ and allows for a one–parameter family of self–similar solutions. Based on a formal large–time analysis LSW predict that a particular one of these self–similar solutions characterizes the long–time behavior of all solutions, which implies universal power laws for statistical quantities. In particular it follows for the mean particle radius $\overline{R}$ that

$$\overline{R} \approx \Big(\frac{4}{9}t\Big)^{1/3}.$$

Unfortunately, the LSW–theory has two serious shortcomings. First, a rigorous analysis of the large–time behavior of the LSW–equation shows, that it is indeed not universal but depends on the contrary very sensitively on the data. We will explain these results in detail in Section 3.

The second problem within the classical LSW–theory is that all self–similar profiles and the corresponding coarsening rates do not agree well with experiments. It is common belief that this discrepancy between theory and experiment is due to the finiteness of the volume fraction ε of the particles. One can find many predictions on the first–order corrections to the LSW–theory in the physics literature; however, there are different approaches which come to quite different conclusions and so far no generally accepted theory seems available.

The first step towards a rigorous study of this issue is to establish the natural range of validity of the LSW–model. In other words, we ask when the LSW–model can be identified as the leading order dynamics of problem (5)-(7) in the limit $N \to \infty$ and $\varepsilon \to 0$.

2.4 The Screening Length

In this section we present the heuristics for a rigorous derivation of homogenization limits of (5)-(7). We do not aim to give the precise mathematical results, which would inevitably involve many technicalities; instead we will elucidate the main ideas.

To establish the range of validity of the LSW–model it is essential to understand the effective range of particle interactions which is limited by screening. This effect will be formally uncovered by the following consideration.

We aim to construct an approximation of the solution u of (5)-(6) and introduce a field $\overline{u}$, which represents an average of u over length scales which are larger than the typical distance between particles. For fixed time t we write

$$u(x) = \sum_i \frac{(1 - R_i\overline{u}(X_i))}{|x - X_i|}. \tag{14}$$

The idea behind this ansatz is, that $\frac{1}{|x-X_i|}$ represents the solution for a single particle. On the other hand, there is an additional "charge" $\frac{R_i\overline{u}(X_i)}{|x-X_i|}$ coming from the presence of the other particles in the system. The notation "charge" comes from an electrostatic analogy: one thinks of the particle as conductors.

Combining now (14) with (6) we obtain

$$0 = -\overline{u}(X_i) + \sum_{j \neq i} \frac{(1 - R_j\overline{u}(X_j)}{|x - X_j|}. \tag{15}$$

By this representation of $\overline{u}$ we can expect that $\overline{u}$ is slowly varying in space. For a continuum description of (15) we denote now by $f(x, R)$ the number density of particles, defined as

$$\int_{R_-}^{R_+} \int_{\Omega} f(x, R)\, dx\, dR \;=\; \#\{\, i \,|\, (X_i, R_i) \in \Omega \times (R_-, R_+)\,\}.$$

We immediately obtain from (7) and (14) that a particle radius evolves as

$$\dot{R}_i = \frac{R_i\overline{u}(X_i) - 1}{R_i^2} \tag{16}$$

which translates into the following evolution law for f:

$$\frac{\partial f}{\partial t}(t, x, R) + \frac{\partial}{\partial R}\Big(\frac{1}{R^2}(R\,\overline{u}(t, x) - 1) f(t, x, R)\Big) = 0. \tag{17}$$

The question is of course, how $\overline{u}$ is determined. We heuristically turn to a continuum version of (15) which is given by

$$\begin{aligned} 0 &= -\overline{u}(x) + \int\!\!\int \frac{1}{|x-y|}\,(1 - R\,\overline{u}(y))\, f(y, R)\, dy\, dR \\ &= -\overline{u}(x) + \int \frac{1}{4\pi\,|x-y|}\,(4\pi\,\rho(y) - \mu(y)\,\overline{u}(y))\, dy, \end{aligned} \tag{18}$$

where ρ and μ are the number resp. capacity density of the particles, i. e.

$$\rho(x) := \int_0^\infty f(x,R)\,dR \quad \text{and} \quad \mu(x) := \int_0^\infty 4\,\pi\,R\,f(x,R)\,dR. \tag{19}$$

We call μ the capacity density since $4\pi R$ is the capacity of a ball of radius R in $\mathbb{R}^3$. We now apply the Laplace operator to the identity (18) and obtain

$$-\Delta\overline{u}(x) + \mu(x)\,\overline{u}(x) \;=\; 4\,\pi\,\rho(x). \tag{20}$$

In the language of electrostatics, this equation for the mesoscopic potential $\overline{u}$ displays the effective screening in the arrangement of charged particles. Indeed, it can be immediately read of from (20) that particles interact only over a finite length, called the screening length, which is related to the average capacity density $\bar{\mu}$ via

$$\text{screening length} := \frac{1}{\sqrt{\bar{\mu}}} \sim \frac{(\text{typical distance})^{3/2}}{(\text{typical radius})^{1/2}} \overset{(4)}{\gg} \text{typical distance}. \tag{21}$$

We see from (21) that the effective interaction range includes indeed many particles in the regime $\varepsilon \ll 1$. Since due to (20) the screening length determines the length scale over which $\overline{u}$ varies, This a posterori legitimates the assumption that $\overline{u}$ is slowly varying in space which allows to pass to the continuum description (18).

2.5 Zero Order Models: Regimes of Validity

In this subsection we review mathematical results which make the formal considerations in subsection 2.4 rigorous. One finds, that there are three different regimes, depending on the ratio between system size and the screening length, as defined in (21).

Case I: The most dilute case is that

$$\text{System size} \quad \ll \quad \text{Screening length} \tag{22}$$

or equivalently, the case that particles have small capacity density.

Since $\overline{u}$ varies on length scales of the order of the screening length, we expect that $\overline{u}(t,x) = \overline{u}(t)$ in Ω. Thus (17), (20) just project on the LSW–model (12), (13) for $f(R) = \int f(x,R)\,dx$.

The first rigorous derivation of the LSW–model and variants of it can be found in [21, 22]. It is derived as a homogenization limit of (5)-(7) in the regime (22), which with (21) is again equivalent to $\varepsilon N^2 \to 0$. The proof is based on (9) and the construction of an approximation of the diffusion field similar as in (14).

Case II: In the case that

$$\text{System size} \quad \sim \quad \text{Screening length} \tag{23}$$

or that the capacity density is finite we expect that the system is to zero order well–described by (17) with (20). Indeed, this has been made rigorous in [24] and corresponding error–estimates can be found in [10]. The proofs in [24] make use of the gradient flow structure of both, the Mullins–Sekerka evolution as well as the reduced model.

Case III: Finally, the case that

$$\text{System size} \quad \gg \quad \text{Screening length} \tag{24}$$

or that the capacity density is large, or $\varepsilon N^2 \to 0$, has been investigated in [28, 29] under general assumption on the particle arrangement. One expects, after rescaling the system with respect to the natural length scale, the screening length, that the evolution is to zero order governed by (17), (20). However, in a rigorous analysis one faces new difficulties, since the size of the rescaled system becomes unbounded. For that reason, new methods compared to the ones used in [21, 24] are developed. One important step in the analysis is to show that the fundamental solution of the homogenization problem decays exponentially on distances larger than the screening length. This is a rigorous formulation of the intuition gained in Section 2.4 on the effective range of particle interactions. In [28] also special care is taken to have the most general assumptions on the initial particle arrangements. (In [10, 21, 24] it is assumed that the minimum distance between particles is of the order of the mean distance.) At the initial time particles are placed i.i.d. in the domain Ω. First, it is shown that the evolution preserves such a statistically uniform distributions over the typical time scales, such that one has a proper definition of "the" screening length. Second, it is established that only a very small fraction of particles can overlap and that this does not affect the macroscopic evolution law for the remaining particles.

In the rest of this section we briefly report on further issues in the theory of Ostwald ripening which we cannot address in detail in the framework of this article.

2.6 Corrections to Zero Order Models

As pointed out in subsection 2.3 one is interested in corrections to the LSW–theory to account for finite volume fraction of particles. One can find several, partially contradicting, theories for the first–order correction in the physics literature. There are

- theories for a finite or periodic system, predicting a correction of order $\varepsilon^{1/3}$ (cf. e.g. [8, 39])
- theories for infinite clusters based on a simple form of the two–particle correlation function predicting a correction of order $\varepsilon^{1/2}$ (cf. e.g. [3, 18, 41])
- theories for infinite clusters which keep track of the evolution of the two–particle correlation function, predicting a correction also of order $\varepsilon^{1/2}$ but of much larger magnitude (cf. e.g. [17, 34])

Recently, a first mathematically rigorous result for finite systems has been presented in [12]. It is established that there is a cross–over of the first–order correction from a $\varepsilon^{1/3}$– to a $\varepsilon^{1/2}$–scaling when the system size becomes larger than the screening length, which confirms numerical experiments in [8].

Despite the large amount of literature, however, the precise form of the first–order correction and the question, whether correlations between particles develop and how the influence the coarsening rate, is still a controversial issue.

2.7 Two Dimensional Ripening

We concentrate in this article for simplicity on the situation in three space dimensions. However, there are many situations where Ostwald ripening takes place in an essentially two–dimensional setting, as e.g. in growth of thin films by chemical vapor deposition. Note that the simple ansatz in the LSW–theory fails in two dimensions due to the divergence of the logarithmic Greens function. It appears that this caused some confusion in the physics literature about the two–dimensional analogue of the LSW–model, which is just (12) coupled with the constraint of constant volume fraction which gives for the mean–field

$$\bar{u}(t) = \frac{\int \frac{1}{R} f \, dR}{\int f \, dR},$$

i.e. in two dimensions the mean–field is just the inverse harmonic radius. This model and the corresponding inhomogeneous extension has been derived rigorously in the regimes I and II above in [23]. The screening length is also in two dimensions related as above to the capacity density, which in dimension two is given by a different formula. The main technical difficulty compared to the three–dimensional case is to make sense of $\int \frac{1}{R} f \, dR$ and to show that $\overline{u}$ is well–defined.

2.8 The Effect of Elasticity

We have so far studied the situation where the last stage of Ostwald ripening is dominated by the reduction of surface energy. However, in most materials of technological importance the effect of elastic stresses due to different lattice parameters in particles and matrix cannot be neglected. For an extensive overview of models and qualitative predictions for phase transformations in alloys with elastic misfit we refer to [9].

It is well–known that the coarsening rate is not affected if the elasticity tensor is homogeneous and isotropic, the misfit (the difference between the equilibrium stresses in the different phases) is isotropic and if boundary effects can be neglected. Extensive numerical simulations for homogeneous anisotropic elasticity have been performed [33], which suggest, that in this case at least the power law for the coarsening rate is not altered. At present neither heuristic predictions for inhomogeneous elasticity exist, nor are there any mathematical results on the influence of elasticity on Ostwald ripening in general.

3 Self–Similarity and Instability in the Mean–Field Theory

This section is devoted to the analysis of the large–time behavior of the homogeneous LSW–model (12), (13). Despite its simplicity, the dynamics of the model turn out to be unusual and interesting.

For what follows it will be convenient to work with the volume of a particle as a variable instead of the radius. Thus, we denote $v = \frac{4}{3}\pi R^3$ and after rescaling time by an appropriate constant we obtain that the volume of a particle satisfies

$$\frac{d}{dt}v = v^{1/3}\overline{u}(t) - 1. \tag{25}$$

For the distribution of particle volume, which we call for simplicity again f, this gives

$$\partial_t f + \partial_v((v^{1/3}\overline{u}(t) - 1)f) = 0, \qquad t > 0, v \in I\!R_+, \tag{26}$$

and $\overline{u}(t)$ is such that the volume fraction is conserved, i.e.

$$\int_0^\infty v\, f(t,v)\, dv = V, \tag{27}$$

which is equivalent to

$$\overline{u}(t) = \frac{\int_0^\infty f(t,v)\, dv}{\int_0^\infty v^{1/3} f(t,v)\, dv}. \tag{28}$$

3.1 Well–Posedness

There are two different approaches to provide an existence– and uniqueness theory for (26). In [14] an L^1–theory of solutions to

$$\partial_t f + \partial_v((k(v)\overline{u}(t) - q(v))f) = 0, \qquad \overline{u}(t) = \frac{\int_0^\infty q(v)\, f(t,v)\, dv}{\int_0^\infty k(v)\, f(t,v)\, dv},$$

is developed, under certain assumptions on the coefficients k, q which in particular ensure that the characteristics leave the domain at $v = 0$. Uniqueness was shown for a smaller class of coefficients, which unfortunately does not allow for the LSW–case $k(v) = v^{1/3}, q(v) = 1$.

In [26] the initial value problem for (26), (28) is treated within a framework that also admits distributions for which a finite fraction of the particles all have the same size, in which case the volume distribution contains Dirac deltas. Existence, uniqueness and continuous dependence on the data is established in [26] with respect to natural metric on the space of probability measures with compact support. It measures the smallest maximum volume change required to rearrange one volume distribution into another and is mathematically speaking just given by the L^∞–Wasserstein metric for the size distributions.

3.2 Basic Results on the Large–Time Behavior

In this and the following subsection we discuss in some detail the results of [25], which show that the large–time dynamics of (26)& (28) depend sensitively on the initial data, more precisely on the largest particles in the system. The idea behind the results is simple and can be described by the following cartoon: the essential effect of the dynamics is to stretch the distribution, then cut off the part which corresponds to the dead particles, and renormalize to satisfy the volume constraint. In other words, one zooms in at the end of the support of the distribution.

In accordance with this intuition, an important role in the analysis of [25] is played by the characteristic corresponding to the maximal particle volume, which is denoted by

$$\bar{v}(t) := \inf \left\{ v \,|\, \int_v^\infty f(t,z)\,dz = 0 \right\}.$$

Since the radius of the largest particle is always larger than the mean radius which by (28) is just the critical radius, it follows that $\bar{v}(t)$ is increasing.

When studying the long–time behavior of solutions one has to distinguish whether the data $f_0(v)$ carries a Dirac distribution at the end of the support or not, in other words whether there is a positive fraction of particles with largest volume or not. Since (26) is a transport equation, one expects that if $f_0(v)$ carries a Dirac a delta then the corresponding solution converges to a Dirac delta for large times. If not, then $\bar{v}(t)$ is unbounded and $\int_{0+}^\infty f(t,z)\,dz \to 0$. Indeed, this behavior is established in Propositions 3.1, 3.2 in [25].

3.3 Long-Time Behavior in Self–Similar Variables

From now on we consider the case that no positive fraction of particles has the maximal volume $\bar{v}$ and we ask whether the convergence occurs in a self–similar manner. In the original work [16, 40] it is assumed that the critical radius is monotonically increasing to infinity and then the system is rescaled with respect to this quantity. In general though, it is not true that the mean radius increases monotonically. However, the maximal particle volume is monotone and thus one can rescale with respect to $\bar{v}(t)$. The same approach has been used in the physics paper [20, 11], where the large–time behavior of the resulting system is studied numerically.

One introduces new variables via

$$\tau = \ln \frac{\bar{v}(t)}{\bar{v}(0)}, \qquad z = 1 - \frac{v}{\bar{v}(t)}, \qquad \psi = \frac{\bar{v}(t)}{V} \int_{v+}^\infty f(t,z)\,dz, \tag{29}$$

where we now use the cumulative volume distribution ψ. For each $\tau \geq 0$, the function $z \mapsto \psi(\tau, z)$ is left continuous and increasing on $[0,1]$ with $\psi(\tau, 0) = 0$. For differentiable initial data ψ satisfies

$$\partial_\tau \psi + (\kappa(\tau) Q(z) - 1) z \partial_z \psi = \psi, \qquad \tau > 0, z \in (0,1), \tag{30}$$

where

$$Q(z) = 3\left(\frac{1-(1-z)^{1/3}}{z}\right)$$

is increasing for $u \in [0,1]$ with $Q(0) = 1$ and $Q(1) = 3$ The rescaled mean–field $\kappa(\tau) > \frac{1}{3}$ is determined from

$$\frac{1}{\bar{v}(t)^{1/3}\overline{u}(t)} = 1 - \frac{1}{3\kappa(\tau)} = \frac{1}{\psi(\tau,1)} \int_0^1 \frac{1}{3}(1-z)^{-2/3}\psi(\tau,z)\,dz \qquad (31)$$

which is equivalent to volume conservation

$$\int_0^1 \psi(\tau,z)\,dz = 1 \qquad \text{for all } t < 0. \qquad (32)$$

Self-similar solutions of (26) now correspond to stationary solutions of (30) and it turns out that there is a one–parameter family of those.

Lemma 3.1 ([25], Lemma 4.2, [20]) *For any $p \in (0,\infty]$, equation (30) has a stationary solution Ψ_p which satisfies*

$$\frac{d}{dz}\ln\Psi_p = \frac{1}{z(\kappa_p Q(z) - 1)},$$

with $\kappa_p = 1 + 1/p$ and is normalized such that $\int_0^1 \Psi_p(z)\,dz = 1$. For $p < \infty$ the profile has the form

$$\Psi_p(z) = \alpha_p(z) z^p$$

where α_p is decreasing and analytic on $[0,1)$. For $p = \infty$ ($\kappa_\infty = 1$) we have $\Psi_\infty(z) = o(z^q)$ as $u \to 0$ for all $q > 0$. (This solution is the one singled out by Lifshitz, Slyozov and Wagner.)

One would naturally like to know under what conditions $\psi(\tau,z)$ converges to a stationary solution as $\tau \to \infty$ and which solution is selected. We shall see that the long–time behavior is a sensitive function of the manner in which the initial data vanish as $z \to 0$. This can be seen already from the following result, whose proof is based on a comparison argument.

Proposition 3.2 *Let ψ be determined by (30)–(32), and suppose $0 < p < \infty$.*

a) If $\inf_{z>0} \psi_0(z)/z^p > 0$, then $\limsup_{\tau\to\infty} \kappa(\tau) \geq 1 + \frac{1}{p}$.
b) If $\sup_{z>0} \psi_0(z)/z^p < \infty$, then $\liminf_{\tau\to\infty} \kappa(\tau) \leq 1 + \frac{1}{p}$.

Next, one shows (cf. [25], Lemma 5.2) that if $\psi(\tau,z)$ converges as $\tau \to \infty$ the limit must be Ψ_p for some $p \in [0,\infty]$. Then the following instability result is just a corollary of Proposition 3.2.

Corollary 3.3 ([25], Corollary 5.4) *Given any $p \in (0,\infty]$, the stationary solution Ψ_p is unstable to perturbations that yield initial data that satisfy*

$$\inf_{z>0} \frac{\psi_0(z)}{z^q} > 0 \text{ for some } q < p.$$

In particular, the LSW solution (with $p = \infty$) is unstable to all perturbations for which $\inf_{z>0} \psi_0(z)/z^q > 0$ for some $q < \infty$.

One naturally would expect now that if $\psi(\tau, z) \to \Psi_p(z)$ for $0 \le z \le 1$ and if $p = 1/(\kappa_* - 1) < \infty$, then $\psi_0(z) \sim cz^p$ as $z \to 0$. This is almost correct. A necessary criterion for convergence in this case is related to the concept of regularly varying function (cf. e.g. [32]).

Definition 3.4 *A positive, measurable function g, defined on some interval of the form $(0, a]$, is called* regularly varying at 0 *(with exponent $p \in I\!R$) if*

$$\lim_{x \to 0+} \frac{g(\lambda x)}{\lambda^p g(x)} = 1 \quad \textit{for all } \lambda > 0. \tag{33}$$

If $p = 0$, we say g is slowly varying *at* 0.

The main result in [25] gives a necessary condition for convergence in the rescaled variables to any one of the stationary solutions Ψ_p with $0 \le p < \infty$.

Theorem 3.5. ([25], Theorem 5.10) *Assume that for some $p \in [0, \infty)$ it holds*

$$\lim_{\tau \to \infty} \psi(\tau, z) = \Psi_p(z) \quad \textit{for all } z \in [0, 1].$$

Then $\psi_0(\cdot)$ is regularly varying at 0 *with exponent p.*

It is conjectured that this condition is also sufficient for convergence towards a stationary solution Ψ_p. This conjecture is supported by numerical simulations and is proved for sufficiently small p for data close to a stationary solution (Theorem 5.12 in [25]). The results of [25] can be summarized as follows:

- The possible limits of the rescaled distribution can be classified by their exponent of vanishing $p \in [0, \infty]$. (For the LSW solution, $p = \infty$.)
- If the distribution function initially satisfies $\psi_0(u) \ge au^p$ for some exponent $p < \infty$ and some $a > 0$, then the scaled distribution function cannot converge to the form predicted by the classical LSW–theory.
- For convergence to a self-similar solution with $p \in [0, \infty)$, it is *necessary* that the initial distribution function is "regularly varying with exponent p" at the end of its support.
- Self-similar solutions with sufficiently small p are stable to perturbations that are small in a certain sense, and for such perturbations the condition of regular variation is both necessary and sufficient for convergence to self-similar form.

The analysis in [25] does not cover the case of data with infinite the support. In fact, the large–time behavior in self–similar variables for the LSW–model (26) for this case is at present unknown. One might expect that for sufficiently nice data, e.g. data with an exponentially decreasing tail, the solution converges to the LSW–solution as claimed by Meerson et al. [20, 11]. Furthermore it seems plausible, that if convergence occurs it should be towards the LSW–solution. However, it is completely unclear, what could be a necessary (and sufficient) condition for convergence towards a self–similar profile, replacing the the condition of regular variation for data with compact support.

A mathematical analysis of a related, but simpler equation can be found in [4] where (25) is replaced by

$$\frac{d}{dt}v = v\,\overline{u}(t) - 1.$$

Here the equation for the characteristics is linear and different methods from the ones in [25] can be employed, to establish as in the LSW–model a sensitive dependence of the large–time behavior on the initial data.

3.4 Conserved Total Mass

In this subsection we briefly describe results for another version of the LSW–model, which is sometimes also referred to as the LS–model. The condition (27) is an approximation of the more general constraint of conservation of total mass. If mass in the diffusion field is taken into account we have that

$$\overline{u}(t) + \int_0^\infty v f(t,v)\,dv = V \tag{34}$$

is conserved in time. Well–posedness theories for this and related models and various type of coefficients can be found in [6, 13, 26], a rigorous derivation in [21]. Since $\overline{u}$ depends linearly on f and is uniformly bounded by V, the constraint (34) simplifies existence– and uniqueness proofs significantly.

Concerning the long–time behavior, however, the situation is more complicated. First, different asymptotic states appear already in the original variables [27, 7]. E.g. sufficiently small initial mean–field can lead to the extinction of all particles in finite time. However, if the particle size distribution carries no Dirac mass at the tip of the support and if the initial maximal volume $\bar{v}(0)$ is sufficiently large, then $\bar{v}(t) \to \infty$ monotonically as $t \to \infty$ and one expects that (34) is well approximated by (8). Self–similar behavior for large times is investigated in [27] (see also [5] for numerical simulations). Here the second difference between (8) and (34) comes into play, which is that (34) does not allow for self–similar solutions. However, there exist exact solutions with constant mean–field which converge towards self–similar profiles. With these asymptotically self–similar solutions at hand analogous results in the self–similar regime can be derived as described in Section 3.3 for the case of conserved volume.

3.5 A Fokker–Planck Approximation

From the discussion in the previous subsections it is clear that to overcome the unphysical weak selection of self–similar solutions one has to go beyond the classical LSW–model. One approach, motivated by a connection with the Becker–Döring equations [31, 15], is to introduce a diffusion term in (26), which leads to

$$\partial_t f + \partial_v((v^{1/3}\overline{u}(t) - 1)f) = \partial_v^2(v^{1/3} f). \tag{35}$$

The interpretation of the additional term is that it accounts for fluctuations. Mathematically speaking, equation (35) is a parabolic equation as compared to the transport equation (26). Consequently, even if the initial data have compact support, the solution has an infinite tail for all positive times. It is shown in [35] that the self–similar solutions described in Section 3.3 with $p < \infty$ cannot occur as asymptotic states in this case. It is also argued in [36, 19] that the solution in fact converges to the LSW–solution, however a rigorous proof is not available. The technical difficulties one faces are similar as in the analysis of the long–time behavior of the LSW–model for data with infinte support.

It has also been speculated that a possible mechanism for a strong selection of an asymptotic state could be finite volume effects. In [17] a diffusion term, however different from the one in (35), has been derived by accounting for screening and correlation between particles. However, it seems that a convincing argument for a strong selection criterion is still missing.

References

1. N. Alikakos, G. Fusco: The equations of Ostwald ripening for dilute systems. J. Stat. Phys., 95, 5/6:851–866, 1999
2. N. Alikakos, G. Fusco: Ostwald ripening for dilute systems under quasistationary dynamics. Commun. Math. Phys., to appear
3. A. D. Brailsford, P. Wynblatt: The dependence of Ostwald ripening kinetics on particle volume fraction. Acta met., 27, 489–497, 1979
4. J. Carr, O. Penrose: Asymptotic behaviour in a simplified Lifshitz–Slyozov equation. Physica D, 124, 166–176, 1998
5. J. Carrillo, T. Goudon: A numerical study on large–time asymptotics of the Lifshitz–Slyozov system. 2001. preprint.
6. J.-F. Collet, T. Goudon: On solutions to the Lifshitz–Slyozov model. Nonlinearity, 13, 1239–1262, 2000
7. J.-F. Collet, T. Goudon, A. Vasseur: Some remarks on large–time asymptotics of the Lifshitz–Slyozov equations. 2001. preprint.
8. V. E. Fradkov, M. E. Glicksman, S. P. Marsh: Coarsening kinetics in finite clusters. Phys. Rev. E, 53, 3925–3932, 1996
9. Peter Fratzl, Oliver Penrose, Joel L. Lebowitz: Modeling of phase separation in alloys with coherent elastic misfit. J. Statist. Phys., 95(5-6), 1429–1503, 1999
10. A. Garroni, B. Niethammer: Correctors and error–estimates in the homogenization of a Mullins–Sekerka problem. IHP Analyse nonlinéaire, 2001. To appear.
11. B. Giron, B. Meerson, P. V. Sasorov: Weak selection and stability of localized distributions in Ostwald ripening. Phys. Rev. E, 58, 4213–4216, 1998
12. A. Hönig, B. Niethammer, F. Otto: On first–order corrections to the LSW–theory. 2001. preprint.
13. P. Laurençot: Weak solutions to the Lifshitz–Slyozov–Wagner equation. Indiana Univ. Math. J., 50, 3, 1319–1346, 2001
14. P. Laurençot: The Lifshitz–Slyozov–Wagner equation with total conserved volume. 2002. To appear.
15. P. Laurençot, S. Mischler: From the Becker–Döring to the Lifshitz–Slyozov–Wagner equations. 2002. To appear in J. Stat. Phys.

16. I. M. Lifshitz, V. V. Slyozov: The kinetics of precipitation from supersaturated solid solutions. J. Phys. Chem. Solids, 19:35–50, 1961
17. M. Marder: Correlations and Ostwald ripening. Phys. Rev. A, 36, 858–874, 1987
18. J. A. Marqusee, J. Ross: Theory of Ostwald ripening: Competitive growth and its dependence on volume fraction. J. Chem. Phys., 80, 536–543, 1984
19. B. Meerson: Fluctuations provide strong selection in Ostwald ripening. Phys. Rev. E, 60, 3, 3072–3075, 1999
20. B. Meerson, P. V. Sasorov: Domain stability, competition, growth and selection in globally constrained bistable systems. Phys. Rev. E, 53, 3491–3494, 1996
21. B. Niethammer: Derivation of the LSW–theory for Ostwald ripening by homogenization methods. Arch. Rat. Mech. Anal., 147, 2, 119–178, 1999
22. B. Niethammer: The LSW–model for Ostwald Ripening with kinetic undercooling. Proc. Roy. Soc. Edinb., 130 A, 1337–1361, 2000
23. B. Niethammer, F. Otto: Domain coarsening in thin films. Comm. Pure Appl. Math., 54, 361–384, 2001
24. B. Niethammer, F. Otto: Ostwald Ripening: The screening length revisited. Calc. Var. and PDE, 13, 1, 33–68, 2001
25. B. Niethammer, R. L. Pego: Non–self–similar behavior in the LSW theory of Ostwald ripening. J. Stat. Phys., 95, 5/6, 867–902, 1999
26. B. Niethammer, R. L. Pego: On the initial-value problem in the Lifshitz-Slyozov-Wagner theory of Ostwald ripening. SIAM J. Math. Anal., 31, 3, 457–485, 2000
27. B. Niethammer, R. L. Pego: The LSW–model for domain coarsening: Asymptotic behavior for total conserved mass. J. Stat. Phys., 104, 5/6, 1113–1144, 2001
28. B. Niethammer, J. J. L. Velázquez: Exponential screening and homogenization in coarsening systems. 2001. Preprint.
29. B. Niethammer, J. J. L. Velázquez: Well–posedness for an inhomogeneous LSW–model in unbounded domains. 2001. Preprint
30. R. L. Pego: Front migration in the nonlinear Cahn–Hilliard equation. Proc. R. Soc. Lond. **A**, 422, 261–278, 1989
31. O. Penrose: The Becker–Döring Equations at Large Times and Their Connection with the LSW Theory of Coarsening. J. Stat. Phys., 89, 1/2, 305–320, 1997
32. E. Seneta: Regular Varying Functions. Springer–Verlag, New York, 1976. Lec. Note in Math. Vol. 508
33. K. Thornton, N. Akaiwa, P. W. Voorhees: Dynamics of Late–Stage Phase Separation in Crystalline Solids. Phys. Rev. Let., 86, 7, 1259–1262, 2001
34. M. Tokuhama, K. Kawasaki: Statistical–mechanical theory of coarsening of spherical droplets. Physica A, 123, 386–411, 1984
35. J. J. L. Velázquez: The Becker–Döring equations and the Lifshitz–Slyozov theory of coarsening. J. Stat. Phys., 92, 195–236, 1998
36. J. J. L. Velázquez: On the effect of stochastic fluctuations in the dynamics of the Lifshitz–Slyozov–Wagner model. J. Stat. Phys., 99, 57–113, 2000
37. P. W. Voorhees: The theory of Ostwald ripening. J. Stat. Phys., 38, 231–252, 1985
38. P. W. Voorhees: Ostwald ripening of two–phase mixtures. Ann. Rev. Mater. Sc, 22, 197–215, 1992
39. P. W. Voorhees, M. E. Glicksman: Solution to the multi–particle diffustion problem with application to Ostwald Ripening – I. Theory. Acta met., 32, 2001–2011, 1984
40. C. Wagner: Theorie der Alterung von Niederschlägen durch Umlösen. Z. Elektrochemie, 65, 581–594, 1961
41. J. H. Yao, K. R. Elder, H. Guo, M. Grant: Theory and simulation of Ostwald ripening. Phys. Rev. B, 47, 14110–14125, 1993

Appendix

Color Plates

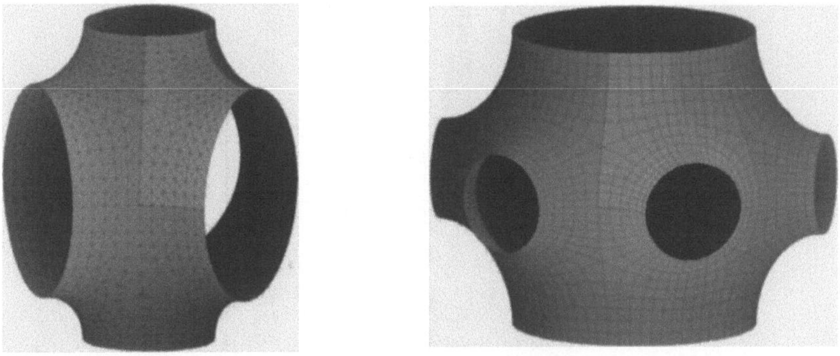

Plate 1. (*cf. Fig. 1, Page 131*) Discrete H-T surface with fundamental patch in a 3-prism, and differently assembled in a 6-prism. In the right figure sets of four coplanar triangles were joined to emphasize the observed alignment of edges with curvature lines.

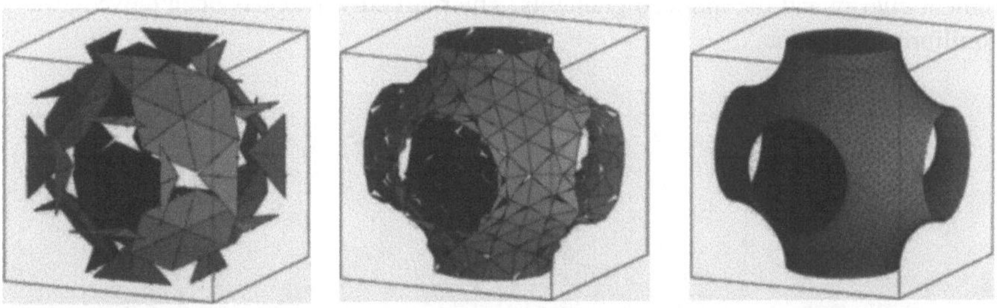

Plate 2. (*cf. Fig. 2, Page 133*) Free-boundary value problem of Schwarz P-surface in a cube solved via the discrete conjugate surface construction. The solution of that method is a non-conforming discrete minimal surface which is shown here at different discretizations.

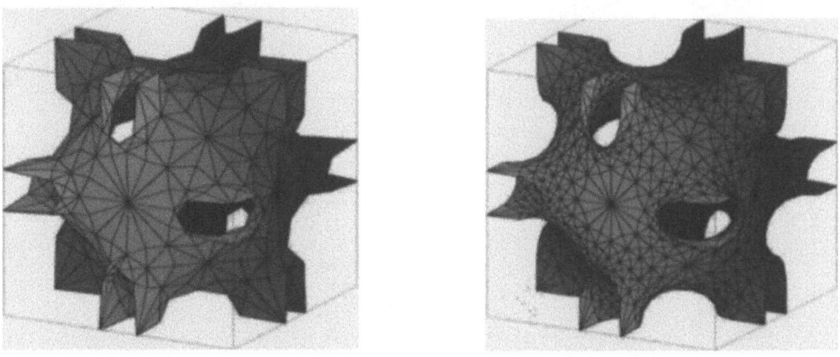

Plate 3. (*cf. Fig. 11, Page 142*) Discrete versions of Neovius surface in a cube. Note the observation that in both minimizers each triangle has a coplanar adjacent triangle which leads to a tessellation with planar quadrilaterals.

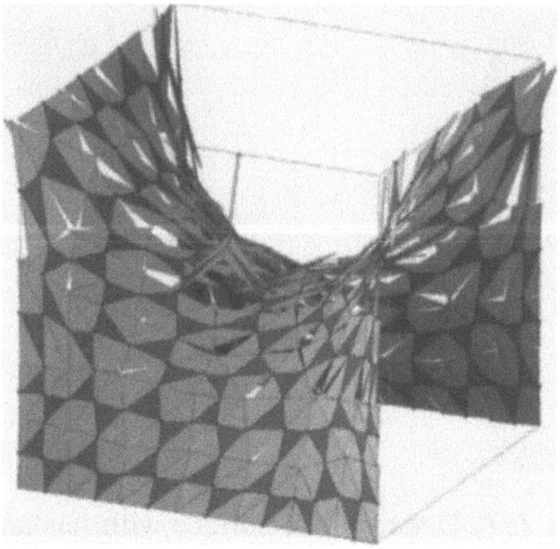

Plate 4. (*cf. Fig. 5, Page 136*) Non-conforming mesh with a stencil of a vertex for the alignment energy and the induced gradients. The marked vertex has been moved away from its equilibrium position.

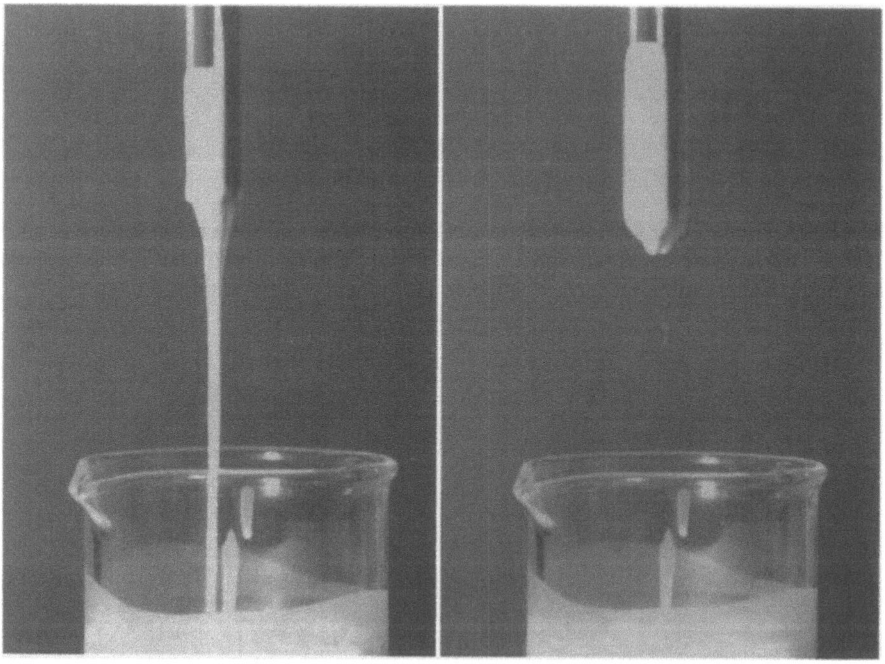

Plate 5. (*cf. Ettwein and Růžička, Pages 591–602*) On the left the image of an electrorheological fluid (in between two electrodes) where no electrical field is applied, on the right the same fluid in case of an applied electrical field

Plate 6. (*cf. Fig. 1, Page 119*) A twisted Karcher-Scherk surface with eight helicoidal ends

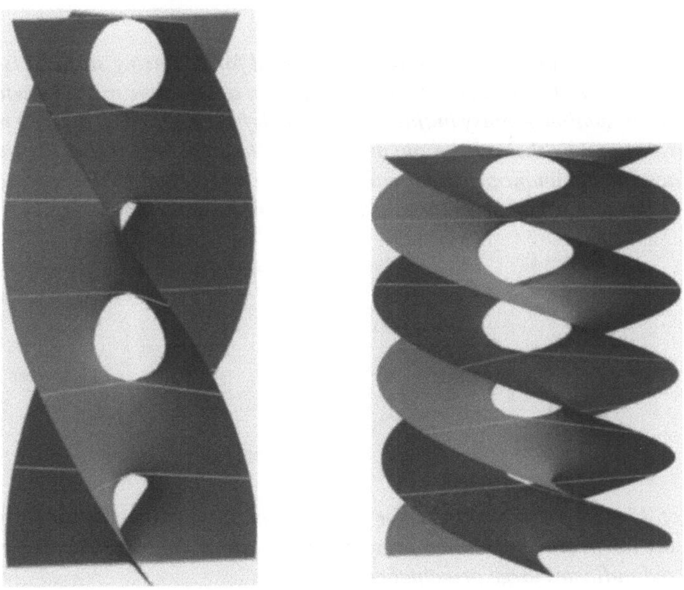

Plate 7. (*cf. Fig. 3, Page 121*) Twisted Karcher-Scherk surface with four helicoidal ends

Plate 8. *(cf. Fig. 1, Page 218) In this figure, a method for denoising surfaces is demonstrated. We compare the isotropic mean curvature flow and an anisotropic mean curvature flow based on curvature evaluation. A noisy initial surface (top left) is evolved by discrete mean curvature flow (top right) and by the anisotropic diffusion method (bottom right). Furthermore for the latter surface the dominant principal curvature - on which the diffusion tensor depends - is color coded (bottom left). The snapshots are taken at the same time-steps.*

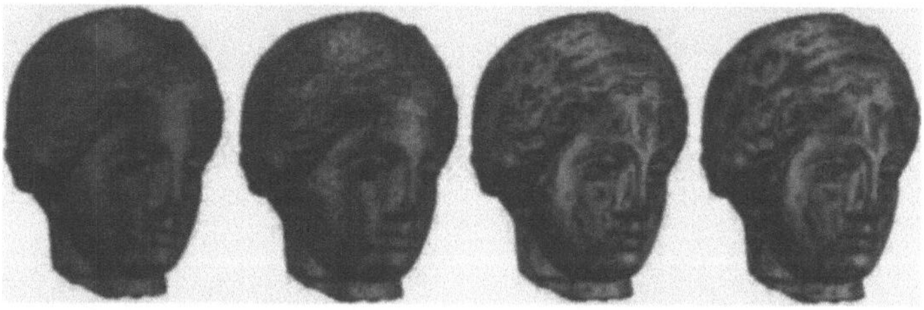

Plate 9. *(cf. Fig. 10, Page 243) Here, for different timesteps of an anisotropic mean curvature evolution, the norm of the dominant principal curvature is color coded, where blue indicates small and red indicates large principal curvature.*

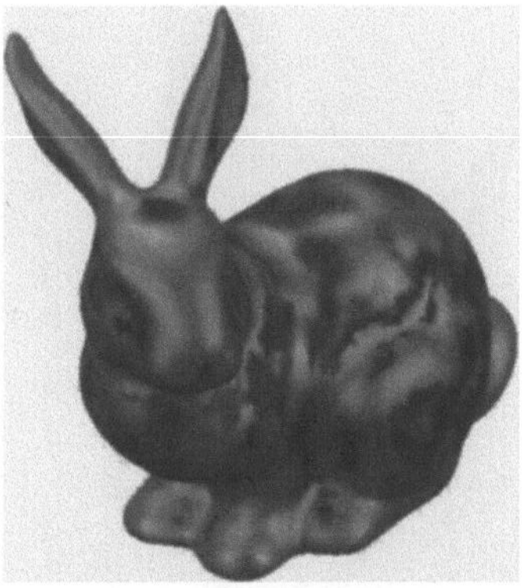

Plate 10. (*cf. Fig. 7, Page 237*) This figure shows a classification scheme based on curvature evaluation. The trace of the classification operator is color coded. Red indicates corners, where areas close to edges are green and smooth domains blue.

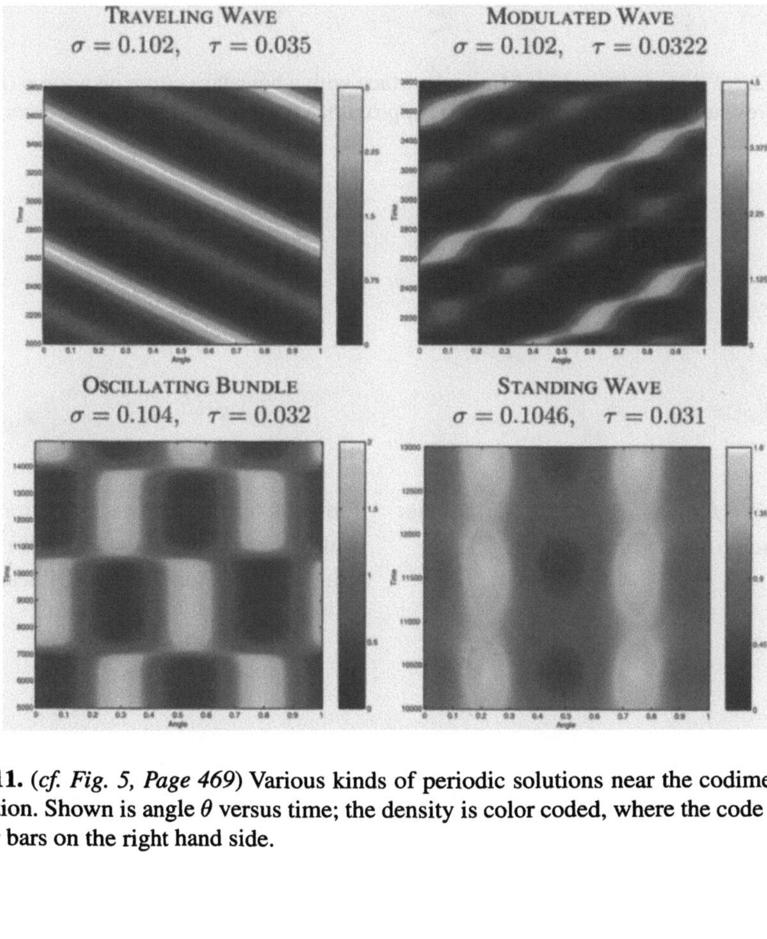

Plate 11. (*cf. Fig. 5, Page 469*) Various kinds of periodic solutions near the codimension 2 bifurcation. Shown is angle θ versus time; the density is color coded, where the code is given in color bars on the right hand side.

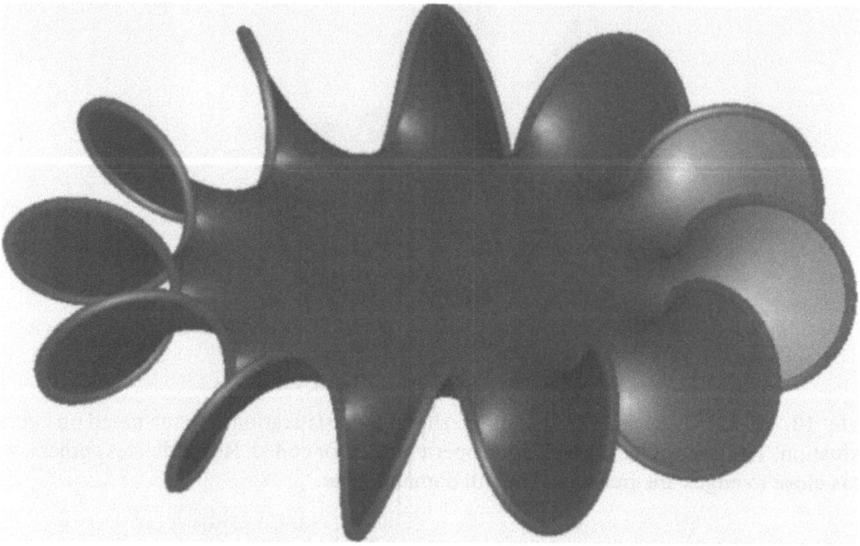

Plate 12. (*cf. Fig. 3, Page 173*) Minimal surface with a boundary curve on a torus (indicated by the grey tube). The minimal surface is approximated using adaptive finite elements.

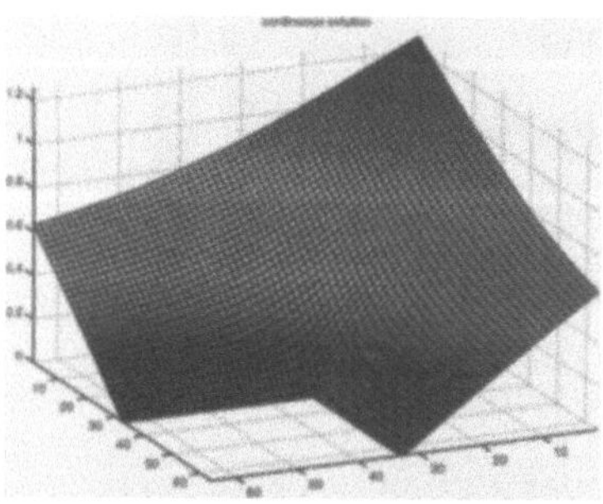

Plate 13. (*cf. Fig. 5, Page 535*) Surface plot of solution $u(r,\theta) = r^{\frac{2}{3}} \sin\left(\frac{2\theta-\pi}{3}\right)$.

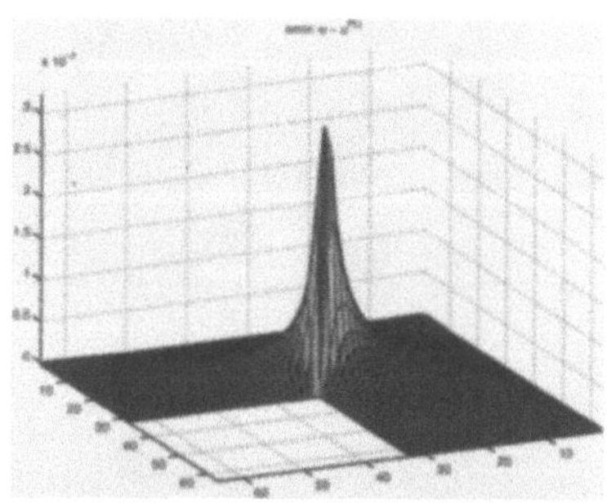

Plate 14. (*cf. Fig. 6, Page 535*) Error $u - u_l^{\mathrm{PU}}$ with $l = 6$ and $u(r,\theta) = r^{\frac{2}{3}} \sin\left(\frac{2\theta-\pi}{3}\right)$ using the h-version PUM with linear polynomials.

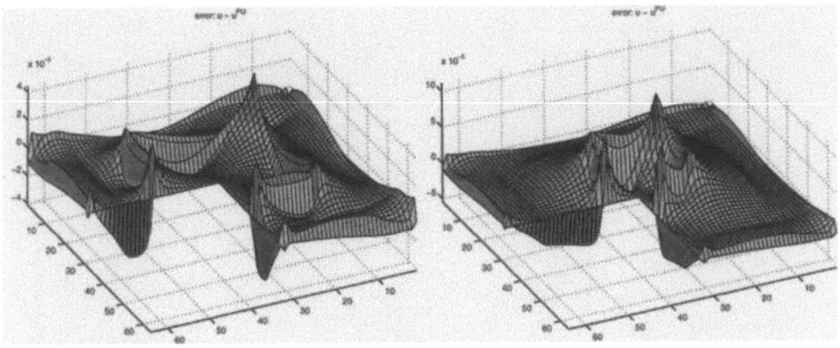

Plate 15. (*cf. Fig. 8, Page 538*) Error $u - u_l^{\mathrm{PU}}$ with $l = 6$ and $u(r,\theta) = r^{\frac{2}{3}} \sin\left(\frac{2\theta-\pi}{3}\right)$ using for the h-version PUM with augmented local spaces (left: $a = 0.5$, right: $a = 0.25$).

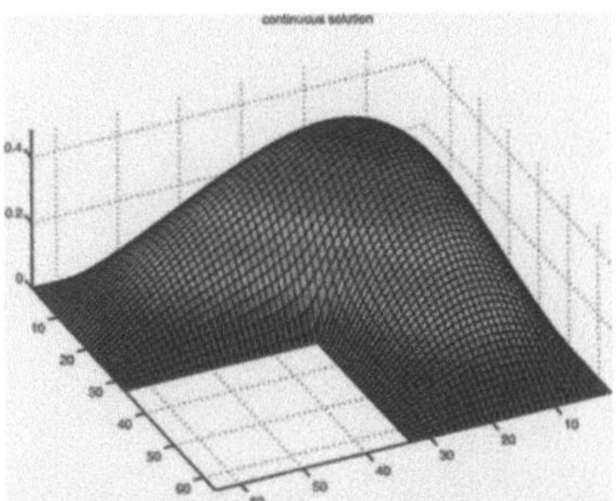

Plate 16. (*cf. Fig. 10, Page 539*) Surface plot of solution $u(r,\theta) = (1-(r\cos(\theta))^2)(1-(r\sin(\theta))^2)r^{\frac{2}{3}} \sin\left(\frac{2\theta-\pi}{3}\right)$.

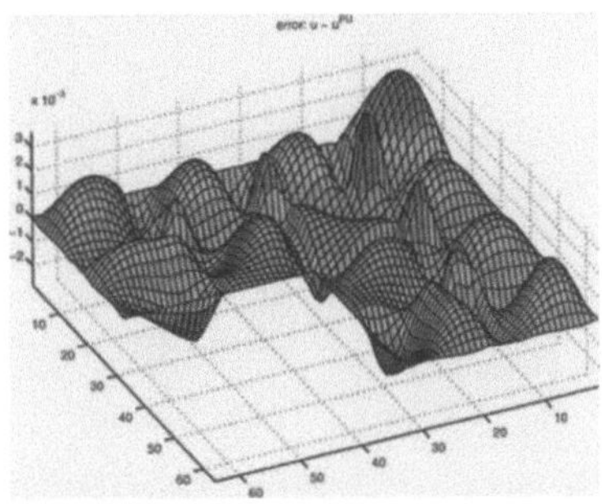

Plate 17. (*cf. Fig. 11, Page 539*) Error $u - u_l^{\mathrm{PU}}$ with $l = 2$, $p = 3$ and $u(r,\theta) = (1-(r\cos(\theta))^2)(1-(r\sin(\theta))^2)r^{\frac{2}{3}} \sin\left(\frac{2\theta-\pi}{3}\right)$.

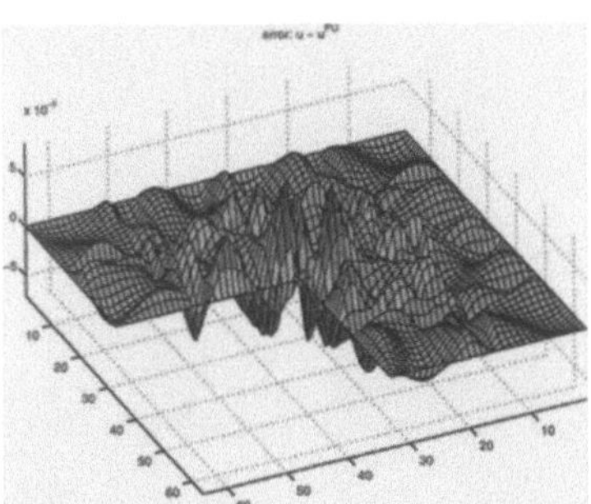

Plate 18. (*cf. Fig. 12, Page 540*) Error $u - u_l^{\mathrm{PU}}$ with $l = 2$, $p = 5$ and $u(r,\theta) = (1-(r\cos(\theta))^2)(1-(r\sin(\theta))^2)r^{\frac{2}{3}} \sin\left(\frac{2\theta-\pi}{3}\right)$.

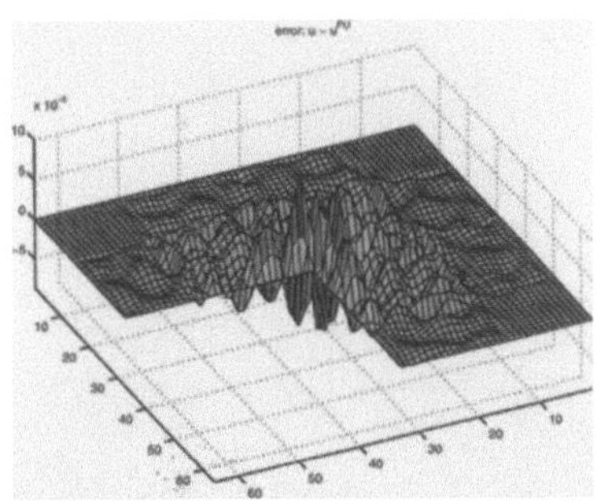

Plate 19. (*cf. Fig. 13, Page 540*) Error $u - u_l^{\mathrm{PU}}$ with $l = 2$, $p = 8$ and $u(r,\theta) = (1-(r\cos(\theta))^2)(1-(r\sin(\theta))^2)r^{\frac{2}{3}} \sin\left(\frac{2\theta-\pi}{3}\right)$.

MIX
Papier aus verantwortungsvollen Quellen
Paper from responsible sources
FSC® C105338

If you have any concerns about our products,
you can contact us on
ProductSafety@springernature.com

In case Publisher is established outside the EU,
the EU authorized representative is:
Springer Nature Customer Service Center GmbH
Europaplatz 3, 69115 Heidelberg, Germany

Printed by Libri Plureos GmbH
in Hamburg, Germany